## Graphs of Common Functions and Models

Constant Function

Identity Function

Absolute Value Function

Square Root Function

Squaring Function

Cubing Function

Exponential Function

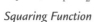

Logarithmic Function

## Algebraic Errors to Avoid

$\dfrac{a}{x+b} \neq \dfrac{a}{x} + \dfrac{a}{b}$ — To see this error, let $a = b = x = 1$.

$\sqrt{x^2 + a^2} \neq x + a$ — To see this error, let $x = 3$ and $a = 4$.

$a - b(x - 1) \neq a - bx - b$ — Remember to distribute negative signs. The equation should be
$$a - b(x - 1) = a - bx + b.$$

$\dfrac{\left(\dfrac{x}{a}\right)}{b} \neq \dfrac{bx}{a}$ — To divide fractions, invert and multiply. The equation should be
$$\frac{\dfrac{x}{a}}{b} = \frac{\dfrac{x}{a}}{\dfrac{b}{1}} = \left(\frac{x}{a}\right)\left(\frac{1}{b}\right) = \frac{x}{ab}.$$

$\sqrt{-x^2 + a^2} \neq -\sqrt{x^2 - a^2}$ — You can't factor a negative sign outside of the square root.

$\dfrac{\not{a} + bx}{\not{a}} \neq 1 + bx$ — This is one of many examples of incorrect cancellation. The equation should be
$$\frac{a + bx}{a} = \frac{a}{a} + \frac{bx}{a} = 1 + \frac{bx}{a}.$$

$(x^2)^3 \neq x^5$ — The equation should be $(x^2)^3 = x^2 x^2 x^2 = x^6$.

# College Algebra

## Concepts and Models

### Third Edition

**Ron Larson**
*The Pennsylvania State University*
*The Behrend College*

**Robert P. Hostetler**
*The Pennsylvania State University*
*The Behrend College*

**Anne V. Hodgkins**
*Phoenix College*

**Houghton Mifflin Company**
Boston   New York

Sponsoring Editor: Jack Shira
Managing Editor: Cathy Cantin
Senior Associate Editor: Maureen Ross
Associate Editor: Laura Wheel
Assistant Editor: Carolyn Johnson
Supervising Editor: Karen Carter
Project Editor: Patty Bergin
Editorial Assistant: Christine E. Lee
Art Supervisor: Gary Crespo
Marketing Manager: Michael Busnach
Composition and Art: Meridian Creative Group

We have included examples and exercises that use real-life data as well as technology output from a variety of software. This would not have been possible without the help of many people and organizations. Our wholehearted thanks goes to all for their time and effort.

Trademark acknowledgments: TI and CBR are registered trademarks of Texas Instruments, Inc.

Printed in the U.S.A.

Library of Congress Catalog Card Number: 99-71987

ISBN: 0-395-97621-9

3456789–DW–03 02 01

# Contents

# A Word from the Authors

Welcome to *College Algebra: Concepts and Models*, Third Edition. In this revision, we continue to focus on developing students' conceptual understanding of college algebra while offering opportunities for them to hone their problem-solving skills. In preparing the Third Edition, we carefully considered all text elements for revision, as well as adding new examples, exercises, and applications.

In response to suggestions from college algebra instructors, we have revised the coverage of some topics in the Third Edition, especially in Chapters P, 3, 4, and 8. To allow greater flexibility at the beginning of the course, we streamlined the algebra review (Chapter P in the Second Edition) and moved it to Appendix A. In Chapter 3, *Transformations of Functions* (Section 3.3) and *The Algebra of Functions* (Section 3.4) are now separate sections. This allows us to cover transformations in greater detail and to expand the discussion of composite functions. We moved several sections from old Chapter 3 to Chapter 4: *Quadratic Functions and Models* (Section 4.1), *Polynomial Functions of Higher Degree* (Section 4.2), and *Rational Functions* (Section 4.7). In addition, we combined old Sections 4.2 and 4.3 as *Real Zeros of Polynomial Functions* (Section 4.4) and addressed the Intermediate Value Theorem in that context. Finally, we omitted the discussion of Descartes's Rule of Signs and the Bounds for Real Zeros from this edition. In Chapter 8, we added a new section, *Mathematical Induction* (8.7), for instructors who want to prove the Binomial Theorem.

We have found that many college algebra students grasp theoretical concepts more easily when they work with them in the context of a real-life situation. Throughout the Third Edition, students now have many more opportunities to collect, analyze, and model real data. We updated all real-data application examples or replaced them with new applications that use current data. In addition, all of the Chapter Opener/Chapter Project applications—many of which ask the student to manipulate real data—are new.

There are many more opportunities in the Third Edition for students to use a graphing utility in the problem-solving process—to visualize and explore theoretical concepts, to analyze real data, and to verify alternative solution methods. However, to accommodate a variety of teaching and learning styles, the use of graphing technology is always optional. At the suggestion of our users, we revised the Technology feature to use only generic references, omitting keystrokes and references to specific calculator models. At the beginning of the Third Edition, we have included "A Brief Introduction to Graphing Utilities." This section addresses the basic functions of graphing calculators that are used in college algebra, including the equation editor and the table, zoom, and trace features.

To encourage mastery and understanding, we have included opportunities for student self-assessment at strategic points throughout each chapter. New to this edition, Learning Objectives at the beginning of each section provide students with a conceptual outline for reference. In the Chapter Summary, these Objectives are linked to the Review Exercises for students who require additional guided practice. Finally, Mid-Chapter Quizzes, Chapter Tests, and Cumulative Tests allow students to test understanding and learning retention at regular intervals.

We hope you will enjoy using the Third Edition in your college algebra class. We think its straightforward, readable style, engaging applications, and study tools will appeal to your students and contribute to their success in this course.


Ron Larson

Robert P. Hostetler

Anne Hodgkins

# Features

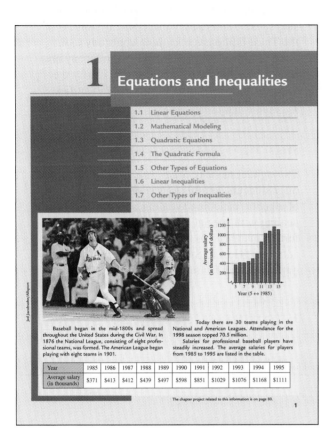

## Chapter Opener

Each chapter begins with a real-life application with current data. At the end of the chapter, students revisit this application as a *Chapter Project* after they have been exposed to the algebra techniques required for solving the problem. A list of section titles outlines the topics covered in the chapter.

## Section Opener

Each section opens with a list of learning *Objectives*, the algebra skills presented in the section. This conceptual outline functions as a useful tool for reference and review for the students and class planning for the instructor. In the *Chapter Summary*, these objectives are linked to *Review Exercises* for additional student practice.

## Definition and Theorem

Definitions, theorems, rules, formulas, guidelines, and properties are highlighted for emphasis and easy reference.

## Historical Note

Notes featuring mathematicians and their work, integrated throughout the text, place algebra in historical context.

---

270    Chapter 4    Polynomial and Rational Functions

### 4.6    The Fundamental Theorem of Algebra

**Objectives**

- Use the Fundamental Theorem of Algebra and the Linear Factorization Theorem to write a polynomial as the product of linear factors.
- Find a polynomial with integer coefficients whose zeros are given.
- Factor a polynomial over the real and complex numbers.
- Find all real and complex zeros of a polynomial function.

### The Fundamental Theorem of Algebra

You have been using the fact than an $n$th-degree polynomial can have at most $n$ real zeros. In the complex number system, this statement can be improved. That is, in the complex number system, every $n$th-degree polynomial function has *precisely* $n$ zeros. This important result is derived from the **Fundamental Theorem of Algebra**, first proved by the famous German mathematician Carl Friedrich Gauss (1777–1855).

**Jean Le Rond d'Alembert**
(1717-1783)
Jean Le Rond d'Alembert worked independently of Carl Gauss trying to prove the Fundamental Theorem of Algebra. His efforts were such that in France, the Fundamental Theorem of Algebra is frequently known as the theorem of d'Alembert.

▶ **The Fundamental Theorem of Algebra**

If $f(x)$ is a polynomial of degree $n$, where $n > 0$, then $f$ has at least one zero in the complex number system.

Using the Fundamental Theorem of Algebra and the equivalence of zeros and factors, you obtain the following theorem.

▶ **Linear Factorization Theorem**

If $f(x)$ is a polynomial of degree $n$

$$f(x) = a_n x^n + a_{n-1} x^{n-1} + \cdots + a_1 x + a_0$$

where $n > 0$, then $f$ has precisely $n$ linear factors

$$f(x) = a_n(x - c_1)(x - c_2) \cdots (x - c_n)$$

where $c_1, c_2, \ldots, c_n$ are complex numbers and $a_n$ is the leading coefficient of $f(x)$.

Note that neither the Fundamental Theorem of Algebra nor the Linear Factorization Theorem tells you *how* to find the zeros or factors of a polynomial. Such theorems are called **existence theorems**. To find the zeros of a polynomial function, you still rely on the techniques developed in the earlier parts of the text.

Remember that the $n$ zeros of a polynomial function can be real or complex, and they may be repeated. Example 1 illustrates several cases.

## Example

A wide variety of *Examples*, many with detailed, step-by-step solutions, enhance the usefulness of the Third Edition as a learning and study tool. Each is labeled for easy reference, and many are accompanied by side comments that clarify the solution method used. The icon identifies *Examples* that reflect real life data.

## Study Tip

Throughout the text, *Study Tips* at point of use offer students helpful suggestions for their study of algebra, including avoiding common errors, dealing with special cases, and developing further insights into theoretical concepts. In addition, "How to Study Algebra" (page xxi) outlines a plan for improving student study skills.

## Technology

Students are encouraged to use a graphing utility as a problem-solving tool. The text offers many opportunities to visualize theoretical concepts, to discover alternative approaches, and to verify the results of other solution methods using technology. However, students are not required to have access to a graphing utility to use this text effectively. As appropriate, the text addresses both the benefits of using technology and its possible misuse or misinterpretation.

## Graphics

The Third Edition has over 1500 figures to help students grasp mathematical concepts through visualization.

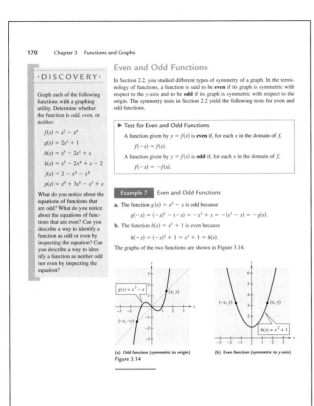

## Discovery

*Discovery* activities offer opportunities for the exploration of selected mathematical concepts. Students are encouraged to use techniques such as visualization and modeling to develop their intuitive understanding of theoretical topics. These optional activities can be omitted at the instructor's discretion without affecting the flow of the material.

**FEATURES**

## Discussing the Concept

Each text section ends with *Discussing the Concept*. Designed to encourage students to think, reason, and write about algebra, these exercises help synthesize the concepts and methods presented in the section. Depending on the nature of the course, these problems can be assigned as individual student work or collaborative projects, or they may form the basis for class discussion.

## Warm Up

Starting with Section 1.1, each exercise set is preceded by ten warm-up exercises. These *Warm Ups* provide students with the opportunity to review and practice previously learned skills necessary to master the new skills presented in the section. Answers to all *Warm Ups* are given in the back of the text.

## Exercises

The Third Edition contains a variety of computational, conceptual, and applied problems to accommodate a range of teaching and learning styles. Designed to build competence, skill, and understanding, each exercise set is graded in difficulty to allow students to gain confidence as they progress. Answers to odd-numbered exercises are given in the back of the text, while detailed solutions are available in the *Study and Solutions Guide*.

There are over 5000 exercises in the Third Edition, including multi-part, exploratory, modeling, data analysis, reasoning, writing, estimation, geometry, and technology-required problems. A rich variety of relevant applications—many of which utilize real data—represent a range of disciplines and clearly demonstrate the applicability of the mathematics to real-world situations. In addition, many section exercises have been labeled in the Instructor's Annotated Edition with the icon ✸✸ to indicate that they are appropriate for group work as well.

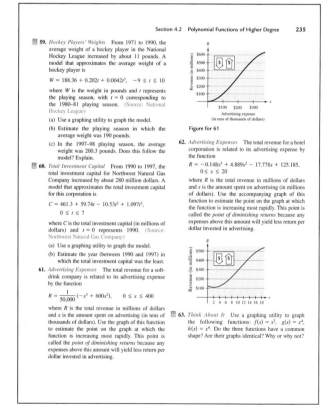

Section 4.1    Quadratic Functions and Models     225

**55.** *Ice Cream* During the mid-1990s, the number of gallons of regular ice cream produced in the United States fell slightly, then rose again according to the model

$$G(x) = 11.25t^2 - 111.25t + 1158.75, \quad 4 \le t \le 7$$

where $G$ is the number of gallons of regular ice cream (in millions) and $t$ represents the calendar year, with $t = 4$ corresponding to 1994. Use a graphing utility to determine the year, from 1994 to 1997, when the number of gallons produced was at its lowest. (Source: U.S. Department of Agriculture)

**56.** *Employment Rate* The annual average employment rate for single women between 16 and 19 years of age from 1990 to 1994 can be approximated by the model

$$P(x) = 0.564t^2 - 2.39t + 51.89, \quad 0 \le t \le 4$$

where $P$ is the percent of single women who were employed and $t$ represents the calendar year, with $t = 0$ corresponding to 1990. Use a graphing utility to determine the year in which the employment rate was at its lowest. (Source: U.S. Bureau of Labor Statistics)

**Math Matters • Circles and Pi**

Pi is a special number that represents a relationship that is found in *all circles*. The relationship is this: The ratio of the circumference of *any* circle to the diameter of the circle always produces the same number—the number denoted by the Greek symbol $\pi$.

Calculation of the decimal representation of the number $\pi$ has consumed the time of many mathematicians (and computers). Because the number $\pi$ is irrational, its decimal representation does not repeat or terminate. Here is a decimal representation of $\pi$ that is accurate to 27 decimal places.

$$\pi \approx 3.141592653589793238462643383.$$

In October 1995, a team of mathematicians in British Columbia discovered a formula for calculating the *n*th digit of $\pi$ without calculating the preceding digits. Run on a computer, the algorithm for the *n*th digit requires that numbers be written in base 16 (hexadecimal representation) instead of base 10 (decimal representation).

The number $\pi$ is used in most formulas that deal with circles. For instance, the area of a circle is given by

$$A = \pi r^2$$

where $r$ is the radius of the circle. Use this formula to determine which of the colored regions inside the circle below has the greater area. Is the area of the red circle larger than that of the blue ring, or do both regions have the same area? (The answer is given in the back of the book.)

Which area is larger?
- A
- B

## Supplementary Lab Assignments

Additional guided discovery activities and applications for selected text topics are available in the supplement *Problem Solving, Modeling, and Data Analysis Labs*. These labs are referenced in the section exercise sets.

## Math Matters

In each chapter, a *Math Matters* feature engages student interest with an historical note or mathematical problem. Answers to those that pose a question are given in the back of the text.

## Review Exercises

*Review Exercises* at the end of each chapter offer students an additional opportunity for practice and review. Answers to odd-numbered review exercises are given in the back of the text.

208     Chapter 3    Functions and Graphs

**REVIEW EXERCISES**

In Exercises 1–4, decide whether the equation represents $y$ as a function of $x$.

**1.** $3x - 4y = 12$     **2.** $y^2 = x^2 - 9$
**3.** $y = \sqrt{x + 3}$     **4.** $x^2 + y^2 - 6x + 8y = 0$

In Exercises 5 and 6, decide whether the set represents a function from $A$ to $B$.

$$A = \{1, 2, 3\} \qquad B = \{-3, -4, -7\}$$

Give reasons for your answer.

**5.** $\{(1, -3), (2, -7), (3, -3)\}$
**6.** $\{(1, -4), (2, -3), (3, -9)\}$

In Exercises 7–10, evaluate the function and simplify the results.

**7.** $f(x) = 3x - 5$
   (a) $f(1)$   (b) $f(-2)$   (c) $f(m)$   (d) $f(x + 1)$
**8.** $f(x) = \sqrt{x + 9} - 3$
   (a) $f(7)$   (b) $f(0)$   (c) $f(-5)$   (d) $f(x + 2)$
**9.** $f(x) = |x| + 5$
   (a) $f(0)$   (b) $f(-3)$   (c) $f(\tfrac{1}{2})$   (d) $f(-x^2)$
**10.** $f(x) = \begin{cases} x^2 + 1, & x \le 2 \\ x + 3, & x > 2 \end{cases}$
   (a) $f(0)$   (b) $f(2)$   (c) $f(3)$   (d) $f(-5)$

In Exercises 11 and 12, find all real values of $x$ such that $f(x) = 0$.

**11.** $f(x) = \dfrac{2x + 7}{3}$     **12.** $f(x) = x^3 - 4x$

In Exercises 13–18, find the domain of the function.

**13.** $f(x) = 3x^2 + 8x + 4$
**14.** $g(t) = \dfrac{5}{t^2 - 9}$
**15.** $h(x) = \sqrt{x + 9}$
**16.** $f(t) = \sqrt[3]{t - 5}$
**17.** $g(t) = \dfrac{\sqrt{t - 3}}{t - 5}$
**18.** $h(x) = \sqrt[4]{9 - x^2}$

**19.** *Reasoning* A student has difficulty understanding why the domains of

$$h(x) = \frac{x^2 - 4}{x} \quad \text{and} \quad k(x) = \frac{x}{x^2 - 4}$$

are different. How would you explain their respective domains algebraically? How could you use a graphing utility to explain their domains?

**20.** *Reasoning* A student has difficulty understanding why the domains of

$$h(x) = \sqrt{x + 1} \quad \text{and} \quad k(x) = \sqrt[3]{x + 1}$$

are different. How would you explain their respective domains algebraically? How could you use a graphing utility to explain their domains?

**21.** *Volume of a Box* An open box is to be made from a square piece of material 20 inches on a side by cutting equal squares from the corners and turning up the sides (see figure).

   (a) Write the volume of the box as a function of $x$.
   (b) What is the domain of this function?
   (c) Use a graphing utility to sketch the graph of this function.

**22.** *Balance in an Account* A person deposits $5000 in an account that pays 6.25% interest compounded quarterly.

   (a) Write the balance of the account in terms of the time $t$ that the principal is left in the account.
   (b) What is the domain of this function?

**23.** *Vertical Motion* The velocity of a ball thrown vertically upward from ground level is given by $v(t) = -32t + 48$, where $t$ is the time in seconds and $v$ is the velocity in feet per second.

   (a) Find the velocity when $t = 1$.
   (b) Find the time when the ball reaches its maximum height. [*Hint:* Find the time when $v(t) = 0$.]
   (c) Find the velocity when $t = 2$.

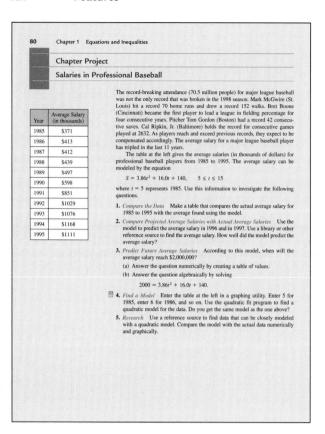

## Chapter Project

*Chapter Projects* are extended applications designed to enhance students' understanding of mathematical concepts. Real data is previewed at the beginning of the chapter and then analyzed in detail in the *Project* at the end of the chapter. Here the student is guided through a set of multi-part exercises using modeling, graphing, and critical thinking skills to analyze the data.

## Chapter Summary

The *Chapter Summary* is another vehicle for student self-assessment. For the student who needs additional practice and review, each section learning *Objective* is keyed to the appropriate *Review Exercises.*

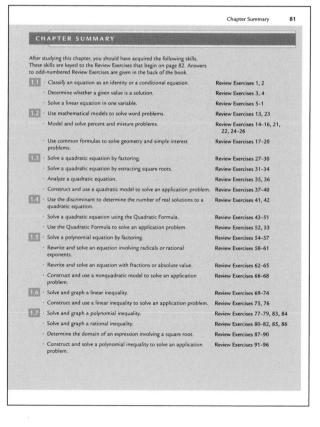

---

## Mid-Chapter Quiz

Take this quiz as you would take a quiz in class. After you are done, check your work against the answers given in the back of the book.

In Exercises 1–4, find the value of $f(x) = 2x^2 - x + 1$.

**1.** $f(3)$  **2.** $f(-1)$  **3.** $f(0)$  **4.** $f(a)$

In Exercises 5 and 6, find all real values of $x$ such that $f(x) = 0$.

**5.** $f(x) = x^3 - 4x$  **6.** $f(x) = 3 - \dfrac{4}{x}$

**7.** A company produces a product for which the variable cost is $11.40 per unit and the fixed costs are $85,000. The product sells for $16.89. Let $x$ be the number of units produced. Write the total profit $P$ as a function of $x$.

In Exercises 8–10, find the domain of the function.

**8.** $f(x) = \dfrac{3}{x^2 - x}$  **9.** $h(x) = \sqrt{5 - x}$  **10.** $g(x) = \dfrac{\sqrt{x-1}}{x+2}$

In Exercises 11–14, use a graphing utility to graph the function. Then estimate the open intervals on which the function is increasing or decreasing.

**11.** $f(x) = -x^2 + 5x + 3$  **12.** $f(x) = 2x^3 - 5x^2$
**13.** $f(x) = x^2 + 2$  **14.** $f(x) = \sqrt{x + 1}$

In Exercises 15 and 16, compare the graphs of $f(x) = \sqrt[3]{x}$ and $g$.

**15.** $g(x) = \sqrt[3]{x - 2}$  **16.** $g(x) = -\sqrt[3]{x + 1}$

**17.** Write the equation for the transformation of $f(x) = x^3$ shown in the figure.

In Exercises 18 and 19, write the equation for the indicated transformation. Verify your result with a graphing utility.

**18.** The graph of $f(x) = x^2$ shifted three units to the right and two units downward.
**19.** The graph of $f(x) = \sqrt{x}$ shifted two units to the left and reflected about the $x$-axis.
**20.** The marketing department of a company estimates that the demand for a product is given by $p = 80 - 0.001x$, where $p$ is the price per unit and $x$ is the number of units. The cost of producing $x$ units is given by $C = 300,000 + 25x$ and the profit for producing $x$ units is given by $P = xp - C$. Write the equation for the profit function, sketch the graph, and estimate the number of units that would produce a maximum profit.

Figure for 17

---

## Chapter Test

Take this test as you would take a test in class. After you are done, check your work against the answers given in the back of the book.

In Exercises 1–4, find the domain of the function.

**1.** $f(x) = 2x^2 - 3x + 8$  **2.** $g(x) = \sqrt{x - 7}$
**3.** $g(t) = \dfrac{\sqrt{t-2}}{t-7}$  **4.** $h(x) = \dfrac{3}{x^2 - 4}$

In Exercises 5–7, decide whether the statement is true or false. Explain.

**5.** The equation $2x - 3y = 5$ identifies $y$ as a function of $x$.
**6.** The equation $y = \pm\sqrt{x^2 - 16}$ identifies $y$ as a function of $x$.
**7.** If $A = \{3, 4, 5\}$ and $B = \{-1, -2, -3\}$, the set $\{(3, -9), (4, -2), (5, -3)\}$ represents a function from $A$ to $B$.

In Exercises 8 and 9, (a) find the range and domain of the function, (b) determine the intervals over which the function is increasing, decreasing, or constant, and (c) determine whether the function is even or odd.

**8.** $f(x) = x^2 + 2$  (See figure.)  **9.** $g(x) = \sqrt{x^2 - 4}$  (See figure.)

**10.** Evaluate the function and simplify the results.

$$g(x) = \begin{cases} x + 1, & x < 0 \\ x^2 + 2, & x \geq 0 \end{cases}$$

(a) $g(-1)$  (b) $g(0)$  (c) $g(3)$

In Exercises 11 and 12, sketch the graph of the function.

**11.** $g(x) = \begin{cases} x + 1, & x < 0 \\ 1, & x = 0 \\ x^2 + 1, & x > 0 \end{cases}$  **12.** $h(x) = (x - 3)^2 + 4$

**13.** Find $f^{-1}(x)$ for the function $f(x) = x^3 - 5$.

In Exercises 14–19, use $f(x) = x^2 + 2$ and $g(x) = 2x - 1$ to find the function.

**14.** $(f + g)(x)$  **15.** $(f - g)(x)$  **16.** $(fg)(x)$
**17.** $\left(\dfrac{f}{g}\right)(x)$  **18.** $(f \circ g)(x)$  **19.** $(g \circ f)(x)$

**20.** A manufacturer determines that the variable costs for a new product are $3.17 per unit and the fixed costs are $75,000. The product is to be sold for $5.25. Let $x$ be the number of units sold.

(a) Write the total cost $C$ as a function of the number of units sold.
(b) Write the total revenue $R$ as a function of the number of units sold.
(c) Write the total profit $P$ as a function of the number of units sold.

Figure for 8

Figure for 9

---

## Cumulative Test: Chapters 1–2

Take this test as you would take a test in class. After you are done, check your work against the answers given in the back of the book.

In Exercises 1–6, solve the equation.

**1.** $4y^2 - 64 = 0$  **2.** $\dfrac{x}{3} + \dfrac{x}{4} = 1$
**3.** $x^4 - 17x^2 + 16 = 0$  **4.** $|2x - 3| = 5$
**5.** $\sqrt{y + 3} + y = 3$  **6.** $2x^2 + x = 5$

**7.** You deposit $5000 in an account that pays 8.5%, compounded monthly. Find the balance after 10 years.

In Exercises 8–10, solve the inequality.

**8.** $\left|\dfrac{2x - 5}{4}\right| \leq 3$  **9.** $(x + 4)^2 \leq 4$  **10.** $\dfrac{3 + 2x}{4 - x} > 2$

**11.** Find the intercepts of the graph of $y = x\sqrt{x + 4}$.
**12.** Describe the symmetry of the graph of $y = -\sqrt{4 - x^2}$.
**13.** Write the equation of the circle in standard form and sketch its graph.

$$x^2 + y^2 - 6x + 4y - 3 = 0$$

**14.** Find the constant $C$ such that $(-2, 3)$ is a solution point of $y = x^3 + C$.
**15.** Find an equation of the line passing through $(3, -2)$ and $(-1, 5)$.
**16.** Find an equation of the line with slope $\frac{2}{3}$ passing through $(2, -1)$.
**17.** Find an equation of the line with zero slope passing through $(-1, -3)$.
**18.** You have accepted a sales job that pays a base salary of $1000 a month plus a 7% commission on monthly sales $x$. Write a linear model that describes your total monthly salary $y$ in terms of your base salary and commission.
**19.** During the second and third quarters of the year, a business had sales of $210,000 and $230,000, respectively. If the sales growth follows a linear pattern, what will the sales be during the fourth quarter?
**20.** The revenue and cost equations for a product are given by

$$R = x(100 - 0.001x) \quad \text{and} \quad C = 20x + 30,000$$

where $R$ and $C$ are measured in dollars and $x$ represents the number of units sold. How many units must be sold for the revenue to equal the cost?

---

## Mid-Chapter Quiz

Each chapter contains a 20-question *Mid-Chapter Quiz*. Appearing midway through the chapter, these are designed as a self-assessment tool for the student. Answers are provided in the back of the text.

## Chapter Test

Each chapter ends with a *Chapter Test*, providing students with another self-assessment opportunity. Answers are provided in the back of the text.

## Cumulative Test

The *Cumulative Tests* that appear after Chapters 2, 5, and 8 give students several opportunities throughout the course to judge their mastery of previously covered material.

## Supplements

*College Algebra: Concepts and Models*, Third Edition, by Larson, Hostetler, and Hodgkins, is accompanied by a comprehensive supplements package with ancillaries for students, for instructors, and for use as classroom resources.

# Printed Resources

### For the Student

**Study and Solutions Guide** by Dianna L. Zook, Purdue University and Indiana University at Fort Wayne, and Anne V. Hodgkins, Phoenix College
(0-395-97623-5)
• Section summaries of key concepts
• Detailed, step-by-step solutions to odd-numbered Section and Review Exercises and to all Mid-Chapter Quizzes, Chapter Tests, and Cumulative Tests

**Graphing Technology Keystroke Guide**
(0-395-97676-6)
• Keystroke instructions for *TI-80, TI-81, TI-82, TI-83, TI-83 Plus, TI-85, TI-86, Casio fx-7700GE, Casio fx-9700GE, Casio CFX-9800G, HP 38G,* and *Sharp EL-9200/9300*
• Examples with step-by-step solutions
• Extensive graphics screen output
• Technology tips

**Problem Solving, Modeling, and Data Analysis Labs** by Wendy Metzger, Palomar College
(0-395-97624-3)
• Guided discovery activities and applications with Answer Key
• Keyed to the text by topic

### For the Instructor

**Instructor's Annotated Edition**
(0-395-97626-X)
• Complete student edition, including student answers section: answers to all odd-numbered exercises, and answers to all Mid-Chapter Quizzes, Chapter Tests, Cumulative Tests, and Math Matters
• Instructor's answers section: answers to all even-numbered exercises, and answers to all Discovery features, Technology features, and Discussing the Concept activities
• Annotations at point of use: teaching strategies, suggestions for implementing features, common student errors, and additional examples, exercises, and group activities

*Complete Solutions Guide* by Dianna L. Zook, Indiana University and Purdue University at Fort Wayne, and Anne V. Hodgkins, Phoenix College (0-395-97625-1)

- Detailed, step-by-step solutions to all Section and Review Exercises; Mid-Chapter Quiz, Chapter Test, and Cumulative Test questions; and Chapter Project Questions

*Instructor's Resource Guide and Test Item File*

(0-395-97622-7)

- Printed test bank with approximately 1700 test items coded by level of difficulty
- Test items include multiple-choice and open-ended format
- Technology-required test items coded for easy reference
- Three tests per chapter
- Two final exams
- Transparency masters
- Lab activities contributed by Phyllis Shaw, Paradise Valley Community College

## Media Resources

*Internet Site*

- Student and Instructor resources
- Graphing calculator programs
- Career Interviews

### For the Student

*Tutor*

(Windows: 0-395-97627-8; Macintosh: 0-395-97638-3)

- Interactive tutorial software keyed to the text by section
- Diagnostic feedback
- Additional practice
- Chapter warm-ups and self-tests
- Glossary

*Videotapes* **by Dana Mosely**

(0-395-97641-3)

- Comprehensive coverage keyed to the text by section
- Detailed explanations of important concepts
- Numerous examples and applications

### For the Instructor

*Computerized Test Bank*

(Windows: 0-395-97639-1; Macintosh: 0-395-97640-5)

- Test-generating software
- Approximately 1700 test items
- Also available as a printed test item file

## Acknowledgments

We would like to thank the many people who have helped us at various stages of this project to prepare the text and supplements package. Their encouragement, criticisms, and suggestions have been invaluable to us.

### Third Edition Reviewers

Judith A. Ahrens, Pellissippi State Technical Community College; Sandra Beken, Horry-Georgetown Technical College; Diane Benjamin, University of Wisconsin—Platteville; Kent Craghead, Colby Community College; Nick Geller, Collin County Community College; Steven Z. Kahn, Anne Arundel Community College; Gael Mericle, Mankato State University; Michael Montano, Riverside Community College; Terrie L. Nichols, Cuyamaca College; Mark Omodt, Anoka-Ramsey Community College

We would also like to thank the staffs of Larson Texts, Inc., and Meridian Creative Group who assisted with proofreading the manuscript; preparing and proofreading the art package; and checking and typesetting supplements.

On a personal level, we are grateful to our families, especially Deanna Gilbert Larson, Eloise Hostetler, and Jay N. Torok, for their love, patience, and support. Also, a special thanks goes to R. Scott O'Neil.

If you have suggestions for improving the text, please feel free to write to us. Over the past two decades, we have received many useful comments from both instructors and students, and we value these very much.

Ron Larson
Robert P. Hostetler
Anne V. Hodgkins

# A Brief Introduction to Graphing Utilities

Graphing utilities such as graphing calculators and computers with graphing software are very valuable tools for visualizing mathematical principles, verifying solutions to equations, exploring mathematical ideas, and developing mathematical models. While graphing utilities are extremely helpful in learning mathematics, their use does not mean that learning algebra is any less important. In fact, the combination of knowledge of mathematics and the use of graphing utilities allows you to explore mathematics more easily and to a greater depth. If you are using a graphing utility in this course, it is up to you to learn its capabilities and to practice using this tool to enhance your mathematical learning.

This introduction gives a brief description of two basic features of your graphing utility: the equation editor and the table feature. If you will be using your graphing utility more extensively in Chapter 1, you may want to review Section 2.3. You may also find it helpful to consult the user's guide for your graphing utility. Additionally, graphing utility keystroke guides are available for most models, and your college library may have a videotape on how to use your graphing utility.

## The Equation Editor

Many graphing utilities are designed to act as "function graphers." In Chapter 3, you will study functions and their graphs in detail. You may recall from previous courses that a function can be thought of as a rule that describes the relationship between two variables. These rules are frequently written in terms of $x$ and $y$. For example, the equation $y = 3x + 5$ represents $y$ as a function of $x$.

Figure 1

Many graphing utilities have an equation editor that requires an equation to be written in "$y =$" form in order to be entered, as shown in Figure 1. (You should note that your equation editor screen may not look like the screen shown in Figure 1.) To determine exactly how to enter an equation into your graphing utility, consult your user's guide.

## The Table Feature

Most graphing utilities have the capability to display a table of values with $x$ and one or more $y$-values. These tables can be used to check solutions of an equation and to generate ordered pairs $(x, y)$ to assist in graphing an equation.

To use the table feature, enter an equation in the equation editor in "$y =$" form. The table may have a setup screen, which will allow you to select the starting $x$-value and the table step or $x$-increment. You may then have an option to automatically generate values for $x$ and $y$ or to build your own table using the ASK mode. In the ASK mode, you enter a value for $x$ and the graphing utility will display the $y$-value.

Figure 2

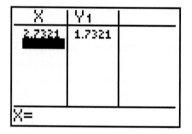

Figure 3

For example, enter the equation

$$y = \frac{3x}{x + 2}$$

in the equation editor as shown in Figure 2. In the table setup screen, set the table to start at $x = -4$ and set the table step to 1. When you view the table, notice that the first $x$-value is $-4$ and each value after it increases by one. Also notice that the $Y_1$ column gives the resulting $y$-value for each $x$-value, as shown in Figure 3. The table shows that the $y$-value when $x = -2$ is ERROR. This means that the variable $x$ may not take on the value $-2$ in this equation. You can use the arrows on the keypad to scroll through the table.

With the same equation in the equation editor, set the table to ASK mode. In this mode you do not need to set the starting $x$-value or the table step, since you are entering any value you choose for $x$. You may enter any real value for $x$— integers, fractions, decimals, irrational numbers, and so forth. If you enter $x = 1 + \sqrt{3}$, the graphing utility may rewrite the number as a decimal approximation, as shown in Figure 4. You can continue to build your own table by entering additional $x$-values in order to generate $y$-values.

If you have several equations in the equation editor, the table may generate $y$-values for each equation.

Before beginning Chapter 1, be sure you are familiar with the use of the equation editor and the table feature. Remember that the screens and names of features and modes may not be the same as your graphing utility, so you should always consult your user's guide for help.

Figure 4

# How to Study Algebra

Studying algebra is a linear process: The material you learn each day builds upon material you learned previously. There are no shortcuts—you must keep up with the coursework every day.

## Preparing for Class

Always refer to your syllabus to make sure you stay on track in the course. In general, you should study two to four hours for every hour you spend in class. If you are prepared for class, you have a very good chance of succeeding. Before each class,

- ❐ Review your notes from the previous class.
- ❐ Review the portion of the text that will be covered in class.
- ❐ Review the *Objectives* to make sure you understand the main ideas of the section.
- ❐ Pay special attention to the definitions, theorems, and rules that are highlighted in the text.
- ❐ Study the *Examples* and their solutions.
- ❐ Make a note of anything you don't understand, and be sure to pay special attention when the instructor addresses that topic in class. If you still have questions, ask the instructor.
- ❐ Try the *Technology* and *Discovery* exercises.

## Keeping Up

- ❐ Always go to class. If you miss a class, get the notes from another student as soon as possible and review them carefully.
- ❐ Take notes during class. As soon as possible after class, review your notes and supplement them, if necessary, to make them clear to you.
- ❐ During class, take time to listen and understand how the instructor thinks and solves problems.
- ❐ Do the homework as soon as possible after class while the concepts are still fresh in your mind.
- ❐ To make sure you don't forget skills you practiced in previous homework assignments, solve the *Warm Ups* at the beginning of each exercise set and check your answers against the ones given in the back of the text.
- ❐ As you do your homework assignments, refer to your notes and the text section. Look for *Examples* that are similar to the *Exercises*. For example, Exercises 11–22 on page 32 are similar to Examples 2 and 3 on page 26.

STUDY PLAN

## Getting Extra Help

There are plenty of resources available to help you with your study of algebra.

❑ If you have trouble with an exercise, check the *Study Tips* and a similar solved *Example* in that text section.

❑ Consult one of the many ancillaries available for this text: *Study and Solutions Guide*, *Tutorial Software*, and *Graphing Technology Keystroke Guide*, as well as additional study resources that are available online at *www.hmco.com*. (Descriptions of each of these can be found on pages xvi–xvii.)

❑ Find a study partner or join a study group so you can work through problems with other algebra students.

❑ Work with a tutor for regular assistance.

## Preparing for an Exam

The best way to prepare for an exam is to keep up with your coursework—cramming seldom works! This text offers you many opportunities to practice your algebra skills.

❑ Take the *Mid-Chapter Quiz* and the *Chapter Test* as if you were in class, and then check your answers against those in the back of the text. Working the *Cumulative Tests* also helps ensure that you don't forget skills learned in earlier chapters.

❑ Review the *Chapter Summary*. If you feel that you need additional practice, work the *Review Exercises* that are keyed to the section *Objectives*.

❑ Rework homework problems, quizzes, and tests. Make sure you understand the methods for solving the types of problems, rather than memorizing the solutions.

❑ Take the exam with a clear mind and a positive attitude. Read the directions carefully and work at a reasonable pace. If you finish early, go over your work.

❑ When you get the exam back, review it carefully and make sure you understand the corrections to any errors that you made so that you can build on that knowledge as you go on to the next topic.

# 1

# Equations and Inequalities

Jed Jacobsohn/Allsport

Baseball began in the mid-1800s and spread throughout the United States during the Civil War. In 1876 the National League, consisting of eight professional teams, was formed. The American League began playing with eight teams in 1901.

Today there are 30 teams playing in the National and American Leagues. Attendance for the 1998 season topped 70.5 million.

Salaries for professional baseball players have steadily increased. The average salaries for players from 1985 to 1995 are listed in the table.

| Year | 1985 | 1986 | 1987 | 1988 | 1989 | 1990 | 1991 | 1992 | 1993 | 1994 | 1995 |
|------|------|------|------|------|------|------|------|------|------|------|------|
| Average salary (in thousands) | $371 | $413 | $412 | $439 | $497 | $598 | $851 | $1029 | $1076 | $1168 | $1111 |

The chapter project related to this information is on page 80.

## 1.1     Linear Equations

### Objectives

- Classify an equation as an identity or a conditional equation.
- Solve a linear equation in one variable.
- Use a linear model to solve an application problem.

## Equations and Solutions

An **equation** is a statement that two algebraic expressions are equal. Some examples of equations in $x$ are

$$3x - 5 = 7, \qquad x^2 - x - 6 = 0, \qquad \text{and} \qquad \sqrt{2x} = 4.$$

To **solve** an equation in $x$ means that you find all values of $x$ for which the equation is **true.** Such values are called **solutions.** For instance, $x = 4$ is a solution of the equation $3x - 5 = 7$, because $3(4) - 5 = 7$ is a true statement.

An equation that is true for *every* real number in the domain of the variable is an **identity.** Two examples of identities are

$$x^2 - 9 = (x + 3)(x - 3) \qquad \text{and} \qquad \frac{x}{3x^2} = \frac{1}{3x}, \qquad x \neq 0.$$

The first equation is an identity because it is a true statement for all real values of $x$. The second is an identity because it is true for all nonzero real values of $x$.

An equation that is true for just *some* (or even none) of the real numbers in the domain of the variable is called a **conditional equation.** For example, the equation $x^2 - 9 = 0$ is conditional because $x = 3$ and $x = -3$ are the only values in the domain that satisfy the equation.

### Example 1     Classifying Equations

Classify each of the following as an identity or a conditional equation.

**a.** $2(x + 3) = 2x + 6$     **b.** $2(x + 3) = x + 6$     **c.** $2(x + 3) = 2x + 3$

#### Solution

**a.** This equation is an identity because it is true for every real value of $x$.

**b.** This equation is a conditional equation because there are real values of $x$ (such as $x = 1$) for which the equation is not true.

**c.** This equation is a conditional equation because there are no real number values of $x$ for which the equation is true.

---

Be sure you understand that equations are used in algebra for two distinct purposes: (1) *identities* are usually used to state mathematical properties and (2) *conditional equations* are usually used to model and solve problems that occur in real life.

# Linear Equations in One Variable

The most common type of conditional equation is a **linear equation.**

---

▶ Definition of a Linear Equation

A **linear equation** in one variable $x$ is an equation that can be written in the standard form

$$ax + b = 0$$

where $a$ and $b$ are real numbers with $a \neq 0$.

---

A linear equation has exactly one solution. To see this, consider the following steps. (Remember that $a \neq 0$.)

| | |
|---|---|
| $ax + b = 0$ | Original equation |
| $ax = -b$ | Subtract $b$ from both sides. |
| $x = -\dfrac{b}{a}$ | Divide both sides by $a$. |

So, the equation $ax + b = 0$ has exactly one solution, $x = -b/a$.

To solve a linear equation in $x$, you should isolate $x$ by a sequence of **equivalent** (and usually simpler) equations, each having the same solution as the original equation. The operations that yield equivalent equations come from the basic rules of algebra reviewed in the appendix.

---

▶ Forming Equivalent Equations

A given equation can be transformed into an equivalent equation by one or more of the following steps.

| | *Given Equation* | *Equivalent Equation* |
|---|---|---|
| **1.** Remove symbols of grouping, combine like terms, or simplify one or both sides of the equation. | $2x - x = 4$ $3(x - 2) = 5$ | $x = 4$ $3x - 6 = 5$ |
| **2.** Add (or subtract) the same quantity to (from) *both* sides of the equation. | $x + 1 = 6$ | $x = 5$ |
| **3.** Multiply (or divide) *both* sides of the equation by the same *nonzero* quantity. | $2x = 6$ | $x = 3$ |
| **4.** Interchange sides of the equation. | $2 = x$ | $x = 2$ |

---

---

| **Example 2** | Solving a Linear Equation |
|---|---|

Solve $3x - 6 = 0$.

**Solution**

$$3x - 6 = 0 \qquad \text{Original equation}$$

$$3x = 6 \qquad \text{Add 6 to both sides.}$$

$$x = 2 \qquad \text{Divide both sides by 3.}$$

---

After solving an equation, you should **check each solution** in the *original* equation. For instance, in Example 2, you can check that 2 is a solution by substituting 2 for $x$ in the original equation $3x - 6 = 0$, as follows.

**Check**

$$3x - 6 = 0 \qquad \text{Original equation}$$

$$3(2) - 6 \overset{?}{=} 0 \qquad \text{Substitute 2 for } x.$$

$$6 - 6 = 0 \qquad \text{Solution checks. } \checkmark$$

## Study Tip

Students sometimes tell us that a solution looks easy when we work it in class, but that they don't see where to begin when trying it alone. Keep in mind that no one—not even great mathematicians—can expect to look at every mathematical problem and immediately know where to begin. Many problems involve some trial and error before a solution is found. To make algebra work for you, you must put in a lot of time, you must expect to try solution methods that end up not working, *and* you must learn from both your successes and your failures.

| **Example 3** | Solving a Linear Equation |
|---|---|

$$6(x - 1) + 4 = 3(7x + 1) \qquad \text{Original equation}$$

$$6x - 6 + 4 = 21x + 3 \qquad \text{Distributive Property}$$

$$6x - 2 = 21x + 3 \qquad \text{Simplify.}$$

$$-15x = 5 \qquad \text{Add 2 and subtract } 21x.$$

$$x = -\tfrac{1}{3} \qquad \text{Divide both sides by } -15.$$

The solution is $-\tfrac{1}{3}$. You can check this as follows.

**Check**

$$6(x - 1) + 4 = 3(7x + 1) \qquad \text{Original equation}$$

$$6\left(-\tfrac{1}{3} - 1\right) + 4 \overset{?}{=} 3\left[7\left(-\tfrac{1}{3}\right) + 1\right] \qquad \text{Substitute } -\tfrac{1}{3} \text{ for } x.$$

$$6\left(-\tfrac{4}{3}\right) + 4 \overset{?}{=} 3\left(-\tfrac{7}{3} + 1\right) \qquad \text{Add fractions.}$$

$$-\tfrac{24}{3} + 4 \overset{?}{=} -\tfrac{21}{3} + 3 \qquad \text{Multiply.}$$

$$-8 + 4 \overset{?}{=} -7 + 3 \qquad \text{Simplify.}$$

$$-4 = -4 \qquad \text{Solution checks. } \checkmark$$

---

# Equations Involving Fractional Expressions

To solve an equation involving fractional expressions, you can multiply every term in the equation by the least common denominator (LCD) of the terms.

**Example 4**    An Equation Involving Fractional Expressions

$$\frac{x}{3} + \frac{3x}{4} = 2 \qquad \text{Original equation}$$

$$(12)\frac{x}{3} + (12)\frac{3x}{4} = (12)2 \qquad \text{Multiply by least common denominator.}$$

$$4x + 9x = 24 \qquad \text{Reduce and multiply.}$$

$$13x = 24 \qquad \text{Combine like terms.}$$

$$x = \frac{24}{13} \qquad \text{Divide both sides by 13.}$$

The solution is $\frac{24}{13}$. Check this in the original equation.

When an equation is multiplied or divided by a *variable* expression, it is possible to introduce an **extraneous** solution—one that does not satisfy the original equation. In such cases a check is especially important.

**Example 5**    An Equation with an Extraneous Solution

Solve $\dfrac{1}{x - 2} = \dfrac{3}{x + 2} - \dfrac{6x}{x^2 - 4}$.

**Solution**

The least common denominator is $x^2 - 4 = (x + 2)(x - 2)$. Multiplying each term by this LCD and reducing produces the following.

$$\frac{1}{x - 2} = \frac{3}{x + 2} - \frac{6x}{x^2 - 4}$$

$$\frac{1}{x - 2}(x + 2)(x - 2) = \frac{3}{x + 2}(x + 2)(x - 2) - \frac{6x}{x^2 - 4}(x + 2)(x - 2)$$

$$x + 2 = 3(x - 2) - 6x, \quad x \neq \pm 2$$

$$x + 2 = 3x - 6 - 6x$$

$$4x = -8$$

$$x = -2$$

After checking $x = -2$, you can see that it yields a denominator of zero. So, $x = -2$ is extraneous, and the equation has *no solution*.

An equation with a *single fraction* on each side can be cleared of denominators by **cross-multiplying,** which is equivalent to multiplying by the least common denominator and then reducing.

Solve $\dfrac{3y - 2}{2y + 1} = \dfrac{6y - 9}{4y + 3}$.

**Solution**

$$\dfrac{3y - 2}{2y + 1} = \dfrac{6y - 9}{4y + 3} \qquad \text{Original equation}$$

$$(3y - 2)(4y + 3) = (6y - 9)(2y + 1) \qquad \text{Cross-multiply.}$$

$$12y^2 + y - 6 = 12y^2 - 12y - 9 \qquad \text{Multiply.}$$

$$13y = -3 \qquad \text{Isolate } y\text{-term on left.}$$

$$y = -\dfrac{3}{13} \qquad \text{Divide both sides by 13.}$$

The solution is $-\frac{3}{13}$. Check this in the original equation.

**TECHNOLOGY**

When using the equation editor on your graphing utility, you may enter

$$y_1 = (3x - 2)/(2x + 1)$$

but not

$$y_1 = (3y - 2)/(2y + 1).$$

To check the solution of an equation not written in terms of $x$, let $x$ "stand in" for the variable. In Example 6, you would enter

$$y_1 = (3x - 2)/(2x + 1)$$

and

$$y_2 = (6x - 9)/(4x + 3).$$

Use the table feature in ASK mode to verify the solution.

**Example 7**    Using a Calculator to Solve an Equation

Solve $\dfrac{1}{9.38} - \dfrac{3}{x} = \dfrac{5}{0.3714}$.

**Solution**

Roundoff error will be minimized if you solve for $x$ before performing any calculations. The least common denominator is $(9.38)(0.3714)(x)$.

$$\dfrac{1}{9.38} - \dfrac{3}{x} = \dfrac{5}{0.3714}$$

$$(9.38)(0.3714)(x)\left(\dfrac{1}{9.38} - \dfrac{3}{x}\right) = (9.38)(0.3714)(x)\left(\dfrac{5}{0.3714}\right)$$

$$0.3714x - 3(9.38)(0.3714) = (9.38)(5)(x), \quad x \neq 0$$

$$[0.3714 - 5(9.38)]x = 3(9.38)(0.3714)$$

$$x = \dfrac{3(9.38)(0.3714)}{0.3714 - 5(9.38)}$$

$$x \approx -0.225 \qquad \text{Round to three places.}$$

Because of roundoff error, a check of a decimal solution may not yield exactly the same values for both sides of the original equation. The difference, however, should be quite small.

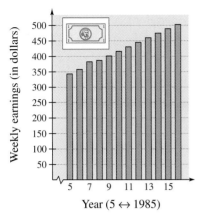

Figure 1.1

# Application

**Example 8**    Weekly Earnings

The average weekly earnings $y$ (in dollars) for workers in the United States from 1985 to 1996 can be modeled by the linear equation $y = 13.4t + 278$ for $t \geq 5$ and where $t = 5$ represents 1985, as shown in Figure 1.1. Use the model to estimate when the average weekly earnings will reach \$600. (Source: U.S. Bureau of Labor Statistics)

**Solution**

$$y = 13.4t + 278 \qquad \text{Given model}$$

$$600 = 13.4t + 278 \qquad \text{Substitute 600 for } y \text{ and solve for } t.$$

$$322 = 13.4t \qquad \text{Subtract 278 from both sides.}$$

$$t = \frac{322}{13.4} \approx 24 \qquad \text{Divide both sides by 13.4.}$$

Because $t = 5$ corresponds to 1985, it follows that $t = 24$ corresponds to 2004. So, you can expect the average weekly earnings to reach \$600 by 2004.

---

| Discussing the Concept | Checking a Solution |
|---|---|

The solution $x = 2$ does not check in either of the following original equations. Why? Is there an error? When a solution does not check, what are the possibilities?

**a.**
$$\frac{2}{x - 1} = \frac{1}{x - 2} - \frac{1}{(x - 2)(x - 1)} \qquad \text{Original equation}$$

$$\frac{2}{x - 1}(x - 2)(x - 1) = \frac{1}{x - 2}(x - 2)(x - 1) -$$
$$\frac{1}{(x - 2)(x - 1)}(x - 2)(x - 1)$$

$$2(x - 2) = (x - 1) - 1$$

$$2x - 4 = x - 2$$

$$x = 2$$

**b.**
$$\frac{2}{x - 1} = \frac{1}{x - 6} - \frac{3}{(x - 1)(x - 6)} \qquad \text{Original equation}$$

$$\frac{2}{x - 1}(x - 1)(x - 6) = \frac{1}{x - 6}(x - 1)(x - 6) -$$
$$\frac{3}{(x - 1)(x - 6)}(x - 1)(x - 6)$$

$$2(x - 6) = (x - 1) - 3$$

$$2x - 6 = x - 4$$

$$x = 2$$

## Warm Up

The following warm-up exercises involve skills that were covered in previous courses. You will use these skills in the exercise set for this section.

In Exercises 1–10, perform the indicated operations and simplify your answer.

**1.** $(2x - 4) - (5x + 6)$

**2.** $(3x - 5) + (2x - 7)$

**3.** $2(x + 1) - (x + 2)$

**4.** $-3(2x - 4) + 7(x + 2)$

**5.** $\dfrac{x}{3} + \dfrac{x}{5}$

**6.** $x - \dfrac{x}{4}$

**7.** $\dfrac{1}{x + 1} - \dfrac{1}{x}$

**8.** $\dfrac{2}{x} + \dfrac{3}{x}$

**9.** $\dfrac{4}{x} + \dfrac{3}{x - 2}$

**10.** $\dfrac{1}{x + 1} - \dfrac{1}{x - 1}$

## 1.1    Exercises

In Exercises 1–6, determine whether the equation is an identity or a conditional equation.

**1.** $2(x - 1) = 2x - 2$

**2.** $3(x + 2) = 3x + 6$

**3.** $2(x - 1) = 3x + 4$

**4.** $3(x + 2) = 2x + 4$

**5.** $2(x + 1) = 2x + 1$

**6.** $3(x + 4) = 3x + 4$

In Exercises 7–14, determine whether the values of $x$ are solutions of the equation.

**7.** $5x - 3 = 3x + 5$

    (a) $x = 0$   (b) $x = -5$   (c) $x = 4$   (d) $x = 10$

**8.** $7 - 3x = 5x - 17$

    (a) $x = -3$   (b) $x = 0$   (c) $x = 8$   (d) $x = 3$

**9.** $3x^2 + 2x - 5 = 2x^2 - 2$

    (a) $x = -3$   (b) $x = 1$   (c) $x = 4$   (d) $x = -5$

**10.** $5x^3 + 2x - 3 = 4x^3 + 2x - 11$

    (a) $x = 2$   (b) $x = -2$   (c) $x = 0$   (d) $x = 10$

**11.** $\dfrac{5}{2x} - \dfrac{4}{x} = 3$

    (a) $x = -\frac{1}{2}$   (b) $x = 4$   (c) $x = 0$   (d) $x = \frac{1}{4}$

**12.** $3 + \dfrac{1}{x + 2} = 4$

    (a) $x = -1$   (b) $x = -2$   (c) $x = 0$   (d) $x = 5$

**13.** $(x + 5)(x - 3) = 20$

    (a) $x = 3$   (b) $x = -2$   (c) $x = 0$   (d) $x = -7$

**14.** $\sqrt[3]{x - 8} = 3$

    (a) $x = 2$   (b) $x = -5$   (c) $x = 35$   (d) $x = 8$

In Exercises 15–52, solve the equation and check your answer. (Some equations have no solution.)

**15.** $x + 10 = 15$

**16.** $7 - x = 18$

**17.** $7 - 2x = 15$

**18.** $7x + 2 = 16$

**19.** $8x - 5 = 3x + 10$

**20.** $7x + 3 = 3x - 13$

**21.** $2(x + 5) - 7 = 3(x - 2)$

**22.** $2(13t - 15) + 3(t - 19) = 0$

**23.** $6[x - (2x + 3)] = 8 - 5x$

**24.** $8(x + 2) - 3(2x + 1) = 2(x + 5)$

**25.** $\dfrac{5x}{4} + \dfrac{1}{2} = x - \dfrac{1}{2}$

**26.** $\dfrac{x}{5} - \dfrac{x}{2} = 3$

**27.** $\frac{3}{2}(z + 5) - \frac{1}{4}(z + 24) = 0$

**28.** $\dfrac{3x}{2} + \dfrac{1}{4}(x - 2) = 10$

**29.** $0.25x + 0.75(10 - x) = 3$

**30.** $0.60x + 0.40(100 - x) = 50$

**31.** $x + 8 = 2(x - 2) - x$

**32.** $3(x + 3) = 5(1 - x) - 1$

**33.** $\dfrac{100 - 4u}{3} = \dfrac{5u + 6}{4} + 6$

**34.** $\dfrac{17 + y}{y} + \dfrac{32 + y}{y} = 100$

**35.** $\dfrac{5x - 4}{5x + 4} = \dfrac{2}{3}$

**36.** $\dfrac{10x + 3}{5x + 6} = \dfrac{1}{2}$

**37.** $10 - \dfrac{13}{x} = 4 + \dfrac{5}{x}$

**38.** $\dfrac{15}{x} - 4 = \dfrac{6}{x} + 3$

**39.** $\dfrac{1}{x - 3} + \dfrac{1}{x + 3} = \dfrac{10}{x^2 - 9}$

**40.** $\dfrac{1}{x - 2} + \dfrac{3}{x + 3} = \dfrac{4}{x^2 + x - 6}$

**41.** $\dfrac{x}{x + 4} + \dfrac{4}{x + 4} + 2 = 0$

**42.** $\dfrac{2}{(x - 4)(x - 2)} = \dfrac{1}{x - 4} + \dfrac{2}{x - 2}$

**43.** $\dfrac{7}{2x + 1} - \dfrac{8x}{2x - 1} = -4$

**44.** $\dfrac{4}{u - 1} + \dfrac{6}{3u + 1} = \dfrac{15}{3u + 1}$

**45.** $\dfrac{3}{x(x - 3)} + \dfrac{4}{x} = \dfrac{1}{x - 3}$

**46.** $3 = 2 + \dfrac{2}{z + 2}$

**47.** $(x + 2)^2 + 5 = (x + 3)^2$

**48.** $(x + 1)^2 + 2(x - 2) = (x + 1)(x - 2)$

**49.** $(x + 2)^2 - x^2 = 4(x + 1)$

**50.** $4(x + 1) - 3x = x + 5$

**51.** $(2x + 1)^2 = 4(x^2 + x + 1)$

**52.** $(2x - 1)^2 = 4(x^2 - x + 6)$

**53.** Explain why a solution of an equation involving fractional expressions may be extraneous.

**54.** What method or methods would you suggest for checking a solution of an equation involving fractional expressions?

In Exercises 55–60, use a calculator to solve the equation. (Round your answer to three decimal places.)

**55.** $0.275x + 0.725(500 - x) = 300$

**56.** $2.763 - 4.5(2.1x - 5.1432) = 6.32x + 5$

**57.** $\dfrac{x}{0.6321} + \dfrac{x}{0.0692} = 1000$

**58.** $\dfrac{2}{7.398} - \dfrac{4.405}{x} = \dfrac{1}{x}$

**59.** $(x + 5.62)^2 + 10.83 = (x + 7)^2$

**60.** $\dfrac{x}{2.625} + \dfrac{x}{4.875} = 1$

**61.** What method or methods would you recommend for checking the solutions to Exercises 55–60 using your graphing utility?

**62.** In Exercises 55–60, your answers are rounded to three decimal places. What effect does rounding have as you check a solution?

In Exercises 63-68, evaluate each expression in two ways. (a) Calculate entirely on your calculator using appropriate parentheses, and then round the answer to two decimal places. (b) Round both the numerator and the denominator to two decimal places before dividing, and then round the final answer to two decimal places. Does the second method introduce an additional roundoff error?

**63.** $\dfrac{1 + 0.73205}{1 - 0.73205}$

**64.** $\dfrac{1 + 0.86603}{1 - 0.86603}$

**65.** $\dfrac{2 - 1.63254}{(2.58)(0.135)}$

**66.** $\dfrac{2 + 0.57735}{1 + 0.57735}$

**67.** $\dfrac{333 + \dfrac{1.98}{0.74}}{4 + \dfrac{6.25}{3.15}}$

**68.** $\dfrac{1.73205 - 1.19195}{3 - (1.73205)(1.19195)}$

**69.** *Legal Services Advertising*  The amount of money spent on advertising for legal services on television from 1990 to 1996 can be approximated by the linear equation

$$y = 10.8t + 92.7$$

where $y$ is the amount in millions of dollars per year and $t$ is the calendar year, with $t = 0$ corresponding to 1990. If this linear pattern continues, when will the amount of advertising for legal services on television reach \$200 million per year? (Source: Television Bureau of Advertising)

**70.** *On-Line Households*   The number of on-line households in the United States from 1995 to 1998 can be approximated by the linear equation

$$y = 6.45t - 23.15, \qquad t \geq 5$$

where $y$ is the number of households in millions, with $t = 5$ corresponding to 1995. If this linear pattern continues, in what year will there be 80 million households on-line in the United States? (Source: *The Wall Street Journal Almanac*)

*Human Height*   In Exercises 71 and 72, use the following information. The relationship between the length of an adult's thigh bone and the height of the adult can be approximated by the linear equations

$$y = 0.432x - 10.44 \qquad \text{Female}$$

$$y = 0.449x - 12.15 \qquad \text{Male}$$

where $y$ is the length of the thigh bone (femur) in inches and $x$ is the height of the adult in inches (see figure).

**71.** An anthropologist discovers a thigh bone belonging to an adult human female. The bone is 16 inches long. How tall would you estimate the female to have been?

**72.** From the foot bone of an adult human male, an anthropologist estimates that the male was 69 inches tall. A few feet away from the site where the foot bone was discovered, the anthropologist discovers an adult male thigh bone that is 19 inches long. Is it possible that both bones came from the same person?

*Social Security Benefits*   In Exercises 73 and 74, use the following information. From 1983 to 1995, the average monthly benefit for retired workers can be approximated by

$$y = 355.3 + 24.3x, \qquad x \geq 3$$

where $y$ is the average monthly benefit in dollars and $x$ is the year, with $x = 3$ corresponding to 1983. (Source: U.S. Social Security Administration)

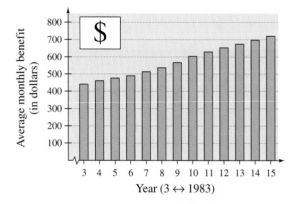

**73.** In which year was the average monthly benefit about $600?

**74.** If the model continues to be accurate, in what year will the average monthly benefit be $1060?

*Probability*   In Exercises 75 and 76, use the equation $p + p' = 1$, which states that the sum of the probability that an event *will* occur $p$ and the probability that it *will not* occur $p'$ is 1.

**75.** If the probability that an event will occur is 0.54, what is the probability that the event will not occur?

**76.** If the probability that an event will not occur is 0.87, what is the probability that the event will occur?

## 1.2    Mathematical Modeling

### Objectives

- Construct a mathematical model from a verbal model.
- Model and solve percent and mixture problems.
- Use common formulas to solve geometry and simple interest problems.
- Develop a general problem-solving strategy.

## Introduction to Problem Solving

In this section, you will study ways of using algebra to solve real-life problems. To do this, you will construct one or more equations that represent the real-life problem. This procedure is called **mathematical modeling.**

A good approach to mathematical modeling is to use two stages. In the first stage, the verbal description of the problem is used to form a *verbal model.* Then, after assigning labels to each of the quantities in the verbal model, you form a *mathematical model* or *algebraic equation.*

| Verbal description | ⇒ | Verbal model | ⇒ | Algebraic equation |
|---|---|---|---|---|

When you are trying to construct a verbal model, it is helpful to look for a *hidden equality*—a statement that two algebraic expressions are equal. For instance, in the following example the hidden equality equates your annual income to 24 paychecks and one bonus check.

### Example 1    Using a Verbal Model

You have accepted a job for which your annual salary will be $24,740. This salary includes a year-end bonus of $500. If you are paid twice a month, what will your gross pay be for each paycheck?

**Solution**

Because there are 12 months in a year and you will be paid twice a month, it follows that you will receive 24 paychecks during the year.

*Verbal Model:*    Income for year  =  24 paychecks  +  Bonus

| *Labels:* | Income for year = 24,740 | (dollars) |
|---|---|---|
| | Amount of each paycheck = $x$ | (dollars) |
| | Bonus = 500 | (dollars) |

*Equation:*    $24{,}740 = 24x + 500$

Using the techniques discussed in Section 1.1, you will find that the solution is $x = \$1010$. Check that your answer is reasonable. Is $1010 a reasonable answer if you are paid twice a month and earn $24,740 a year?

► Translating Key Words and Phrases

| Key Words and Phrases | Verbal Description | Algebraic Statement |
|---|---|---|
| **Consecutive** | | |
| Next, subsequent | Consecutive integers | $n, n + 1$ |
| **Addition** | | |
| Sum, plus, greater, increased by, more than, exceeds, total of | The sum of 5 and $x$ <br> Seven more than $y$ | $5 + x$ <br> $y + 7$ |
| **Subtraction** | | |
| Difference, minus, less than, decreased by, subtracted from, reduced by, the remainder | Four decreased by $b$ <br> Three less than $z$ | $4 - b$ <br> $z - 3$ |
| **Multiplication** | | |
| Product, multiplied by, twice, times, percent of | Two times $x$ | $2x$ |
| **Division** | | |
| Quotient, divided by, per | The quotient of $x$ and 8 | $\dfrac{x}{8}$ |

**Example 2** Constructing Mathematical Models

**a.** A salary of \$22,400 is increased by 9%. What is the new salary?

*Verbal Model:* New salary = 9%(salary) + Salary

*Labels:*
Salary = 22,400 (dollars)
New salary = $S$ (dollars)
Percent = 0.09 (percent in decimal form)

*Equation:* $S = 0.09(22{,}400) + 22{,}400$

**b.** All items of clothing in a store are reduced by 20%. Find the original price of a suit selling for \$140.

*Verbal Model:* Original price − 20%(original price) = Sale price

*Labels:*
Original price = $p$ (dollars)
Sale price = 140 (dollars)
Percent = 0.2 (percent in decimal form)

*Equation:* $p - 0.2p = 140$

**Study Tip**

In Example 2, notice that part of the labeling process is to list the unit of measure for each labeled quantity. If you develop this habit, you will find that it is a big help in checking the validity of your verbal model.

## Using Mathematical Models

Study the next several examples carefully. Your goal should be to develop a *general problem-solving strategy.*

| Example 3 | Finding the Percent of a Raise |
|---|---|

You accept a job that pays $8 an hour. You are told that after a 2-month probationary period, your hourly wage will be increased to $9 an hour. What percent raise will you receive after the 2-month period?

**Solution**

*Verbal Model:*    Raise = Percent · Old wage

*Labels:*    Old wage = 8                                                    (dollars)
Raise = 1                                                       (dollar)
Percent = $r$                        (percent in decimal form)

*Equation:*    $1 = r \cdot 8$

By solving this equation, you will find that you will receive a raise of $\frac{1}{8} = 0.125$ or 12.5%.

| Example 4 | Finding the Percent of a Benefit Package |
|---|---|

Your annual salary is $22,000. In addition to your salary, your employer also provides the following benefits. The total of this benefit package represents what percent of your annual salary?

| Social security (employer's portion): | 6.20% of salary | $1364 |
| Workmen's compensation: | 0.5% of salary | $110 |
| Unemployment compensation: | 0.75% of salary | $165 |
| Medical insurance: | $2240 per year | $2240 |
| Retirement contribution: | 5% of salary | $1100 |

**Solution**

*Verbal Model:*    Benefit package = Percent · Salary

*Labels:*    Salary = 22,000                                                (dollars)
Benefit package = 4979                                    (dollars)
Percent = $r$                        (percent in decimal form)

*Equation:*    $4979 = r \cdot 22{,}000$

By solving this equation, you will find that your benefit package is $r = 4979/22{,}000$ or about 22.6% of your salary.

*In 1995, 15.4% of the population of the United States had no health insurance.* (Source: U.S. Bureau of the Census)

### Example 5    Finding the Measurements of a Room

A rectangular family room is twice as long as it is wide, and its perimeter is 84 feet. Find the measurements of the family room.

#### Solution

For this problem, it helps to sketch a picture (see Figure 1.2).

*Verbal Model:*    $2 \cdot$ **Length** $+ 2 \cdot$ **Width** $=$ **Perimeter**

*Labels:*    Perimeter $= 84$          (feet)
Width $= w$              (feet)
Length $= l = 2w$       (feet)

*Equation:*    $2(2w) + 2w = 84$

$$4w + 2w = 84$$

$$6w = 84$$

$$w = 14 \text{ feet}$$

$$l = 2w = 28 \text{ feet}$$

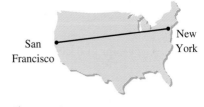

**Figure 1.2**

The measurements of the room are 14 feet by 28 feet.

### Example 6    A Distance Problem

A plane is flying nonstop from New York to San Francisco, a distance of about 2700 miles, as shown in Figure 1.3. After 1.5 hours in the air, the plane flies over Chicago (a distance of 800 miles from New York). How long will it take the plane to fly from New York to San Francisco? (Assume that the plane flies at a constant speed during the entire flight.)

#### Solution

To solve this problem, use the formula that relates distance, rate, and time. That is, (distance) = (rate)(time). Because it took the plane 1.5 hours to travel a distance of 800 miles, you can conclude that its rate (or speed) must have been

$$\text{Rate} = \frac{\text{distance}}{\text{time}} = \frac{800 \text{ miles}}{1.5 \text{ hours}} \approx 533.33 \text{ miles per hour.}$$

Because the entire trip is 2700 miles, the time for the entire trip is

$$\text{Time} = \frac{\text{distance}}{\text{rate}} = \frac{2700 \text{ miles}}{533.33 \text{ miles per hour}} \approx 5.06 \text{ hours.}$$

Because 0.06 hours represents about 4 minutes, you can conclude that the trip will take about 5 hours and 4 minutes.

**Figure 1.3**

Another way to solve the distance problem in Example 6 is to use the concept of **ratio and proportion.** To do this, let $x$ represent the time required to fly from New York to San Francisco, and set up the following proportion.

$$\frac{\text{Time to San Francisco}}{\text{Time to Chicago}} = \frac{\text{Distance to San Francisco}}{\text{Distance to Chicago}}$$

$$\frac{x}{1.5} = \frac{2700}{800}$$

$$x = 1.5 \cdot \frac{2700}{800}$$

$$x \approx 5.06$$

Notice how a ratio and proportion are used with a result from geometry to solve the problem in the following example.

> **Example 7**    An Application Involving Similar Triangles

To measure the height of the twin towers of the World Trade Center (in New York City), you measure the shadow cast by one of the towers to be 170.25 feet long, as shown in Figure 1.4. Then you measure the shadow cast by a 4-foot post and find that its shadow is 6 inches long. How can this information be used to determine the height of the tower?

### Solution

To find the height of the tower, you can use a result from geometry that states that the ratios of corresponding sides of similar triangles are equal.

*Verbal Model:*    $\dfrac{\text{Height of building}}{\text{Length of shadow}} = \dfrac{\text{Height of post}}{\text{Length of shadow}}$

*Labels:*    
Height of building $= x$    (feet)
Length of building's shadow $= 170.25$    (feet)
Height of post $= 48$    (inches)
Length of post's shadow $= 6$    (inches)

*Equation:*    $\dfrac{x}{170.25} = \dfrac{48}{6}$

$$x = 8 \cdot 170.25$$

$$x = 1362 \text{ feet}$$

The tower is about 1362 feet high. (This is the building from which King Kong fell in the remake of the movie *King Kong.* In the original movie, King Kong fell from the Empire State Building, which has a height of 1250 feet.)

$x$ ft

48 in.

6 in.

170.25 ft

Figure 1.4

## Mixture Problems

The next example is called a **mixture problem** because it involves two different unknown quantities that are *mixed* in a specific way. Watch for a *hidden product* in the verbal model.

| Example 8 | A Simple Interest Problem  |

You have received an inheritance of $10,000 that has been invested in two ways. Part of the money was invested at $9\frac{1}{2}\%$ simple interest and the remainder was invested at 11%. After 1 year, the two investments have returned a combined interest of $1038.50. How much was invested in each type of account?

### Solution

Simple interest problems are based on the formula $I = Prt$, where $I$ is the interest, $P$ is the principal, $r$ is the annual percentage rate (in decimal form), and $t$ is the time in years.

*Verbal Model:*  | Interest from $9\frac{1}{2}\%$ | $+$ | Interest from 11% | $=$ | Total interest |

*Labels:*

| | |
|---|---|
| Amount invested at $9\frac{1}{2}\%$ $= x$ | (dollars) |
| Amount invested at 11% $= 10,000 - x$ | (dollars) |
| Interest from $9\frac{1}{2}\%$ $= Prt = (x)(0.095)(1)$ | (dollars) |
| Interest from 11% $= Prt = (10,000 - x)(0.11)(1)$ | (dollars) |
| Total interest $= 1038.50$ | (dollars) |

*Equation:*

$$0.095x + 0.11(10,000 - x) = 1038.5$$
$$0.095x + 1100 - 0.11x = 1038.5$$
$$-0.015x = -61.5$$
$$x = \$4100 \text{ at } 9\frac{1}{2}\%$$
$$10,000 - x = \$5900 \text{ at } 11\%$$

The amount invested at $9\frac{1}{2}\%$ is $4100 and the amount invested at 11% is $5900. Check these results in the original statement of the problem.

In Example 8, did you recognize the hidden products in the two terms on the left side of the equation? Both hidden products come from the common formula

| Interest | $=$ | Principal | $\cdot$ | Rate | $\cdot$ | Time |

$$I = Prt.$$

## Common Formulas

Many common types of geometric, scientific, and investment problems use ready-made equations, called **formulas.** Knowing formulas such as those in the following lists will help you translate and solve a wide variety of real-life problems involving perimeter, area, volume, temperature, interest, and principal.

▶ **Common Formulas for Area, Perimeter, and Volume**

| *Square* | *Rectangle* | *Circle* | *Triangle* |
|---|---|---|---|
| $A = s^2$ | $A = lw$ | $A = \pi r^2$ | $A = \frac{1}{2}bh$ |
| $P = 4s$ | $P = 2l + 2w$ | $C = 2\pi r$ | $P = a + b + c$ |

| *Cube* | *Rectangular Solid* | *Circular Cylinder* | *Sphere* |
|---|---|---|---|
| $V = s^3$ | $V = lwh$ | $V = \pi r^2 h$ | $V = \frac{4}{3}\pi r^3$ |

▶ **Miscellaneous Common Formulas**

*Temperature:*    $F$ = degrees Fahrenheit, $C$ = degrees Celsius

$$F = \frac{9}{5}C + 32$$

*Simple interest:*  $I$ = interest, $P$ = principal, $r$ = interest rate, $t$ = time

$$I = Prt$$

*Distance:*    $d$ = distance traveled, $r$ = rate, $t$ = time

$$d = rt$$

When working with applied problems, you often need to rewrite one of the common formulas. For instance, the formula $P = 2l + 2w$ for the perimeter of a rectangle can be rewritten or solved for $w$ to produce $w = \frac{1}{2}(P - 2l)$.

**Example 9**    Using a Formula

A cylindrical can has a volume of 300 cubic centimeters and a radius of 3 centimeters, as shown in Figure 1.5. Find the height of the can.

Solution

The formula for the *volume of a cylinder* is $V = \pi r^2 h$. To find the height of the can, solve for $h$. Then, using $V = 300$ centimeters$^3$ and $r = 3$ centimeters, find the height.

$$h = \frac{V}{\pi r^2} = \frac{300 \text{ centimeters}^3}{\pi(3 \text{ centimeters})^2} = \frac{300 \text{ centimeters}^3}{9\pi \text{ centimeters}^2} \approx 10.61 \text{ centimeters}$$

3 cm

$h$

Figure 1.5

▶ **Strategy for Solving Word Problems**

1. *Search* for the hidden equality—two expressions said to be equal or known to be equal. A sketch may be helpful.

2. *Write* a verbal model that equates these two expressions.

3. *Assign* numbers to the known quantities and letters (or algebraic expressions) to the variable quantities.

4. *Rewrite* the verbal model as an algebraic equation using the assigned labels.

5. *Solve* the resulting algebraic equation.

6. *Check* to see that the answer satisfies the word problem as stated. (Remember that "solving for $x$" or some other variable may not completely answer the question.)

Discussing the Concept    **Red Herrings**

Applied problems in textbooks usually give precisely the right amount of information that is necessary to solve a given problem. In real life, however, you must often sort through the given information and discard information that is irrelevant to the problem. Such irrelevant information is called a **red herring.** Find the red herrings in the following problem.

From 100 to 200 feet beneath the surface of the ocean, pressure changes at a rate of approximately 4.4 pounds per square inch for every 10-foot change in depth. A diver takes 30 minutes to ascend 25 feet from a depth of 150 feet. What change in pressure does the diver experience?

# Warm Up

The following warm-up exercises involve skills that were covered in previous courses and earlier sections. You will use these skills in the exercise set for this section.

In Exercises 1–10, solve the equation (if possible) and check your answer.

**1.** $3x - 42 = 0$

**2.** $64 - 16x = 0$

**3.** $2 - 3x = 14 + x$

**4.** $7 + 5x = 7x - 1$

**5.** $5[1 + 2(x + 3)] = 6 - 3(x - 1)$

**6.** $2 - 5(x - 1) = 2[x + 10(x - 1)]$

**7.** $\dfrac{x}{3} + \dfrac{x}{2} = \dfrac{1}{3}$

**8.** $\dfrac{2}{x} + \dfrac{2}{5} = 1$

**9.** $1 - \dfrac{2}{z} = \dfrac{z}{z + 3}$

**10.** $\dfrac{x}{x + 1} - \dfrac{1}{2} = \dfrac{4}{3}$

## 1.2    Exercises

*Creating a Mathematical Model*   In Exercises 1–10, write an algebraic expression for the verbal expression.

**1.** The sum of two consecutive natural numbers

**2.** The product of two natural numbers whose sum is 25

**3.** *Distance Traveled*   The distance traveled in $t$ hours by a car traveling at 50 miles per hour

**4.** *Travel Time*   The travel time for a plane that is traveling at a rate of $r$ miles per hour for 200 miles

**5.** *Acid Solution*   The amount of acid in $x$ gallons of a 20% acid solution

**6.** *Discount*   The sale price for an item that is discounted by 20% of its list price $L$

**7.** *Geometry*   The perimeter of a rectangle whose width is $x$ and whose length is twice the width

**8.** *Geometry*   The area of a triangle whose base is 20 inches and whose height is $h$ inches

**9.** *Total Cost*   The total cost of producing $x$ units for which the fixed costs are $1200 and the cost per unit is $25

**10.** *Total Revenue*   The total revenue obtained by selling $x$ units at $3.59 per unit

*Using a Mathematical Model*   In Exercises 11–16, write a mathematical model for the number problem, and solve the problem.

**11.** The sum of two consecutive natural numbers is 525. Find the two numbers.

**12.** Find three consecutive natural numbers whose sum is 804.

**13.** One positive number is five times another positive number. The difference between the two numbers is 148. Find the numbers.

**14.** One number is one-fifth of another number. The difference between the two numbers is 76. Find the numbers.

**15.** Find two consecutive integers whose product is 5 less than the square of the smaller number.

**16.** Find two consecutive natural numbers such that the difference of their reciprocals is one-fourth the reciprocal of the smaller number.

**17.** *Weekly Paycheck*   Your weekly paycheck is 15% *more* than your coworker's. Your two paychecks total $645. Find the amount of each paycheck.

**18.** *Monthly Profit*   The total profit for a company in February was 20% *higher* than it was in January. The total profit for the 2 months was $157,498. Find the profit for each month.

**19.** *Weekly Paycheck*   Your weekly paycheck is 15% *less* than your coworker's. Your two paychecks total $645. Find the amount of each paycheck.

**20.** *Monthly Profit*   The total profit for a company in February was 20% *lower* than it was in January. The total profit for the 2 months was $157,498. Find the profit for each month.

*Price Comparison*   In Exercises 21–24, the prices of different items are given for 1917 and 1994. Find the percent increase or decrease for each item.   (Source: *Everything Has Its Price*)

| | *Item* | *1917* | *1994* |
|---|---|---|---|
| **21.** | Three-minute phone call from New York to Los Angeles | $20.70 | $0.75 |
| **22.** | Chocolate bar | $0.02 | $0.45 |
| **23.** | College tuition | $20.00 | $3880.00 |
| **24.** | Movie ticket | $0.15 | $5.45 |

**25.** *Average Age of a Car*   In 1980 the average age of a passenger car in the United States was 6.6 years. By 1990 the average age had increased by 18.18%, and by 1995 the average age had increased (over the 1990 age) by an additional 8.97%.   (Source: American Automobile Manufacturers Association)

(a) Find the average age in 1990.

(b) Find the average age in 1995.

**26.** *Highest Football Salaries*   In 1970 the highest paid professional football player (Joe Namath) had a salary of $150,000 per year. By 1980 the highest paid player (Walter Payton) had a salary that was 216.67% higher. By 1990 the highest paid player (Joe Montana) had a salary that was 742.11% higher than the 1980 high. By 1996 the highest paid player (Troy Aikman) had a salary that was 122.82% higher than the 1990 high. Find the salaries of Walter Payton in 1980, Joe Montana in 1990, and Troy Aikman in 1996.   (Source: *Sport Magazine, The Wall Street Journal Almanac*)

**27.** *Retirement Nest Eggs*   Based on a survey of 740 companies, the most common types of investments by corporate retirement plans are those shown in the bar graph. How many respondents in the survey used each type of investment? (Most used more than one type of investment.)   (Source: Wyatt Company)

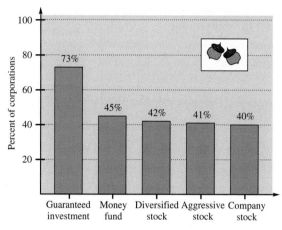

**28.** *School Lunch Snacks*   Based on a survey of 426 mothers, the most common types of snacks put in children's school lunches are those shown in the bar graph. How many respondents put each type of snack in their children's lunches? (Most put in more than one type of snack.)   (Source: Nabisco Brands)

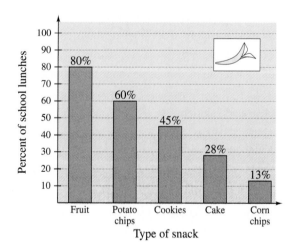

29. *Enjoying Nature*   The Defenders of Wildlife organization conducted a survey in which 300 adults were asked how they enjoyed nature. (Respondents were allowed to answer in more than one category). Of the 300 people, 189 answered "feeding birds," 186 answered "watching wildlife in neighborhood," 129 answered "driving to wildlife site," 108 answered "camping, biking, photography," and 105 answered "hiking." Suppose that one of the 300 people was chosen at random. What is the probability that the person said that feeding birds was a way that he or she enjoyed nature?   (Source: Defenders of Wildlife)

30. *Business Headaches*   In a survey, 414 small-business owners were asked what their biggest business headache was. (Respondents were allowed to answer in only one category.) The responses were as follows.

| Government regulations | 128 |
| Employee relations | 83 |
| Cash flow | 46 |
| Insurance | 37 |
| Paperwork | 25 |
| Taxes | 25 |
| Other | 70 |

Suppose that one of the 414 people was chosen at random. What is the probability that the person said that government regulations was his or her biggest headache?   (Source: MasterCard)

31. *Geometry*   A room is 1.5 times as long as it is wide, and its perimeter is 75 feet (see figure). Find the measurements of the room.

32. *Geometry*   A picture frame has a total perimeter of 3 feet (see figure). The width of the frame is 0.62 times its length. Find the measurements of the frame.

33. *Simple Interest*   You invest $2500 at 7% simple interest. How many years will it take for the investment to earn $1000 in interest?

34. *Simple Interest*   An investment earns $3200 interest over a 7-year period. What is the rate of interest on a $4800 principal investment?

35. *Course Grade*   To get an A in a course you must have an average of at least 90% on four tests that have 100 points each. If your scores on the first three tests were 87, 92, and 84, what must you score on the fourth test to get an A for the course?

36. *Course Grade*   Suppose you are taking a course that has four tests. The first three tests have 100 points each and the fourth test has 200 points. To get an A in the course you must have an average of at least 90% on the four tests. Your scores on the first three tests were 87, 92, and 84. What must you score on the fourth test to get an A for the course?

37. *Loan Payments*   A family has annual loan payments totaling $13,077.75, or 58.6% of its annual income. What is the family's income?

38. *Food Budget*   Suppose your annual budget for food (eating at home and eating at restaurants) is 23.2% of your annual income. During the year, you spent $4832.12 on food. What is your annual income?

39. *List Price*   The price of a swimming pool has been discounted 16.5%. The sale price is $849. Find the original list price of the pool.

40. *List Price*   The price of a compact disc player has been discounted 27.5%. The sale price is $628. Find the original list price of the player.

41. *Discount Rate*   The price of a stereo has been discounted by $150. The sale price is $245. What percent is the discount of the original list price?

**42.** *Discount Rate*    The price of a shirt has been discounted by $20. The sale price is $29.95. What percent is the discount of the original list price?

*Weekly Salary*    In Exercises 43 and 44, use the following information to write a mathematical model and solve. Due to economic factors, your employer has reduced your weekly wage by 15%. Before the reduction, your weekly salary was $425.

**43.** What is your reduced salary?

**44.** What percent raise must you receive to bring your weekly salary back up to $425? Explain why the percent raise is different from the percent reduction.

**45.** *Travel Time*    Suppose you are driving on a freeway to another town that is 150 miles from your home. After 30 minutes, you pass a freeway exit that you know is 25 miles from your home. Assuming that you continue at the same constant speed, how long will it take for the entire trip?

**46.** *Travel Time*    A plane is flying from Orlando to Denver, a distance of about 1950 miles. After 1 hour and 15 minutes, the plane flies over a town that is 600 miles from Orlando. How long will it take the plane to fly from Orlando to Denver? (Assume that the plane flies at a constant speed the entire flight.)

**47.** *Travel Time*    Two cars start at a given point and travel in the same direction at average speeds of 40 miles per hour and 55 miles per hour. How much time must elapse before the cars are 5 miles apart?

**48.** *Catch-Up Time*    Students are traveling in two cars to a football game 135 miles away. The first car leaves on time and travels at an average speed of 45 miles per hour. The second car starts $\frac{1}{2}$ hour later and travels at an average speed of 55 miles per hour. At these speeds, how long will it take the second car of students to catch up to the first car?

**49.** *Travel Time*    Two families meet at a park for a picnic. At the end of the day, one family travels east at an average speed of 42 miles per hour and the other family travels west at an average speed of 50 miles per hour. Both families have approximately 160 miles to travel. Find the time it takes each family to get home.

**50.** *Average Speed*    A truck traveled at an average speed of 55 miles per hour on a 200-mile trip to pick up a load of freight. On the return trip (with the truck fully loaded), the average speed was 40 miles per hour. Find the average speed for the round trip.

**51.** *Radio Waves*    Radio waves travel at the same speed as light, $3.0 \times 10^8$ meters per second. Find the time required for a radio wave to travel from mission control in Houston to NASA astronauts on the surface of the moon $3.86 \times 10^8$ meters away.

**52.** *Distance to a Star*    Find the distance to a star that is 50 light years (distance traveled by light in 1 year) away. (Light travels at 186,000 miles per second.)

**53.** *Height of a Tree*    Suppose that you want to measure the height of a tree that is in your yard. To do this, you measure the tree's shadow and find that it is 25 feet long. You also measure the shadow of a 5-foot lamppost and find its shadow to be 2 feet long (see figure). How tall is the tree?

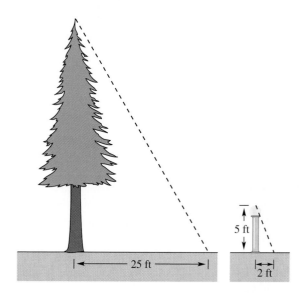

**54.** *Height of a Building*    Suppose that you want to measure the height of a building. To do this, you measure the building's shadow and find that it is 50 feet long. You also measure the shadow of a 4-foot stake and find its shadow to be $3\frac{1}{2}$ feet long. How tall is the building?

**55.** *Projected Expenses* From January through May, a company's expenses have totaled $234,980. If the monthly expenses continue at this rate, what will the total expenses for the year be?

**56.** *Projected Revenue* From January through August, a company's revenues have totaled $345,950. If the monthly revenue continues at this rate, what will the total revenue for the year be?

**57.** *Investment Mix* Suppose you invest $12,000 in two funds paying $10\frac{1}{2}\%$ and 13% simple interest. The total annual interest is $1447.50. How much is invested in each fund?

**58.** *Investment Mix* Suppose you invest $25,000 in two funds paying 11% and $12\frac{1}{2}\%$ simple interest. The total annual interest is $2975.00. How much is invested in each fund?

**59.** *Comparing Investment Returns* You invest $12,000 in a fund paying $9\frac{1}{2}\%$ simple interest and $8000 in a fund for which the interest rate is variable. At the end of the year you receive notification that the total interest for both funds is $2054.40. Find the equivalent simple interest rate on the variable rate fund.

**60.** *Comparing Investment Returns* You have $10,000 on deposit earning simple interest that is linked to the prime rate. When the prime rate dropped, your rate dropped by $1\frac{1}{2}\%$ for the last quarter of the year. Your total annual interest was $1112.50. What was your interest rate for the first three quarters and for the last quarter?

**61.** *Production Limit* A company has fixed costs of $10,000 per month and variable costs of $8.50 per unit manufactured. The company has $85,000 available to cover the monthly costs. How many units can the company manufacture? (*Fixed costs* are those that occur regardless of the level of production. *Variable costs* depend on the level of production.)

**62.** *Production Limit* A company has fixed costs of $10,000 per month and variable costs of $9.30 per unit manufactured. The company has $85,000 available to cover the monthly costs. How many units can the company manufacture? (*Fixed costs* are those that occur regardless of the level of production. *Variable costs* depend on the level of production.)

**63.** *Length of a Tank* The diameter of a cylindrical propane gas tank is 4 feet (see figure). The total volume of the tank is 603.2 cubic feet. Find the length of the tank.

**64.** *Water Depth* A trough is 12 feet long, 3 feet deep, and 3 feet wide (see figure). Find the depth of the water when the trough contains 70 gallons. (1 gallon $\approx$ 0.13368 cubic foot.)

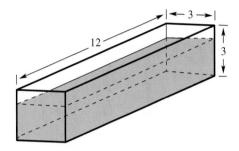

**65.** *Mixture* A 55-gallon barrel contains a mixture with a concentration of 40%. How much of this mixture must be withdrawn and replaced by 100% concentrate to bring the mixture up to 75% concentration? (See figure.)

**66.** *Mixture* A farmer mixed gasoline and oil to make 2 gallons of mixture for his two-cycle chain saw engine. This mixture was 32 parts gasoline and 1 part two-cycle oil. How much gasoline must be added to bring the mixture to 40 parts gasoline and 1 part oil?

**67.** Sometimes, when developing a mathematical model for a problem, the most accurate equation is not used as the model. Explain why and include factors that influence a mathematical model's selection.

*New York City Marathon* In Exercises 68 and 69, find the average speed of the record-holding runners in the New York City Marathon. The length of the course is 26 miles, 385 yards. (Note that 1 mile = 5280 feet = 1760 yards.)

**68.** Men's record time: 2 hours, 8 minutes

**69.** Women's record time: 2 hours, $24\frac{2}{5}$ minutes

In Exercises 70–83, solve for the variable.

**70.** *Area* Solve for $h$ in $A = \frac{1}{2}bh$.

**71.** *Perimeter* Solve for $l$ in $P = 2l + 2w$.

**72.** *Volume* Solve for $l$ in $V = lwh$.

**73.** *Volume* Solve for $h$ in $V = \pi r^2 h$.

**74.** *Markup* Solve for $C$ in $S = C + RC$.

**75.** *Discount* Solve for $L$ in $S = L - RL$.

**76.** *Interest* Solve for $r$ in $A = P + Prt$.

**77.** *Interest* Solve for $P$ in $A = P\left(1 + \dfrac{r}{n}\right)^{nt}$.

**78.** *Area* Solve for $b$ in $A = \frac{1}{2}(a + b)h$.

**79.** *Area* Solve for $\theta$ in $A = \dfrac{\pi r^2 \theta}{360}$.

**80.** *Sequence* Solve for $n$ in $L = a + (n - 1)d$.

**81.** *Series* Solve for $r$ in $S = \dfrac{rL - a}{r - 1}$.

**82.** *Lateral Surface Area* Solve for $h$ in $A = 2\pi rh$.

**83.** *Surface Area* Solve for $r$ in $A = 4\pi r^2$.

## 1.3   Quadratic Equations

### Objectives

- Solve a quadratic equation by factoring.
- Solve a quadratic equation by extracting square roots.
- Construct and use a quadratic model to solve an application problem.

## Solving Quadratic Equations by Factoring

In the first two sections of this chapter, you studied linear equations in one variable. In this and the next section, you will study quadratic equations.

> ▶ **Definition of a Quadratic Equation**
>
> A **quadratic equation** in $x$ is an equation that can be written in the standard form
>
> $$ax^2 + bx + c = 0$$
>
> where $a$, $b$, and $c$ are real numbers with $a \neq 0$. Another name for a quadratic equation in $x$ is a **second-degree polynomial equation in $x$.**

There are three basic techniques for solving quadratic equations: factoring, extracting square roots, and the Quadratic Formula. (The Quadratic Formula is discussed in the next section.) The first technique is based on the Zero-Factor Property given in the appendix.

If $ab = 0$, then $a = 0$ or $b = 0$.         Zero-Factor Property

To use this property, rewrite the left side of the standard form of a quadratic equation as the product of two linear factors. Then find the solutions of the quadratic equation by setting each factor equal to zero.

### Study Tip

The Zero-Factor Property applies *only* to equations written in standard form (in which one side of the equation is zero). So, be sure that all terms are collected on one side *before* factoring. For instance, in the equation

$$(x - 5)(x + 2) = 8$$

it is *incorrect* to set each factor equal to 8. Can you solve this equation correctly?

---

**Example 1**   Solving a Quadratic Equation

Solve $x^2 - 3x - 10 = 0$.

**Solution**

| | | |
|---|---|---|
| $x^2 - 3x - 10 = 0$ | | Original equation |
| $(x - 5)(x + 2) = 0$ | | Factor. |
| $x - 5 = 0$ | $x = 5$ | Set 1st factor equal to 0. |
| $x + 2 = 0$ | $x = -2$ | Set 2nd factor equal to 0. |

The solutions are 5 and $-2$. Check these in the original equation.

| Example 2 | Solving a Quadratic Equation by Factoring |

Solve $6x^2 - 3x = 0$.

**Solution**

$$6x^2 - 3x = 0 \qquad \text{Original equation}$$

$$3x(2x - 1) = 0 \qquad \text{Factor.}$$

$$3x = 0 \implies x = 0 \qquad \text{Set 1st factor equal to 0.}$$

$$2x - 1 = 0 \implies x = \tfrac{1}{2} \qquad \text{Set 2nd factor equal to 0.}$$

The solutions are 0 and $\frac{1}{2}$. Check these by substituting in the original equation, as follows.

**Check**

$$6x^2 - 3x = 0 \qquad \text{Original equation}$$

$$6(0)^2 - 3(0) \stackrel{?}{=} 0 \qquad \text{Substitute 0 for } x.$$

$$0 - 0 = 0 \qquad \text{First solution checks. } \checkmark$$

$$6\left(\tfrac{1}{2}\right)^2 - 3\left(\tfrac{1}{2}\right) \stackrel{?}{=} 0 \qquad \text{Substitute } \tfrac{1}{2} \text{ for } x.$$

$$\tfrac{6}{4} - \tfrac{3}{2} = 0 \qquad \text{Second solution checks. } \checkmark$$

If the two factors of a quadratic expression are the same, the corresponding solution is a **double** or **repeated** solution.

## • TECHNOLOGY •

To check the solution in Example 3 with your graphing utility, you should first write the equation in standard form.

$$9x^2 - 6x + 1 = 0$$

Then enter the expression $9x^2 - 6x + 1$ into $y_1$ of the equation editor. Now you can use the ASK mode of the table feature of your graphing utility to check the solution.

| Example 3 | A Quadratic Equation with a Repeated Solution |

Solve $9x^2 - 6x = -1$.

**Solution**

$$9x^2 - 6x = -1 \qquad \text{Original equation}$$

$$9x^2 - 6x + 1 = 0 \qquad \text{Write in standard form.}$$

$$(3x - 1)^2 = 0 \qquad \text{Factor.}$$

$$3x - 1 = 0 \qquad \text{Set repeated factor equal to 0.}$$

$$x = \tfrac{1}{3} \qquad \text{Solution}$$

The only solution is $\frac{1}{3}$. Check this in the original equation.

## Extracting Square Roots

There is a nice shortcut for solving equations of the form $u^2 = d$, where $d > 0$. By factoring, you can see that this equation has two solutions.

| | |
|---|---|
| $u^2 = d$ | Original equation |
| $u^2 - d = 0$ | Standard form |
| $(u + \sqrt{d})(u - \sqrt{d}) = 0$ | Factor. |
| $u + \sqrt{d} = 0 \quad \Longrightarrow \quad u = -\sqrt{d}$ | Set 1st factor equal to 0. |
| $u - \sqrt{d} = 0 \quad \Longrightarrow \quad u = \sqrt{d}$ | Set 2nd factor equal to 0. |

Solving an equation of the form $u^2 = d$ without going through the steps of factoring is called **extracting square roots.**

---

▶ **Extracting Square Roots**

The equation $u^2 = d$, where $d > 0$, has exactly two solutions:

$$u = \sqrt{d} \quad \text{and} \quad u = -\sqrt{d}.$$

These solutions can also be written as $u = \pm\sqrt{d}$.

---

**Example 4**   Extracting Square Roots

Solve $4x^2 = 12$.

**Solution**

| | |
|---|---|
| $4x^2 = 12$ | Original equation |
| $x^2 = 3$ | Divide both sides by 4. |
| $x = \pm\sqrt{3}$ | Extract square roots. |

The solutions are $\sqrt{3}$ and $-\sqrt{3}$. Check these in the original equation.

**Example 5**   Extracting Square Roots

| | |
|---|---|
| $(x - 3)^2 = 7$ | Original equation |
| $x - 3 = \pm\sqrt{7}$ | Extract square roots. |
| $x = 3 \pm \sqrt{7}$ | Add 3 to both sides. |

The solutions are $3 \pm \sqrt{7}$. Check these in the original equation.

## Ancient Forms of Quadratic Equations

In ancient and medieval times, quadratic equations were classified in three categories.

$$x^2 + px = q$$
$$x^2 = px + q$$
$$x^2 + q = px$$

Solutions of these types of quadratic equations have been found in Babylonian texts over 4000 years old. Quadratic equations of the form $x^2 + px + q = 0$, where $p$ and $q$ are positive, were not solved because such equations have no positive solutions.

# Applications

Quadratic equations often occur in problems dealing with area. Here is a simple example.

> A square room has an area of 144 square feet. Find the measurements of the room.

To solve this problem, you can let $x$ represent the length of each side of the room. Then, by solving the equation

$$x^2 = 144$$

you can conclude that each side of the room is 12 feet long. Note that although the equation $x^2 = 144$ has two solutions, $-12$ and $12$, the negative solution makes no sense (for this problem), so you should choose the positive solution.

## Example 6  Finding the Measurements of a Room

A bedroom is 3 feet longer than it is wide (see Figure 1.6) and has an area of 154 square feet. Find the measurements of the room.

### Solution

You can begin by using the same type of problem-solving strategy that was presented in Section 1.2.

*Verbal Model:*  Width of room · Length of room = Area of room

*Labels:*
Area of room = 154 (square feet)
Width of room = $w$ (feet)
Length of room = $w + 3$ (feet)

*Equation:*
$$w(w + 3) = 154$$
$$w^2 + 3w - 154 = 0$$
$$(w - 11)(w + 14) = 0$$
$$w - 11 = 0 \implies w = 11$$
$$w + 14 = 0 \implies w = -14$$

Choosing the positive value, you can conclude that the width is 11 feet and the length is $w + 3 = 14$ feet. You can check this in the original statement of the problem as follows.

### Check

The length of 14 feet is 3 feet more than the width of 11 feet. ✓
The area of the room is $(11)(14) = 154$ square feet. ✓

Figure 1.6

Another application of quadratic equations involves an object that is falling (or vertically projected into the air). The equation that gives the height of such an object is called a **position equation,** and on earth's surface it has the form

$$s = -16t^2 + v_0t + s_0.$$

In this equation, $s$ represents the height of the object (in feet), $v_0$ represents the initial velocity of the object (in feet per second), $s_0$ represents the initial height of the object (in feet), and $t$ represents the time (in seconds).

> **Example 7**    **Falling Object**

A construction worker is working on the 24th floor of a building project, as shown in Figure 1.7. The worker accidentally drops a wrench and immediately yells "Look out below!" Could a person at ground level hear this warning in time to get out of the way of the falling wrench?

**Solution**

Let's assume that each floor of the building is 10 feet high, so that the wrench was dropped from a height of 240 feet. Because sound travels at about 1100 feet per second, you can assume that a person at ground level hears the warning within 1 second of the time the wrench is dropped. To set up a mathematical model for the height of the wrench, you can use the position equation

$$s = -16t^2 + v_0t + s_0.$$

Because the object is dropped (rather than thrown), you can conclude that the initial velocity is $v_0 = 0$. Moreover, because the initial height is $s_0 = 240$ feet, you have the following model.

$$s = -16t^2 + 240$$

After falling for 1 second, the wrench is at a height of $-16(1^2) + 240 = 224$ feet. After falling for 2 seconds, the wrench is at a height of $-16(2^2) + 240 = 176$ feet. To find the number of seconds it takes the wrench to hit the ground, let the height $s$ be zero and solve the resulting equation for $t$.

| | |
|---|---|
| $s = -16t^2 + 240$ | Position equation |
| $0 = -16t^2 + 240$ | Set height equal to 0. |
| $16t^2 = 240$ | Add $16t^2$ to both sides. |
| $t^2 = 15$ | Divide both sides by 16. |
| $t = \sqrt{15} \approx 3.87$ | Extract positive square root. |

The wrench will take about 3.87 seconds to hit the ground. If the person heard the warning 1 second after the wrench was dropped, the person would still have almost 3 seconds to get out of the way.

240 ft

**Figure 1.7**

A third type of application that often involves a quadratic equation is one dealing with the hypotenuse of a right triangle. Recall from geometry that the sides of a right triangle are related by a formula called the **Pythagorean Theorem.** This theorem states that if $a$ and $b$ are the lengths of the sides of the triangle and $c$ is the length of the hypotenuse (see Figure 1.8),

$$a^2 + b^2 = c^2. \qquad \text{Pythagorean Theorem}$$

Notice how this formula is used in the next example.

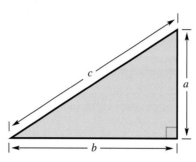

Figure 1.8

**Example 8** Cutting Across the Lawn

Your house is on a large corner lot. Several children in the neighborhood cut across your lawn in such a way that the distance across the lawn is 32 feet, as shown in Figure 1.9. Assuming that Figure 1.9 is drawn to scale, how many feet does a person save by walking across the lawn instead of walking on the side-walk?

**Solution**

In Figure 1.9, let $x$ represent the length of the shorter part of the sidewalk. Using a ruler, it appears that the length of the longer part of the sidewalk is twice the shorter, so you can represent its length by $2x$.

$$x^2 + (2x)^2 = 32^2 \qquad \text{Pythagorean Theorem}$$
$$5x^2 = 1024 \qquad \text{Add like terms.}$$
$$x^2 = 204.8 \qquad \text{Divide both sides by 5.}$$
$$x = \sqrt{204.8} \qquad \text{Extract positive square root.}$$

The total distance on the sidewalk is

$$x + 2x = 3x = 3\sqrt{204.8} \approx 42.9 \text{ feet.}$$

So, cutting across the lawn saves a person about 10.9 feet of walking.

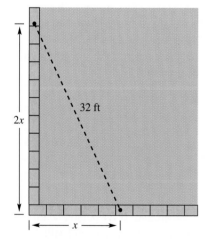

Figure 1.9

A fourth type of common application of a quadratic equation is one in which a quantity $y$ is changing over time $t$ according to a quadratic model.

### Example 9    The Cost of a New Car

From 1987 to 1995, the average cost of a new import car can be modeled by

$$\text{Cost} = 57.4t^2 + 11{,}352, \qquad t \geq 7$$

where $t = 7$ represents 1987. (See Figure 1.10.) If the average cost of a new import car continued to increase according to this model, in what year would the average cost reach \$35,000? (Source: Commerce Department, American Autodatum)

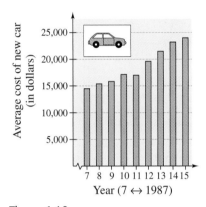

**Figure 1.10**

### Solution

To solve this problem, let the cost be \$35,000 and solve the equation for $t$.

| | |
|---|---|
| $57.4t^2 + 11{,}352 = 35{,}000$ | Set cost equal to 35,000. |
| $57.4t^2 = 23{,}648$ | Subtract 11,352 from both sides. |
| $t^2 \approx 411.986$ | Divide both sides by 57.4. |
| $t \approx \sqrt{411.986}$ | Extract positive square root. |

The solution is $t \approx 20.3$. However, because $t$ is greater than 20, you can round up to $t = 21$ years. Because $t = 7$ represents 1987, you can conclude that the average cost of a new import car would reach \$35,000 in 2001.

---

### Discussing the Concept    **Geometry**

A **Pythagorean triple** is a set of three positive integers $(a, b, c)$ that satisfy the Pythagorean Theorem for right triangles, $a^2 + b^2 = c^2$. One such Pythagorean triple is $(3, 4, 5)$, because $3^2 + 4^2 = 5^2$. How many other Pythagorean triples can you find? (Remember that each member of the triple must be an integer.) Make a list of these triples, and discuss the method you used to find them.

# Warm Up

The following warm-up exercises involve skills that were covered in previous courses and earlier sections. You will use these skills in the exercise set for this section.

In Exercises 1–4, simplify the expression.

**1.** $\sqrt{\frac{7}{50}}$

**2.** $\sqrt{32}$

**3.** $\sqrt{7^2 + 3 \cdot 7^2}$

**4.** $\sqrt{\frac{1}{4} + \frac{3}{8}}$

In Exercises 5–10, factor the expression.

**5.** $3x^2 + 7x$

**6.** $4x^2 - 25$

**7.** $16 - (x - 11)^2$

**8.** $x^2 + 7x - 18$

**9.** $10x^2 + 13x - 3$

**10.** $6x^2 - 73x + 12$

## 1.3   Exercises

In Exercises 1–10, write the quadratic equation in standard form.

**1.** $2x^2 = 3 - 5x$

**2.** $4x^2 - 2x = 9$

**3.** $x^2 = 25x$

**4.** $10x^2 = 90$

**5.** $(x - 3)^2 = 2$

**6.** $12 - 3(x + 7)^2 = 0$

**7.** $x(x + 2) = 3x^2 + 1$

**8.** $x^2 + 1 = \dfrac{x - 3}{2}$

**9.** $\dfrac{3x^2 - 10}{5} = 12x$

**10.** $x(x + 5) = 2(x + 5)$

In Exercises 11–22, solve the quadratic equation by factoring.

**11.** $6x^2 + 3x = 0$

**12.** $9x^2 - 1 = 0$

**13.** $x^2 - 2x - 8 = 0$

**14.** $x^2 - 10x + 9 = 0$

**15.** $x^2 + 10x + 25 = 0$

**16.** $16x^2 + 56x + 49 = 0$

**17.** $3 + 5x - 2x^2 = 0$

**18.** $2x^2 = 19x + 33$

**19.** $x^2 + 4x = 12$

**20.** $x^2 + 4x = 21$

**21.** $-x^2 - 7x = 10$

**22.** $-x^2 + 8x = 12$

In Exercises 23–36, solve the equation by extracting square roots. List both the exact answer *and* a decimal answer that has been rounded to two decimal places.

**23.** $x^2 = 16$

**24.** $x^2 = 144$

**25.** $x^2 = 7$

**26.** $x^2 = 27$

**27.** $3x^2 = 36$

**28.** $9x^2 = 25$

**29.** $(x - 12)^2 = 18$

**30.** $(x + 13)^2 = 21$

**31.** $(x + 2)^2 = 12$

**32.** $(x + 5)^2 = 20$

**33.** $12x^2 = 300$

**34.** $6x^2 = 250$

**35.** $3x^2 + 2(x^2 - 4) = 15$

**36.** $x^2 + 3(x^2 - 5) = 10$

In Exercises 37–58, solve the equation by any convenient method.

**37.** $x^2 = 64$

**38.** $7x^2 = 32$

**39.** $x^2 - 2x + 1 = 0$

**40.** $x^2 - 6x + 5 = 0$

**41.** $16x^2 - 9 = 0$

**42.** $11x^2 + 33x = 0$

**43.** $4x^2 - 12x + 9 = 0$

**44.** $x^2 - 14x + 49 = 0$

**45.** $(x + 3)^2 = 81$

**46.** $(x - 5)^2 = 8$

**47.** $4x = 4x^2 - 3$

**48.** $80 + 6x = 9x^2$

**49.** $50 + 5x = 3x^2$

**50.** $144 - 73x + 4x^2 = 0$

**51.** $12x = x^2 + 27$        **52.** $26x = 8x^2 + 15$

**53.** $50x^2 - 60x + 10 = 0$   **54.** $9x^2 + 12x + 3 = 0$

**55.** $(x + 3)^2 - 4 = 0$       **56.** $(x - 2)^2 - 9 = 0$

**57.** $(x + 1)^2 = x^2$          **58.** $(x + 1)^2 = 4x^2$

**59.** Consider the expression $(x + 2)^2$. How would you convince someone in your class that $(x + 2)^2 \neq x^2 + 4$? Give an argument based on the rules of algebra. Give an argument using your graphing utility.

**60.** Consider the expression $\sqrt{a^2 + b^2}$. How would you convince someone in your class that $\sqrt{a^2 + b^2} \neq a + b$? Give an argument based on the rules of algebra or geometry. Give an argument using your graphing utility.

**61.** *Geometry*   A one-story building is 14 feet longer than it is wide (see figure). The building has 1632 square feet of floor space. What are the measurements of the building?

**62.** *Geometry*   A billboard is 10 feet longer than it is high (see figure). The billboard has 336 square feet of advertising space. What are the measurements of the billboard?

**63.** *Geometry*   A triangular sign has a height that is equal to its base. The area of the sign is 4 square feet. Find the base and height of the sign.

**64.** *Geometry*   The building lot shown in the figure has an area of 8000 square feet. What are the measurements of the lot?

In Exercises 65–68, assume that air resistance is negligible, which implies that the position equation discussed in this section is a reasonable model.

**65.** *Royal Gorge Bridge*   The Royal Gorge Bridge near Canon City, Colorado is the highest suspension bridge in the world. The bridge is 1053 feet above the Arkansas river. If a rock were dropped from the bridge, how long would it take the rock to hit the water?

**66.** *Falling Object*   A rock is dropped from the top of a 200-foot cliff that overlooks the ocean. How long will it take for the rock to hit the water?

**67.** *Olympic Diver*   The high-dive platform in the Olympics is 10 meters above the water. A diver wants to perform a double flip into the water. How long will the diver be in the air? [*Hint:* 1 meter = 3.2808 feet]

**68.** *The Owl and the Mouse*   An owl is circling a field and sees a mouse. The owl folds its wings and begins to dive. If the owl starts its dive from a height of 100 feet, how long does the mouse have to escape?

**69.** *Geometry*   The hypotenuse of an isosceles right triangle is 5 centimeters long. How long are the sides? (An isosceles triangle is one that has two sides of equal length.)

**70.** *Geometry* An equilateral triangle has a height of 1 foot. How long are each of its sides? (*Hint:* Use the height of the triangle to partition the triangle into two right triangles of the same size.)

**71.** *Flying Distance* The cities of Chicago, Atlanta, and Buffalo approximately form the vertices of a right triangle. The distance from Atlanta to Buffalo is about 650 miles and the distance from Atlanta to Chicago is about 565 miles. Use the map in the figure to approximate the flying distance from Atlanta to Buffalo *by way of* Chicago.

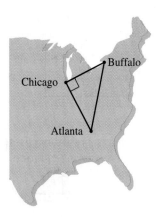

**72.** *Depth of a Submarine* The sonar of a navy cruiser detects a submarine that is 3000 feet from the cruiser. The angle formed by the ocean surface and a line from the cruiser to the submarine is 45° (see figure). How deep is the submarine?

**73.** *Salvage Depth* A marine salvage ship detects a wreck 2000 feet off the bow. If the angle between the ocean surface and a line from the ship to the wreck is 45°, how deep is the wreck?

**74.** *Golf Winnings* The leading golf winnings (in dollars) from 1992 to 1996 in the United States can be approximated by the model

$$\text{Winnings} = 13{,}000t^2 + 1{,}317{,}556, \qquad t \geq 2$$

where $t = 2$ represents 1992 (see figure). Assuming that the leading winnings continue to follow the pattern given by this quadratic model, in what year would the leading winnings exceed \$3,500,000? (Source: Professional Golfers' Association)

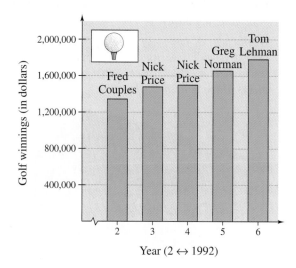

Year (2 ↔ 1992)

**75.** *Total Revenue* The demand equation for a certain product is $p = 20 - 0.0002x$, where $p$ is the price per unit and $x$ is the number of units sold. The total revenue for selling $x$ units is given by

$$\text{Revenue} = xp = x(20 - 0.0002x).$$

How many units must be sold to produce a revenue of \$500,000?

**76.** *Total Revenue* The demand equation for a certain product is $p = 30 - 0.0005x$, where $p$ is the price per unit and $x$ is the number of units sold. The total revenue for selling $x$ units is given by

$$\text{Revenue} = xp = x(30 - 0.0005x).$$

How many units must be sold to produce a revenue of \$450,000?

**77.** *Production Cost*    A company determines that the average monthly cost of raw materials for manufacturing a product line can be modeled by

$$\text{Cost} = 20.75t^2 + 5104, \qquad 0 \leq t$$

where $t = 0$ represents 1990. If the average monthly cost of raw materials continues to increase according to this model, in what year will the average monthly cost reach \$10,000?

**78.** *Monthly Cost*    A company determines that the average monthly cost for staffing temporary positions can be modeled by

$$\text{Cost} = 115.35t^2 + 12,072, \qquad 0 \leq t$$

where $t = 0$ represents 1990. If the average monthly cost for staffing temporary positions continues to increase according to this model, in what year will the average monthly cost reach \$20,000?

**79.** *U.S. Population*    The population of the United States from 1800 to 1890 can be approximated by the model

$$\text{Population} = 694.2t^2 + 6183.3, \qquad 0 \leq t \leq 9$$

where the population is given in thousands of people and the time $t$ represents the calendar year, with $t = 0$ corresponding to 1800, $t = 1$ corresponding to 1810, and so on (see figure). If this model had continued to be valid up through the present time, in what year would the population of the United States have reached 250,000,000? Judging from the given figure, would you say this model was a good representation of the population through 1890? How about through 1990? (Source: U.S. Bureau of the Census)

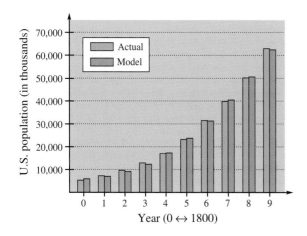

Year (0 ↔ 1800)

**80.** *U.S. Population*    The population of the United States from 1900 to 1990 can be approximated by the model

$$\text{Population} = 2046.7t^2 + 95,401.6, \qquad 0 \leq t \leq 9$$

where the population is given in thousands of people and the time $t$ represents the calendar year, with $t = 0$ corresponding to 1900, $t = 1$ corresponding to 1910, and so on (see figure). If this model continues to be valid, in what year will the population of the United States reach 300,000,000? (Source: U.S. Bureau of the Census)

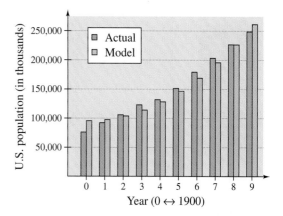

Year (0 ↔ 1900)

**81.** The U.S. Bureau of the Census predicts that the population in 2050 will be 390,000,000 people. Does the model in Exercise 80 appear to be a valid model for the year 2050?

**82.** *A Short Deck*    A standard deck of cards is missing several cards. One of the cards in the deck, however, is known to be the ace of hearts. A card is drawn from the deck at random, and then replaced, and then a card is drawn again. The probability that both drawings produce the ace of hearts is given by

$$p^2 \approx 0.000567.$$

How many cards are in the deck? (*Hint:* Write $p$ in the form $1/n$, where $n$ is the number of cards in the deck.)

**83.** *Airline Flight*    Suppose that you are going on a *round-trip* airline flight. You are told that the probability that both flights will be late is

$$p^2 = 0.2016.$$

What is the probability $p$ that you will be late on the return trip?

## Objectives

- Develop the Quadratic Formula by completing the square.
- Use the discriminant to determine the number of real solutions to a quadratic equation.
- Solve a quadratic equation using the Quadratic Formula.
- Use the Quadratic Formula to solve an application problem.

## Development of the Quadratic Formula

In Section 1.3 you studied two methods for solving quadratic equations. These two methods are efficient for special quadratic equations that are factorable or that can be solved by extracting square roots.

There are, however, many quadratic equations that cannot be solved efficiently by either of these two techniques. Fortunately, there is a general formula that can be used to solve *any* quadratic equation. It is called the **Quadratic Formula.** You can develop this important formula by a process called **completing the square,** as follows.

$$ax^2 + bx + c = 0 \qquad \text{Standard form, } a \neq 0$$

$$ax^2 + bx = -c \qquad \text{Subtract } c \text{ from both sides.}$$

$$x^2 + \frac{b}{a}x = -\frac{c}{a} \qquad \text{Divide both sides by } a.$$

$$x^2 + \frac{b}{a}x + \left(\frac{b}{2a}\right)^2 = -\frac{c}{a} + \left(\frac{b}{2a}\right)^2 \qquad \text{Add } \left(\frac{b}{2a}\right)^2 \text{ to both sides.}$$

$$\left(x + \frac{b}{2a}\right)^2 = \frac{b^2 - 4ac}{4a^2} \qquad \text{Simplify.}$$

$$x + \frac{b}{2a} = \pm\sqrt{\frac{b^2 - 4ac}{4a^2}} \qquad \text{Extract square roots.}$$

$$x = \frac{-b \pm \sqrt{b^2 - 4ac}}{2a} \qquad \text{Solutions}$$

This result is summarized below.

### Study Tip

The Quadratic Formula is one of the most important formulas in algebra, and you should memorize it. We have found that it helps to try to memorize a verbal statement of the rule. For instance, you might try to remember the following verbal statement of the Quadratic Formula: "The opposite of $b$, plus or minus the square root of $b$ squared minus $4ac$, all divided by $2a$."

---

▶ **The Quadratic Formula**

The solutions of

$$ax^2 + bx + c = 0, \quad a \neq 0$$

are given by the **Quadratic Formula,**

$$x = \frac{-b \pm \sqrt{b^2 - 4ac}}{2a}.$$

## The Discriminant

In the Quadratic Formula, the quantity under the radical sign, $b^2 - 4ac$, is called the **discriminant** of the quadratic expression $ax^2 + bx + c$.

$b^2 - 4ac$     Discriminant

It can be used to determine the nature of the solutions of a quadratic equation.

---

▶ **Solutions of a Quadratic Equation**

The solutions of a quadratic equation

$$ax^2 + bx + c = 0, \quad a \neq 0$$

can be classified by the discriminant, $b^2 - 4ac$, as follows.

**1.** If $b^2 - 4ac > 0$, the equation has *two* distinct real solutions.

**2.** If $b^2 - 4ac = 0$, the equation has *one* repeated real solution.

**3.** If $b^2 - 4ac < 0$, the equation has *two* distinct imaginary solutions. (In Section 4.5, you will study imaginary numbers.)

---

In case 3 above, the equation has no real solutions.

**Example 1**   Using the Discriminant

Use the discriminant to determine the number of real solutions of each of the following quadratic equations.

**a.** $4x^2 - 20x + 25 = 0$     **b.** $13x^2 + 7x + 1 = 0$     **c.** $5x^2 = 8x$

Solution

**a.** Using $a = 4$, $b = -20$, and $c = 25$, the discriminant is

$$b^2 - 4ac = (-20)^2 - 4(4)(25) = 400 - 400 = 0.$$

So, there is *one* repeated real solution.

**b.** Using $a = 13$, $b = 7$, and $c = 1$, the discriminant is

$$b^2 - 4ac = (7)^2 - 4(13)(1) = 49 - 52 = -3.$$

So, there are *no* real solutions.

**c.** In standard form, this equation is $5x^2 - 8x = 0$, with $a = 5$, $b = -8$, and $c = 0$, which implies that the discriminant is

$$b^2 - 4ac = (-8)^2 - 4(5)(0) = 64.$$

So, there are *two* distinct real solutions.

## Using the Quadratic Formula

When using the Quadratic Formula, remember that *before* the formula can be applied, you must first write the quadratic equation in standard form.

**Example 2**    Two Distinct Solutions

Solve $x^2 + 3x = 9$.

**Solution**

$$x^2 + 3x = 9 \qquad \text{Original equation}$$

$$x^2 + 3x - 9 = 0 \qquad \text{Write in standard form.}$$

$$x = \frac{-3 \pm \sqrt{(3)^2 - 4(1)(-9)}}{2(1)} \qquad \text{Quadratic Formula}$$

$$x = \frac{-3 \pm \sqrt{45}}{2} \qquad \text{Simplify.}$$

$$x = \frac{-3 \pm 3\sqrt{5}}{2} \qquad \text{Simplify.}$$

The solutions are

$$x = \frac{-3 + 3\sqrt{5}}{2} \qquad \text{and} \qquad x = \frac{-3 - 3\sqrt{5}}{2}.$$

Check these in the original equation.

**Example 3**    One Repeated Solution

Solve $8x^2 - 24x + 18 = 0$.

**Solution**

Begin by dividing both sides by the common factor 2.

$$8x^2 - 24x + 18 = 0 \qquad \text{Original equation}$$

$$4x^2 - 12x + 9 = 0 \qquad \text{Divide both sides by 2.}$$

$$x = \frac{-(-12) \pm \sqrt{(-12)^2 - 4(4)(9)}}{2(4)} \qquad \text{Quadratic Formula}$$

$$x = \frac{12 \pm \sqrt{0}}{8} \qquad \text{Simplify.}$$

$$x = \frac{3}{2} \qquad \text{Repeated solution}$$

The only solution is $\frac{3}{2}$. Check this in the original equation.

The discriminant in Example 3 is a perfect square (zero in this case), and you could have factored the quadratic as

$$4x^2 - 12x + 9 = 0$$

$$(2x - 3)^2 = 0$$

to conclude that the solution is $x = \frac{3}{2}$. Because factoring is easier than applying the Quadratic Formula, try factoring first when solving a quadratic equation. If, however, factors cannot be easily found, then use the Quadratic Formula. For instance, try solving the quadratic equation

$$x^2 - x - 12 = 0$$

in two ways—by factoring and by the Quadratic Formula—to see that you get the same solutions either way.

When using a calculator with the Quadratic Formula, you should get in the habit of using the memory key to store intermediate steps. This will save steps and minimize roundoff error.

---

| Example 4 | Using a Calculator with the Quadratic Formula |

Solve $16.3x^2 - 197.6x + 7.042 = 0$.

**Solution**

In this case, $a = 16.3$, $b = -197.6$, $c = 7.042$, and you have

$$x = \frac{-(-197.6) \pm \sqrt{(-197.6)^2 - 4(16.3)(7.042)}}{2(16.3)}.$$

To evaluate these solutions, begin by calculating the square root of the discriminant, as follows.

*Scientific Calculator Steps*

197.6 $\boxed{+/-}$ $\boxed{x^2}$ $\boxed{-}$ 4 $\boxed{\times}$ 16.3 $\boxed{\times}$ 7.042 $\boxed{=}$ $\boxed{\sqrt{}}$

*Graphing Calculator Steps*

$\boxed{\sqrt{}}$ $\boxed{(}$ $\boxed{(}$ $\boxed{(-)}$ 197.6 $\boxed{)}$ $\boxed{x^2}$ $\boxed{-}$ 4 $\boxed{\times}$ 16.3 $\boxed{\times}$ 7.042 $\boxed{)}$ $\boxed{\text{ENTER}}$

In either case, the result is 196.43478. Storing this result and using the recall key, you can find the following two solutions.

$$x \approx \frac{197.6 + 196.43478}{2(16.3)} \approx 12.087 \qquad \text{Add stored value.}$$

$$x \approx \frac{197.6 - 196.43478}{2(16.3)} \approx 0.036 \qquad \text{Subtract stored value.}$$

---

**·TECHNOLOGY·**

Try to calculate the value of $x$ in Example 4 by using additional parentheses instead of storing the intermediate result, 196.43478, in your calculator.

## Applications

In Section 1.3, you studied four basic types of applications involving quadratic equations: area, falling bodies, Pythagorean Theorem, and quadratic models. The solution of each of these types of problems can involve the Quadratic Formula. For instance, Example 5 shows how the Quadratic Formula can be used to analyze a quadratic model for school expenditures.

**Example 5**     School Expenditures

From 1990 to 1996, the total school expenditures (elementary through university) in the United States closely followed the quadratic model

$$\text{Expenditures} = 77.0t^2 + 9958.3t + 452{,}234, \quad t \geq 0$$

where the expenditures are measured in millions of dollars and the time $t = 0$ represents 1990. These expenses are shown graphically in Figure 1.11. Assuming that this model continues to be valid, in what year will the total school expenditures reach \$620,000,000,000 per year? (Source: U.S. National Center for Education Statistics)

### Solution

Because the expenditures are measured in millions of dollars, you need to solve the equation

$$77.0t^2 + 9958.3t + 452{,}234 = 620{,}000.$$

To begin, write the equation in standard form as

$$77.0t^2 + 9958.3t - 167{,}766 = 0.$$

Then apply the Quadratic Formula with $a = 77.0$, $b = 9958.3$, and $c = -167{,}766$ to obtain

$$t = \frac{-9958.3 \pm \sqrt{(9958.3)^2 - 4(77.0)(-167{,}766)}}{2(77.0)}.$$

Choosing the positive solution, you will find that

$$t = \frac{-9958.3 + \sqrt{(9958.3)^2 - 4(77.0)(-167{,}766)}}{2(77.0)}$$

$$\approx 15.$$

Because $t = 0$ corresponds to 1990, it follows that $t = 15$ must correspond to 2005. So, from the model you can conclude that the total school expenditures will reach \$620,000,000,000 in 2005. A table that shows the expenditures for several years can help confirm this result.

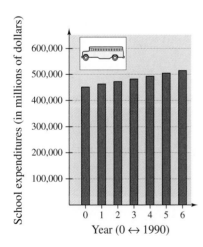

Figure 1.11

In Section 1.3, you learned that the position equation for a falling object is of the form

$$s = -16t^2 + v_0 t + s_0$$

where $s$ is the height (in feet), $v_0$ is the initial velocity (in feet per second), $t$ is the time (in seconds), and $s_0$ is the initial height (in feet). This equation is valid only for free-falling objects near earth's surface. Because of differences in gravitational force, position equations are different on other planets or moons. For instance, the next example looks at a position equation on our moon.

---

## •TECHNOLOGY•

Use the recall expression feature of your graphing calculator to find the time in the air on earth for the rock in Example 6. Simply replace $-2.7$ with $-16$ in the expression for $t$.

---

**Example 6**   Throwing an Object on the Moon

An astronaut standing on the surface of the moon throws a rock into space, as shown in Figure 1.12. The height of the rock is given by

$$s = -2.7t^2 + 27t + 6.$$

How much time will elapse before the rock strikes the lunar surface?

### Solution

Because $s$ gives the height of the rock at any time $t$, you can find the time that the rock hits the surface of the moon by setting $s$ equal to zero and solving for $t$.

$$-2.7t^2 + 27t + 6 = 0$$

$$t = \frac{-27 \pm \sqrt{(27)^2 - 4(-2.7)(6)}}{2(-2.7)} \qquad \text{Quadratic Formula}$$

$$\approx 10.2 \text{ seconds} \qquad \text{Choose positive solution.}$$

So, about 10.2 seconds will elapse before the rock hits the surface.

**Figure 1.12**

---

| Discussing the Concept | Methods for Solving a Quadratic Equation |
|---|---|

List all of the methods that can be used to solve a quadratic equation. Try not to refer to your book or notes. Identify the types of quadratic equations that are more easily solved by each method. Give an example of each type of equation and use the method to solve it.

# Warm Up

The following warm-up exercises involve skills that were covered in previous courses and earlier sections. You will use these skills in the exercise set for this section.

In Exercises 1–4, simplify the expression.

1. $\sqrt{9 - 4(3)(-12)}$
2. $\sqrt{36 - 4(2)(3)}$
3. $\sqrt{12^2 - 4(3)(4)}$
4. $\sqrt{15^2 + 4(9)(12)}$

In Exercises 5–10, solve the quadratic equation by factoring.

5. $x^2 - x - 2 = 0$
6. $2x^2 + 3x - 9 = 0$
7. $x^2 - 4x = 5$
8. $2x^2 + 13x = 7$
9. $x^2 = 5x - 6$
10. $x(x - 3) = 4$

## 1.4　Exercises

In Exercises 1–8, use the discriminant to determine the number of real solutions of the quadratic equation.

1. $4x^2 - 4x + 1 = 0$
2. $2x^2 - x - 1 = 0$
3. $3x^2 + 4x + 1 = 0$
4. $x^2 + 2x + 4 = 0$
5. $2x^2 - 5x + 5 = 0$
6. $3x^2 - 6x + 3 = 0$
7. $\frac{1}{5}x^2 + \frac{6}{5}x - 8 = 0$
8. $\frac{1}{3}x^2 - 5x + 25 = 0$

In Exercises 9–30, use the Quadratic Formula to solve the equation.

9. $2x^2 + x - 1 = 0$
10. $2x^2 - x - 1 = 0$
11. $16x^2 + 8x - 3 = 0$
12. $25x^2 - 20x + 3 = 0$
13. $2 + 2x - x^2 = 0$
14. $x^2 - 10x + 22 = 0$
15. $x^2 + 14x + 44 = 0$
16. $6x = 4 - x^2$
17. $x^2 + 8x - 4 = 0$
18. $4x^2 - 4x - 4 = 0$
19. $12x - 9x^2 = -3$
20. $16x^2 + 22 = 40x$
21. $36x^2 + 24x = 7$
22. $3x + x^2 - 1 = 0$
23. $4x^2 + 4x = 7$
24. $16x^2 - 40x + 5 = 0$
25. $28x - 49x^2 = 4$
26. $9x^2 + 24x + 16 = 0$
27. $8t = 5 + 2t^2$
28. $25h^2 + 80h + 61 = 0$
29. $(y - 5)^2 = 2y$
30. $(x + 6)^2 = -2x$

In Exercises 31–36, use a calculator to solve the equation. (Round your answer to three decimal places.)

31. $5.1x^2 - 1.7x - 3.2 = 0$
32. $10.4x^2 + 8.6x + 1.2 = 0$
33. $7.06x^2 - 4.85x + 0.50 = 0$
34. $-0.005x^2 + 0.101x - 0.193 = 0$
35. $422x^2 - 506x - 347 = 0$
36. $2x^2 - 2.50x - 0.42 = 0$

In Exercises 37–46, solve the equation by any convenient method.

**37.** $3x + 4 = 2x - 7$

**38.** $x^2 - 2x + 5 = x^2 - 5$

**39.** $4x^2 - 15 = 25$

**40.** $4x^2 + 2x + 4 = 2x - 8$

**41.** $x^2 + 3x + 1 = 0$

**42.** $x^2 + 3x - 4 = 0$

**43.** $(x - 1)^2 = 9$

**44.** $2x^2 - 4x - 6 = 0$

**45.** $100x^2 - 400 = 0$

**46.** $2x^2 + 4x - 9 = 2(x - 1)^2$

*Writing Real-Life Problems*  In Exercises 47–50, solve the number problem *and* write a real-life problem that could be represented by this verbal model. For instance, an applied problem that could be represented by Exercise 47 is as follows.

*The sum of the length and width of a one-story house is 100 feet. The house has 2500 square feet of floor space. What are the length and width of the house?*

**47.** Find two numbers whose sum is 100 and whose product is 2500.

**48.** One number is 1 more than another number. The product of the two numbers is 72. Find the numbers.

**49.** One number is 1 more than another number. The sum of their squares is 113. Find the numbers.

**50.** One number is 2 more than another number. The product of the two numbers is 440. Find the numbers.

*Cost Equation*  In Exercises 51–54, use the cost equation to find the number of units $x$ that a manufacturer can produce for the cost $C$. (Round your answer to the nearest positive integer.)

**51.** $C = 0.125x^2 + 20x + 5000$ $\quad$ $C = \$14,000$

**52.** $C = 0.5x^2 + 15x + 5000$ $\quad$ $C = \$11,500$

**53.** $C = 800 + 0.04x + 0.002x^2$ $\quad$ $C = \$1680$

**54.** $C = 800 - 10x + \dfrac{x^2}{4}$ $\quad$ $C = \$896$

**55.** *Seating Capacity*  A rectangular classroom seats 72 students. If the seats were rearranged with three more seats in each row, the classroom would have two fewer rows. Find the original number of seats in each row.

**56.** *Measurements of a Corral*  A rancher has 200 feet of fencing to enclose two adjacent rectangular corrals (see figure). Find the measurements such that the enclosed area will be 1400 square feet.

$4x + 3y = 200$

**57.** *Geometry*  An open box is to be made from a square piece of material by cutting 2-inch squares from the corners and turning up the sides (see figure). The volume of the finished box is to be 200 cubic inches. Find the size of the original piece of material.

**58.** *Geometry*  An open box (see figure) is to be constructed from 108 square inches of material. Find the measurements of the square base.

**59.** *On the Moon*  An astronaut on the moon throws a rock straight up into space. The initial velocity is 40 feet per second and the initial height is 5 feet. How long will it take the rock to hit the surface? If the rock had been thrown with the same initial velocity and height on earth, how long would it remain in the air? (See Example 6.)

**60.** *Hot Air Balloon*    Two people are riding in a hot air balloon that is 200 feet above the ground. One person drops a sack of sand and the balloon starts to rise. One second later, the balloon is 20 feet higher and the other person throws a sack of sand toward the ground (see figure). The position equation for the first sack is $s = -16t^2 + 200$, and (using an initial velocity of $-20$ feet per second) the position equation for the second sack is

$$s = -16t^2 - 20t + 220.$$

Which sack of sand will hit the ground first? (*Hint:* Remember that the first sack was dropped one second before the second sack.)

*Falling Objects*   In Exercises 61 and 62, use the following information. The position equation for falling objects on earth is of the form

$$s = -16t^2 + v_0t + s_0$$

where $s$ is the height (in feet), $v_0$ is the initial velocity (in feet per second), $t$ is the time (in seconds), and $s_0$ is the initial height (in feet).

**61.** If a rock were thrown on the surface of the earth with an initial velocity of 27 feet per second from an initial height of 6 feet, would it take a longer or a shorter period of time to reach the ground than it would on the surface of the moon? (See Example 6.)

**62.** Use the Quadratic Formula to find the actual time that the rock would remain in the air.

**63.** *Flying Distance*    A small commuter airline flies to three cities whose locations form the vertices of a right triangle (see figure). The total flight distance (from City A to City B to City C and back to City A) is 1400 miles. It is 600 miles between the two cities that are farthest apart. Find the other two distances between cities.

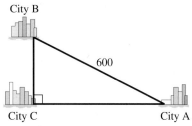

Figure for 63

**64.** *Distance from the Dock*    A windlass is used to tow a boat to a dock. The rope is attached to the boat at a point 15 feet below the level of the windlass (see figure). Find the distance from the boat to the dock when there is 75 feet of rope out.

**65.** *Recreational Spending*    The total number of dollars (in billions) spent on recreation in the United States from 1985 to 1995 can be approximated by the model $S = 134 + 7.9t + 0.64t^2$ where $t = 5$ represents 1985. The figure shows the actual spending and the spending represented by the model. (Source: U.S. Bureau of Economic Analysis)

(a) During which year was $310 billion spent on recreation?

(b) From the model, in what year will the total annual recreational spending reach $550 billion?

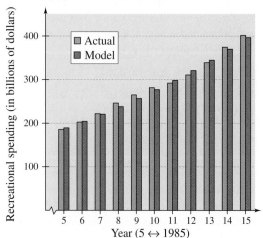

**66.** *U.S. Inventories*    The total amount of inventory (in billions of dollars) for manufacturing and trade in the United States from 1990 to 1996 can be approximated by the model

Inventory $= 5.93t^2 - 3.7t + 829$

where $t = 0$ represents 1990. The figure shows the actual spending and the spending represented by the model. From this model, in what year will total inventory reach $1,500,000,000,000? (Source: U.S. Bureau of the Census)

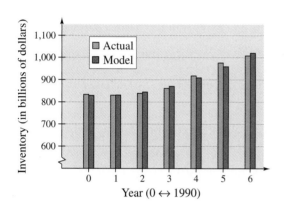

**67.** *Computers in the Classroom*    The number of personal computers (in millions) used in grades K through 12 in the United States from 1992 to 1996 can be modeled by

Number $= 0.071t^2 + 0.09t + 3.58$

where $t = 2$ represents 1992. From this model, in what year will the number of computers in use in grades K through 12 reach 15 million? (Source: Future Computing/Data Inc.)

**68.** *Physicians' Services*    The amount of money spent annually for physicians' services in the United States from 1960 to 1995 is approximated by the model

Amount $= 0.248t^2 - 3.241t + 15.21$

where the amount spent is in billions of dollars and $t$ represents the year, with $t = 0$ corresponding to 1960. From this model, in what year will the amount reach $350 billion? (Source: U.S. Health Care Financing Administration)

**69.** *Flying Speed*    Two planes leave simultaneously from the same airport, one flying due east and the other due south (see figure). The eastbound plane is flying 50 miles per hour faster than the southbound plane. After 3 hours the planes are 2440 miles apart. Find the speed of each plane.

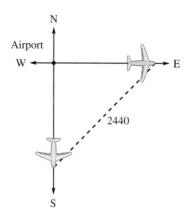

**70.** *Flying Speed*    Two planes leave simultaneously from the same airport, one flying due east and the other due south. The eastbound plane is flying 100 miles per hour faster than the southbound plane. After 2 hours the planes are 1500 miles apart. Find the speed of each plane.

**71.** *Total Revenue*    The demand equation for a certain product is

$p = 60 - 0.0004x$

where $p$ is the price per unit and $x$ is the number of units sold. The total revenue for selling $x$ units is given by $R = xp$. How many units must be sold to produce a revenue of $220,000?

**72.** *Total Revenue*    The demand equation for a certain product is

$p = 50 - 0.0005x$

where $p$ is the price per unit and $x$ is the number of units sold. The total revenue for selling $x$ units is given by $R = xp$. How many units must be sold to produce a revenue of $250,000?

**73.** When solving some problems, such as the problem in Example 5 on page 40, using the Quadratic Formula, why is only one solution used?

## Mid-Chapter Quiz

Take this quiz as you would take a quiz in class. After you are done, check your work against the answers given in the back of the book.

In Exercises 1–4, solve the equation and check the solution.

**1.** $2(x + 3) - 5(2x - 3) = 5$

**2.** $\dfrac{3x + 3}{5x - 2} = \dfrac{3}{4}$

**3.** $\dfrac{2}{x(x - 1)} + \dfrac{1}{x} = \dfrac{1}{x - 4}$

**4.** $(x + 3)^2 - x^2 = 6(x + 2)$

In Exercises 5 and 6, solve the equation. (Round to three decimal places.)

**5.** $\dfrac{x}{3.057} + \dfrac{x}{4.392} = 100$

**6.** $0.378x + 0.757(500 - x) = 215$

**7.** Describe how you can check your answers to Exercises 1–4, using your graphing utility.

In Exercises 8 and 9, write an algebraic equation for the verbal description. Find the solution if possible and check.

**8.** *Production Limit*   A company has fixed costs of $20,000 per month and variable costs of $7.50 per unit manufactured. The company has $80,000 to cover monthly costs. How many units can the company manufacture?

**9.** *Total Revenue*   The demand equation for a certain product is $p = 60 - 0.0004x$, where $p$ is the price per unit and $x$ is the number of units sold. The total revenue for selling $x$ units is given by

$$\text{Revenue} = xp = x(60 - 0.0004x).$$

How many units must be sold to produce revenue of $250,000?

In Exercises 10–15, solve the equation.

**10.** *Factoring:* $3x^2 + 13x = 10$

**11.** *Extracting roots:* $3x^2 = 15$

**12.** *Extracting roots:* $(x + 3)^2 = 17$

**13.** *Quadratic Formula:* $3x^2 + 7x - 2 = 0$

**14.** *Quadratic Formula:* $2x + x^2 = 5$

**15.** *Quadratic Formula:* $3x^2 - 4.50x - 0.32 = 0$

In Exercises 16 and 17, use the discriminant to determine the number of real solutions.

**16.** $x^2 + 3x - 5 = 0$

**17.** $2x^2 - 4x + 9 = 0$

**18.** Describe how you would convince a fellow student that $(x + 3)^2 = x^2 + 6x + 9$.

**19.** A fire retardant is dropped from a height of 250 feet. With no initial velocity, what is the minimum time it will take the fire retardant to hit the ground? Explain.

**20.** An open box 5 inches deep and 180 cubic inches in volume is to be constructed. Find the measurements of the square base.

## 1.5    Other Types of Equations

### Objectives

- Solve a polynomial equation by factoring.
- Rewrite and solve an equation involving radicals or rational exponents.
- Rewrite and solve an equation with fractions or absolute values.
- Construct and use a nonquadratic model to solve an application problem.

## Polynomial Equations

In this section you will extend the techniques for solving equations to nonlinear and nonquadratic equations. At this point in the text, you have only three basic methods for solving nonlinear equations—*factoring, extracting roots,* and the *Quadratic Formula.* So the main goal of this section is to learn to *rewrite* nonlinear equations in a form to which you can apply one of these methods.

**Example 1**    Solving a Polynomial Equation by Factoring

Solve $3x^4 = 48x^2$.

### Solution

The basic approach here is first to write the polynomial equation in standard form with zero on one side, to factor the other side, and then to set each factor equal to zero.

| | |
|---|---|
| $3x^4 = 48x^2$ | Original equation |
| $3x^4 - 48x^2 = 0$ | Collect terms on left side. |
| $3x^2(x^2 - 16) = 0$ | Common monomial factor |
| $3x^2(x + 4)(x - 4) = 0$ | Difference of two squares |
| $3x^2 = 0 \implies x = 0$ | Set 1st factor equal to 0. |
| $x + 4 = 0 \implies x = -4$ | Set 2nd factor equal to 0. |
| $x - 4 = 0 \implies x = 4$ | Set 3rd factor equal to 0. |

You can check these solutions by substituting in the original equation, as follows.

### Check

| | |
|---|---|
| $3x^4 = 48x^2$ | Original equation |
| $3(0)^4 = 48(0)^2$ | 0 checks. ✓ |
| $3(-4)^4 = 48(-4)^2$ | −4 checks. ✓ |
| $3(4)^4 = 48(4)^2$ | 4 checks. ✓ |

After checking, you can conclude that the solutions are 0, −4, and 4.

### Study Tip

A common mistake that is made in solving an equation such as that in Example 1 is dividing both sides of the equation by the variable factor $x^2$. This loses the solution $x = 0$. When using factoring to solve an equation, be sure to set each factor equal to zero. Don't divide both sides of an equation by a variable factor in an attempt to simplify the equation.

Example 2    Solving a Polynomial Equation by Factoring

Solve $x^3 - 3x^2 - 3x + 9 = 0$.

**Solution**

$$x^3 - 3x^2 - 3x + 9 = 0 \qquad \text{Original equation}$$

$$x^2(x - 3) - 3(x - 3) = 0 \qquad \text{Group terms.}$$

$$(x - 3)(x^2 - 3) = 0 \qquad \text{Factor by grouping.}$$

$$x - 3 = 0 \quad \Longrightarrow \quad x = 3 \qquad \text{Set 1st factor equal to 0.}$$

$$x^2 - 3 = 0 \quad \Longrightarrow \quad x = \pm\sqrt{3} \qquad \text{Set 2nd factor equal to 0.}$$

The solutions are $3$, $\sqrt{3}$, and $-\sqrt{3}$. Check these in the original equation.

·DISCOVERY·

What do you observe about the degree of the polynomials in Examples 1, 2, and 3 and the possible number of solutions of the equations? Does your observation apply to the quadratic equations in Sections 1.3 and 1.4?

Occasionally, mathematical models involve equations that are of **quadratic type**. In general, an equation is of quadratic type if it can be written in the form

$$au^2 + bu + c = 0$$

where $a \neq 0$ and $u$ is an algebraic expression.

Example 3    Solving an Equation of Quadratic Type

Solve $x^4 - 3x^2 + 2 = 0$.

**Solution**

This equation is of quadratic type with $u = x^2$.

$$(x^2)^2 - 3(x^2) + 2 = 0$$

To solve this equation, you can factor the left side of the equation as the product of two second-degree polynomials.

$$x^4 - 3x^2 + 2 = 0 \qquad \text{Original equation}$$

$$(x^2 - 1)(x^2 - 2) = 0 \qquad \text{Partially factor.}$$

$$(x + 1)(x - 1)(x^2 - 2) = 0 \qquad \text{Completely factor.}$$

$$x + 1 = 0 \quad \Longrightarrow \quad x = -1 \qquad \text{Set 1st factor equal to 0.}$$

$$x - 1 = 0 \quad \Longrightarrow \quad x = 1 \qquad \text{Set 2nd factor equal to 0.}$$

$$x^2 - 2 = 0 \quad \Longrightarrow \quad x = \pm\sqrt{2} \qquad \text{Set 3rd factor equal to 0.}$$

The solutions are $-1$, $1$, $\sqrt{2}$, and $-\sqrt{2}$. Check these in the original equation.

## Solving Equations Involving Radicals

The steps involved in solving the remaining equations in this section will often introduce *extraneous solutions,* as discussed in Section 1.1. Operations such as squaring both sides of an equation, raising both sides of an equation to a rational power, or multiplying both sides by a variable quantity all have this potential danger. So, when you use any of these operations, a check is crucial.

### Example 4    An Equation Involving a Rational Exponent

Solve $4x^{3/2} - 8 = 0$.

**Solution**

| | |
|---|---|
| $4x^{3/2} - 8 = 0$ | Original equation |
| $4x^{3/2} = 8$ | Add 8 to both sides. |
| $x^{3/2} = 2$ | Isolate $x^{3/2}$. |
| $x = 2^{2/3}$ | Raise both sides to $\frac{2}{3}$ power. |
| $x \approx 1.578$ | Round to three decimal places. |

**Check**

| | |
|---|---|
| $4x^{3/2} - 8 = 0$ | Original equation |
| $4(2^{2/3})^{3/2} \overset{?}{=} 8$ | Substitute $2^{2/3}$ for $x$. |
| $4(2) \overset{?}{=} 8$ | Property of exponents |
| $8 = 8$ | Solution checks. ✓ |

### Study Tip

The basic technique used in Example 4 is to isolate the factor with the rational exponent, and raise both sides to the *reciprocal power.* In Example 5, this is equivalent to isolating the square root and squaring both sides.

### Example 5    An Equation Involving a Radical

| | |
|---|---|
| $\sqrt{2x + 7} - x = 2$ | Original equation |
| $\sqrt{2x + 7} = x + 2$ | Isolate the square root. |
| $2x + 7 = x^2 + 4x + 4$ | Square both sides. |
| $0 = x^2 + 2x - 3$ | Standard form |
| $0 = (x + 3)(x - 1)$ | Factor. |
| $x + 3 = 0 \implies x = -3$ | Set 1st factor equal to 0. |
| $x - 1 = 0 \implies x = 1$ | Set 2nd factor equal to 0. |

By checking these values, you can determine that the only solution is 1.

## Equations with Fractions or Absolute Values

To solve an equation involving fractions, multiply both sides of the equation by the least common denominator (LCD) of each term in the equation. This procedure will "clear the equation of fractions." For instance, in the equation

$$\frac{2}{x^2 + 1} + \frac{1}{x} = \frac{2}{x}$$

you can multiply both sides of the equation by $x(x^2 + 1)$. Try doing this and solving the resulting equation. You should obtain one solution: $x = 1$.

### Example 6  An Equation Involving Fractions

Solve $\frac{2}{x} = \frac{3}{x - 2} - 1$.

#### Solution

For this equation, the LCD of the three terms is $x(x - 2)$, so we begin by multiplying each term in the equation by this expression.

$$\frac{2}{x} = \frac{3}{x - 2} - 1$$

$$x(x - 2)\frac{2}{x} = x(x - 2)\frac{3}{x - 2} - x(x - 2)(1)$$

$$2(x - 2) = 3x - x(x - 2)$$

$$2x - 4 = -x^2 + 5x$$

$$x^2 - 3x - 4 = 0$$

$$(x - 4)(x + 1) = 0$$

$$x - 4 = 0 \quad\Longrightarrow\quad x = 4$$

$$x + 1 = 0 \quad\Longrightarrow\quad x = -1$$

#### Check

$$\frac{2}{x} = \frac{3}{x - 2} - 1 \qquad \text{Original equation}$$

$$\frac{2}{4} \overset{?}{=} \frac{3}{4 - 2} - 1 \qquad \text{Substitute 4 for } x.$$

$$\frac{1}{2} = \frac{3}{2} - 1 \qquad \text{4 checks. } \checkmark$$

$$\frac{2}{-1} \overset{?}{=} \frac{3}{-1 - 2} - 1 \qquad \text{Substitute } -1 \text{ for } x.$$

$$-2 = -1 - 1 \qquad -1 \text{ checks. } \checkmark$$

The solutions are 4 and $-1$.

To solve an equation involving an absolute value, remember that the expression inside the absolute value signs can be positive or negative. This results in *two* separate equations, each of which must be solved. For instance, the equation

$$|x - 2| = 3$$

results in the two equations

$$x - 2 = 3 \quad \text{and} \quad -(x - 2) = 3$$

which implies that the equation has two solutions: 5 and $-1$.

### Example 7    An Equation Involving Absolute Value

Solve $|x^2 - 3x| = -4x + 6$.

### Solution

Because the variable expression inside the absolute value signs can be positive or negative, you must solve the following two equations.

*First Equation*

| | |
|---|---|
| $x^2 - 3x = -4x + 6$ | Use positive expression. |
| $x^2 + x - 6 = 0$ | Standard form |
| $(x + 3)(x - 2) = 0$ | Factor. |
| $x + 3 = 0 \implies x = -3$ | Set 1st factor equal to 0. |
| $x - 2 = 0 \implies x = 2$ | Set 2nd factor equal to 0. |

**Study Tip**

When solving an equation with absolute value, such as in Example 7, write the two equations to be solved. Then solve each equation independently. Be sure to check your solution.

*Second Equation*

| | |
|---|---|
| $-(x^2 - 3x) = -4x + 6$ | Use negative expression. |
| $x^2 - 7x + 6 = 0$ | Standard form |
| $(x - 1)(x - 6) = 0$ | Factor. |
| $x - 1 = 0 \implies x = 1$ | Set 1st factor equal to 0. |
| $x - 6 = 0 \implies x = 6$ | Set 2nd factor equal to 0. |

### Check

| | |
|---|---|
| $|(-3)^2 - 3(-3)| = -4(-3) + 6$ | $-3$ checks. ✓ |
| $|(2)^2 - 3(2)| \neq -4(2) + 6$ | 2 does not check. ✗ |
| $|(1)^2 - 3(1)| = -4(1) + 6$ | 1 checks. ✓ |
| $|(6)^2 - 3(6)| \neq -4(6) + 6$ | 6 does not check. ✗ |

The solutions are $-3$ and 1.

## Applications

It would be impossible to categorize the many different types of applications that involve nonlinear and nonquadratic models. However, from the few examples and exercises that are given below, we hope that you will gain some appreciation for the variety of applications that can occur.

**Example 8**    Reduced Rates

A ski club chartered a bus for a ski trip at a cost of $480. In an attempt to lower the bus fare per skier, the club invited nonmembers to go along. When five non-members joined the trip, the fare per skier decreased by $4.80. How many club members are going on the trip?

### Solution

Begin the solution by creating a verbal model and assigning labels, as follows.

*Verbal Model:*   $\boxed{\text{Cost per skier}} \cdot \boxed{\text{Number of skiers}} = \boxed{\text{Cost of trip}}$

*Labels:*

Cost of trip = 480                                      (dollars)

Number of ski club members = $x$                       (people)

Number of skiers = $x + 5$                             (people)

Original cost per member = $\dfrac{480}{x}$            (dollars per person)

Cost per skier = $\dfrac{480}{x} - 4.80$              (dollars per person)

*Equation:*

$$\left(\frac{480}{x} - 4.80\right)(x + 5) = 480$$

$$\left(\frac{480 - 4.8x}{x}\right)(x + 5) = 480$$

$$(480 - 4.8x)(x + 5) = 480x$$

$$480x - 4.8x^2 - 24x + 2400 = 480x$$

$$-4.8x^2 - 24x + 2400 = 0$$

$$x^2 + 5x - 500 = 0$$

$$(x + 25)(x - 20) = 0$$

$$x + 25 = 0 \implies x = -25$$

$$x - 20 = 0 \implies x = 20$$

Choosing the positive value of $x$, you can conclude that 20 ski club members are going on the trip. Check this in the original statement of the problem.

Interest in a savings account is calculated by one of three basic methods: simple interest, interest compounded $n$ times per year, and interest compounded continuously. The next example uses the formula for interest that is compounded $n$ times per year.

$$A = P\left(1 + \frac{r}{n}\right)^{nt}$$

In this formula, $A$ is the balance in the account, $P$ is the principal (or original deposit), $r$ is the annual percentage rate (in decimal form), $n$ is the number of compoundings per year, and $t$ is the time in years. Later, in Chapter 5, you will study the derivation of this formula for compound interest.

**Example 9**   Compound Interest

Suppose that when you were born, your grandparents deposited $5000 in a long-term investment in which the interest was compounded quarterly. On your 25th birthday the balance in the account is $25,062.59. What was the annual percentage rate for this investment?

Solution

*Formula:*    $A = P\left(1 + \frac{r}{n}\right)^{nt}$

*Labels:*    Amount $= A = 25{,}062.59$                               (dollars)
Principal $= P = 5000.00$                               (dollars)
Time $= t = 25$                                         (years)
Compoundings per year $= n = 4$            (compoundings)
Annual rate $= r$                  (percent in decimal form)

*Equation:*      $25{,}062.59 = 5000\left(1 + \dfrac{r}{4}\right)^{4(25)}$

$$\frac{25{,}062.59}{5000} = \left(1 + \frac{r}{4}\right)^{100}$$

$$5.0125 \approx \left(1 + \frac{r}{4}\right)^{100}$$

$$(5.0125)^{1/100} \approx 1 + \frac{r}{4}$$

$$1.01625 \approx 1 + \frac{r}{4}$$

$$0.01625 \approx \frac{r}{4}$$

$$0.065 \approx r$$

The annual percentage rate is $0.065 = 6.5\%$. Check this in the original statement of the problem.

**Example 10**    Market Research

The marketing department at a publisher is asked to determine the price of a book. The department determines that the demand for the book depends on the price of the book according to the formula

$$p = 40 - \sqrt{0.0001x + 1}, \quad 0 \le x$$

where $p$ is the price per book in dollars and $x$ is the number of books sold at the given price. For instance, in Figure 1.13, note that if the price were $39 then (according to the model) no one would be willing to buy the book. On the other hand, if the price were $17.60, 5 million copies could be sold. If the publisher set the price at $12.95, how many copies would be sold?

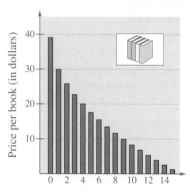

Number of books sold (in millions)

**Figure 1.13**

**Solution**

| | | |
|---|---|---|
| $p = 40 - \sqrt{0.0001x + 1}$ | Given model |
| $12.95 = 40 - \sqrt{0.0001x + 1}$ | Set price at $12.95. |
| $\sqrt{0.0001x + 1} = 27.05$ | Isolate radical. |
| $0.0001x + 1 = 731.7025$ | Square both sides. |
| $0.0001x = 730.7025$ | Subtract 1 from both sides. |
| $x = 7,307,025$ | Divide both sides by 0.0001. |

So, by setting the book's price at $12.95, the publisher can expect to sell about 7.3 million copies.

**Discussing the Concept    Compound Interest**

Suppose that on the day you were born a family member deposited money in an investment that earned interest compounded monthly at the rate of 7% per year. How much money needed to be deposited for the account to be worth $100,000 on your next birthday? How much money needed to be deposited for the account to be worth $1,000,000 on your next birthday?

# Warm Up

The following warm-up exercises involve skills that were covered in previous courses and earlier sections. You will use these skills in the exercise set for this section.

In Exercises 1–10, find the real solutions of the equation.

**1.** $x^2 - 22x + 121 = 0$

**2.** $x(x - 20) + 3(x - 20) = 0$

**3.** $(x + 20)^2 = 625$

**4.** $5x^2 + x = 0$

**5.** $3x^2 + 4x - 4 = 0$

**6.** $12x^2 + 8x - 55 = 0$

**7.** $x^2 + 4x - 5 = 0$

**8.** $4x^2 + 4x - 15 = 0$

**9.** $x^2 - 3x + 1 = 0$

**10.** $x^2 - 4x + 2 = 0$

## 1.5   Exercises

In Exercises 1–58, find the real solutions of the equation. Check your solutions.

**1.** $4x^4 - 18x^2 = 0$

**2.** $20x^3 - 125x = 0$

**3.** $x^3 - 2x^2 - 3x = 0$

**4.** $2x^4 - 15x^3 + 18x^2 = 0$

**5.** $x^4 - 81 = 0$

**6.** $x^6 - 64 = 0$

**7.** $5x^3 + 30x^2 + 45x = 0$

**8.** $9x^4 - 24x^3 + 16x^2 = 0$

**9.** $x^3 - 3x^2 - x + 3 = 0$

**10.** $x^3 + 2x^2 + 3x + 6 = 0$

**11.** $x^4 - x^3 + x - 1 = 0$

**12.** $x^4 + 2x^3 - 8x - 16 = 0$

**13.** $x^4 - 10x^2 + 9 = 0$

**14.** $x^4 - 29x^2 + 100 = 0$

**15.** $x^4 + 5x^2 - 36 = 0$

**16.** $x^4 - 4x^2 + 3 = 0$

**17.** $4x^4 - 65x^2 + 16 = 0$

**18.** $36t^4 + 29t^2 - 7 = 0$

**19.** $x^6 + 7x^3 - 8 = 0$

**20.** $x^6 + 3x^3 + 2 = 0$

**21.** $\sqrt{2x} - 10 = 0$

**22.** $4\sqrt{x} - 3 = 0$

**23.** $\sqrt{x - 10} - 4 = 0$

**24.** $\sqrt{5 - x} - 3 = 0$

**25.** $\sqrt[3]{2x + 5} + 3 = 0$

**26.** $\sqrt[3]{3x + 1} - 5 = 0$

**27.** $2x + 9\sqrt{x} - 5 = 0$

**28.** $6x - 7\sqrt{x} - 3 = 0$

**29.** $x = \sqrt{11x - 30}$

**30.** $2x - \sqrt{15 - 4x} = 0$

**31.** $-\sqrt{26 - 11x} + 4 = x$

**32.** $x + \sqrt{31 - 9x} = 5$

**33.** $\sqrt{x + 1} - 3x = 1$

**34.** $\sqrt{2x + 1} + x = 7$

**35.** $(x - 5)^{2/3} = 16$

**36.** $(x + 3)^{4/3} = 16$

**37.** $(x + 3)^{3/2} = 8$

**38.** $(x^2 + 2)^{2/3} = 9$

**39.** $(x^2 - 5)^{2/3} = 16$

**40.** $(x^2 - x - 22)^{4/3} = 16$

**41.** $\dfrac{1}{x} - \dfrac{1}{x + 1} = 3$

**42.** $\dfrac{x}{x^2 - 4} + \dfrac{1}{x + 2} = 3$

**43.** $\dfrac{20 - x}{x} = x$

**44.** $\dfrac{4}{x} - \dfrac{5}{3} = \dfrac{x}{6}$

**45.** $x = \dfrac{3}{x} + \dfrac{1}{2}$

**46.** $4x + 1 = \dfrac{3}{x}$

**47.** $\dfrac{1}{x} = \dfrac{4}{x - 1} + 1$

**48.** $x + \dfrac{9}{x + 1} = 5$

**49.** $\dfrac{4}{x + 1} - \dfrac{3}{x + 2} = 1$

**50.** $\dfrac{x + 1}{3} - \dfrac{x + 1}{x + 2} = 0$

**51.** $|x + 1| = 2$

**52.** $|x - 2| = 3$

**53.** $|2x - 1| = 5$

**54.** $|3x + 2| = 7$

**55.** $|x| = x^2 + x - 3$

**56.** $|x^2 + 6x| = 3x + 18$

**57.** $|x - 10| = x^2 - 10x$

**58.** $|x + 1| = x^2 - 5$

**59.** *Error Analysis*   Find the error(s) in the following solution.

**60.** *Error Analysis*    Find the error(s) in the solution.

$$\sqrt{6 - 2x} - 3 = 0$$
$$6 - 2x + 9 = 0$$
$$-2x = -15$$
$$x = \frac{15}{2}$$

In Exercises 61–64, use a calculator to find the real solutions of the equation. (Round your answers to three decimal places.)

**61.** $3.2x^4 - 1.5x^2 - 2.1 = 0$

**62.** $7.08x^6 + 4.15x^3 - 9.6 = 0$

**63.** $1.8x - 6\sqrt{x} - 5.6 = 0$

**64.** $4x + 8\sqrt{x} + 3.6 = 0$

**65.** *Sharing the Cost*    A college charters a bus for $1700 to take a group of students to the Fiesta Bowl. When six more students join the trip, the cost per student drops by $7.50. How many students were in the original group?

**66.** *Sharing the Cost*    Three students plan to rent an apartment and share equally on the rent. By adding a fourth person, each person could save $75 a month. How much is the monthly rent?

**67.** *Compound Interest*    A deposit of $2500 reaches a balance of $3544.06 after 5 years. The interest on the account is compounded monthly. What is the annual percentage rate for this investment?

**68.** *Compound Interest*    A sales representative describes a "guaranteed investment fund" that is offered to new investors. You are told that if you deposit $10,000 in the fund you will be guaranteed a return of at least $25,000 after 20 years. (a) If after 20 years you received the minimum guarantee, what annual percentage rate did you receive? (b) If after 20 years you received $35,000, what annual percentage rate did you receive? (Assume that the interest in the fund is compounded quarterly.)

**69.** *Borrowing Money*    Suppose you borrow $100 from a friend and agree to pay the money back, plus $10 in interest, after 6 months. Assuming that the interest is compounded monthly, what annual percentage rate are you paying?

**70.** *Cash Advance*    You take out a cash advance of $500 on a credit card. After 2 months, you owe $515.75. The interest is compounded monthly. What is the annual percentage rate for this loan?

**71.** *Airline Passengers*    An airline offers daily flights between Chicago and Denver. The total monthly cost of these flights is modeled by

$$C = \sqrt{0.2x + 1}$$

where $C$ is measured in millions of dollars and $x$ is measured in thousands of passengers (see figure). The total cost of the flights for a month is 2.5 million dollars. How many passengers flew that month?

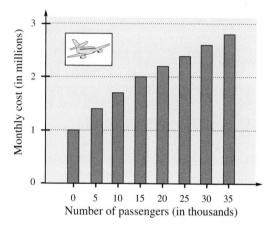

**72.** *Grocery Stores*    The number of grocery stores $G$ (in thousands) in the United States for the period from 1980 to 1995 can be approximated by the model

$$G = \sqrt{31,764 + 39t - 25t^2}$$

where $t = 0$ represents 1980. (Source: U.S. Department of Agriculture)

(a) Complete the table.

| $t$ | 0 | 5 | 10 | 15 |
|---|---|---|---|---|
| $G$ | | | | |

(b) From this model, in what year will there be 150,000 grocery stores in the United States?

**73.** *Life Expectancy*    The life expectancy table (for ages 58–75) used by the U.S. National Center for Health Statistics is modeled by

$$y = \sqrt{0.6632x^2 - 110.55x + 4680.24}$$

where $x$ represents a person's current age and $y$ represents the average number of years the person is expected to live (see figure). If a person's life expectancy is estimated to be 20 years, how old is the person? (Source: U.S. National Center for Health Statistics)

Figure for 73

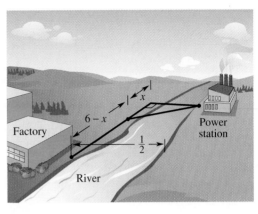

Figure for 76

**74.** *Federal Deficit*    The deficit (in millions of dollars) for the federal government from 1992 to 1997 can be modeled by

$$\text{Deficit} = -142,292,330 + 584,557t + \frac{7,774,119,500}{t}$$

where $t = 92$ represents 1992 (see figure). Predict the year that the deficit will reach 6 trillion dollars. (Source: U.S. Office of Management and Budget)

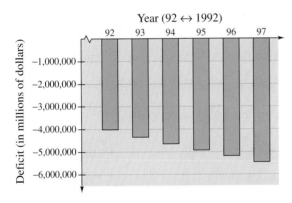

**75.** *Market Research*    The demand equation for a certain product is modeled by

$$p = 30 - \sqrt{0.0001x + 1}$$

where $x$ is the number of units demanded per day and $p$ is the price per unit. Find the demand if the price is set at $13.95.

**76.** *Power Line*    A power station is on one side of a river that is $\frac{1}{2}$ mile wide. A factory is 6 miles downstream on the other side of the river. It costs $18 per foot to run power lines over land and $24 per foot to run them under water. The project's cost is $616,877.27. Find the length $x$ as labeled in the figure.

**77.** *Sailboat Stays*    Two stays for the mast on a sailboat are attached to the boat at two points, as shown in the figure. One point is 10 feet from the mast and the other point is 15 feet from the mast. The total length of the two stays is 35 feet. How high on the mast are the stays attached?

**78.** *Batting Average*    A softball player has been at bat 120 times and has hit the ball safely 20 times. The player's batting average is $20/120 \approx .167$. How many consecutive times must the player hit the ball safely to obtain a batting average of .200? (*Hint:* (Batting average) = (total hits) ÷ (total times at bat))

**79.** *Work Rate*    With only the cold water valve open, it takes 8 minutes to fill the tub of a washing machine. With both the hot and cold water valves open, it takes only 5 minutes. The time it takes for the tub to fill with only the hot water valve open can be modeled by the equation

$$\frac{1}{8} + \frac{1}{t} = \frac{1}{5}$$

where $t$ is the time (in minutes) that it takes the tub to fill. How long does it take for the tub of the washing machine to fill with only the hot water valve open?

## 1.6    Linear Inequalities

### Objectives

- Write bounded and unbounded intervals using inequalities or interval notation.
- Solve and graph a linear inequality.
- Construct and use a linear inequality to solve an application problem.

## Introduction

Simple inequalities are used to *order* the real numbers. Inequality symbols $<, \le, >$, and $\ge$ are used to compare two numbers and to denote subsets of real numbers. For instance, the simple inequality

$$x \ge 3$$

denotes all real numbers $x$ that are greater than or equal to 3.

In this section you will expand your work with inequalities to include more involved statements such as

$$5x - 7 < 3x + 9 \quad \text{and} \quad -3 \le 6x - 1 < 3.$$

As with an equation, you **solve an inequality** in the variable $x$ by finding all values of $x$ for which the inequality is true. Such values are **solutions** and are said to **satisfy** the inequality. The set of all real numbers that are solutions of an inequality is the **solution set** of the inequality.

The set of all points on the real number line that represent the solution set is the **graph** of the inequality. Graphs of many types of inequalities consist of intervals on the real number line. The four different types of **bounded** intervals are summarized below.

---

▶ **Bounded Intervals on the Real Number Line**

Let $a$ and $b$ be real numbers such that $a < b$. The following intervals on the real number line are **bounded.** The numbers $a$ and $b$ are the **endpoints** of each interval.

| Notation | Interval Type | Inequality | Graph |
|----------|---------------|------------|-------|
| $[a, b]$ | Closed | $a \le x \le b$ | |
| $(a, b)$ | Open | $a < x < b$ | |
| $[a, b)$ | Half-open | $a \le x < b$ | |
| $(a, b]$ | Half-open | $a < x \le b$ | |

---

Note that a closed interval contains both of its endpoints, a half-open interval contains only one of its endpoints, and an open interval does not contain either of its endpoints. Often, the solution of an inequality is an interval on the real line that is **unbounded.** For instance, the interval consisting of all positive numbers is unbounded.

---

▶ Unbounded Intervals on the Real Number Line

Let $a$ and $b$ be real numbers. The following intervals on the real number line are **unbounded.**

| Notation | Interval Type | Inequality | Graph |
|---|---|---|---|
| $[a, \infty)$ | Half-open | $a \leq x$ | |
| $(a, \infty)$ | Open | $a < x$ | |
| $(-\infty, b]$ | Half-open | $x \leq b$ | |
| $(-\infty, b)$ | Open | $x < b$ | |
| $(-\infty, \infty)$ | Entire real line | | |

---

The symbols $\infty$ (**positive infinity**) and $-\infty$ (**negative infinity**) do not represent real numbers. They are simply convenient symbols used to describe the unboundedness of an interval such as $(1, \infty)$.

---

**Example 1**   Intervals and Inequalities

Write an inequality to represent each of the following intervals and state whether the interval is bounded or unbounded.

**a.** $(-3, 5]$     **b.** $(-3, \infty)$     **c.** $[0, 2]$

**Solution**

**a.** $(-3, 5]$ corresponds to $-3 < x \leq 5$.     Bounded

**b.** $(-3, \infty)$ corresponds to $-3 < x$.     Unbounded

**c.** $[0, 2]$ corresponds to $0 \leq x \leq 2$.     Bounded

## Properties of Inequalities

The procedures for solving linear inequalities in one variable are much like those for solving linear equations. To isolate the variable, you can make use of **properties of inequalities.** These properties are similar to the properties of equality, but there are two important exceptions. When both sides of an inequality are multiplied or divided by a negative number, the direction of the inequality symbol must be reversed. Here is an example.

$$-2 < 5 \qquad \text{Original inequality}$$

$$(-3)(-2) > (-3)(5) \qquad \text{Multiply both sides by } -3 \text{ and reverse the inequality.}$$

$$6 > -15 \qquad \text{Solution of inequality}$$

Two inequalities that have the same solution set are **equivalent.** For instance, the inequalities

$$x + 2 < 5 \quad \text{and} \quad x < 3$$

are equivalent. The following list describes operations that can be used to create equivalent inequalities.

---

▶ **Properties of Inequalities**

Let $a$, $b$, $c$, and $d$ be real numbers.

**1.** *Transitive Property*

$$a < b \text{ and } b < c \quad \Longrightarrow \quad a < c$$

**2.** *Addition of Inequalities*

$$a < b \text{ and } c < d \quad \Longrightarrow \quad a + c < b + d$$

**3.** *Addition of a Constant*

$$a < b \quad \Longrightarrow \quad a + c < b + c$$

**4.** *Multiplying by a Constant*

$$\text{For } c > 0, a < b \quad \Longrightarrow \quad ac < bc$$

$$\text{For } c < 0, a < b \quad \Longrightarrow \quad ac > bc$$

---

Each of the properties above is true if the symbol $<$ is replaced by $\leq$. For instance, another form of the multiplication property would be as follows.

$$\text{For } c > 0, a \leq b \quad \Longrightarrow \quad ac \leq bc.$$

$$\text{For } c < 0, a \leq b \quad \Longrightarrow \quad ac \geq bc.$$

## Solving a Linear Inequality

The simplest type of inequality to solve is a **linear inequality** in a single variable. For instance, $2x + 3 > 4$ is a linear inequality in $x$.

As you read through the following examples, pay special attention to the steps in which the inequality symbol is reversed. Remember that when you multiply or divide by a negative number, you must reverse the inequality symbol.

**Example 2**    Solving a Linear Inequality

Solve $5x - 7 > 3x + 9$.

**Solution**

| | |
|---|---|
| $5x - 7 > 3x + 9$ | Original inequality |
| $5x > 3x + 16$ | Add 7 to both sides. |
| $5x - 3x > 16$ | Subtract $3x$ from both sides. |
| $2x > 16$ | Combine like terms. |
| $x > 8$ | Divide both sides by 2. |

The solution set is all real numbers that are greater than 8, which is denoted by $(8, \infty)$. The graph is shown in Figure 1.14.

**Figure 1.14**    *Solution Interval:* $(8, \infty)$

Checking the solution set of an inequality is not as simple as checking the solutions of an equation. You can, however, get an indication of the validity of a solution set by substituting a few convenient values of $x$ to see whether the original inequality is satisfied.

**Example 3**    Solving a Linear Inequality

Solve $1 - \dfrac{3x}{2} \geq x - 4$.

**Solution**

| | |
|---|---|
| $1 - \dfrac{3x}{2} \geq x - 4$ | Original inequality |
| $2 - 3x \geq 2x - 8$ | Multiply both sides by 2. |
| $-3x \geq 2x - 10$ | Subtract 2 from both sides. |
| $-5x \geq -10$ | Subtract $2x$ from both sides. |
| $x \leq 2$ | Divide both sides by $-5$ and reverse inequality. |

The solution set is all real numbers that are less than or equal to 2, which is denoted by $(-\infty, 2]$. The graph is shown in Figure 1.15.

**Figure 1.15**    *Solution Interval:* $(-\infty, 2]$

Sometimes it is convenient to write two inequalities as a **double inequality.** For instance, you can write the two inequalities $-4 \le 5x - 2$ and $5x - 2 < 7$ more simply as $-4 \le 5x - 2 < 7$.

### Example 4    Solving a Double Inequality

To solve the following double inequality, we isolate $x$ as the middle term.

| | |
|---|---|
| $-3 \le 6x - 1 < 3$ | Original inequality |
| $-2 \le 6x < 4$ | Add 1 to all three parts. |
| $-\dfrac{1}{3} \le x < \dfrac{2}{3}$ | Divide by 6 and reduce. |

The solution set is all real numbers that are greater than or equal to $-\frac{1}{3}$ and less than $\frac{2}{3}$. The interval notation for this solution set is

$$\left[-\tfrac{1}{3}, \tfrac{2}{3}\right).$$     Solution set

The graph of this solution set is shown in Figure 1.16.

**Figure 1.16**    *Solution interval:* $\left[-\tfrac{1}{3}, \tfrac{2}{3}\right)$

The double inequality in Example 4 could have been solved in two parts as follows.

| | |
|---|---|
| $-3 \le 6x - 1$    and | $6x - 1 < 3$ |
| $-2 \le 6x$ | $6x < 4$ |
| $-\dfrac{1}{3} \le x$ | $x < \dfrac{2}{3}$ |

The solution set consists of all real numbers that satisfy *both* inequalities. In other words, the solution set is the set of all values of $x$ for which $-\frac{1}{3} \le x < \frac{2}{3}$.

When combining two inequalities to form a double inequality, be sure that the inequalities satisfy the Transitive Property. For instance, it is *incorrect* to combine the inequalities $3 < x$ and $x \le -1$ as $3 < x \le -1$. This "inequality" is obviously wrong because 3 is not less than $-1$.

# Inequalities Involving Absolute Value

▶ Solving an Absolute Value Inequality

Let $x$ be a variable or an algebraic expression and let $a$ be a real number such that $a \geq 0$.

1. The solutions of $|x| < a$ are all values of $x$ that lie between $-a$ and $a$.

$$|x| < a \qquad \text{if and only if } -a < x < a.$$

2. The solutions of $|x| > a$ are all values of $x$ that are less than $-a$ or greater than $a$.

$$|x| > a \qquad \text{if and only if } x < -a \text{ or } x > a.$$

These rules are also valid if $<$ is replaced by $\leq$ and $>$ is replaced by $\geq$.

---

| **Example 5** | Solving an Absolute Value Inequality |

Solve $|x - 5| < 2$.

**Solution**

$$|x - 5| < 2 \qquad \text{Original inequality}$$
$$-2 < x - 5 < 2 \qquad \text{Equivalent inequalities}$$
$$-2 + 5 < x - 5 + 5 < 2 + 5 \qquad \text{Add 5 to all three parts.}$$
$$3 < x < 7 \qquad \text{Solution set}$$

Figure 1.17

The solution set consists of all real numbers that are greater than 3 and less than 7, which is denoted by $(3, 7)$. The graph is shown in Figure 1.17.

---

| **Example 6** | Solving an Absolute Value Inequality |

Solve $|x + 3| \geq 7$.

**Solution**

$$|x + 3| \geq 7 \qquad \text{Original inequality}$$
$$x + 3 \leq -7 \qquad \text{or} \qquad x + 3 \geq 7 \qquad \text{Equivalent inequalities}$$
$$x + 3 - 3 \leq -7 - 3 \qquad x + 3 - 3 \geq 7 - 3 \qquad \text{Subtract 3 from both sides.}$$
$$x \leq -10 \qquad \qquad x \geq 4 \qquad \text{Solution set}$$

Figure 1.18

The solution set is all real numbers that are less than or equal to $-10$ *or* greater than or equal to 4, which is denoted by $(-\infty, -10] \cup [4, \infty)$ (see Figure 1.18). The symbol $\cup$ (union) means *or*.

## Applications

| Example 7 | Comparative Shopping |  |

A subcompact car can be rented from Company A for $180 per week with no extra charge for mileage. A similar car can be rented from Company B for $100 per week, plus 20 cents for each mile driven. How many miles must you drive in a week to make the rental fee for Company B more than that for Company A?

**Solution**

*Verbal Model:*

$$\boxed{\text{Weekly cost for Company B}} \; > \; \boxed{\text{Weekly cost for Company A}}$$

*Labels:*     Miles driven in one week $= m$                                  (miles)
Weekly cost for Company A $= 180$                          (dollars)
Weekly cost for Company B $= 100 + 0.20m$                (dollars)

*Inequality:*    $100 + 0.2m > 180$

$$0.2m > 80$$

$$m > 400 \text{ miles}$$

If you drive more than 400 miles in a week, Company A is cheaper.

| Example 8 | Exercise Program |  |

A man begins an exercise and diet program that is designed to reduce his weight by at least 2 pounds per week. At the beginning of the diet the man weighs 225 pounds. Find the maximum number of weeks before the man's weight will reach (or fall below) his goal of 192 pounds.

**Solution**

*Verbal Model:*

$$\boxed{\text{Desired weight}} \; \leq \; \boxed{\text{Current weight}} \; - \; \boxed{\text{2 pounds per week}} \; \cdot \; \boxed{\text{Number of weeks}}$$

*Labels:*     Desired weight $= 192$                              (pounds)
Current weight $= 225$                              (pounds)
Number of weeks $= x$                               (weeks)

*Inequality:*    $192 \leq 225 - 2x$

$$-33 \leq -2x$$

$$16.5 \geq x \text{ weeks}$$

Losing at least 2 pounds per week, it will take at most $16\frac{1}{2}$ weeks to reach his goal.

**Example 9**    Accuracy of a Measurement

You go to a candy store to buy chocolates that cost \$9.89 per pound. The scale that is used in the store has a state seal of approval that indicates the scale is accurate to within half an ounce. According to the scale, your purchase weighs one-half pound and costs \$4.95. How much might you have been undercharged or overcharged due to an error in the scale?

### Solution

To solve this problem, let $x$ represent the *true* weight of the candy. Because the state seal indicates that the scale is accurate to within half an ounce (or $\frac{1}{32}$ of a pound), you can conclude that the difference between the exact weight $(x)$ and the scale weight $\left(\frac{1}{2}\right)$ is less than or equal to $\frac{1}{32}$ of a pound. That is,

$$\left| x - \frac{1}{2} \right| \leq \frac{1}{32}.$$

You can solve this inequality as follows.

$$-\frac{1}{32} \leq x - \frac{1}{2} \leq \frac{1}{32}$$

$$\frac{15}{32} \leq x \leq \frac{17}{32}$$

$$0.46875 \leq x \leq 0.53125$$

In other words, your "one-half" pound of candy could have weighed as little as 0.46875 pound (which would have cost \$4.64) or as much as 0.53125 pound (which would have cost \$5.25). So, you could have been undercharged by as much as \$0.30 or overcharged by as much as \$0.31.

---

| Discussing the Concept | Absolute Value Inequalities |
| --- | --- |

Describe the solutions of

$$|3x + 4| > -2.$$

Give an analytical argument. Describe the solutions of

$$|3x + 4| < -2.$$

Give an analytical argument. Explain the difference between the inequalities.

## Warm Up

The following warm-up exercises involve skills that were covered in previous courses and earlier sections. You will use these skills in the exercise set for this section.

In Exercises 1–4, determine which of the two numbers is larger.

**1.** $-\frac{1}{2}, -7$

**2.** $-\frac{1}{3}, -\frac{1}{6}$

**3.** $-\pi, -3$

**4.** $-6, -\frac{13}{2}$

In Exercises 5–8, use inequality notation to denote the statement.

**5.** $x$ is nonnegative.

**6.** $z$ is strictly between $-3$ and $10$.

**7.** $P$ is no more than 2.

**8.** $W$ is at least 200.

In Exercises 9 and 10, evaluate the expression for the values of $x$.

**9.** $|x - 10|, x = 12, x = 3$

**10.** $|2x - 3|, x = \frac{3}{2}, x = 1$

## 1.6    Exercises

In Exercises 1 and 2, determine whether the values of $x$ are solutions of the inequality.

**1.** $5x - 12 > 0$

  (a) $x = 3$    (b) $x = -3$    (c) $x = \frac{5}{2}$    (d) $x = \frac{3}{2}$

**2.** $x + 1 < \dfrac{2x}{3}$

  (a) $x = 0$    (b) $x = 4$    (c) $x = -4$    (d) $x = -3$

In Exercises 3–10, match the inequality with its graph. [The graphs are labeled (a), (b), (c), (d), (e), (f), (g), and (h).]

(a)

(b)

(c)

(d)

(e)

(f)

(g)

(h)

**3.** $x < 4$

**4.** $x \geq 6$

**5.** $-2 < x \leq 5$

**6.** $0 \leq x \leq \frac{7}{2}$

**7.** $|x| < 4$

**8.** $|x| > 3$

**9.** $|x - 5| > 2$

**10.** $|x + 6| < 3$

In Exercises 11–46, solve the inequality and sketch its graph.

**11.** $3x \geq 9$

**12.** $5x \leq 20$

**13.** $\dfrac{x}{3} < 4$

**14.** $\dfrac{2}{5}x > 7$

**15.** $-10x < 40$

**16.** $-6x > 15$

**17.** $x - 5 \geq 7$

**18.** $x + 7 \leq 12$

**19.** $4(x + 1) < 2x + 3$

**20.** $2x + 7 < 3$

**21.** $2x - 1 \geq 0$

**22.** $3x + 1 \geq 2$

**23.** $3 \leq 2x - 1 < 7$

**24.** $3 > 1 - \dfrac{x}{2} > -3$

**25.** $1 < 2x + 3 < 9$

**26.** $-8 \leq 1 - 3(x - 2) < 13$

**27.** $-4 < \dfrac{2x - 3}{3} < 4$

**28.** $0 \leq \dfrac{x + 3}{2} < 5$

**29.** $\dfrac{3}{4} > x + 1 > \dfrac{1}{4}$

**30.** $-1 < -\dfrac{x}{3} < 1$

**31.** $|x + 3| < 5$

**32.** $\left|\dfrac{2x + 1}{2}\right| < 6$

**33.** $\left|\dfrac{x}{2}\right| > 3$

**34.** $|5x| > 10$

**35.** $|x - 20| \leq 4$

**36.** $|x - 7| < 6$

**37.** $|2x - 5| > 6$

**38.** $2|5 - 3x| + 7 < 21$

**39.** $\left|\dfrac{x - 3}{2}\right| \geq 5$

**40.** $|1 - 2x| < 5$

**41.** $|9 - 2x| - 2 < -1$

**42.** $\left|1 - \dfrac{2x}{3}\right| < 1$

**43.** $2|x + 10| \geq 9$

**44.** $3|4 - 5x| \leq 9$

**45.** $|x - 5| < 0$

**46.** $|x - 5| \geq 0$

In Exercises 47–54, use absolute value notation to define the solution set.

**47.**

**48.**

**49.**

**50.**

**51.** All real numbers at most 10 units of 12

**52.** All real numbers at least 5 units from 8

**53.** All real numbers whose distances from $-3$ are more than 5

**54.** All real numbers whose distances from $-6$ are no more than 7

**55.** *Comparative Shopping*   You can rent a midsize car from Company A for $250 per week with no extra charge for mileage. A similar car can be rented from Company B for $150 per week, plus 25 cents for each mile driven. How many miles must you drive in a week to make the rental fee for Company B greater than that for Company A?

**56.** *Comparative Shopping*   Your department sends its copying to the photocopy center of your company. The photocopy center bills your department $0.10 per page. You are considering buying a departmental copier for $3000. With your own copier the cost per page would be $0.03. The expected life of the copier is 4 years. How many copies must you make in the 4-year period to justify purchasing the copier?

**57.** *Simple Interest*   For $1000 to grow to more than $1250 in 2 years, what must the simple interest rate be?

**58.** *Simple Interest*   For $1000 to grow to more than $1500 in 2 years, what must the simple interest rate be?

**59.** *Weight Loss Program*   A person enrolls in a diet program that guarantees a loss of at least $1\frac{1}{2}$ pounds per week. The person's weight at the beginning of the program is 164 pounds. Find the maximum number of weeks that the person must be in the program before attaining a weight of 128 pounds.

**60.** *Salary Increase*  You accept a new job with a starting salary of $18,500. When you are employed, you are told that you will receive an annual raise of at least $1250. What is the maximum number of years that you must work before your annual salary will be $23,500?

**61.** *Break-Even Analysis*  The revenue for selling $x$ units of a product is $R = 115.95x$. The cost of producing $x$ units is $C = 95x + 750$. In order to obtain a profit, the revenue must be greater than the cost.

(a) Complete the table.

| $x$ | 10 | 20 | 30 | 40 | 50 |
|-----|----|----|----|----|----|
| $R$ |    |    |    |    |    |
| $C$ |    |    |    |    |    |

(b) For what values of $x$ will this product return a profit?

**62.** *Break-Even Analysis*  The revenue for selling $x$ units of a product is $R = 24.55x$. The cost of producing $x$ units is

$$C = 15.4x + 150,000.$$

In order to obtain a profit, the revenue must be greater than the cost. For what values of $x$ will this product return a profit?

**63.** *Annual Operating Cost*  A utility company has a fleet of vans. The annual operating cost per van is

$$C = 0.32m + 2300$$

where $m$ is the number of miles traveled by a van in a year. What number of miles will yield an annual operating cost that is less than $10,000?

**64.** *Daily Sales*  A doughnut shop at a shopping mall sells a dozen doughnuts for $2.95. Beyond the fixed costs (rent, utilities, and insurance) of $150 per day, it costs $1.45 for enough materials (flour, sugar, and so on) and labor to produce a dozen doughnuts. If the daily profit varies between $50 and $200, between what levels (in dozens) do the daily sales vary?

**65.** *Cable Television Subscribers*  The number of cable television subscribers (in thousands) in the United States between 1990 and 1996 can be modeled by

$$\text{Subscribers} = 1744t + 49,664, \qquad t \geq 0$$

where $t = 0$ represents 1990. According to this model, in what year will the number of subscribers exceed 70 million (see figure)?  (Source: *Television and Cable Fact Book*)

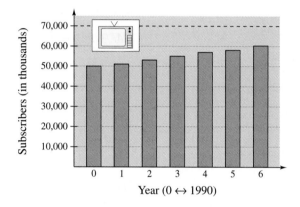

**66.** *Teachers' Salaries*  The average salary for elementary and secondary school teachers in the United States from 1985 to 1997 can be modeled by

$$\text{Salary} = 1345t + 17,479, \qquad t \geq 0$$

where $t = 5$ represents 1985 (see figure). According to the model, in what year will the average salary exceed $45,000?  (Source: National Education Association)

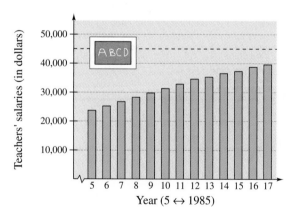

**67.** *Accuracy of Measurement*   You buy six T-bone steaks that cost $3.98 per pound. The weight that is listed on the package is 5.72 pounds. If the scale that weighed the package is accurate to within $\frac{1}{2}$ ounce, how much money might you have been undercharged or overcharged?

**68.** *Accuracy of Measurement*   You buy a bag of oranges that cost $0.69 per pound. The weight that is listed on the bag is 4.65 pounds. If the scale that weighed the bag is only accurate to within 1 ounce, how much money might you have been undercharged or overcharged?

**69.** *Human Height*   The heights $h$ of two-thirds of a population satisfy the inequality

$$|h - 68.5| \le 2.7$$

where $h$ is measured in inches. Determine the interval on the real number line in which these heights lie.

**70.** *Humidity Control*   An electronic device is to be operated in a room with relative humidity $h$ defined by

$$|h - 50| \le 30.$$

What are the minimum and maximum relative humidities for the operation of this device?

## Math Matters • Man and Mouse

A man who falls from a height of 2000 feet (without a parachute) will strike the ground with lethal force. But a mouse can fall from the same height and simply get up and walk away. Why?

The answer is that the speed at which a falling object hits the ground depends partly on the air resistance of the object, which in turn depends on the object's weight and surface area. If the ratio of an object's surface area to its weight is large, then its air resistance will be large. On the other hand, if the ratio of an object's surface area to its weight is small, then its air resistance will be small. This is why a parachute works—a person with a parachute has a much larger surface area (for approximately the same weight) than a person without a parachute. So how does this relate to the falling man and mouse? The falling mouse has a much greater air resistance than the falling man because the ratio of the mouse's surface area to its weight is greater than the ratio of the man's surface area to his weight. To convince yourself that the mouse's ratio is greater than the man's, try the following experiment. Find the ratio of the surface area and weight for the cubes described below.

Notice that as the cube becomes larger, the ratio of its surface area to its weight becomes smaller. (The answers are given in the back of the text.)

| Length of Side | Surface Area | Volume | Density | Weight |
|---|---|---|---|---|
| 1 ft | 6 ft² | 1 ft³ | 1 lb/ft³ | 1 lb |
| 2 ft | 24 ft² | 8 ft³ | 1 lb/ft³ | 8 lb |
| 3 ft | 54 ft² | 27 ft³ | 1 lb/ft³ | 27 lb |
| 4 ft | 96 ft² | 64 ft³ | 1 lb/ft³ | 64 lb |

## 1.7 Other Types of Inequalities

### Objectives

- Use critical numbers to determine test intervals for a polynomial inequality.
- Solve and graph a polynomial inequality.
- Solve and graph a rational inequality.
- Determine the domain of an expression involving a square root.
- Construct and use a polynomial inequality to solve an application problem.

## Polynomial Inequalities

To solve a polynomial inequality such as

$$x^2 - 2x - 3 < 0$$

you can use the fact that a polynomial can change signs only at its zeros (the $x$-values that make the polynomial equal to zero). Between two consecutive zeros a polynomial must be entirely positive or entirely negative. This means that when the real zeros of a polynomial are put in order, they divide the real number line into intervals in which the polynomial has no sign changes. These zeros are the **critical numbers** of the inequality, and the resulting intervals are the **test intervals** for the inequality. For example, the polynomial

$$x^2 - 2x - 3 = (x + 1)(x - 3)$$

has two zeros, $x = -1$ and $x = 3$, and these zeros divide the real number line into three test intervals:

$$(-\infty, -1), \quad (-1, 3), \quad \text{and} \quad (3, \infty).$$

So, to solve the inequality $x^2 - 2x - 3 < 0$, you need only test one value from each of these test intervals. You can use the same basic approach to determine the test intervals for any polynomial.

> ▶ **Finding Test Intervals for a Polynomial**
>
> To determine the intervals on which the values of a polynomial are entirely negative or entirely positive, use the following steps.
>
> **1.** Find all real zeros of the polynomial, and arrange the zeros in increasing order (from smallest to largest). These zeros are the **critical numbers** of the polynomial.
>
> **2.** Use the critical numbers of the polynomial to determine its **test intervals.**
>
> **3.** Choose one representative $x$-value in each test interval and evaluate the polynomial at that value. If the value of the polynomial is negative, the polynomial will have negative values for *every* $x$-value in the interval. If the value of the polynomial is positive, the polynomial will have positive values for *every* $x$-value in the interval.

| Example 1 | Finding Test Intervals for a Polynomial |

Solve $x^2 - x - 6 < 0$.

**Solution**

By factoring the quadratic as

$$x^2 - x - 6 = (x + 2)(x - 3)$$

you can see that the critical numbers are $x = -2$ and $x = 3$. So, the polynomial's test intervals are

$$(-\infty, -2), \quad (-2, 3), \quad \text{and} \quad (3, \infty). \qquad \text{Test intervals}$$

In each test interval, choose a representative $x$-value and evaluate the polynomial.

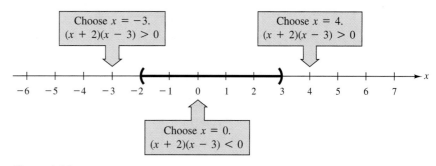

| Interval | $x$-Value | Polynomial Value | Conclusion |
|---|---|---|---|
| $(-\infty, -2)$ | $x = -3$ | $(-3)^2 - (-3) - 6 = 6$ | Positive |
| $(-2, 3)$ | $x = 0$ | $(0)^2 - (0) - 6 = -6$ | Negative |
| $(3, \infty)$ | $x = 4$ | $(4)^2 - (4) - 6 = 6$ | Positive |

From this, you can conclude that the polynomial is positive for all $x$-values in $(-\infty, -2)$ and $(3, \infty)$, and is negative for all $x$-values in $(-2, 3)$. This implies that the solution of the inequality $x^2 - x - 6 < 0$ is the interval $(-2, 3)$, as shown in Figure 1.19.

Choose $x = -3$.
$(x + 2)(x - 3) > 0$

Choose $x = 4$.
$(x + 2)(x - 3) > 0$

Choose $x = 0$.
$(x + 2)(x - 3) < 0$

**Figure 1.19**

As with linear inequalities, you can check the reasonableness of a solution by substituting $x$-values into the original inequality. For instance, to check the solution found in Example 1, try substituting several $x$-values from the interval $(-2, 3)$ into the inequality

$$x^2 - x - 6 < 0.$$

Regardless of which $x$-values you choose, the inequality should be satisfied.

**Study Tip**

The critical numbers act as boundaries between the real numbers that satisfy the inequality and the real numbers that do not satisfy the inequality.

In Example 1, the polynomial inequality was given in standard form. Whenever this is not the case, you should begin the solution process by writing the inequality in standard form—with the polynomial on one side and zero on the other.

### Example 2　Solving a Polynomial Inequality

Solve $x^3 - 3x^2 > 10x$.

### Solution

Begin by writing the inequality in standard form.

$$x^3 - 3x^2 > 10x \qquad \text{Original inequality}$$

$$x^3 - 3x^2 - 10x > 0 \qquad \text{Write in standard form.}$$

$$x(x - 5)(x + 2) > 0 \qquad \text{Factor.}$$

*Critical numbers:* $\quad x = -2, x = 0, x = 5$

*Test intervals:* $\quad (-\infty, -2), (-2, 0), (0, 5), (5, \infty)$

*Test:* $\quad$ Is $x(x - 5)(x + 2) > 0$?

After testing these intervals, as shown in Figure 1.20, you can see that $x^3 - 3x^2 - 10x$ is positive in the open intervals $(-2, 0)$ and $(5, \infty)$. So, the solution set consists of all real numbers in the intervals $(-2, 0)$ and $(5, \infty)$.

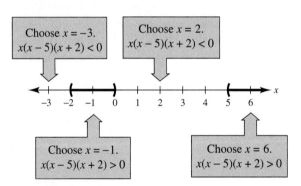

**Figure 1.20**

When solving a polynomial inequality, be sure you have accounted for the particular type of inequality symbol given in the inequality. For instance, in Example 2, note that the solution consisted of two open intervals because the original inequality contained a "greater than" symbol. If the original inequality had been $x^3 - 3x^2 \geq 10x$, the solution would have consisted of the two intervals $[-2, 0]$ and $[5, \infty)$.

Each of the polynomial inequalities in Examples 1 and 2 has a solution set that consists of a single interval or the union of two intervals. When solving the exercises for this section, you should watch for some unusual solution sets, as illustrated in Example 3.

**Example 3**    Unusual Solution Sets

**a.** The solution set of the following inequality consists of the entire set of real numbers, $(-\infty, \infty)$.

$$x^2 + 2x + 4 > 0$$

**b.** The solution set of the following inequality consists of the single real number $\{-1\}$.

$$x^2 + 2x + 1 \le 0$$

**c.** The solution set of the following inequality is empty.

$$x^2 + 3x + 5 < 0$$

**d.** The solution set of the following inequality consists of all real numbers except the number 2.

$$x^2 - 4x + 4 > 0$$

**·TECHNOLOGY·**

**Graphs of Inequalities and Graphing Utilities**    Most graphing utilities can sketch the graph of an inequality. Consult the user's guide to determine the steps for your graphing utility. Once you know how to graph an inequality, you may check solutions by graphing. For example, the solution to

$$x^2 - 5x < 0$$

is the interval $(0, 5)$. When graphed, the solution occurs as an interval above the horizontal axis on the graphing utility, as shown at the right. The graph does not indicate whether 0 and/or 5 are part of the solution. You must determine whether the endpoints are part of the solution based on the inequality sign and the type of inequality.

## Rational Inequalities

The concepts of critical numbers and test intervals can be extended to inequalities involving rational expressions. To do this, use the fact that the value of a rational expression can change sign only at its *zeros* (the *x*-values for which its numerator is zero) and its *undefined values* (the *x*-values for which its denominator is zero). These two types of numbers make up the **critical numbers** of a rational inequality.

**Example 4**     Solving a Rational Inequality

Solve $\dfrac{2x - 7}{x - 5} \leq 3$.

**Solution**

$$\frac{2x - 7}{x - 5} \leq 3 \qquad \text{Original inequality}$$

$$\frac{2x - 7}{x - 5} - 3 \leq 0 \qquad \text{Write in standard form.}$$

$$\frac{2x - 7 - 3x + 15}{x - 5} \leq 0 \qquad \text{Add fractions.}$$

$$\frac{-x + 8}{x - 5} \leq 0 \qquad \text{Simplify.}$$

*Critical numbers:*     $x = 5, x = 8$

*Test intervals:*     $(-\infty, 5), (5, 8), (8, \infty)$

*Test:*     Is $\dfrac{-x + 8}{x - 5} \leq 0$?

After testing these intervals, as shown in Figure 1.21, you can see that the rational expression $(-x + 8)/(x - 5)$ is negative in the open intervals $(-\infty, 5)$ and $(8, \infty)$. Moreover, because $(-x + 8)(x - 5) = 0$ when $x = 8$, you can conclude that the solution set consists of all real numbers in the intervals

$$(-\infty, 5) \cup [8, \infty). \qquad \text{Solution set}$$

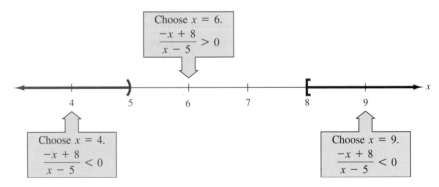

Choose $x = 6$.
$\dfrac{-x + 8}{x - 5} > 0$

Choose $x = 4$.
$\dfrac{-x + 8}{x - 5} < 0$

Choose $x = 9$.
$\dfrac{-x + 8}{x - 5} < 0$

**Figure 1.21**

## Applications

One common application of inequalities comes from business and involves profit, revenue, and cost. The formula that relates these three quantities is

> Profit  =  Revenue  −  Cost
>
> $P = R - C.$

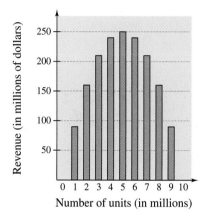

Revenue (in millions of dollars) vs Number of units (in millions)

**Figure 1.22**

**Example 5**    Increasing the Profit for a Product

The marketing department of a calculator manufacturer has determined that the demand for a new model of calculator is given by

$$p = 100 - 0.00001x, \qquad 0 \le x \le 10{,}000{,}000 \qquad \text{Demand equation}$$

where the price per calculator is in dollars and $x$ represents the number of calculators sold. (If this model is accurate, no one would be willing to pay $100 for the calculator. At the other extreme, the company couldn't *give* away more than 10 million calculators.) The revenue for selling $x$ calculators is given by

$$R = xp = x(100 - 0.00001x) \qquad \text{Revenue equation}$$

as shown in Figure 1.22. The total cost of producing $x$ calculators is $10 per calculator plus a development cost of $2,500,000. So, the total cost is

$$C = 10x + 2{,}500{,}000. \qquad \text{Cost equation}$$

What price should the company charge per calculator to obtain a profit of at least $190,000,000?

### Solution

*Verbal Model:*    Profit  =  Revenue  −  Cost

*Equation:*    $P = R - C$

$$P = 100x - 0.00001x^2 - (10x + 2{,}500{,}000)$$

$$P = -0.00001x^2 + 90x - 2{,}500{,}000$$

To answer the question, you must now solve the inequality

$$-0.00001x^2 + 90x - 2{,}500{,}000 \ge 190{,}000{,}000.$$

Using the techniques described in this section, you can find the solution to be $3{,}500{,}000 \le x \le 5{,}500{,}000$, as shown in Figure 1.23. The prices that correspond to these $x$-values are given by

$$\$45.00 \le p \le \$65.00.$$

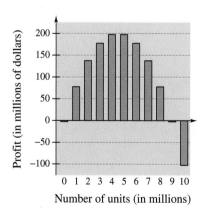

Profit (in millions of dollars) vs Number of units (in millions)

**Figure 1.23**

Another common application of inequalities is finding the domain of an expression that involves a square root, as shown in Example 6.

### Example 6 Finding the Domain of an Expression

Find the domain of the expression $\sqrt{64 - 4x^2}$.

#### Solution

Remember that the domain of an expression is the set of all $x$-values for which the expression is defined. Because $\sqrt{64 - 4x^2}$ is defined (has real values) only if $64 - 4x^2$ is nonnegative, the domain is given by $64 - 4x^2 \geq 0$.

$$64 - 4x^2 \geq 0 \qquad \text{Standard form}$$

$$16 - x^2 \geq 0 \qquad \text{Divide both sides by 4.}$$

$$(4 - x)(4 + x) \geq 0 \qquad \text{Factor.}$$

So, the inequality has two critical numbers: $-4$ and $4$. You can use these two numbers to test the inequality as follows.

*Critical numbers:* $\qquad x = -4, x = 4$

*Test intervals:* $\qquad (-\infty, -4), (-4, 4), (4, \infty)$

*Test:* $\qquad$ Is $(4 - x)(4 + x) \geq 0$?

A test shows that $64 - 4x^2$ is greater than or equal to 0 in the *closed interval* $[-4, 4]$. So, the domain of the expression $\sqrt{64 - 4x^2}$ is the interval $[-4, 4]$, as shown in Figure 1.24.

**Figure 1.24**

### Discussing the Concept    Profit Analysis

Consider the relationship $P = R - C$ described on page 75. Discuss why it might be beneficial to solve $P < 0$ if you owned a business. Use the situation described in Example 5 to illustrate your reasoning.

# Warm Up

The following warm-up exercises involve skills that were covered in previous courses and earlier sections. You will use these skills in the exercise set for this section.

In Exercises 1–10, solve the inequality.

**1.** $-\dfrac{y}{3} > 2$

**2.** $-6z < 27$

**3.** $-3 \le 2x + 3 < 5$

**4.** $-3x + 5 \ge 20$

**5.** $10 > 4 - 3(x + 1)$

**6.** $3 < 1 + 2(x - 4) < 7$

**7.** $2|x| \le 7$

**8.** $|x - 3| > 1$

**9.** $|x + 4| > 2$

**10.** $|2 - x| \le 4$

## 1.7   Exercises

In Exercises 1–28, solve the inequality and sketch its graph.

**1.** $x^2 \le 9$

**2.** $x^2 < 5$

**3.** $x^2 > 4$

**4.** $(x - 3)^2 \ge 1$

**5.** $(x + 2)^2 < 25$

**6.** $(x + 6)^2 \le 8$

**7.** $x^2 + 4x + 4 \ge 9$

**8.** $x^2 - 6x + 9 < 16$

**9.** $x^2 + x < 6$

**10.** $x^2 + 2x > 3$

**11.** $3(x - 1)(x + 1) > 0$

**12.** $6(x + 2)(x - 1) < 0$

**13.** $x^2 + 2x - 3 < 0$

**14.** $x^2 - 4x - 1 > 0$

**15.** $4x^3 - 6x^2 < 0$

**16.** $4x^3 - 12x^2 > 0$

**17.** $x^3 - 4x \ge 0$

**18.** $2x^3 - x^4 \le 0$

**19.** $\dfrac{1}{x} > x$

**20.** $\dfrac{1}{x} < 4$

**21.** $\dfrac{x + 6}{x + 1} < 2$

**22.** $\dfrac{x + 12}{x + 2} \ge 3$

**23.** $\dfrac{3x - 5}{x - 5} > 4$

**24.** $\dfrac{5 + 7x}{1 + 2x} < 4$

**25.** $\dfrac{4}{x + 5} > \dfrac{1}{2x + 3}$

**26.** $\dfrac{5}{x - 6} > \dfrac{3}{x + 2}$

**27.** $\dfrac{1}{x - 3} \le \dfrac{9}{4x + 3}$

**28.** $\dfrac{1}{x} \ge \dfrac{1}{x + 3}$

In Exercises 29–36, find the domain of the expression.

**29.** $\sqrt[4]{4 - x^2}$

**30.** $\sqrt{x^2 - 4}$

**31.** $\sqrt{x^2 - 7x + 12}$

**32.** $\sqrt{144 - 9x^2}$

**33.** $\sqrt{12 - x - x^2}$

**34.** $\sqrt{x^2 + 4}$

**35.** $\sqrt{x^2 - 3x + 3}$

**36.** $\sqrt[4]{-x^2 + 2x - 2}$

In Exercises 37 and 38, consider the domains of the expressions $\sqrt[3]{x^2 - 7x + 12}$ and $\sqrt{x^2 - 7x + 12}$.

**37.** Explain why the domain of $\sqrt[3]{x^2 - 7x + 12}$ consists of all real numbers.

**38.** Explain why the domain of $\sqrt{x^2 - 7x + 12}$ is different from the domain of $\sqrt[3]{x^2 - 7x + 12}$.

In Exercises 39–44, use a calculator to solve the inequality. (Round each number in your answer to two decimal places.)

**39.** $0.4x^2 + 5.26 < 10.2$

**40.** $-1.3x^2 + 3.78 > 2.12$

**41.** $-0.5x^2 + 12.5x + 1.6 > 0$

**42.** $1.2x^2 + 4.8x + 3.1 < 5.3$

**43.** $\dfrac{1}{2.3x - 5.2} > 3.4$

**44.** $\dfrac{2}{3.1x - 3.7} > 5.8$

**45.** *Height of a Projectile*   A projectile is fired straight upward from ground level with an initial velocity of 160 feet per second. During what time period will its height exceed 384 feet?

**46.** *Height of a Projectile*   A projectile is fired straight upward from ground level with an initial velocity of 128 feet per second. During what time period will its height be less than 128 feet?

**47.** *Geometry*   A rectangular playing field (see figure) with a perimeter of 100 meters is to have an area of at least 500 square meters. Within what bounds must the length lie?

**48.** *Geometry*   A rectangular room with a perimeter of 50 feet is to have an area of at least 120 square feet. Within what bounds must the length lie?

**49.** *Company Profits*   The revenue and cost equations for a product are given by

$$R = x(50 - 0.0002x)$$

$$C = 12x + 150,000$$

where $R$ and $C$ are measured in dollars and $x$ represents the number of units sold (see figure).

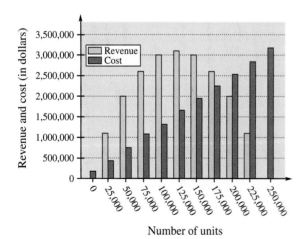

Number of units

(a)  How many units must be sold to obtain a profit of at least $1,650,000?

(b)  The demand equation for the product is

$$p = 50 - 0.0002x$$

where $p$ is the price per unit. What price per unit will produce a profit of at least $1,650,000?

**50.** *Company Profits*   The revenue and cost equations for a product are given by

$$R = x(75 - 0.0005x)$$

$$C = 30x + 250,000$$

where $R$ and $C$ are measured in dollars and $x$ represents the number of units sold (see figure).

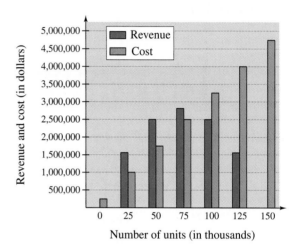

Number of units (in thousands)

(a)  How many units must be sold to obtain a profit of at least $750,000?

(b)  The demand equation for the product is

$$p = 75 - 0.0005x$$

where $p$ is the price per unit. What price per unit will produce a profit of at least $750,000?

**51.** *Compound Interest*   $P$ dollars, invested at interest rate $r$ compounded annually, increases to an amount

$$A = P(1 + r)^2$$

in 2 years. If an investment of $1000 is to increase to an amount greater than $1200 in 2 years, the interest rate must be greater than what percent?

**52.** *Compound Interest*   $P$ dollars, invested at interest rate $r$ compounded annually, increases to an amount

$$A = P(1 + r)^3$$

in 3 years. If an investment of $500 is to increase to an amount greater than $600 in 3 years, the interest rate must be greater than what percent?

**53.** *World Population*    The world population (in millions) from 1985 to 1996 can be modeled by

Population $= 4407 + 90.7t - 0.34t^2$,    $t \geq 5$

where $t = 5$ represents 1985 (see figure). According to this model, in what year will the world population exceed 6,300,000,000?    (Source: U.S. Bureau of the Census)

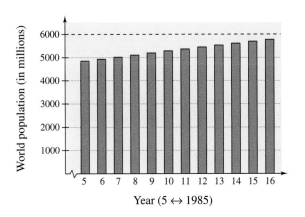

Year (5 ↔ 1985)

**54.** *Percent of College Graduates*    The percent of the American population that graduated from college from 1960 to 1995 can be modeled by

Percent of graduates $= 7.28 + 0.42t + 0.001t^2$,

$t \geq 0$

where $t = 0$ represents 1960. According to this model, in what year will the percent of college graduates exceed 25%?    (Source: U.S. Bureau of the Census)

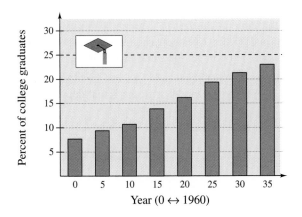

Year (0 ↔ 1960)

**55.** *Vehicle Registration*    The number of motor vehicle registrations (in thousands) in the United States from 1991 to 1995 can be modeled by

Registrations $= (429 + 3.9t)^2$,    $t \geq 1$

where $t = 1$ represents 1991 (see figure). According to this model, in what year will the number of motor vehicle registrations exceed 225,000,000? (Source: U.S. Federal Highway Administration)

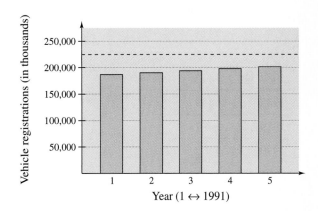

Year (1 ↔ 1991)

**56.** *Vehicle Registration*    The number of motor vehicle registrations (in millions) in the world from 1990 to 1996 can be modeled by

Registrations $= 583 + 4.7t + 1.66t^2$,    $t \geq 0$

where $t = 0$ represents 1990 (see figure). According to this model, in what year will the number of motor vehicle registrations exceed 925,000,000? (Source: American Automobile Manufacturers Association)

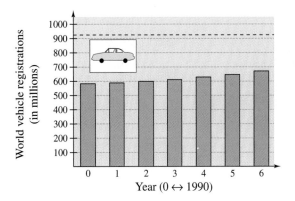

Year (0 ↔ 1990)

## Chapter Project

## Salaries in Professional Baseball

| Year | Average Salary (in thousands) |
|------|-------------------------------|
| 1985 | $371 |
| 1986 | $413 |
| 1987 | $412 |
| 1988 | $439 |
| 1989 | $497 |
| 1990 | $598 |
| 1991 | $851 |
| 1992 | $1029 |
| 1993 | $1076 |
| 1994 | $1168 |
| 1995 | $1111 |

The record-breaking attendance (70.5 million people) for major league baseball was not the only record that was broken in the 1998 season. Mark McGwire (St. Louis) hit a record 70 home runs and drew a record 152 walks. Bret Boone (Cincinnati) became the first player to lead a league in fielding percentage for four consecutive years. Pitcher Tom Gordon (Boston) had a record 42 consecutive saves. Cal Ripkin, Jr. (Baltimore) holds the record for consecutive games played at 2632. As players reach and exceed previous records, they expect to be compensated accordingly. The average salary for a major league baseball player has tripled in the last 11 years.

The table at the left gives the average salaries (in thousands of dollars) for professional baseball players from 1985 to 1995. The average salary can be modeled by the equation

$$S = 3.86t^2 + 16.0t + 140, \qquad 5 \leq t \leq 15$$

where $t = 5$ represents 1985. Use this information to investigate the following questions.

1. *Compare the Data*    Make a table that compares the actual average salary for 1985 to 1995 with the average found using the model.

2. *Compare Projected Average Salaries with Actual Average Salaries*    Use the model to predict the average salary in 1996 and in 1997. Use a library or other reference source to find the average salary. How well did the model predict the average salary?

3. *Predict Future Average Salaries*    According to this model, when will the average salary reach $2,000,000?

   (a) Answer the question numerically by creating a table of values.

   (b) Answer the question algebraically by solving

   $$2000 = 3.86t^2 + 16.0t + 140.$$

4. *Find a Model*    Enter the table at the left in a graphing utility. Enter 5 for 1985, enter 6 for 1986, and so on. Use the quadratic fit program to find a quadratic model for the data. Do you get the same model as the one above?

5. *Research*    Use a reference source to find data that can be closely modeled with a quadratic model. Compare the model with the actual data numerically and graphically.

# CHAPTER SUMMARY

After studying this chapter, you should have acquired the following skills.
These skills are keyed to the Review Exercises that begin on page 82. Answers
to odd-numbered Review Exercises are given in the back of the book.

**1.1** · Classify an equation as an identity or a conditional equation.    Review Exercises 1, 2

· Determine whether a given value is a solution.    Review Exercises 3, 4

· Solve a linear equation in one variable.    Review Exercises 5–1

**1.2** · Use mathematical models to solve word problems.    Review Exercises 13, 23

· Model and solve percent and mixture problems.    Review Exercises 14–16, 21, 22, 24–26

· Use common formulas to solve geometry and simple interest problems.    Review Exercises 17–20

**1.3** · Solve a quadratic equation by factoring.    Review Exercises 27–30

· Solve a quadratic equation by extracting square roots.    Review Exercises 31–34

· Analyze a quadratic equation.    Review Exercises 35, 36

· Construct and use a quadratic model to solve an application problem.    Review Exercises 37–40

**1.4** · Use the discriminant to determine the number of real solutions to a quadratic equation.    Review Exercises 41, 42

· Solve a quadratic equation using the Quadratic Formula.    Review Exercises 43–51

· Use the Quadratic Formula to solve an application problem.    Review Exercises 52, 53

**1.5** · Solve a polynomial equation by factoring.    Review Exercises 54–57

· Rewrite and solve an equation involving radicals or rational exponents.    Review Exercises 58–61

· Rewrite and solve an equation with fractions or absolute value.    Review Exercises 62–65

· Construct and use a nonquadratic model to solve an application problem.    Review Exercises 66–68

**1.6** · Solve and graph a linear inequality.    Review Exercises 69–74

· Construct and use a linear inequality to solve an application problem.    Review Exercises 75, 76

**1.7** · Solve and graph a polynomial inequality.    Review Exercises 77–79, 83, 84

· Solve and graph a rational inequality.    Review Exercises 80–82, 85, 86

· Determine the domain of an expression involving a square root.    Review Exercises 87–90

· Construct and solve a polynomial inequality to solve an application problem.    Review Exercises 91–96

## REVIEW EXERCISES

In Exercises 1 and 2, determine whether the equation is an identity or a conditional equation.

**1.** $5(x - 3) = 2x + 9$    **2.** $3(x + 2) = 3x + 6$

In Exercises 3 and 4, determine whether each given value of $x$ is a solution of the equation.

**3.** $3x^2 + 7x + 5 = x^2 + 9$

   (a) $x = 0$    (b) $x = \frac{1}{2}$    (c) $x = -4$    (d) $x = -1$

**4.** $6 + \dfrac{3}{x - 4} = 5$

   (a) $x = 5$    (b) $x = 0$    (c) $x = -2$    (d) $x = 7$

In Exercises 5–10, solve the equation (if possible) and check your answer.

**5.** $4(x + 3) - 3 = 2(4 - 3x) - 4$

**6.** $\dfrac{3x - 2}{5x - 1} = \dfrac{3}{4}$

**7.** $(x + 3) + 2(x - 4) = 5(x + 3)$

**8.** $\dfrac{3}{x - 4} + \dfrac{8}{2x + 5} = \dfrac{11}{2x^2 - 3x - 20}$

**9.** $\dfrac{x}{x + 3} - \dfrac{4}{x + 3} + 2 = 0$

**10.** $7 - \dfrac{3}{x} = 8 + \dfrac{5}{x}$

In Exercises 11 and 12, use a calculator to solve the equation for $x$. (Round your answer to three decimal places.)

**11.** $0.375x - 0.75(300 - x) = 200$

**12.** $\dfrac{x}{0.0645} + \dfrac{x}{0.098} = 2$

**13.** Three consecutive even integers have a sum of 42. Find the smallest of these integers.

**14.** Suppose your annual salary is $24,500. You receive a 6% raise. What is your new annual salary?

**15.** *Monthly Profit*    The total profit for a company in June was 15% less than it was in May. The total profit for the 2 months was $225,392. Find the profit for each month.

**16.** *Study Habits*    Based on a survey of 320 math students, the most common study techniques are shown in the graph. How many students in the survey used each type of study technique? (Most students used more than one type of technique.)

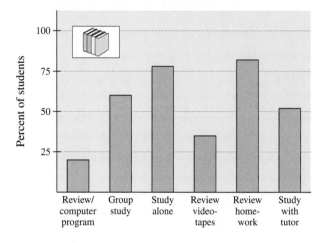

**17.** *Geometry*    A volleyball court is twice as long as it is wide, and its perimeter is 177 feet. Find the measurements of the volleyball court.

**18.** *Geometry*    A room is 1.25 times as long as it is wide, and its perimeter is 90 feet. Find the measurements of the room.

**19.** *Simple Interest*    You deposit $400 in a savings account earning 3% annually. How much interest will you have earned after 1 year?

**20.** *Simple Interest*    You deposit $750 in a money market account. One year later the account balance is $783.75. What was the interest rate for 1 year?

**21.** *List Price*    The price of a microwave oven has been discounted 15%. The sale price is $339.15. Find the original price of the microwave.

**22.** *Discount Rate*    The price of a VCR has been discounted by $85. The sale price is $257. What was the percent discount?

**23.** *Travel Time*   Two cars start at the same time at a given point and travel in the same direction at average speeds of 45 miles per hour and 50 miles per hour. How much time must elapse before the two cars are 10 miles apart?

**24.** *Projected Revenue*   From February through July, a company's revenues have totaled $435,112. If the monthly revenues continue at this rate, what will the total revenue for the year be?

**25.** *Mixture*   A car radiator contains 10 quarts of a 30% antifreeze solution. How many quarts will have to be replaced with pure antifreeze if the resulting 10-quart solution is to be 50% antifreeze?

**26.** *Mixture*   A 3-gallon acid solution contains 5% boric acid. How many gallons of 20% boric acid solution should be added to make a final solution that is 12% boric acid?

In Exercises 27–30, solve the quadratic equation by factoring. Check your solutions by substituting in the original equation.

**27.** $6x^2 = 5x + 4$

**28.** $-x^2 = 15x + 36$

**29.** $x^2 - 11x + 24 = 0$

**30.** $15 + x - 2x^2 = 0$

In Exercises 31–34, solve the equation by extracting square roots. List both the exact answer and a decimal answer that has been rounded to two decimal places.

**31.** $x^2 = 11$

**32.** $16x^2 = 25$

**33.** $(x + 4)^2 = 18$

**34.** $(x - 1)^2 = 5$

**35.** Describe at least two ways you can use a graphing utility to check a solution to a quadratic equation.

**36.** *Error Analysis*   A student solves Exercise 31 by extracting square roots and states that the two solutions are $x = \sqrt{11}$ and $x \approx 3.32$. What are some of the possible errors the student may have made? Give an analytical argument to persuade the student that there are two *different* solutions to Exercise 31.

**37.** *Geometry*   A billboard is 12 feet longer than it is high. The billboard has 405 square feet of advertising space. What are the measurements of the billboard? Use a diagram to help answer the question.

**38.** *Grand Canyon*   The Grand Canyon is 6000 feet deep at the deepest part. If a rock was dropped over the deepest part of the canyon, how long would the rock take to hit the water in the Colorado River below?

**39.** *Total Revenue*   The demand equation for a certain product is

$$p = 50 - 0.0001x$$

where $p$ is the price per unit and $x$ is the number of units sold. The total revenue for selling $x$ units is given by

$$\text{Revenue} = xp = x(50 - 0.0001x).$$

How many units must be sold to produce a revenue of $6,000,000?

**40.** *Depth of an Underwater Cable*   A ship's sonar locates a cable 2000 feet from the ship (see figure). The angle between the surface of the water and a line from the ship to the cable is 45 degrees. How deep is the cable?

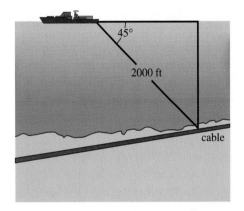

In Exercises 41 and 42, use the discriminant to determine the number of real solutions of the quadratic equation.

**41.** $x^2 + 11x + 24 = 0$

**42.** $x^2 + 5x + 12 = 0$

In Exercises 43–48, use the Quadratic Formula to solve the equation. Check your solutions by substituting in the original equation.

**43.** $x^2 - 12x + 30 = 0$

**44.** $5x^2 + 16x - 12 = 0$

**45.** $(y + 7)^2 = -5y$

**46.** $6x = 7 - 2x^2$

**47.** $x^2 + 6x - 3 = 0$

**48.** $10x^2 - 11x = 2$

**49.** *Error Analysis*   Using the Quadratic Formula, a student finds the solution of Exercise 48 to be

$$x = \frac{11 \pm \sqrt{(-11)^2 - 4(10)(2)}}{2(10)} = \frac{11 \pm \sqrt{41}}{20}.$$

Identify the error in the solution. How would you help the student correct the error?

In Exercises 50 and 51, use a calculator to solve the equation. (Round your answers to three decimal places.)

**50.** $3.6x^2 - 5.7x - 1.9 = 0$

**51.** $34x^2 - 296x + 47 = 0$

**52.** *On the Moon*  An astronaut standing on the edge of a cliff on the moon drops a rock over the cliff. If the cliff is 100 feet high, how long will it take the rock to hit the lunar surface? If the rock were dropped off a similar cliff on earth, how long would it remain in the air?

**53.** *Geometry*  An open box is to be made from a square piece of material by cutting 3-inch squares from the corners and turning up the sides (see figure). The volume of the finished box is to be 363 cubic inches. Find the size of the original piece of material.

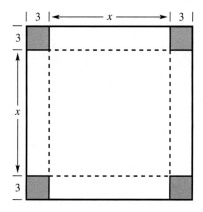

In Exercises 54–65, find all the solutions of the equation. (Check your solutions.)

**54.** $3x^3 - 9x^2 - 12x = 0$     **55.** $x^4 - 5x^2 + 4 = 0$

**56.** $x^4 + 3x^3 - 5x - 15 = 0$

**57.** $x^6 + 26x^3 - 27 = 0$

**58.** $2\sqrt{x} - 5 = 0$     **59.** $\sqrt{3x - 2} + x = 4$

**60.** $(x^2 - 5)^{2/3} = 9$     **61.** $(x^2 - 5x - 6)^{4/3} = 16$

**62.** $|5x + 4| = 11$     **63.** $|x^2 + 4x| - 2x = 8$

**64.** $\dfrac{5}{x + 1} + \dfrac{3}{x + 3} = 1$     **65.** $x + \dfrac{3}{x + 2} = 2$

**66.** *Sharing the Cost*  Three students are planning to share the expense of renting a condominium at a resort for 2 weeks. By adding a fourth person to the group, each person could save $50 in rental fees. How much is the rent for the 2-week period?

**67.** *Cash Advance*  Suppose you took out a cash advance of $400 on a credit card. After 3 months, the amount you owe is $421.56. What is the annual percentage rate for this loan? (Assume that the interest is compounded monthly and that you made no payments after the first and second months.)

**68.** *Market Research*  The demand equation for a product is given by

$$p = 42 - \sqrt{0.001x + 2}$$

where $x$ is the number of units demanded per day and $p$ is the price per unit. Find the demand if the price is set at $29.95.

In Exercises 69–74, solve the inequality and sketch its graph.

**69.** $3(x - 1) < 2x + 8$     **70.** $-5 \le 2 - 4(x + 2) \le 6$

**71.** $-3 < \dfrac{2x + 1}{4} < 3$     **72.** $-1 \le -5 - 3x < 4$

**73.** $|x + 10| + 3 < 5$     **74.** $|2x - 3| - 4 > 2$

**75.** *Break-Even Analysis*  The revenue for selling $x$ units of a product is

$$R = 125.95x.$$

The cost of producing $x$ units is

$$C = 92x + 1200.$$

In order to obtain a profit, the revenue must be greater than the cost. For what values of $x$ will this product return a profit?

**76.** *Accuracy of Measurement*  Suppose you buy an 18-inch gold chain that costs $8.95 per inch. If the chain is measured accurately to within $\frac{1}{16}$ of an inch, how much money might you have been undercharged or overcharged?

In Exercises 77–82, solve the inequality and sketch its graph.

**77.** $5(x + 1)(x - 3) < 0$

**78.** $(x + 4)^2 \le 4$

**79.** $x^3 - 9x < 0$

**80.** $\dfrac{x + 5}{x + 8} \ge 2$

**81.** $\dfrac{2 + 3x}{4 - x} < 2$

**82.** $\dfrac{1}{x + 1} \ge \dfrac{1}{x + 5}$

In Exercises 83–86, use a calculator to solve the inequality. (Round to two decimal places.)

**83.** $-1.2x^2 + 4.76 > 1.32$

**84.** $3.5x^2 + 4.9x - 6.1 < 2.4$

**85.** $\dfrac{1}{3.7x - 6.1} > 2.9$

**86.** $\dfrac{3}{5.4x - 2.7} < 8.9$

In Exercises 87–90, find the domain of the expression.

**87.** $\sqrt{x - 10}$

**88.** $\sqrt[4]{2x + 5}$

**89.** $\sqrt{x^2 - 15x + 54}$

**90.** $\sqrt{81 - 4x^2}$

**91.** *Height of a Projectile*   A projectile is fired straight upward from ground level with an initial velocity of 134 feet per second. During what time period will its height exceed 276 feet?

**92.** *Height of a Flare*   A flare is fired straight upward from ground level with an initial velocity of 100 feet per second. During what time period will its height exceed 150 feet?

**93.** *Compound Interest*   P dollars invested at interest rate r compounded annually increased to an amount

$$A = P(1 + r)^5$$

in 5 years. If an investment of $1000 is to increase to an amount greater than $1400 in 5 years, the interest rate must be greater than what percent?

**94.** *Compound Interest*   P dollars invested at an interest rate r compounded semiannually increased to an amount

$$A = P(1 + r/2)^{2 \cdot 8}$$

in 8 years. If an investment of $2000 is to increase to an amount greater than $4200 in 8 years, the interest rate must be greater than what percent?

**95.** *Company Profits*   The revenue and cost equations for a product are given by

$$R = x(80 - 0.0005x)$$

$$C = 20x + 300,000$$

where R and C are measured in dollars and x represents the number of units sold. How many units must be sold to obtain a profit of at least $900,000?

**96.** *Price of a Product*   In Exercise 95, the revenue equation is

$$R = x(80 - 0.0005x)$$

which implies that the demand equation is

$$p = 80 - 0.0005x$$

where p is the price per unit. What price per unit should the company set to obtain a profit of at least $1,200,000?

## Chapter Test

Take this test as you would take a test in class. After you are done, check your work against the answers given in the back of the book.

1. Solve the equation $3(x + 2) - 8 = 4(2 - 5x) + 7$.

2. Find the domain of the expression $\sqrt{2x - 9}$.

3. In May the total profit for a company was 20% less than it was in April. The total profit for the 2 months was \$315,655.20. Find the profit for each month.

In Exercises 4–13, solve the equation.

4. *Factoring:* $6x^2 + 7x = 5$

5. *Factoring:* $12 + 5x - 2x^2 = 0$

6. *Extracting roots:* $x^2 - 5 = 10$

7. *Quadratic Formula:* $(x + 5)^2 = -3x$

8. *Quadratic Formula:* $3x^2 - 11x = 2$

9. *Quadratic Formula:* $5.4x^2 - 3.2x - 2.5 = 0$

10. $|3x + 2| = 8$

11. $\sqrt{x - 3} + x = 5$

12. $x^4 - 10x^2 + 9 = 0$

13. $(x^2 - 9)^{2/3} = 9$

14. The demand equation for a product is $p = 40 - 0.0001x$, where $p$ is the price per unit and $x$ is the number of units sold. The total revenue for selling $x$ units is given by $R = xp$. How many units must be sold to produce a revenue of \$2,000,000? Explain your reasoning.

In Exercises 15–18, solve the inequality and sketch its graph.

15. $\dfrac{3x + 1}{5} < 2$

16. $|4 - 5x| \geq 24$

17. $\dfrac{x + 3}{x + 7} > 2$

18. $3x^3 - 12x \leq 0$

19. The revenue and cost equations for a product are given by

$$R = x(100 - 0.0005x) \quad \text{and} \quad C = 30x + 200,000$$

where $R$ and $C$ are measured in dollars and $x$ represents the number of units sold. How many units must be sold to obtain a profit of at least \$500,000?

20. $P$ dollars, invested at interest rate $r$ compounded annually, increases to an amount $A = P(1 + r)^{10}$ in 10 years. If an investment of \$2000 is to increase to an amount greater than \$5000 in 10 years, the interest rate must be greater than what percent?

# 2 The Cartesian Plane and Graphs

Demography is the statistical study of human populations with reference to size, density, distribution, and vital statistics. The United States Bureau of the Census tracks the demographics of the nation and uses these findings to forecast future trends. For example, the Census Bureau predicts that the U.S. population will have increased to about 392 million by the year 2050.

The table below gives the average sizes of U.S. households from 1960 to 1995. These data are also shown in the graph above. (Source: U.S. Bureau of the Census)

Although the overall population of the United States is increasing, from the graph you can see that the average household size is decreasing. Can you think of any possible explanations for this?

| Year | 1960 | 1965 | 1970 | 1975 | 1980 | 1985 | 1990 | 1995 |
|------|------|------|------|------|------|------|------|------|
| People per household | 3.35 | 3.32 | 3.14 | 2.94 | 2.76 | 2.69 | 2.63 | 2.65 |

The chapter project related to this information is on page 143.

## 2.1 The Cartesian Plane

### Objectives

- Plot points in the Cartesian plane.
- Find the distance between two points in the plane.
- Use the Distance Formula to solve distance and geometry problems.
- Find the midpoint of a line segment.
- Use the Midpoint Formula to solve an application problem.

## The Cartesian Plane

**René Descartes**

1596–1650

The Cartesian coordinate plane named after René Descartes was developed independently by another French mathematician, Pierre de Fermat. Fermat's *Introduction to Loci,* written about 1629, was clearer and more systematic than Descartes's *La géométrie.* However, Fermat's work was not published during his life-time. Consequently, Descartes received the credit for the development of the coordinate plane with the now familiar *x* and *y* axes.

Just as you can represent real numbers by points on the real number line, you can represent ordered pairs of real numbers by points in a **rectangular coordinate plane.** This plane is called the **Cartesian plane,** after the French mathematician René Descartes (1596–1650).

The Cartesian plane is formed by two real lines intersecting at right angles, as shown in Figure 2.1. The horizontal number line is the **x-axis** and the vertical number line is the **y-axis.** (The plural of axis is *axes.*) The point of intersection of the two axes is the **origin,** and the axes separate the plane into four **quadrants.**

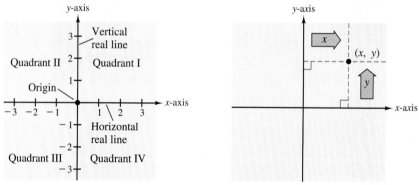

**Figure 2.1** *The Cartesian Plane*      **Figure 2.2**

Each point in the plane corresponds to an **ordered pair** $(x, y)$ of real numbers $x$ and $y$, called the **coordinates** of the point. The first number (or **x-coordinate**) tells how far to the left or right the point is from the vertical axis, and the second number (or **y-coordinate**) tells how far up or down the point is from the horizontal axis, as shown in Figure 2.2.

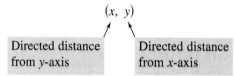

The notation $(a, b)$ to denote an open interval on the real number line and the notation $(x, y)$ to denote a point in the plane are similar but have different meanings. When these notations are used in this text, the nature of the specific problem will show which of the two is being discussed.

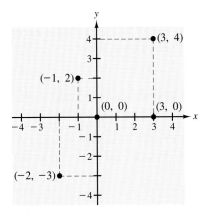

**Figure 2.3**

### Example 1    Plotting Points in the Cartesian Plane

Plot the points $(-1, 2)$, $(3, 4)$, $(0, 0)$, $(3, 0)$, and $(-2, -3)$ in the Cartesian plane.

#### Solution

To plot the point $(-1, 2)$ you can envision a vertical line through $-1$ on the $x$-axis and a horizontal line through 2 on the $y$-axis. The intersection of these two lines is the point $(-1, 2)$, as shown in Figure 2.3.

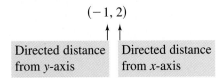

The other four points can be plotted in a similar way.

### Example 2    Shifting Points in the Plane

The triangle shown in Figure 2.4(a) has vertices at the points $(-1, 2)$, $(1, -4)$, and $(2, 3)$. Shift the triangle three units to the right and two units up and find the vertices of the shifted triangle.

#### Solution

To shift the vertices three units to the right, add 3 to each $x$-coordinate. To shift the vertices two units up, add 2 to each $y$-coordinate. See Figure 2.4(b).

| *Original Vertices* | *Shifted Vertices* |
|---|---|
| $(-1, 2)$ | $(-1 + 3, 2 + 2) = (2, 4)$ |
| $(1, -4)$ | $(1 + 3, -4 + 2) = (4, -2)$ |
| $(2, 3)$ | $(2 + 3, 3 + 2) = (5, 5)$ |

**(a)**

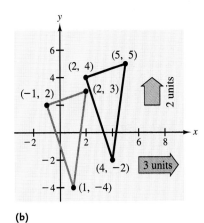

**(b)**

**Figure 2.4**

The value of the rectangular coordinate system is that it allows you to visualize relationships between variables $x$ and $y$. Today, Descartes's ideas are in common use in virtually every scientific and business-related field.

| Example 3 | Number of Doctorates in Mathematics |  |

The numbers of doctorates in mathematics granted to United States citizens by universities in the United States in the years 1982 to 1997 are given in the table. Plot these points on a rectangular coordinate system. (Source: American Mathematical Society)

*In 1971, nearly 25,000 bachelors degrees in mathematics were earned in the United States. Of these, 38% were earned by women. In 1995, the total number of bachelors degrees in mathematics had fallen to 13,851. Of these, 46.9% were earned by women.*

| Year | 1982 | 1983 | 1984 | 1985 | 1986 | 1987 | 1988 | 1989 |
|------|------|------|------|------|------|------|------|------|
| Degrees | 519 | 455 | 433 | 396 | 386 | 362 | 363 | 411 |

| Year | 1990 | 1991 | 1992 | 1993 | 1994 | 1995 | 1996 | 1997 |
|------|------|------|------|------|------|------|------|------|
| Degrees | 401 | 461 | 430 | 526 | 469 | 567 | 493 | 516 |

**Solution**

The points are shown in Figure 2.5. Note that the break in the $x$-axis indicates that the numbers for the years prior to 1982 have been omitted.

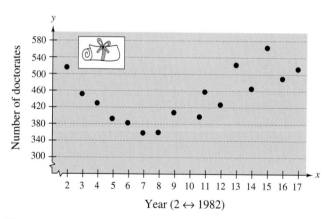

Figure 2.5

# The Distance Between Two Points in the Plane

The distance $d$ between two points $a$ and $b$ on the real number line is simply

$$d = |b - a|.$$

The same "absolute value rule" is used to find the distance between two points that lie on the same *vertical* or *horizontal* line in the plane.

### Example 4    Finding Horizontal and Vertical Distances

**a.** Find the distance between the points $(1, -1)$ and $(1, 4)$.

**b.** Find the distance between the points $(-3, -1)$ and $(1, -1)$.

### Solution

**a.** Because the $x$-coordinates are equal, you can envision a vertical line through the points $(1, -1)$ and $(1, 4)$, as shown in Figure 2.6. The distance between these two points is given by the absolute value of the difference of their $y$-coordinates. That is,

$$\text{Vertical distance} = |4 - (-1)| \qquad \text{Subtract } y\text{-coordinates.}$$
$$= 5. \qquad \text{Simplify.}$$

So, the points are five units apart.

**b.** Because the $y$-coordinates are equal, you can envision a horizontal line through the points $(-3, -1)$ and $(1, -1)$, as shown in Figure 2.6. The distance between these two points is given by the absolute value of the difference of their $x$-coordinates. That is,

$$\text{Horizontal distance} = |1 - (-3)| \qquad \text{Subtract } x\text{-coordinates.}$$
$$= 4. \qquad \text{Simplify.}$$

So, the points are four units apart.

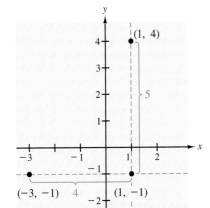

Figure 2.6

The technique used in Example 4 can be used to develop a general formula for finding the distance between two points in the plane. This general formula will work for any two points, even if they do not lie on the same vertical or horizontal line. To develop the formula, you can use the Pythagorean Theorem, which states that for a right triangle, the hypotenuse $c$ and sides $a$ and $b$ are related by the formula

$$a^2 + b^2 = c^2 \qquad \text{Pythagorean Theorem}$$

as shown in Figure 2.7. (The converse is also true. That is, if $a^2 + b^2 = c^2$, then the triangle is a right triangle.)

Figure 2.7    *Pythagorean Theorem*

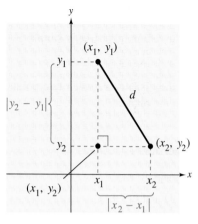

**Figure 2.8**

To develop a general formula for the distance between two points, let $(x_1, y_1)$ and $(x_2, y_2)$ represent two points in the plane that do not lie on the same horizontal or vertical line. With these two points, a right triangle can be formed, as shown in Figure 2.8. Note that the third vertex of the triangle is $(x_1, y_2)$. Because $(x_1, y_1)$ and $(x_1, y_2)$ lie on the same vertical line, the length of the vertical side of the triangle is $|y_2 - y_1|$. Similarly, the length of the horizontal side is $|x_2 - x_1|$. So, by the Pythagorean Theorem, the distance between $(x_1, y_1)$ and $(x_2, y_2)$ is given by

$$d^2 = |x_2 - x_1|^2 + |y_2 - y_1|^2.$$

Because the distance $d$ must be positive, choose the positive square root and write

$$d = \sqrt{|x_2 - x_1|^2 + |y_2 - y_1|^2}.$$

Finally, replacing $|x_2 - x_1|^2$ and $|y_2 - y_1|^2$ by the equivalent expressions $(x_2 - x_1)^2$ and $(y_2 - y_1)^2$ gives the following formula for the distance between two points in a rectangular coordinate plane.

---

▶ **The Distance Formula**

The distance $d$ between two points $(x_1, y_1)$ and $(x_2, y_2)$ in the coordinate plane is

$$d = \sqrt{(x_2 - x_1)^2 + (y_2 - y_1)^2}.$$

---

**Example 5**　**Finding the Distance Between Two Points**

Find the distance between the points $(-2, 1)$ and $(3, 4)$.

**Solution**

Let $(x_1, y_1) = (-2, 1)$ and $(x_2, y_2) = (3, 4)$, and apply the Distance Formula.

$$d = \sqrt{[3 - (-2)]^2 + (4 - 1)^2} \qquad \text{Distance Formula}$$

$$= \sqrt{5^2 + 3^2} \qquad \text{Simplify.}$$

$$= \sqrt{25 + 9} \qquad \text{Simplify.}$$

$$= \sqrt{34} \qquad \text{Simplify.}$$

$$\approx 5.83 \qquad \text{Use a calculator.}$$

See Figure 2.9.

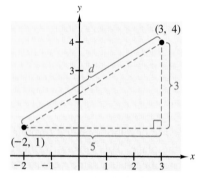

**Figure 2.9**　$d = \sqrt{34} \approx 5.83$

In Example 5, try letting $(x_1, y_1) = (3, 4)$ and $(x_2, y_2) = (-2, 1)$ to see that the Distance Formula yields the same result.

## Example 6    Using the Distance Formula in Geometry

Show that the points $(2, 1)$, $(4, 0)$, and $(5, 7)$ are vertices of a right triangle.

### Solution

The three points are plotted in Figure 2.10. Using the Distance Formula, you can find the lengths of the three sides of the triangle.

$$d_1 = \sqrt{(5 - 2)^2 + (7 - 1)^2} = \sqrt{9 + 36} = \sqrt{45}$$

$$d_2 = \sqrt{(4 - 2)^2 + (0 - 1)^2} = \sqrt{4 + 1} = \sqrt{5}$$

$$d_3 = \sqrt{(5 - 4)^2 + (7 - 0)^2} = \sqrt{1 + 49} = \sqrt{50}$$

Because

$$d_1^2 + d_2^2 = 45 + 5$$
$$= 50$$
$$= d_3^2$$

you can conclude from the Pythagorean Theorem that the triangle is a right triangle.

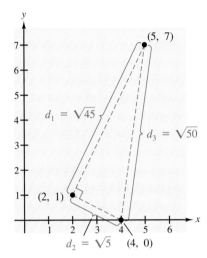

**Figure 2.10**

The next example shows how the Distance Formula can be used to solve a real-life problem.

## Example 7    The Length of a Football Pass

In a football game, a quarterback throws a pass from the 5-yard line, 20 yards from the sideline. The pass is caught by a wide receiver on the 45-yard line, 50 yards from the same sideline, as shown in Figure 2.11. How long was the pass?

### Solution

Using the Distance Formula, you can find the distance to be

$$d = \sqrt{(50 - 20)^2 + (45 - 5)^2} \qquad \text{Distance Formula}$$
$$= \sqrt{900 + 1600} \qquad \text{Simplify.}$$
$$= \sqrt{2500} \qquad \text{Simplify.}$$
$$= 50 \text{ yards} \qquad \text{Simplify.}$$

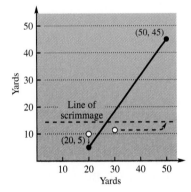

**Figure 2.11**

In Example 7, note that a scale was added to the goal line. When you use coordinate geometry to solve real-life problems, you are free to place the coordinate system in any way that is convenient in the problem.

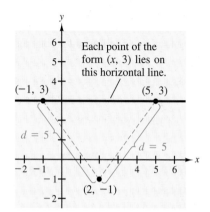

Figure 2.12

### Example 8    Using the Distance Formula

Find $x$ such that the distance between $(x, 3)$ and $(2, -1)$ is 5.

### Solution

Using the Distance Formula, you can write the following.

| | |
|---|---|
| $\sqrt{(x - 2)^2 + (3 + 1)^2} = 5$ | Distance Formula |
| $(x^2 - 4x + 4) + 16 = 25$ | Square both sides. |
| $x^2 - 4x - 5 = 0$ | Standard form |
| $(x - 5)(x + 1) = 0$ | Factor. |
| $x - 5 = 0 \implies x = 5$ | Set 1st factor equal to 0. |
| $x + 1 = 0 \implies x = -1$ | Set 2nd factor equal to 0. |

So, you can conclude that each of the points $(5, 3)$ and $(-1, 3)$ lies five units from the point $(2, -1)$, as shown in Figure 2.12.

## The Midpoint Formula

The following formula shows how to find the midpoint of the line segment that joins two points.

> ### ▶ The Midpoint Formula
>
> The **midpoint** of the line segment joining the points $(x_1, y_1)$ and $(x_2, y_2)$ in the coordinate plane is
> $$\left(\frac{x_1 + x_2}{2}, \frac{y_1 + y_2}{2}\right).$$

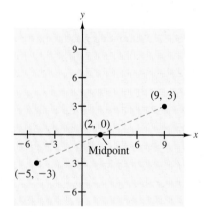

Figure 2.13

### Example 9    Finding the Midpoint of a Line Segment

Find the midpoint of the line segment joining the points $(-5, -3)$ and $(9, 3)$.

### Solution

Figure 2.13 shows the two given points and their midpoint. By the Midpoint Formula, you have

$$\text{Midpoint} = \left(\frac{-5 + 9}{2}, \frac{-3 + 3}{2}\right) = (2, 0).$$

| Example 10 | Retail Sales |  |

A business had annual retail sales of $240,000 in 1994 and $312,000 in 2000. Find the sales for 1997. (Assume that the annual increase in sales followed a linear pattern. That is, assume that the points representing the sales lie on a straight line, as shown in Figure 2.14.)

### Solution

To make the computations simpler, let $t = 4$ represent the year 1994 and let $t = 10$ represent the year 2000. Then, if you measure the retail sales in thousands of dollars, the sales for 1994 and 2000 are represented by the points

$$(4, 240) \quad \text{and} \quad (10, 312).$$

Because 1997 is midway between 1994 and 2000 and because the growth pattern is linear, you can use the Midpoint Formula to find the 1997 sales, as follows.

$$\text{Midpoint} = \left( \frac{4 + 10}{2}, \frac{240 + 312}{2} \right)$$

$$= (7, 276)$$

So, the 1997 sales were approximately $276,000, as indicated in Figure 2.14.

**Figure 2.14**

---

## Discussing the Concept        A Misleading Graph

Although graphs can help us visualize relationships between two variables, they can also be used to mislead people. The graphs shown below represent the same data points. Which of the two graphs is misleading, and why? Discuss other ways in which graphs can be misleading. Try to find another example of a misleading graph in a newspaper or magazine. Why is it misleading? Why would it be beneficial for someone to use a misleading graph?

## Warm Up

The following warm-up exercises involve skills that were covered in earlier sections. You will use these skills in the exercise set for this section.

In Exercises 1–6, simplify the expression.

1. $\sqrt{(2 - 6)^2 + [1 - (-2)]^2}$
2. $\sqrt{(1 - 4)^2 + (-2 - 1)^2}$
3. $\dfrac{4 + (-2)}{2}$
4. $\dfrac{-1 + (-3)}{2}$
5. $\sqrt{18} + \sqrt{45}$
6. $\sqrt{12} + \sqrt{44}$

In Exercises 7–10, solve the equation.

7. $\sqrt{(4 - x)^2 + (5 - 2)^2} = \sqrt{58}$
8. $\sqrt{(8 - 6)^2 + (y - 5)^2} = 2\sqrt{5}$
9. $\dfrac{x + 3}{2} = 7$
10. $\dfrac{-2 + y}{2} = 1$

## 2.1    Exercises

In Exercises 1–4, sketch the polygon with the indicated vertices.

1. Triangle: $(-3, 4), (1, -1), (-4, -2)$
2. Triangle: $(0, 3), (-1, -2), (4, 8)$
3. Square: $(2, 4), (5, 1), (2, -2), (-1, 1)$
4. Parallelogram: $(5, 2), (7, 0), (1, -2), (-1, 0)$

*Shifting a Graph*    In Exercises 5 and 6, the figure is shifted in the plane. Find the vertices of the shifted figure.

5.

6.

In Exercises 7–10, plot the points and find the distance between them.

7. $(8, -1), (8, 5)$
8. $(-4, 2), (-4, 9)$
9. $(5, -7), (-6, -7)$
10. $(-3, 3), (1, 3)$

11. Find two points in Quadrant I that are two units apart.

12. Find two points in Quadrant II that are three units apart.

In Exercises 13–16, find the length of the hypotenuse in two ways: (a) use the Pythagorean Theorem and (b) use the Distance Formula.

13.

14.

**15.**

**16.**

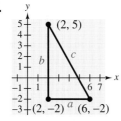

In Exercises 35 and 36, find $x$ such that the distance between the points is 15.

**35.** $(3, -4), (x, 5)$        **36.** $(x, 8), (-9, -4)$

In Exercises 37 and 38, find $y$ such that the distance between the points is 20.

**37.** $(-15, y), (-3, -7)$     **38.** $(6, -1), (-10, y)$

In Exercises 17–28, (a) plot the points, (b) find the distance between the points, and (c) find the midpoint of the line segment joining the points.

**17.** $(2, -5), (-6, 1)$

**18.** $(1, 12), (6, 0)$

**19.** $(3, -11), (-12, -3)$

**20.** $(-7, 3), (2, -9)$

**21.** $(-1, 2), (5, 4)$

**22.** $(2, 10), (10, 2)$

**23.** $\left(\frac{1}{2}, 1\right), \left(-\frac{5}{2}, \frac{4}{3}\right)$

**24.** $\left(-\frac{1}{3}, -\frac{1}{3}\right), \left(-\frac{1}{6}, -\frac{1}{2}\right)$

**25.** $(1.7, 8.5), (-3.2, 5.3)$

**26.** $(21.7, 10.2), (7.9, -2.3)$

**27.** $(-36, -18), (48, -72)$

**28.** $(1.451, 3.051), (5.906, 11.360)$

In Exercises 29 and 30, use the Midpoint Formula to estimate the sales for 1998.

**29.**

| Year | 1996 | 2000 |
|------|------|------|
| Sales | $520,000 | $740,000 |

**30.**

| Year | 1996 | 2000 |
|------|------|------|
| Sales | $4,200,000 | $5,650,000 |

In Exercises 31–34, show that the points form the vertices of the indicated polygon.

**31.** Right triangle: $(4, 0), (2, 1), (-1, -5)$

**32.** Isosceles triangle: $(1, -3), (3, 2), (-2, 4)$

**33.** Rhombus: $(0, 0), (1, 2), (2, 1), (3, 3)$

**34.** Parallelogram: $(0, 1), (3, 7), (4, 4), (1, -2)$

In Exercises 39–42, state the quadrant in which $(x, y)$ lies.

**39.** $x > 0$ and $y < 0$      **40.** $x < 0$ and $y < 0$

**41.** $x > 0$ and $y > 0$      **42.** $x < 0$ and $y > 0$

**43.** *Football Pass*    In a football game, the quarterback throws a pass from the 20-yard line, 15 yards from the sideline. The pass is caught on the 50-yard line, 30 yards from the same sideline. How long was the pass?

**44.** *Driving Distance*    Two trucks leave a warehouse at the same time. One truck travels east at a rate of 55 miles per hour and the other travels south at a rate of 65 miles per hour. How far apart are the two trucks after 2 hours? (Assume that both trucks are traveling in a straight line.)

*Average Life Expectancy*    In Exercises 45 and 46, use the figure, which shows the average life expectancy for Americans from 1920 to 1995.    (Source: U.S. National Center for Health Statistics)

**45.** Estimate the average life expectancy in 1920.

**46.** Estimate the average life expectancy in 1995.

*Gold Prices*    In Exercises 47 and 48, use the figure, which shows the average price of gold from 1975 to 1996.   (Source: U.S. Bureau of Mines)

Year (5 ↔ 1975)

**47.** What is the highest price of gold shown in the graph? When did this occur?

**48.** What is the lowest price of gold shown in the graph? When did this occur?

*Fuel Efficiency*    In Exercises 49 and 50, use the figure, which shows the average fuel efficiency for automobiles in the United States from 1970 to 1994.   (Source: U.S. Federal Highway Administration)

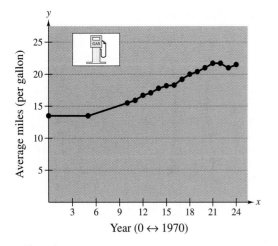

Year (0 ↔ 1970)

**49.** Estimate the percent increase in fuel efficiency from 1975 to 1988.

**50.** Estimate the percent increase in fuel efficiency from 1970 to 1994.

*Football Attendance*    In Exercises 51 and 52, use the figure, which shows the average paid attendance at NFL football games from 1982 to 1993.   (Source: National Football League)

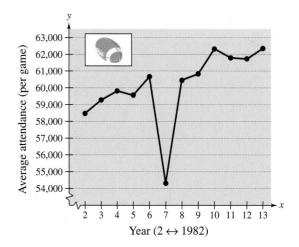

Year (2 ↔ 1982)

**51.** Estimate the increase in attendance from 1982 to 1993.

**52.** Estimate the decrease in attendance from 1986 to 1987.

## 2.2    Graphs of Equations

### Objectives

- Determine whether a point lies on the graph of an equation.
- Sketch the graph of an equation using a table of values.
- Find the *x*- and *y*-intercepts of the graph of an equation.
- Determine the symmetry of a graph.
- Write the equation of a circle in standard form.

## The Graph of an Equation

Frequently, the relationship between two quantities is expressed in the form of an equation. In this section you will study a basic procedure for sketching the graph of an equation. For an equation in variables $x$ and $y$, a point $(a, b)$ is a **solution point** if the substitution $x = a$ and $y = b$ satisfies the equation.

### Example 1    Solution of an Equation

Determine if $(-1, 0)$ is a solution of the equation $y = 2x^2 - 4x - 6$.

**Solution**

$$y = 2x^2 - 4x - 6 \qquad \text{Original equation}$$

$$0 = 2(-1)^2 - 4(-1) - 6 \qquad \text{Substitute } -1 \text{ for } x \text{ and } 0 \text{ for } y.$$

$$0 = 0 \qquad \text{Simplify.}$$

Both sides of the equation are equivalent, so the point $(-1, 0)$ is a solution.

---

Most equations have *infinitely* many solution points. The set of all solution points of an equation is its **graph.**

### Example 2    Sketching the Graph of an Equation

Sketch the graph of $3x + y = 5$ by rewriting the equation as $y = 5 - 3x$ with $y$ isolated on the left. Next, construct a table of values by choosing several values of $x$ and calculating the corresponding values of $y$.

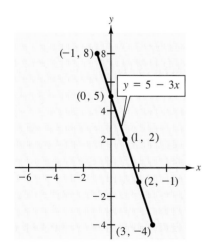

Figure 2.15

| $x$ | $-1$ | 0 | 1 | 2 | 3 |
|---|---|---|---|---|---|
| $y = 5 - 3x$ | 8 | 5 | 2 | $-1$ | $-4$ |

Finally, plot the points given in the table and connect them as shown in Figure 2.15. It appears that the graph is a straight line.

---

---

▶ **The Point-Plotting Method of Graphing**

**1.** If possible, isolate one of the variables.

**2.** Construct a table of several solution points.

**3.** Plot these points in the coordinate plane.

**4.** Connect the points with a smooth curve.

---

Figure 2.16

Step 4 of the point-plotting method can be difficult. For instance, how would you connect the four points in Figure 2.16? Without further information about the equation, any one of the three graphs in Figure 2.17 would be reasonable. These graphs show that with too few solution points, you can grossly misrepresent the graph of an equation. Throughout this course, you will study many ways to improve your graphing techniques. For now, we suggest that you plot enough points to reveal the essential behavior of the graph.

**Study Tip**

When constructing a table, use negative, zero, and positive values for $x$.

 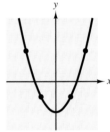

Figure 2.17

**Example 3** Sketching the Graph of an Equation

Sketch the graph of

$$y = x^2 - 2.$$

**Solution**

First, construct a table of values by choosing several convenient values of $x$ and calculating the corresponding values of $y$.

| $x$ | $-2$ | $-1$ | 0 | 1 | 2 | 3 |
|---|---|---|---|---|---|---|
| $y = x^2 - 2$ | 2 | $-1$ | $-2$ | $-1$ | 2 | 7 |

Next, plot the corresponding solution points. Finally, connect the points with a smooth curve, as shown in Figure 2.18.

Figure 2.18

$y = x^2 - 2$

## Intercepts of a Graph

When you are sketching a graph, two types of points that are especially useful are those for which either the $y$-coordinate or the $x$-coordinate is zero.

> ▶ Definition of Intercepts
>
> 1. The **$x$-intercepts** of a graph are the points at which the graph intersects the $x$-axis. To find the $x$-intercepts, let $y$ be zero and solve for $x$.
>
> 2. The **$y$-intercepts** of a graph are the points at which the graph intersects the $y$-axis. To find the $y$-intercepts, let $x$ be zero and solve for $y$.

Some graphs have no intercepts and some have several. For instance, consider the three graphs in Figure 2.19.

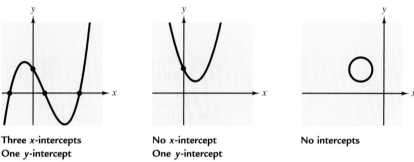

**Three $x$-intercepts**          **No $x$-intercept**
**One $y$-intercept**             **One $y$-intercept**          **No intercepts**
Figure 2.19

**Example 4**    Finding $x$- and $y$-Intercepts

Find the $x$- and $y$-intercepts of the graph of

$$y^2 - 3 = x.$$

### Solution

To find the $x$-intercept, let $y = 0$. This produces $-3 = x$, which implies that the graph has one $x$-intercept, which occurs at

$$(-3, 0). \qquad \text{\footnotesize $x$-intercept}$$

To find the $y$-intercept, let $x = 0$. This produces $y^2 - 3 = 0$, which has two solutions: $y = \pm\sqrt{3}$. So, the equation has two $y$-intercepts, which occur at

$$\left(0, \sqrt{3}\right) \quad \text{and} \quad \left(0, -\sqrt{3}\right). \qquad \text{\footnotesize $y$-intercepts}$$

See Figure 2.20.

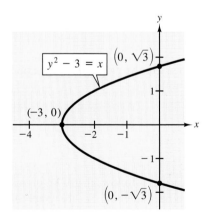

Figure 2.20

## Symmetry

Symmetry with respect to the $x$-axis means that if the Cartesian plane were folded along the $x$-axis, the portion of the graph above the $x$-axis would coincide with the portion below the $x$-axis. Symmetry with respect to the $y$-axis or the origin can be described in a similar manner. Symmetry with respect to the origin means that if the Cartesian plane were rotated $180°$ about the origin, the portion of the graph to the right of the origin would coincide with the portion to the left of the origin. (See Figure 2.21.)

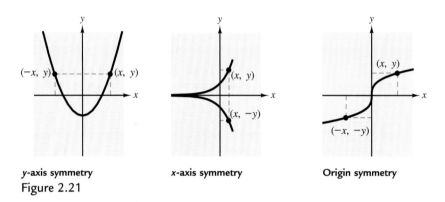

**y-axis symmetry**          **x-axis symmetry**          **Origin symmetry**

Figure 2.21

---

▶ **Definition of Symmetry**

1. A graph is **symmetric with respect to the $y$-axis** if, whenever $(x, y)$ is on the graph, $(-x, y)$ is also on the graph.

2. A graph is **symmetric with respect to the $x$-axis** if, whenever $(x, y)$ is on the graph, $(x, -y)$ is also on the graph.

3. A graph is **symmetric with respect to the origin** if, whenever $(x, y)$ is on the graph, $(-x, -y)$ is also on the graph.

---

Suppose you apply this definition of symmetry to the graph of the equation $y = x^2 - 1$. Replacing $x$ with $-x$ produces the following.

$$y = x^2 - 1 \qquad \text{Original equation}$$

$$y = (-x)^2 - 1 \qquad \text{Replace } x \text{ with } -x.$$

$$y = x^2 - 1 \qquad \text{Replacement yields equivalent equation.}$$

Because the substitution did not change the equation, it follows that if $(x, y)$ is a solution of the equation, then $(-x, y)$ must also be a solution. So, the graph of $y = x^2 - 1$ is symmetric with respect to the $y$-axis, as shown in Figure 2.22.

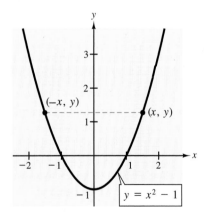

Figure 2.22    *y-Axis Symmetry*

> ▶ **Tests for Symmetry**
>
> **1.** The graph of an equation is symmetric with respect to the *y-axis* if replacing $x$ with $-x$ yields an equivalent equation.
>
> **2.** The graph of an equation is symmetric with respect to the *x-axis* if replacing $y$ with $-y$ yields an equivalent equation.
>
> **3.** The graph of an equation is symmetric with respect to the *origin* if replacing $x$ with $-x$ *and* $y$ with $-y$ yields an equivalent equation.

| Example 5 | **Using Symmetry as a Sketching Aid** |

Describe the symmetry of the graph of $x - y^2 = 1$.

**Solution**

Of the three tests for symmetry, the only one that is satisfied by this equation is the test for *x*-axis symmetry.

$$x - y^2 = 1 \qquad \text{Original equation}$$

$$x - (-y)^2 = 1 \qquad \text{Replace } y \text{ with } -y.$$

$$x - y^2 = 1 \qquad \text{Replacement yields equivalent equation.}$$

So, the graph is symmetric with respect to the *x*-axis. To sketch the graph, plot the points above the *x*-axis and use symmetry to complete the graph, as shown in Figure 2.23.

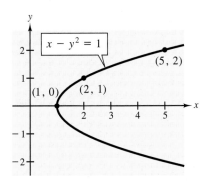

Figure 2.23

# The Equation of a Circle

So far in this section you have studied the point-plotting method and two additional concepts (intercepts and symmetry) that can be used to streamline the graphing procedure. Another graphing aid is *equation recognition,* which is the ability to recognize the general shape of a graph simply by looking at its equation. A circle is one type of graph that is easily recognized.

Figure 2.24 shows a circle of radius $r$ with center at the point $(h, k)$. The point $(x, y)$ is on this circle if and only if its distance from the center $(h, k)$ is $r$. This means that a **circle** in the plane consists of all points $(x, y)$ that are a given positive distance $r$ from a fixed point $(h, k)$. Using the Distance Formula, you can conclude that the point $(x, y)$ lies on the circle if and only if

$$\sqrt{(x - h)^2 + (y - k)^2} = r.$$

By squaring both sides of this equation, you obtain the **standard form of the equation of a circle.**

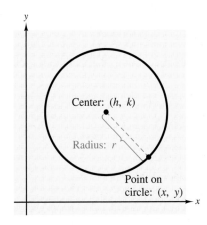

Figure 2.24

> ▶ **Standard Form of the Equation of a Circle**
>
> The **standard form of the equation of a circle** is
>
> $$(x - h)^2 + (y - k)^2 = r^2.$$
>
> The point $(h, k)$ is called the **center** of the circle and the positive number $r$ is called the **radius** of the circle.

The standard form of the equation of a circle whose center is the *origin* is simply

$$x^2 + y^2 = r^2.$$

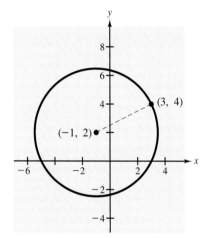

**Figure 2.25**

**Example 6**    Finding an Equation of a Circle

The point $(3, 4)$ lies on a circle whose center is at $(-1, 2)$, as shown in Figure 2.25. Find an equation for the circle.

**Solution**

The radius $r$ of the circle is the distance between $(-1, 2)$ and $(3, 4)$. So, you have

$$r = \sqrt{[3 - (-1)]^2 + (4 - 2)^2}$$
$$= \sqrt{16 + 4} = \sqrt{20}.$$

So, the center of the circle is $(h, k) = (-1, 2)$, the radius is $r = \sqrt{20}$, and you can write the standard form of the equation of the circle as follows.

$$(x - h)^2 + (y - k)^2 = r^2 \qquad \text{Standard form}$$
$$[x - (-1)]^2 + (y - 2)^2 = \left(\sqrt{20}\right)^2 \qquad \text{Let } h = -1, k = 2, \text{ and } r = \sqrt{20}.$$
$$(x + 1)^2 + (y - 2)^2 = 20 \qquad \text{Equation of circle}$$

If you remove the parentheses in the standard equation in Example 6, you obtain the following.

$$(x + 1)^2 + (y - 2)^2 = 20 \qquad \text{Standard form}$$
$$x^2 + 2x + 1 + y^2 - 4y + 4 = 20 \qquad \text{Expand terms.}$$
$$x^2 + y^2 + 2x - 4y - 15 = 0 \qquad \text{General form}$$

The last equation is in the **general form of the equation of a circle.**

$$Ax^2 + Ay^2 + Dx + Ey + F = 0, \qquad A \neq 0$$

The general form of the equation of a circle is less useful than the standard form. For instance, it is not immediately apparent from the general equation shown above that the center is $(-1, 2)$ and the radius is $\sqrt{20}$. To graph the equation of a circle, it is best to write the equation in standard form. You can do this by **completing the square,** as demonstrated in Example 7.

**Study Tip**

To complete the square, add the square of half the coefficient of the linear term.

| Example 7 | Completing the Square to Sketch a Circle |

Describe the circle given by

$$4x^2 + 4y^2 + 20x - 16y + 37 = 0.$$

**Solution**

Begin by writing the given equation in standard form by completing the square for both the $x$-terms *and* the $y$-terms.

$$4x^2 + 4y^2 + 20x - 16y + 37 = 0 \qquad \text{General form}$$

$$x^2 + y^2 + 5x - 4y + \frac{37}{4} = 0 \qquad \text{Divide by 4.}$$

$$\left(x^2 + 5x + \boxed{\phantom{xx}}\right) + \left(y^2 - 4y + \boxed{\phantom{xx}}\right) = -\frac{37}{4} \qquad \text{Group terms.}$$

$$\left[x^2 + 5x + \left(\frac{5}{2}\right)^2\right] + (y^2 - 4y + 2^2) = -\frac{37}{4} + \frac{25}{4} + 4 \qquad \text{Complete square.}$$

$$\left(x + \frac{5}{2}\right)^2 + (y - 2)^2 = 1 \qquad \text{Standard form}$$

So, the center of the circle is $\left(-\frac{5}{2}, 2\right)$ and the radius of the circle is 1. Using this information, you can sketch the circle shown in Figure 2.26.

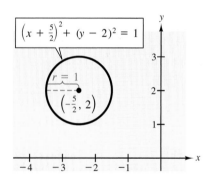

Figure 2.26

**Discussing the Concept    Interpreting an Intercept**

For the graph of each of the following real-life situations, find the indicated intercept and explain its practical significance.

**a.** (*y-intercept*) The trade-in value $y$ of a used car $x$ years after its purchase is $y = 9500 - 2400x$.

**b.** (*x-intercept*) The number of families $y$ demanding child care services each week is related to the price $x$ per hour of care by $y = 7850 - 20x^2$, $x \geq 0$.

Find an example of a real-life situation in which it makes sense for there to be no x-intercept.

# Warm Up

The following warm-up exercises involve skills that were covered in earlier sections. You will use these skills in the exercise set for this section.

In Exercises 1 and 2, solve for $y$ in terms of $x$.

**1.** $3x - 5y = 2$

**2.** $x^2 - 4x + 2y - 5 = 0$

In Exercises 3–6, solve the equation.

**3.** $x^2 - 4x + 4 = 0$

**4.** $(x - 1)(x + 5) = 0$

**5.** $x^3 - 9x = 0$

**6.** $x^4 - 8x^2 + 16 = 0$

In Exercises 7–10, simplify the equation.

**7.** $-y = (-x)^3 + 4(-x)$

**8.** $(-x)^2 + (-y)^2 = 4$

**9.** $y = 4(-x)^2 + 8$

**10.** $(-y)^2 = 3(-x) + 4$

## 2.2 Exercises

In Exercises 1–6, determine whether the points lie on the graph of the equation.

| Equation | Points | |
|---|---|---|
| **1.** $y = 2x^2 - 3$ | (a) $(-1, -1)$ | (b) $(2, 5)$ |
| **2.** $y = 5 - x - x^2$ | (a) $(0, 5)$ | (b) $(-2, -1)$ |
| **3.** $4y - 3x + 7 = 0$ | (a) $(1, -2)$ | (b) $(4, 3)$ |
| **4.** $x^2 - y^2 = 13$ | (a) $(5, 3)$ | (b) $(-7, 6)$ |
| **5.** $y = \sqrt{x - 8}$ | (a) $(9, 1)$ | (b) $(17, 3)$ |
| **6.** $y = \dfrac{2x}{6 - x}$ | (a) $(-3, 0)$ | (b) $(-4, -8)$ |

In Exercises 7–14, find the intercepts of the graph.

**7.** $y = 2x - 1$

**8.** $y = (x - 4)(x + 2)$

**9.** $y = x^2 + x - 2$

**10.** $y = 4 - x^2$

**11.** $y = x\sqrt{x + 2}$

**12.** $y = x\sqrt{x + 3}$

**13.** $2y - xy + 3x = 4$

**14.** $x^2y - x^2 + 4y = 0$

**15.** Discuss how you would use your knowledge of the Cartesian plane and intercepts to explain why you let $y$ equal zero when you are finding the $x$-intercepts and let $x$ equal zero when you are finding the $y$-intercepts.

**16.** Is it possible for a graph to have no $x$-intercepts? Is it possible for a graph to have no $y$-intercepts? Is it possible for a graph to have no $x$-intercepts and no $y$-intercepts? Give examples to support your answers.

In Exercises 17–24, check for symmetry.

**17.** $x^4 - 2y = 0$

**18.** $x - y^2 = 0$

**19.** $y = \sqrt{4 - x^2}$

**20.** $y = \sqrt{9 - x^2}$

**21.** $x^3y = 1$

**22.** $xy = 4$

**23.** $y = \dfrac{x}{x^2 + 1}$

**24.** $y = x^4 - x^2 + 3$

In Exercises 25–28, use symmetry to complete the graph.

**25.** $y$-axis symmetry

$y = -x^2 + 4$

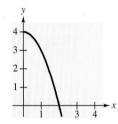

**26.** $x$-axis symmetry

$y^2 = -x + 4$

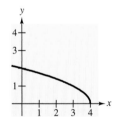

**27.** Origin symmetry

$y = -x^3 + x$

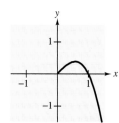

**28.** $y$-axis symmetry

$y = |x| - 2$

In Exercises 29–34, match the equation with its graph. [The graphs are labeled (a), (b), (c), (d), (e), and (f).]

(a)

(b)

(c)

(d)

(e)

(f)

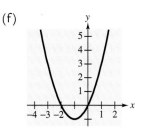

**29.** $y = 4 - x$     **30.** $y = x^2 + 2x$

**31.** $y = \sqrt{4 - x^2}$     **32.** $y = \sqrt{x}$

**33.** $y = x^3 - x$     **34.** $y = -\sqrt{4 - x^2}$

In Exercises 35–54, sketch the graph of the equation. Identify any intercepts and test for symmetry.

**35.** $y = 5 - 3x$     **36.** $y = 2x - 3$

**37.** $y = 1 - x^2$     **38.** $y = x^2 - 1$

**39.** $y = x^2 - 4x + 3$     **40.** $y = -x^2 - 4x$

**41.** $y = x^3 + 2$     **42.** $y = x^3 - 1$

**43.** $y = \dfrac{8}{x^2 + 4}$     **44.** $y = \dfrac{4}{x^2 + 1}$

**45.** $y = \sqrt{x + 1}$     **46.** $y = \sqrt{1 - x}$

**47.** $y = \sqrt[3]{x}$     **48.** $y = \sqrt[3]{x + 1}$

**49.** $y = |x - 4|$     **50.** $y = |x| - 3$

**51.** $x = y^2 - 1$     **52.** $x = y^2 - 4$

**53.** $x^2 + y^2 = 4$     **54.** $x^2 + y^2 = 16$

In Exercises 55–62, find the standard form of the equation of the specified circle.

**55.** Center: $(0, 0)$;   radius: 3

**56.** Center: $(0, 0)$;   radius: 5

**57.** Center: $(-4, 1)$;   radius: $\sqrt{2}$

**58.** Center: $\left(0, \frac{1}{3}\right)$;   radius: $\frac{1}{3}$

**59.** Center: $(-1, 2)$;   solution point: $(0, 0)$

**60.** Center: $(3, -2)$;   solution point: $(-1, 1)$

**61.** Endpoints of a diameter: $(-3, 4)$, $(5, -2)$

**62.** Endpoints of a diameter: $(-4, -1)$, $(4, 1)$

In Exercises 63–70, write the equation of the circle in standard form. Then sketch the circle.

**63.** $x^2 + y^2 - 6x + 4y - 3 = 0$

**64.** $x^2 + y^2 - 2x + 6y - 15 = 0$

**65.** $x^2 + y^2 - 2x + 6y + 10 = 0$

**66.** $5x^2 + 5y^2 + 10x + 1 = 0$

**67.** $2x^2 + 2y^2 - 2x - 2y - 3 = 0$

**68.** $4x^2 + 4y^2 - 4x + 2y - 1 = 0$

**69.** $16x^2 + 16y^2 + 16x + 40y - 7 = 0$

**70.** $x^2 + y^2 - 4x + 2y + 3 = 0$

In Exercises 71–74, find the constant $C$ such that the ordered pair is a solution point of the equation.

**71.** $y = C - 4x^2$      $(2, -10)$

**72.** $y = Cx^3$      $(-4, 8)$

**73.** $y = C\sqrt{x + 1}$      $(3, 8)$

**74.** $x + C(y + 2) = 0$      $(4, 3)$

In Exercises 75 and 76, an equation of a circle is written in standard form. Indicate the coordinates of the center of the circle and determine the radius of the circle. Rewrite the equation of the circle in general form.

**75.** $(x - 2)^2 + (y + 3)^2 = 16$

**76.** $\left(x + \frac{1}{2}\right)^2 + (y + 1)^2 = 5$

In Exercises 77 and 78, (a) create a table to compare the data to the values found using the model, (b) sketch a graph comparing the data and the model for that data, and (c) use the model to predict $y$ for the year 2005.

**77.** *Purchasing Power of the Dollar*   The table gives the purchasing power of the dollar for *consumers* from 1989 to 1996. The base year for comparison is 1982, in which the purchasing power of the dollar for *producers* is $1.00.   (Source: U.S. Bureau of Labor Statistics)

| Year | 1989 | 1990 | 1991 | 1992 |
|------|------|------|------|------|
| Purchasing power | 0.807 | 0.766 | 0.734 | 0.713 |

| Year | 1993 | 1994 | 1995 | 1996 |
|------|------|------|------|------|
| Purchasing power | 0.692 | 0.675 | 0.656 | 0.638 |

A mathematical model that approximates the purchasing power of the dollar is

$$y = \frac{0.768 + 0.034t}{1 + 0.087t}, \quad -1 \le t \le 6$$

where $y$ represents the purchasing power of the dollar and $t = 0$ represents 1990.

**78.** *Life Expectancy*   The table gives the life expectancy of a child (at birth) for selected years from 1970 to 1995.   (Source: U.S. National Center for Health Statistics)

| Year | 1970 | 1975 | 1980 |
|------|------|------|------|
| Life expectancy | 70.8 | 72.6 | 73.7 |

| Year | 1985 | 1990 | 1995 |
|------|------|------|------|
| Life expectancy | 74.7 | 75.4 | 75.8 |

A mathematical model for the life expectancy during this period is

$$y = \frac{70.801 + 3.751t}{1 + 0.047t}, \quad 0 \le t \le 25$$

where $y$ represents the life expectancy and $t = 0$ represents 1970.

**79.** *Earnings per Share*   The earnings per share $y$ (in dollars) for Bristol-Myers Squibb Corporation from 1990 to 1997 can be modeled by

$$y = 0.016t^2 + 0.08t + 1.77, \quad 0 \le t \le 7$$

where $t = 0$ represents 1990. Sketch the graph of this equation.   (Source: *Value Line*)

**80.** *Earnings per Share*   The earnings per share $y$ (in dollars) for Amgen, Inc. from 1990 to 1997 can be modeled by

$$y = 0.33t + 0.31, \quad 0 \le t \le 7$$

where $t = 0$ represents 1990. Sketch the graph of this equation.   (Source: *Value Line*)

## 2.3    Graphing Utilities

### Objectives

- Sketch the graph of an equation using a graphing utility.
- Determine an appropriate viewing window for the graph of an equation.
- Use a square setting to obtain a true geometric perspective.

## Introduction

In Section 2.2 you studied the point-plotting method for sketching the graph of an equation. One of the disadvantages of the point-plotting method is that to get a good idea about the shape of a graph you need to plot *many* points. By plotting only a few points, you can badly misrepresent the graph.

For instance, consider the equation $y = \frac{1}{30}x(x^4 - 10x^2 + 39)$. To graph this equation, suppose you calculated only the following five points.

| $x$ | $-3$ | $-1$ | $0$ | $1$ | $3$ |
|---|---|---|---|---|---|
| $y = \frac{1}{30}x(x^4 - 10x^2 + 39)$ | $-3$ | $-1$ | $0$ | $1$ | $3$ |

By plotting these five points, as shown in Figure 2.27(a), you might assume that the graph of the equation is a straight line. This, however, is not correct. By plotting several more points, as shown in Figure 2.27(b), you can see that the actual graph is not straight at all.

### Blaise Pascal

1623–1662

Calculators have interested mathematicians for hundreds of years. In 1642, at the age of 18, Blaise Pascal began designing a calculating machine. Over the next few years, Pascal built and sold approximately 50 machines.

**(a)**

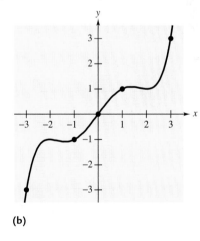

**(b)**

Figure 2.27

So, the point-plotting method leaves you with a dilemma. On the one hand, the method can be very inaccurate if only a few points are plotted. But, on the other hand, it is very time consuming to plot a dozen (or more) points. Technology can help you resolve this dilemma. Plotting several points (even hundreds of points) on a rectangular coordinate system is something that a graphing utility can do easily.

## Using a Graphing Utility

There are many different graphing utilities: some are graphing packages for computers and some are hand-held graphing calculators. The steps used to graph an equation with a graphing utility are similar for all types of utilities and are summarized below.

---

▶ Graphing an Equation with a Graphing Utility

Before performing the following steps, set your graphing utility so that all the standard defaults are active.

1. Set the viewing window for the graph.

2. Rewrite the equation so that $y$ is isolated on the left side of the equation.

3. Using the equation editor, enter the right side of the equation on the first line of the display. (The first line may be labeled $Y_1=$.)

4. Press the appropriate key to graph the equation.

---

Graphing utilities are equipped with a *window* feature. This feature allows you to determine the portion of the coordinate plane that you want to view. You can also determine the distance between consecutive tick marks (scale) on each axis. If your calculator is set to the standard graphing defaults, the screen should show the following values.

| | |
|---|---|
| Xmin=-10 | The minimum $x$-value is $-10$. |
| Xmax=10 | The maximum $x$-value is 10. |
| Xscl=1 | The $x$-scale is one unit per tick mark. |
| Ymin=-10 | The minimum $y$-value is $-10$. |
| Ymax=10 | The maximum $y$-value is 10. |
| Yscl=1 | The $y$-scale is one unit per tick mark. |

These settings are summarized visually in Figure 2.28. This viewing window is usually called the **standard viewing window.**

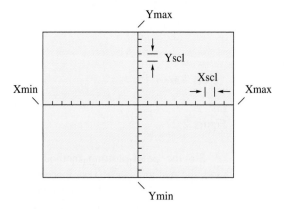

Figure 2.28

The following examples use the standard viewing window. Notice how the window contains all of the intercepts of the graph.

**Example 1** Graphing an Equation

Sketch the graph of $2y + x^3 = 4x$.

**Solution**

To begin, solve the given equation for $y$ in terms of $x$.

$$2y + x^3 = 4x \qquad \text{Original equation}$$

$$2y = -x^3 + 4x \qquad \text{Subtract } x^3 \text{ from both sides.}$$

$$y = -\frac{1}{2}x^3 + 2x \qquad \text{Divide both sides by 2.}$$

Enter the equation in the equation editor of your graphing utility. The equation may use the notation

$$Y_1 = -(1/2)X^\wedge 3 + 2X.$$

Now graph using the standard viewing window. Your screen should look like the one shown in Figure 2.29.

**Figure 2.29**

The next example illustrates how to graph an absolute value equation. Most graphing utilities use the abbreviation "ABS" or "abs" for absolute value. The expression within the absolute value signs should be enclosed in parentheses in the graphing utility. For example, the equation

$$y = \frac{1}{2}|3x + 4|$$

should appear in the equation editor as

$$Y_1 = (1/2)\,\text{abs}\,(3X+4).$$

**Example 2** Graphing an Equation Involving Absolute Value

Sketch the graph of $y = |x - 3|$.

**Solution**

This equation is already written with $y$ isolated on the left side of the equation. Enter the equation in the equation editor of your graphing utility. The equation may use the notation

$$Y_1 = \text{abs}\,(X-3).$$

Now graph using the standard viewing window. Your screen should look like the one shown in Figure 2.30.

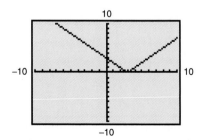

**Figure 2.30**

## Using a Graphing Utility's Special Features

To use your graphing utility to its best advantage, you must learn how to set the viewing window, as illustrated in the next example.

**Example 3**   Setting the Viewing Window

Sketch the graph of $y = x^2 + 12$.

### Solution

Enter the equation in the equation editor of your graphing utility and graph. If your graphing utility is set to the standard viewing window, nothing will appear on the screen. The reason for this is that the lowest point on the graph of $y = x^2 + 12$ occurs at the point $(0, 12)$. Using the standard viewing window, you obtain a screen whose largest $y$-value is 10. In other words, none of the graph is visible on a screen whose $y$-values vary between $-10$ and 10, as shown in Figure 2.31(a). To change these settings, enter the following values.

| | |
|---|---|
| Xmin=-10 | The minimum $x$-value is $-10$. |
| Xmax=10 | The maximum $x$-value is 10. |
| Xscl=1 | The $x$-scale is one unit per tick mark. |
| Ymin=-10 | The minimum $y$-value is $-10$. |
| Ymax=30 | The maximum $y$-value is 30. |
| Yscl=5 | The $y$-scale is five units per tick mark. |

Graph the equation with the new viewing window and you will obtain the graph shown in Figure 2.31(b). On this graph, note that each tick mark on the $y$-axis represents five units because you changed the $y$-scale to 5. Also note that the highest point on the $y$-axis is now 30 because you changed the maximum value of $y$ to 30.

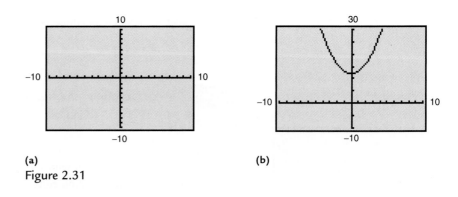

(a)                                         (b)

Figure 2.31

A graphing utility is especially useful for graphing more complicated equations such as

$$y = -0.166 + 0.02x - 0.0953x^2.$$

You must be careful to select a viewing window that shows all of the important features of the graph.

Most graphing utilities have a feature that will change the standard viewing window to a viewing window with a square setting. That is, the tick marks on the x-axis will appear the same distance apart as the tick marks on the y-axis. Look for the square setting in the *zoom* feature of your graphing utility.

## Study Tip

A square setting is useful anytime you want to sketch a graph that has a true geometric perspective. For instance, if you want the graph of $x^2 + y^2 = 25$ to look like a circle, you need to use a square setting. You also need to enter the top and bottom portions of the circle separately, as follows.

$y = \sqrt{25 - x^2}$   Top half

$y = -\sqrt{25 - x^2}$   Bottom half

In Figure 2.33(a) the graph of the circle appears stretched in a standard viewing window, whereas in Figure 2.33(b) the square setting shows the circle in true geometric perspective.

(a)

(b)
Figure 2.33

### Example 4   Using a Square Setting

Sketch the graph of $y = x$. The graph of this equation is a straight line that makes a 45° angle with the x-axis and y-axis. From the graph on your calculator, does the angle appear to be 45°?

### Solution

Graph the equation $y = x$ in a standard viewing window and you will obtain the graph shown in Figure 2.32(a). Note that the angle the line makes with the x-axis doesn't appear to be 45°. The reason for this is that the screen is wider than it is tall. This has the effect of making the tick marks on the x-axis farther apart than the tick marks on the y-axis. To obtain the same distance between tick marks on both axes, you can change the graphing settings from "standard" to "square." The screen should look like that shown in Figure 2.32(b). Note in this figure that the square setting has changed the viewing window so that the x-values vary between −15 and 15.

(a)

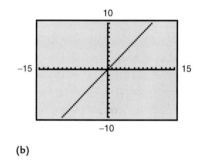

(b)

Figure 2.32

There are many possible square settings on a graphing calculator. To create a square setting, you need the following ratio to be $\frac{2}{3}$.

$$\frac{\text{Ymax} - \text{Ymin}}{\text{Xmax} - \text{Xmin}}$$

For instance, the setting in Example 4 is square because $(\text{Ymax} - \text{Ymin}) = 20$ and $(\text{Xmax} - \text{Xmin}) = 30$.

**Example 5**    Sketching More than One Graph on the Same Screen

Sketch the graphs of the following equations on the same screen.

$$y = -x + 5, \qquad y = -x, \qquad \text{and} \qquad y = -x - 5$$

Solution

Enter all three equations on the first three lines, $Y_1$, $Y_2$, and $Y_3$, of the equation editor. The display may be as follows.

$Y_1 = $ -X + 5

$Y_2 = $ -X

$Y_3 = $ -X − 5

The graphs of the equations are shown in Figure 2.34. Note that the graph of each equation is a straight line, and that the lines are parallel to each other.

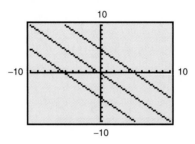

Figure 2.34

Most graphing utilities have other convenient graphing aides. You may be able to *box* a portion of the viewing window and enlarge that area of the graph. *Zoom in* and *zoom out* enlarge or shrink the area of the graph centered at the cursor. Additional features may also be available. Check the user's guide for your graphing utility.

---

**Discussing the Concept    Graphing with Technology**

For each of the following equations, describe what you must do before you can graph the equation using a graphing utility.

**a.** $-4y + 32 = 2x$       **b.** $y = \frac{1}{3}(x^3 - 10x)$

**c.** $x^2 + 2y^2 = 16$       **d.** $x^2 = y + 12x$

Now use a graphing utility and a standard viewing setting to graph each equation. Enlarge your view of the graph by zooming out or increasing the boundaries of your viewing window several times. For which equations are all of the *x*-intercepts shown in the standard viewing window?

## Warm Up

The following warm-up exercises involve skills that were covered in earlier sections. You will use these skills in the exercise set for this section.

In Exercises 1–10, solve for $y$ in terms of $x$.

**1.** $3x + y = 4$

**2.** $x - y = 0$

**3.** $2x + 3y = 2$

**4.** $4x - 5y = -2$

**5.** $3x + 4y - 5 = 0$

**6.** $-2x - 3y + 6 = 0$

**7.** $x^2 + y - 4 = 0$

**8.** $-2x^2 + 3y + 2 = 0$

**9.** $x^2 + y^2 = 4$

**10.** $x^2 - y^2 = 9$

## 2.3   Exercises

In Exercises 1–10, use a graphing utility to match the equation with its graph. [The graphs are labeled (a), (b), (c), (d), (e), (f), (g), (h), (i), and (j).]

**1.** $y = x$

**2.** $y = -x$

**3.** $y = x^2$

**4.** $y = -x^2$

**5.** $y = x^3$

**6.** $y = -x^3$

**7.** $y = |x|$

**8.** $y = -|x|$

**9.** $y = \sqrt{x}$

**10.** $y = -\sqrt{x}$

(a)

(b)

(c)

(d)

(e)

(f)

(g)

(h)

(i)

(j)

In Exercises 11–30, use a graphing utility to graph the equation. Use a standard setting for each graph.

**11.** $y = x - 5$

**12.** $y = -x + 4$

**13.** $y = -\frac{1}{2}x + 3$

**14.** $y = \frac{2}{3}x + 1$

**15.** $2x - 3y = 4$

**16.** $x + 2y = 3$

**17.** $y = \frac{1}{2}x^2 - 1$

**18.** $y = -x^2 + 6$

**19.** $y = x^2 - 4x - 5$

**20.** $y = x^2 - 3x + 2$

**21.** $y = -x^2 + 2x + 1$

**22.** $y = -x^2 + 4x - 1$

**23.** $2y = x^2 + 2x - 3$

**24.** $3y = -x^2 - 4x + 5$

**25.** $y = |x + 5|$

**26.** $y = \frac{1}{2}|x - 6|$

**27.** $y = \sqrt{x^2 + 1}$

**28.** $y = 2\sqrt{x^2 + 2} - 4$

**29.** $y = \frac{1}{5}(-x^3 + 16x)$

**30.** $y = \frac{1}{8}(x^3 + 8x^2)$

In Exercises 31–34, use a graphing utility to graph the equation. Does the setting give a good representation of the graph? Explain.

**31.** $y = -2x^2 + 12x + 14$   **32.** $y = -x^2 + 5x + 6$

| Xmin = -5 |
|---|
| Xmax = 10 |
| Xscl = 1 |
| Ymin = -5 |
| Ymax = 35 |
| Yscl = 5 |

| Xmin = -8 |
|---|
| Xmax = 4 |
| Xscl = 1 |
| Ymin = -5 |
| Ymax = 15 |
| Yscl = 5 |

**33.** $y = x^3 + 6x^2$   **34.** $y = -x^3 + 16x$

| Xmin = -10 |
|---|
| Xmax = 5 |
| Xscl = 1 |
| Ymin = -4 |
| Ymax = 36 |
| Yscl = 4 |

| Xmin = -6 |
|---|
| Xmax = 6 |
| Xscl = 1 |
| Ymin = -25 |
| Ymax = 25 |
| Yscl = 5 |

In Exercises 35 and 36, find a setting on a graphing utility such that the viewing window will be square. In the square viewing window, the following ratio must be $\frac{2}{3}$.

$$\frac{Ymax - Ymin}{Xmax - Xmin}$$

**35.**

| Xmin = ? |
|---|
| Xmax = 30 |
| Xscl = 2 |
| Ymin = -5 |
| Ymax = 35 |
| Yscl = 5 |

**36.**

| Xmin = -4 |
|---|
| Xmax = 11 |
| Xscl = 1 |
| Ymin = -2 |
| Ymax = ? |
| Yscl = 1 |

In Exercises 37–40, find a setting on a graphing utility such that the graph of the equation agrees with the graph shown.

**37.** $y = -x^2 - 4x + 20$   **38.** $y = x^2 + 12x - 8$

**39.** $y = -x^3 + x^2 + 2x$   **40.** $y = x^3 + 3x^2 - 2x$

In Exercises 41–44, use a graphing utility to find the number of $x$-intercepts of the equation.

**41.** (a) $y = \frac{1}{8}(4x^2 - 32x + 65)$

(b) $y = \frac{1}{8}(4x^2 + 32x + 63)$

**42.** (a) $y = \frac{1}{4}(-4x^2 + 16x - 15)$

(b) $y = \frac{1}{4}(-4x^2 + 16x - 17)$

**43.** $y = 4x^3 - 20x^2 - 4x + 61$

**44.** $y = \frac{1}{4}(2x^3 + 6x^2 - 4x + 1)$

**45.** What do you observe about the graphs of the equation in Exercises 41(b) and 42(a)? 41(a) and 42(b)?

**46.** How are the graphs in Exercises 43 and 44 alike? How are the graphs different?

In Exercises 47–50, use a graphing utility to graph the equations on the same screen. Use a square setting that shows a good representation of the graphs. What geometrical shape is bounded by the graphs?

**47.** $y = |x| - 4$, $y = -|x| + 4$

**48.** $y = x + |x| - 4$, $y = x - |x| + 4$

**49.** $y = -\sqrt{25 - x^2}$, $y = \sqrt{25 - x^2}$

**50.** $y = 6$, $y = -\sqrt{3}x - 4$, $y = \sqrt{3}x - 4$

**51.** A mathematical model that approximates the purchasing power of the dollar is given by

$$y = \frac{1.13 - 0.268t + 0.0165t^2}{1 - 0.192t + 0.0037t^2 + 0.00068t^3},$$

$2 \le t \le 16$

where $y$ represents the purchasing power of the dollar and $t = 2$ represents 1982. (Source: U.S. Bureau of Labor Statistics)

(a) Use this model to find the purchasing power of the dollar in 1993.

(b) The purchasing power of the dollar in 1994 was 0.675. What does this tell you about the model?

**52.** Describe how you could use the table feature of your graphing utility to help select an appropriate viewing window for the equation in Exercise 51. If your graphing utility does not have a table feature, is there another way you can obtain ideas for the size of a viewing window?

**53.** A friend uses a graphing utility to graph the equation in Exercise 51. No graph appears on the viewing screen. What might have happened? How would you help your friend find an appropriate viewing window?

**54.** A mathematical model for the life expectancy from 1970 to 1995 is

$$y = \sqrt{5008.5 + 66.00t - 2.412t^2 + 0.0384t^3},$$

$$0 \le t \le 25$$

where $y$ represents the life expectancy and $t = 0$ represents 1970.   (Source: U.S. Bureau of the Census)

(a) Using this model, predict the life expectancy in 2010 and in 2020.

(b) Are your answers reasonable? What cautions would you give when using a mathematical model for a time span that is greatly different from the period in which the data for the model were collected?

**Ever Been Married?**   In Exercises 55–58, use the following models, which relate ages to the percents $y$ of American males and females who have never been married.

$$y = \frac{47.07 - 0.7487x}{1 - 0.0738x + 0.00205x^2} \qquad \text{Males}$$

$$y = \frac{7.28 + 0.1864x}{1 - 0.09853x + 0.00275x^2} \qquad \text{Females}$$

In these models, $x$ is the age of the person, and each model is valid for $20 \le x \le 50$.   (Source: U.S. Bureau of the Census)

**55.** Use a graphing utility to graph both models on the same screen.

**56.** Write a paragraph describing the relationship between the two graphs that were plotted in Exercise 55.

**57.** Suppose that an American male is chosen at random from the population. If the person is 25 years old, what is the probability that he has never been married?

**58.** Suppose that an American female is chosen at random from the population. If the person is 25 years old, what is the probability that she has never been married?

**Earnings and Dividends**  In Exercises 59–62, use the following model, which approximates the relationship between dividends $y$ (in dollars) per share and earnings $x$ (in dollars) per share for Paychex, Inc. between 1990 and 1997.   (Source: Value Line)

$$y = 0.20x + 0.27x^2, \quad 0.06 \le x \le 0.47$$

**59.** Use a graphing utility to graph the model.

**60.** According to the model, what size dividend would Paychex, Inc. pay if the earnings per share were $0.35?

**61.** Use the trace key on your graphing utility to estimate the earnings per share that would produce a dividend per share of $0.10.

**62.** The *payout ratio* for a stock is the ratio of the dividend per share to earnings per share. Use the model to find the payout ratio for an earnings per share of

(a) $0.20     (b) $0.35     (c) $0.45

## Mid-Chapter Quiz

Take this quiz as you would take a quiz in class. After you are done, check your work against the answers given in the back of the book.

In Exercises 1 and 2, (a) plot the points, (b) find the distance between the points, and (c) find the midpoint of the line segment joining the points.

**1.** $(-3, 2)$, $(4, -5)$

**2.** $(1.3, -4.5)$, $(-3.7, 0.7)$

**3.** A business had sales of $1,330,000 in 1995 and $1,890,000 in 1999. Estimate the sales in 1997. Explain your reasoning.

**4.** One plane is 200 miles due north of a transmitter and another plane is 150 miles due east of the transmitter. What is the distance between the planes?

In Exercises 5 and 6, describe the figure with the given vertices.

**5.** $(-1, -1)$, $(10, 7)$, $(2, 18)$

**6.** $(-3, -1)$, $(0, 2)$, $(4, 2)$, $(1, -1)$

In Exercises 7–12, sketch the graph of the equation. Identify any intercepts and symmetry.

**7.** $y = 9 - x^2$      **8.** $y = x\sqrt{x + 4}$      **9.** $xy = 9$

**10.** $y = \sqrt{16 - x^2}$      **11.** $x^2 + y^2 = 9$      **12.** $y = |x - 3|$

In Exercises 13 and 14, find the standard form of the equation of the circle.

**13.** Center: $(2, -3)$;    radius: 4

**14.** Center: $\left(0, -\frac{1}{2}\right)$;    radius: 2

**15.** Write the equation $x^2 + y^2 - 2x + 4y - 4 = 0$ in standard form. Then sketch the circle.

In Exercises 16 and 17, use a graphing utility to find the number of $x$-intercepts of the equation.

**16.** $y = \frac{1}{4}(-4x^2 + 16x - 14)$

**17.** $y = \frac{1}{8}(4x^2 - 32x + 63)$

In Exercises 18–20, use the following model. The earnings per share $y$ (in dollars) for Amgen, Inc. from 1990 to 1997 can be modeled by $y = 0.33t + 0.31$, $0 \le t \le 7$, where $t = 0$ represents 1990.   (Source: *Value Line*)

**18.** Determine a convenient graphing utility viewing window.

**19.** Use a graphing utility to graph this model.

**20.** Use the trace feature to predict the earnings per share in 2000.

## 2.4 Lines in the Plane

### Objectives

- Determine the slope of a line passing through two points.
- Use the point-slope form to find the equation of a line.
- Use the slope-intercept form to sketch the graph of a line.
- Determine if lines are parallel or perpendicular using slope.
- Write the equation of a line parallel or perpendicular to a given line.

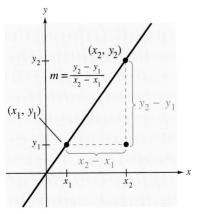

**Figure 2.35**

## The Slope of a Line

The **slope** of a nonvertical line is a measure of the steepness of the line. Specifically, the slope represents the number of units the line rises or falls vertically for each unit of horizontal change from left to right. For instance, consider the two points $(x_1, y_1)$ and $(x_2, y_2)$ on the line shown in Figure 2.35. As you move from left to right along this line, a change of $y_2 - y_1$ units in the vertical direction corresponds to a change of $x_2 - x_1$ units in the horizontal direction. That is,

$$y_2 - y_1 = \text{the change in } y$$

and

$$x_2 - x_1 = \text{the change in } x.$$

The slope of the line is defined as the quotient of these two changes.

> ▶ **Definition of the Slope of a Line**
>
> The **slope** $m$ of the nonvertical line passing through the points $(x_1, y_1)$ and $(x_2, y_2)$ is
>
> $$m = \frac{y_2 - y_1}{x_2 - x_1} = \frac{\text{change in } y}{\text{change in } x}$$
>
> where $x_1 \neq x_2$.

The change in $x$ is sometimes called the "run" and the change in $y$ is sometimes called the "rise."

When this formula is used, the *order of subtraction* is important. Given two points on a line, you are free to label either one of them as $(x_1, y_1)$ and the other as $(x_2, y_2)$. However, once this is done, you must form the numerator and denominator using the same order of subtraction.

$$m = \frac{y_2 - y_1}{x_2 - x_1} \qquad m = \frac{y_1 - y_2}{x_1 - x_2} \qquad m = \frac{y_2 - y_1}{x_1 - x_2}$$

  Correct          Correct          Incorrect

## Study Tip

When working with the slope of a line, it helps to remember the following interpretations of slope.

1. A line with positive slope *rises* from left to right.

2. A line with negative slope *falls* from left to right.

3. A line with zero slope is *horizontal*.

4. A line with undefined slope is *vertical*.

### Example 1    Finding the Slope of a Line Through Two Points

Find the slopes of the lines passing through the following pairs of points.

**a.** $(-2, 0)$ and $(3, 1)$    **b.** $(-1, 2)$ and $(2, 2)$    **c.** $(0, 4)$ and $(1, -1)$

**Solution**

**a.** Let $(x_1, y_1) = (-2, 0)$ and $(x_2, y_2) = (3, 1)$.

$$m = \frac{y_2 - y_1}{x_2 - x_1} \qquad \Leftarrow \text{Difference in } y\text{-values}$$
$$\Leftarrow \text{Difference in } x\text{-values}$$

$$= \frac{1 - 0}{3 - (-2)}$$

$$= \frac{1}{5}$$

**b.** The slope of the line through $(-1, 2)$ and $(2, 2)$ is

$$m = \frac{2 - 2}{2 - (-1)} = \frac{0}{3} = 0.$$

**c.** The slope of the line through $(0, 4)$ and $(1, -1)$ is

$$m = \frac{-1 - 4}{1 - 0} = \frac{-5}{1} = -5.$$

The graphs of the three lines are shown in Figure 2.36.

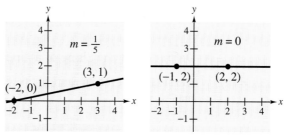

(a) If *m* is positive, the line rises.

(b) If *m* is zero, the line is horizontal.

(c) If *m* is negative, the line falls.

**Figure 2.36**

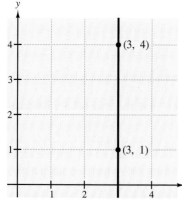

**Figure 2.37**    *If the line is vertical, the slope is undefined.*

The definition of slope does not apply to vertical lines. For instance, consider the points $(3, 4)$ and $(3, 1)$ on the vertical line shown in Figure 2.37. Applying the formula for slope, you have

$$\frac{1 - 4}{3 - 3} = \frac{-3}{0}. \qquad \text{Undefined division by zero}$$

Because division by zero is not defined, the slope of a vertical line is not defined.

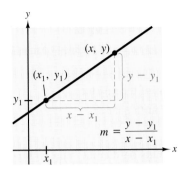

**Figure 2.38**  *Any two points on a line can be used to determine the slope of the line.*

# The Point-Slope Form

If you know the slope of a line and you also know the coordinates of one point on the line, you can find an equation for the line. For instance, in Figure 2.38, let $(x_1, y_1)$ be a given point on the line whose slope is $m$. If $(x, y)$ is any *other* point on the line, it follows that

$$\frac{y - y_1}{x - x_1} = m.$$

This equation in variables $x$ and $y$ can be rewritten to produce the following **point-slope form** of the equation of a line.

> ▶ **Point-Slope Form of the Equation of a Line**
>
> The **point-slope form** of the equation of the line that passes through the point $(x_1, y_1)$ and has a slope of $m$ is
>
> $$y - y_1 = m(x - x_1).$$

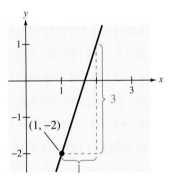

**Figure 2.39**

▮ **Example 2**   The Point-Slope Form of the Equation of a Line

Find an equation of the line that passes through $(1, -2)$ and has a slope of 3.

**Solution**

Use the point-slope form with $(x_1, y_1) = (1, -2)$ and $m = 3$.

$$\begin{aligned}
y - y_1 &= m(x - x_1) && \text{Point-slope form} \\
y - (-2) &= 3(x - 1) && \text{Substitute } y_1 = -2, x_1 = 1, \text{ and } m = 3. \\
y + 2 &= 3x - 3 && \text{Simplify.} \\
y &= 3x - 5 && \text{Equation of line}
\end{aligned}$$

The graph of this line is shown in Figure 2.39.

---

The point-slope form can be used to find the equation of a line passing through two points $(x_1, y_1)$ and $(x_2, y_2)$. First, use the formula for the slope of the line passing through two points. Then, use the point-slope form to obtain the equation

$$y - y_1 = \frac{y_2 - y_1}{x_2 - x_1}(x - x_1).$$

This is sometimes called the **two-point form** of the equation of a line.

**Example 3** A Linear Model for Sales Prediction

During the first two quarters of the year, a company had sales of $3.4 million and $3.7 million, respectively.

**a.** Write a linear equation giving the sales $y$ in terms of the quarter $x$.

**b.** Use the equation to predict the sales during the fourth quarter.

Solution

**a.** In Figure 2.40 let $(1, 3.4)$ and $(2, 3.7)$ be two points on the line representing the total sales. The slope of the line passing through these two points is

$$m = \frac{3.7 - 3.4}{2 - 1} = 0.3.$$

By the point-slope form, the equation of the line is as follows.

| | |
|---|---|
| $y - y_1 = m(x - x_1)$ | Point-slope form |
| $y - 3.4 = 0.3(x - 1)$ | Substitute for $y_1$, $m$, and $x_1$. |
| $y = 0.3x - 0.3 + 3.4$ | Simplify. |
| $y = 0.3x + 3.1$ | Equation of line |

**b.** Using the equation from part (a), the fourth quarter sales $(x = 4)$ should be

$$y = 0.3(4) + 3.1 = 4.3 \text{ million dollars.}$$

Figure 2.40

The estimation method illustrated in Example 3 is called **linear extrapolation.** Note in Figure 2.41(a), that for linear extrapolation, the estimated point lies to the right of the given points. When the estimated point lies *between* two given points, the procedure is called **linear interpolation,** as shown in Figure 2.41(b).

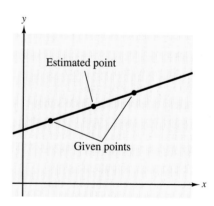

**(a) Linear extrapolation**

**(b) Linear interpolation**

Figure 2.41

## Sketching Graphs of Lines

You have seen that to *find the equation of a line* it is convenient to use the point-slope form. This formula, however, is not particularly useful for *sketching the graph of a line*. The form that is better suited to graphing linear equations is the **slope-intercept form** of the equation of a line. To derive the slope-intercept form, write the following.

$$y - y_1 = m(x - x_1) \qquad \text{Point-slope form}$$

$$y = mx - mx_1 + y_1 \qquad \text{Solve for } y.$$

$$y = mx + b \qquad \text{Slope-intercept form}$$

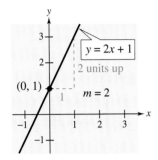

(a) When *m* is positive, the line rises.

> ### ► Slope-Intercept Form of the Equation of a Line
>
> The graph of the equation
>
> $$y = mx + b$$
>
> is a line whose slope is *m* and whose *y*-intercept is $(0, b)$.

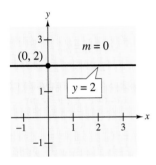

(b) When *m* is zero, the line is horizontal.

**Example 4**   Sketching the Graph of a Linear Equation

Sketch the graph of each of the following linear equations.

**a.** $y = 2x + 1$    **b.** $y = 2$    **c.** $x + y = 2$

Solution

**a.** Because $b = 1$, the *y*-intercept is $(0, 1)$. Moreover, because the slope is $m = 2$, this line *rises* two units for each unit the line moves to the right, as shown in Figure 2.42(a).

**b.** By writing the equation $y = 2$ in the form

$$y = (0)x + 2$$

you can see that the *y*-intercept is $(0, 2)$ and the slope is zero. A zero slope implies that the line is horizontal, as shown in Figure 2.42(b).

**c.** By writing the equation $x + y = 2$ in slope-intercept form,

$$y = -x + 2$$

you can see that the *y*-intercept is $(0, 2)$. Moreover, because the slope is $m = -1$, this line *falls* one unit for each unit the line moves to the right, as shown in Figure 2.42(c).

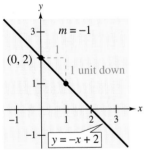

(c) When *m* is negative, the line falls.

**Figure 2.42**

From the slope-intercept form of the equation of a line, you can see that a horizontal line ($m = 0$) has an equation of the form

$$y = (0)x + b \quad \text{or} \quad y = b. \qquad \text{Horizontal line}$$

This is consistent with the fact that each point on a horizontal line through $(0, b)$ has a $y$-coordinate of $b$, as shown in Figure 2.43.

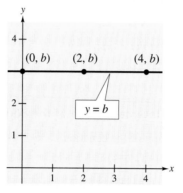

**Figure 2.43**  *Horizontal Line*          **Figure 2.44**  *Vertical Line*

Similarly, each point on a vertical line through $(a, 0)$ has an $x$-coordinate of $a$, as shown in Figure 2.44. Hence, a vertical line has an equation of the form

$$x = a. \qquad \text{Vertical line}$$

This equation cannot be written in the slope-intercept form because the slope of a vertical line is undefined. However, *every* line has an equation that can be written in the **general form**

$$Ax + By + C = 0 \qquad \text{General form}$$

where $A$ and $B$ are not *both* zero. If $A = 0$ (and $B \neq 0$), the equation can be reduced to the form $y = b$, which represents a horizontal line. If $B = 0$ (and $A \neq 0$), the general equation can be reduced to the form $x = a$, which represents a vertical line.

---

▶ **Summary of Equations of Lines**

1. General form:          $Ax + By + C = 0$

2. Vertical line:          $x = a$

3. Horizontal line:          $y = b$

4. Slope-intercept form:          $y = mx + b$

5. Point-slope form:          $y - y_1 = m(x - x_1)$

# Parallel and Perpendicular Lines

The slope of a line is a convenient tool for determining whether two lines are parallel or perpendicular.

> ▶ **Parallel Lines**
>
> Two distinct nonvertical lines are **parallel** if and only if their slopes are equal.

### Example 5    Equations of Parallel Lines

Find an equation of the line that passes through the point $(2, -1)$ and is parallel to the line $2x - 3y = 5$, as shown in Figure 2.45.

### Solution

Writing the given equation in slope-intercept form produces the following.

$$2x - 3y = 5 \qquad \text{Original equation}$$

$$3y = 2x - 5 \qquad \text{Isolate } y\text{-term.}$$

$$y = \frac{2}{3}x - \frac{5}{3} \qquad \text{Slope-intercept form}$$

So, the given line has a slope of $m = \frac{2}{3}$. Because any line parallel to the given line must also have a slope of $\frac{2}{3}$, the required line through $(2, -1)$ has the following equation.

$$y - (-1) = \frac{2}{3}(x - 2) \qquad \text{Point-slope form}$$

$$y = \frac{2}{3}x - \frac{4}{3} - 1 \qquad \text{Solve for } y.$$

$$y = \frac{2}{3}x - \frac{7}{3} \qquad \text{Slope-intercept form}$$

Notice the similarity between the slope-intercept form of the original equation and the slope-intercept form of the parallel equation.

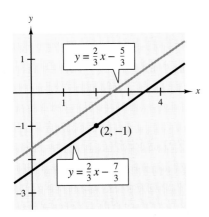

$$y = \frac{2}{3}x - \frac{5}{3}$$

$$(2, -1)$$

$$y = \frac{2}{3}x - \frac{7}{3}$$

**Figure 2.45**

You have seen that two nonvertical lines are parallel if and only if they have the same slope. Two nonvertical lines are *perpendicular* if and only if their slopes are negative reciprocals of each other. For instance, the lines $y = 2x$ and $y = -\frac{1}{2}x$ are perpendicular because one has a slope of 2 and the other has a slope of $-\frac{1}{2}$.

·DISCOVERY·

Use a graphing utility to graph $y_1 = \frac{2}{3}x + \frac{5}{2}$ and $y_2 = -\frac{3}{2}x + 2$. When you examine the graphs with a square setting, what do you observe? What do you notice about the slopes of the two lines?

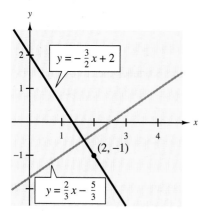

**Figure 2.46**

▶ **Perpendicular Lines**

Two nonvertical lines are **perpendicular** if and only if their slopes are negative reciprocals of each other. That is,

$$m_1 = -\frac{1}{m_2}.$$

**Example 6**   Equations of Perpendicular Lines

Find an equation of the line that passes through the point $(2, -1)$ and is perpendicular to the line $2x - 3y = 5$.

**Solution**

By writing the given line in the form $y = \frac{2}{3}x - \frac{5}{3}$, you can see that the line has a slope of $\frac{2}{3}$. Hence, any line that is perpendicular to this line must have a slope of $-\frac{3}{2}$ (because $-\frac{3}{2}$ is the negative reciprocal of $\frac{2}{3}$). So, the required line through the point $(2, -1)$ has the following equation.

$$y - (-1) = -\frac{3}{2}(x - 2) \qquad \text{Point-slope form}$$

$$y = -\frac{3}{2}x + 3 - 1 \qquad \text{Solve for } y.$$

$$y = -\frac{3}{2}x + 2 \qquad \text{Slope-intercept form}$$

The graphs of both equations are shown in Figure 2.46.

| Discussing the Concept | Interpreting Slope |
|---|---|

You think the relationship between daily high temperature $y$ (in degrees Fahrenheit) and time $x$ (in days) is linear. You observe the high temperature on day 0 to be 60°F and the high temperature 15 days later to be 68°F.

**a.** Find the slope of the line passing through these two points.

**b.** Interpret the meaning of the slope in this situation.

**c.** Use linear extrapolation to predict the high temperatures on days 30, 60, 105, and 180. Are all of these predictions reasonable? Why or why not?

# Warm Up

The following warm-up exercises involve skills that were covered in earlier sections. You will use these skills in the exercise set for this section.

In Exercises 1–4, simplify the expression.

**1.** $\dfrac{4 - (-5)}{-3 - (-1)}$

**2.** $\dfrac{-5 - 8}{0 - (-3)}$

**3.** Find $-1/m$ for $m = 4/5$.

**4.** Find $-1/m$ for $m = -2$.

In Exercises 5–10, solve for $y$ in terms of $x$.

**5.** $2x - 3y = 5$

**6.** $4x + 2y = 0$

**7.** $y - (-4) = 3[x - (-1)]$

**8.** $y - 7 = \frac{2}{3}(x - 3)$

**9.** $y - (-1) = \dfrac{3 - (-1)}{2 - 4}(x - 4)$

**10.** $y - 5 = \dfrac{3 - 5}{0 - 2}(x - 2)$

## 2.4   Exercises

In Exercises 1–6, estimate the slope of the line from its graph.

**1.**

**2.**

**3.**

**4.**

**5.**

**6.**
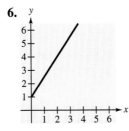

In Exercises 7 and 8, sketch the lines through the given point with the given slopes. Sketch all four lines on the same set of coordinate axes.

|  | Point | | Slopes |
|---|---|---|---|

**7.** $(5, -1)$   (a) $-2$   (b) $1$   (c) $0$   (d) $\frac{3}{2}$

**8.** $(-4, -3)$   (a) $-1$   (b) $2$   (c) $\frac{1}{4}$   (d) undefined

In Exercises 9–14, plot the points and find the slope of the line passing through the points.

**9.** $(6, 9), (-4, -1)$   **10.** $(2, 4), (4, -4)$

**11.** $(-6, -1), (-6, 4)$   **12.** $(0, -10), (-4, 0)$

**13.** $\left(-\frac{1}{3}, 1\right), \left(-\frac{2}{3}, \frac{5}{6}\right)$   **14.** $\left(\frac{7}{8}, \frac{3}{4}\right), \left(\frac{5}{4}, -\frac{1}{4}\right)$

In Exercises 15–20, determine if the lines $L_1$ and $L_2$ passing through the pairs of points are parallel, perpendicular, or neither.

**15.** $L_1$: $(0, -1), (5, 9)$;   $L_2$: $(0, 3), (4, 1)$

**16.** $L_1$: $(3, 6), (-6, 0)$;   $L_2$: $(0, -1), \left(5, \frac{7}{3}\right)$

**17.** $L_1$: $(-2, -1), (1, 5)$;   $L_2$: $(1, 3), (5, -5)$

**18.** $L_1$: $(4, 8), (-4, 2)$;   $L_2$: $(3, -5), \left(-1, \frac{1}{3}\right)$

**19.** $L_1$: $(-1, 7), (-6, 4)$;   $L_2$: $(0, 1), (5, 4)$

**20.** $L_1$: $(-1, 3), (2, -5)$;   $L_2$: $(3, 0), (2, -7)$

In Exercises 21–26, use the point on the line and the slope of the line to find three additional points that the line passes through. (Each problem has more than one correct answer.)

| | Point | Slope |
|---|---|---|
| **21.** | $(5, -2)$ | $m = 0$ |
| **22.** | $(-1, 3)$ | $m$ is undefined |
| **23.** | $(5, -6)$ | $m = 1$ |
| **24.** | $(10, -6)$ | $m = -1$ |
| **25.** | $(-6, -1)$ | $m = \frac{1}{2}$ |
| **26.** | $(7, -5)$ | $m = -\frac{2}{3}$ |

In Exercises 27–32, find the slope and $y$-intercept (if possible) of the line specified by the equation.

**27.** $4x - y - 6 = 0$     **28.** $2x + 3y - 9 = 0$

**29.** $8 - 3x = 0$     **30.** $3y + 5 = 0$

**31.** $7x + 6y - 30 = 0$     **32.** $x - y - 10 = 0$

In Exercises 33–40, find an equation of the line passing through the points and use a graphing utility to verify your answer.

**33.** $(7, -4), (-7, 3)$     **34.** $(4, 3), (-4, -4)$

**35.** $(-9, 11), (-9, 14)$     **36.** $(-1, 4), (6, 4)$

**37.** $\left(2, \frac{1}{2}\right), \left(\frac{1}{2}, \frac{5}{4}\right)$     **38.** $(1, 1), \left(6, -\frac{2}{3}\right)$

**39.** $(1, 0.6), (-2, -0.6)$     **40.** $(-8, 0.6), (2, -2.4)$

In Exercises 41–50, find an equation of the line that passes through the point and has the indicated slope. Sketch the line.

| | Point | Slope |
|---|---|---|
| **41.** | $(7, 0)$ | $m = 1$ |
| **42.** | $(-2, 0)$ | $m = -4$ |
| **43.** | $(-3, 6)$ | $m = -2$ |
| **44.** | $(-8, 3)$ | $m = -\frac{1}{2}$ |
| **45.** | $(4, 0)$ | $m = -\frac{1}{3}$ |
| **46.** | $(-2, -5)$ | $m = \frac{3}{4}$ |
| **47.** | $(6, -1)$ | $m$ is undefined |
| **48.** | $(-10, 4)$ | $m = 0$ |
| **49.** | $\left(4, \frac{5}{2}\right)$ | $m = \frac{4}{3}$ |
| **50.** | $\left(-\frac{1}{2}, \frac{3}{2}\right)$ | $m = -3$ |

**51.** A fellow student does not understand why the slope of a vertical line is undefined. Describe how you would help this student understand the concept of undefined slope.

**52.** Another student overhears your conversation in Exercise 51 and states, "I do not understand why a horizontal line has zero slope and why that is different from undefined or no slope." Describe how you would explain the concepts of zero slope and undefined slope and why they are different from each other.

In Exercises 53–56, use the *intercept form* of the equation of a line

$$\frac{x}{a} + \frac{y}{b} = 1, \quad a \neq 0, \quad b \neq 0$$

to find the equation of the line with the given intercepts $(a, 0)$ and $(0, b)$.

**53.** $x$-intercept: $(1, 0)$
  $y$-intercept: $(0, -4)$

**54.** $x$-intercept: $(-3, 0)$
  $y$-intercept: $(0, 4)$

**55.** $x$-intercept: $\left(-\frac{1}{6}, 0\right)$
  $y$-intercept: $\left(0, -\frac{2}{3}\right)$

**56.** $x$-intercept: $\left(-\frac{2}{3}, 0\right)$
  $y$-intercept: $\left(0, \frac{1}{2}\right)$

In Exercises 57–62, write an equation of the line through the point (a) parallel to the line and (b) perpendicular to the line. Use a graphing utility to verify your answer.

| | Point | Line |
|---|---|---|
| **57.** | $(6, 2)$ | $y - 2x = -1$ |
| **58.** | $(-5, 4)$ | $x + y = 8$ |
| **59.** | $\left(\frac{1}{4}, -\frac{2}{3}\right)$ | $2x - 3y = 5$ |
| **60.** | $\left(\frac{7}{8}, \frac{3}{4}\right)$ | $5x + 3y = 0$ |
| **61.** | $(-1, 0)$ | $y = -3$ |
| **62.** | $(2, 5)$ | $x = 4$ |

**63.** *Temperature*   Find an equation of the line that gives the relationship between the temperature in degrees Celsius, $C$, and degrees Fahrenheit, $F$. Remember that water freezes at $0°$ Celsius ($32°$ Fahrenheit) and boils at $100°$ Celsius ($212°$ Fahrenheit).

**64.** *Temperature*    Use the result of Exercise 63 to complete the table. Is there a temperature for which the Fahrenheit reading is the same as the Celsius reading? If so, what is it?

| $C$ |  | $-10°$ | $10°$ |  |  | $177°$ |
|---|---|---|---|---|---|---|
| $F$ | $0°$ |  |  | $68°$ | $90°$ |  |

**65.** *Simple Interest*    A person deposits $P$ dollars in an account that pays simple interest. After 2 months the balance in the account is $759, and after 3 months the balance in the account is $763.50. Find an equation that gives the relationship between the balance $A$ and the time $t$ in months.

**66.** *Simple Interest*    Use the result of Exercise 65 to complete the table.

| $A$ |  |  | $759.00 | $763.50 |  |  |  |
|---|---|---|---|---|---|---|---|
| $t$ | 0 | 1 |  |  | 4 | 5 | 6 |

**67.** *Fourth Quarter Sales*    During the first and second quarters of the year, a business had sales of $145,000 and $152,000, respectively. If the growth of the sales follows a linear pattern, what will the sales be during the fourth quarter?

**68.** *College Enrollment*    A small college had 2546 students in 1997 and 2702 students in 1999. If the enrollment follows a linear growth pattern, how many students will the college have in 2004?

**69.** *Annual Salary*    Suppose that your salary was $28,500 in 1996 and $31,800 in 1999. If your salary follows a linear growth pattern, what salary will you be making in 2003?

**70.** *Endowment Income*    In 1989, the total amount of money endowed to symphony orchestras was $46.8 million. By 1992 the total amount had increased to $55.3 million. If these contributions continued to increase in a linear pattern, how much money would have been endowed to symphony orchestras in 1995? (Source: American Symphony Orchestra League, Inc.)

**71.** *Time Warner Revenue*    In 1993, Time Warner, Inc. had a total revenue of $6581 million. By 1995, the total revenue had increased to $8067 million. If the total revenue continued to increase in a linear pattern, how much would the revenue have been in 1997? The actual revenue in 1997 was $13,294 million. Do you think the increase in revenue was approximately linear?    (Source: Time Warner, Inc.)

**72.** *Shopping Centers*    In 1991, the number of shopping centers in the United States was 37,975. By 1994, the number had increased to 40,368. If the number of shopping centers continued to increase in a linear pattern, what would the number have been in 1996? The actual number of shopping centers in 1996 was 42,130. Do you think the increase in the number of shopping centers was approximately linear?    (Source: National Research Bureau)

**73.** *Apartment Rent*    The rent for a two-bedroom apartment was $625 in 1995 and $800 in 1998. If the rent for a two-bedroom apartment follows a linear growth pattern, what is the rate of change per year for the rent? What will the rent be in 2005?

**74.** *Scuba Diving*    Pressure increases at the rate of 1 atmosphere (the weight of air pressure on a body at sea level) for each additional 33 feet of depth under water. At sea level the pressure is 1 atmosphere. At 66 feet below the surface, the total pressure is 3 atmospheres. Write a linear equation to describe the pressure $p$ with respect to depth $d$ below the surface of the sea. What is the rate of change of pressure with depth?    (Source: PADI Open Water Diver Manual)

## Math Matters • The Pythagorean Theorem

The Pythagorean Theorem is one of the most famous theorems in mathematics. Its name comes from Pythagoras of Samos (c. 580–500 B.C.), a Greek mathematician who founded a school of mathematics and philosophy at Crotona in what is now southern Italy. The Pythagorean Theorem states a relationship among the three sides of a right triangle. Specifically, the theorem states that the sum of the squares of the two legs of a right triangle is equal to the square of the hypotenuse. This result is usually written as

$$a^2 + b^2 = c^2.$$    Pythagorean Theorem

One way to prove the Pythagorean Theorem is to use a geometrical argument, as indicated in the series of figures below. In the figure at the left, the areas of the three squares

represent the squares of the sides of a right triangle. To prove that $a^2 + b^2 = c^2$, you need to show that the sum of the areas of the two smaller squares is equal to the area of the larger square. One way to do this is to make four copies of the given right triangle, and use the four triangles to form a square whose sides have lengths of $a + b$. By doing this in two different ways, you can conclude that $a^2 + b^2 = c^2$. Do you see why this is true?

According to the *Guinness Book of World Records,* 1995, the Pythagorean Theorem is the "most proven theorem" in mathematics. This publication claims that "*The Pythagorean Proposition* published in 1940 contained 370 different proofs of Pythagoras' theorem, including one by President James Garfield."

## 2.5    Linear Modeling

### Objectives

- Construct a linear model to relate quantities that vary directly.
- Construct and use a linear model with slope as the rate of change.
- Use a scatter plot to find a linear model that fits a set of data.

## Introduction

The primary objective of applied mathematics is to find equations or **mathematical models** that describe real-world phenomena. In developing a mathematical model to represent actual data, you should strive for two (often conflicting) goals—accuracy and simplicity. That is, you want the model to be simple enough to be workable, yet accurate enough to produce meaningful results.

### Example 1    A Mathematical Model

The total annual amounts of advertising expenses (in billions of dollars) in the United States from 1991 to 1996 are given in the table.    (Source: McCann-Erickson, Inc.)

| Year | 1 | 2 | 3 | 4 | 5 | 6 |
|------|-----|-----|-----|-----|-----|-----|
| $y$ | \$126.4 | \$131.3 | \$138.1 | \$150.0 | \$160.9 | \$173.2 |

A linear model that approximates these data is

$$y = 113.18 + 9.56t, \quad 1 \le t \le 6$$

where $y$ represents the advertising expenses (in billions of dollars) and $t$ represents the year, with $t = 1$ corresponding to 1991. Plot the actual data *and* the model on the same graph. How closely does the model represent the data?

### Solution

The actual data are plotted in Figure 2.47, along with the graph of the linear model. From the figure, it appears that the model is a "good fit" for the actual data. You can see how well the model fits by comparing the actual values of $y$ with the values of $y$ given by the model (these values are labeled $y^*$ in the table below).

Figure 2.47

| $t$ | 1 | 2 | 3 | 4 | 5 | 6 |
|------|-----|-----|-----|-----|-----|-----|
| $y$ | \$126.4 | \$131.3 | \$138.1 | \$150.0 | \$160.9 | \$173.2 |
| $y^*$ | \$122.7 | \$132.3 | \$141.9 | \$151.4 | \$161.0 | \$170.5 |

## Direct Variation

There are two basic types of linear models. The more general model has a $y$-intercept that is nonzero: $y = mx + b$, $b \neq 0$. The simpler model, $y = mx$, has a $y$-intercept that is zero. In the simpler model, $y$ is said to **vary directly** as $x$, or be **proportional** to $x$.

---

▶ **Direct Variation**

The following statements are equivalent.

**1.** $y$ **varies directly** as $x$.

**2.** $y$ is **directly proportional** to $x$.

**3.** $y = mx$ for some nonzero constant $m$.

$m$ is the **constant of variation** or the **constant of proportionality.**

---

**Example 2**    State Income Tax

In Pennsylvania, the state income tax is directly proportional to *gross income*. Suppose you were working in Pennsylvania and your state income tax deduction was $43.50 for a gross monthly income of $1500.00. Find a mathematical model that gives the Pennsylvania state income tax in terms of gross income.

**Solution**

| *Verbal Model:* | $\dfrac{\text{State}}{\text{income tax}} = m \cdot \dfrac{\text{Gross}}{\text{income}}$ | |
|---|---|---|

*Labels:*    State income tax $= y$                            (dollars)
Gross income $= x$                            (dollars)
Income tax rate $= m$            (percent in decimal form)

*Equation:*    $y = mx$

To solve for $m$, substitute the given information into the equation $y = mx$, and then solve for $m$.

$$y = mx \qquad \text{Direct variation model}$$

$$43.50 = m(1500) \qquad \text{Substitute } y = 43.50 \text{ and } x = 1500.$$

$$0.029 = m \qquad \text{Income tax rate}$$

So, the equation (or model) for state income tax in Pennsylvania is $y = 0.029x$. In other words, Pennsylvania has a state income tax rate of 2.9% of gross income. The graph of this equation is shown in Figure 2.48.

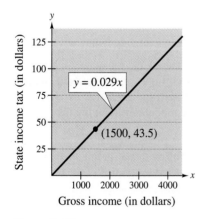

**Figure 2.48**

Most measurements in the English system and metric system are directly proportional. The next example shows how to use a direct proportion to convert between miles per hour and kilometers per hour.

**Example 3**    The English and Metric Systems

You are traveling at a rate of 64 miles per hour. You switch your speedometer reading to metric units and notice that the speed is 103 kilometers per hour. Use this information to find a mathematical model that relates miles per hour to kilometers per hour.

### Solution

If you let $y$ represent the speed in miles per hour and $x$ represent the speed in kilometers per hour, you know that $y$ and $x$ are related by the equation

$$y = mx.$$

You are given that $y = 64$ when $x = 103$. By substituting these values into the equation $y = mx$, you can find the value of $m$.

Figure 2.49

| | |
|---|---|
| $y = mx$ | Direct variation model |
| $64 = m(103)$ | Substitute $y = 64$ and $x = 103$. |
| $\dfrac{64}{103} = m$ | Divide both sides by 103. |
| $0.62136 \approx m$ | Use a calculator. |

So, the conversion factor from kilometers per hour to miles per hour is approximately 0.62136, and the model is

$$y = 0.62136x.$$

The graph of this equation is shown in Figure 2.49.

---

Once you have found a model that converts speeds from kilometers per hour to miles per hour, you can use the model to convert other speeds from the metric system to the English system, as shown in the table.

| Kilometers per hour | 20.0 | 40.0 | 60.0 | 80.0 | 100.0 | 120.0 |
|---|---|---|---|---|---|---|
| Miles per hour | 12.4 | 24.9 | 37.3 | 49.7 | 62.1 | 74.6 |

The conversion equation $y = 0.62136x$ can be approximated by the simpler equation $y = \frac{5}{8}x$. For instance, to convert 40 kilometers per hour, divide by 8 and multiply by 5 to obtain 25 miles per hour.

## Rates of Change

A second common type of linear model is one that involves a known rate of change. In the linear equation

$$y = mx + b$$

you know that $m$ represents the slope of the line. In real-life problems, the slope can often be interpreted as the **rate of change** of $y$ with respect to $x$. Rates of change should always be listed in appropriate units of measure.

**Example 4**   Height of a Mountain Climber

A mountain climber is climbing up a 500-foot cliff. By 1 P.M., the mountain climber has climbed 115 feet up the cliff. By 4 P.M., the climber has reached a height of 280 feet, as shown in Figure 2.50. Find the average rate of change of the climber and use this rate of change to find the equation that relates the height of the climber to the time. Use the model to estimate the time when the climber will reach the top of the cliff.

### Solution

Let $y$ represent the height of the climber and let $t$ represent the time. Then the two points that represent the climber's two positions are

$$(t_1, y_1) = (1, 115) \quad \text{and} \quad (t_2, y_2) = (4, 280).$$

So, the average rate of change of the climber is

$$\text{Average rate of change} = \frac{y_2 - y_1}{t_2 - t_1}$$

$$= \frac{280 - 115}{4 - 1}$$

$$= 55 \text{ feet per hour.}$$

An equation that relates the height of the climber to the time is

$$y - y_1 = m(t - t_1) \qquad \text{Point-slope form}$$

$$y - 115 = 55(t - 1) \qquad \text{Substitute } y_1 = 115,\, t_1 = 1,\text{ and } m = 55.$$

$$y = 55t + 60. \qquad \text{Linear model}$$

To find the time when the climber reaches the top of the cliff, let $y = 500$ and solve for $t$ to obtain

$$t = 8.$$

So, continuing at the same rate, the climber will reach the top of the cliff at 8 P.M.

**Figure 2.50**

**Example 5**    Population of Omaha, Nebraska

Between 1980 and 1990, the population of the city of Omaha, Nebraska increased at an average rate of approximately 2200 people per year. In 1980, the population was 314,000. Find a mathematical model that gives the population of Omaha in terms of the year, and use the model to estimate the population in 1994. (Source: U.S. Bureau of the Census)

**Solution**

Let $y$ represent the population of Omaha, and let $t$ represent the calendar year, with $t = 0$ corresponding to 1980. Letting $t = 0$ correspond to 1980 is convenient because you were given the population in 1980. Now, using the rate of change of 2200 people per year, you have

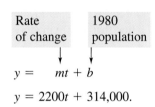

$$y = 2200t + 314,000.$$

Using this model, you can predict the 1994 population to be

$$(1994 \text{ population}) = 2200(14) + 314,000 = 344,800.$$

The graph is shown in Figure 2.51. (In this particular example, the linear model is quite good—the actual population of Omaha, Nebraska in 1994 was 345,000.)

**Figure 2.51**

**Example 6**    Straight-Line Depreciation

Your company has purchased a $15,000 machine that has a useful life of 10 years. The salvage value at the end of the 10 years is $3000. Write a linear equation that describes the book value of the machine each year.

**Solution**

Let $V$ represent the value of the machine at the end of the year $t$. You can represent the initial value of the machine by the ordered pair $(0, 15,000)$ and the salvage value by the ordered pair $(10, 3000)$. The slope of the line is

$$m = \frac{3000 - 15,000}{10 - 0} = -1200$$

which represents the annual depreciation in *dollars per year.* Using the point-slope form, you can write the equation of the line as follows.

$$V - 15,000 = -1200(t - 0) \qquad \text{Point-slope form}$$

$$V = -1200t + 15,000 \qquad \text{Slope-intercept form}$$

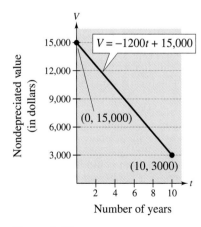

**Figure 2.52**

The graph of the equation is shown in Figure 2.52.

## Scatter Plots

Another type of linear modeling is a graphical approach that is commonly used in statistics. To find a mathematical model that approximates a set of actual data points, plot the points on a rectangular coordinate system. This collection of points is called a **scatter plot.** Once the points have been plotted, try to find the line that most closely represents the plotted points. (In this section, we will rely on a visual technique for fitting a line to a set of points. If you take a course in statistics, you will encounter *regression analysis* formulas that can fit a line to a set of points.)

### Example 7    Fitting a Line to a Set of Points

The scatter plot in Figure 2.53(a) shows 35 different points in the plane. Find the equation of a line that approximately fits these points.

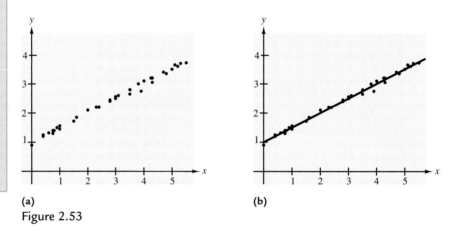

(a)               (b)

Figure 2.53

### Solution

From Figure 2.53(a), you can see that there is no line that *exactly* fits the given points. The points, however, do appear to resemble a linear pattern. Figure 2.53(b) shows a line that appears to best describe the given points. (Notice that about as many points lie above the line as below it.) From this figure, you can see that the "best-fitting line" has a $y$-intercept at about $(0, 1)$ and has a slope of about $\frac{1}{2}$. So, the equation of the line is

$$y = \frac{1}{2}x + 1.$$

If you had been given the coordinates of the 35 points, you could have checked the accuracy of this model by constructing a table that compared the actual $y$-values with the $y$-values given by the model.

## Study Tip

Use a scatter plot to first determine whether a line is a reasonable model for a given set of data.

| Example 8 | Prize Money at the Indianapolis 500 |

The total prize money $p$ (in millions of dollars) awarded at the Indianapolis 500 race from 1990 to 1998 is given in the following table. Construct a scatter plot that represents the data and find a linear model that approximates the data. (Source: Indianapolis Motor Speedway Hall of Fame)

| Year | 1990 | 1991 | 1992 | 1993 | 1994 | 1995 | 1996 | 1997 | 1998 |
|------|------|------|------|------|------|------|------|------|------|
| $p$ | $6.63 | $7.01 | $7.53 | $7.68 | $7.86 | $8.06 | $8.11 | $8.62 | $8.72 |

### Solution

Let $t = 0$ represent 1990. The scatter plot for the points is shown in Figure 2.54. From the scatter plot, draw a line that approximates the data. Then, to find the equation of the line, approximate two points *on the line*: $(1, 7)$ and $(5, 8)$. The slope of this line is

$$m \approx \frac{p_2 - p_1}{t_2 - t_1} = \frac{8 - 7}{5 - 1} = 0.25.$$

Using the point-slope form, you can determine that the equation of the line is

$$p = 0.25t + 6.75.$$

To check this model, compare the actual $p$-values with the $p$-values given by the model (these values are labeled $p^*$ in the table below).

Figure 2.54

| $t$ | 0 | 1 | 2 | 3 | 4 | 5 | 6 | 7 | 8 |
|-----|---|---|---|---|---|---|---|---|---|
| $p$ | $6.63 | $7.01 | $7.53 | $7.68 | $7.86 | $8.06 | $8.11 | $8.62 | $8.72 |
| $p^*$ | $6.75 | $7.00 | $7.25 | $7.50 | $7.75 | $8.00 | $8.25 | $8.50 | $8.75 |

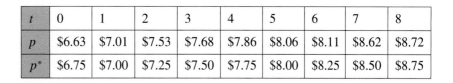

| Discussing the Concept | **Gathering and Analyzing Data** |

Measure the height ($h$) and forearm ($f$) of each person in the class with a tape measure. Gather the data in the form $(h, f)$ and plot the points on a set of coordinate axes. Do the points appear to follow a linear model? Find an equation for a line that approximately represents the points.

# Warm Up

The following warm-up exercises involve skills that were covered in earlier sections. You will use these skills in the exercise set for this section.

In Exercises 1–4, sketch the graph of the line.

**1.** $y = 2x$                **2.** $y = \frac{1}{2}x$

**3.** $y = 2x + 1$         **4.** $y = \frac{1}{2}x + 1$

In Exercises 5 and 6, find an equation of the line that has the given slope and $y$-intercept.

**5.** Slope: 1; $y$-intercept: $(0, 2)$       **6.** Slope: $\frac{3}{2}$; $y$-intercept: $(0, 3)$

In Exercises 7–10, find an equation of the line that passes through the two points.

**7.** $(1, 3)$ and $(6, 8)$          **8.** $(0, 4)$ and $(7, 10)$

**9.** $(1, 5.2)$ and $(5, 4.7)$      **10.** $(2, 6.5)$ and $(8, 3.6)$

---

## 2.5    Exercises

LAB   2

1. *Employment*   The total numbers of employed (in thousands) in the United States from 1991 to 1996 are given by the following ordered pairs.

   (1991, 117,718)
   (1992, 118,492)
   (1993, 120,259)
   (1994, 123,060)
   (1995, 124,900)
   (1996, 126,708)

   A linear model that approximates these data is

   $$y = 115{,}158.7 + 1913.6t, \quad 1 \le t \le 6$$

   where $y$ represents the number of employed (in thousands) and $t = 1$ represents 1991. Plot the actual data *and* the model on the same graph. How closely does the model represent the data? (Source: U. S. Bureau of Labor Statistics)

2. *Olympic Swimming*   The winning times (in minutes) in the women's 400-meter freestyle swimming event in the Olympics from 1948 to 1996 are given by the following ordered pairs.

   (1948, 5.30), (1952, 5.20)
   (1956, 4.91), (1960, 4.84)
   (1964, 4.72), (1968, 4.53)
   (1972, 4.32), (1976, 4.16)
   (1980, 4.15), (1984, 4.12)
   (1988, 4.06), (1992, 4.12)
   (1996, 4.12)

   A linear model that approximates these data is

   $$y = 5.4 - 0.03t, \quad 8 \le t \le 56$$

   where $y$ represents the winning time in minutes and $t = 8$ represents 1948. Plot the actual data *and* the model on the same graph. How closely does the model represent the data? (Source: *The Universal Almanac 1997*)

3. *Olympic Swimming Revisited*   In Exercise 2, the winning times (in minutes) for the years 1948 to 1964 are relatively high compared with the winning times from 1968 to 1996. Consider the winning times for the years 1968 to 1996 only and use your graphing utility to make a scatter plot. Then use the linear regression program to find a best-fitting line for the data. Graph the line on the scatter plot. Use the equation of the line to predict the winning times in this event at the 2000 Summer Olympics. Is your prediction reasonable?

4. *Olympic Swimming Revisited*   Use the winning times for the years 1948 to 1964 from Exercise 2 and your graphing utility to make a scatter plot. Then use the linear regression program to find a best-fitting line for the data. Graph the line on the scatter plot. Use the equation of the line to predict the winning times in this event at the 2000 Summer Olympics. Is your prediction reasonable? How does this prediction compare with the one you found in Exercise 3?

*Direct Variation*    In Exercises 5–10, *y* is proportional to *x*. Use the *x*- and *y*-values to find a linear model that relates *y* and *x*.

**5.** $x = 8$, $y = 3$    **6.** $x = 5$, $y = 9$

**7.** $x = 15$, $y = 300$    **8.** $x = 12$, $y = 204$

**9.** $x = 7$, $y = 3.2$    **10.** $x = 11$, $y = 1.5$

**11.** *Simple Interest*    The simple interest that a person receives for an investment is directly proportional to the amount of the investment. By investing $2500 in a bond issue, you obtained an interest payment of $187.50 at the end of 1 year. Find a mathematical model that gives the interest *I* at the end of 1 year in terms of the amount invested *P*.

**12.** *Simple Interest*    The simple interest that a person receives for an investment is directly proportional to the amount of the investment. By investing $5000 in a municipal bond, you obtained interest of $337.50 at the end of 1 year. Find a mathematical model that gives the interest *I* for the municipal bond at the end of 1 year in terms of the amount invested *P*.

**13.** *Property Tax*    Your property tax is based on the assessed value of your property. (The assessed value is often lower than the actual value of the property.) A house that has an assessed value of $50,000 has a property tax of $1840.

(a) Find a mathematical model that gives the amount of property tax *y* in terms of the assessed value *x* of the property.

(b) Use the model to find the property tax on a house that has an assessed value of $85,000.

**14.** *State Sales Tax*    An item that sells for $145.99 has a sales tax of $10.22.

(a) Find a mathematical model that gives the amount of sales tax *y* in terms of the retail price *x*.

(b) Use the model to find the sales tax on a purchase that has a retail price of $540.50.

**15.** *Centimeters and Inches*    On a yardstick, you notice that 13 inches is the same length as 33 centimeters.

(a) Use this information to find a mathematical model that relates centimeters to inches.

(b) Use the model to complete the table.

| Inches | 5 | 10 | 20 | 25 | 30 |
|---|---|---|---|---|---|
| Centimeters | | | | | |

**16.** *Liters and Gallons*    You are buying gasoline and notice that 14 gallons of gasoline is the same as 53 liters.

(a) Use this information to find a linear model that relates gallons to liters.

(b) Use the model to complete the table.

| Gallons | 5 | 10 | 20 | 25 | 30 |
|---|---|---|---|---|---|
| Liters | | | | | |

In Exercises 17–22, you are given the 1998 value of a product *and* the rate at which the value is expected to change during the next 5 years. Use this information to write a linear equation that gives the dollar value of the product in terms of the year. (Let $t = 0$ represent 1998.)

| | 1998 Value | Rate |
|---|---|---|
| **17.** | $2540 | $125 increase per year |
| **18.** | $156 | $4.50 increase per year |
| **19.** | $20,400 | $2000 decrease per year |
| **20.** | $45,000 | $2800 decrease per year |
| **21.** | $154,000 | $12,500 increase per year |
| **22.** | $245,000 | $5600 increase per year |

**23.** *Straight-Line Depreciation*    A business purchases a piece of equipment for $875. After 5 years the equipment will have no value. Write a linear equation giving the value *V* of the equipment during the 5 years.

**24.** *Straight-Line Depreciation*    A business purchases a piece of equipment for $25,000. The equipment will be replaced in 10 years, at which time its salvage value is expected to be $2000. Write a linear equation giving the value *V* of the equipment during the 10 years.

**25.** *Sales Price and List Price*    A store is offering a 15% discount on all items. Write a linear equation giving the sale price *S* for an item with a list price *L*.

**26.** *Sales Price and List Price*    A store is offering a 25% discount on all shirts. Write a linear equation giving the sale price *S* for a shirt with a list price *L*.

**27.** *Hourly Wages*    A manufacturer pays its assembly line workers $11.50 per hour. In addition, workers receive a piecework rate of $0.75 per unit produced. Write a linear equation for the hourly wages *W* in terms of the number of units *x* produced per hour.

**28.** *Sales Commission*   A salesperson receives a monthly salary of $2500 plus a commission of 7% of sales. Write a linear equation for the salesperson's monthly wage $W$ in terms of the person's monthly sales $S$.

**29.** *Height of a Parachutist*   After opening the parachute, the descent of a parachutist follows a linear model. At 2:08 P.M. the height of the parachutist is 7000 feet. At 2:10 P.M. the height is 4600 feet.

(a) Write a linear equation that gives the height of the parachutist in terms of the time $t$. (Let $t = 0$ represent 2:08 P.M. and let $t$ be measured in seconds.)

(b) Use the equation in part (a) to find the time when the parachutist will reach the ground.

**30.** *Distance Traveled by a Car*   You are driving at a constant speed on a freeway. At 4:30 P.M. you drive by a sign that gives the distance to Montgomery, Alabama as 84 miles. At 4:59 P.M. you drive by another sign that says the distance to Montgomery is 56 miles.

(a) Write a linear equation that gives your distance from Montgomery in terms of $t$. (Let $t = 0$ represent 4:30 P.M. and let $t$ be measured in minutes.)

(b) Use the equation in part (a) to find the time when you will reach Montgomery.

In Exercises 31–34, match the description with one of the graphs. Also find the slope and describe how it is interpreted in the real-life situation. [The graphs are labeled (a), (b), (c), and (d).]

**31.** A person is paying $10 per week to a friend to repay a $100 loan.

**32.** An employee is paid $12.50 per hour plus $1.50 for each unit produced per hour.

**33.** A sales representative receives $20 per day for food, plus $0.25 for each mile traveled.

**34.** A typewriter that was purchased for $600 depreciates $100 per year.

**35.** *High School Enrollment*   A high school had an enrollment of 1200 students in 1990. During the next 10 years the enrollment increased by approximately 50 students per year.

(a) Write a linear equation giving the enrollment $N$ in terms of the year $t$. (Let $t = 0$ correspond to the year 1990.)

(b) If this constant rate of growth continues, predict the enrollment in the year 2010.

(c) Use a graphing utility to confirm graphically the result in part (b).

**36.** *Cost of Renting*   The monthly rent in an office building is related to the office size. The rent for a 600-square-foot office is $750. The rent for a 900-square-foot office is $1150.

(a) Write a linear equation giving the monthly rent $r$ in terms of the square footage $x$.

(b) Use the equation in part (a) to find the monthly rent for an office that has 1300 square feet of floor space.

(c) Use a graphing utility to confirm graphically the result in part (b).

In Exercises 37–40, can the data be approximated by a linear model?

(a)

(b)

(c)

(d)

**37.**

**38.**

**39.**

**40.**

In Exercises 41–44, sketch the line that you think best approximates the data. Find an equation of the line.

**41.**

**42.**

**43.**

**44.**

**45.** *Average Weekly Overtime*   The average numbers of weekly overtime hours worked by production workers nationwide in the manufacturing industry in the month of July from 1988 to 1998 are given in the table.   (Source: U.S. Bureau of Labor Statistics)

| Year | 1988 | 1989 | 1990 | 1991 |
|------|------|------|------|------|
| Avg. Weekly Overtime | 3.8 | 3.7 | 3.7 | 3.6 |

| Year | 1992 | 1993 | 1994 | 1995 |
|------|------|------|------|------|
| Avg. Weekly Overtime | 3.8 | 4.1 | 4.7 | 4.3 |

| Year | 1996 | 1997 | 1998 |
|------|------|------|------|
| Avg. Weekly Overtime | 4.5 | 4.8 | 4.8 |

(a) Use your graphing utility to draw a scatter plot for these data. (Let $t = 8$ represent 1988.) Do the data appear linear?

(b) Use a linear regression program to develop a model for predicting average weekly overtime hours.

(c) Use your linear model to predict the average weekly overtime hours for July 1999 and July 2000. Are your predictions reasonable?

**46.** *Think About It*   A linear mathematical model for predicting prize winnings in a race is based on data for only 3 years. Discuss the potential accuracy and inaccuracy of such a model.

**47.** *Discus Throw*   The lengths of the winning men's discus throws in the Olympics from 1900 to 1996 are shown in the scatter plot.   (Source: *The Universal Almanac 1997*)

(a) Find the equation of the line that you think best fits these data. (Let $y$ represent the length of the winning discus throw in feet and let $t = 0$ represent 1900.)

(b) Use the equation to predict the length of the winning men's discus throw in 2000.

(c) Use your school's library or some other reference source to analyze the accuracy of the prediction in part (b).

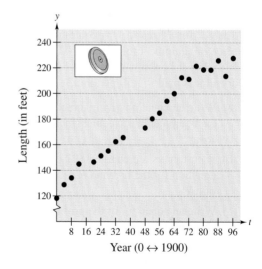

Year (0 ↔ 1900)

**48.** *Total Sales* The total sales for Microsoft Corporation from 1995 to 1998 (in millions of dollars) are shown in the scatter plot. (Source: *Value Line*)

(a) Find the equation of the line that you think best fits the data. (Let $y$ represent the total sales in millions of dollars and let $t = 5$ represent 1995.)

(b) Use a graphing utility to predict the sales of Microsoft in 1999. Compare the result with the prediction given by the equation in part (a).

(c) Use your school's library or some other reference source to analyze the accuracy of the prediction in part (b).

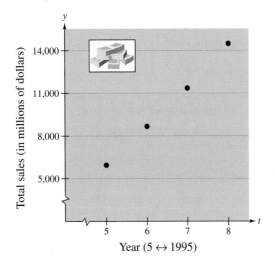

Year (5 $\leftrightarrow$ 1995)

**49.** *Contracting Purchase* A contractor purchases a piece of equipment for $36,500. The operating cost is $5.25 per hour for fuel and maintenance, and the operator is paid $11.50 per hour.

(a) Write an equation giving the cost $C$ of operating the equipment for $t$ hours. (Include the purchase cost.)

(b) If customers are charged $27 per hour, write an equation for the revenue $R$ derived from $t$ hours of use.

(c) Write an equation for the profit $P$ derived from $t$ hours of use.

(d) *Break-Even Point* Use the result of part (c) to find the number of hours this equipment must be used to yield a profit of 0 dollars.

**50.** *Contracting Purchase* A contractor purchases a piece of equipment for $55,000. The operating cost is $5.95 per hour for fuel and maintenance, and the operator is paid $12.25 per hour.

(a) Write an equation giving the cost $C$ of operating the equipment for $t$ hours. (Include the purchase cost.)

(b) If customers are charged $35 per hour, write an equation for the revenue $R$ derived from $t$ hours of use.

(c) Write an equation for the profit $P$ derived from $t$ hours of use.

(d) *Break-Even Point* Use the result of part (c) to find the number of hours this equipment must be used to yield a profit of 0 dollars.

**51.** *Real Estate Purchase* A real estate office handles an apartment complex with 50 units. When the rent per unit is $380 per month, all 50 units are occupied. However, when the rent is $425 per month, the average number of occupied units drops to 47. Assume that the relationship between the monthly rent $p$ and the demand $x$ is linear.

(a) Write the equation of the line giving the demand $x$ in terms of the rent $p$.

(b) Use the equation in part (a) to predict the number of units occupied if the rent is raised to $455.

## Chapter Project

## Demographics

The United States Bureau of the Census tracks statistics on the social, political, and economic organization of the nation. These data are published in *The Statistical Abstract of the United States,* and are also available on the Internet. The Census Bureau gathers the information by taking surveys and accessing other government and private statistical publications. Governments, businesses, research groups, companies, organizations, and individuals use the information to construct mathematical models that allow them to make predictions about future trends.

The table at the left shows the average number of people in U.S. households from 1960 to 1995. It is not always easy to see from a table what type of pattern, if any, is developing in the data. Mathematical modeling is a way to find an equation that represents a set of real-life data. In this chapter you studied one of the simplest mathematical models to use—the line.

1. *Inspecting the Data*   Look at the table at the left. What possible pattern do you see in the data?

2. *Graphing the Data*   Make a scatter plot of the data. How does the scatter plot compare with your answer to Question 1? Do the points appear to resemble a linear pattern?

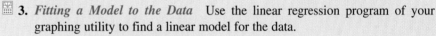 3. *Fitting a Model to the Data*   Use the linear regression program of your graphing utility to find a linear model for the data.

   (a) Write the linear model.

   (b) What is the slope of the line? What does the slope of the line tell you about the average size of U.S. households?

   (c) What is the *x*-intercept of the line? Does an *x*-intercept make sense in the context of the data? Explain.

4. *Comparing Actual Data with the Model*   Enter the linear model from Question 3(a) into the equation editor of your graphing utility. Use the table feature to evaluate the values that your model gives for average household sizes for different years. How do the model values compare with the actual values from the table?

5. *Extending the Exploration*   Collect a set of data by yourself or with a group of students from your class. For example, you could ask students in your class how much time they spent studying for an exam and their scores on the exam. Another source of data could be the Census Bureau's publications or website. Analyze the data you have collected by making a scatter plot and using the statistical and table features of your graphing utility. Can you find a linear model that fits the data reasonably well? What conclusions or predictions can you make about the subject of your investigation?

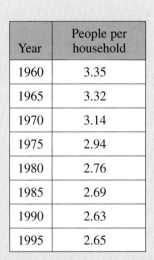

| Year | People per household |
|------|----------------------|
| 1960 | 3.35 |
| 1965 | 3.32 |
| 1970 | 3.14 |
| 1975 | 2.94 |
| 1980 | 2.76 |
| 1985 | 2.69 |
| 1990 | 2.63 |
| 1995 | 2.65 |

## CHAPTER SUMMARY

After studying this chapter, you should have acquired the following skills.
These skills are keyed to the Review Exercises that begin on page 145.
Answers to odd-numbered Review Exercises are given in the back of the book.

**2.1** · Plot points in the Cartesian plane.    Review Exercises 1, 2, 5–8

· Find the distance between two points in the plane.    Review Exercises 1–8

· Use the Distance Formula to solve distance and geometry problems.    Review Exercises 9–14

· Find the midpoint of a line segment.    Review Exercises 5–8

· Use the Midpoint Formula to solve an application problem.    Review Exercises 15, 16

**2.2** · Determine whether a point lies on the graph of an equation.    Review Exercises 17, 18, 49, 50

· Find the x- and y-intercepts of the graph of an equation.    Review Exercises 19–22, 27–32

· Determine the symmetry of a graph.    Review Exercises 23–32

· Sketch the graph of an equation using a table of values.    Review Exercises 27–32

· Write the equation of a circle in standard form.    Review Exercises 33–38

**2.3** · Sketch the graph of an equation using a graphing utility.    Review Exercises 39–48, 51, 52

· Use a square setting to obtain a true geometric perspective.    Review Exercises 53, 54

**2.4** · Determine the slope of a line passing through two points.    Review Exercises 55, 56

· Use the point-slope form to find the equation of a line.    Review Exercises 57–64

· Use the slope-intercept form to sketch the graph of a line.    Review Exercises 65, 66

· Determine if lines are parallel or perpendicular using slope.    Review Exercises 67–70

· Write the equation of a line parallel or perpendicular to a given line.    Review Exercises 71, 72

**2.5** · Construct a linear model to relate quantities that vary directly.    Review Exercises 73–80

· Construct and use a linear model with slope as the rate of change.    Review Exercises 81–88

# REVIEW EXERCISES

In Exercises 1 and 2, plot the points and find the distance between the points.

**1.** $(5, -2), (-8, -2)$  **2.** $(3, 7), (3, 11)$

In Exercises 3 and 4, (a) find the lengths of the two sides of the right triangle and use the Pythagorean Theorem to find the length of the hypotenuse, and (b) use the Distance Formula to find the length of the hypotenuse of the triangle.

**3.**    **4.**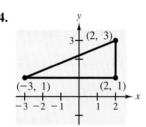

In Exercises 5-8, (a) plot the points, (b) find the distance between the points, and (c) find the midpoint of the line segment joining the points.

**5.** $(-9, 3), (5, 7)$  **6.** $(3, 2), (-3, -5)$
**7.** $(-6.7, -3.9), (5.1, 8.2)$
**8.** $(3.45, 6.55), (-1.06, -3.87)$

In Exercises 9 and 10, show that the points form the vertices of the indicated polygon.

**9.** Parallelogram: $(1, 1), (8, 2), (9, 5), (2, 4)$
**10.** Right triangle: $(-1, -1), (10, 7), (2, 18)$

In Exercises 11 and 12, find $x$ such that the distance between the points is 25.

**11.** $(10, 10), (x, -5)$
**12.** $(x, -5), (-15, 10)$

In Exercises 13 and 14, find $y$ such that the distance between the points is 13.

**13.** $(6, y), (-6, -2)$  **14.** $(5, 1), (17, y)$

In Exercises 15 and 16, use the Midpoint Formula to estimate the sales of a company for 1998 given the sales in 1996 and 2000.

**15.**

| Year | 1996 | 2000 |
|------|------|------|
| Sales | $640,000 | $810,000 |

**16.**

| Year | 1996 | 2000 |
|------|------|------|
| Sales | $3,250,000 | $5,690,000 |

In Exercises 17 and 18, determine whether the points lie on the graph of the equation.

**17.** $y = x^2 - 7x + 10$  (a) $(1, 4)$  (b) $(2, 0)$
**18.** $y = \sqrt{25 - x^2}$  (a) $(3, -2)$  (b) $(0, 5)$

In Exercises 19-22, find the $x$- and $y$-intercepts of the graph of the equation.

**19.** $y = (x + 3)(x - 2)$  **20.** $y = x^2 - 9$
**21.** $y = x\sqrt{x + 3}$
**22.** $xy + x - 2y + 4 = 0$

In Exercises 23-26, check for symmetry with respect to both axes and the origin.

**23.** $y = \dfrac{x}{x^2 - 2}$  **24.** $x^2 + y^2 = 9$
**25.** $y = x^3 + 1$  **26.** $y^2 = x$

In Exercises 27-32, sketch the graph of the equation. Identify any intercepts and test for symmetry.

**27.** $y = x^2 + 3$  **28.** $x^2 + y^2 = 25$
**29.** $y = 3x - 4$  **30.** $y = \sqrt{9 - x}$
**31.** $y = x^3 + 1$  **32.** $y = |x - 3|$

In Exercises 33–36, find the standard form of the equation of the specified circle.

**33.** Center: $(1, 3)$; radius: $5$

**34.** Center: $(-1, 0)$; solution point: $(2, 4)$

**35.** Endpoints of a diameter: $(-1, -2), (3, 1)$

**36.** Center: $\left(\frac{1}{2}, -\frac{1}{3}\right)$; radius: $\frac{3}{2}$

In Exercises 37 and 38, write the equation of the circle in standard form and sketch its graph.

**37.** $x^2 + y^2 - 6x + 4y - 3 = 0$

**38.** $4x^2 + 4y^2 + 8x - 16y - 44 = 0$

 In Exercises 39–44, use a graphing utility to match the equation with its graph. [The graphs are labeled (a), (b), (c), (d), (e), and (f).]

(a)

(b)

(c)

(d)

(e)

(f)
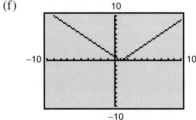

**39.** $y = x^3$      **40.** $y = -x^3$

**41.** $y = \sqrt{x}$      **42.** $y = -\sqrt{x}$

**43.** $y = |x - 1|$      **44.** $y = -|x - 1|$

In Exercises 45–48, use a graphing utility to sketch the graph of the equation. Use a standard setting for each graph.

**45.** $y = -2x + 3$      **46.** $4y = x^2 + 5x - 2$

**47.** $y = |x + 4|$      **48.** $y = \frac{1}{3}(x^3 - 3x)$

In Exercises 49 and 50, find the constant $C$ such that the ordered pair is a solution point of the equation.

**49.** $y = Cx^2$          $(2, 12)$

**50.** $x + C(y + 3) = 0$    $(-3, 4)$

In Exercises 51 and 52, use a graphing utility to find the number of $x$-intercepts of the equation.

**51.** $y = \frac{1}{3}(3x^2 - 27x + 55)$

**52.** $y = \frac{1}{4}(-8x^2 + 32x - 9)$

In Exercises 53 and 54, use a graphing utility to sketch the graphs on the same screen. Using a square setting, what geometrical shape is bounded by the graphs?

**53.** $y = -\sqrt{16 - x^2}, y = \sqrt{16 - x^2}$

**54.** $y = |x| - 6, y = -|x| + 6$

In Exercises 55 and 56, plot the points and find the slope of the line passing through the two points.

**55.** $(1, 5), (-3, -6)$

**56.** $(0, -5), (3, 2)$

In Exercises 57–60, find an equation of the line passing through the points and sketch the line.

**57.** $(3, 7), (2, -1)$

**58.** $(3, -2), (-1, -2)$

**59.** $(3, 4), (3, -2)$

**60.** $(-1, 5), (2, -3)$

In Exercises 61–64, find an equation of the line that passes through the point and has the indicated slope. Sketch the graph of the line.

**61.** $(0, -5)$      $m = \frac{3}{2}$

**62.** $(-2, 6)$      $m = 0$

**63.** $(3, 0)$      $m = -\frac{2}{3}$

**64.** $(5, 4)$      $m$ is undefined

In Exercises 65 and 66, find the slope and $y$-intercept (if possible) of the line specified by the equation and sketch the line.

**65.** $5x - 4y + 11 = 0$

**66.** $3y - 2 = 0$

In Exercises 67–70, determine whether the lines $L_1$ and $L_2$ passing through the pairs of points are parallel, perpendicular, or neither.

**67.** $L_1$: $(0, 3), (-2, 1)$;  $L_2$: $(-8, -3), (4, 9)$

**68.** $L_1$: $(-3, -1), (2, 5)$;  $L_2$: $(2, 1), (8, 6)$

**69.** $L_1$: $(3, 6), (-1, -5)$;  $L_2$: $(-2, 3), (4, 7)$

**70.** $L_1$: $(-1, 2), (-1, 4)$;  $L_2$: $(7, 3), (4, 7)$

In Exercises 71 and 72, write an equation of the line through the point (a) parallel to the given line and (b) perpendicular to the line.

**71.** $(3, -2)$      $5x - 4y = 8$

**72.** $(-8, 3)$      $2x + 3y = 5$

*Direct Variation*   In Exercises 73–76, assume that $y$ is proportional to $x$. Use the $x$-value and $y$-value to find a linear model that relates $y$ and $x$.

**73.** $x = 3$      $y = 7$

**74.** $x = 5$      $y = 3.8$

**75.** $x = 10$      $y = 3480$

**76.** $x = 14$      $y = 1.95$

**77.** *Simple Interest*   The simple interest that a person receives for an investment is directly proportional to the amount of the investment. Suppose that by investing $7000 in a municipal bond, you obtained an interest of $438.50 at the end of 1 year. Find a mathematical model that gives the amount of interest $I$ for this municipal bond at the end of 1 year in terms of the amount invested.

**78.** *Property Tax*   The property tax in a certain city is based on the assessed value of the property. A house that has an assessed value of $40,000 has a property tax of $1620. Find a mathematical model that gives the amount of property tax $y$ in terms of the assessed value of the property $x$. Use the model to find the property tax on a house that has an assessed value of $65,000.

**79.** *Grams and Ounces*   Suppose that you are buying cherry pie filling and notice that 21 ounces of filling is the same as 595 grams. Use this information to find a linear model that relates ounces to grams. Use the model to complete the table.

| Ounces | 6 | 8 | 12 | 16 | 24 |
|--------|---|---|----|----|----|
| Grams  |   |   |    |    |    |

**80.** *Kilometers and Miles*   Suppose you are driving on the highway and you notice that a billboard indicates that it is 2.5 miles or 4 kilometers to the next restaurant of a national fast food chain. Use this information to find a linear model that relates miles to kilometers. Use the model to complete the table.

| Miles      | 2 | 5 | 10 | 12 |
|------------|---|---|----|----|
| Kilometers |   |   |    |    |

**81.** *Fourth Quarter Sales*   During the second and third quarters of the year, a business had sales of $160,000 and $185,000, respectively. If the growth of the sales follows a linear pattern, what will sales be during the fourth quarter?

**82.** *Annual Salary*   Suppose that your annual salary was $22,000 in 1998 and $24,500 in 2000. If your salary follows a linear growth pattern, what salary will you be making in 2004?

**83.** *Dollar Value*   The dollar value of a product in 2000 is $85 and the item is expected to increase in value at a rate of $3.75 per year. Write a linear equation that gives the dollar value of the product in terms of the year. Use this model to estimate the dollar value of the product in 2005. (Let $t = 0$ represent 2000.)

**84.** *Straight-Line Depreciation*   A small business purchases a piece of equipment for $30,000. After 10 years, the equipment will have to be replaced. Its salvage value at that time is expected to be $2500. Write a linear equation giving the value $V$ of the equipment during the 10 years it will be used.

**85.** *Cost of Renting*   The monthly rent in an office building is related to the size of the office that is rented. The monthly rent for an office that has 800 square feet of floor space is $975. The monthly rent for an office that has 1100 square feet of floor space is $1350. Write a linear equation giving the monthly rent $r$ in terms of the number of square feet of floor space $x$. Use the result to find the monthly rent for an office that has 1400 square feet of floor space.

**86.** *Hourly Wages*   A manufacturer pays its assembly line workers $9.50 per hour. In addition, workers receive a piecework rate of $0.60 per unit produced. Write a linear equation for the hourly wages $W$ in terms of the number of units $x$ produced per hour.

**87.** *U.S. Cigarette Exports*   The numbers of cigarettes exported (in billions) by the United States in selected years are listed in the table.

| Year                              | 1960 | 1970 | 1980 |
|-----------------------------------|------|------|------|
| Cigarettes exported (in billions) | 20.2 | 29.2 | 82   |

| Year                              | 1990  | 1996  |
|-----------------------------------|-------|-------|
| Cigarettes exported (in billions) | 164.3 | 243.9 |

Graph a scatter plot using the data from the table. (Let $t = 0$ represent 1960.) Use a linear regression program to find a mathematical model. Use the model to predict the number of cigarettes exported in 2000. (Source: U.S. Department of Agriculture)

**88.** *Contracting Purchase*   A contractor purchases a piece of equipment for $24,500. The equipment requires an average expenditure of $5.75 per hour for fuel and maintenance, and the operator is paid $9.50 per hour.

(a) Write a linear equation giving the total cost $C$ of operating this equipment for $t$ hours. (Include the purchase price of the equipment.)

(b) If customers are charged $35 per hour of machine use, write an equation for the revenue $R$ derived from $t$ hours of use.

(c) Use the formula for profit $(P = R - C)$ to write an equation for the profit derived from $t$ hours of use.

(d) *Break-Even Point*   Use the result of part (c) to find the number of hours this equipment must be used to yield a profit of 0 dollars.

## Chapter Test

Take this test as you would take a test in class. After you are done, check your work against the answers given in the back of the book.

(a)

(b)

(c)

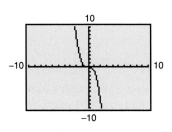

(d)

Figures for 10–13

In Exercises 1 and 2, find the distance between the points and the midpoint of the line segment connecting the points.

**1.** $(-3, 2), (5, -2)$     **2.** $(3.25, 7.05), (-2.37, 1.62)$

**3.** Find the intercepts of the graph of $y = (x + 5)(x - 3)$.

**4.** Find the intercepts of the graph of $y = x\sqrt{x - 2}$.

**5.** Describe the symmetry of the graph of $y = \dfrac{x}{x^2 - 4}$.

**6.** Find an equation of the line through $(3, 4)$ and $(-2, 4)$.

**7.** Find an equation of the line through $(4, -3)$ with a slope of $\frac{3}{4}$.

**8.** Find an equation of the vertical line through $(2, -7)$.

**9.** Find an equation of the line with intercepts $(3, 0)$ and $(0, -4)$.

In Exercises 10–13, match the equation with its graph. [The graphs are labeled (a), (b), (c), and (d).]

**10.** $y = \sqrt{x}$     **11.** $y = -\sqrt{x - 2}$

**12.** $y = |x - 2|$     **13.** $y = -x^3$

**14.** Write the equation of the circle in standard form and sketch its graph.

$$x^2 + y^2 - 4x - 2y - 4 = 0$$

**15.** During the first and second quarters of the year, a business had sales of $150,000 and $185,000, respectively. If the sales follow a linear pattern, what will sales be in the third quarter?

**16.** An assembly line worker is paid $8.75 per hour, plus a piecework rate of $0.55 per unit. Write a linear equation for the hourly wages $W$ in terms of the number of units $x$ produced per hour.

**17.** A business purchases a piece of equipment for $25,000. After 5 years, the equipment will be worth only $5000. Write a linear equation that gives the value $V$ of the equipment during the 5 years.

**18.** A store is offering a 30% discount on all summer apparel. Write a linear equation that gives the sale price $S$ of an item in terms of its list price $L$.

**19.** The numbers of complaints (in thousands) submitted to the U.S. Department of Labor, Wage and Hours Division, for the years 1990 to 1997 are listed in the table. Graph a scatter plot and find a linear model for the data. (Source: U.S. Department of Labor)

| Year | 1990 | 1991 | 1992 | 1993 | 1994 | 1995 | 1996 | 1997 |
|---|---|---|---|---|---|---|---|---|
| Complaints (in thousands) | 60.5 | 54.1 | 47.9 | 46.1 | 40.5 | 37.5 | 34.7 | 34.7 |

## Cumulative Test: Chapters 1–2

Take this test as you would take a test in class. After you are done, check your work against the answers given in the back of the book.

In Exercises 1–6, solve the equation.

**1.** $4y^2 - 64 = 0$

**2.** $\dfrac{x}{3} + \dfrac{x}{4} = 1$

**3.** $x^4 - 17x^2 + 16 = 0$

**4.** $|2x - 3| = 5$

**5.** $\sqrt{y + 3} + y = 3$

**6.** $2x^2 + x = 5$

**7.** You deposit $5000 in an account that pays 8.5%, compounded monthly. Find the balance after 10 years.

In Exercises 8–10, solve the inequality.

**8.** $\left| \dfrac{2x - 5}{4} \right| \leq 3$

**9.** $(x + 4)^2 \leq 4$

**10.** $\dfrac{3 + 2x}{4 - x} > 2$

**11.** Find the intercepts of the graph of $y = x\sqrt{x + 4}$.

**12.** Describe the symmetry of the graph of $y = -\sqrt{4 - x^2}$.

**13.** Write the equation of the circle in standard form and sketch its graph.

$$x^2 + y^2 - 6x + 4y - 3 = 0$$

**14.** Find the constant $C$ such that $(-2, 3)$ is a solution point of $y = x^3 + C$.

**15.** Find an equation of the line passing through $(3, -2)$ and $(-1, 5)$.

**16.** Find an equation of the line with slope $\frac{2}{3}$ passing through $(2, -1)$.

**17.** Find an equation of the line with zero slope passing through $(-1, -3)$.

**18.** You have accepted a sales job that pays a base salary of $1000 a month plus a 7% commission on monthly sales $x$. Write a linear model that describes your total monthly salary $y$ in terms of your base salary and commission.

**19.** During the second and third quarters of the year, a business had sales of $210,000 and $230,000, respectively. If the sales growth follows a linear pattern, what will the sales be during the fourth quarter?

**20.** The revenue and cost equations for a product are given by

$$R = x(100 - 0.001x) \quad \text{and} \quad C = 20x + 30,000$$

where $R$ and $C$ are measured in dollars and $x$ represents the number of units sold. How many units must be sold for the revenue to equal the cost?

Steven Lunetta/PhotoEdit

Year (0 ↔ 1970)

When computers started to become common in the workplace and later in the home, some experts believed that use of paper would diminish. Based on that prediction, you would expect that the consumption of paper and paper products would decrease, but this has not happened. The table below gives the consumption of paper and paperboard in millions of metric tons for North and Central America. A model for this data is

$$C = 0.026t^2 + 1.09t + 54.9, \quad 0 \le t \le 24$$

where $C$ is the total consumption of paper and paperboard in metric tons and $t$ is time, with $t = 0$ corresponding to the year 1970. The increase in the consumption of paper and paperboard is shown in the graph of the model.   (Source: *State of the World's Forests 1997,* United Nations Conference on Environment and Development (UNCED))

| Year | 1970 | 1980 | 1990 | 1994 |
|---|---|---|---|---|
| Consumption (in millions of metric tons) | 55.0 | 68.1 | 87.9 | 95.9 |

The chapter project related to this information is on page 206.

## Objectives

- Determine if an equation or a set of ordered pairs represents a function.
- Use function notation.
- Evaluate a function.
- Find the domain of a function.
- Write a function that relates quantities in an application problem.

## Introduction to Functions

Many everyday phenomena involve two quantities that are related to each other by some rule of correspondence. Here are some examples.

1. The simple interest $I$ earned on $1000 for 1 year is related to the annual percentage rate $r$ by the formula $I = 1000r$.

2. The distance $d$ traveled on a bicycle in 2 hours is related to the speed $s$ of the bicycle by the formula $d = 2s$.

3. The area $A$ of a circle is related to its radius $r$ by the formula $A = \pi r^2$.

Not all correspondences between two quantities have simple mathematical formulas. For instance, people commonly match up NFL starting quarterbacks with touchdown passes and hours of the day with temperature. In each of these cases, however, there is some rule of correspondence that matches each item from one set with exactly one item from a different set. Such a rule of correspondence is called a **function.**

Hours of the day

Celsius temperature

▶ Definition of a Function

A **function** $f$ from a set $A$ to a set $B$ is a rule of correspondence that assigns to each element $x$ in the set $A$ exactly one element $y$ in the set $B$. The set $A$ is the **domain** (or set of inputs) of the function $f$, and the set $B$ contains the **range** (or set of outputs).

To get a better idea of this definition, look at the function illustrated in Figure 3.1. This function can be represented by the following set of ordered pairs.

$$\{(1, 9°), (2, 13°), (3, 15°), (4, 15°), (5, 12°), (6, 4°)\}$$

In each ordered pair, the first coordinate is the input and the second coordinate is the output. In this example, note the following characteristics of a function.

Set $A$ is the domain.
Input: 1, 2, 3, 4, 5, 6

Set $B$ contains the range.
Output: 4°, 9°, 12°, 13°, 15°

1. Each element in $A$ must be matched with an element of $B$.

2. Some elements in $B$ may not be matched with any element in $A$.

3. Two or more elements of $A$ may be matched with the same element of $B$.

**Figure 3.1** *Function from Set A to Set B*

The converse of the third statement is not true. That is, an element of $A$ (the domain) cannot be matched with two different elements of $B$.

In the following two examples, you are asked to decide whether different correspondences are functions. To do this, you must decide whether each element in the domain $A$ is matched with exactly one element in the range $B$. If any element in $A$ is matched with two or more elements in $B$, the correspondence is not a function.

### Example 1   Testing for Functions

Let $A = \{a, b, c\}$ and $B = \{1, 2, 3, 4, 5\}$. Which of the following sets of ordered pairs or figures represent functions from set $A$ to set $B$?

**a.** $\{(a, 2), (b, 3), (c, 4)\}$

**b.** $\{(a, 4), (b, 5)\}$

**c.**

**d.**

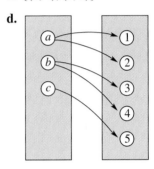

### Solution

**a.** This collection of ordered pairs *does* represent a function from $A$ to $B$. Each element of $A$ is matched with exactly one element of $B$.

**b.** This collection of ordered pairs *does not* represent a function from $A$ to $B$. Not every element of $A$ is matched with an element of $B$.

**c.** This figure *does* represent a function from $A$ to $B$. It does not matter that each element of $A$ is matched with the same element of $B$.

**d.** This figure *does not* represent a function from $A$ to $B$. The element $a$ in $A$ is matched with *two* elements, 1 and 2, of $B$. This is also true of the element $b$.

Representing functions by sets of ordered pairs is a common practice in *discrete mathematics*. In algebra, however, it is more common to represent functions by equations or formulas involving two variables. For instance, the equation

$$y = x^2 \qquad \text{\small $y$ is a function of $x$.}$$

represents the variable $y$ as a function of the variable $x$. In this equation, $x$ is the **independent variable** and $y$ is the **dependent variable.** The domain of the function is the set of all values taken on by the independent variable $x$, and the range of the function is the set of all values taken on by the dependent variable $y$.

**Example 2**    Testing for Functions Represented by Equations

Which of the following equations represent $y$ as a function of $x$?

**a.** $x^2 + y = 1$     **b.** $-x + y^2 = 1$

**Solution**

To determine whether $y$ is a function of $x$, try to solve for $y$ in terms of $x$.

**a.** Solving for $y$ yields the following.

$$x^2 + y = 1 \qquad \text{Original equation}$$

$$y = 1 - x^2 \qquad \text{Solve for } y.$$

To each value of $x$ there corresponds exactly one value of $y$. So, $y$ *is* a function of $x$.

**b.** Solving for $y$ yields the following.

$$-x + y^2 = 1 \qquad \text{Original equation}$$

$$y^2 = 1 + x \qquad \text{Add } x \text{ to both sides.}$$

$$y = \pm\sqrt{1 + x} \qquad \text{Solve for } y.$$

The $\pm$ indicates that to a given value of $x$ there correspond two values of $y$. So, $y$ *is not* a function of $x$.

## Function Notation

When an equation is used to represent a function, it is convenient to name the function so that it can be referenced easily. For example, you know that the equation $y = 1 - x^2$ describes $y$ as a function of $x$. Suppose you give this function the name "$f$." Then you can use the following **function notation.**

| Input | Output | Equation |
|-------|--------|----------|
| $x$ | $f(x)$ | $f(x) = 1 - x^2$ |

The symbol $f(x)$ is read as the **value of $f$ at $x$** or simply **$f$ of $x$.** The symbol $f(x)$ corresponds to the $y$-value for a given $x$. So, you can write $y = f(x)$. Keep in mind that $f$ is the *name* of the function, whereas $f(x)$ is the *value* of the function at $x$. For instance, the function given by

$$f(x) = 3 - 2x$$

has *function values* denoted by $f(-1)$, $f(0)$, $f(2)$, and so on. To find these values, substitute the specified input values into the given equation.

For $x = -1$,    $f(-1) = 3 - 2(-1) = 3 + 2 = 5.$

For $x = 0$,       $f(0) = 3 - 2(0) = 3 - 0 = 3.$

For $x = 2$,       $f(2) = 3 - 2(2) = 3 - 4 = -1.$

Although $f$ is often used as a convenient function name and $x$ is often used as the independent variable, you can use other letters. For instance,

$$f(x) = x^2 - 4x + 7, \quad f(t) = t^2 - 4t + 7, \quad \text{and} \quad g(s) = s^2 - 4s + 7$$

all define the same function. In fact, the role of the independent variable in a function is simply that of a "placeholder." Consequently, the function above could be described by the form

$$f(\;\;\;) = (\;\;\;)^2 - 4(\;\;\;) + 7.$$

### Example 3    Evaluating a Function

Let $g(x) = -x^2 + 4x + 1$ and find the following.

**a.** $g(2)$      **b.** $g(t)$      **c.** $g(x + 2)$

**Solution**

**a.** Replacing $x$ with 2 in $g(x) = -x^2 + 4x + 1$ yields the following.

$$g(2) = -(2)^2 + 4(2) + 1 = -4 + 8 + 1 = 5$$

**b.** Replacing $x$ with $t$ yields the following.

$$g(t) = -(t)^2 + 4(t) + 1 = -t^2 + 4t + 1$$

**c.** Replacing $x$ with $x + 2$ yields the following.

$$\begin{aligned} g(x + 2) &= -(x + 2)^2 + 4(x + 2) + 1 \\ &= -(x^2 + 4x + 4) + 4x + 8 + 1 \\ &= -x^2 - 4x - 4 + 4x + 8 + 1 \\ &= -x^2 + 5 \end{aligned}$$

### Study Tip

In Example 3(c), note that $g(x + 2)$ is not equal to $g(x) + g(2)$. In general, $g(u + v) \ne g(u) + g(v)$.

A function defined by two or more equations over a specified domain is called a *piecewise-defined* function.

### Example 4    A Piecewise-Defined Function

Evaluate the following function when $x = -1, 0,$ and 1.

$$f(x) = \begin{cases} x^2 + 1, & x < 0 \\ x - 1, & x \ge 0 \end{cases}$$

**Solution**

Because $x = -1$ is less than 0, use $f(x) = x^2 + 1$ to obtain

$$f(-1) = (-1)^2 + 1 = 2.$$

For $x = 0$, use $f(x) = x - 1$ to obtain

$$f(0) = (0) - 1 = -1.$$

For $x = 1$, use $f(x) = x - 1$ to obtain $f(1) = (1) - 1 = 0$. The graph of the function is shown in Figure 3.2.

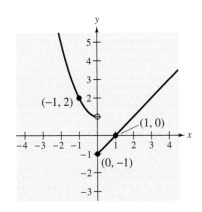

**Figure 3.2**

# Finding the Domain of a Function

The domain of a function can be described explicitly or it can be *implied* by the expression used to define the function. The **implied domain** is the set of all real numbers for which the expression is defined. For instance, the function given by

$$f(x) = \frac{1}{x^2 - 4}$$

has an implied domain that consists of all real $x$ other than $x = \pm 2$. These two values are excluded from the domain because division by zero is undefined. Another common type of implied domain is that used to avoid even roots of negative numbers. For example, the function given by

$$f(x) = \sqrt{x}$$

is defined only for $x \geq 0$. So, its implied domain is the interval $[0, \infty)$. In general, the domain of a function *excludes* values that would cause division by zero *or* result in the even root of a negative number.

---

**Example 5**   Finding the Domain of a Function

Find the domain of each of the following functions.

**a.** $f$: $\{(-3, 0), (-1, 4), (0, 2), (2, 2), (4, -1)\}$    **b.** $g(x) = \dfrac{1}{x + 5}$

**c.** Volume of a sphere: $V = \frac{4}{3}\pi r^3$    **d.** $h(x) = \sqrt{4 - x^2}$

**e.** $r(x) = \sqrt[3]{x + 3}$

**Solution**

**a.** The domain of $f$ consists of all first coordinates in the set of ordered pairs.

Domain $= \{-3, -1, 0, 2, 4\}$

**b.** Excluding $x$-values that yield zero in the denominator, the domain of $g$ is the set of all real numbers $x \neq -5$.

**c.** Because this function represents the volume of a sphere, the values of the radius $r$ must be positive. So, the domain is the set of all real numbers $r$ such that $r \geq 0$.

**d.** This function is defined only for $x$-values for which $4 - x^2 \geq 0$. Using the methods described in Section 1.7, you can conclude that $-2 \leq x \leq 2$. So, the domain is the interval $[-2, 2]$.

**e.** Because the cube root of a number may be positive, zero, or negative, the expression under the radical may be positive, zero, or negative. So, the domain of $r$ is the set of all real numbers, or $(-\infty, \infty)$.

---

In Example 5(c), note that the domain of a function may be implied by the physical context. For instance, from the equation $V = \frac{4}{3}\pi r^3$, you would have no reason to restrict $r$ to nonnegative values, but the physical context implies that a sphere cannot have a negative radius.

## Applications

Figure 3.3

**Example 6**    The Measurements of a Container

You work in the marketing department of a soft-drink company and are experimenting with a new soft-drink can that is slightly narrower and taller than a standard can. For your experimental can, the ratio of the height to the radius is 4, as shown in Figure 3.3.

**a.** Express the volume of the can as a function of the radius $r$.

**b.** Express the volume of the can as a function of the height $h$.

**Solution**

**a.** $V = \pi r^2 h = \pi r^2(4r) = 4\pi r^3$ 　　　　$V$ is a function of $r$.

**b.** $V = \pi\left(\dfrac{h}{4}\right)^2 h = \dfrac{\pi h^3}{16}$ 　　　　$V$ is a function of $h$.

**Example 7**    The Path of a Baseball

A baseball is hit 3 feet above ground at a velocity of 100 feet per second and at an angle of 45°. The path of the baseball is given by the function

$$y = -0.0032x^2 + x + 3$$

where $y$ and $x$ are measured in feet, as shown in Figure 3.4. Will the baseball clear a 10-foot fence located 300 feet from home plate?

**Solution**

When $x = 300$, the height of the baseball is given by

$$y = -0.0032(300^2) + 300 + 3 = 15 \text{ feet.}$$

So, the ball will clear the fence.

Figure 3.4

*In 1987, Americans made about 39.4 million trips to foreign countries. By 1995, this number had increased to about 50.8 million. The most visited country was Mexico.*

| Example 8 | U.S. Travelers Abroad |  |

The amount of money spent by U.S. travelers in other countries increased in a linear pattern from 1987 to 1990, as shown in Figure 3.5. Then, in 1991, the amount of money spent decreased and until 1995 increased in a *different* linear pattern. These two patterns can be approximated by the function

$$C = \begin{cases} 11{,}440.7 + 2541.9t, & 7 \le t \le 10 \\ 6660.4 + 2629.6t, & 11 \le t \le 15 \end{cases}$$

where $C$ represents the amount of money spent (in millions of dollars) and $t$ represents the calendar year, with $t = 7$ corresponding to 1987. Use this function to approximate the total amount of money spent by U.S. travelers abroad between 1987 and 1995.   (Source: U.S. Bureau of Economic Analysis)

### Solution

From 1987 to 1990, use the formula $C = 11{,}440.7 + 2541.9t$ to approximate the money spent.

| Year | 1987 | 1988 | 1989 | 1990 |
|---|---|---|---|---|
| $t$ | 7 | 8 | 9 | 10 |
| Money spent | $29,234 | $31,776 | $34,318 | $36,860 |

From 1991 to 1995, use the formula $C = 6660.4 + 2629.6t$ to approximate the money spent.

| Year | 1991 | 1992 | 1993 | 1994 | 1995 |
|---|---|---|---|---|---|
| $t$ | 11 | 12 | 13 | 14 | 15 |
| Money spent | $35,586 | $38,216 | $40,845 | $43,475 | $46,104 |

To find the total amount spent, add the amounts for each of the years, as shown below.

$$29{,}234 + 31{,}776 + 34{,}318 + 36{,}860 + 35{,}586 + 38{,}216 +$$
$$40{,}845 + 43{,}475 + 46{,}104 = 336{,}414$$

Because the money spent is measured in millions of dollars, you can conclude that the total amount spent by U.S. travelers abroad from 1987 through 1995 was approximately

$$\text{Total spent} \approx \$336{,}414{,}000{,}000.$$

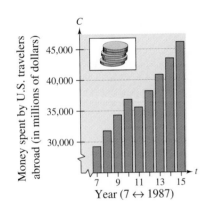

Figure 3.5

▶ Summary of Function Terminology

*Function:*  A **function** is a relationship between two variables such that to each value of the independent variable there corresponds exactly one value of the dependent variable.

*Function Notation:*    $y = f(x)$

$f$ is the **name** of the function.

$y$ is the **dependent variable.**

$x$ is the **independent variable.**

$f(x)$ is the **value of the function at $x$.**

*Domain:*  The **domain** of a function is the set of all values (inputs) of the independent variable for which the function is defined. If $x$ is in the domain of $f$, we say that $f$ is **defined** at $x$. If $x$ is not in the domain of $f$, we say that $f$ is **undefined** at $x$.

*Range:*  The **range** of a function is the set of all values (outputs) assumed by the dependent variable (that is, the set of all function values).

*Implied Domain:*  If $f$ is defined by an algebraic expression and the domain is not specified, the **implied domain** consists of all real numbers for which the expression is defined.

## Discussing the Concept    Everyday Functions

In groups of two or three, identify common real-life functions. Consider everyday activities, events, and expenses, such as long distance telephone calls and car insurance. Here are a couple of examples.

**a.** The statement, "Your happiness is a function of the grade you receive in this course" *is not* a correct mathematical use of the word "function." The word "happiness" is ambiguous.

**b.** The statement, "Your federal income tax is a function of your adjusted gross income" *is* a correct mathematical use of the word "function." Once you have determined your adjusted gross income, your income tax can be determined.

Describe your function in words. Avoid using ambiguous words. Can you find an example of a piecewise-defined function?

## Warm Up

The following warm-up exercises involve skills that were covered in earlier sections. You will use these skills in the exercise set for this section.

In Exercises 1–4, simplify the expression.

**1.** $2(-3)^3 + 4(-3) - 7$

**2.** $4(-1)^2 - 5(-1) + 4$

**3.** $(x + 1)^2 + 3(x + 1) - 4 - (x^2 + 3x - 4)$

**4.** $(x - 2)^2 - 4(x - 2) - (x^2 - 4)$

In Exercises 5 and 6, solve for $y$ in terms of $x$.

**5.** $2x + 5y - 7 = 0$

**6.** $y^2 = x^2$

In Exercises 7–10, solve the inequality.

**7.** $x^2 - 4 \geq 0$

**8.** $9 - x^2 \geq 0$

**9.** $x^2 + 2x + 1 \geq 0$

**10.** $x^2 - 3x + 2 \geq 0$

## 3.1    Exercises

In Exercises 1–4, evaluate the function.

**1.** $f(x) = 6 - 4x$

  (a) $f(3) = 6 - 4(\quad)$

  (b) $f(-7) = 6 - 4(\quad)$

  (c) $f(t) = 6 - 4(\quad)$

  (d) $f(c + 1) = 6 - 4(\quad)$

**2.** $f(s) = \dfrac{1}{s + 1}$

  (a) $f(4) = \dfrac{1}{(\quad) + 1}$

  (b) $f(0) = \dfrac{1}{(\quad) + 1}$

  (c) $f(4x) = \dfrac{1}{(\quad) + 1}$

  (d) $f(x + 1) = \dfrac{1}{(\quad) + 1}$

**3.** $g(x) = \dfrac{1}{x^2 - 2x}$

  (a) $g(1) = \dfrac{1}{(\quad)^2 - 2(\quad)}$

  (b) $g(-3) = \dfrac{1}{(\quad)^2 - 2(\quad)}$

  (c) $g(t) = \dfrac{1}{(\quad)^2 - 2(\quad)}$

  (d) $g(t + 1) = \dfrac{1}{(\quad)^2 - 2(\quad)}$

**4.** $f(t) = \sqrt{25 - t^2}$

  (a) $f(3) = \sqrt{25 - (\quad)^2}$

  (b) $f(5) = \sqrt{25 - (\quad)^2}$

  (c) $f(x + 5) = \sqrt{25 - (\quad)^2}$

  (d) $f(2x) = \sqrt{25 - (\quad)^2}$

In Exercises 5–16, evaluate the function and simplify the results.

**5.** $f(x) = 2x - 3$

(a) $f(1)$        (b) $f(-3)$

(c) $f(x - 1)$        (d) $f\left(\frac{1}{4}\right)$

**6.** $g(y) = 7 - 3y$

(a) $g(0)$        (b) $g\left(\frac{7}{3}\right)$

(c) $g(s)$        (d) $g(s + 2)$

**7.** $h(t) = t^2 - 2t$

(a) $h(2)$        (b) $h(-1)$

(c) $h(x + 2)$        (d) $h(1.5)$

**8.** $V(r) = \frac{4}{3}\pi r^3$

(a) $V(3)$        (b) $V(0)$

(c) $V\left(\frac{3}{2}\right)$        (d) $V(2r)$

**9.** $f(y) = 3 - \sqrt{y}$

(a) $f(4)$        (b) $f(100)$

(c) $f(4x^2)$        (d) $f(0.25)$

**10.** $f(x) = \sqrt{x + 8} + 2$

(a) $f(-8)$        (b) $f(1)$

(c) $f(x - 8)$        (d) $f(x + 8)$

**11.** $q(x) = \dfrac{1}{x^2 - 9}$

(a) $q(4)$        (b) $q(0)$

(c) $q(3)$        (d) $q(y + 3)$

**12.** $q(t) = \dfrac{2t^2 + 3}{t^2}$

(a) $q(2)$        (b) $q(0)$

(c) $q(x)$        (d) $q(-x)$

**13.** $f(x) = \dfrac{|x|}{x}$

(a) $f(2)$        (b) $f(-2)$

(c) $f(x^2)$        (d) $f(x - 1)$

**14.** $f(x) = |x| + 4$

(a) $f(2)$        (b) $f(-2)$

(c) $f(x^2)$        (d) $f(x + 2)$

**15.** $f(x) = \begin{cases} 2x + 1, & x < 0 \\ 2x + 2, & x \geq 0 \end{cases}$

(a) $f(-1)$        (b) $f(0)$

(c) $f(1)$        (d) $f(2)$

**16.** $f(x) = \begin{cases} x^2 + 2, & x \leq 1 \\ 2x^2 + 2, & x > 1 \end{cases}$

(a) $f(-2)$        (b) $f(0)$

(c) $f(1)$        (d) $f(2)$

In Exercises 17–22, find all real values of $x$ such that $f(x) = 0$.

**17.** $f(x) = 15 - 3x$        **18.** $f(x) = \dfrac{3x - 4}{5}$

**19.** $f(x) = x^2 - 9$        **20.** $f(x) = x^3 - x$

**21.** $f(x) = \dfrac{3}{x - 1} + \dfrac{4}{x - 2}$

**22.** $f(x) = 2 + \dfrac{3}{x}$

In Exercises 23–34, find the domain of the function.

**23.** $g(x) = 1 - 2x^2$        **24.** $f(x) = 5x^2 + 2x - 1$

**25.** $h(t) = \dfrac{4}{t}$        **26.** $s(y) = \dfrac{3y}{y + 5}$

**27.** $g(y) = \sqrt[3]{y - 10}$        **28.** $f(t) = \sqrt[3]{t + 4}$

**29.** $f(x) = \sqrt[4]{1 - x^2}$        **30.** $g(x) = \sqrt{x + 1}$

**31.** $g(x) = \dfrac{1}{x} - \dfrac{3}{x + 2}$        **32.** $h(x) = \dfrac{10}{x^2 - 2x}$

**33.** $f(x) = \dfrac{\sqrt{x + 1}}{x - 2}$        **34.** $f(s) = \dfrac{\sqrt{s - 1}}{s - 4}$

**35.** Consider $f(x) = \sqrt{x - 2}$ and $g(x) = \sqrt[3]{x - 2}$. Why are the domains of $f$ and $g$ different?

**36.** A student says that the domain of

$$f(x) = \dfrac{\sqrt{x + 1}}{x - 3}$$

is all real numbers except $x = 3$. Is the student correct? Describe how you would explain what the domain of $f$ is analytically.

In Exercises 37–46, decide whether the equation represents $y$ as a function of $x$.

**37.** $x^2 + y^2 = 4$        **38.** $x = y^2$

**39.** $x^2 + y = 4$        **40.** $x + y^2 = 4$

**41.** $2x + 3y = 4$

**42.** $x^2 + y^2 - 2x - 4y + 1 = 0$

**43.** $y^2 = x^2 - 1$        **44.** $y = \sqrt{x + 5}$

**45.** $x^2y - x^2 + 4y = 0$        **46.** $xy - y - x - 2 = 0$

In Exercises 47–50, decide whether the set of ordered pairs represents a function from $A$ to $B$.

$A = \{0, 1, 2, 3\}$ and $B = \{-2, -1, 0, 1, 2\}$

Give reasons for your answers.

**47.** $\{(0, 1), (1, -2), (2, 0), (3, 2)\}$

**48.** $\{(0, -1), (2, 2), (1, -2), (3, 0), (1, 1)\}$

**49.** $\{(0, 0), (1, 0), (2, 0), (3, 0)\}$

**50.** $\{(0, 2), (3, 0), (1, 1)\}$

In Exercises 51–54, decide whether the set of ordered pairs represents a function from $A$ to $B$.

$A = \{a, b, c\}$ and $B = \{0, 1, 2, 3\}$

Give reasons for your answers.

**51.** $\{(a, 1), (c, 2), (c, 3), (b, 3)\}$

**52.** $\{(a, 1), (b, 2), (c, 3)\}$

**53.** $\{(1, a), (0, a), (2, c), (3, b)\}$

**54.** $\{(c, 0), (b, 0), (a, 3)\}$

In Exercises 55–58, decide whether the set of ordered pairs represents a function from $A$ to $B$.

$A = \{a, b, c\}$ and $B = \{1, 2, 3, 4\}$

Give reasons for your answers.

**55.**

**56.**

**57.**

**58.**

In Exercises 59–62, the domain of $f$ is the set

$A = \{-2, -1, 0, 1, 2\}$.

Write the function as a set of ordered pairs.

**59.** $f(x) = x^2$

**60.** $f(x) = \dfrac{2x}{x^2 + 1}$

**61.** $f(x) = \sqrt{x + 2}$

**62.** $f(x) = |x + 1|$

**63.** *Volume of a Box*  An open box is to be made from a square piece of material 12 inches on a side by cutting equal squares from the corners and turning up the sides (see figure).

(a) Write the volume $V$ as a function of $x$.

(b) What is the domain of this function?

**64.** *Volume of a Package*  A rectangular package to be sent by a postal service can have a maximum combined length and girth (perimeter of a cross section) of 108 inches (see figure).

(a) Write the volume as a function of $x$.

(b) What is the domain of this function?

**65.** *Height of a Balloon*    A balloon carrying a transmitter ascends vertically from a point 2000 feet from the receiving station (see figure). Let $d$ be the distance between the balloon and the receiving station. Express the height of the balloon as a function of $d$. What is the domain of this function?

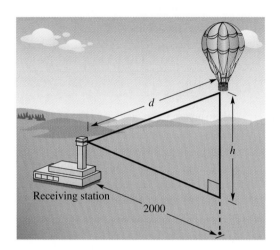

**66.** *Price of Mobile Homes*    The average price of a new mobile home in the United States from 1985 to 1996 can be approximated by the model

$$p = \begin{cases} 20{,}568 + 9t^3, & 5 \le t \le 9 \\ 233{,}800 - 48{,}335t + 3636t^2 - 86t^3, & 10 \le t \le 16 \end{cases}$$

where $p$ is the price in dollars and $t$ is the year, with $t = 5$ corresponding to 1985 (see figure). Find the average prices of mobile homes in 1987 and in 1994. (Source: U.S. Bureau of the Census)

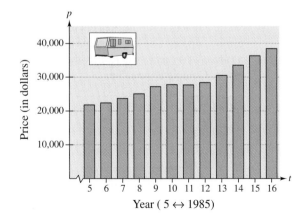

Year ( 5 ↔ 1985)

**67.** *Balance in an Account*    A person deposits $2000 in an account that pays 6% interest compounded quarterly. (See page 53 for the compound interest formula.)

(a) Write the balance $A$ of the account in terms of the number of years $t$ that the principal is left in the account.

(b) What is the domain of this function?

**68.** *Unemployment Rates*    The average annual unemployment rate in the United States from 1987 to 1996 is given by the function

$$r = \begin{cases} 29.0 - 5.40t + 0.307t^2, & 7 \le t \le 11 \\ 29.24 - 2.75t + 0.079t^2, & 12 \le t \le 16 \end{cases}$$

where $r$ is the unemployment rate (as a percent) and $t = 7$ represents 1987. Use this function to find the average annual unemployment rates in 1990 and in 1995.    (Source: U.S. Bureau of Labor Statistics)

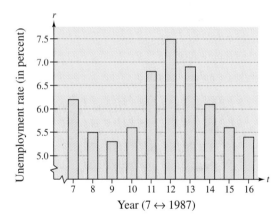

Year (7 ↔ 1987)

**69.** *Revenue, Cost, and Profit*    A company produces a product for which the variable cost is $12.30 per unit and the fixed costs are $98,000. The product sells for $17.98. Let $x$ be the number of units produced.

(a) Write the total cost $C$ as a function of the number of units produced.

(b) Write the revenue $R$ as a function of the number of units produced.

(c) Write the profit $P$ as a function of the number of units produced.

(d) Find the number of units that must be produced to break even. That is, for what values of $x$ will the profit function be equal to zero?

**70.** *Revenue, Cost, and Profit*    A company produces a product for which the variable cost is $15.95 per unit and the fixed costs are $110,000. The product sells for $23.79. Let $x$ be the number of units produced.

(a) Write the total cost $C$ as a function of the number of units produced.

(b) Write the revenue $R$ as a function of the number of units produced.

(c) Write the profit $P$ as a function of the number of units produced.

(d) Find the number of units that must be produced to break even. That is, for what values of $x$ will the profit function be equal to zero?

**71.** *Average Cost*    The inventor of a new game believes that the variable cost for producing the game is $0.95 per unit and the fixed costs are $6000. The inventor sells each game for $1.69. Let $x$ be the number of games sold.

(a) Write the total cost $C$ as a function of the number of games sold.

(b) Write the average cost per unit $\overline{C} = C/x$ as a function of $x$.

(c) Complete the table.

| $x$ | 100 | 1000 | 10,000 | 100,000 |
|-----|-----|------|--------|---------|
| $\overline{C}$ | | | | |

(d) Write a paragraph analyzing the data in the table. What do you observe about the average cost per unit as $x$ gets larger?

**72.** *Average Cost*    A manufacturer determines that the variable cost for a new product is $1.15 per unit and the fixed costs are $35,000. The product is to be sold for $2.19. Let $x$ be the number of units sold.

(a) Write the total cost $C$ as a function of the number of units sold.

(b) Write the average cost per unit $\overline{C} = C/x$ as a function of $x$.

(c) Complete the table.

| $x$ | 100 | 1000 | 10,000 | 100,000 |
|-----|-----|------|--------|---------|
| $\overline{C}$ | | | | |

(d) Write a paragraph analyzing the data in the table. What do you observe about the average cost per unit as $x$ gets larger?

**73.** *Charter Bus Fares*    For groups of 80 or more people, a charter bus company determines the rate per person as Rate $= 8 - 0.05(n - 80)$, $n \geq 80$, where the rate is given in dollars and $n$ is the number of people.

(a) Express the total revenue $R$ for the bus company as a function of $n$.

(b) Complete the table.

| $n$ | 90 | 100 | 110 | 120 | 130 | 140 | 150 |
|-----|-----|-----|-----|-----|-----|-----|-----|
| $R$ | | | | | | | |

(c) Write a paragraph analyzing the data in the table.

**74.** *Apartment Rentals*    An apartment complex finds that apartment occupancy depends on the monthly rent, according to the model Rent $= 1700 - 10x$, $0 \leq x \leq 150$, where rent is given in dollars and $x$ is the number of rented apartments.

(a) Express the total revenue $R$ for the apartment complex as a function of $x$.

(b) Complete the table.

| $x$ | 25 | 50 | 75 | 100 | 125 |
|-----|-----|-----|-----|-----|-----|
| $R$ | | | | | |

(c) Write a paragraph analyzing the data in the table.

**75.** *Ripples in a Pond*    A stone is thrown into the middle of a calm pond, causing ripples to form in concentric circles. The radius of the outermost ripple increases at the rate of 0.8 foot per second.

(a) Write a function for the radius of the circle formed by the outermost ripple in terms of time $t$.

(b) Write a function for the area enclosed by the outermost ripple. Use the formula for the area of a circle, $A = \pi r^2$, where $r$ is the radius of the circle. Complete the table.

| Time $t$ | 1 | 2 | 3 | 4 | 5 |
|----------|---|---|---|---|---|
| Radius $r$ | | | | | |
| Area $A$ | | | | | |

(c) Compare the ratios $A(2)/A(1)$ and $A(4)/A(2)$. What do you observe? Based on your observation, predict the area when $t = 8$. Verify by checking $t = 8$ in the area function.

**3.2**    Graphs of Functions

### Objectives

- Find the domain and range using the graph of a function.
- Identify the graph of a function using the Vertical Line Test.
- Describe the increasing and decreasing behavior of a function.
- Classify a function as even or odd.
- Identify six common graphs and use them to sketch the graph of a function.

## The Graph of a Function

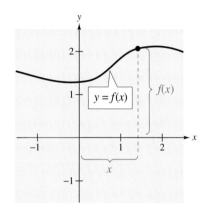

Figure 3.6

In Section 3.1 you studied functions from an algebraic point of view. In this section, you will study functions from a geometric perspective. The **graph of a function $f$** is the collection of ordered pairs $(x, f(x))$ such that $x$ is in the domain of $f$. As you study this section, remember that

$$x = \text{the directed distance from the } y\text{-axis}$$

$$f(x) = \text{the directed distance from the } x\text{-axis}$$

as shown in Figure 3.6. If the graph of a function has an $x$-intercept at $(a, 0)$, then $a$ is a **zero** of the function. In other words, the zeros of a function are the values of $x$ for which $f(x) = 0$. For instance, the function given by $f(x) = x^2 - 4$ has two zeros: $-2$ and $2$.

The **range** of a function (the set of values assumed by the dependent variable) is often more easily determined graphically than algebraically. This technique is illustrated in Example 1.

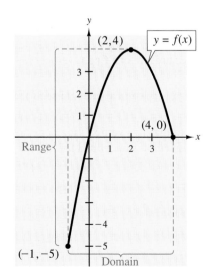

Figure 3.7

**Example 1**    Finding the Domain and Range of a Function

Use the graph of the function $f$, shown in Figure 3.7, to find the following.

**a.** The domain of $f$.

**b.** The function values $f(-1)$ and $f(2)$.

**c.** The range of $f$.

### Solution

**a.** Because the graph does not extend beyond $x = -1$ (on the left) and $x = 4$ (on the right), the domain of $f$ is all $x$ in the interval $[-1, 4]$.

**b.** Because $(-1, -5)$ is a point on the graph of $f$, it follows that

$$f(-1) = -5.$$

Similarly, because $(2, 4)$ is a point on the graph of $f$, it follows that

$$f(2) = 4.$$

**c.** Because the graph does not extend below $f(-1) = -5$ nor above $f(2) = 4$, the range of $f$ is the interval $[-5, 4]$.

By the definition of a function, at most one *y*-value corresponds to a given *x*-value. It follows, then, that a vertical line can intersect the graph of a function at most once. This observation provides a convenient visual test for functions.

▶ **Vertical Line Test for Functions**

A set of points in a coordinate plane is the graph of *y* as a function of *x* if and only if no vertical line intersects the graph at more than one point.

**Example 2** Vertical Line Test for Functions

Use the Vertical Line Test to decide whether the graphs in Figure 3.8 represent *y* as a function of *x*.

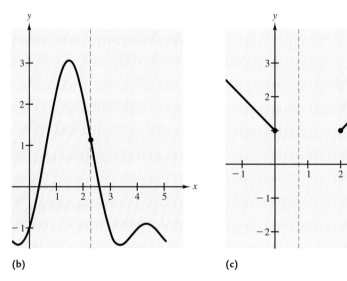

(a)

(b)

(c)

Figure 3.8

**Solution**

**a.** This *is not* a graph of *y* as a function of *x* because we can find a vertical line that intersects the graph twice.

**b.** This *is* a graph of *y* as a function of *x* because every vertical line intersects the graph at most once.

**c.** This *is* a graph of *y* as a function of *x*. (Note that if a vertical line does not intersect the graph, it simply means that the function is undefined for that particular value of *x*.)

# Increasing and Decreasing Functions

The more you know about the graph of a function, the more you know about the function itself. Consider the graph shown in Figure 3.9. As you move from *left to right*, this graph decreases, then is constant, and then increases.

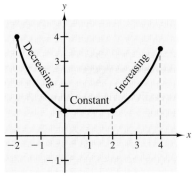

**Figure 3.9**

> ▶ **Increasing, Decreasing, and Constant Functions**
>
> A function $f$ is **increasing** on an interval if, for any $x_1$ and $x_2$ in the interval, $x_1 < x_2$ implies $f(x_1) < f(x_2)$.
>
> A function $f$ is **decreasing** on an interval if, for any $x_1$ and $x_2$ in the interval, $x_1 < x_2$ implies $f(x_1) > f(x_2)$.
>
> A function $f$ is **constant** on an interval if, for any $x_1$ and $x_2$ in the interval, $f(x_1) = f(x_2)$.

**Example 3**    Increasing and Decreasing Functions

Describe the increasing or decreasing behavior of each function in Figure 3.10.

Solution

**a.** This function is increasing over the entire real line.

**b.** This function is increasing on the interval $(-\infty, -1)$, decreasing on the interval $(-1, 1)$, and increasing on the interval $(1, \infty)$.

**c.** This function is increasing on the interval $(-\infty, 0)$, constant on the interval $(0, 2)$, and decreasing on the interval $(2, \infty)$.

(a)
**Figure 3.10**

(b)

(c)

To be able to sketch an accurate graph of a function, it is important that you know the points at which the function changes its increasing, decreasing, or constant behavior. These points often identify *maximum* or *minimum* values of the function.

| Example 4 | The Price of One-Family Houses |  |

During the 1990s, the average price of a new privately owned one-family house decreased and then increased according to the model

$$C = -147.22t^3 + 1967.88t^2 - 3673.02t + 122,540.48, \quad 0 \le t \le 6$$

where $C$ is the average price in dollars and $t$ represents the calendar year, with $t = 0$ corresponding to 1990. According to this model, during which years was the price of one-family houses decreasing? During which years was the price of one-family houses increasing? Approximate the minimum price of a one-family house between 1990 and 1996. (Source: U.S. Bureau of the Census, U.S. Department of Housing and Urban Development)

### Solution

To solve this problem, sketch an accurate graph of the function, as shown in Figure 3.11. From the graph, you can see that the price of a one-family house decreased from 1990 to early 1991. Then, from early 1991 to 1996, the price increased. The minimum price during the 7-year period was approximately $120,000.

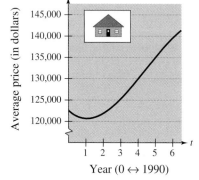

Figure 3.11

---

## •TECHNOLOGY•

There are several ways you can use a graphing utility to find the minimum and maximum values of a function. Begin by graphing the function from Example 4

$$C = -147.22t^3 + 1967.88t^2 -$$

$$3673.02t + 122,540.48, \quad 0 \le t \le 6$$

as shown below. One way to find the minimum value is to use the trace feature. A second way to find the minimum is to use the minimum value feature. Using either method, you can determine that the minimum value occurs when $x \approx 1.1$. Check your user's manual for information on the minimum value and maximum value features of your graphing utility.

## Step Functions

| Example 5 | The Greatest Integer Function |
|---|---|

The **greatest integer function** is denoted by $[\![x]\!]$ and is defined by

$$[\![x]\!] = \text{the greatest integer less than or equal to } x.$$

The graph of this function is shown in Figure 3.12. Note that the graph of the greatest integer function jumps vertically one unit at each integer and is constant (a horizontal line segment) between each pair of consecutive integers. Because of the jumps in its graph, the greatest integer function is an example of a category of functions called **step functions.** Some values of the greatest integer function are as follows.

$$[\![-1]\!] = -1 \qquad [\![-0.5]\!] = -1$$
$$[\![0]\!] = 0 \qquad [\![0.5]\!] = 0$$
$$[\![1]\!] = 1 \qquad [\![1.5]\!] = 1$$

The range of the greatest integer function is the set of all integers.

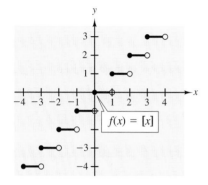

Figure 3.12    *Greatest Integer Function*

If you use a graphing utility to sketch a step function, you should set the utility to *Dot* mode rather than *Connected* mode.

| Example 6 | The Price of a Telephone Call |  |
|---|---|---|

The cost of a telephone call between Los Angeles and San Francisco is $0.50 for the first minute and $0.36 for each additional minute (or portion of a minute). The greatest integer function can be used to create a model for the cost of this call, as follows.

$$C = 0.50 + 0.36[\![t]\!], \quad t > 0$$

where $C$ is the total cost of the call in dollars and $t$ is the length of the call in minutes. Sketch the graph of this function.

### Solution

For calls up to 1 minute, the cost is $0.50. For calls between 1 and 2 minutes, the cost is $0.86, and so on.

| Length of call | $0 \leq t < 1$ | $1 \leq t < 2$ | $2 \leq t < 3$ | $3 \leq t < 4$ | $4 \leq t < 5$ |
|---|---|---|---|---|---|
| Cost of call | $0.50 | $0.86 | $1.22 | $1.58 | $1.94 |

Using these values, you can sketch the graph shown in Figure 3.13.

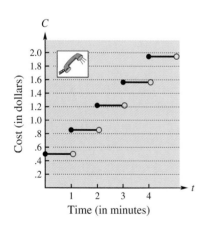

Figure 3.13

# Even and Odd Functions

Graph each of the following functions with a graphing utility. Determine whether the function is *odd, even,* or *neither.*

$$f(x) = x^2 - x^4$$

$$g(x) = 2x^3 + 1$$

$$h(x) = x^5 - 2x^3 + x$$

$$k(x) = x^5 - 2x^4 + x - 2$$

$$j(x) = 2 - x^6 - x^8$$

$$p(x) = x^9 + 3x^5 - x^3 + x$$

What do you notice about the equations of functions that are odd? What do you notice about the equations of functions that are even? Can you describe a way to identify a function as odd or even by inspecting the equation? Can you describe a way to identify a function as neither odd nor even by inspecting the equation?

In Section 2.2, you studied different types of symmetry of a graph. In the terminology of functions, a function is said to be **even** if its graph is symmetric with respect to the $y$-axis and to be **odd** if its graph is symmetric with respect to the origin. The symmetry tests in Section 2.2 yield the following tests for even and odd functions.

▶ **Test for Even and Odd Functions**

A function given by $y = f(x)$ is **even** if, for each $x$ in the domain of $f$,

$$f(-x) = f(x).$$

A function given by $y = f(x)$ is **odd** if, for each $x$ in the domain of $f$,

$$f(-x) = -f(x).$$

**Example 7**    Even and Odd Functions

**a.** The function $g(x) = x^3 - x$ is odd because

$$g(-x) = (-x)^3 - (-x) = -x^3 + x = -(x^3 - x) = -g(x).$$

**b.** The function $h(x) = x^2 + 1$ is even because

$$h(-x) = (-x)^2 + 1 = x^2 + 1 = h(x).$$

The graphs of the two functions are shown in Figure 3.14.

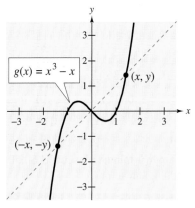

(a) **Odd function (symmetric to origin)**

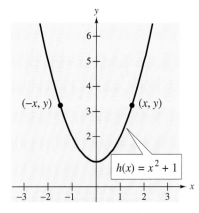

(b) **Even function (symmetric to y-axis)**

**Figure 3.14**

## Common Graphs

Figure 3.15 shows the graphs of six common functions. You need to be familiar with these graphs. They can be used as an aid to sketching other graphs. For instance, the graph of $f(x) = |x - 2|$, an absolute value function, will be $\vee$ -shaped.

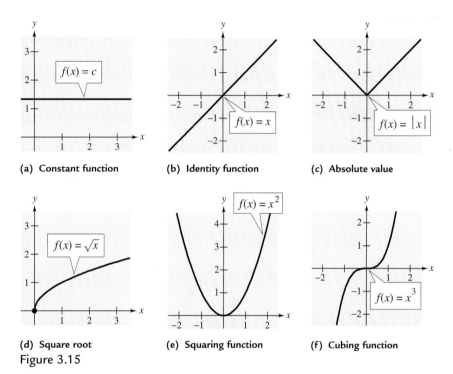

(a)  **Constant function**   (b)  **Identity function**   (c)  **Absolute value**

(d)  **Square root**   (e)  **Squaring function**   (f)  **Cubing function**

Figure 3.15

| Discussing the Concept | Increasing and Decreasing Functions |
|---|---|

Use your school's library or some other reference source to find examples of three different functions that represent the behaviors of quantities between 1989 and 1999. Find one quantity that decreased during the decade, one that increased, and one that was constant. For instance, the value of the dollar decreased, the population of the United States increased, and the land size of the United States remained constant. Can you find three other examples? If the examples you find appear to represent *linear growth or decline,* use the methods described in Section 2.5 to find a linear function $f(x) = a + bx$ that approximates the data.

**1.** Find $f(2)$ for $f(x) = -x^3 + 5x$.

**2.** Find $f(6)$ for $f(x) = x^2 - 6x$.

**3.** Find $f(-x)$ for $f(x) = 3/x$.

**4.** Find $f(-x)$ for $f(x) = x^2 + 3$.

In Exercises 5 and 6, solve the equation.

**5.** $x^3 - 16x = 0$

**6.** $2x^2 - 3x + 1 = 0$

In Exercises 7–10, find the domain of the function.

**7.** $g(x) = 4(x - 4)^{-1}$

**8.** $f(x) = 2x/(x^2 - 9x + 20)$

**9.** $h(t) = \sqrt[4]{5 - 3t}$

**10.** $f(t) = t^3 + 3t - 5$

## 3.2   Exercises

In Exercises 1–8, find the domain and range of the function.

**1.** $f(x) = \sqrt{x - 1}$

**2.** $f(x) = 4 - x^2$

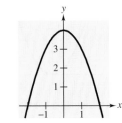

**3.** $f(x) = \sqrt{x^2 - 4}$

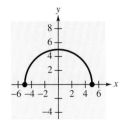

**4.** $f(x) = |x - 2|$

**5.** $f(x) = \sqrt{25 - x^2}$

**6.** $f(x) = \dfrac{|x|}{x}$

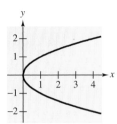

**7.** $f(x) = x^3 - 1$

**8.** $f(x) = \sqrt{x^2 - 9}$

In Exercises 9–14, use the Vertical Line Test to decide whether $y$ is a function of $x$.

**9.** $y = x^2$

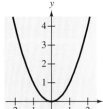

**10.** $y = x^3 - 1$

**11.** $x - y^2 = 0$

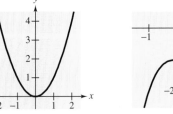

**12.** $x^2 + y^2 = 9$

**13.** $x^2 = xy - 1$

**14.** $x = |y|$

In Exercises 15–22, describe the increasing and decreasing behavior of the function. Find the point or points where the behavior of the function changes.

**15.** $f(x) = 2x$

**16.** $f(x) = x^2 - 2x$

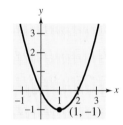

**17.** $f(x) = x^3 - 3x^2$

**18.** $f(x) = \sqrt{x^2 - 4}$

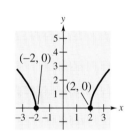

**19.** $f(x) = 3x^4 - 6x^2$

**20.** $f(x) = x^{2/3}$

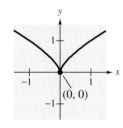

**21.** $y = x\sqrt{x + 3}$

**22.** $y = |x + 1| + |x - 1|$

In Exercises 23–28, decide whether the function is even, odd, or neither.

**23.** $f(x) = x^6 - 2x^2 + 3$   **24.** $h(x) = x^3 - 5$

**25.** $g(x) = x^3 - 5x$   **26.** $f(x) = x\sqrt{1 - x^2}$

**27.** $f(t) = t^2 + 2t - 3$   **28.** $g(s) = 4s^{2/3}$

In Exercises 29–40, sketch the graph of the function and determine whether the function is even, odd, or neither.

**29.** $f(x) = 3$   **30.** $g(x) = x$

**31.** $f(x) = 5 - 3x$   **32.** $h(x) = x^2 - 4$

**33.** $g(s) = \dfrac{s^3}{4}$   **34.** $f(t) = -t^4$

**35.** $f(x) = \sqrt{1 - x}$   **36.** $f(x) = x^{3/2}$

**37.** $g(t) = \sqrt[3]{t - 1}$   **38.** $f(x) = |x + 2|$

**39.** $f(x) = \begin{cases} x + 3, & x \le 0 \\ 3, & 0 < x \le 2 \\ 2x - 1, & x > 2 \end{cases}$

**40.** $f(x) = \begin{cases} 2x + 1, & x \le -1 \\ x^2 - 2, & x > -1 \end{cases}$

In Exercises 41–54, sketch the graph of the function.

**41.** $f(x) = 4 - x$   **42.** $f(x) = 4x + 2$

**43.** $f(x) = x^2 - 9$   **44.** $f(x) = x^2 - 4x$

**45.** $f(x) = 1 - x^4$   **46.** $f(x) = \sqrt{x + 2}$

**47.** $f(x) = x^2 + 1$   **48.** $f(x) = -1(1 + |x|)$

**49.** $f(x) = -5$   **50.** $f(x) = \frac{1}{2}(2 + |x|)$

**51.** $f(x) = -[\![x]\!]$   **52.** $f(x) = 2[\![x]\!]$

**53.** $f(x) = [\![x - 1]\!]$   **54.** $f(x) = [\![x + 1]\!]$

In Exercises 55–58, use a graphing utility to graph the function and then estimate the open intervals on which the function is increasing or decreasing.

**55.** $f(x) = x^2 - 4x + 1$   **56.** $f(x) = -x^2 + 6x + 3$

**57.** $f(x) = x^3 - 3x^2$   **58.** $f(x) = -x^3 + 3x + 1$

**59.** *Population of Oklahoma*   The population of the state of Oklahoma fell during the latter part of the 1980s and then began to increase. A model that approximates the population of Oklahoma is

$$P = -0.706t^3 + 27.54t^2 - 327.9t + 4387,$$

$$6 \le t \le 16$$

where $P$ represents the population in thousands and $t = 6$ represents 1986. Use the figure to estimate the years that the population was decreasing and the years that the population was increasing.   (Source: U.S. Bureau of the Census)

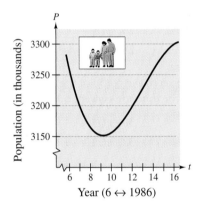

Year ($6 \leftrightarrow 1986$)

**60.** *Population of Rhode Island*   The population of the state of Rhode Island increased during the latter part of the 1980s and then began to decrease. A model that approximates the population of Rhode Island is

$$P = 799 + 48.6t - 3.67t^2 + 0.086t^3, \quad 6 \le t \le 16$$

where $P$ represents the population in thousands and $t = 6$ represents 1986. Use the figure to estimate the years that the population was increasing and the years that the population was decreasing.   (Source: U.S. Bureau of the Census)

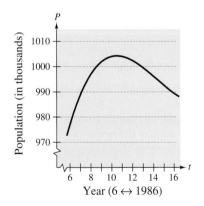

Year ($6 \leftrightarrow 1986$)

**61.** *Company Revenue*   A company determines that the total revenue for the years from 1990 to 2000 can be approximated by the function

$$R = -0.035t^3 - 0.03t^2 + 3.25t + 14.41,$$

$$0 \le t \le 10$$

where $R$ is the total revenue in hundreds of thousands of dollars and $t = 0$ represents 1990. Graph the revenue function with a graphing utility and use the trace feature to estimate the years in which the revenue was increasing and the years in which the revenue was decreasing. Find the maximum revenue for the years 1990 to 2000. Find the minimum revenue for the years 1990 to 2000.

**62.** *Radio Stations*   From 1990 to 1996, the total number of radio stations that operated with a country format can be approximated by the function

$$R = 2443 + 20.4t + 20.8t^2 - 3.75t^3, \quad 0 \le t \le 6$$

where $R$ is the number of radio stations and $t = 0$ represents 1990. Use the figure to estimate the years in which the number of country stations was increasing and the years in which the number of country stations was decreasing. Estimate the maximum number of country stations from 1990 to 1996. Estimate the minimum number of country stations from 1990 to 1996.   (Source: M Street Corporation)

Year ($0 \leftrightarrow 1990$)

**63.** *Reasoning*   When finding a maximum or minimum value for Exercises 61 and 62, why should you also check the endpoints of the function?

**64.** *Maximum Profit*   The marketing department of a company estimates that the demand for a product is given by $p = 100 - 0.0001x$, where $p$ is the price per unit and $x$ is the number of units. The cost of producing $x$ units is given by $C = 350,000 + 30x$, and the profit for producing $x$ units is given by

$$P = R - C = xp - C.$$

Sketch the graph of the profit function and estimate the number of units that would produce a maximum profit.

**65.** *Maximum Profit*   The marketing department of a company estimates that the demand for a product is given by $p = 120 - 0.0001x$, where $p$ is the price per unit and $x$ is the number of units. The cost of producing $x$ units is given by $C = 450,000 + 50x$, and the profit for producing $x$ units is given by

$$P = R - C = xp - C.$$

Sketch the graph of the profit function and estimate the number of units that would produce a maximum profit.

**66.** *Cost of a Phone Call*   Suppose that the cost of a telephone call between Denver and Boise is $0.60 for the first minute and $0.42 for each additional minute (or portion of a minute). A model for the total cost of the phone call is

$$C = 0.60 + 0.42[\![x]\!], \quad 0 < x$$

where $C$ is the total cost of the call in dollars and $x$ is the length of the call in minutes. Sketch the graph of this function.

**67.** *Cost of a Phone Call*   Suppose the cost of a telephone call between Dallas and New York City is $0.75 for the first minute and $0.46 for each additional minute (or portion of a minute). A model for the total cost of the phone call is

$$C = 0.75 + 0.46[\![x]\!], \quad 0 < x$$

where $C$ is the total cost of the call in dollars and $x$ is the length of the call in minutes. Sketch the graph of this function.

**68.** *Cost of Overnight Delivery*   Suppose that the cost of sending an overnight package from New York to Atlanta is $9.80 for the first pound and $2.50 for each additional pound (or portion of a pound). A model for the total cost of sending the package is

$$C = 9.8 + 2.5[\![x]\!], \quad 0 < x$$

where $C$ is the total cost in dollars and $x$ is the weight of the package in pounds. Sketch the graph of this function.

**69.** *Cost of Overnight Delivery*   Suppose that the cost of sending an overnight package from Los Angeles to Miami is $10.75 for the first pound and $3.95 for each additional pound (or portion of a pound). A model for the total cost of sending the package is

$$C = 10.75 + 3.95[\![x]\!], \quad 0 < x$$

where $C$ is the total cost in dollars and $x$ is the weight in pounds. Sketch the graph of this function.

**70.** *Research and Development*   The total amount spent on basic research and development in the United States from 1960 to 1996 can be approximated by the model

$$C = \begin{cases} 12.70 + 1.45t, & 0 \le t \le 15.9 \\ -87.66 + 7.76t, & 15.9 < t \le 36 \end{cases}$$

where $C$ is the amount in billions of dollars and $t$ is the year, with $t = 0$ corresponding to 1960. Sketch the graph of this function.   (Source: National Science Foundation)

**71.** *Grade Level Salaries*   The 1998 average salary for federal employees can be approximated by the model

$$S = \begin{cases} 9472.27 + 2263.39x, & x = 1, 2, \ldots, 10 \\ -63,127.2 + 8961.2x, & x = 11, \ldots, 15 \end{cases}$$

where $S$ is the salary in dollars and $x$ represents the "GS" level. Sketch a *bar graph* that represents this function.   (Source: U.S. Office of Personnel Management)

## 3.3 Transformations of Functions

### Objectives

- Identify rigid and nonrigid transformations.
- Sketch the graph of a function using transformations and common graphs.
- Write the equation of a function using transformations and common graphs.

## Vertical and Horizontal Shifts

Many functions have graphs that are simple transformations of the common graphs that are summarized on page 171. For example, you can obtain the graph of $h(x) = x^2 + 2$ by shifting the graph of $f(x) = x^2$ *up* two units, as shown in Figure 3.16. In function notation, $h$ and $f$ are related as follows.

$$h(x) = x^2 + 2 = f(x) + 2 \qquad \text{Upward shift of 2}$$

Similarly, you can obtain the graph of $g(x) = (x - 2)^2$ by shifting the graph of $f(x) = x^2$ to the *right* two units, as shown in Figure 3.17. In this case, the functions $g$ and $f$ have the following relationship.

$$g(x) = (x - 2)^2 = f(x - 2) \qquad \text{Right shift of 2}$$

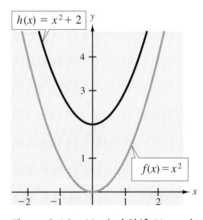

Figure 3.16 *Vertical Shift Upward* Figure 3.17 *Horizontal Shift to the Right*

---

▶ **Vertical and Horizontal Shifts**

Let $c$ be a positive real number. **Vertical** and **horizontal shifts** in the graph of $y = f(x)$ are represented as follows.

1. Vertical shift $c$ units **upward:** $\qquad h(x) = f(x) + c$

2. Vertical shift $c$ units **downward:** $\qquad h(x) = f(x) - c$

3. Horizontal shift $c$ units to the **right:** $\quad h(x) = f(x - c)$

4. Horizontal shift $c$ units to the **left:** $\qquad h(x) = f(x + c)$

---

In items 3 and 4, be sure you see that $h(x) = f(x - c)$ corresponds to a *right* shift and $h(x) = f(x + c)$ corresponds to a *left* shift.

Some graphs can be obtained from a combination of vertical and horizontal shifts. This is demonstrated in Example 1(b).

## Study Tip

Vertical and horizontal shifts generate a *family of functions,* each with the same shape but at different locations in the plane. The functions $f$, $g$, and $h$ in Example 1 belong to the family of cubic functions.

### Example 1    Shifts in the Graph of a Function

Use the graph of $f(x) = x^3$ to sketch the graphs of the following functions.

**a.** $g(x) = x^3 + 1$

**b.** $h(x) = (x + 2)^3 + 1$

**Solution**

Relative to the graph of $f(x) = x^3$, the graph of $g(x) = x^3 + 1$ is an upward shift of one unit, and the graph of $h(x) = (x + 2)^3 + 1$ involves a left shift of two units *and* an upward shift of one unit. The graphs of both functions are compared with the graph of $f(x) = x^3$ in Figure 3.18.

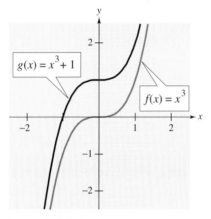

(a)  Vertical shift: 1 up

(b)  Horizontal shift: 2 left; vertical shift: 1 up

Figure 3.18

## ·DISCOVERY·

The point $(2, 4)$ is on the graph of $f(x) = x^2$. Predict the location of that point if the following transformations are performed.

**a.** $f(x - 4)$

**b.** $f(x) + 1$

**c.** $f(x + 1) - 2$

Use a graphing utility to verify your prediction. Can you find a general description to represent an ordered pair that has been shifted horizontally? vertically?

## ·TECHNOLOGY·

Graphing utilities are ideal tools for exploring translations of functions. Try to predict how the graphs of $g$ and $h$ relate to the graph of $f$. Graph $f$, $g$, and $h$ on the same screen to check your prediction.

**a.** $f(x) = x^2$,  $g(x) = (x - 4)^2$,  $h(x) = (x - 4)^2 + 3$

**b.** $f(x) = x^2$,  $g(x) = (x + 1)^2$,  $h(x) = (x + 1)^2 - 2$

**c.** $f(x) = x^2$,  $g(x) = (x + 4)^2$,  $h(x) = (x + 4)^2 + 2$

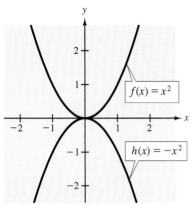

**Figure 3.19**  *Reflection*

# Reflections

The second common type of transformation is a **reflection.** For instance, if you consider the $x$-axis to be a mirror, the graph of $h(x) = -x^2$ is the mirror image (or reflection) of the graph of $f(x) = x^2$, as shown in Figure 3.19.

> ▶ **Reflections in the Coordinate Axes**
>
> Reflections in the coordinate axes of the graph of $y = f(x)$ are represented as follows.
>
> 1. **Reflection in the $x$-axis:** $h(x) = -f(x)$
>
> 2. **Reflection in the $y$-axis:** $h(x) = f(-x)$

---

**Example 2**  Reflections of the Graph of a Function

Compare the graphs of the following with the graph of $f(x) = \sqrt{x}$.

**a.** $g(x) = -\sqrt{x}$

**b.** $h(x) = \sqrt{-x}$

**Solution**

**a.** The graph of $g$ is a reflection of the graph of $f$ in the $x$-axis because

$$g(x) = -\sqrt{x} = -f(x).$$

**b.** The graph of $h$ is a reflection of the graph of $f$ in the $y$-axis because

$$h(x) = \sqrt{-x} = f(-x).$$

The graphs of both functions are shown in Figure 3.20.

**(a)** Reflection in $x$-axis

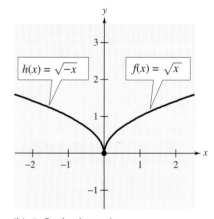

**(b)** Reflection in $y$-axis

Figure 3.20

---

**·TECHNOLOGY·**

You will find programs for a variety of calculator models that will give you practice working with reflections, horizontal shifts, and vertical shifts at our website *www.hmco.com*. These programs will sketch a graph of the function

$$y = R(x + H)^2 + V$$

where $R = \pm 1$, $H$ is an integer between $-6$ and 6, and $V$ is an integer between $-3$ and 3. Each time you run the program, different values of $R$, $H$, and $V$ are possible. From the graph, you should be able to determine the values of $R$, $H$, and $V$.

When sketching the graph of a function involving square roots, remember that the domain must be restricted to exclude negative numbers inside the radical. For instance, here are the domains of the functions in Example 2.

Domain of $g(x) = -\sqrt{x}$:    $x \geq 0$

Domain of $h(x) = \sqrt{-x}$:    $x \leq 0$

**Example 3**    Reflections and Shifts

Use the graph of $f(x) = x^2$ to sketch the graph of each function.

**a.** $g(x) = -(x - 3)^2$

**b.** $h(x) = -x^2 + 2$

**Solution**

**a.** To sketch the graph of $g(x) = -(x - 3)^2$, first shift the graph of $f(x) = x^2$ to the right three units. Then reflect the result in the x-axis.

**b.** To sketch the graph of $h(x) = -x^2 + 2$, first reflect the graph of $f(x) = x^2$ in the x-axis. Then shift the result upward two units.

The graphs of both functions are shown in Figure 3.21.

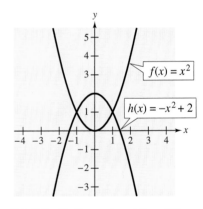

(a) Shift and then reflect in the x-axis    (b) Reflect in the x-axis and then shift

Figure 3.21

**Example 4**    Finding Equations from Graphs

The graphs labeled g and h in Figure 3.22 are transformations of the graph of $f(x) = x^4$. Find an equation for each function.

**Solution**

The graph of g is a reflection in the x-axis *followed* by a downward shift of two units of the graph of $f(x) = x^4$. Thus, the equation for g is $g(x) = -x^4 - 2$. The graph of h is a horizontal shift of one unit to the left *followed* by a reflection in the x-axis of the graph $f(x) = x^4$. Thus, the equation for h is $h(x) = -(x + 1)^4$.

Figure 3.22

## Nonrigid Transformations

Horizontal shifts, vertical shifts, and reflections are **rigid** transformations because the basic shape of the graph is unchanged. These transformations change only the *position* of the graph in the $xy$-plane. A **nonrigid** transformation is one that causes a *distortion*—a change in the shape of the original graph. For instance, a nonrigid transformation of the graph of $y = f(x)$ is represented by $g(x) = cf(x)$, where the transformation is a **vertical stretch** if $c > 1$ and a **vertical shrink** if $0 < c < 1$.

### Example 5    Nonrigid Transformations

Compare the graphs of the following with the graph of $f(x) = |x|$.

**a.** $h(x) = 3|x|$

**b.** $g(x) = \dfrac{1}{3}|x|$

### Solution

**a.** Relative to the graph of $f(x) = |x|$, the graph of

$$h(x) = 3|x| = 3f(x)$$

is a vertical stretch (multiply each $y$-value by 3) of the graph of $f$.

**b.** Similarly, the equation

$$g(x) = \frac{1}{3}|x| = \frac{1}{3}f(x)$$

indicates that the graph of $g$ is a vertical shrink of the graph of $f$. The graphs of both functions are shown in Figure 3.23.

(a)

(b)

Figure 3.23

**Figure 3.24**

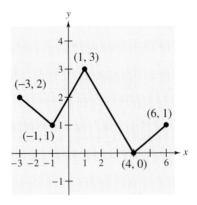

## Example 6    Rigid and Nonrigid Transformations

Use the graph of $f$ in Figure 3.24 to sketch the following graphs.

**a.** $g(x) = f(x - 2) + 1$

**b.** $h(x) = \frac{1}{2}f(x)$

### Solution

**a.** The graph of $g(x)$ is a horizontal shift to the right two units and a vertical shift upward one unit of the graph of $f(x)$. The graph of $g(x)$ is shown in Figure 3.25(a).

**b.** The graph of $h(x)$ is a vertical shrink of the graph of $f(x)$.

For $x = -5$, $h(-5) = \frac{1}{2}f(-5) = \frac{1}{2}(1) = \frac{1}{2}$.

For $x = -3$, $h(-3) = \frac{1}{2}f(-3) = \frac{1}{2}(0) = 0$.

For $x = -1$, $h(-1) = \frac{1}{2}f(-1) = \frac{1}{2}(2) = 1$.

For $x = 2$, $h(2) = \frac{1}{2}f(2) = \frac{1}{2}(-1) = -\frac{1}{2}$.

For $x = 4$, $h(4) = \frac{1}{2}f(4) = \frac{1}{2}(0) = 0$.

The graph of $h(x)$ is shown in Figure 3.25(b).

**(a)**

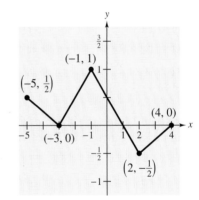

**(b)**

**Figure 3.25**

---

## Discussing the Concept    Constructing Transformations

Use a graphing utility to graph $f(x) = 3 - x^3$. Describe how to alter this function to produce each of the following transformation descriptions. Graph each transformation on the same screen with $f$; confirm that the transformation moves $f$ as described.

**a.** The graph of $f$ shifted upward four units

**b.** The graph of $f$ shifted to the left two units

**c.** The graph of $f$ shifted downward two units and to the right one unit

**d.** The graph of $f$ reflected in the $x$-axis

# Warm Up

The following warm-up exercises involve skills that were covered in earlier sections. You will use these skills in the exercise set for this section.

**1.** Find $f(3)$ for $f(x) = x^2 - 4x + 15$.

**2.** Find $f(-x)$ for $f(x) = 2x/(x - 3)$.

In Exercises 3 and 4, solve the equation.

**3.** $-x^3 + 10x = 0$

**4.** $3x^2 + 2x - 8 = 0$

In Exercises 5–10, sketch the graph of the function.

**5.** $f(x) = -2$

**6.** $f(x) = -x$

**7.** $f(x) = x + 5$

**8.** $f(x) = 2 - x$

**9.** $f(x) = 3x - 4$

**10.** $f(x) = 9x + 10$

## 3.3 Exercises

LAB I

In Exercises 1–8, use the graph of $f(x) = x^2$ to sketch the graph of the function by hand. Verify with a graphing utility.

**1.** $g(x) = x^2 + 3$

**2.** $g(x) = x^2 - 2$

**3.** $g(x) = (x + 3)^2$

**4.** $g(x) = (x - 4)^2$

**5.** $g(x) = (x - 2)^2 + 2$

**6.** $g(x) = (x + 1)^2 - 3$

**7.** $g(x) = -x^2 + 1$

**8.** $g(x) = -(x - 2)^2$

In Exercises 9–14, use the graph of $f(x) = |x|$ to sketch the graph of the function by hand. Verify with a graphing utility.

**9.** $g(x) = |x| + 2$

**10.** $g(x) = |x - 1|$

**11.** $g(x) = -|x| + 3$

**12.** $g(x) = |x + 2| - 3$

**13.** $g(x) = 4 - |x - 2|$

**14.** $g(x) = |x - 2| + 2$

In Exercises 15–22, use the graph of $f(x) = \sqrt{x}$ to sketch the graph of the function by hand. Verify with a graphing utility.

**15.** $y = \sqrt{x - 2}$

**16.** $y = \sqrt{x + 3}$

**17.** $y = \sqrt{x - 3} + 1$

**18.** $y = \sqrt{x + 5} - 2$

**19.** $y = 2 - \sqrt{x - 4}$

**20.** $y = \sqrt{2x}$

**21.** $y = \sqrt{-x} + 1$

**22.** $y = \sqrt{2x} - 5$

In Exercises 23–30, use the graph of $f(x) = \sqrt[3]{x}$ to sketch the graph of the function by hand. Verify with a graphing utility.

**23.** $y = \sqrt[3]{x} - 1$

**24.** $y = \sqrt[3]{x + 1}$

**25.** $y = 2 - \sqrt[3]{x + 1}$

**26.** $y = -\sqrt[3]{x - 1} - 4$

**27.** $y = \sqrt[3]{x + 1} - 1$

**28.** $y = \frac{1}{2}\sqrt[3]{x}$

**29.** $y = \frac{1}{2}\sqrt[3]{x} - 3$

**30.** $y = 2\sqrt[3]{x - 2} + 1$

In Exercises 31–36, identify the common function and the transformation shown in the graph. Write the equation for the graphed function.

**31.**

**32.**

**33.**

**34.**

**35.**

**36.**

⊞ **37.** Use a graphing utility to graph $f$ for $c = -2, 0$, and 2 on the same screen.

(a) $f(x) = \frac{1}{2}x + c$

(b) $f(x) = \frac{1}{2}(x - c)$

(c) $f(x) = \frac{1}{2}(cx)$

In each case, compare the three graphs with the graph of $y = \frac{1}{2}x$.

⊞ **38.** Use a graphing utility to graph $f$ for $c = -2, 0$, and 2 on the same screen.

(a) $f(x) = x^3 + c$

(b) $f(x) = (x - c)^3$

(c) $f(x) = (x - 2)^3 + c$

In each case, compare the three graphs with the graph of $y = x^3$.

**39.** Use the graph of $f(x) = x^2$ to write equations for the functions whose graphs are shown.

(a)

(b)

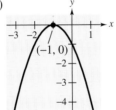

**40.** Use the graph of $f(x) = x^3$ to write equations for the functions whose graphs are shown.

(a)

(b)

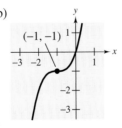

**41.** Use the graph of $f$ (see figure) to sketch the graphs.

(a) $y = f(x) + 2$     (b) $y = -f(x)$

(c) $y = f(x - 2)$     (d) $y = f(x + 3)$

(e) $y = 2f(x)$     (f) $y = f(-x)$

Figure for 41

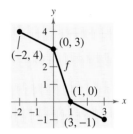

Figure for 42

**42.** Use the graph of $f$ (see figure) to sketch the graphs.

(a) $y = f(x) - 1$     (b) $y = f(x + 1)$

(c) $y = f(x - 1)$     (d) $y = -f(x - 2)$

(e) $y = f(-x)$     (f) $y = \frac{1}{2}f(x)$

In Exercises 43–46, consider the graph of $f(x) = |x|$. Use your knowledge of rigid and nonrigid transformations to write the equation for the following. Verify with a graphing utility.

**43.** The graph of $f(x)$ is shifted two units to the right and one unit upward.

**44.** The graph of $f(x)$ is reflected about the $x$-axis, shifted one unit to the left, and shifted two units upward.

**45.** The graph of $f(x)$ is vertically stretched by a factor of 3 and reflected about the $x$-axis.

**46.** The graph of $f(x)$ is vertically shrunk by a factor of $\frac{1}{2}$ and shifted two units to the right.

In Exercises 47–50, consider the graph of $g(x) = \sqrt{x}$. Use your knowledge of rigid and nonrigid transformations to write the equation for the following. Verify with a graphing utility.

**47.** The graph of $g(x)$ is shifted three units to the left and two units upward.

**48.** The graph of $g(x)$ is reflected about the $x$-axis, shifted two units to the right, and shifted one unit downward.

**49.** The graph of $g(x)$ is vertically stretched by a factor of 4 and reflected about the $x$-axis.

**50.** The graph of $g(x)$ is vertically shrunk by a factor of $\frac{1}{2}$ and shifted three units to the left.

In Exercises 51 and 52, use the graph of $f(x) = x^3 - 3x^2$ to write an equation for the function $g$ shown in the graph.

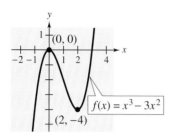

$f(x) = x^3 - 3x^2$

(0, 0)

(2, −4)

**51.**

(2, 5)

(0, 1)

**52.**

(2, 1)

(4, −3)

**53.** The point $(3, 9)$ on the graph of $f(x) = x^2$ has been shifted to the point $(4, 7)$ after a rigid transformation. Identify the shift and write the new function $g(x)$ in terms of $f(x)$.

**54.** The point $(8, 2)$ on the graph of $f(x) = \sqrt[3]{x}$ has been shifted to the point $(5, 0)$ after a rigid transformation. Identify the shift and write the new function $h(x)$ in terms of $f(x)$.

**55.** *Profit*     The profit $P$ per week on a certain product is given by the model

$$P(x) = 80 + 20x - 0.5x^2, \qquad 0 \le x \le 20$$

where $x$ is the amount spent on advertising. In this model, $x$ and $P$ are both measured in hundreds of dollars.

(a) Use a graphing utility to graph the profit function.

(b) The business estimates that taxes and operating costs will increase by an average of $2500 per week during the next year. Rewrite the profit equation to reflect this expected decrease in profits. Identify the type of transformation applied to the graph of the equation.

(c) Rewrite the profit equation so that $x$ measures advertising expenditures in dollars. [Find $P(x/100)$.] Identify the type of transformation applied to the graph of the profit function.

**56.** *Automobile Aerodynamics*     The number of horsepower $H$ required to overcome wind drag on a certain automobile is approximated by

$$H(x) = 0.002x^2 + 0.005x - 0.029, \quad 10 \le x \le 100$$

where $x$ is the speed of the car in miles per hour.

(a) Use a graphing utility to graph the power function.

(b) Rewrite the power function so that $x$ represents the speed in kilometers per hour. [Find $H(1.6x)$.] Identify the type of transformation applied to the graph of the power function.

**57.** *Exploration*     Use a graphing utility to graph each function. Describe any similarities and differences you observe among the graphs.

(a) $y = x$          (b) $y = x^2$

(c) $y = x^3$        (d) $y = x^4$

(e) $y = x^5$        (f) $y = x^6$

**58.** *Conjecture*     Use the results of Exercise 57 to make a conjecture about the shapes of the graphs of $y = x^7$ and $y = x^8$. Use a graphing utility to verify your conjecture.

**59.** Use the results of Exercise 57 to sketch the graph of $y = (x - 3)^3$ by hand. Use a graphing utility to verify your graph.

**60.** Use the results of Exercise 57 to sketch the graph of $y = (x + 1)^2$ by hand. Use a graphing utility to verify your graph.

## Mid-Chapter Quiz

Take this quiz as you would take a quiz in class. After you are done, check your work against the answers given in the back of the book.

In Exercises 1–4, find the value of $f(x) = 2x^2 - x + 1$.

**1.** $f(3)$        **2.** $f(-1)$        **3.** $f(0)$        **4.** $f(a)$

In Exercises 5 and 6, find all real values of $x$ such that $f(x) = 0$.

**5.** $f(x) = x^3 - 4x$        **6.** $f(x) = 3 - \dfrac{4}{x}$

**7.** A company produces a product for which the variable cost is \$11.40 per unit and the fixed costs are \$85,000. The product sells for \$16.89. Let $x$ be the number of units produced. Write the total profit $P$ as a function of $x$.

In Exercises 8–10, find the domain of the function.

**8.** $f(x) = \dfrac{3}{x^2 - x}$        **9.** $h(x) = \sqrt{5 - x}$        **10.** $g(x) = \dfrac{\sqrt{x - 1}}{x + 2}$

In Exercises 11–14, use a graphing utility to graph the function. Then estimate the open intervals on which the function is increasing or decreasing.

**11.** $f(x) = -x^2 + 5x + 3$        **12.** $f(x) = 2x^3 - 5x^2$

**13.** $f(x) = x^2 + 2$        **14.** $f(x) = \sqrt{x + 1}$

In Exercises 15 and 16, compare the graphs of $f(x) = \sqrt[3]{x}$ and $g$.

**15.** $g(x) = \sqrt[3]{x - 2}$        **16.** $g(x) = -\sqrt[3]{x + 1}$

**17.** Write the equation for the transformation of $f(x) = x^3$ shown in the figure.

In Exercises 18 and 19, write the equation for the indicated transformation. Verify your result with a graphing utility.

**18.** The graph of $f(x) = x^2$ shifted three units to the right and two units downward.

**19.** The graph of $f(x) = \sqrt{x}$ shifted two units to the left and reflected about the $x$-axis.

**20.** The marketing department of a company estimates that the demand for a product is given by $p = 80 - 0.001x$, where $p$ is the price per unit and $x$ is the number of units. The cost of producing $x$ units is given by $C = 300,000 + 25x$ and the profit for producing $x$ units is given by $P = xp - C$. Write the equation for the profit function, sketch the graph, and estimate the number of units that would produce a maximum profit.

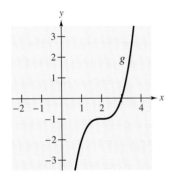

Figure for 17

## 3.4  The Algebra of Functions

### Objectives

- Find the sum, difference, product, and quotient of functions.
- Form the composition of two functions and determine the domain.
- Identify a function as the composite of two functions.
- Use combinations and composition of functions to solve an application problem.

## Arithmetic Combinations of Functions

Just as two real numbers can be combined by the operations of addition, subtraction, multiplication, and division to form other real numbers, two *functions* can be combined to create new functions. For example, the functions $f(x) = 2x - 3$ and $g(x) = x^2 - 1$ can be combined as follows.

$$f(x) + g(x) = (2x - 3) + (x^2 - 1) = x^2 + 2x - 4 \qquad \text{Sum}$$

$$f(x) - g(x) = (2x - 3) - (x^2 - 1) = -x^2 + 2x - 2 \qquad \text{Difference}$$

$$f(x)g(x) = (2x - 3)(x^2 - 1) = 2x^3 - 3x^2 - 2x + 3 \qquad \text{Product}$$

$$\frac{f(x)}{g(x)} = \frac{2x - 3}{x^2 - 1}, \quad x \neq \pm 1 \qquad \text{Quotient}$$

The domain of an arithmetic combination of functions $f$ and $g$ consists of all real numbers that are common to the domains of $f$ and $g$. In the case of the quotient $f(x)/g(x)$, there is the further restriction that $g(x) \neq 0$.

---

▶ **Sum, Difference, Product, and Quotient of Functions**

Let $f$ and $g$ be two functions with overlapping domains. Then, for all $x$ common to both domains, the **sum, difference, product,** and **quotient** of $f$ and $g$ are defined as follows.

1. *Sum:* $\qquad\qquad (f + g)(x) = f(x) + g(x)$

2. *Difference:* $\qquad (f - g)(x) = f(x) - g(x)$

3. *Product:* $\qquad\quad (fg)(x) = f(x) \cdot g(x)$

4. *Quotient:* $\qquad\quad \left(\dfrac{f}{g}\right)(x) = \dfrac{f(x)}{g(x)}, \quad g(x) \neq 0$

---

**Example 1**  Finding the Sum of Two Functions

Given $f(x) = 2x + 1$ and $g(x) = x^2 + 2x - 1$, find $(f + g)(x)$.

**Solution**

$$(f + g)(x) = f(x) + g(x)$$

$$= (2x + 1) + (x^2 + 2x - 1) = x^2 + 4x$$

### Example 2    Finding the Difference of Two Functions

Given $f(x) = 2x + 1$ and $g(x) = x^2 + 2x - 1$, find $(f - g)(x)$. Then evaluate the difference when $x = 2$.

#### Solution

The difference of the functions $f$ and $g$ is given by

$$(f - g)(x) = f(x) - g(x)$$
$$= (2x + 1) - (x^2 + 2x - 1)$$
$$= -x^2 + 2.$$

When $x = 2$, the value of this difference is

$$(f - g)(2) = -(2)^2 + 2 = -2.$$

In Examples 1 and 2, both $f$ and $g$ have domains that consist of all real numbers. Thus, the domains of $(f + g)$ and $(f - g)$ are also the set of all real numbers. Remember that any restrictions on the domains of $f$ and $g$ must be considered when forming the sum, difference, product, or quotient of $f$ and $g$.

### Example 3    The Quotient of Two Functions

Find the domains of $(f/g)(x)$ and $(g/f)(x)$ for the functions

$$f(x) = \sqrt{x} \quad \text{and} \quad g(x) = \sqrt{4 - x^2}.$$

#### Solution

The quotient of $f$ and $g$ is given by

$$\left(\frac{f}{g}\right)(x) = \frac{f(x)}{g(x)} = \frac{\sqrt{x}}{\sqrt{4 - x^2}}$$

and the quotient of $g$ and $f$ is given by

$$\left(\frac{g}{f}\right)(x) = \frac{g(x)}{f(x)} = \frac{\sqrt{4 - x^2}}{\sqrt{x}}.$$

The domain of $f$ is $[0, \infty)$ and the domain of $g$ is $[-2, 2]$. The intersection of these two domains is $[0, 2]$, which implies that the domains of $f/g$ and $g/f$ are as follows.

Domain of $\dfrac{f}{g}$: $[0, 2)$

Domain of $\dfrac{g}{f}$: $(0, 2]$

Can you see why these two domains differ slightly?

## Composition of Functions

Another way of combining two functions is to form the **composition** of one with the other. For instance, if $f(x) = x^2$ and $g(x) = x + 1$, the composition of $f$ with $g$ is given by

$$f(g(x)) = f(x + 1) = (x + 1)^2.$$

This composition is denoted as $f \circ g$.

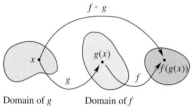

$f \circ g$

$g(x)$

$x$

$f(g(x))$

$g$

$f$

Domain of $g$        Domain of $f$

**Figure 3.26**

---

▶ **Definition of Composition of Two Functions**

The **composition** of the functions $f$ and $g$ is given by

$$(f \circ g)(x) = f(g(x)).$$

The domain of $(f \circ g)$ is the set of all $x$ in the domain of $g$ such that $g(x)$ is in the domain of $f$. (See Figure 3.26.)

---

**Example 4**    **Composition of Functions**

Given $f(x) = x + 2$ and $g(x) = 4 - x^2$, find the following.

**a.** $(f \circ g)(x)$

**b.** $(g \circ f)(x)$

**Solution**

**a.** The composition of $f$ with $g$ is as follows.

$$(f \circ g)(x) = f(g(x)) \qquad \text{Definition of } f \circ g$$
$$= f(4 - x^2) \qquad \text{Definition of } g(x)$$
$$= (4 - x^2) + 2 \qquad \text{Definition of } f(x)$$
$$= -x^2 + 6 \qquad \text{Simplify.}$$

**b.** The composition of $g$ with $f$ is as follows.

$$(g \circ f)(x) = g(f(x)) \qquad \text{Definition of } g \circ f$$
$$= g(x + 2) \qquad \text{Definition of } f(x)$$
$$= 4 - (x + 2)^2 \qquad \text{Definition of } g(x)$$
$$= 4 - (x^2 + 4x + 4) \qquad \text{Expand.}$$
$$= -x^2 - 4x \qquad \text{Simplify.}$$

Note that, in this case, $(f \circ g)(x) \neq (g \circ f)(x)$.

**Example 5**    Finding the Domain of a Composite Function

Find the composition $(f \circ g)(x)$ for the functions

$$f(x) = x^2 - 9 \quad \text{and} \quad g(x) = \sqrt{9 - x^2}.$$

Then find the domain of $(f \circ g)$.

**Solution**

The composition of the functions is as follows.

$$(f \circ g)(x) = f(g(x))$$
$$= f\left(\sqrt{9 - x^2}\right)$$
$$= \left(\sqrt{9 - x^2}\right)^2 - 9$$
$$= 9 - x^2 - 9$$
$$= -x^2$$

From this, it might appear that the domain of the composition is the set of all real numbers. This, however, is not true because the domain of $g$ is $[-3, 3]$. So, the domain of $f \circ g$ is $[-3, 3]$.

In Examples 4 and 5, you formed the composition of two functions. To "decompose" a composite function, look for an "inner" function and an "outer" function. For instance, the function $h$ given by

$$h(x) = (3x - 5)^3$$

is the composition of $f$ with $g$, where $f(x) = x^3$ and $g(x) = 3x - 5$. That is,

$$h(x) = (3x - 5)^3 = [g(x)]^3 = f(g(x)).$$

In the function $h$, $g(x) = 3x - 5$ is the inner function and $f(x) = x^3$ is the outer function.

**Example 6**    Identifying a Composite Function

Express the function

$$h(x) = \frac{1}{(x - 2)^2}$$

as a composition of two functions.

**Solution**

One way to write $h$ as a composition of two functions is to take the inner function to be $g(x) = x - 2$ and the outer function to be

$$f(x) = \frac{1}{x^2} = x^{-2}.$$

Then you can write

$$h(x) = \frac{1}{(x - 2)^2} = (x - 2)^{-2} = f(x - 2) = f(g(x)).$$

· **TECHNOLOGY** ·

In Example 5, the domain of the composite function is $-3 \le x \le 3$. To convince yourself of this, use a graphing utility to graph

$$y = \left(\sqrt{9 - x^2}\right)^2 - 9$$

as shown in the figure below. Notice that the graphing utility does not extend the graph to the left of $x = -3$ or to the right of $x = 3$.

## Applications

**Example 7** Political Makeup of the U.S. Senate

Consider three functions $R$, $D$, and $I$ that represent the numbers of Republicans, Democrats, and Independents in the U.S. Senate from 1963 to 1997. Sketch the graphs of $R$, $D$, and $I$, and the sum of $R$, $D$, and $I$, in the same coordinate plane. The numbers of Republicans and Democrats in the Senate are shown below.

| Year | Republicans | Democrats |
|------|-------------|-----------|
| 1963 | 33 | 67 |
| 1965 | 32 | 68 |
| 1967 | 36 | 64 |
| 1969 | 43 | 57 |
| 1971 | 44 | 54 |
| 1973 | 42 | 56 |
| 1975 | 37 | 60 |
| 1977 | 38 | 61 |
| 1979 | 41 | 58 |

| Year | Republicans | Democrats |
|------|-------------|-----------|
| 1981 | 53 | 46 |
| 1983 | 54 | 46 |
| 1985 | 53 | 47 |
| 1987 | 45 | 55 |
| 1989 | 45 | 55 |
| 1991 | 44 | 56 |
| 1993 | 43 | 57 |
| 1995 | 52 | 48 |
| 1997 | 54 | 46 |

### Solution

The graphs of $R$, $D$, and $I$ are shown in Figure 3.27. Note that the sum of $R$, $D$, and $I$ is the constant function $R + D + I = 100$. This follows from the fact that the number of senators in the United States is 100 (two from each state).

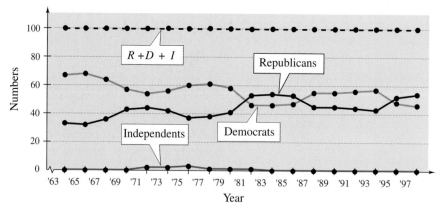

**Figure 3.27** *Number of U.S. Senators (by political party)*

Example 8    Bacteria Count

The number of bacteria in a refrigerated food is given by

$$N(T) = 20T^2 - 80T + 500, \qquad 2 \le T \le 14$$

where $T$ is the temperature of the food. When the food is removed from refrigeration, the temperature is given by

$$T(t) = 4t + 2, \qquad 0 \le t \le 3$$

where $t$ is the time in hours. Find (a) the composite $N(T(t))$, (b) the number of bacteria in the food when $t = 2$ hours, and (c) the time when the bacteria count reaches 2000.

Solution

**a.** $N(T(t)) = 20(4t + 2)^2 - 80(4t + 2) + 500$

$$= 20(16t^2 + 16t + 4) - 320t - 160 + 500$$

$$= 320t^2 + 320t + 80 - 320t - 160 + 500$$

$$= 320t^2 + 420$$

**b.** When $t = 2$, the number of bacteria is

$$N = 320(2)^2 + 420 = 1280 + 420 = 1700.$$

**c.** The bacteria count will reach $N = 2000$ when $320t^2 + 420 = 2000$. By solving this equation, you can determine that the bacteria count will reach 2000 when

$$t \approx 2.2 \text{ hours.}$$

---

| Discussing the Concept | Composition of Functions |
|---|---|

The suggested retail price of a new car is $p$ dollars. The dealer has advertised a factory rebate of \$1500 and a 7% discount.

**a.** Write a function $R$, in terms of $p$, giving the cost of the car after receiving the rebate.

**b.** Write a function $D$, in terms of $p$, giving the cost of the car after receiving the discount.

**c.** Form the composite functions $(R \circ D)(p)$ and $(D \circ R)(p)$. Explain what each composite function represents.

**d.** Find $(R \circ D)(21{,}000)$ and $(D \circ R)(21{,}000)$. Which function yields the lower price for the car? Explain.

**e.** Research the rebates and discounts offered by several car dealers in your area. How do they compare with one another?

In Exercises 1–10, perform the indicated operations and simplify the result.

1. $\dfrac{1}{x} + \dfrac{1}{1-x}$

2. $\dfrac{2}{x+3} - \dfrac{2}{x-3}$

3. $\dfrac{3}{x-2} - \dfrac{2}{x(x-2)}$

4. $\dfrac{x}{x-5} + \dfrac{1}{3}$

5. $(x-1)\left(\dfrac{1}{\sqrt{x^2-1}}\right)$

6. $\left(\dfrac{x}{x^2-4}\right)\left(\dfrac{x^2-x-2}{x^2}\right)$

7. $(x^2-4) \div \left(\dfrac{x+2}{5}\right)$

8. $\left(\dfrac{x}{x^2+3x-10}\right) \div \left(\dfrac{x^2+3x}{x^2+6x+5}\right)$

9. $\dfrac{(1/x)+5}{3-(1/x)}$

10. $\dfrac{(x/4)-(4/x)}{x-4}$

## 3.4 Exercises

In Exercises 1–4, use the graphs of $f$ and $g$ to graph $h(x) = (f+g)(x)$.

**1.**

**2.**

**3.**

**4.**

In Exercises 5–12, find (a) $(f+g)(x)$, (b) $(f-g)(x)$, (c) $(fg)(x)$, and (d) $(f/g)(x)$. What is the domain of $f/g$?

5. $f(x) = x + 1$, $\quad g(x) = x - 1$

6. $f(x) = 2x - 5$, $\quad g(x) = 1 - x$

7. $f(x) = x^2$, $\quad g(x) = 1 - x$

8. $f(x) = 2x - 5$, $\quad g(x) = 5$

9. $f(x) = x^2 + 5$, $\quad g(x) = \sqrt{1-x}$

10. $f(x) = \sqrt{x^2 - 4}$, $\quad g(x) = \dfrac{x^2}{x^2+1}$

11. $f(x) = \dfrac{1}{x}$, $\quad g(x) = \dfrac{1}{x^2}$

12. $f(x) = \dfrac{x}{x+1}$, $\quad g(x) = x^3$

In Exercises 13–24, evaluate the function for $f(x) = x^2 + 1$ and $g(x) = x - 4$.

13. $(f+g)(3)$

14. $(f-g)(-2)$

15. $(f-g)(2t)$

16. $(f+g)(t-1)$

17. $(fg)(4)$

18. $(fg)(-6)$

19. $\left(\dfrac{f}{g}\right)(5)$

20. $\left(\dfrac{f}{g}\right)(0)$

21. $(f-g)(0)$

22. $(f+g)(1)$

23. $\left(\dfrac{f}{g}\right)(-1) - g(3)$

24. $(2f)(5)$

In Exercises 25–28, find (a) $f \circ g$ and (b) $f \circ f$.

25. $f(x) = x^2$, $\quad g(x) = x - 1$

26. $f(x) = 3x$, $\quad g(x) = 2x + 1$

27. $f(x) = 3x + 5$, $\quad g(x) = 5 - x$

28. $f(x) = x^3$, $\quad g(x) = \dfrac{1}{x}$

In Exercises 29–36, find (a) $f \circ g$ and (b) $g \circ f$.

**29.** $f(x) = \sqrt{x + 4}, \qquad g(x) = x^2$

**30.** $f(x) = \sqrt[3]{x - 1}, \qquad g(x) = x^3 + 1$

**31.** $f(x) = \frac{1}{3}x - 3, \qquad g(x) = 3x + 1$

**32.** $f(x) = x^4, \qquad g(x) = x^4$

**33.** $f(x) = \sqrt{x}, \qquad g(x) = \sqrt{x}$

**34.** $f(x) = 2x - 3, \qquad g(x) = 2x - 3$

**35.** $f(x) = |x|, \qquad g(x) = x + 6$

**36.** $f(x) = x^{2/3}, \qquad g(x) = x^6$

In Exercises 37–40, use the graphs of $f$ and $g$ to evaluate the functions.

(a)

(b)

**37.** (a) $(f + g)(3)$      (b) $\left(\dfrac{f}{g}\right)(2)$

**38.** (a) $(f - g)(1)$      (b) $(fg)(4)$

**39.** (a) $(f \circ g)(2)$      (b) $(g \circ f)(2)$

**40.** (a) $(f \circ g)(1)$      (b) $(g \circ f)(3)$

In Exercises 41–44, determine the domain of (a) $f$, (b) $g$, and (c) $f \circ g$.

**41.** $f(x) = \sqrt{x}, \qquad g(x) = x^2 + 1$

**42.** $f(x) = \dfrac{1}{x}, \qquad g(x) = x + 3$

**43.** $f(x) = \dfrac{3}{x^2 - 1}, \qquad g(x) = x + 1$

**44.** $f(x) = 2x + 3, \qquad g(x) = \dfrac{x}{2}$

In Exercises 45–52, find two functions $f$ and $g$ such that $(f \circ g)(x) = h(x)$. (There are many correct answers.)

**45.** $h(x) = (2x + 1)^2$

**46.** $h(x) = (1 - x)^3$

**47.** $h(x) = \sqrt[3]{x^2 - 4}$

**48.** $h(x) = \sqrt{9 - x}$

**49.** $h(x) = \dfrac{1}{x + 2}$

**50.** $h(x) = \dfrac{4}{(5x + 2)^2}$

**51.** $h(x) = (x + 4)^2 + 2(x + 4)$

**52.** $h(x) = (x + 3)^{3/2}$

**53.** *Stopping Distance* While traveling in a car at $x$ miles per hour, you are required to stop quickly to avoid an accident. The distance the car travels during your reaction time is given by $R(x) = \frac{3}{4}x$. The distance traveled while you are braking is given by

$$B(x) = \frac{1}{15}x^2.$$

Find the function giving total stopping distance $T$. (*Hint:* $T = R + B$.) Graph the functions $R$, $B$, and $T$ on the same set of coordinate axes for $0 \le x \le 60$.

**54.** *Comparing Profits* A company has two manufacturing plants, one in New Jersey and the other in California. From 1995 to 2000, the profits for the manufacturing plant in New Jersey have been decreasing according to the function

$$P_1 = 16.97 - 0.43t, \qquad t = 5, 6, 7, 8, 9, 10$$

where $P_1$ represents the profit in millions of dollars and $t = 5$ represents 1995. On the other hand, the profits for the manufacturing plant in California have been increasing according to the function

$$P_2 = 13.36 + 0.56t, \qquad t = 5, 6, 7, 8, 9, 10.$$

Write a function that represents the overall company profits during the 6-year period. Use the *stacked bar graph* in the figure, which represents the total profit for the company during this 6-year period, to determine whether the overall company profits have been increasing or decreasing.

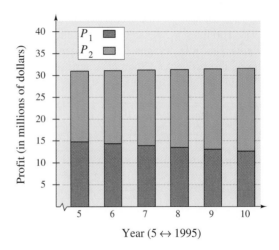

**55. *Comparing Sales*** You own two fast-food restaurants. During the years 1995 to 2000, the sales for the first restaurant have been decreasing according to the function

$$R_1 = 500 - 0.8t^2, \quad t = 5, 6, 7, 8, 9, 10$$

where $R_1$ represents the sales for the first restaurant (in thousands of dollars) and $t = 5$ represents 1995. During the same 6-year period, the sales for the second restaurant have been increasing according to the function

$$R_2 = 253.9 + 0.78t, \quad t = 5, 6, 7, 8, 9, 10.$$

Write a function that represents the total sales for the two restaurants. Use the *stacked bar graph* in the figure, which represents the total sales during this 6-year period, to determine whether the total sales have been increasing or decreasing.

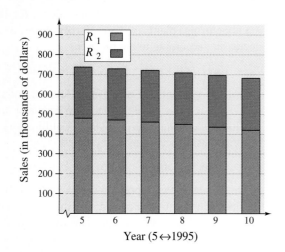

Year (5 ↔ 1995)

**56. *Operating Costs*** Use the table, which gives the variable costs for operating an automobile in the United States for the years 1991 through 1996. The variables $y_1, y_2,$ and $y_3$ represent the costs in cents per mile for gas and oil, maintenance, and tires, respectively. (Source: American Automobile Manufacturers Association)

| Year | 1991 | 1992 | 1993 | 1994 | 1995 | 1996 |
|------|------|------|------|------|------|------|
| $y_1$ | 6.70 | 6.00 | 6.00 | 5.60 | 6.00 | 5.90 |
| $y_2$ | 2.20 | 2.20 | 2.40 | 2.50 | 2.60 | 2.80 |
| $y_3$ | 0.90 | 0.90 | 0.90 | 1.10 | 1.40 | 1.40 |

(a) Create a stacked bar graph for the data.

(b) Mathematical models for the data are given by

$$y_1 = 7.80 - 1.41t + 0.32t^2 - 0.02t^3$$

$$y_2 = 2.02 + 0.12t$$

$$y_3 = 0.68 + 0.12t$$

where $t = 1$ represents 1991. Use a graphing utility to graph $y_1, y_2, y_3,$ and $y_1 + y_2 + y_3$ in the same viewing window. Use the model to estimate the total variable cost per mile in the year 2000.

**57. *Bacteria Count*** The number of bacteria in a refrigerated food product is given by

$$N(T) = 10T^2 - 20T + 600, \quad 1 \le T \le 20$$

where $T$ is the temperature of the food. When the food is removed from the refrigerator, the temperature is given by

$$T(t) = 3t + 1$$

where $t$ is the time in hours. Find (a) the composite $N(T(t))$ and (b) the time when the bacteria count reaches 1500.

**58. *Bacteria Count*** The number of bacteria in a refrigerated food product is given by

$$N(T) = 25T^2 - 50T + 300, \quad 2 \le T \le 20$$

where $T$ is the temperature of the food. When the food is removed from the refrigerator, the temperature is given by

$$T(t) = 2t + 1$$

where $t$ is the time in hours. Find (a) the composite $N(T(t))$ and (b) the time when the bacteria count reaches 750.

*Price-Earnings Ratio* In Exercises 59 and 60, the average annual price-earnings ratio for a corporation's stock is defined as the average price of the stock divided by the earnings per share. The average price of a corporation's stock is given as the function $P$ and the earnings per share is given as the function $E$. Find the price-earnings ratio, $P/E$, for the years from 1989 to 1997.

**59.** *McDonald's Corporation*

| Year | P | E |
|------|------|------|
| 1989 | $14.41 | $0.98 |
| 1990 | $15.51 | $1.10 |
| 1991 | $16.76 | $1.18 |
| 1992 | $22.10 | $1.30 |
| 1993 | $26.13 | $1.46 |
| 1994 | $28.73 | $1.68 |
| 1995 | $37.43 | $1.97 |
| 1996 | $47.52 | $2.21 |
| 1997 | $48.09 | $2.29 |

**60.** *Maytag Corporation*

| Year | P | E |
|------|------|------|
| 1989 | $21.34 | $1.27 |
| 1990 | $15.51 | $0.94 |
| 1991 | $14.01 | $0.87 |
| 1992 | $16.31 | $0.62 |
| 1993 | $15.41 | $0.48 |
| 1994 | $17.53 | $1.38 |
| 1995 | $17.01 | $1.35 |
| 1996 | $20.32 | $1.60 |
| 1997 | $26.86 | $1.84 |

**61.** *Troubled Waters*   A pebble is dropped into a calm pond, causing ripples in the form of concentric circles (see figure). The radius (in feet) of the outer ripple is given by $r(t) = 0.6t$, where $t$ is time in seconds after the pebble strikes the water. The area of the circle is given by the function $A(r) = \pi r^2$. Find and interpret $(A \circ r)(t)$.

**62.** *Cost*   The weekly cost of producing $x$ units in a manufacturing process is given by the function

$$C(x) = 60x + 750.$$

The number of units produced in $t$ hours is given by $x(t) = 50t$. Find and interpret $(C \circ x)(t)$.

**63.** *Cost*   The weekly cost of producing $x$ units in a manufacturing process is given by the function

$$C(x) = 70x + 375.$$

The number of units produced in $t$ hours is given by $x(t) = 40t$. Find and interpret $(C \circ x)(t)$.

**64.** Find the domain of $(f/g)(x)$ and $(g/f)(x)$ for the functions

$$f(x) = \sqrt{x} \quad \text{and} \quad g(x) = \sqrt{9 - x^2}.$$

Why do the two domains differ?

## 3.5    Inverse Functions

### Objectives

- Determine if a function has an inverse.
- Find the inverse of a function.
- Graph a function and its inverse.

## The Inverse of a Function

Recall from Section 3.1 that a function can be represented by a set of ordered pairs. For instance, the function $f(x) = x + 4$ from the set $A = \{1, 2, 3, 4\}$ to the set $B = \{5, 6, 7, 8\}$ can be written as follows.

$$f(x) = x + 4: \ \{(1, 5), (2, 6), (3, 7), (4, 8)\}$$

By interchanging the first and second coordinates of each of these ordered pairs, you can form the **inverse function** of $f$, which is denoted by $f^{-1}$. It is a function from the set $B$ to the set $A$, and can be written as follows.

$$f^{-1}(x) = x - 4: \ \{(5, 1), (6,2), (7, 3), (8, 4)\}$$

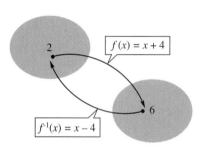

Figure 3.28

Note that the domain of $f$ is equal to the range of $f^{-1}$, and vice versa, as shown in Figure 3.28. Also note that the functions $f$ and $f^{-1}$ have the effect of "undoing" each other. In other words, when you form the composition of $f$ with $f^{-1}$ or the composition of $f^{-1}$ with $f$, you obtain identity functions, as follows.

$$f(f^{-1}(x)) = f(x - 4) = (x - 4) + 4 = x$$
$$f^{-1}(f(x)) = f^{-1}(x + 4) = (x + 4) - 4 = x$$

### Example 1    Finding Inverse Functions Informally

Find the inverse of $f(x) = 4x$. Then verify that both $f(f^{-1}(x))$ and $f^{-1}(f(x))$ are equal to the identity function.

#### Solution

The given function *multiplies* each input by 4. To "undo" this function, you need to *divide* each input by 4. Thus, the inverse function of $f(x) = 4x$ is

$$f^{-1}(x) = \frac{x}{4}.$$

You can verify that both $f(f^{-1}(x))$ and $f^{-1}(f(x))$ are equal to the identity function as follows.

$$f(f^{-1}(x)) = f\left(\frac{x}{4}\right) = 4\left(\frac{x}{4}\right) = x$$

$$f^{-1}(f(x)) = f^{-1}(4x) = \frac{4x}{4} = x$$

Example 2    Finding Inverse Functions Informally

Find the inverse of $f(x) = x - 6$. Then verify that both $f(f^{-1}(x))$ and $f^{-1}(f(x))$ are equal to the identity function.

### Solution

The given function *subtracts* 6 from each input. To "undo" this function, you need to *add* 6 to each input. Thus, the inverse function of $f(x) = x - 6$ is

$$f^{-1}(x) = x + 6.$$

You can verify that both $f(f^{-1}(x))$ and $f^{-1}(f(x))$ are equal to the identity function as follows.

$$f(f^{-1}(x)) = f(x + 6) = (x + 6) - 6 = x$$
$$f^{-1}(f(x)) = f^{-1}(x - 6) = (x - 6) + 6 = x$$

---

The formal definition of the inverse of a function is as follows.

---

▶ **Definition of the Inverse of a Function**

Let $f$ and $g$ be two functions such that

$$f(g(x)) = x \qquad \text{for every } x \text{ in the domain of } g$$

and

$$g(f(x)) = x \qquad \text{for every } x \text{ in the domain of } f.$$

Under these conditions, the function $g$ is the **inverse** of the function $f$. The function $g$ is denoted by $f^{-1}$ (read "$f$-inverse"). Thus,

$$f(f^{-1}(x)) = x \qquad \text{and} \qquad f^{-1}(f(x)) = x.$$

The domain of $f$ must be equal to the range of $f^{-1}$, and the range of $f$ must be equal to the domain of $f^{-1}$.

---

Don't be confused by the use of $-1$ to denote the inverse function $f^{-1}$. In this text, whenever we write $f^{-1}$, we will *always* be referring to the inverse of the function $f$ and *not* to the reciprocal of $f(x)$. That is,

$$f^{-1}(x) \neq \frac{1}{f(x)}.$$

If the function $g$ is the inverse of the function $f$, it must also be true that the function $f$ is the inverse of the function $g$. For this reason, you can say that the functions $f$ and $g$ are *inverses of each other*.

### Example 3 Verifying Inverse Functions

Show that the following functions are inverses of each other.

$$f(x) = 2x^3 - 1 \quad \text{and} \quad g(x) = \sqrt[3]{\frac{x+1}{2}}$$

**Solution**

$$f(g(x)) = f\left(\sqrt[3]{\frac{x+1}{2}}\right) = 2\left(\sqrt[3]{\frac{x+1}{2}}\right)^3 - 1$$

$$= 2\left(\frac{x+1}{2}\right) - 1$$

$$= x + 1 - 1$$

$$= x$$

$$g(f(x)) = g(2x^3 - 1) = \sqrt[3]{\frac{(2x^3 - 1) + 1}{2}}$$

$$= \sqrt[3]{\frac{2x^3}{2}}$$

$$= \sqrt[3]{x^3}$$

$$= x$$

**·DISCOVERY·**

Graph the equations from Example 3 and the equation $y = x$ on your graphing utility using a square viewing window.

$$y_1 = 2x^3 - 1$$

$$y_2 = \sqrt[3]{\frac{x+1}{2}}$$

$$y_3 = x$$

What do you observe about the graphs of $y_1$ and $y_2$?

### Example 4 Verifying Inverse Functions

Which of the functions

$$g(x) = \frac{x-2}{5} \quad \text{and} \quad h(x) = \frac{5}{x} + 2$$

is the inverse of the function $f(x) = \dfrac{5}{x-2}$?

**Solution**

By forming the composition of $f$ with $g$, you can see that

$$f(g(x)) = f\left(\frac{x-2}{5}\right) = \frac{5}{[(x-2)/5] - 2} = \frac{25}{x - 12} \neq x.$$

Because this composition is not equal to the identity function $x$, it follows that $g$ *is not* the inverse of $f$. By forming the composition of $f$ with $h$, you have

$$f(h(x)) = f\left(\frac{5}{x} + 2\right) = \frac{5}{(5/x) + 2 - 2} = \frac{5}{5/x} = x.$$

So, it appears that $h$ is the inverse of $f$. You can confirm this by showing that the composition of $h$ with $f$ is also equal to the identity function. (Try doing this.)

## Finding the Inverse of a Function

For simple functions (such as the ones in Examples 1 and 2), you can find inverse functions by inspection. For more complicated functions, however, it is best to use the following guidelines. The key step in these guidelines is switching the roles of $x$ and $y$. This step corresponds to the fact that inverse functions have ordered pairs with the coordinates reversed.

### Study Tip

Note in Step 3 of the guidelines for finding the inverse of a function that it is possible that a function has no inverse. For instance, the function $f(x) = x^2$ has no inverse function.

▶ **Finding the Inverse of a Function**

1. In the equation for $f(x)$, replace $f(x)$ by $y$.
2. Interchange the roles of $x$ and $y$.
3. If the new equation does not represent $y$ as a function of $x$, the function $f$ does not have an inverse function. If the new equation does represent $y$ as a function of $x$, solve the new equation for $y$.
4. Replace $y$ by $f^{-1}(x)$.
5. Verify that $f$ and $f^{-1}$ are inverses of each other by showing that the domain of $f$ is equal to the range of $f^{-1}$, the range of $f$ is equal to the domain of $f^{-1}$, and $f(f^{-1}(x)) = x = f^{-1}(f(x))$.

**Example 5**    Finding the Inverse of a Function

Find the inverse of $f(x) = \dfrac{5 - 3x}{2}$.

**Solution**

$$f(x) = \frac{5 - 3x}{2} \qquad \text{Given function}$$

$$y = \frac{5 - 3x}{2} \qquad \text{Replace } f(x) \text{ by } y.$$

$$x = \frac{5 - 3y}{2} \qquad \text{Interchange } x \text{ and } y.$$

$$2x = 5 - 3y \qquad \text{Multiply both sides by 2.}$$

$$3y = 5 - 2x \qquad \text{Isolate the } y\text{-term.}$$

$$y = \frac{5 - 2x}{3} \qquad \text{Solve for } y.$$

$$f^{-1}(x) = \frac{5 - 2x}{3} \qquad \text{Replace } y \text{ by } f^{-1}(x).$$

Note that both $f$ and $f^{-1}$ have domains and ranges that consist of the entire set of real numbers. Check that $f(f^{-1}(x)) = x$ and $f^{-1}(f(x)) = x$.

# The Graph of the Inverse of a Function

The graphs of a function $f$ and its inverse $f^{-1}$ are related to each other in the following way. If the point $(a, b)$ lies on the graph of $f$, then the point $(b, a)$ must lie on the graph of $f^{-1}$, and vice versa. This means that the graph of $f^{-1}$ is a *reflection* of the graph of $f$ in the line $y = x$, as shown in Figure 3.29.

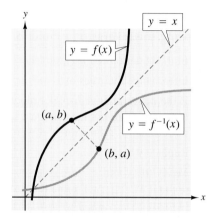

Figure 3.29  *The graph of $f^{-1}$ is a reflection of the graph of $f$ in the line $y = x$.*

**Example 6**  The Graphs of $f$ and $f^{-1}$

Sketch the graphs of the inverse functions $f(x) = 2x - 3$ and $f^{-1}(x) = \frac{1}{2}(x + 3)$ on the same rectangular coordinate system and show that the graphs are reflections of each other in the line $y = x$.

### Solution

The graphs of $f$ and $f^{-1}$ are shown in Figure 3.30. Visually, it appears that the graphs are reflections of each other in the line $y = x$. You can further verify this reflective property by testing a few points on each graph. Note in the following list that if the point $(a, b)$ is on the graph of $f$, then the point $(b, a)$ is on the graph of $f^{-1}$.

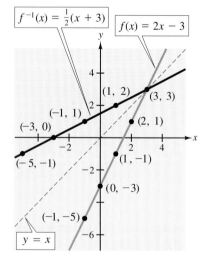

Figure 3.30

| $f(x) = 2x - 3$ | $f^{-1}(x) = \frac{1}{2}(x + 3)$ |
|:---:|:---:|
| $(-1, -5)$ | $(-5, -1)$ |
| $(0, -3)$ | $(-3, 0)$ |
| $(1, -1)$ | $(-1, 1)$ |
| $(2, 1)$ | $(1, 2)$ |
| $(3, 3)$ | $(3, 3)$ |

In the study tip on page 199, we mentioned that the function

$$f(x) = x^2$$

has no inverse. What we really meant is that *assuming the domain of f is the entire real line*, the function $f(x) = x^2$ has no inverse. If, however, we restrict the domain of $f$ to the nonnegative real numbers, then $f$ does have an inverse, as demonstrated in Example 7.

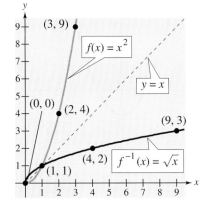

Figure 3.31

---

**Example 7**   The Graphs of *f* and *f*⁻¹

Sketch the graphs of the inverse functions

$$f(x) = x^2, \quad x \geq 0, \quad \text{and} \quad f^{-1}(x) = \sqrt{x}$$

on the same rectangular coordinate system and show that the graphs are reflections of each other in the line $y = x$.

**Solution**

The graphs of $f$ and $f^{-1}$ are shown in Figure 3.31. Visually, it appears that the graphs are reflections of each other in the line $y = x$. You can further verify this reflective property by testing a few points on each graph. Note in the following list that if the point $(a, b)$ is on the graph of $f$, then the point $(b, a)$ is on the graph of $f^{-1}$.

| $f(x) = x^2, \quad x \geq 0$ | $f^{-1}(x) = \sqrt{x}$ |
|:---:|:---:|
| $(0, 0)$ | $(0, 0)$ |
| $(1, 1)$ | $(1, 1)$ |
| $(2, 4)$ | $(4, 2)$ |
| $(3, 9)$ | $(9, 3)$ |

Try showing that $f(f^{-1}(x)) = x$ and $f^{-1}(f(x)) = x$.

---

The guidelines for finding the inverse of a function include an *algebraic* test for determining whether a function has an inverse. The reflective property of the graphs of inverse functions gives you a nice *geometric* test for determining whether a function has an inverse. This test is called the **Horizontal Line Test** for inverse functions.

▶ **Horizontal Line Test for Inverse Functions**

A function $f$ has an inverse function if and only if no *horizontal* line intersects the graph of $f$ at more than one point.

| Example 8 | Applying the Horizontal Line Test |

**a.** The graph of the function $f(x) = x^3 - 1$ is shown in Figure 3.32(a). Because no horizontal line intersects the graph of $f$ at more than one point, you can conclude that $f$ *does* have an inverse function.

**b.** The graph of the function $f(x) = x^2 - 1$ is shown in Figure 3.32(b). Because it is possible to find a horizontal line that intersects the graph of $f$ at more than one point, you can conclude that $f$ *does not* have an inverse function.

(a)

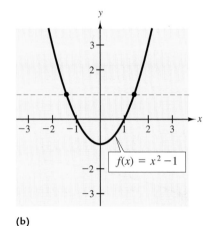

(b)

Figure 3.32

## Discussing the Concept    Visualizing Inverses

Sketch the graphs of $f(x) = x^2$ and $g(x) = x$ on graph paper. Fold the graph paper on the line $g(x) = x$ and sketch the reflection of $f(x) = x^2$ about $g(x) = x$. What do you notice about the resulting graph?

Does $f(x)$ pass the Vertical Line Test for functions? That is, is $f(x)$ a function? Does $f(x)$ pass the Horizontal Line Test for inverse functions? That is, is $f^{-1}(x)$ a function? Does $f^{-1}(x)$ pass the Vertical Line Test for functions?

Write a short paragraph discussing what information can be derived from the Vertical Line Test for functions and the Horizontal Line Test for inverse functions.

# Warm Up

The following warm-up exercises involve skills that were covered in earlier sections. You will use these skills in the exercise set for this section.

In Exercises 1–4, find the domain of the function.

**1.** $f(x) = \sqrt[3]{x + 1}$

**2.** $f(x) = \sqrt{x + 1}$

**3.** $g(x) = \dfrac{2}{x^2 - 2x}$

**4.** $h(x) = \dfrac{x}{3x + 5}$

In Exercises 5–8, simplify the expression.

**5.** $2\left(\dfrac{x + 5}{2}\right) - 5$

**6.** $7 - 10\left(\dfrac{7 - x}{10}\right)$

**7.** $\sqrt[3]{2\left(\dfrac{x^3}{2} - 2\right) + 4}$

**8.** $\sqrt[5]{(x + 2)^5} - 2$

In Exercises 9 and 10, solve for $x$ in terms of $y$.

**9.** $y = \dfrac{2x - 6}{3}$

**10.** $y = \sqrt[3]{2x - 4}$

## 3.5    Exercises

In Exercises 1–10, show that $f$ and $g$ are inverses algebraically *and* using a graphing utility.

**1.** $f(x) = 2x$,    $g(x) = \dfrac{x}{2}$

**2.** $f(x) = x - 5$,    $g(x) = x + 5$

**3.** $f(x) = 5x + 1$,    $g(x) = \dfrac{x - 1}{5}$

**4.** $f(x) = 3 - 4x$,    $g(x) = \dfrac{3 - x}{4}$

**5.** $f(x) = x^3$,    $g(x) = \sqrt[3]{x}$

**6.** $f(x) = \dfrac{1}{x}$,    $g(x) = \dfrac{1}{x}$

**7.** $f(x) = \sqrt{x - 4}$,    $g(x) = x^2 + 4$,    $x \geq 0$

**8.** $f(x) = 9 - x^2$,    $x \geq 0$,
$g(x) = \sqrt{9 - x}$,    $x \leq 9$

**9.** $f(x) = 1 - x^3$,    $g(x) = \sqrt[3]{1 - x}$

**10.** $f(x) = \dfrac{1}{1 + x}$,    $x \geq 0$

$g(x) = \dfrac{1 - x}{x}$,    $0 < x \leq 1$

In Exercises 11–20, use the graph of $y = f(x)$, $y = g(x)$, or $y = h(x)$ to determine whether the function has an inverse.

**11.**

**12.**

**13.**

**14.**
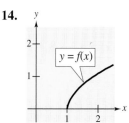

**15.** $g(x) = \dfrac{4 - x}{6}$

**16.** $f(x) = 10$

**17.** $h(x) = |x + 4|$

**18.** $g(x) = (x + 5)^3$

**19.** $f(x) = -\sqrt{16 - x^2}$

**20.** $f(x) = (x + 2)^2$

*Error Analysis* In Exercises 21 and 22, a student has handed in the answer to a problem on a quiz. Find the error(s) in each solution and discuss how to explain each error to your student.

**21.** Find the inverse $f^{-1}$ of $y = \sqrt{2x - 5}$.

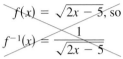

$$f(x) = \sqrt{2x - 5}, \text{ so}$$

$$f^{-1}(x) = \frac{1}{\sqrt{2x - 5}}$$

**22.** Find the inverse $f^{-1}$ of $y = \frac{3}{5}x + \frac{1}{3}$.

$$f(x) = \frac{3}{5}x + \frac{1}{3}, \text{ so}$$

$$f^{-1}(x) = \frac{5}{3}x - 3$$

In Exercises 23–32, find the inverse of the function $f$. Then, using a graphing utility, graph both $f$ and $f^{-1}$ on the same screen.

**23.** $f(x) = 2x - 3$

**24.** $f(x) = 3x$

**25.** $f(x) = x^5$

**26.** $f(x) = x^3 + 1$

**27.** $f(x) = \sqrt{x}$

**28.** $f(x) = x^2, \ x \ge 0$

**29.** $f(x) = \sqrt{4 - x^2}, 0 \le x \le 2$

**30.** $f(x) = \dfrac{4}{x}$

**31.** $f(x) = \sqrt[3]{x - 1}$

**32.** $f(x) = x^{3/5}$

In Exercises 33–48, determine whether the function has an inverse. If it does, find its inverse.

**33.** $f(x) = x^4$

**34.** $f(x) = \dfrac{1}{x^2}$

**35.** $g(x) = \dfrac{x}{8}$

**36.** $f(x) = 3x + 5$

**37.** $p(x) = -4$

**38.** $f(x) = \dfrac{3x + 4}{5}$

**39.** $f(x) = (x + 3)^2, \ x \ge -3$

**40.** $q(x) = (x - 5)^2$

**41.** $h(x) = \dfrac{1}{x}$

**42.** $f(x) = |x - 2|, \ x \le 2$

**43.** $f(x) = \sqrt{2x + 3}$

**44.** $f(x) = \sqrt{x - 2}$

**45.** $g(x) = x^2 - x^4$

**46.** $f(x) = \dfrac{x^2}{x^2 + 1}$

**47.** $f(x) = 25 - x^2, \ x \le 0$

**48.** $f(x) = 36 + x^2, \ x \le 0$

In Exercises 49 and 50, use the graph of $f$ to complete the table and to sketch the graph of $f^{-1}$.

**49.**

| $x$ | 0 | 1 | 2 | 3 | 4 |
|------|---|---|---|---|---|
| $f^{-1}(x)$ | | | | | |

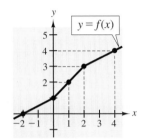

**50.**

| $x$ | 0 | 2 | 4 | 6 |
|------|---|---|---|---|
| $f^{-1}(x)$ | | | | |

In Exercises 51–54, use the functions

$$f(x) = \frac{1}{8}x - 3 \text{ and } g(x) = x^3$$

to find the value.

**51.** $(f^{-1} \circ g^{-1})(1)$

**52.** $(g^{-1} \circ f^{-1})(-3)$

**53.** $(f^{-1} \circ f^{-1})(6)$

**54.** $(g^{-1} \circ g^{-1})(-4)$

In Exercises 55–58, use the functions

$$f(x) = x + 4 \text{ and } g(x) = 2x - 5$$

to find the composition of functions.

**55.** $g^{-1} \circ f^{-1}$

**56.** $f^{-1} \circ g^{-1}$

**57.** $(f \circ g)^{-1}$

**58.** $(g \circ f)^{-1}$

**59.** *Cellular Phones* From 1987 to 1996, the average local phone bill for cellular phones decreased. A function that approximates the average phone bill per month is

$$y = 141.919 - 6.05503t, \qquad 7 \le t \le 16$$

where $y$ represents the amount of the bill (in dollars) and $t = 7$ represents 1987 (see figure). Find the inverse of this function, and use $f^{-1}$ to find the year with the average monthly bill of \$51.00. (Source: Cellular Telecommunications Industry Association)

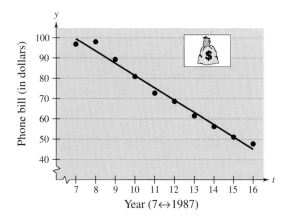

Year (7↔1987)

**60.** *Reasoning* You are teaching another student how to find the inverse function of a one-to-one function. The student you are helping states that interchanging the roles of $x$ and $y$ is "cheating." Discuss how you would use the graphs of $f(x) = x^2$, $x \ge 0$ and $f^{-1}(x) = \sqrt{x}$ to justify that particular step in the process of finding an inverse function.

**61.** *Automated Teller Machines (ATMs)* From 1985 to 1996, the annual number of ATM transactions in the United States increased according to the function

$$y = 2594 + 31.036t^2, \qquad 5 \le t \le 16$$

where $y$ is the number of transactions (in millions) and $t = 5$ represents 1985. Solve the equation for $t$ in terms of $y$ and use the result to find the year in which there were 7,000,000,000 transactions. (Source: *Bank Network News*)

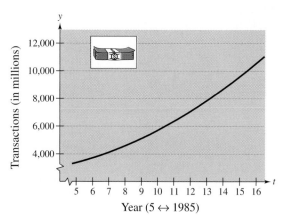

Year (5 ↔ 1985)

**Figure for 61**

**62.** *Median Family Income* The median family income in the United States increased from 1970 to 1995 according to the function

$$y = (94 + 3.77t)^2, \qquad 0 \le t \le 25$$

where $y$ is the median income in dollars and $t = 0$ represents 1970. Solve the equation for $t$ in terms of $y$ and use the result to find the year in which the median family income was \$29,553. (Source: U.S. Bureau of the Census)

**63.** *Earnings-Dividend Ratio* From 1990 to 1997, the earnings per share for the Coca-Cola Corporation were approximately related to the dividends per share by the function

$$f(x) = \sqrt{-0.086 + 0.239x}, \qquad 0.51 \le x \le 1.64$$

where $f(x)$ represents the dividends per share (in dollars) and $x$ represents the earnings per share (in dollars). In 1996 Coca-Cola paid dividends of \$0.50 per share. What were the earnings per share in 1996? (Source: Coca-Cola)

## Chapter Project

## World Paper Production and Consumption

With the advent of the computer age, experts in the late 1970s predicted that less paper would be used. It was expected that many office documents and messages would be sent electronically (via electronic mail and the Internet) and that future offices would become "paperless." The table at the left gives the amount of paper and paperboard (in millions of metric tons) produced, $P$, and consumed, $C$, in North and Central America for the given years. Use the data in the table to investigate the following questions.

| Year | $P$ | $C$ |
|------|------|------|
| 1970 | 58.1 | 55.0 |
| 1980 | 72.4 | 68.1 |
| 1990 | 91.5 | 87.9 |
| 1994 | 102.1 | 95.9 |

1. *Fitting a Model to Data*  Using a graphing utility, enter the data for the amount of paper and paperboard produced. Let $t = 0$ represent 1970. Use the quadratic fit feature to find an equation that models the production of paper and paperboard in North and Central America.

2. *Comparing Actual Data with a Model Numerically*  Create a table comparing the actual production of paper and paperboard with the values found using the model in Question 1. How does the model compare with the actual data?

3. *Comparing Actual Data with a Model Graphically*  Graph a scatter plot of the actual data and its model equation for production in the same viewing window. How well does the model fit the actual data?

4. *Fitting a Model to Data*  Using a graphing utility, enter the data for the amount of paper and paperboard consumed. Let $t = 0$ represent 1970. Use the quadratic fit feature to find an equation that models the consumption of paper and paperboard in North and Central America.

5. *Comparing Actual Data with a Model Numerically*  Create a table comparing the actual consumption of paper and paperboard with the values found using the model in Question 4. How does the model compare with the actual data?

6. *Comparing Actual Data with a Model Graphically*  Graph a scatter plot of the actual data and its model equation for consumption in the same viewing window. How well does the model fit the actual data?

7. *Discussion*  Graph the model for paper and paperboard production and the model for paper and paperboard consumption in the same viewing window.

   (a) What observations can you make about the two graphs?

   (b) What do the two graphs indicate is happening with the production and consumption of paper and paperboard?

   (c) Despite the use of electronic mail and the Internet, the production and consumption of paper and paperboard are increasing. Discuss in a paragraph why the two may continue to increase. Would it be possible for both to continue to increase? What might cause one or both to decrease?

## CHAPTER SUMMARY

After studying this chapter, you should have acquired the following skills.
These skills are keyed to the Review Exercises that begin on page 208.
Answers to odd-numbered Review Exercises are given in the back of the book.

| | | |
|---|---|---|
| **3.1** | · Determine if an equation or a set of ordered pairs represents a function. | Review Exercises 1–6 |
| | · Use function notation. | Review Exercises 7–12 |
| | · Evaluate a function. | Review Exercises 7–10 |
| | · Find the domain of a function. | Review Exercises 13–18 |
| | · Write a function that relates quantities in an application problem. | Review Exercises 19–24 |
| **3.2** | · Find the domain and range using the graph of a function. | Review Exercises 25–28 |
| | · Describe the increasing and decreasing behavior of a function. | Review Exercises 25–28, 40 |
| | · Classify a function as even or odd. | Review Exercises 25–28 |
| | · Identify six common graphs and use them to sketch the graph of a function. | Review Exercises 29–39 |
| | · Identify the graph of a function using the Vertical Line Test. | Review Exercises 41–46 |
| **3.3** | · Sketch the graph of a function using transformations and common graphs. | Review Exercises 47–58 |
| | · Write the equation of a function using transformations and common graphs. | Review Exercises 59–62 |
| **3.4** | · Find the sum, difference, product, and quotient of functions. | Review Exercises 63–70 |
| | · Form the composition of two functions and determine the domain. | Review Exercises 71–76 |
| | · Identify a function as the composition of two functions. | Review Exercises 77–80 |
| | · Use combinations and compositions of functions to solve an application problem. | Review Exercises 81–86 |
| **3.5** | · Determine if a function has an inverse. | Review Exercises 87–94 |
| | · Find the inverse of a function. | Review Exercises 89–94, 102 |
| | · Graph a function and its inverse. | Review Exercises 89–100, 101 |

## REVIEW EXERCISES

In Exercises 1–4, decide whether the equation represents $y$ as a function of $x$.

**1.** $3x - 4y = 12$        **2.** $y^2 = x^2 - 9$

**3.** $y = \sqrt{x + 3}$       **4.** $x^2 + y^2 - 6x + 8y = 0$

In Exercises 5 and 6, decide whether the set represents a function from $A$ to $B$.

$A = \{1, 2, 3\}$     $B = \{-3, -4, -7\}$

Give reasons for your answer.

**5.** $\{(1, -3), (2, -7), (3, -3)\}$

**6.** $\{(1, -4), (2, -3), (3, -9)\}$

In Exercises 7–10, evaluate the function and simplify the results.

**7.** $f(x) = 3x - 5$

    (a) $f(1)$   (b) $f(-2)$   (c) $f(m)$   (d) $f(x + 1)$

**8.** $f(x) = \sqrt{x + 9} - 3$

    (a) $f(7)$   (b) $f(0)$   (c) $f(-5)$   (d) $f(x + 2)$

**9.** $f(x) = |x| + 5$

    (a) $f(0)$   (b) $f(-3)$   (c) $f\left(\frac{1}{2}\right)$   (d) $f(-x^2)$

**10.** $f(x) = \begin{cases} x^2 + 1, & x \le 2 \\ x + 3, & x > 2 \end{cases}$

    (a) $f(0)$   (b) $f(2)$   (c) $f(3)$   (d) $f(-5)$

In Exercises 11 and 12, find all real values of $x$ such that $f(x) = 0$.

**11.** $f(x) = \dfrac{2x + 7}{3}$       **12.** $f(x) = x^3 - 4x$

In Exercises 13–18, find the domain of the function.

**13.** $f(x) = 3x^2 + 8x + 4$

**14.** $g(t) = \dfrac{5}{t^2 - 9}$

**15.** $h(x) = \sqrt{x + 9}$

**16.** $f(t) = \sqrt[3]{t - 5}$

**17.** $g(t) = \dfrac{\sqrt{t - 3}}{t - 5}$

**18.** $h(x) = \sqrt[4]{9 - x^2}$

**19.** *Reasoning*   A student has difficulty understanding why the domains of

$$h(x) = \frac{x^2 - 4}{x} \quad \text{and} \quad k(x) = \frac{x}{x^2 - 4}$$

are different. How would you explain their respective domains algebraically? How could you use a graphing utility to explain their domains?

**20.** *Reasoning*   A student has difficulty understanding why the domains of

$$h(x) = \sqrt{x + 1} \quad \text{and} \quad k(x) = \sqrt[3]{x + 1}$$

are different. How would you explain their respective domains algebraically? How could you use a graphing utility to explain their domains?

**21.** *Volume of a Box*   An open box is to be made from a square piece of material 20 inches on a side by cutting equal squares from the corners and turning up the sides (see figure).

    (a) Write the volume of the box as a function of $x$.

    (b) What is the domain of this function?

    (c) Use a graphing utility to sketch the graph of this function.

**22.** *Balance in an Account*   A person deposits $5000 in an account that pays 6.25% interest compounded quarterly.

    (a) Write the balance of the account in terms of the time $t$ that the principal is left in the account.

    (b) What is the domain of this function?

**23.** *Vertical Motion*   The velocity of a ball thrown vertically upward from ground level is given by $v(t) = -32t + 48$, where $t$ is the time in seconds and $v$ is the velocity in feet per second.

    (a) Find the velocity when $t = 1$.

    (b) Find the time when the ball reaches its maximum height. [*Hint:* Find the time when $v(t) = 0$.]

    (c) Find the velocity when $t = 2$.

**24.** *Cost and Profit*   A company produces a product for which the variable cost is $5.35 per unit and the fixed costs are $16,000. The company sells the product for $8.20 and can sell all that it produces.

(a) Find the total cost as a function of $x$, the number of units produced.

(b) Find the profit as a function of $x$.

In Exercises 25–28, (a) determine the range and domain of the function, (b) determine the intervals over which the function is increasing, decreasing, or constant, and (c) determine if the function is even, odd, or neither.

**25.** $f(x) = x^2 + 1$       **26.** $f(x) = \sqrt{x^2 - 9}$

**27.** $f(x) = x^3 - 4x^2$     **28.** $f(x) = |x - 2|$

      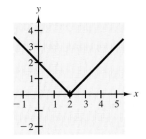

In Exercises 29–38, sketch the graph of the function.

**29.** $f(x) = 3 - 2x$           **30.** $f(x) = |x + 3|$

**31.** $g(x) = \sqrt{x^2 - 16}$        **32.** $h(x) = [\![x]\!]$

**33.** $h(x) = 2[\![x]\!] + 1$          **34.** $f(x) = 3$

**35.** $g(x) = \begin{cases} x + 2, & x < 0 \\ 2, & x = 0 \\ x^2 + 2, & x > 0 \end{cases}$

**36.** $g(x) = \begin{cases} 3x + 1, & x < -1 \\ x^2 - 3, & x \ge -1 \end{cases}$

**37.** $h(x) = x^2 - 3x$       **38.** $f(x) = \sqrt{9 - x^2}$

**39.** *Cost of a Phone Call*   The cost of a phone call between Dallas and Philadelphia is $0.70 for the first minute and $0.38 for each additional minute (or portion of a minute). A model for the total cost of the phone call is

$$C = 0.70 + 0.38[\![x]\!], \qquad 0 < x$$

where $C$ is the total cost of the call in dollars and $x$ is the length of the call in minutes. Sketch the graph of this function.

**40.** *Revenue*   A company determines that the total revenue for the years 1988 to 1999 can be approximated by the function

$$R = -0.035t^3 + 0.85t^2 - 1.2t + 4.75, \\ 8 \le t \le 19$$

where $R$ is the total revenue in hundreds of thousands of dollars and $t = 8$ represents 1988. Graph the revenue function using a graphing utility and use the trace feature to estimate the years in which the revenue was increasing and the years in which the revenue was decreasing.

In Exercises 41–46, use the Vertical Line Test to determine if $y$ is a function of $x$.

**41.** $y = \frac{1}{2}x^2$                 **42.** $y = \frac{1}{4}x^3$

**43.** $x - y^2 = 1$              **44.** $x^2 + y^2 = 25$

      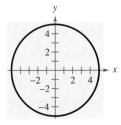

**45.** $x^2 = 2xy - 1$            **46.** $x = |y + 2|$

      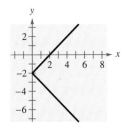

In Exercises 47 and 48, use the graph of $f(x) = x^2$ to sketch the graph of the function.

**47.** $g(x) = -(x - 1)^2 - 2$     **48.** $g(x) = -x^2 + 3$

In Exercises 49 and 50, use the graph of $f(x) = \sqrt{x}$ to sketch the graph of the function.

**49.** $h(x) = -\sqrt{x - 2}$     **50.** $h(x) = \sqrt{x} + 2$

In Exercises 51 and 52, use the graph of $f(x) = \sqrt[3]{x}$ to sketch the graph of the function.

**51.** $y = \sqrt[3]{x + 2}$     **52.** $y = 2\sqrt[3]{x}$

In Exercises 53–58, use the graph of $f(x)$ (see figure) to sketch the graph of the specified function.

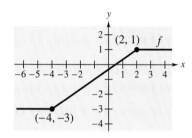

**53.** $f(x - 4)$     **54.** $f(x + 2)$
**55.** $f(x) + 4$     **56.** $f(x) - 1$
**57.** $2f(x)$     **58.** $\frac{1}{2}f(x)$

In Exercises 59–62, identify the common function and the transformation shown in the graph. Write the equation for the graphed function.

**59.**

**60.**

**61.**

**62.**

In Exercises 63–66, find $(f + g)(x)$, $(f - g)(x)$, $(fg)(x)$, and $(f/g)(x)$. What is the domain of $f/g$?

**63.** $f(x) = x + 2$,     $g(x) = x - 3$
**64.** $f(x) = x^2 + 1$,     $g(x) = 8 - x$
**65.** $f(x) = \sqrt{x^2 - 4}$,     $g(x) = 3$
**66.** $f(x) = 3x + 2$,     $g(x) = x^2 - 4x$

In Exercises 67–70, evaluate the function for $f(x) = x^2 - 1$ and $g(x) = x + 2$.

**67.** $(f + g)(2)$     **68.** $(f - g)(-1)$

**69.** $(fg)(3)$     **70.** $\left(\dfrac{f}{g}\right)(0)$

In Exercises 71–76, find (a) $f \circ g$ and (b) $g \circ f$.

**71.** $f(x) = x^2$,     $g(x) = x + 3$
**72.** $f(x) = 2x - 5$,     $g(x) = x^2 + 2$
**73.** $f(x) = |x| + 2$,     $g(x) = x - 5$
**74.** $f(x) = \dfrac{1}{x}$,     $g(x) = 3x + x^2$
**75.** $f(x) = x^2$,     $g(x) = x - 1$
**76.** $f(x) = \dfrac{1}{x}$,     $g(x) = x^3$

In Exercises 77–80, express $h(x)$ as a composition of two functions, $f(x)$ and $g(x)$.

**77.** $h(x) = \dfrac{1}{(x - 2)^2}$     **78.** $h(x) = \sqrt{x^2 + 3}$

**79.** $h(x) = x - 6\sqrt{x} + 9$     **80.** $h(x) = x^2 - 8x + 16$

**81.** *Comparing Sales*   You own two dry cleaning establishments. From 1995 to 1999, the sales for one of the establishments have been decreasing according to the function

$$R_1 = 399.7 - 0.6t - 0.3t^2, \qquad t = 5, 6, 7, 8, 9$$

where $R_1$ represents the sales (in thousands of dollars) and $t = 5$ represents 1995. During the same 5-year period, the sales for the second establishment have been increasing according to the function

$$R_2 = 200.82 + 0.82t, \qquad t = 5, 6, 7, 8, 9.$$

Write a function that represents the total sales for the two establishments. Make a stacked bar graph to represent the total sales during this 5-year period. Are total sales increasing or decreasing?

**82.** *Cost, Revenue, and Profit*   A manufacturer determines that the variable cost for a new product is $2.25 per unit and the fixed costs are $55,000. The product is to be sold for $3.95. Let $x$ be the number of units sold.

(a) Write the total cost function $C$ as a function of the number of units sold.

(b) Write the total revenue function $R$ as a function of the number of units sold.

(c) Write the total profit function $P$ as a function of the number of units sold. (*Hint*: $P = R - C$)

(d) Write the average cost per unit $\overline{C} = C/x$ as a function of $x$.

(e) Complete the table. What do you observe about $\overline{C}$ as $x$ increases?

| $x$ | 100 | 1000 | 10,000 |
|-----|-----|------|--------|
| $\overline{C}$ | | | |

| $x$ | 100,000 | 1,000,000 |
|-----|---------|-----------|
| $\overline{C}$ | | |

**83.** *Area*   A square concrete foundation was prepared as a base for a large cylindrical gasoline tank (see figure).

(a) Express the radius $r$ of the tank as a function of the length $x$ of the sides of the square.

(b) Express the area $A$ of the circular base of the tank as a function of the radius $r$.

(c) Find and interpret $(A \circ r)(x)$.

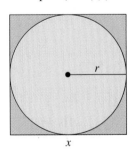

$x$

**84.** *Think About It*   You are a sales representative for an automobile manufacturer. You are paid an annual salary plus a bonus of 3% of your sales over $500,000. Consider the two functions $f(x) = x - 500,000$ and $g(x) = 0.03x$. If $x$ is greater than $500,000$, which of the following represents your bonus? Explain.

(a) $f(g(x))$     (b) $g(f(x))$

**85.** *Cost*   The weekly cost $C$ of producing $x$ units in a manufacturing process is given by $C(x) = 50x + 900$. The number of units produced in $t$ hours is given by $x(t) = 100t$. Find and interpret $(C \circ x)(t)$.

**86.** *Cost*   The daily cost of producing $x$ units in a manufacturing process is given by the function $C(x) = 30x + 800$. The number of units produced in $t$ hours is given by $x(t) = 150t$. Find and interpret $(C \circ x)(t)$.

In Exercises 87 and 88, show that $f$ and $g$ are inverses of each other.

**87.** $f(x) = 3x + 5, \qquad g(x) = \dfrac{x - 5}{3}$

**88.** $f(x) = \sqrt[3]{x - 3}, \qquad g(x) = x^3 + 3$

In Exercises 89–94, determine whether the function has an inverse. If it does, find the inverse and graph $f$ and $f^{-1}$ in the same coordinate plane.

**89.** $f(x) = 3x^2$            **90.** $f(x) = \sqrt[3]{x + 1}$

**91.** $f(x) = \dfrac{1}{x}$            **92.** $f(x) = x^4$

**93.** $f(x) = \dfrac{2x - 5}{3}$        **94.** $f(x) = \dfrac{x^2}{x^2 - 9}$

In Exercises 95–100, (a) find $f^{-1}$, (b) sketch the graphs of $f$ and $f^{-1}$ on the same coordinate system, and (c) verify that $f^{-1}(f(x)) = x = f(f^{-1}(x))$.

**95.** $f(x) = \frac{1}{2}x - 3$        **96.** $f(x) = 5x - 7$

**97.** $f(x) = \sqrt{x + 1}$        **98.** $f(x) = x^3 + 2$

**99.** $f(x) = x^2, \quad x \ge 0$       **100.** $f(x) = \sqrt[3]{x - 1}$

**101.** *Hourly Wage*   Your wage is $8.00 per hour plus $0.75 for each unit produced per hour. Thus, your hourly wage $y$ in terms of the number of units produced is given by $y = 8 + 0.75x$.

(a) Determine the inverse of the function. What does each variable in the inverse function represent?

(b) Use a graphing utility to graph the function and its inverse.

(c) Use the trace feature to find the hourly wage if 10 units are produced per hour.

(d) Use the trace feature to find the number of units produced when your hourly wage is $22.25.

**102.** *Company Income*   The income $y$ for a company can be modeled by $y = (965 + 37.5t)^2$, $0 \le t \le 10$, where $t = 0$ represents 1990. Solve the model for $t$ and use the result to find the year in which the income was $1,620,000.

## Chapter Test

Take this test as you would take a test in class. After you are done, check your work against the answers given in the back of the book.

In Exercises 1–4, find the domain of the function.

**1.** $f(x) = 2x^2 - 3x + 8$

**2.** $g(x) = \sqrt{x - 7}$

**3.** $g(t) = \dfrac{\sqrt{t - 2}}{t - 7}$

**4.** $h(x) = \dfrac{3}{x^2 - 4}$

In Exercises 5–7, decide whether the statement is true or false. Explain.

**5.** The equation $2x - 3y = 5$ identifies $y$ as a function of $x$.

**6.** The equation $y = \pm\sqrt{x^2 - 16}$ identifies $y$ as a function of $x$.

**7.** If $A = \{3, 4, 5\}$ and $B = \{-1, -2, -3\}$, the set $\{(3, -9), (4, -2), (5, -3)\}$ represents a function from $A$ to $B$.

In Exercises 8 and 9, (a) find the range and domain of the function, (b) determine the intervals over which the function is increasing, decreasing, or constant, and (c) determine whether the function is even or odd.

**8.** $f(x) = x^2 + 2$   (See figure.)

**9.** $g(x) = \sqrt{x^2 - 4}$   (See figure.)

**10.** Evaluate the function and simplify the results.

$$g(x) = \begin{cases} x + 1, & x < 0 \\ x^2 + 2, & x \geq 0 \end{cases}$$

(a) $g(-1)$   (b) $g(0)$   (c) $g(3)$

In Exercises 11 and 12, sketch the graph of the function.

**11.** $g(x) = \begin{cases} x + 1, & x < 0 \\ 1, & x = 0 \\ x^2 + 1, & x > 0 \end{cases}$

**12.** $h(x) = (x - 3)^2 + 4$

**13.** Find $f^{-1}(x)$ for the function $f(x) = x^3 - 5$.

In Exercises 14–19, use $f(x) = x^2 + 2$ and $g(x) = 2x - 1$ to find the function.

**14.** $(f + g)(x)$

**15.** $(f - g)(x)$

**16.** $(fg)(x)$

**17.** $\left(\dfrac{f}{g}\right)(x)$

**18.** $(f \circ g)(x)$

**19.** $(g \circ f)(x)$

**20.** A manufacturer determines that the variable costs for a new product are $3.17 per unit and the fixed costs are $75,000. The product is to be sold for $5.25. Let $x$ be the number of units sold.

(a) Write the total cost $C$ as a function of the number of units sold.

(b) Write the total revenue $R$ as a function of the number of units sold.

(c) Write the total profit $P$ as a function of the number of units sold.

Figure for 8

Figure for 9

# 4 Polynomial and Rational Functions

Julian Herbert/Allsport

When a ball is dropped and bounces on a surface, the bounces become smaller and smaller until the ball comes to a state of rest. The path of each bounce can be modeled by a quadratic function.

The table below shows data that was collected when a soccer ball was dropped from a height of 5 feet. The table gives the actual heights of the ball at several times during one of its bounces. This data can be modeled by the function

$$f(t) = -13.44t^2 + 51.97t - 47.44,$$

where $t$ is the time in seconds after the ball has been dropped and $f(t)$ is the height of the ball in feet. The graph above shows the model with the scatter plot of the data.

| Time (sec)  | 1.5  | 1.6  | 1.7  | 1.8  | 1.9  | 2.0  | 2.1  | 2.2  | 2.3  | 2.4  |
|-------------|------|------|------|------|------|------|------|------|------|------|
| Height (ft) | 0.31 | 1.28 | 2.01 | 2.68 | 2.87 | 2.78 | 2.48 | 1.92 | 0.67 | 0.13 |

The chapter project related to this information is on page 290.

**213**

## 4.1 Quadratic Functions and Models

### Objectives

- Sketch the graph of a quadratic function and identify its vertex and intercepts.
- Find a quadratic function given its vertex and a point on the graph.
- Construct and use a quadratic model to solve an application problem.

## The Graph of a Quadratic Function

In this and the next section, you will study the graphs of polynomial functions.

> ▶ **Definition of a Polynomial Function**
>
> Let $n$ be a nonnegative integer and let $a_n, a_{n-1}, \ldots, a_2, a_1, a_0$ be real numbers with $a_n \neq 0$. The function given by
>
> $$f(x) = a_n x^n + a_{n-1} x^{n-1} + \cdots + a_2 x^2 + a_1 x + a_0$$
>
> is called a **polynomial function of $x$ with degree $n$.**

Polynomial functions are classified by degree. For instance, the polynomial function

$$f(x) = a, \quad a \neq 0 \qquad \text{Constant function}$$

has degree 0 and is called a **constant function.** In Chapter 2, you learned that the graph of this type of function is a horizontal line. The polynomial function

$$f(x) = ax + b, \quad a \neq 0 \qquad \text{Linear function}$$

has degree 1 and is called a **linear function.** In Chapter 2, you learned that the graph of the linear function $f(x) = ax + b$ is a line whose slope is $a$ and whose $y$-intercept is $(0, b)$. In this section you will study second-degree polynomial functions, which are called **quadratic functions.**

> ▶ **Definition of a Quadratic Function**
>
> Let $a$, $b$, and $c$ be real numbers with $a \neq 0$. The function of $x$ given by
>
> $$f(x) = ax^2 + bx + c \qquad \text{Quadratic function}$$
>
> is called a **quadratic function.**

The graph of a quadratic function is called a **parabola.** It is "∪"-shaped and can open upward or downward.

All parabolas are symmetric with respect to a line called the **axis of symmetry,** or simply the **axis** of the parabola. The point where the axis intersects the parabola is the **vertex** of the parabola, as shown in Figure 4.1. If the leading coefficient is positive, the graph of $f(x) = ax^2 + bx + c$ is a parabola that opens upward, and if the leading coefficient is negative, the graph of $f(x) = ax^2 + bx + c$ is a parabola that opens downward.

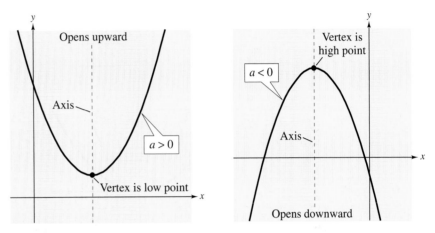

**Figure 4.1**

The simplest type of quadratic function is $f(x) = ax^2$. Its graph is a parabola whose vertex is $(0, 0)$. If $a > 0$, the vertex is the *minimum* point on the graph, and if $a < 0$, the vertex is the *maximum* point on the graph, as shown in Figure 4.2.

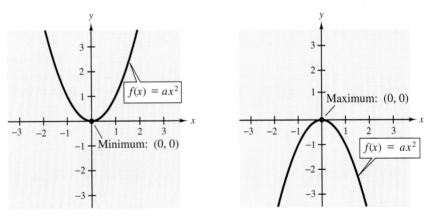

*a* > 0: Parabola opens upward          *a* < 0: Parabola opens downward

**Figure 4.2**

When sketching the graph of $f(x) = ax^2$, it is helpful to use the graph of $y = x^2$ as a reference, as discussed in Section 3.3. There you saw that when $a > 1$, the graph of $y = af(x)$ is a vertical stretch of the graph of $y = f(x)$. When $0 < a < 1$, the graph of $y = af(x)$ is a vertical shrink of the graph of $y = f(x)$. This is demonstrated again in Example 1.

Example 1 **Sketching the Graph of a Quadratic Function**

**a.** Compared with $y = x^2$, each output of $f(x) = \frac{1}{3}x^2$ "shrinks" by a factor of $\frac{1}{3}$, creating the broader parabola shown in Figure 4.3(a).

**b.** Compared with $y = x^2$, each output of $g(x) = 2x^2$ "stretches" by a factor of 2, creating the narrower parabola shown in Figure 4.3(b).

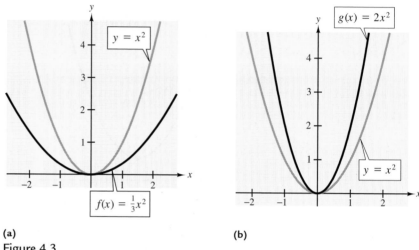

(a)     (b)

Figure 4.3

In Example 1, note that the coefficient $a$ determines how widely the parabola given by $f(x) = ax^2$ opens. If $|a|$ is small, the parabola opens more widely than if $|a|$ is large.

Recall from Section 3.3 that the graphs of $y = f(x \pm c)$, $y = f(x) \pm c$, and $y = -f(x)$ are rigid transformations of the graph of $y = f(x)$. For instance, in Figure 4.4, notice how the graph of $y = x^2$ can be transformed to produce the graphs of $f(x) = -x^2 + 1$ and $g(x) = (x + 2)^2 - 3$.

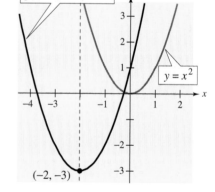

Figure 4.4

# The Standard Form of a Quadratic Function

The **standard form** of a quadratic function is

$$f(x) = a(x - h)^2 + k.$$

This form is especially convenient for sketching a parabola because it identifies the vertex of the parabola.

> ▶ **Standard Form of a Quadratic Function**
>
> The quadratic function
>
> $$f(x) = a(x - h)^2 + k, \qquad a \neq 0$$
>
> is said to be in **standard form.** The graph of $f$ is a parabola whose axis is the vertical line $x = h$ and whose vertex is the point $(h, k)$. If $a > 0$, the parabola opens upward, and if $a < 0$, the parabola opens downward.

To write a quadratic function in standard form, you can use the process of *completing the square*, as illustrated in Example 2.

### Example 2    Writing a Quadratic Function in Standard Form

Sketch the graph of $f(x) = 2x^2 + 8x + 7$ and identify the vertex.

#### Solution

Begin by writing the quadratic function in standard form. Notice that the first step in completing the square is to factor out any coefficient of $x^2$ that is not 1.

$$
\begin{aligned}
f(x) &= 2x^2 + 8x + 7 && \text{Given form} \\
&= 2(x^2 + 4x) + 7 && \text{Factor 2 out of } x \text{ terms.} \\
&= 2(x^2 + 4x + 4 - 4) + 7 && \text{Add and subtract 4 within parentheses.} \\
&\qquad\quad \underset{2^2}{\underline{\quad\quad}} \\
&= 2(x^2 + 4x + 4) - 2(4) + 7 && \text{Regroup terms.} \\
&= 2(x^2 + 4x + 4) - 8 + 7 && \text{Simplify.} \\
&= 2(x + 2)^2 - 1 && \text{Standard form}
\end{aligned}
$$

From this form, you can see that the graph of $f$ is a parabola that opens upward with vertex $(-2, -1)$. This corresponds to a left shift of two units and a downward shift of one unit relative to the graph of $y = 2x^2$, as shown in Figure 4.5.

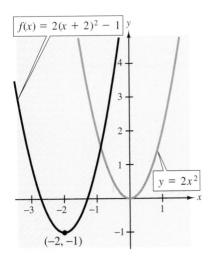

$f(x) = 2(x + 2)^2 - 1$

$y = 2x^2$

$(-2, -1)$

**Figure 4.5**

Figure 4.6

**Example 3**    Writing a Quadratic Function in Standard Form

Sketch the graph of $f(x) = -x^2 + 6x - 8$ and identify the vertex.

**Solution**

As in Example 2, we begin by writing the quadratic function in standard form.

$$f(x) = -x^2 + 6x - 8 \qquad \text{Given form}$$

$$= -(x^2 - 6x) - 8 \qquad \text{Factor } -1 \text{ out of } x \text{ terms.}$$

$$= -(x^2 - 6x + 9 - 9) - 8 \qquad \text{Add and subtract 9 within parentheses.}$$

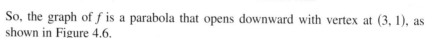

$$3^2$$

$$= -(x^2 - 6x + 9) - (-9) - 8 \qquad \text{Regroup terms.}$$

$$= -(x - 3)^2 + 1 \qquad \text{Standard form}$$

So, the graph of $f$ is a parabola that opens downward with vertex at $(3, 1)$, as shown in Figure 4.6.

**Example 4**    Finding the Equation of a Parabola

Find an equation for the parabola whose vertex is $(1, 2)$ and that passes through the point $(0, 0)$, as shown in Figure 4.7.

**Solution**

Because the parabola has a vertex at $(h, k) = (1, 2)$, the equation must have the form

$$f(x) = a(x - 1)^2 + 2. \qquad \text{Standard form}$$

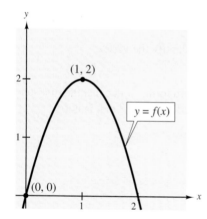

Figure 4.7

Because the parabola passes through the point $(0, 0)$, it follows that $f(0) = 0$. So, you obtain

$$0 = a(0 - 1)^2 + 2 \qquad \Longrightarrow \qquad a = -2$$

which implies that the equation is

$$f(x) = -2(x - 1)^2 + 2 = -2x^2 + 4x.$$

To find the $x$-intercepts of the graph of $f(x) = ax^2 + bx + c$, you must solve the equation

$$ax^2 + bx + c = 0.$$

If $ax^2 + bx + c$ does not factor, you can use the Quadratic Formula to find the $x$-intercepts. Remember, however, that a parabola may have no $x$-intercepts.

# Applications

Many applications involve finding the maximum or minimum value of a quadratic function. By writing the quadratic function $f(x) = ax^2 + bx + c$ in standard form, you can determine that the vertex occurs when $x = -b/2a$.

| Example 5 | The Maximum Height of a Baseball |  |

A baseball is hit 3 feet above ground at a velocity of 100 feet per second and at an angle of 45° with respect to the ground. The path of the baseball is given by the function

$$f(x) = -0.0032x^2 + x + 3$$

where $f(x)$ is the height of the baseball (in feet) and $x$ is the distance from home plate (in feet). What is the maximum height reached by the baseball?

### Solution

For this quadratic function, you have

$$f(x) = ax^2 + bx + c = -0.0032x^2 + x + 3.$$

So, $a = -0.0032$ and $b = 1$. Because the function has a maximum when $x = -b/2a$, you can conclude that the baseball reaches its maximum height when it is

$$x = -\frac{b}{2a} = -\frac{1}{2(-0.0032)} = 156.25 \text{ feet}$$

from home plate. At this distance, the maximum height is

$$f(156.25) = -0.0032(156.25)^2 + 156.25 + 3 = 81.125 \text{ feet}.$$

The path of the baseball is shown in Figure 4.8.

Figure 4.8

In Section 2.5 you plotted data points in the coordinate plane and estimated the best fit line. Fitting a quadratic model by this same process would be difficult. Most graphing utilities have a built-in statistical program that easily calculates the best-fitting quadratic for a set of data points. Refer to the user's guide of your graphing utility for the required steps.

Courtesy of Daimler Chrysler

*The fuel efficiency is one of many specifications consumers are interested in when buying a new car. The fuel efficiency estimates for the Chrysler 300M, named Motor Trends Car of the Year for 1999, are 18 miles per gallon for city driving and 27 miles per gallon for highway driving.*
(Source: *www.edmunds.com*)

**Example 6    Fitting a Quadratic Function to Data**

A study was done to compare the speed $x$ (in miles per hour) with the mileage $y$ (in miles per gallon) of an automobile. The results are shown in the table. Use a graphing utility to plot the data and find the quadratic model that best fits the data. Estimate the speed for which the miles per gallon is greatest.    (Source: Federal Highway Administration)

| Speed $x$ | 15 | 20 | 25 | 30 | 35 | 40 | 45 |
|-----------|------|------|------|------|------|------|------|
| Mileage $y$ | 22.3 | 25.5 | 27.5 | 29.0 | 28.8 | 30.0 | 29.9 |

| Speed $x$ | 50 | 55 | 60 | 65 | 70 | 75 |
|-----------|------|------|------|------|------|------|
| Mileage $y$ | 30.2 | 30.4 | 28.8 | 27.4 | 25.3 | 23.3 |

### Solution

Begin by entering the data into your graphing utility and displaying the scatter plot. From the scatter plot in Figure 4.9(a) you can see that the points have a parabolic trend. Use the quadratic regression feature to find the quadratic function that best fits the data. The quadratic equation that best fits the data is

$$y = -0.008x^2 + 0.75x + 13.5.$$

Graph the data and the equation in the same viewing window as shown in Figure 4.9(b). From the graph you can see that the vertex of the graph is approximately $(47, 31)$. So the speed at which the mileage is greatest is 47 miles per hour.

(a)

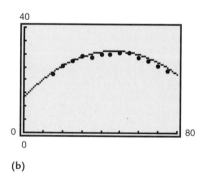

(b)

Figure 4.9

| Example 7 | Charitable Contributions |  |

According to a survey conducted by *Independent Sector*, the percent of income that Americans give to charities is related to the amount of income. For families with annual incomes of $100,000 or less, the percent can be modeled by

$$P(x) = 0.0014x^2 - 0.1529x + 5.855, \quad 5 \le x \le 100$$

where $x$ is the annual income in thousands of dollars. According to this model, what income level corresponds to the minimum percent of charitable contributions?

### Solution

There are two ways to answer this question. One is to sketch the graph of the quadratic function, as shown in Figure 4.10. From this graph, it appears that the minimum percent corresponds to an income level of about $55,000. The other way to answer the question is to use the fact that the minimum point of the parabola occurs when $x = -b/2a$.

$$x = -\frac{b}{2a} = -\frac{-0.1529}{2(0.0014)} \approx 54.6.$$

From this $x$-value, you can conclude that the minimum percent corresponds to an income level of about $54,600.

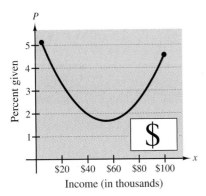

**Figure 4.10**

---

| Discussing the Concept | Approximating Intercepts and Extreme Points |

A graphing utility with zoom and trace features can be a useful tool in approximating the intercepts and vertex coordinates of quadratic functions to acceptable degrees of accuracy. Use a graphing utility to locate and approximate the intercepts and vertices of the following quadratic functions to two decimal places. Confirm the coordinates of the vertices algebraically. How do the estimates and the algebraic results compare?

**a.** $y = x^2 - 3x + 4$      **b.** $y = x^2 - 17x + 64$

**c.** $y = 7.52x^2 + 4.16x$      **d.** $y = -\frac{1}{2}x^2 + 5x - \frac{13}{2}$

**e.** $y = \frac{x^2}{19} + \frac{13x}{3} + \frac{4}{21}$      **f.** $y = 14x^2 - 5x + 2$

# Warm Up

The following warm-up exercises involve skills that were covered in earlier sections. You will use these skills in the exercise set for this section.

In Exercises 1–4, solve the equation by factoring.

**1.** $2x^2 + 11x - 6 = 0$    **2.** $5x^2 - 12x - 9 = 0$

**3.** $3 + x - 2x^2 = 0$    **4.** $x^2 + 20x + 100 = 0$

In Exercises 5–10, use the Quadratic Formula to solve the equation.

**5.** $x^2 - 6x + 4 = 0$    **6.** $x^2 + 4x + 1 = 0$

**7.** $2x^2 - 16x + 25 = 0$    **8.** $3x^2 + 30x + 74 = 0$

**9.** $x^2 + 3x + 1 = 0$    **10.** $x^2 + 3x - 3 = 0$

## 4.1    Exercises

LAB **3**

In Exercises 1–8, match the quadratic function with its graph. [The graphs are labeled (a), (b), (c), (d), (e), (f), (g), and (h).]

**1.** $f(x) = -(x - 3)^2$    **2.** $f(x) = (x + 5)^2$

**3.** $f(x) = x^2 - 4$    **4.** $f(x) = 5 - x^2$

**5.** $f(x) = (x + 3)^2 - 2$    **6.** $f(x) = (x - 1)^2 - 5$

**7.** $f(x) = -(x + 1)^2 + 3$    **8.** $f(x) = (x - 2)^2 - 4$

In Exercises 9–14, find an equation of the parabola.

(a)

(b)

(c)

(d)

(e)

(f)

(g)

(h)

**9.**

**10.**

**11.**

**12.**

**13.**

**14.**
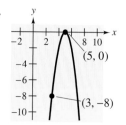

In Exercises 15–32, sketch the graph of the quadratic function. Identify the vertex and intercepts.

**15.** $f(x) = x^2 + 2$

**16.** $f(x) = x^2 - 4$

**17.** $f(x) = 16 - x^2$

**18.** $h(x) = 25 - x^2$

**19.** $f(x) = (x + 5)^2 - 6$

**20.** $f(x) = (x - 6)^2 + 3$

**21.** $h(x) = x^2 - 4x + 2$

**22.** $g(x) = x^2 + 2x + 1$

**23.** $f(x) = -(x^2 + 2x - 3)$

**24.** $f(x) = -(x^2 + 6x - 3)$

**25.** $f(x) = x^2 - x + \frac{5}{4}$

**26.** $f(x) = x^2 + 3x + \frac{1}{4}$

**27.** $f(x) = -x^2 + 2x + 5$

**28.** $f(x) = -x^2 - 4x + 1$

**29.** $h(x) = 4x^2 - 4x + 21$

**30.** $f(x) = 2x^2 - x + 1$

**31.** $f(x) = \frac{1}{4}(x^2 - 16x + 32)$

**32.** $g(x) = \frac{1}{2}(x^2 + 4x - 2)$

In Exercises 33–36, find the quadratic function that has the indicated vertex and whose graph passes through the given point.

**33.** Vertex: $(2, -1)$; point: $(4, -3)$

**34.** Vertex: $(-3, 5)$; point: $(-6, -1)$

**35.** Vertex: $(5, 12)$; point: $(7, 15)$

**36.** Vertex: $(-2, -2)$; point: $(-1, 0)$

In Exercises 37–42, find two quadratic functions whose graphs have the given x-intercepts. Find one function that has a graph that opens upward and another that has a graph that opens downward. (Each exercise has many correct answers.)

**37.** $(2, 0), (-1, 0)$

**38.** $(-4, 0), (0, 0)$

**39.** $(0, 0), (10, 0)$

**40.** $(4, 0), (8, 0)$

**41.** $(-3, 0), \left(-\frac{1}{2}, 0\right)$

**42.** $\left(-\frac{5}{2}, 0\right), (2, 0)$

**43.** *Maximum Area*    The perimeter of a rectangle is 100 feet. Let $x$ represent the width of the rectangle and write a quadratic function that expresses the area of the rectangle in terms of its width. Of all possible rectangles with perimeters of 100 feet, what are the measurements of the one that has the greatest area?

**44.** *Maximum Area*    The perimeter of a rectangle is 400 feet. Let $x$ represent the width of the rectangle and write a quadratic function that expresses the area of the rectangle in terms of its width. Of all possible rectangles with perimeters of 400 feet, what are the measurements of the one that has the greatest area?

**45.** *Maximum Area*    A rancher has 200 feet of fencing with which to enclose two adjacent rectangular corrals (see figure). What measurements will produce a maximum enclosed area?

**46.** *Maximum Area*    An indoor physical-fitness room consists of a rectangular region with a semicircle on each end (see figure). The perimeter of the room is to be a 200-meter running track. What measurements will produce a maximum area of the rectangle?

*Maximum Revenue*    In Exercises 47 and 48, find the number of units that produces a maximum revenue. The revenue $R$ is measured in dollars and $x$ is the number of units produced.

**47.** $R = 900x - 0.01x^2$

**48.** $R = 100x - 0.0002x^2$

**49.** *Minimum Cost*    A manufacturer of lighting fixtures has daily production costs of

$$C(x) = 800 - 10x + 0.25x^2$$

where $C$ is the total cost in dollars and $x$ is the number of units produced. How many fixtures should be produced each day to yield a minimum cost?

**50. *Maximum Profit*** The profit for a company is given by

$$P(x) = -0.0002x^2 + 140x - 250{,}000$$

where $x$ is the number of units produced. What production level will yield a maximum profit

**51. *Maximum Height of a Diver*** The path of a diver is given by

$$y = -\frac{4}{9}x^2 + \frac{24}{9}x + 10$$

where $y$ is the height in feet and $x$ is the horizontal distance from the end of the diving board in feet (see figure). Use a graphing utility and the trace feature to find the maximum height of the diver.

**52. *Maximum Height*** The winning women's shot put in the 1996 Summer Olympics was recorded by Astrid Kumbernuss of Germany. The path of her winning toss was approximately given by

$$y = -0.01707x^2 + 1.0775x + 5$$

where $y$ is the height of the shot in feet and $x$ is the horizontal distance in feet. Use a graphing utility and the trace feature to find how long the winning toss was and the maximum height of the shot.

**53. *Regression Problem*** Let $x$ be the number of units (in tens of thousands) that a company produces and let $p(x)$ be the profit (in hundreds of thousands of dollars). The following table gives the profit for different levels of production.

| $x$ | 2 | 4 | 6 | 8 | 10 |
|---|---|---|---|---|---|
| $p(x)$ | 270.5 | 307.8 | 320.1 | 329.2 | 325 |

| $x$ | 12 | 14 | 16 | 18 | 20 |
|---|---|---|---|---|---|
| $p(x)$ | 311.2 | 287.8 | 254.8 | 212.2 | 160 |

(a) Make a scatter plot using the data in the table.

(b) Use a quadratic regression program to find a mathematical model for $p(x)$.

(c) Use your graphing utility to graph your model for $p(x)$ with the scatter plot for the data.

(d) Use the model to approximate the level of production that will yield a maximum profit.

(e) With these data and this model, the profit begins to decrease. Discuss how it is possible for production to increase and profit to decrease.

**54. *Regression Problem*** Let $x$ be the angle (in degrees) at which a baseball is hit with no spin at an initial speed of 40 miles per hour and let $d(x)$ be the distance the ball travels. The following table gives the distances for the different angles the ball is hit. (Source: *The Physics of Sports*)

| $x$ | 10° | 15° | 30° | 36° |
|---|---|---|---|---|
| $d(x)$ | 58.3 | 79.7 | 126.9 | 136.6 |

| $x$ | 42° | 43° | 44° | 45° |
|---|---|---|---|---|
| $d(x)$ | 140.6 | 140.8 | 140.9 | 140.9 |

| $x$ | 48° | 54° | 60° |
|---|---|---|---|
| $d(x)$ | 139.3 | 132.5 | 120.5 |

(a) Make a scatter plot using the data in the table.

(b) Use a quadratic regression program to find a mathematical model for $d(x)$.

(c) Use your graphing utility to graph your model for $d(x)$ with the scatter plot for the data.

(d) Use the model to approximate the angle that will result in a maximum distance.

**55.** *Ice Cream*    During the mid-1990s, the number of gallons of regular ice cream produced in the United States fell slightly, then rose again according to the model

$$G(x) = 11.25t^2 - 111.25t + 1158.75, \quad 4 \leq t \leq 7$$

where $G$ is the number of gallons of regular ice cream (in millions) and $t$ represents the calendar year, with $t = 4$ corresponding to 1994. Use a graphing utility to determine the year, from 1994 to 1997, when the number of gallons produced was at its lowest.    (Source: U.S. Department of Agriculture)

**56.** *Employment Rate*    The annual average employment rate for single women between 16 and 19 years of age from 1990 to 1994 can be approximated by the model

$$P(x) = 0.564t^2 - 2.39t + 51.89, \quad 0 \leq t \leq 4$$

where $P$ is the percent of single women who were employed and $t$ represents the calendar year, with $t = 0$ corresponding to 1990. Use a graphing utility to determine the year in which the employment rate was at its lowest.    (Source: U.S. Bureau of Labor Statistics)

## Math Matters • Circles and Pi

Pi is a special number that represents a relationship that is found in *all circles*. The relationship is this: The ratio of the circumference of *any* circle to the diameter of the circle always produces the same number—the number denoted by the Greek symbol $\pi$.

Calculation of the decimal representation of the number $\pi$ has consumed the time of many mathematicians (and computers). Because the number $\pi$ is irrational, its decimal representation does not repeat or terminate. Here is a decimal representation of $\pi$ that is accurate to 27 decimal places.

$$\pi = 3.141592653589793238462643383.$$

In October 1995, a team of mathematicians in British Columbia discovered a formula for calculating the $n$th digit of $\pi$ without calculating the preceding digits. Run on a computer, the algorithm for the $n$th digit requires that numbers be written in base 16 (hexadecimal representation) instead of base 10 (decimal representation).

The number $\pi$ is used in most formulas that deal with circles. For instance, the area of a circle is given by

$$A = \pi r^2$$

where $r$ is the radius of the circle. Use this formula to determine which of the colored regions inside the circle below has the greater area. Is the area of the red circle larger than that of the blue ring, or do both regions have the same area? (The answer is given in the back of the book.)

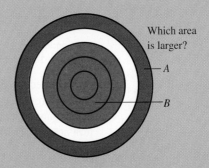

Which area is larger?

— A

— B

## 4.2    Polynomial Functions of Higher Degree

### Objectives

- Identify the characteristics of polynomial functions.
- Apply the Leading Coefficient Test to determine right and left behavior of the graph of a polynomial function.
- Find the real zeros of a polynomial function.
- Sketch the graph of a polynomial function.
- Use a polynomial model to solve an application problem.

## Graphs of Polynomial Functions

### Study Tip

The graphs of polynomial functions of degree greater than 2 are more difficult to analyze than graphs of polynomials of degree 0, 1, or 2. However, using the characteristics presented in this section, together with point plotting, intercepts, and symmetry, you should be able to make reasonably accurate sketches *by hand.* Of course, if you have a graphing utility, the task is easier.

In this section, you will study basic characteristics of the graphs of polynomial functions. The first characteristic is that the graph of a polynomial function is **continuous.** Essentially, this means that the graph of a polynomial function has no breaks, as shown in Figure 4.11(a).

(a)  **Continuous**

(b)  **Discontinuous**

Figure 4.11

The second characteristic is that the graph of a polynomial function has only smooth, rounded turns, as shown in Figure 4.12(a). A polynomial function cannot have a sharp turn. For instance, the function $f(x) = |x|$, which has a sharp turn at the point $(0, 0)$, as shown in Figure 4.12(b), is not a polynomial function.

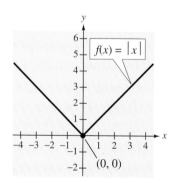

(a)  **Polynomial functions have smooth, rounded graphs.**

(b)

Figure 4.12

The polynomial functions that have the simplest graphs are monomials of the form $f(x) = x^n$, where $n$ is an integer greater than zero. From Figure 4.13, you can see that when $n$ is *even* the graph is similar to the graph of $f(x) = x^2$ and when $n$ is *odd* the graph is similar to the graph of $f(x) = x^3$. Moreover, the greater the value of $n$, the flatter the graph near the origin.

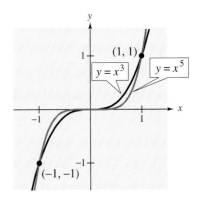

(a) If *n* is even, the graph of $y = x^n$ touches the axis at the *x*-intercept.

(b) If *n* is odd, the graph of $y = x^n$ crosses the axis at the *x*-intercept.

Figure 4.13

### Example 1    Sketching Transformations of Monomial Functions

**a.** Because the degree of $f(x) = -x^5$ is odd, its graph is similar to the graph of $y = x^3$. In Figure 4.14(a), note that the negative coefficient has the effect of reflecting the graph about the *x*-axis.

**b.** The graph of $h(x) = (x + 1)^4$ is a left shift, by one unit, of the graph of $y = x^4$, as shown in Figure 4.14(b).

(a)

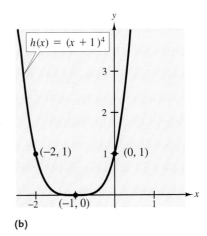

(b)

Figure 4.14

## The Leading Coefficient Test

For each function, identify the degree of the function and whether it is even or odd. Identify the leading coefficient, and whether the leading coefficient is greater than 0 or less than 0. Use a graphing utility to graph each function. Describe the relationship between the degree and leading coefficient of the function and the behavior of the ends of the graph of the function.

(a) $y = x^3 - 2x^2 - x + 1$
(b) $y = 2x^5 + 2x^2 - 5x + 1$
(c) $y = -2x^5 - x^2 + 5x + 3$
(d) $y = -x^3 + 5x - 2$
(e) $y = 2x^2 + 3x - 4$
(f) $y = x^4 - 3x^2 + 2x - 1$
(g) $y = x^2 + 3x + 2$
(h) $y = -x^6 - x^2 - 5x + 4$

In Example 1, note that both graphs eventually rise or fall without bound as $x$ moves to the right. Whether the graph of a polynomial eventually rises or falls can be determined by the function's degree (even or odd) and by its leading coefficient, as indicated in the **Leading Coefficient Test.**

▶ **Leading Coefficient Test**

As $x$ moves without bound to the left or to the right, the graph of the polynomial function $f(x) = a_n x^n + \cdots + a_1 x + a_0$ eventually rises or falls in the following manner.

**1.** When $n$ is *odd*:

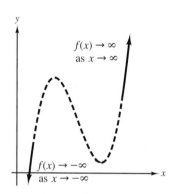

If the leading coefficient is positive $(a_n > 0)$, the graph falls to the left and rises to the right.

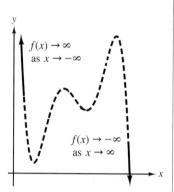

If the leading coefficient is negative $(a_n < 0)$, the graph rises to the left and falls to the right.

**2.** When $n$ is *even*:

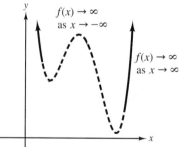

If the leading coefficient is positive $(a_n > 0)$, the graph rises to the left and right.

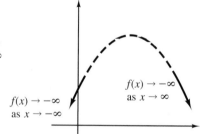

If the leading coefficient is negative $(a_n < 0)$, the graph falls to the left and right.

The dashed portions of the graphs indicate that the test determines *only* the right and left behavior of the graph.

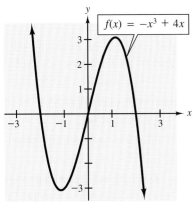

$f(x) = -x^3 + 4x$

**Figure 4.15**

**Example 2**    Applying the Leading Coefficient Test

Describe the right and left behavior of the graph of

$$f(x) = -x^3 + 4x.$$

**Solution**

Because the degree is odd and the leading coefficient is negative, the graph rises to the left and falls to the right, as shown in Figure 4.15.

In Example 2, note that the Leading Coefficient Test tells you only whether the graph *eventually* rises or falls to the right or left. Other characteristics of the graph, such as intercepts and minimum and maximum points, must be determined by means of other tests.

**Example 3**    Applying the Leading Coefficient Test

Describe the right and left behavior of the graphs of the following functions.

**a.**  $f(x) = x^4 - 5x^2 + 4$      **b.**  $f(x) = x^5 - x$

**Solution**

**a.** Because the degree is even and the leading coefficient is positive, the graph rises to the left and right, as shown in Figure 4.16(a).

**b.** Because the degree is odd and the leading coefficient is positive, the graph falls to the left and rises to the right, as shown in Figure 4.16(b).

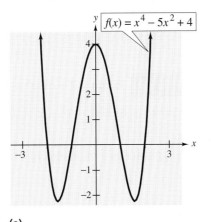

$f(x) = x^4 - 5x^2 + 4$

**(a)**

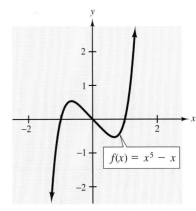

$f(x) = x^5 - x$

**(b)**

**Figure 4.16**

## Real Zeros of Polynomial Functions

It can be shown that for a polynomial function $f$ of degree $n$, the following statements are true. (Remember that the **zeros** of a function are the $x$-values for which the function is zero.)

1. The graph of $f$ has, at most, $n - 1$ turning points. (Turning points are points at which the graph changes from increasing to decreasing or vice versa.)

2. The function $f$ has, at most, $n$ real zeros. (You will study this result in detail in Section 4.6 on the Fundamental Theorem of Algebra.)

Finding the zeros of polynomial functions is one of the most important problems in algebra. There is a strong interplay between graphical and algebraic approaches to this problem. Sometimes you can use information about the graph of a function to help find its zeros, and in other cases you can use information about the zeros of a function to help sketch its graph.

---

▶ **Real Zeros of Polynomial Functions**

If $f$ is a polynomial function and $a$ is a real number, the following statements are equivalent.

1. $x = a$ is a *zero* of the function $f$.

2. $x = a$ is a *solution* of the polynomial equation $f(x) = 0$.

3. $(x - a)$ is a *factor* of the polynomial $f(x)$.

4. $(a, 0)$ is an *x-intercept* of the graph of $f$.

---

In the equivalent statements above, notice that finding zeros of polynomial functions is closely related to factoring and finding $x$-intercepts.

---

**Example 4**    Finding Zeros of a Polynomial Function

Find all real zeros of $f(x) = x^3 - x^2 - 2x$.

**Solution**

By factoring, you obtain the following.

$$f(x) = x^3 - x^2 - 2x \qquad \text{Original function}$$

$$= x(x^2 - x - 2) \qquad \text{Remove common monomial factor.}$$

$$= x(x - 2)(x + 1) \qquad \text{Factor completely.}$$

So, the real zeros are $x = 0$, $x = 2$, and $x = -1$, and the corresponding $x$-intercepts are $(0, 0)$, $(2, 0)$, and $(-1, 0)$, as shown in Figure 4.17. In the figure, note that the graph has two turning points. This is consistent with the fact that a third-degree polynomial can have *at most* two turning points.

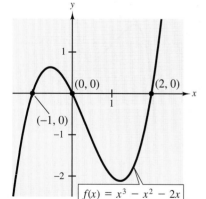

Figure 4.17

### Example 5    Finding Zeros of a Polynomial Function

Find all real zeros of $f(x) = -2x^4 + 2x^2$.

#### Solution

In this case, the polynomial factors as follows.

$$f(x) = -2x^4 + 2x^2 \qquad \text{Original function}$$

$$= -2x^2(x^2 - 1) \qquad \text{Remove common monomial factor.}$$

$$= -2x^2(x - 1)(x + 1) \qquad \text{Factor completely.}$$

So, the real zeros are $x = 0$, $x = 1$, and $x = -1$, and the corresponding $x$-intercepts are $(0, 0)$, $(1, 0)$, and $(-1, 0)$, as shown in Figure 4.18. Note in the figure that the graph has three turning points, which is consistent with the fact that a fourth-degree polynomial can have at most three turning points.

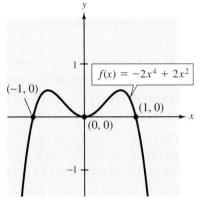

**Figure 4.18**

In Example 5, the real zero arising from $-2x^2 = 0$ is called a **repeated zero.** In general, a factor $(x - a)^k$ yields a repeated zero $x = a$ of **multiplicity** $k$. If $k$ is odd, the graph *crosses* the $x$-axis at $x = a$. If $k$ is even, the graph *touches* the $x$-axis (but does not cross the $x$-axis) at $x = a$. Note how this occurs in Figure 4.18.

### Example 6    Charitable Contributions Revisited

Example 7 in Section 4.1 discussed the model

$$P(x) = 0.0014x^2 - 0.1529x + 5.855, \qquad 5 \le x \le 100$$

where $P$ is the percent of annual income given and $x$ is the annual income in thousands of dollars. Note that this model gives the charitable contributions as a *percent* of annual income. To find the average *amount* that a family gives to charity, you can multiply the given model by the income $1000x$ (and divide by 100 to change from percent to decimal form) to obtain

$$A(x) = 0.014x^3 - 1.529x^2 + 58.55x, \qquad 5 \le x \le 100$$

where $A$ represents the amount of charitable contributions in dollars. Sketch the graph of this function and use the graph to estimate the annual salary for a family that gives $1000 a year to charities.

#### Solution

The graph of this function is shown in Figure 4.19. From the graph you see that an annual contribution of $1000 corresponds to an annual income of about $59,000.

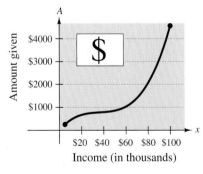

**Figure 4.19**

Example 7 shows how the Leading Coefficient Test and zeros of polynomial functions can be used as sketching aids.

## • TECHNOLOGY •

Example 7 uses an "algebraic approach" to describe the graph of the function. A graphing utility is a valuable complement to this approach. Remember that the most important part of using a graphing utility is to find a viewing window that shows all important parts of the graph. For instance, the graph below shows the important parts of the graph of the function in Example 7.

### Example 7  Sketching the Graph of a Polynomial Function

Sketch the graph of $f(x) = 3x^4 - 4x^3$.

**Solution**

Because the leading coefficient is positive and the degree is even, you know that the graph eventually rises to the left and to the right, as shown in Figure 4.20(a). By factoring

$$f(x) = 3x^4 - 4x^3 = x^3(3x - 4)$$

you can see that the zeros of $f$ are $x = 0$ and $x = \frac{4}{3}$ (both of odd multiplicity). So the x-intercepts occur at $(0, 0)$ and $\left(\frac{4}{3}, 0\right)$. To sketch the graph by hand, find a few additional points, as shown in the table. Then plot the points and complete the graph, as shown in Figure 4.20(b).

| $x$ | $-1$ | $0.5$ | $1$ | $1.5$ |
|---|---|---|---|---|
| $f(x)$ | $7$ | $-0.3125$ | $-1$ | $1.6875$ |

(a)

(b)

Figure 4.20

---

## Discussing the Concept    Modeling Polynomial Functions

Suppose you are writing a quiz for your algebra class and want to make up several polynomial functions for your students to investigate. Discuss how you could find polynomial functions that have reasonably simple zeros. Then use the methods you have discussed to find polynomial functions that have the following zeros.

**a.** $(2, 4, -3)$        **b.** $\left(3, -\frac{2}{3}, \frac{3}{4}\right)$

**c.** $(-3, -3, -3, -3)$        **d.** $(1, -2, 3, -4)$

# Warm Up

The following warm-up exercises involve skills that were covered in earlier sections. You will use these skills in the exercise set for this section.

In Exercises 1–6, factor the expression completely.

**1.** $12x^2 + 7x - 10$      **2.** $25x^3 - 60x^2 + 36x$

**3.** $12z^4 + 17z^3 + 5z^2$      **4.** $y^3 + 125$

**5.** $x^3 + 3x^2 - 4x - 12$      **6.** $x^3 + 2x^2 + 3x + 6$

In Exercises 7–10, find all real solutions of the equation.

**7.** $5x^2 + 8 = 0$      **8.** $x^2 - 6x + 4 = 0$

**9.** $4x^2 + 4x - 11 = 0$      **10.** $x^4 - 18x^2 + 81 = 0$

## 4.2    Exercises

In Exercises 1–8, match the polynomial function with its graph. [The graphs are labeled (a), (b), (c), (d), (e), (f), (g), and (h).]

(a)

(b)

(c)

(d)

(e)

(f)

(g)

(h)

**1.** $f(x) = \frac{1}{2}(x^3 + 2x^2 - 3x)$      **2.** $f(x) = x^2 - 2x$

**3.** $f(x) = \frac{1}{3}x^4 - x^2$      **4.** $f(x) = -3x^4 - 4x^3$

**5.** $f(x) = 3x^3 - 9x + 1$

**6.** $f(x) = x^5 - 5x^3 + 4x$

**7.** $f(x) = -\frac{1}{3}x^3 + x - \frac{2}{3}$

**8.** $f(x) = -x^5 + 5x^3 - 4x$

In Exercises 9–18, determine the right-hand and left-hand behavior of the graph of the function.

**9.** $f(x) = -x^3 + 1$

**10.** $f(x) = \frac{1}{3}x^3 + 5x$

**11.** $g(x) = 6 - 4x^2 + x - 3x^5$

**12.** $f(x) = 2x^5 - 5x + 7.5$

**13.** $f(x) = 4x^8 - 2$

**14.** $h(x) = 1 - x^6$

**15.** $f(x) = 2 + 5x - x^2 - x^3 + 2x^4$

**16.** $f(x) = \dfrac{3x^4 - 2x + 5}{4}$

**17.** $h(t) = -\frac{2}{3}(t^2 - 5t + 3)$

**18.** $f(s) = -\frac{7}{8}(s^3 + 5s^2 - 7s + 1)$

*Algebraic and Graphical Approaches*  In Exercises 19–34, find all real zeros of the function algebraically. Then use a graphing utility to confirm your results.

**19.** $f(x) = x^2 - 16$

**20.** $f(x) = 64 - x^2$

**21.** $h(t) = t^2 + 8t + 16$

**22.** $f(x) = x^2 - 12x + 36$

**23.** $f(x) = \frac{1}{3}x^2 + \frac{1}{3}x - \frac{2}{3}$

**24.** $f(x) = \frac{1}{2}x^2 + \frac{5}{2}x - \frac{3}{2}$

**25.** $f(x) = 2x^2 + 4x + 6$

**26.** $g(x) = -5(x^2 + 2x - 4)$

**27.** $f(t) = t^3 - 4t^2 + 4t$

**28.** $f(x) = x^4 - x^3 - 20x^2$

**29.** $g(t) = \frac{1}{2}t^4 - \frac{1}{2}$

**30.** $f(x) = x^5 + x^3 - 6x$

**31.** $f(x) = 2x^4 - 2x^2 - 40$

**32.** $g(t) = t^5 - 6t^3 + 9t$

**33.** $f(x) = x^3 - 3x^2 + 2x - 6$

**34.** $f(x) = x^3 - 4x^2 - 25x + 100$

In Exercises 35–38, use the graph of $y = x^3$ to sketch the graph of the function.

**35.** $f(x) = x^3 + 4$

**36.** $f(x) = (x - 3)^3$

**37.** $f(x) = (x - 2)^3 - 2$

**38.** $f(x) = -(x + 1)^3 + 3$

In Exercises 39–42, use the graph of $y = x^4$ to sketch the graph of the function.

**39.** $f(x) = (x + 3)^4$      **40.** $f(x) = x^4 - 3$

**41.** $f(x) = 4 - x^4$        **42.** $f(x) = \frac{1}{2}(x - 1)^4$

*Analyzing a Graph*  In Exercises 43–54, analyze the graph of the function algebraically and use the results to sketch the graph *by hand*. Then use a graphing utility to confirm your sketch.

**43.** $f(x) = -\frac{3}{2}$

**44.** $h(x) = \frac{1}{3}x - 3$

**45.** $f(t) = \frac{1}{2}(t^2 - 4t - 1)$

**46.** $g(x) = -x^2 + 10x - 16$

**47.** $f(x) = 4x^2 - x^3$

**48.** $f(x) = 1 - x^3$

**49.** $f(x) = x^3 - 9x$

**50.** $f(x) = \frac{1}{4}x^4 - 2x^2$

**51.** $g(t) = -\frac{1}{4}(t - 2)^2(t + 2)^2$

**52.** $f(x) = x(x - 2)^2(x + 1)$

**53.** $f(x) = 1 - x^6$

**54.** $g(x) = 1 - (x + 1)^6$

**55.** *Modeling Polynomials*  Sketch the graph of a polynomial function that is fourth degree, has a root of multiplicity 2, and has a negative leading coefficient. Sketch another graph under the same conditions but with a positive leading coefficient.

**56.** *Modeling Polynomials*  Sketch the graph of a polynomial function that is fifth degree, has a root of multiplicity 2, and has a negative leading coefficient. Sketch another graph under the same conditions but with a positive leading coefficient.

**57.** *Modeling Polynomials*  Determine the equation of the fourth-degree polynomial function $f(x)$ shown below.

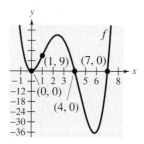

**58.** *Modeling Polynomials*  Determine the equation of the third-degree polynomial function $g(x)$ shown below.

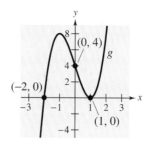

**59.** *Hockey Players' Weights* From 1971 to 1990, the average weight of a hockey player in the National Hockey League increased by about 11 pounds. A model that approximates the average weight of a hockey player is

$$W = 188.36 + 0.202t + 0.0042t^3, \quad -9 \le t \le 10$$

where $W$ is the weight in pounds and $t$ represents the playing season, with $t = 0$ corresponding to the 1980–81 playing season. (Source: National Hockey League)

(a) Use a graphing utility to graph the model.

(b) Estimate the playing season in which the average weight was 190 pounds.

(c) In the 1997–98 playing season, the average weight was 200.3 pounds. Does this follow the model? Explain.

**60.** *Total Investment Capital* From 1990 to 1997, the total investment capital for Northwest Natural Gas Company increased by about 280 million dollars. A model that approximates the total investment capital for this corporation is

$$C = 461.3 + 59.74t - 10.53t^2 + 1.097t^3,$$
$$0 \le t \le 7$$

where $C$ is the total investment capital (in millions of dollars) and $t = 0$ represents 1990. (Source: Northwest Natural Gas Company)

(a) Use a graphing utility to graph the model.

(b) Estimate the year (between 1990 and 1997) in which the total investment capital was the least.

**61.** *Advertising Expenses* The total revenue for a soft-drink company is related to its advertising expense by the function

$$R = \frac{1}{50,000}(-x^3 + 600x^2), \quad 0 \le x \le 400$$

where $R$ is the total revenue in millions of dollars and $x$ is the amount spent on advertising (in tens of thousands of dollars). Use the graph of this function to estimate the point on the graph at which the function is increasing most rapidly. This point is called the *point of diminishing returns* because any expenses above this amount will yield less return per dollar invested in advertising.

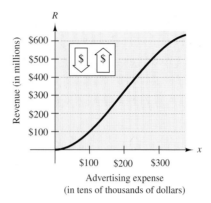

Figure for 61

**62.** *Advertising Expenses* The total revenue for a hotel corporation is related to its advertising expense by the function

$$R = -0.148x^3 + 4.889x^2 - 17.778x + 125.185,$$
$$0 \le x \le 20$$

where $R$ is the total revenue in millions of dollars and $x$ is the amount spent on advertising (in millions of dollars). Use the accompanying graph of this function to estimate the point on the graph at which the function is increasing most rapidly. This point is called the *point of diminishing returns* because any expenses above this amount will yield less return per dollar invested in advertising.

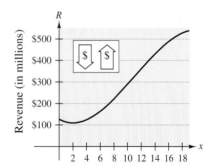

**63.** *Think About It* Use a graphing utility to graph the following functions: $f(x) = x^2$, $g(x) = x^4$, $h(x) = x^6$. Do the three functions have a common shape? Are their graphs identical? Why or why not?

## 4.3    Polynomial Division

### Objectives

- Divide one polynomial by a second polynomial using long division.
- Simplify a rational expression using long division.
- Use synthetic division to divide two polynomials.
- Use the Remainder Theorem and synthetic division to evaluate a polynomial.
- Use the Factor Theorem to factor a polynomial function.

## Long Division of Polynomials

In this section, you will study two procedures for *dividing* polynomials. These procedures are especially valuable in factoring and finding the zeros of polynomial functions. To begin, suppose you are given the graph of

$$f(x) = 6x^3 - 19x^2 + 16x - 4.$$

Notice that a zero of $f$ occurs at $x = 2$, as shown in Figure 4.21. Because $x = 2$ is a zero of the polynomial function $f$, you know that $(x - 2)$ is a factor of $f(x)$. This means that there exists a second-degree polynomial $q(x)$ such that

$$f(x) = (x - 2) \cdot q(x).$$

To find $q(x)$, you can use **long division,** as illustrated in Example 1.

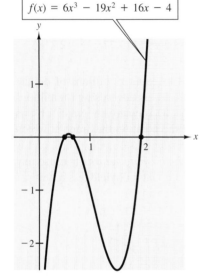

$f(x) = 6x^3 - 19x^2 + 16x - 4$

**Figure 4.21**

### Example 1    Long Division of Polynomials

Divide the polynomial $6x^3 - 19x^2 + 16x - 4$ by $x - 2$, and use the result to factor the polynomial completely.

**Solution**

$$
\require{enclose}
\begin{array}{r}
6x^2 - 7x + 2 \\[-3pt]
x - 2 \enclose{longdiv}{6x^3 - 19x^2 + 16x - 4}
\end{array}
$$

| | |
|---|---|
| $\underline{6x^3 - 12x^2}$ | Multiply: $6x^2(x - 2)$. |
| $-7x^2 + 16x$ | Subtract. |
| $\underline{-7x^2 + 14x}$ | Multiply: $-7x(x - 2)$. |
| $2x - 4$ | Subtract. |
| $\underline{2x - 4}$ | Multiply: $2(x - 2)$. |
| $0$ | Subtract. |

From this division, you can conclude that

$$6x^3 - 19x^2 + 16x - 4 = (x - 2)(6x^2 - 7x + 2)$$

and by factoring the quadratic $6x^2 - 7x + 2$, you have

$$6x^3 - 19x^2 + 16x - 4 = (x - 2)(2x - 1)(3x - 2).$$

Note that the factorization shown in Example 1 agrees with the graph shown in Figure 4.21 in that the three $x$-intercepts occur at $x = 2$, $x = \frac{1}{2}$, and $x = \frac{2}{3}$.

In Example 1, $x - 2$ is a factor of the polynomial $6x^3 - 19x^2 + 16x - 4$, and the long division process produces a remainder of zero. Often, long division will produce a nonzero remainder. For instance, if you divide $x^2 + 3x + 5$ by $x + 1$, you obtain the following.

$$
\begin{array}{r}
x + 2 \quad\longleftarrow \text{ Quotient} \\
x + 1 \overline{) \; x^2 + 3x + 5} \quad\longleftarrow \text{ Dividend} \\
\underline{x^2 + \;x} \\
2x + 5 \\
\underline{2x + 2} \\
3 \quad\longleftarrow \text{ Remainder}
\end{array}
$$

Divisor $\Longrightarrow$

In fractional form, you can write this result as follows.

$$
\underbrace{\frac{x^2 + 3x + 5}{x + 1}}_{\substack{\text{Dividend} \\ \text{Divisor}}} = \overbrace{x + 2}^{\text{Quotient}} + \underbrace{\frac{\overset{\text{Remainder}}{3}}{x + 1}}_{\text{Divisor}}
$$

This example illustrates the following well-known theorem called the **Division Algorithm.**

---

▶ **The Division Algorithm**

If $f(x)$ and $d(x)$ are polynomials such that $d(x) \neq 0$, and the degree of $d(x)$ is less than or equal to the degree of $f(x)$, there exist unique polynomials $q(x)$ and $r(x)$ such that

$$
f(x) = d(x)q(x) + r(x)
$$

Dividend $\quad$ Quotient

Divisor $\quad$ Remainder

where $r(x) = 0$ *or* the degree of $r(x)$ is less than the degree of $d(x)$. If the remainder $r(x)$ is zero, $d(x)$ **divides evenly** into $f(x)$.

---

The Division Algorithm can also be written as

$$
\frac{f(x)}{d(x)} = q(x) + \frac{r(x)}{d(x)}.
$$

In the Division Algorithm, the rational expression $f(x)/d(x)$ is **improper** because the degree of $f(x)$ is greater than or equal to the degree of $d(x)$. On the other hand, the rational expression $r(x)/d(x)$ is **proper** because the degree of $r(x)$ is less than the degree of $d(x)$.

## Study Tip

Be sure to include a place-holder for each missing power of $x$. For example, to set up $x^4 - 9x^2 + 10$ divided by $x - 1$, you would write the following.

$$x - 1 \,\overline{)\, x^4 + 0x^3 - 9x^2 + 0x + 10}$$

**Example 2**    Long Division of Polynomials

Divide $x^3 - 1$ by $x - 1$.

### Solution

Because there is no $x^2$-term or $x$-term in the dividend, you need to line up the subtraction by using zero coefficients (or leaving spaces) for the missing terms.

$$
\begin{array}{r}
x^2 + \phantom{x}x + 1 \\
x - 1 \,\overline{)\, x^3 + 0x^2 + 0x - 1} \\
\underline{x^3 - \phantom{x}x^2} \phantom{+ 0x - 1} \\
x^2 \phantom{+ 0x - 1} \\
\underline{x^2 - \phantom{x}x} \phantom{- 1} \\
x - 1 \\
\underline{x - 1} \\
0
\end{array}
$$

So, $x - 1$ divides evenly into $x^3 - 1$ and you can write

$$\frac{x^3 - 1}{x - 1} = x^2 + x + 1.$$

You can check the result of a division problem by multiplying. For instance, in Example 2, try checking that $(x - 1)(x^2 + x + 1) = x^3 - 1$.

**Example 3**    Long Division of Polynomials

Divide $2x^4 + 4x^3 - 5x^2 + 3x - 2$ by $x^2 + 2x - 3$.

### Solution

$$
\begin{array}{r}
2x^2 \phantom{+ 3x} + 1 \\
x^2 + 2x - 3 \,\overline{)\, 2x^4 + 4x^3 - 5x^2 + 3x - 2} \\
\underline{2x^4 + 4x^3 - 6x^2} \phantom{+ 3x - 2} \\
x^2 + 3x - 2 \\
\underline{x^2 + 2x - 3} \\
x + 1
\end{array}
$$

Note that the first subtraction eliminated two terms from the dividend. When this happens, the quotient skips a term. So, you can write

$$\frac{2x^4 + 4x^3 - 5x^2 + 3x - 2}{x^2 + 2x - 3} = 2x^2 + 1 + \frac{x + 1}{x^2 + 2x - 3}.$$

## Synthetic Division

There is a nice shortcut for long division by polynomials of the form $x - k$. The shortcut is called **synthetic division.** We summarize the pattern for synthetic division of a cubic polynomial as follows. (The pattern for higher-degree polynomials is similar.)

---

▶ **Synthetic Division (for a Cubic Polynomial)**

To divide $ax^3 + bx^2 + cx + d$ by $x - k$, use the following pattern.

*Vertical pattern:* Add terms.

*Diagonal pattern:* Multiply by $k$.

---

Synthetic division works *only* for divisors of the form $x - k$. You cannot use synthetic division to divide a polynomial by a quadratic, a cubic, or any other higher-degree polynomial.

**Example 4**    Using Synthetic Division

Use synthetic division to divide $x^4 - 10x^2 - 2x + 4$ by $x + 3$.

**Solution**

You should set up the array as follows. Note that a zero is included for each missing term in the dividend.

Divisor: $x + 3$        Dividend: $x^4 - 10x^2 - 2x + 4$

$$
\begin{array}{r|rrrrr}
-3 & 1 & 0 & -10 & -2 & 4 \\
   &   & -3 & 9 & 3 & -3 \\
\hline
   & 1 & -3 & -1 & 1 & 1 \quad\leftarrow \text{Remainder: } 1
\end{array}
$$

Quotient: $x^3 - 3x^2 - x + 1$

So, you have

$$\frac{x^4 - 10x^2 - 2x + 4}{x + 3} = x^3 - 3x^2 - x + 1 + \frac{1}{x + 3}.$$

## Remainder and Factor Theorems

The remainder obtained in the synthetic division process has an important interpretation found in the Remainder Theorem.

---

▶ **The Remainder Theorem**

If a polynomial $f(x)$ is divided by $x - k$, the remainder is

$$r = f(k).$$

---

The Remainder Theorem tells you that synthetic division can be used to evaluate a polynomial function. That is, to evaluate a polynomial function $f(x)$ when $x = k$, divide $f(x)$ by $x - k$. The remainder will be $f(k)$, as illustrated in Example 5.

---

**Example 5**     Using the Remainder Theorem

Use the Remainder Theorem to evaluate the following function at $x = -2$.

$$f(x) = 3x^3 + 8x^2 + 5x - 7$$

**Solution**

Using synthetic division, you obtain the following.

$$
\begin{array}{r|rrrr}
-2 & 3 & 8 & 5 & -7 \\
   &   & -6 & -4 & -2 \\
\hline
   & 3 & 2 & 1 & -9
\end{array}
$$

Because the remainder is $r = -9$, you can conclude that

$$f(-2) = -9.$$

This means that $(-2, -9)$ is a point on the graph of $f$. Try checking this by substituting $x = -2$ in the original function.

---

•**TECHNOLOGY**•

Remember, you can also evaluate a function with your graphing utility by entering the function in the equation editor and using the table feature in ASK mode.

---

Another important theorem is the Factor Theorem, which is stated below.

---

▶ **Factor Theorem**

A polynomial $f(x)$ has a factor $(x - k)$ if and only if $f(k) = 0$.

---

Example 6   **Factoring a Polynomial**

Show that $(x - 2)$ and $(x + 3)$ are factors of the polynomial

$$f(x) = 2x^4 + 7x^3 - 4x^2 - 27x - 18.$$

Then find the remaining factors of $f(x)$.

### Solution

Using synthetic division with 2 and $-3$ *successively*, you obtain the following.

```
2 |  2    7   -4  -27  -18
   |       4   22   36   18        0 remainder
   ---------------------------     (x − 2) is a factor.
      2   11   18    9    0
```

```
-3 |  2   11   18    9
   |      -6  -15   -9             0 remainder
   ----------------------         (x + 3) is a factor.
      2    5    3    0
```

Because the resulting quadratic expression factors as

$$2x^2 + 5x + 3 = (2x + 3)(x + 1)$$

the complete factorization of $f(x)$ is

$$f(x) = (x - 2)(x + 3)(2x + 3)(x + 1).$$

Note that this factorization implies that $f$ has four real zeros:

$$2, -3, -\tfrac{3}{2}, \text{ and } -1.$$

This is confirmed by the graph of $f$, which is shown in Figure 4.22.

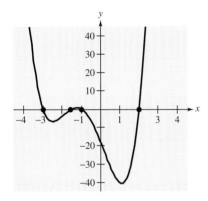

**Figure 4.22**

---

> ▶ **Results of Synthetic Division**
>
> The remainder $r$, obtained in the synthetic division of $f(x)$ by $x - k$, provides the following information.
>
> **1.** The remainder $r$ gives the value of $f$ at $x = k$. That is, $r = f(k)$.
>
> **2.** If $r = 0$, $(x - k)$ is a factor of $f(x)$.
>
> **3.** If $r = 0$, $(k, 0)$ is an $x$-intercept of the graph of $f$.

Throughout this text, we have emphasized the importance of developing several problem-solving strategies. In the exercises for this section, try using more than one strategy to solve several of the exercises. For instance, if you find that $x - k$ divides evenly into $f(x)$, try sketching the graph of $f$. You should find that $(k, 0)$ is an $x$-intercept of the graph.

**(a)**

Gross income (in dollars)

**(b)**

**Figure 4.23**

## Application

Example 7 Take-Home Pay

The 1996 yearly take-home pay for an employee who was married and claimed four deductions is given by the function

$$y = -0.00000108016x^2 + 0.90939x + 2075.911, \quad 25{,}000 \le x \le 75{,}000$$

where $y$ represents the take-home pay and $x$ represents the yearly salary. Find a function that gives the take-home pay as a *percent* of the yearly salary. (Source: U.S. Department of the Treasury)

### Solution

Because the yearly salary is given by $x$ and the take-home pay is given by $y$, the percent of yearly salary that the person takes home is

$$P = \frac{y}{x}$$

$$= \frac{-0.00000108016x^2 + 0.90939x + 2075.911}{x}$$

$$= -0.00000108016x + 0.90939 + \frac{2075.911}{x}.$$

The graphs of these functions are shown in Figure 4.23(a) and (b). Note in Figure 4.23(b) that as a person's yearly salary increases, the percent that he or she takes home decreases.

| Discussing the Concept | Understanding the Remainder Theorem |
| --- | --- |

The Remainder Theorem states: "If a polynomial $f(x)$ is divided by $x - k$, the remainder is $r = f(k)$." Given $f(x) = x^3 + 6x^2 + 5x - 12$ with the real zeros $-4$, $-3$, and $1$, explain algebraically and graphically what it means if $r = 0$. Explain algebraically and graphically what it means if $r \ne 0$. What is the relationship between the Remainder Theorem and the $x$-intercepts and factors of $f(x)$?

# Warm Up

The following warm-up exercises involve skills that were covered in earlier sections. You will use these skills in the exercise set for this section.

In Exercises 1–4, write the expression in standard polynomial form.

**1.** $(x - 1)(x^2 + 2) + 5$     **2.** $(x^2 - 3)(2x + 4) + 8$

**3.** $(x^2 + 1)(x^2 - 2x + 3) - 10$     **4.** $(x + 6)(2x^3 - 3x) - 5$

In Exercises 5 and 6, factor the polynomial.

**5.** $x^2 - 4x + 3$     **6.** $4x^3 - 10x^2 + 6x$

In Exercises 7–10, find a polynomial function that has the given zeros.

**7.** $0, 3, 4$     **8.** $-6, 1$

**9.** $-3, 1 + \sqrt{2}, 1 - \sqrt{2}$     **10.** $1, -2, 2 + \sqrt{3}, 2 - \sqrt{3}$

## 4.3    Exercises

In Exercises 1–14, divide by long division.

| | *Dividend* | *Divisor* |
|---|---|---|
| **1.** | $3x^2 - 7x + 4$ | $x - 1$ |
| **2.** | $5x^2 - 17x - 12$ | $x - 4$ |
| **3.** | $2x^3 + 6x^2 - x - 3$ | $2x^2 - 1$ |
| **4.** | $3x^3 - 12x^2 - 2x + 8$ | $3x^2 - 2$ |
| **5.** | $x^4 + 5x^3 + 6x^2 - x - 2$ | $x + 2$ |
| **6.** | $x^4 + 2x^3 - 3x^2 - 8x - 4$ | $x^2 - 4$ |
| **7.** | $7x + 3$ | $x + 2$ |
| **8.** | $8x - 5$ | $2x + 1$ |
| **9.** | $6x^3 + 10x^2 + x + 8$ | $2x^2 + 1$ |
| **10.** | $x^3 - 9$ | $x^2 + 1$ |
| **11.** | $x^3 - 4x^2 + 5x - 2$ | $x - 1$ |
| **12.** | $x^3 - x^2 + 2x - 8$ | $x - 2$ |
| **13.** | $2x^5 - 8x^3 + 4x - 1$ | $x^2 - 2x + 1$ |
| **14.** | $x^5 + 7$ | $x^3 - 1$ |

In Exercises 15–32, divide by synthetic division.

| | *Dividend* | *Divisor* |
|---|---|---|
| **15.** | $2x^3 + 5x^2 - 7x + 20$ | $x + 4$ |
| **16.** | $3x^3 - 23x^2 - 12x + 32$ | $x - 8$ |
| **17.** | $4x^3 - 9x + 8x^2 - 18$ | $x + 2$ |
| **18.** | $9x^3 - 16x - 18x^2 + 32$ | $x - 2$ |
| **19.** | $-x^3 + 75x - 250$ | $x + 10$ |
| **20.** | $3x^3 - 16x^2 - 72$ | $x - 6$ |
| **21.** | $5x^3 - 6x^2 + 8$ | $x - 4$ |
| **22.** | $6x^4 - 15x^3 - 11x$ | $x + 2$ |
| **23.** | $10x^4 - 50x^3 - 800$ | $x - 6$ |
| **24.** | $x^5 - 13x^4 - 120x + 80$ | $x + 3$ |
| **25.** | $2x^5 - 30x^3 - 37x + 13$ | $x - 4$ |
| **26.** | $5x^3$ | $x + 3$ |
| **27.** | $-3x^4$ | $x - 2$ |
| **28.** | $-3x^4$ | $x + 2$ |
| **29.** | $5 - 3x + 2x^2 - x^3$ | $x + 1$ |
| **30.** | $180x - x^4$ | $x - 6$ |
| **31.** | $4x^3 + 16x^2 - 23x - 15$ | $x + \frac{1}{2}$ |
| **32.** | $3x^3 - 4x^2 + 5$ | $x - \frac{3}{2}$ |

In Exercises 33–38, use synthetic division to show that $x$ is a solution of the third-degree polynomial equation, and use the result to factor the polynomial completely.

**33.** $x^3 - 12x - 16 = 0$, $\quad x = -2$

**34.** $x^3 - 28x - 48 = 0$, $\quad x = -4$

**35.** $2x^3 + 3x^2 - 17x + 12 = 0$, $\quad x = \frac{3}{2}$

**36.** $4x^3 + x^2 - 16x - 4 = 0$, $\quad x = -\frac{1}{4}$

**37.** $x^3 + 2x^2 - 3x - 6 = 0$, $\quad x = \sqrt{3}$

**38.** $x^3 + 2x^2 - 2x - 4 = 0$, $\quad x = \sqrt{2}$

In Exercises 39–42, express the function in the form

$$f(x) = (x - k)q(x) + r$$

for the given value of $k$, and demonstrate that $f(k) = r$.

**39.** $f(x) = 2x^3 - 9x^2 + 10x - 9$, $\quad k = \frac{1}{2}$

**40.** $f(x) = \frac{1}{3}(15x^4 + 10x^3 - 6x^2 + 17x + 14)$,
$\quad k = -\frac{2}{3}$

**41.** $f(x) = x^3 + 3x^2 - 2x - 14$, $\quad k = \sqrt{2}$

**42.** $f(x) = 4x^3 - 6x^2 - 12x - 4$, $\quad k = 1 - \sqrt{3}$

In Exercises 43–48, use synthetic division to find the required function values.

**43.** $f(x) = 2x^5 - 3x^2 - 4x - 1$
   (a) $f(-2)$ (b) $f(-5)$
   (c) $f(1)$ (d) $f(3)$

**44.** $g(x) = x^6 - 4x^4 + 3x^2 + 2$
   (a) $g(2)$ (b) $g(-4)$
   (c) $g(3)$ (d) $g(-1)$

**45.** $f(x) = 1.2x^3 - 0.5x^2 - 2.1x - 2.4$
   (a) $f(2)$ (b) $f(-6)$
   (c) $f\left(\frac{2}{3}\right)$ (d) $f(1)$

**46.** $f(x) = 0.4x^4 - 1.6x^3 + 0.7x^2 - 2$
   (a) $f(1)$ (b) $f(-2)$
   (c) $f(5)$ (d) $f(-10)$

**47.** $f(x) = x^3 - 2x^2 - 11x + 52$
   (a) $f(5)$ (b) $f(-4)$
   (c) $f(1.2)$ (d) $f(2)$

**48.** $g(x) = x^3 - x^2 + 25x - 25$
   (a) $g(5)$ (b) $g\left(\frac{1}{5}\right)$
   (c) $g(-1.5)$ (d) $g(-1)$

In Exercises 49–52, match the function with its graph and use the result to find all real solutions of $f(x) = 0$. [The graphs are labeled (a), (b), (c), and (d).]

(a)

(b)

(c)

(d)

**49.** $f(x) = x^3 - 2x^2 - 7x + 12$

**50.** $f(x) = x^3 - x^2 - 5x + 2$

**51.** $f(x) = x^3 + 5x^2 + 6x + 2$

**52.** $f(x) = x^3 - 5x^2 + 2x + 12$

**53.** *Modeling Polynomials* A third-degree polynomial function $f(x)$ has real zeros $-1$, $2$, and $\frac{10}{3}$. Find two different polynomial functions, one with a positive leading coefficient and one with a negative leading coefficient, that could be $f(x)$. How many different polynomial functions are there for $f(x)$?

**54.** *Modeling Polynomials* A fourth-degree polynomial function $g(x)$ has real zeros $-2$, $0$, $1$, and $5$. Find two different polynomial functions, one with a positive leading coefficient and one with a negative leading coefficient, that could be $g(x)$. How many different polynomial functions could there be for $g(x)$?

In Exercises 55–60, simplify the rational function.

**55.** $\dfrac{4x^3 - x^2 - 13x + 10}{4x - 5}$

**56.** $\dfrac{2x^3 + 7x^2 - 10x + 3}{2x - 1}$

**57.** $\dfrac{3x^3 - 7x^2 - 13x - 28}{x - 4}$

**58.** $\dfrac{2x^3 + 3x^2 - 3x - 2}{x - 1}$

**59.** $\dfrac{x^4 + 6x^3 + 11x^2 + 6x}{x^2 + 3x + 2}$

**60.** $\dfrac{x^4 + 9x^3 - 5x^2 - 36x + 4}{x^2 - 4}$

**61.** *Measurements of a Room*   A rectangular room has a volume of

$$x^3 + 11x^2 + 34x + 24$$

cubic feet. The height of the room is $x + 1$ (see figure). Find the number of square feet of floor space in the room.

**62.** *Measurements of a House*   A rectangular house has a volume of

$$x^3 + 55x^2 + 650x + 2000$$

cubic feet (the space in the attic is not counted). The height of the house is $x + 5$ (see figure). Find the number of square feet of floor space *on the first floor* of the house.

**63.** *Profit*   A company making compact discs estimates that the profit from selling a particular disc is

$$P = -153.6x^3 + 5760x^2 - 100,000,$$
$$0 \le x \le 35$$

where $P$ is the profit in dollars and $x$ is the advertising expense (in tens of thousands of dollars). For this disc, the advertising expense was $300,000 ($x = 30$) and the profit was $936,800.

(a) From the graph shown in the figure, it appears that the company could have obtained the same profit by spending less on advertising. Use the graph to estimate another amount that the company could have spent that would have produced the same profit.

(b) Use synthetic division with $x = 30$ to confirm algebraically the result of part (a).

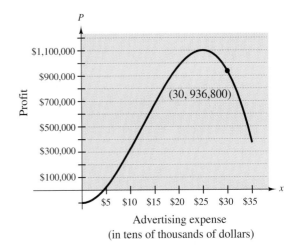

Advertising expense
(in tens of thousands of dollars)

**64.** *Profit*   A company that produces calculators estimates that the profit from selling a particular model of calculator is

$$P = -125x^3 + 6450x^2 - 150,000, \quad 0 \le x \le 50$$

where $P$ is the profit in dollars and $x$ is the advertising expense (in tens of thousands of dollars). For this model of calculator, the advertising expense was $400,000 ($x = 40$) and the profit was $2,170,000.

(a) Use a graphing utility to graph the profit function.

(b) Could the company have obtained the same profit by spending less on advertising? Explain your reasoning.

## 4.4     Real Zeros of Polynomial Functions

### Objectives

- Find all possible rational zeros of a function using the Rational Zero Test.
- Approximate the real zeros of a polynomial function using the Intermediate Value Theorem.
- Approximate the real zeros of a polynomial function using a graphing utility.
- Apply techniques for approximating real zeros to solve an application problem.

## The Rational Zero Test

The **Rational Zero Test** relates the possible rational zeros of a polynomial (having integer coefficients) to the leading coefficient and to the constant term of the polynomial.

> ▶ **The Rational Zero Test**
>
> If the polynomial $f(x) = a_n x^n + a_{n-1} x^{n-1} + \cdots + a_2 x^2 + a_1 x + a_0$ has *integer* coefficients, then every rational zero of $f$ has the form
>
> $$\text{Rational zero} = \frac{p}{q}$$
>
> where $p$ and $q$ have no common factors other than 1, and
>
> $p =$ a factor of the constant term $a_0$
>
> $q =$ a factor of the leading coefficient $a_n$.

**Study Tip**

When the leading coefficient is 1, the possible rational zeros are simply the factors of the constant term.

To use the Rational Zero Test, you should first list all rational numbers whose numerators are factors of the constant term and whose denominators are factors of the leading coefficient.

$$\text{Possible rational zeros} = \frac{\text{factors of constant term}}{\text{factors of leading coefficient}}$$

Having formed this list of *possible rational zeros*, use a trial-and-error method to determine which, if any, are actual zeros of the polynomial.

**Example 1**     Rational Zero Test with Leading Coefficient of 1

Find the rational zeros of $f(x) = x^3 + x + 1$. (See Figure 4.24.)

**Solution**

The only factors of the leading coefficient are 1 and $-1$. By testing these possible zeros, you can see that neither works.

$$f(1) = (1)^3 + 1 + 1 = 3$$
$$f(-1) = (-1)^3 + (-1) + 1 = -1$$

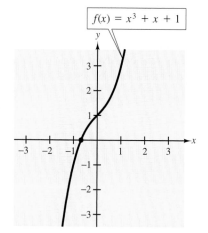

$f(x) = x^3 + x + 1$

Figure 4.24

So, you can conclude that the given polynomial has *no* rational zeros. Note from the graph of $f$ in Figure 4.24 that $f$ does have one real zero (between $-1$ and $0$). By the Rational Zero Test, you know that this real zero is *not* a rational number.

**Example 2**    Rational Zero Test with Leading Coefficient of 1

Find the rational zeros of

$$f(x) = x^4 - x^3 + x^2 - 3x - 6.$$

**Solution**

Because the leading coefficient is 1, the possible rational zeros are the factors of the constant term.

*Possible rational zeros:*  $\pm 1, \pm 2, \pm 3, \pm 6$

A test of these possible zeros shows that $x = -1$ and $x = 2$ are the only two that work. Check the others to be sure.

___

If the leading coefficient of a polynomial is not 1, the list of possible rational zeros can increase dramatically. In such cases the search can be shortened in several ways: (1) a programmable calculator can be used to speed up the calculations; (2) a graph, either by hand or with a graphing utility, can give a good estimate of the locations of the zeros; and (3) synthetic division can be used to test the possible rational zeros.

To see how to use synthetic division to test the possible rational zeros, let's take another look at the function

$$f(x) = x^4 - x^3 + x^2 - 3x - 6$$

given in Example 2. To test that $x = -1$ and $x = 2$ are zeros of $f$, you can apply synthetic division, as follows.

$$
\begin{array}{r|rrrrr}
-1 & 1 & -1 & 1 & -3 & -6 \\
   &   & -1 & 2 & -3 & 6 \\
\hline
   & 1 & -2 & 3 & -6 & 0 \\
\end{array}
$$

$$
\begin{array}{r|rrrr}
2 & 1 & -2 & 3 & -6 \\
  &   & 2  & 0 & 6 \\
\hline
  & 1 & 0 & 3 & 0 \\
\end{array}
$$

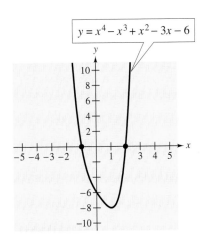

$$y = x^4 - x^3 + x^2 - 3x - 6$$

So, you have

$$f(x) = (x + 1)(x - 2)(x^2 + 3).$$

Because the factor $(x^2 + 3)$ produces no real zeros, you can conclude that $x = -1$ and $x = 2$ are the only *real* zeros of $f$. This is verified in the graph of $f$ shown in Figure 4.25.

**Figure 4.25**

Finding the first zero is often the hardest part. After that, the search is simplified by using the lower-degree polynomial obtained in synthetic division. Once the lower-degree polynomial is quadratic, either factoring or the Quadratic Formula can be used to find the remaining zeros.

### Example 3   Using the Rational Zero Test

Find the rational zeros of $f(x) = 2x^3 + 3x^2 - 8x + 3$.

**Solution**

The leading coefficient is 2 and the constant term is 3.

$$\text{Possible rational zeros: } \frac{\text{Factors of 3}}{\text{Factors of 2}} = \frac{\pm 1, \pm 3}{\pm 1, \pm 2} = \pm 1, \pm 3, \pm \frac{1}{2}, \pm \frac{3}{2}$$

By synthetic division, you can determine that $x = 1$ is a zero.

$$
\begin{array}{c|cccc}
1 & 2 & 3 & -8 & 3 \\
  &   & 2 & 5 & -3 \\
\hline
  & 2 & 5 & -3 & 0
\end{array}
$$

So, $f(x)$ factors as

$$f(x) = (x - 1)(2x^2 + 5x - 3) = (x - 1)(2x - 1)(x + 3)$$

and you can conclude that the zeros of $f$ are $x = 1$, $x = \frac{1}{2}$, and $x = -3$.

### Example 4   Using the Rational Zero Test

Find all the real zeros of $f(x) = 10x^3 - 15x^2 - 16x + 12$.

**Solution**

The leading coefficient is 10 and the constant term is 12.

$$\text{Possible rational zeros: } \frac{\text{Factors of 12}}{\text{Factors of 10}} = \frac{\pm 1, \pm 2, \pm 3, \pm 4, \pm 6, \pm 12}{\pm 1, \pm 2, \pm 5, \pm 10}$$

With so many possibilities (32, in fact), it is worth your time to stop and sketch a graph. From Figure 4.26, it looks like three reasonable choices would be $x = -\frac{6}{5}$, $x = \frac{1}{2}$, and $x = 2$. Testing these by synthetic division shows that only $x = 2$ works. So, you have

$$f(x) = (x - 2)(10x^2 + 5x - 6).$$

Using the Quadratic Formula, you find that the two additional zeros are irrational numbers.

$$x = \frac{-5 + \sqrt{265}}{20} \approx 0.5639 \quad \text{and} \quad x = \frac{-5 - \sqrt{265}}{20} \approx -1.0639$$

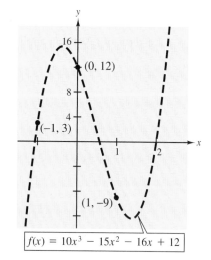

$f(x) = 10x^3 - 15x^2 - 16x + 12$

**Figure 4.26**

# The Intermediate Value Theorem

The next theorem, called the **Intermediate Value Theorem,** tells you of the existence of real zeros of polynomial functions. The theorem implies that if $(a, f(a))$ and $(b, f(b))$ are two points on the graph of a polynomial such that $f(a) \neq f(b)$, then for any number $d$ between $f(a)$ and $f(b)$ there must be a number $c$ between $a$ and $b$ such that $f(c) = d$. (See Figure 4.27.)

▶ **Intermediate Value Theorem**

Let $a$ and $b$ be real numbers such that $a < b$. If $f$ is a polynomial function such that $f(a) \neq f(b)$, then, in the interval $[a, b]$, $f$ takes on every value between $f(a)$ and $f(b)$.

The Intermediate Value Theorem helps you locate the real zeros of a polynomial function in the following way. If you can find a value $x = a$ where a polynomial function is positive, and another value $x = b$ where it is negative, you can conclude that the function has at least one real zero between these two values. For example, the function

$$f(x) = x^3 + x^2 + 1$$

is negative when $x = -2$ and positive when $x = -1$. So, it follows from the Intermediate Value Theorem that $f$ must have a real zero somewhere between $-2$ and $-1$, as shown in Figure 4.28.

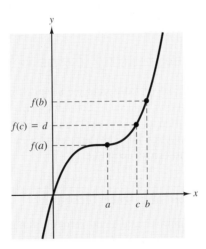

**Figure 4.27**   *If d lies between f(a) and f(b), there exists c between a and b such that f(c) = d.*

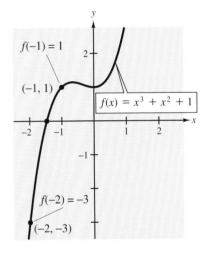

**Figure 4.28**   *f has a zero between $-2$ and $-1$.*

| Example 5 | Approximating a Zero of a Polynomial Function |

Use the Intermediate Value Theorem to approximate the real zero of

$$f(x) = x^3 - x^2 + 1.$$

**Solution**

Begin by computing a few function values, as follows.

| $x$ | $-2$ | $-1$ | 0 | 1 |
|------|------|------|---|---|
| $f(x)$ | $-11$ | $-1$ | 1 | 1 |

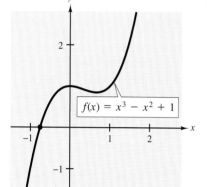

$f(x) = x^3 - x^2 + 1$

**Figure 4.29** *f has a zero between* $-0.8$ *and* $-0.7$.

Because $f(-1)$ is negative and $f(0)$ is positive, you can apply the Intermediate Value Theorem to conclude that the function has a zero between $-1$ and 0. To pinpoint this zero more closely, divide the interval $[-1, 0]$ into tenths and evaluate the function at each point. When you do this, you will find that

$$f(-0.8) = -0.152$$

and

$$f(-0.7) = 0.167.$$

So, $f$ must have a zero between $-0.8$ and $-0.7$, as shown in Figure 4.29. By continuing this process you can approximate this zero to any desired accuracy.

## Approximating Zeros of Polynomial Functions

There are several different techniques for approximating the zeros of a polynomial function. All such techniques are better suited to computers or graphing utilities than they are to "hand calculations." In this section, you will study two techniques that can be used with a graphing utility. The first is called the **zoom-and-trace** technique.

---

▶ **Zoom-and-Trace Technique**

To approximate a real zero of a function with a graphing utility, use the following steps.

1. Graph the function so that the real zero you want to approximate appears as an $x$-intercept on the screen.

2. Move the cursor near the $x$-intercept and use the zoom feature to zoom in to get a better look at the intercept.

3. Use the trace feature to find the $x$-values that occur just before and just after the $x$-intercept. If the difference in these values is sufficiently small, use their average as the approximation. If not, continue zooming in until the approximation reaches the desired accuracy.

To help you visually determine when you have zoomed in enough times to reach the desired accuracy, you can set the X-scale of the viewing window to the accuracy you need and zoom in repeatedly.

- To approximate the zero to the nearest hundredth, set the X-scale to 0.01.
- To approximate the zero to the nearest thousandth, set the X-scale to 0.001.
- To approximate the zero to the nearest ten-thousandth, set the X-scale to 0.0001.

The amount that a graphing utility zooms in is determined by the *zoom factor*. The zoom factor is an integer that gives the ratio of the larger screen to the smaller screen. For instance, if you zoom in with a zoom factor of 2, you will obtain a screen in which the $x$- and $y$-values are half their original values. In this text, we use a zoom factor of 4.

## Example 6   Approximating the Zeros of a Polynomial Function

Approximate the real zeros of the polynomial function

$$f(x) = x^3 + 4x + 2.$$

Use an accuracy of 0.001.

### Solution

To begin, use a graphing utility to graph the function, as shown in Figure 4.30(a). Set the X-scale to 0.001 and zoom in several times until the tick marks on the $x$-axis become visible. The final screen should be similar to the one shown in Figure 4.30(b). At this point, you can use the trace feature to determine that the $x$-values just to the left and right of the $x$-intercept are

$$x \approx -0.4735 \quad \text{and} \quad x \approx -0.4733.$$

So, to the nearest thousandth, you can approximate the zero of the function to be

$$x \approx -0.473.$$

To check this, try substituting $-0.473$ into the function. You should obtain a result that is approximately zero.

(a)

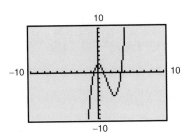

(b)
Figure 4.30

In Example 6, the cubic polynomial function has only one real zero. Remember that functions can have two or more real zeros. In such cases, you can use the zoom-and-trace technique for each zero separately. For instance, the function

$$f(x) = x^3 - 4x^2 + x + 2$$

has three real zeros, as shown in Figure 4.31. Using a zoom-and-trace approach for each real zero, you can approximate the real zeros to be

$$-0.562, \quad 1.000, \quad \text{and} \quad 3.562.$$

Figure 4.31

The second technique that can be used with some graphing utilities is to use the graphing utility's **solve, root,** or **zero key.** The name of this key or feature differs with different calculators. Consult your user's guide to determine if this feature is available.

**Example 7** Approximating the Zeros of a Polynomial Function

Approximate the real zeros of $f(x) = x^3 - 2x^2 - x + 1$.

**Solution**

To begin, use a graphing utility to graph the function, as shown in Figure 4.32. Notice that the graph has three $x$-intercepts. To approximate the leftmost intercept, find an appropriate viewing window and use the root feature. After this has been done, the calculator should display an approximation of

$$x \approx -0.8019377$$

which is accurate to seven decimal places. By repeating this process, you can determine that the other two zeros are $x \approx 0.556$ and $x \approx 2.247$.

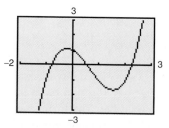

*Find an appropriate viewing window, then use the root feature.*

*Cursor to the left of the intercept and press "Enter."*

*Cursor to the right of the intercept and press "Enter."*
Figure 4.32

*Cursor near the intercept and press "Enter."*

You may be wondering why we spend so much time in algebra trying to find the *zeros* of a function. The reason is that if you have a technique that will allow you to solve the equation $f(x) = 0$, you can use the same technique to solve the more general equation

$$f(x) = c$$

where $c$ is any real number. This procedure is demonstrated in Example 8.

**Example 8**    **Solving the Equation $f(x) = c$**

Find a value of $x$ such that $f(x) = 30$ for the function

$$f(x) = x^3 - 4x + 4.$$

### Solution

The graph of $f(x) = x^3 - 4x + 4$ is shown in Figure 4.33. Note from the graph that $f(x) = 30$ when $x$ is about 3.5. To use the zoom-and-trace technique to approximate this $x$-value more closely, consider the equation

$$x^3 - 4x + 4 = 30$$

$$x^3 - 4x - 26 = 0.$$

So, the *solutions* of the equation $f(x) = 30$ are precisely the same $x$-values as the *zeros* of $g(x) = x^3 - 4x - 26$. Using the graph of $g$, as shown in Figure 4.34, you can approximate the zero of $g$ to be

$$x \approx 3.41.$$

You can check this value by substituting $x = 3.41$ into the original function.

$$f(3.41) = (3.41)^3 - 4(3.41) + 4$$

$$\approx 30.01 \checkmark$$

Remember that with decimal approximations, a check will usually not produce an exact value.

Figure 4.33

Figure 4.34

## Application

### Example 9    Profit and Advertising Expenses

A company that produces sports clothes estimates that the profit from selling a particular line of sportswear is given by

$$P = -140x^3 + 7520x^2 - 400{,}000, \qquad 0 \le x \le 50$$

where $P$ is the profit (in tens of thousands of dollars) and $x$ is the advertising expense (in tens of thousands of dollars). According to this model, how much money should the company spend to obtain a profit of $2,750,000?

### Solution

From Figure 4.35, it appears that there are two different values of $x$ between 0 and 50 that will produce a profit of $2,750,000. However, because of the context of the problem, it is clear that the better answer is the smaller of the two numbers. So, to solve the equation

$$-140x^3 + 7520x^2 - 400{,}000 = 2{,}750{,}000$$

$$-140x^3 + 7520x^2 - 3{,}150{,}000 = 0$$

you can find the zeros of the function $g(x) = -140x^3 + 7520x^2 - 3{,}150{,}000$. Using the zoom-and-trace technique, you can find that the leftmost zero is $x \approx 32.8$. You can check this solution by substituting $x = 32.8$ into the original function.

$$P = -140(32.8)^3 + 7520(32.8)^2 - 400{,}000 \approx \$2{,}750{,}059$$

So, the company should plan to spend about $328,000 in advertising for the line of sportswear.

Figure 4.35

---

| Discussing the Concept | Comparing Real Zeros and Rational Zeros |
|---|---|

Discuss the meanings of *real zeros* and *rational zeros* of a polynomial function and compare these two types of zeros. Then answer the following questions.

**a.** Is it possible for a polynomial function to have no rational zeros but to have real zeros? If so, give an example.

**b.** If a polynomial function has three real zeros, and only one of them is a rational number, must the other two zeros be irrational numbers?

**c.** Consider a cubic polynomial function $f(x) = ax^3 + bx^2 + cx + d$ where $a \ne 0$. Is it possible that $f$ has no real zeros? Is it possible that $f$ has no rational zeros?

**d.** Is it possible that a second-degree polynomial function with integer coefficients has one rational zero and one irrational zero? If so, give an example.

# Warm Up

The following warm-up exercises involve skills that were covered in earlier sections. You will use these skills in the exercise set for this section.

In Exercises 1 and 2, find a polynomial function with integer coefficients having the given zeros.

**1.** $-1, \frac{2}{3}, 3$

**2.** $-2, 0, \frac{3}{4}, 2$

In Exercises 3 and 4, divide by synthetic division.

**3.** $\dfrac{x^5 - 9x^3 + 5x + 18}{x + 3}$

**4.** $\dfrac{3x^4 + 17x^3 + 10x^2 - 9x - 8}{x + \frac{2}{3}}$

In Exercises 5–8, use the given zero to find all the real zeros of $f$.

**5.** $f(x) = 2x^3 + 11x^2 + 2x - 4, \ x = \frac{1}{2}$

**6.** $f(x) = 6x^3 - 47x^2 - 124x - 60, \ x = 10$

**7.** $f(x) = 4x^3 - 13x^2 - 4x + 6, \ x = -\frac{3}{4}$

**8.** $f(x) = 10x^3 + 51x^2 + 48x - 28, \ x = \frac{2}{5}$

In Exercises 9 and 10, find all real solutions.

**9.** $x^4 - 3x^2 + 2 = 0$

**10.** $x^4 - 7x^2 + 12 = 0$

## 4.4    Exercises

LAB **4**

In Exercises 1–4, use the Rational Zero Test to list all possible rational zeros of $f$. Then use a graphing utility to graph the function. Use the graph to help determine which of the possible rational zeros are actual zeros of the function.

**1.** $f(x) = x^3 + x^2 - 4x - 4$

**2.** $f(x) = -4x^3 + 15x^2 - 8x - 3$

**3.** $f(x) = 4x^4 - 17x^2 + 4$

**4.** $f(x) = -2x^4 + 13x^3 - 21x^2 + 2x + 8$

In Exercises 5–12, find the real zeros of the function.

**5.** $f(x) = x^3 - 6x^2 + 11x - 6$

**6.** $g(x) = x^3 - 4x^2 - x + 4$

**7.** $h(t) = t^3 + 12t^2 + 21t + 10$

**8.** $f(x) = x^3 - 4x^2 + 5x - 2$

**9.** $C(x) = 2x^3 + 3x^2 - 1$

**10.** $f(x) = 3x^3 - 19x^2 + 33x - 9$

**11.** $f(x) = x^4 - 3x^2 + 2$

**12.** $P(t) = t^4 - 7t^2 + 12$

In Exercises 13–18, find all real solutions of the polynomial equation.

**13.** $z^4 - z^3 - 2z - 4 = 0$

**14.** $x^4 - 13x^2 - 12x = 0$

**15.** $2y^4 + 7y^3 - 26y^2 + 23y - 6 = 0$

**16.** $2x^4 - 11x^3 - 6x^2 + 64x + 32 = 0$

**17.** $x^5 - x^4 - 3x^3 + 5x^2 - 2x = 0$

**18.** $x^5 - 7x^4 + 10x^3 + 14x^2 - 24x = 0$

In Exercises 19 and 20, (a) list the possible rational zeros of $f$, (b) sketch the graph of $f$ so that some of the possible zeros in part (a) can be discarded, and then (c) determine all real zeros of $f$.

**19.** $f(x) = 32x^3 - 52x^2 + 17x + 3$

**20.** $f(x) = 4x^3 + 7x^2 - 11x - 18$

In Exercises 21 and 22, find the rational zeros of the polynomial function.

**21.** $P(x) = x^4 - \frac{25}{4}x^2 + 9$

**22.** $f(x) = x^3 - \frac{3}{2}x^2 - \frac{23}{2}x + 6$

In Exercises 23–26, follow the procedure given in Example 5 to approximate the zero of $f(x)$ in the interval $[a, b]$. Give your approximation to the nearest tenth. (If you have a graphing utility, use it to help approximate the zero.)

**23.** $f(x) = x^3 + x - 1,$     $[0, 1]$

**24.** $f(x) = x^5 + x + 1,$     $[-1, 0]$

**25.** $f(x) = x^4 - 10x^2 - 11,$     $[3, 4]$

**26.** $f(x) = -x^3 + 3x^2 + 9x - 2,$     $[4, 5]$

In Exercises 27–32, match the function with its graph. Then approximate the real zeros of the function to three decimal places. [The graphs are labeled (a), (b), (c), (d), (e), and (f).]

(a)

(b)

(c)

(d)

(e)

(f)

**27.** $f(x) = x^3 - 2x + 2$

**28.** $f(x) = x^5 + x - 1$

**29.** $f(x) = 2x^3 - 6x^2 + 6x - 1$

**30.** $f(x) = 5x^3 - 20x^2 + 20x - 4$

**31.** $f(x) = -x^3 + 3x^2 - x + 1$

**32.** $f(x) = -x^3 + 4x + 2$

In Exercises 33–40, use a graphing utility to approximate the real zeros of the function. Use an accuracy of 0.001.

**33.** $f(x) = x^4 - x - 3$

**34.** $f(x) = x^3 - 3.9x^2 + 4.79x - 1.881$

**35.** $f(x) = 4x^3 + 14x - 8$

**36.** $f(x) = -x^3 + 2x^2 + 4x + 5$

**37.** $f(x) = x^4 + x - 3$

**38.** $f(x) = -x^4 + 2x^3 + 4$

**39.** $f(x) = 7x^4 - 42x^3 + 43x^2 + 216x - 324$

**40.** $f(x) = 3x^4 - 12x^3 + 27x^2 + 4x - 4$

**41.** *Measurements of a Box*   An open box is to be made from a rectangular piece of material, 12 inches by 10 inches, by cutting equal squares from the corners and turning up the sides.

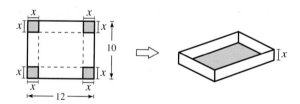

(a) Use the figure to write the volume $V$ of the box as a function of $x$. Determine the domain of the function.

(b) Sketch the graph of the function and approximate the measurements of the box that yield a maximum volume.

(c) Find values of $x$ such that $V = 96$. Which of these values is a physical impossibility in the construction of the box? Explain.

**42.** *Measurements of a Box*    An open box is to be made from a rectangular piece of material, 9 inches by 5 inches, by cutting equal squares from the corners and turning up the sides.

(a) Use the figure to write the volume $V$ of the box as a function of $x$. Determine the domain of the function.

(b) Sketch the graph of the function and approximate the measurements of the box that yield a maximum volume.

(c) Find values of $x$ such that $V = 56$. Which of these values is a physical impossibility in the construction of the box? Explain.

**43.** *Measurements of a Package*    A rectangular package to be sent by a postal service can have a maximum combined length and girth (perimeter of a cross section) of 108 inches (see figure). Find the measurements of the package, given that the volume is 11,664 cubic inches.

**44.** *Measurements of a Package*    A rectangular package to be sent by a postal service can have a maximum combined length and girth (perimeter of a cross section) of 120 inches. Find the measurements of the package, given that the volume is 16,000 cubic inches.

**45.** *Measurements of a Box*    An open box is to be made from a rectangular piece of material, 10 inches by 8 inches, by cutting equal squares from the corners and turning up the sides. Find the measurements of the box if the volume is to be 52.5 cubic inches.

**46.** *Demand Function*    A company that produces cellular pagers estimates that the demand for a particular model of pager is

$$D = -x^3 + 54x^2 - 140x - 3000, \quad 10 \le x \le 50$$

where $D$ is the demand for the pager and $x$ is the price of the pager.

(a) Use a graphing utility to graph $D$. Use the trace feature to determine for what values of $x$ the demand will be 14,400 pagers.

(b) You may also determine the price $x$ of the pagers that will yield a demand of 14,400 by setting $D$ equal to 14,400 and solving for $x$ with a graphing utility. Discuss this alternative method of solution. Of the actual solutions, what price would you recommend the company charge for the pagers?

**47.** *Imports to the United States*    The values (in billions of dollars) of goods imported into the United States from 1987 to 1996 are given in the table. (Source: U.S. Bureau of the Census)

| Year | 1987 | 1988 | 1989 | 1990 | 1991 |
|---|---|---|---|---|---|
| Imports | 402 | 437 | 468 | 491 | 483 |

| Year | 1992 | 1993 | 1994 | 1995 | 1996 |
|---|---|---|---|---|---|
| Imports | 525 | 575 | 658 | 740 | 790 |

(a) Use a graphing utility to make a scatter plot of the data. Let $x = 7$ correspond to 1987.

(b) Use a regression program to find a cubic model for the data. Let $x = 7$ correspond to 1987.

(c) Graph the data and the model in the same viewing window. Does the model appear to be a good fit for the data?

(d) According to the model, when did the imports reach 650 billion dollars?

**48.** *Cost of Dental Care*   The values of the consumer price index for dental care in the United States from 1987 to 1996 are given in the table.   (Source: U.S. Bureau of Labor Statistics)

| Year | 1987 | 1988 | 1989 | 1990 | 1991 |
|------|------|------|------|------|------|
| Index | 128.8 | 137.5 | 146.1 | 155.8 | 167.4 |

| Year | 1992 | 1993 | 1994 | 1995 | 1996 |
|------|------|------|------|------|------|
| Index | 178.7 | 188.1 | 197.1 | 206.8 | 216.5 |

(a) Use a graphing utility to make a scatter plot of the data. Let $x = 7$ correspond to 1987.

(b) Use a regression program to find a cubic model for the data. Let $x = 7$ correspond to 1987.

(c) Graph the data and the model in the same viewing window. Does the model appear to be a good fit for the data?

(d) According to the model, when did the consumer price index for dental care reach 215?

**49.** *Medicine*   The concentration of a certain chemical in the bloodstream $t$ hours after injection into muscle tissue is given by

$$C = \frac{3t^2 + t}{t^3 + 50}, \qquad 0 \le t.$$

The concentration is greatest when

$$3t^4 + 2t^3 - 300t - 50 = 0.$$

Approximate this time to the nearest tenth of an hour.

**50.** *Transportation Cost*   The transportation cost $C$ of the components used in manufacturing a certain product is given by

$$C = 100\left(\frac{200}{x^2} + \frac{x}{x + 30}\right), \qquad 1 \le x$$

where $C$ is measured in thousands of dollars and $x$ is the order size in hundreds. The cost is a minimum when

$$3x^3 - 40x^2 - 2400x - 36{,}000 = 0.$$

Approximate the optimal order size to the nearest hundred units.

**51.** *Advertising Costs*   A company that produces portable cassette players estimates that the profit from selling a particular model is given by

$$P = -76x^3 + 4830x^2 - 320{,}000, \qquad 0 \le x \le 60$$

where $P$ is the profit (in dollars) and $x$ is the advertising expense (in tens of thousands of dollars). According to this model, how much money should the company spend to obtain a profit of $2,500,000?

**52.** *Advertising Costs*   A company that manufactures bicycles estimates that the profit from selling a particular model is given by

$$P = -45x^3 + 2500x^2 - 275{,}000, \qquad 0 \le x \le 50$$

where $P$ is the profit (in dollars) and $x$ is the advertising expense (in tens of thousands of dollars). According to this model, how much money should the company spend to obtain a profit of $800,000?

# Mid-Chapter Quiz

Take this quiz as you would take a quiz in class. After you are done, check your work against the answers given in the back of the book.

In Exercises 1 and 2, sketch the graph of the quadratic function. Identify the vertex and the intercepts.

**1.** $f(x) = (x + 1)^2 - 2$

**2.** $f(x) = 16 - x^2$

In Exercises 3 and 4, determine the right and left behavior of the graph of the function. Verify with a graphing utility.

**3.** $f(x) = 5x^3 - 7x^2 + 2$

**4.** $f(x) = -x^4 + 5x^2 - 4$

**5.** Completely factor the polynomial $2x^3 - x^2 - 13x - 6$ using long division. One factor is $(x + 2)$.

**6.** Use synthetic division to evaluate $f(x) = 3x^3 - 5x^2 + 9$ at $x = 2$.

In Exercises 7 and 8, express the function in the form $f(x) = (x - k)q(x) + r$ for the given value of $k$, and demonstrate that $f(k) = r$.

**7.** $f(x) = x^4 - 5x^2 + 4$,  $k = 1$

**8.** $f(x) = x^3 + 5x^2 - 2x - 24$,  $k = -3$

**9.** Simplify: $\dfrac{2x^4 + 9x^3 - 32x^2 - 99x + 180}{x^2 + 2x - 15}$

**10.** The profit for a company is $P = -95x^3 + 5650x^2 - 250{,}000$, $0 \le x \le 55$, where $x$ is the advertising expense (in tens of thousands of dollars). What is the profit for an advertising expense of \$450,000? Use a graphing utility to approximate another advertising expense that would yield the same profit.

In Exercises 11–14, find all real zeros of the function.

**11.** $f(x) = -2x^3 - 7x^2 + 10x + 35$

**12.** $f(x) = 4x^4 - 37x^2 + 9$

**13.** $f(x) = 3x^4 + 4x^3 - 3x - 4$

**14.** $f(x) = 2x^3 - 3x^2 + 2x - 3$

In Exercises 15 and 16, use a graphing utility to approximate the real zeros of the function.

**15.** $f(x) = 3x^3 + 4x^2 + 1$

**16.** $f(x) = x^4 - x^3 + x^2 + 2x - 6$

In Exercises 17–20, an open box is to be made from a 12-inch-by-14-inch piece of material by cutting equal squares from the corners and turning up the sides. The volume of the box is to be 160 cubic inches.

**17.** Write an equation that will allow you to determine the measurements of the box.

**18.** Find the measurements of the box algebraically.

**19.** Use a graphing utility to find the measurements of the box.

**20.** When this problem is solved algebraically or graphically, not all possible zeros are used. Explain.

### 4.5     Complex Numbers

## Objectives

- Perform operations with complex numbers and write the results in standard form.
- Find the complex conjugate of a complex number.
- Solve a polynomial equation.
- Plot a complex number in the complex plane.
- Determine if a complex number is in the Mandelbrot Set.

## The Imaginary Unit $i$

In Section 1.4, you learned that some quadratic equations have no real solutions. For instance, the quadratic equation

$$x^2 + 1 = 0 \qquad \text{Equation with no real solution}$$

has no real solution because there is no real number $x$ that can be squared to produce $-1$. To overcome this deficiency, mathematicians created an expanded system of numbers using the **imaginary unit $i$,** defined as

$$i = \sqrt{-1} \qquad \text{Imaginary unit}$$

where $i^2 = -1$. By adding real numbers to real multiples of this imaginary unit, we obtain the set of **complex numbers.** Each complex number can be written in the **standard form, $a + bi$.**

> ▶ **Definition of a Complex Number**
>
> If $a$ and $b$ are real numbers, the number $a + bi$ is a **complex number,** and it is said to be written in **standard form.** If $b = 0$, the number $a + bi = a$ is a real number. If $b \neq 0$, the number $a + bi$ is called an **imaginary number.** A number of the form $bi$, where $b \neq 0$, is called a **pure imaginary number.**

Complex numbers

| Real numbers | Imaginary numbers |
|---|---|
| $3, -\frac{1}{2},$ | $-2 + i$ |
| $\sqrt{2}, \ 0$ | Pure imaginary numbers $3i$ |

**Figure 4.36**

The set of real numbers is a subset of the set of complex numbers, as shown in Figure 4.36. This is true because every real number $a$ can be written as a complex number using $b = 0$. That is, for every real number $a$, we can write $a = a + 0i$.

> ▶ **Equality of Complex Numbers**
>
> Two complex numbers $a + bi$ and $c + di$, written in standard form, are **equal** to each other
>
> $$a + bi = c + di \qquad \text{Equality of two complex numbers}$$
>
> if and only if $a = c$ and $b = d$.

# Operations with Complex Numbers

To add (or subtract) two complex numbers, you add (or subtract) the real and imaginary parts of the numbers separately.

---

▶ **Addition and Subtraction of Complex Numbers**

If $a + bi$ and $c + di$ are two complex numbers written in standard form, their sum and difference are defined as follows.

$$\text{Sum: } (a + bi) + (c + di) = (a + c) + (b + d)i$$
$$\text{Difference: } (a + bi) - (c + di) = (a - c) + (b - d)i$$

---

The **additive identity** in the complex number system is zero (the same as in the real number system). Furthermore, the **additive inverse** of the complex number $a + bi$ is

$$-(a + bi) = -a - bi. \qquad \text{Additive inverse}$$

So, you have

$$(a + bi) + (-a - bi) = 0 + 0i = 0.$$

---

**Example 1**    Adding and Subtracting Complex Numbers

**a.** $(3 - i) + (2 + 3i) = 3 - i + 2 + 3i$    Remove parentheses.

$$= 3 + 2 - i + 3i \qquad \text{Group real and imaginary terms.}$$
$$= (3 + 2) + (-1 + 3)i$$
$$= 5 + 2i \qquad \text{Standard form}$$

**b.** $2i + (-4 - 2i) = 2i - 4 - 2i$    Remove parentheses.

$$= -4 + 2i - 2i \qquad \text{Group real and imaginary terms.}$$
$$= -4 \qquad \text{Standard form}$$

**c.** $3 - (-2 + 3i) + (-5 + i) = 3 + 2 - 3i - 5 + i$

$$= 3 + 2 - 5 - 3i + i$$
$$= 0 - 2i$$
$$= -2i$$

---

Note in Example 1(b) that the sum of two imaginary numbers can be a real number.

Many of the properties of real numbers are valid for complex numbers as well. Here are some examples.

*Associative Property of Addition and Multiplication*

*Commutative Property of Addition and Multiplication*

*Distributive Property of Multiplication Over Addition*

Notice how these properties are used when two complex numbers are multiplied.

$$(a + bi)(c + di) = a(c + di) + bi(c + di) \quad \text{Distributive}$$
$$= ac + (ad)i + (bc)i + (bd)i^2 \quad \text{Distributive}$$
$$= ac + (ad)i + (bc)i + (bd)(-1) \quad \text{Definition of } i$$
$$= ac - bd + (ad)i + (bc)i \quad \text{Commutative}$$
$$= (ac - bd) + (ad + bc)i \quad \text{Associative}$$

Rather than trying to memorize this multiplication rule, we suggest that you simply remember how the distributive property is used to multiply two complex numbers. The procedure is similar to multiplying two polynomials and combining like terms (as in the FOIL Method).

### Example 2   Multiplying Complex Numbers

**a.** $(i)(-3i) = -3i^2$     Multiply.

$\qquad\qquad = -3(-1)$     $i^2 = -1$

$\qquad\qquad = 3$     Simplify.

**b.** $(2 - i)(4 + 3i) = 8 + 6i - 4i - 3i^2$     Product of binomials

$\qquad\qquad\qquad = 8 + 6i - 4i - 3(-1)$     $i^2 = -1$

$\qquad\qquad\qquad = 8 + 3 + 6i - 4i$     Collect terms.

$\qquad\qquad\qquad = 11 + 2i$     Standard form

**c.** $(3 + 2i)(3 - 2i) = 9 - 6i + 6i - 4i^2$     Product of binomials

$\qquad\qquad\qquad = 9 - 4(-1)$     $i^2 = -1$

$\qquad\qquad\qquad = 9 + 4$     Simplify.

$\qquad\qquad\qquad = 13$     Simplify.

**d.** $(3 + 2i)^2 = 9 + 6i + 6i + 4i^2$     Product of binomials

$\qquad\qquad = 9 + 4(-1) + 12i$     $i^2 = -1$

$\qquad\qquad = 9 - 4 + 12i$     Simplify.

$\qquad\qquad = 5 + 12i$     Simplify.

## Complex Conjugates

Notice in Example 2(c) that the product of two imaginary numbers can be a real number. This occurs with pairs of complex numbers of the form $a + bi$ and $a - bi$, called **complex conjugates.** In general, the product of two complex conjugates can be written as follows.

$$(a + bi)(a - bi) = a^2 - abi + abi - b^2 i^2$$
$$= a^2 - b^2(-1)$$
$$= a^2 + b^2$$

Complex conjugates can be used to divide one complex number by another. That is, to find the quotient of $a + bi$ and $c + di$, multiply the numerator and denominator by the conjugate of the denominator to obtain

$$\frac{a + bi}{c + di} = \frac{a + bi}{c + di}\left(\frac{c - di}{c - di}\right) = \frac{(ac + bd) + (bc - ad)i}{c^2 + d^2}.$$

**•TECHNOLOGY•**

Some graphing utilities can perform operations with complex numbers. Check the user's guide to determine if your graphing utility has this capability.

---

**Example 3**    Dividing Complex Numbers

$$\frac{1}{1 + i} = \frac{1}{1 + i}\left(\frac{1 - i}{1 - i}\right) \qquad \text{Multiply by conjugate.}$$

$$= \frac{1 - i}{1^2 - i^2} \qquad \text{Product of conjugates}$$

$$= \frac{1 - i}{1 - (-1)} \qquad i^2 = -1$$

$$= \frac{1 - i}{2} \qquad \text{Simplify.}$$

$$= \frac{1}{2} - \frac{1}{2}i \qquad \text{Standard form}$$

---

**Example 4**    Dividing Complex Numbers

$$\frac{2 + 3i}{4 - 2i} = \frac{2 + 3i}{4 - 2i}\left(\frac{4 + 2i}{4 + 2i}\right) \qquad \text{Multiply by conjugate.}$$

$$= \frac{8 + 4i + 12i + 6i^2}{16 - 4i^2} \qquad \text{Expand.}$$

$$= \frac{8 - 6 + 16i}{16 + 4} \qquad i^2 = -1$$

$$= \frac{2 + 16i}{20} \qquad \text{Simplify.}$$

$$= \frac{1}{10} + \frac{4}{5}i \qquad \text{Standard form}$$

## Complex Solutions

When using the Quadratic Formula to solve a quadratic equation, you often obtain a result such as $\sqrt{-3}$, which you know is not a real number. By factoring out $i = \sqrt{-1}$, you can write this number in standard form.

$$\sqrt{-3} = \sqrt{3(-1)} = \sqrt{3}\sqrt{-1} = \sqrt{3}\,i$$

The number $\sqrt{3}\,i$ is called the principal square root of $-3$.

---

▶ **Principal Square Root of a Negative Number**

If $a$ is a positive number, the **principal square root** of the negative number $-a$ is defined as

$$\sqrt{-a} = \sqrt{a}\,i.$$

---

### Study Tip

The definition of principal square root uses the rule

$$\sqrt{ab} = \sqrt{a}\sqrt{b}$$

for $a > 0$ and $b < 0$. This rule is not valid if *both* $a$ and $b$ are negative. For example,

$$\sqrt{-5}\sqrt{-5} = 5i^2 = -5$$

whereas

$$\sqrt{(-5)(-5)} = \sqrt{25} = 5.$$

To avoid problems with multiplying square roots of negative numbers, be sure to convert to standard form *before* multiplying.

---

**Example 5**    Writing Complex Numbers in Standard Form

**a.** $\sqrt{-3}\sqrt{-12} = \sqrt{3}\,i\sqrt{12}\,i = \sqrt{36}\,i^2 = 6(-1) = -6$

**b.** $\sqrt{-48} - \sqrt{-27} = \sqrt{48}\,i - \sqrt{27}\,i = 4\sqrt{3}\,i - 3\sqrt{3}\,i = \sqrt{3}\,i$

**c.** $\left(-1 + \sqrt{-3}\right)^2 = \left(-1 + \sqrt{3}\,i\right)^2$

$$= (-1)^2 - 2\sqrt{3}\,i + \left(\sqrt{3}\right)^2(i^2)$$

$$= 1 - 2\sqrt{3}\,i + 3(-1)$$

$$= -2 - 2\sqrt{3}\,i$$

---

**Example 6**    Complex Solutions of a Quadratic Equation

Solve the equation $3x^2 - 2x + 5 = 0$.

**Solution**

$$x = \frac{-(-2) \pm \sqrt{(-2)^2 - 4(3)(5)}}{2(3)} \qquad \text{Quadratic Formula}$$

$$= \frac{2 \pm \sqrt{-56}}{6} \qquad \text{Simplify.}$$

$$= \frac{2 \pm 2\sqrt{14}\,i}{6} \qquad \text{Write in } i\text{-form.}$$

$$= \frac{1}{3} \pm \frac{\sqrt{14}}{3}\,i \qquad \text{Standard form}$$

---

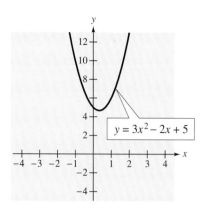

Figure 4.37

The graph of $f(x) = 3x^2 - 2x + 5$, shown in Figure 4.37, does not touch or cross the $x$-axis. This confirms that Example 6 has no real solutions.

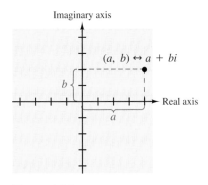

Figure 4.38

# Applications

Most applications involving complex numbers are either theoretical (see the next section) or very technical, and so are not appropriate for inclusion in this text. However, to give you some idea of how complex numbers can be used in applications, we give a general description of their use in **fractal geometry.**

To begin, consider a coordinate system called the **complex plane.** Just as every real number corresponds to a point on the real line, every complex number corresponds to a point in the complex plane, as shown in Figure 4.38. In this figure, note that the vertical axis is the **imaginary axis** and the horizontal axis is the **real axis.** The point that corresponds to the complex number $a + bi$ is $(a, b)$.

| Complex number $a + bi$ | $\Longrightarrow$ | Ordered pair $(a, b)$ |
|---|---|---|

**Example 7**   Plotting Complex Numbers in the Complex Plane

Plot the following complex numbers in the complex plane.

**a.** $2 + 3i$

**b.** $-1 + 2i$

**c.** $4$

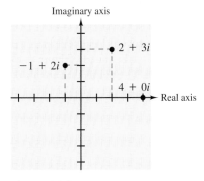

Figure 4.39

### Solution

**a.** To plot the complex number $2 + 3i$, move (from the origin) two units to the right on the real axis and then three units up, as shown in Figure 4.39. In other words, plotting the complex number $2 + 3i$ in the complex plane is comparable to plotting the point $(2, 3)$ in the Cartesian plane.

**b.** The complex number $-1 + 2i$ corresponds to the point $(-1, 2)$, as shown in Figure 4.39.

**c.** The complex number $4$ corresponds to the point $(4, 0)$, as shown in Figure 4.39.

---

In the hands of a person who understands "fractal geometry," the complex plane can become an easel on which stunning pictures, called **fractals,** can be drawn. The most famous such picture is called the **Mandelbrot Set,** named after the Polish-born mathematician Benoit Mandelbrot. To draw the Mandelbrot Set, consider the following sequence of numbers.

$$c, \quad c^2 + c, \quad (c^2 + c)^2 + c, \quad [(c^2 + c)^2 + c]^2 + c, \quad \ldots$$

The behavior of this sequence depends on the value of the complex number $c$. For some values of $c$ this sequence is **bounded,** and for other values it is **unbounded.** If the sequence is bounded, the complex number $c$ is in the Mandelbrot Set, and if the sequence is unbounded, the complex number $c$ is not in the Mandelbrot Set.

### Example 8    Members of the Mandelbrot Set

**a.** The complex number $-2$ is in the Mandelbrot Set because for $c = -2$, the corresponding Mandelbrot sequence is $-2, 2, 2, 2, 2, 2, \ldots$, which is bounded.

**b.** The complex number $i$ is also in the Mandelbrot Set because for $c = i$, the corresponding Mandelbrot sequence is

$$i, \qquad -1 + i, \qquad -i, \qquad -1 + i, \qquad -i, \qquad -1 + i, \qquad \ldots$$

which is bounded.

**c.** The complex number $1 + i$ is *not* in the Mandelbrot Set because for $c = 1 + i$, the corresponding Mandelbrot sequence is

$$1 + i, \qquad 1 + 3i, \qquad -7 + 7i, \qquad 1 - 97i, \qquad -9407 - 193i,$$

$$88454401 + 3631103i, \ldots$$

which is unbounded.

---

With this definition, a picture of the Mandelbrot Set would have only two colors. One color for points that are in the set (the sequence is bounded) and one color for points that are outside the set (the sequence is unbounded). Figure 4.40 shows a black and yellow picture of the Mandelbrot Set. The points that are black are in the Mandelbrot Set and the points that are yellow are not.

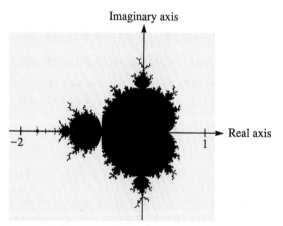

**Figure 4.40**    *Mandelbrot Set*

To add more interest to the picture, computer scientists discovered that the points that are not in the Mandelbrot Set can be assigned a variety of colors, depending on "how quickly" their sequences diverge. Figure 4.41 shows three different appendages of the Mandelbrot Set using a spectrum of colors. (The colored portions of the picture represent points that are *not* in the Mandelbrot Set.)

American Mathematical Society

**Figure 4.41**

Figures 4.42, 4.43, and 4.44 show other types of fractal sets. From these pictures, you can see why fractals have fascinated people since their discovery (around 1980).

Alfred Pasieka/SPL/Photo Researchers, Inc.

Gregory Sams/Photo Researchers, Inc.

Gregory Sams/Photo Researchers, Inc.

**Figure 4.42**          **Figure 4.43**          **Figure 4.44**

| Discussing the Concept | **Building Your Own Fractal** |
|---|---|

On a large sheet of paper, use a ruler to draw a large equilateral triangle. Find the midpoint of each side. Connect the midpoints as shown in the figure on the left. Repeat the process for each of the smaller upward-pointing triangles, as shown in the figure on the right. Continue the process, always connecting the midpoints of the upward-pointing triangles. This fractal triangle is called the Sierpinski triangle.

# Warm Up

The following warm-up exercises involve skills that were covered in earlier sections. You will use these skills in the exercise set for this section.

In Exercises 1–8, simplify the expression.

1. $\sqrt{12}$

2. $\sqrt{500}$

3. $\sqrt{20} - \sqrt{5}$

4. $\sqrt{27} - \sqrt{243}$

5. $\sqrt{24}\sqrt{6}$

6. $2\sqrt{18}\sqrt{32}$

7. $\dfrac{1}{\sqrt{3}}$

8. $\dfrac{2}{\sqrt{2}}$

In Exercises 9 and 10, solve the quadratic equation.

9. $x^2 + x - 1 = 0$

10. $x^2 + 2x - 1 = 0$

## 4.5    Exercises

1. Write out the first 16 positive integer powers of $i$ $(i, i^2, i^3, \ldots, i^{16})$, and express each as $i$, $-i$, 1, or $-1$.

2. Express each of the powers of $i$ as $i$, $-i$, 1, or $-1$.

   (a) $i^{28}$    (b) $i^{37}$    (c) $i^{127}$    (d) $i^{82}$

In Exercises 3–6, find the real numbers $a$ and $b$ such that the equation is true.

3. $a + bi = 3 + 12i$

4. $a + bi = -2 - 5i$

5. $(a + 3) + (b - 1)i = 7 - 4i$

6. $(a + 6) + 2bi = 6 - 5i$

In Exercises 7–18, write the complex number in standard form and find its complex conjugate.

7. $9 + \sqrt{-16}$

8. $2 + \sqrt{-25}$

9. $-3 - \sqrt{-12}$

10. $1 + \sqrt{-8}$

11. $-21$

12. $45$

13. $-6i + i^2$

14. $4i^2 - 2i^3$

15. $-5i^5$

16. $(-i)^3$

17. $\left(\sqrt{-6}\right)^2 + 3$

18. $\left(\sqrt{-4}\right)^2 - 5$

In Exercises 19–56, perform the indicated operation and write the result in standard form.

19. $(-4 + 3i) + (6 - 2i)$

20. $(13 - 2i) + (-5 + 6i)$

21. $(12 + 5i) - (7 - i)$

22. $(3 + 2i) - (6 + 13i)$

23. $\left(-2 + \sqrt{-8}\right) + \left(5 - \sqrt{-50}\right)$

24. $\left(8 + \sqrt{-18}\right) - \left(4 + 3\sqrt{2}\,i\right)$

25. $-\left(\frac{3}{2} + \frac{5}{2}i\right) + \left(\frac{5}{3} + \frac{11}{3}i\right)$

26. $(1.6 + 3.2i) + (-5.8 + 4.3i)$

27. $\sqrt{-3} \cdot \sqrt{-8}$

28. $\sqrt{-5} \cdot \sqrt{-10}$

29. $\left(\sqrt{-10}\right)^2$

30. $\left(\sqrt{-75}\right)^3$

31. $(1 + i)(3 - 2i)$

32. $(9 - 4i)(1 - i)$

33. $(4 + 5i)(4 - 5i)$

34. $(6 + 7i)(6 - 7i)$

35. $7i(6 - 3i)$

36. $-8i(9 + 4i)$

37. $\left(\sqrt{5} - \sqrt{3}\,i\right)\left(\sqrt{5} + \sqrt{3}\,i\right)$

38. $\left(\sqrt{14} + \sqrt{10}\,i\right)\left(\sqrt{14} - \sqrt{10}\,i\right)$

39. $\left(2 - \sqrt{-8}\right)\left(8 + \sqrt{-6}\right)$

40. $\left(3 + \sqrt{-5}\right)\left(7 - \sqrt{-10}\right)$

**41.** $(4 + 5i)^2$

**42.** $(2 - 3i)^3$

**43.** $\dfrac{2 + i}{2 - i}$

**44.** $\dfrac{3}{1 - i}$

**45.** $\dfrac{5}{4 - 2i}$

**46.** $\dfrac{8 - 7i}{1 - 2i}$

**47.** $\dfrac{6 - 7i}{i}$

**48.** $\dfrac{8 + 20i}{2i}$

**49.** $\dfrac{1}{(2i)^3}$

**50.** $\dfrac{1}{(4 - 5i)^2}$

**51.** $\dfrac{5}{(1 + i)^3}$

**52.** $\dfrac{(2 - 3i)(5i)}{2 + 3i}$

**53.** $\dfrac{(21 - 7i)(4 + 3i)}{2 - 5i}$

**54.** $\dfrac{1}{i^3}$

**55.** $(2 + 3i)^2 + (2 - 3i)^2$    **56.** $(1 - 2i)^2 - (1 + 2i)^2$

*Error Analysis*    In Exercises 57 and 58, a student has handed in the following problem. Find the error(s) and discuss how to explain each error to the student.

**57.** Write $\dfrac{5}{3 - 2i}$ in standard form.

$$\frac{5}{3 - 2i} \cdot \frac{3 + 2i}{3 + 2i} = \frac{15 + 10i}{9 - 4} = 3 + 2i$$

**58.** Multiply $\left(\sqrt{-4} + 3\right)\left(i - \sqrt{-3}\right)$.

$$\left(\sqrt{-4} + 3\right)\left(i - \sqrt{-3}\right)$$
$$= i\sqrt{-4} - \sqrt{-4}\sqrt{-3} + 3i - 3\sqrt{-3}$$
$$= -2i - \sqrt{12} + 3i - 3i\sqrt{3}$$
$$= \left(1 - 3\sqrt{3}\right)i - 2\sqrt{3}$$

In Exercises 59–66, solve the equation.

**59.** $x^2 - 2x + 2 = 0$

**60.** $x^2 + 6x + 10 = 0$

**61.** $4x^2 + 16x + 17 = 0$

**62.** $9x^2 - 6x + 37 = 0$

**63.** $4x^2 + 16x + 15 = 0$

**64.** $9x^2 - 6x + 35 = 0$

**65.** $16t^2 - 4t + 3 = 0$

**66.** $5s^2 + 6s + 3 = 0$

In Exercises 67–72, plot the complex number.

**67.** $-2 + i$

**68.** $i$

**69.** $3$

**70.** $-2 - 3i$

**71.** $1 - 2i$

**72.** $-2i$

In Exercises 73–78, decide whether the number is in the Mandelbrot Set. Explain your reasoning.

**73.** $c = 0$

**74.** $c = 2$

**75.** $c = \frac{1}{2}i$

**76.** $c = -i$

**77.** $c = 1$

**78.** $c = -1$

## Math Matters • Acceptance of Imaginary Numbers

Imaginary numbers were given the name "imaginary" because many mathematicians had a difficult time accepting such numbers. The following excerpt, written by the famous mathematician Carl Gauss (1777–1855), shows that, even in the early 1800s, some people were hesitant to accept the use of imaginary numbers.

"Our general arithmetic, which far surpasses the extent of the geometry of the ancients, is entirely the creation of modern times. Starting with the notion of whole numbers, it has gradually enlarged its domain. To whole numbers have been added fractions; to rational numbers have been added irrational; to positive numbers have been added negative; and to real numbers have been added imaginary numbers.

This advance, however, has always been made with timid steps. The early algebraists called negative solutions of equations false solutions (and this is indeed the case when the problem to which they relate has been stated in such a way that negative solutions have no meaning). The reality of negative numbers is sufficiently justified since there are many cases where they have meaningful interpretations. This has long been admitted, but the imaginary numbers (formerly and occasionally now called impossible numbers) are still rather tolerated than fully accepted."

## 4.6 The Fundamental Theorem of Algebra

### Objectives

- Use the Fundamental Theorem of Algebra and the Linear Factorization Theorem to write a polynomial as the product of linear factors.
- Find a polynomial with integer coefficients whose zeros are given.
- Factor a polynomial over the real and complex numbers.
- Find all real and complex zeros of a polynomial function.

### Jean Le Rond d'Alembert

(1717–1783)

Jean Le Rond d'Alembert worked independently of Carl Gauss trying to prove the Fundamental Theorem of Algebra. His efforts were such that in France, the Fundamental Theorem of Algebra is frequently known as the theorem of d'Alembert.

## The Fundamental Theorem of Algebra

You have been using the fact than an $n$th-degree polynomial can have at most $n$ real zeros. In the complex number system, this statement can be improved. That is, in the complex number system, every $n$th-degree polynomial function has *precisely n* zeros. This important result is derived from the **Fundamental Theorem of Algebra,** first proved by the famous German mathematician Carl Friedrich Gauss (1777–1855).

▶ **The Fundamental Theorem of Algebra**

If $f(x)$ is a polynomial of degree $n$, where $n > 0$, then $f$ has at least one zero in the complex number system.

Using the Fundamental Theorem of Algebra and the equivalence of zeros and factors, you obtain the following theorem.

▶ **Linear Factorization Theorem**

If $f(x)$ is a polynomial of degree $n$

$$f(x) = a_n x^n + a_{n-1} x^{n-1} + \cdots + a_1 x + a_0$$

where $n > 0$, then $f$ has precisely $n$ linear factors

$$f(x) = a_n(x - c_1)(x - c_2) \cdots (x - c_n)$$

where $c_1, c_2, \ldots, c_n$ are complex numbers and $a_n$ is the leading coefficient of $f(x)$.

Note that neither the Fundamental Theorem of Algebra nor the Linear Factorization Theorem tells you *how* to find the zeros or factors of a polynomial. Such theorems are called **existence theorems.** To find the zeros of a polynomial function, you still rely on the techniques developed in the earlier parts of the text.

Remember that the $n$ zeros of a polynomial function can be real or complex, and they may be repeated. Example 1 illustrates several cases.

Example 1   Zeros of Polynomial Functions

**a.** The first-degree polynomial $f(x) = x - 2$ has exactly *one* zero: $x = 2$.

**b.** Counting multiplicity, the second-degree polynomial function

$$f(x) = x^2 - 6x + 9 = (x - 3)(x - 3)$$

has exactly *two* zeros: $x = 3$ and $x = 3$.

**c.** The third-degree polynomial function

$$f(x) = x^3 + 4x = x(x - 2i)(x + 2i)$$

has exactly *three* zeros: $x = 0$, $x = 2i$, and $x = -2i$.

**d.** The fourth-degree polynomial function

$$f(x) = x^4 - 1 = (x - 1)(x + 1)(x - i)(x + i)$$

has exactly *four* zeros: $x = 1$, $x = -1$, $x = i$, and $x = -i$.

## •TECHNOLOGY•

Remember that when you use a graphing utility to locate the zeros of a function, the only zeros that appear as *x*-intercepts are the *real zeros*. Compare the graphs below with the four polynomial functions in Example 1. Which zeros appear on the graphs?

Example 2 shows how you can use the methods described in Sections 4.3 and 4.4 (Rational Zero Test, synthetic division, and factoring) to find all the zeros of a polynomial function, including complex zeros.

### Example 2    Finding the Zeros of a Polynomial Function

Write the polynomial function $f(x) = x^5 + x^3 + 2x^2 - 12x + 8$ as the product of linear factors and list all of its zeros.

**Solution**

From the Rational Zero Test, the possible rational zeros are $\pm 1, \pm 2, \pm 4,$ and $\pm 8$. Synthetic division produces the following.

$$
\begin{array}{r|rrrrrr}
1 & 1 & 0 & 1 & 2 & -12 & 8 \\
  &   & 1 & 1 & 2 & 4 & -8 \\
\hline
  & 1 & 1 & 2 & 4 & -8 & 0
\end{array}
\quad\Longrightarrow\quad \text{1 is a zero.}
$$

$$
\begin{array}{r|rrrrr}
1 & 1 & 1 & 2 & 4 & -8 \\
  &   & 1 & 2 & 4 & 8 \\
\hline
  & 1 & 2 & 4 & 8 & 0
\end{array}
\quad\Longrightarrow\quad \text{1 is a repeated zero.}
$$

$$
\begin{array}{r|rrrr}
-2 & 1 & 2 & 4 & 8 \\
   &   & -2 & 0 & -8 \\
\hline
   & 1 & 0 & 4 & 0
\end{array}
\quad\Longrightarrow\quad -2 \text{ is a zero.}
$$

So, you have

$$f(x) = x^5 + x^3 + 2x^2 - 12x + 8$$

$$= (x - 1)(x - 1)(x + 2)(x^2 + 4).$$

By factoring $x^2 + 4$ as

$$x^2 - (-4) = \left(x - \sqrt{-4}\right)\left(x + \sqrt{-4}\right) = (x - 2i)(x + 2i)$$

you obtain

$$f(x) = (x - 1)(x - 1)(x + 2)(x - 2i)(x + 2i)$$

which gives the following five zeros of $f$.

$$1, \quad 1, \quad -2, \quad 2i, \quad \text{and} \quad -2i$$

Note from the graph of $f$ shown in Figure 4.45 that the *real* zeros are the only ones that appear as $x$-intercepts.

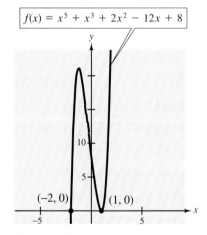

$f(x) = x^5 + x^3 + 2x^2 - 12x + 8$

$(-2, 0)$    $(1, 0)$

**Figure 4.45**

## Conjugate Pairs

In Example 2, note that the two imaginary zeros are **conjugates.** That is, they are of the form $a + bi$ and $a - bi$.

> ▶ **Complex Zeros Occur in Conjugate Pairs**
>
> Let $f(x)$ be a polynomial function that has real coefficients. If $a + bi$, where $b \neq 0$, is a zero of the function, then the conjugate $a - bi$ is also a zero of the function.

Be sure you see that this result is true only if the polynomial function has *real coefficients.* For instance, the result applies to the function $f(x) = x^2 + 1$, but not to the function $g(x) = x - i$.

**Example 3**    Finding a Polynomial with Given Zeros

Find a *fourth-degree* polynomial function, with real coefficients, that has $-1$, $-1$, and $3i$ as zeros.

### Solution

Because $3i$ is a zero *and* the polynomial is stated to have real coefficients, you know that the conjugate $-3i$ must also be a zero. So, from the Linear Factorization Theorem, $f(x)$ can be written as

$$f(x) = a(x + 1)(x + 1)(x - 3i)(x + 3i).$$

For simplicity, let $a = 1$, and obtain

$$f(x) = (x^2 + 2x + 1)(x^2 + 9)$$
$$= x^4 + 2x^3 + 10x^2 + 18x + 9.$$

## Factoring a Polynomial

The Linear Factorization Theorem shows that you can write any $n$th-degree polynomial as the product of $n$ linear factors.

$$f(x) = a(x - c_1)(x - c_2)(x - c_3) \cdots (x - c_n)$$

However, this result includes the possibility that some of the values of $c_i$ are complex. The following result implies that even if you do not want to get involved with "imaginary factors," you can still write $f(x)$ as the product of linear and/or quadratic factors.

> ▶ Factors of a Polynomial
>
> Every polynomial of degree $n > 0$ with real coefficients can be written as the product of linear and quadratic factors with real coefficients, where the quadratic factors have no real zeros.

A quadratic factor with no real zeros is said to be **irreducible over the reals.** Be sure you see that this is not the same as being *irreducible over the rationals.* For example, the quadratic

$$x^2 + 1 = (x - i)(x + i)$$

is irreducible over the reals (and therefore over the rationals). On the other hand, the quadratic

$$x^2 - 2 = \left(x - \sqrt{2}\right)\left(x + \sqrt{2}\right)$$

is irreducible over the rationals, but it is *reducible* over the reals.

### Example 4   Factoring a Polynomial

Write the polynomial $f(x) = x^4 - x^2 - 20$

**a.** as the product of factors that are irreducible over the *rationals,*

**b.** as the product of linear factors and quadratic factors that are irreducible over the *reals,* and

**c.** in completely factored form.

### Solution

**a.** Begin by factoring the polynomial into the product of two quadratic polynomials.

$$x^4 - x^2 - 20 = (x^2 - 5)(x^2 + 4)$$

Both of these factors are irreducible over the rationals.

**b.** By factoring over the reals, you have

$$x^4 - x^2 - 20 = \left(x + \sqrt{5}\right)\left(x - \sqrt{5}\right)(x^2 + 4)$$

where the quadratic factor is irreducible over the reals.

**c.** In completely factored form, you have

$$x^4 - x^2 - 20 = \left(x + \sqrt{5}\right)\left(x - \sqrt{5}\right)(x - 2i)(x + 2i).$$

**Example 5**    Finding the Zeros of a Polynomial Function

Find all the zeros of $f(x) = x^4 - 3x^3 + 6x^2 + 2x - 60$, given that $1 + 3i$ is a zero of $f$.

### Solution

Because imaginary zeros occur in conjugate pairs, you know that $1 - 3i$ is also a zero of $f$. This means that both

$$[x - (1 + 3i)] \quad \text{and} \quad [x - (1 - 3i)]$$

are factors of $f(x)$. Multiplying these two factors produces

$$[x - (1 + 3i)][x - (1 - 3i)] = [(x - 1) - 3i][(x - 1) + 3i]$$
$$= (x - 1)^2 - 9i^2$$
$$= x^2 - 2x + 10.$$

Using long division, you can divide $x^2 - 2x + 10$ into $f(x)$ to obtain the following.

$$
\require{enclose}
\begin{array}{r}
x^2 - \phantom{x}x - \phantom{x}6 \phantom{0000000} \\
x^2 - 2x + 10 \enclose{longdiv}{x^4 - 3x^3 + \phantom{x}6x^2 + \phantom{x}2x - 60} \\
\underline{x^4 - 2x^3 + 10x^2} \phantom{00000000000} \\
-x^3 - \phantom{x}4x^2 + \phantom{x}2x \phantom{000000} \\
\underline{-x^3 + \phantom{x}2x^2 - 10x} \phantom{000000} \\
-6x^2 + 12x - 60 \\
\underline{-6x^2 + 12x - 60} \\
0
\end{array}
$$

So, you have

$$f(x) = (x^2 - 2x + 10)(x^2 - x - 6) = (x^2 - 2x + 10)(x - 3)(x + 2)$$

and you can conclude that the zeros of $f$ are $1 + 3i$, $1 - 3i$, $3$, and $-2$.

---

**Discussing the Concept**    **Factoring a Polynomial**

Compile a list of all the various techniques that can be used to find the zeros of a polynomial with real coefficients. Include the techniques that make use of a graphing utility. Give an example illustrating each technique and discuss when the use of each technique is appropriate.

# Warm Up

The following warm-up exercises involve skills that were covered in earlier sections. You will use these skills in the exercise set for this section.

In Exercises 1–4, write the complex number in standard form and give its complex conjugate.

**1.** $4 - \sqrt{-29}$

**2.** $-5 - \sqrt{-144}$

**3.** $-1 + \sqrt{-32}$

**4.** $6 + \sqrt{-1/4}$

In Exercises 5–10, perform the indicated operations and write your answer in standard form.

**5.** $(-3 + 6i) - (10 - 3i)$

**6.** $(12 - 4i) + 20i$

**7.** $(4 - 2i)(3 + 7i)$

**8.** $(2 - 5i)(2 + 5i)$

**9.** $\dfrac{1 + i}{1 - i}$

**10.** $(3 + 2i)^3$

## 4.6   Exercises

In Exercises 1–26, find all the zeros of the function and write the polynomial as a product of linear factors.

**1.** $f(x) = x^2 + 16$

**2.** $f(x) = x^2 + 36$

**3.** $h(x) = x^2 - 5x + 5$

**4.** $g(x) = x^2 + 10x + 23$

**5.** $f(x) = x^4 - 256$

**6.** $f(t) = t^4 - 16$

**7.** $g(x) = x^3 + 5x$

**8.** $h(x) = x^3 - 3x^2 + 4x - 2$

**9.** $g(x) = x^3 - 6x^2 + 13x - 10$

**10.** $f(x) = x^3 - 2x^2 - 11x + 52$

**11.** $f(t) = t^3 - 3t^2 - 15t + 125$

**12.** $f(x) = x^3 + 8x^2 + 20x + 13$

**13.** $f(x) = x^3 + 24x^2 + 214x + 740$

**14.** $f(s) = 2s^3 - 5s^2 + 12s - 5$

**15.** $f(x) = 16x^3 - 20x^2 - 4x + 15$

**16.** $f(x) = 9x^3 - 15x^2 + 11x - 5$

**17.** $h(x) = x^3 - x + 6$

**18.** $h(x) = x^3 + 9x^2 + 27x + 35$

**19.** $f(x) = 5x^3 - 9x^2 + 28x + 6$

**20.** $g(x) = 3x^3 - 4x^2 + 8x + 8$

**21.** $g(x) = x^4 - 4x^3 + 8x^2 - 16x + 16$

**22.** $h(x) = x^4 + 6x^3 + 10x^2 + 6x + 9$

**23.** $f(x) = x^4 + 10x^2 + 9$

**24.** $f(x) = x^4 + 29x^2 + 100$

**25.** $f(t) = t^5 + 5t^4 - 7t^3 - 43t^2 - 8t - 48$

**26.** $g(x) = x^5 - 8x^4 + 28x^3 - 56x^2 + 64x - 32$

In Exercises 27–36, find a polynomial with integer coefficients that has the given zeros.

**27.** $-3, 6i, -6i$

**28.** $3i, -3i$

**29.** $1, 2 + i, 2 - i$

**30.** $6, -5 + 2i, -5 - 2i$

**31.** $-4, 3i, -3i, 2i, -2i$

**32.** $2, 2, 2, 4i, -4i$

**33.** $-5, -5, 1 + \sqrt{3}\,i$

**34.** $\frac{2}{3}, -1, 3 + \sqrt{2}\,i$

**35.** $\frac{3}{4}, -2, -\frac{1}{2} + i$

**36.** $0, 0, 4, 1 + i$

In Exercises 37–40, write the polynomial (a) as the product of factors that are irreducible over the *rationals*, (b) as the product of linear and quadratic factors that are irreducible over the *reals*, and (c) in completely factored form.

**37.** $f(x) = x^4 - 7x^2 - 8$

**38.** $f(x) = x^4 - 4x^3 - 2x^2 + 32x - 48$
(*Hint:* One factor is $x^2 - 8$.)

**39.** $f(x) = x^4 - 4x^3 + 5x^2 - 2x - 6$
(*Hint:* One factor is $x^2 - 2x - 2$.)

**40.** $f(x) = x^4 - 3x^3 - x^2 - 12x - 20$
(*Hint:* One factor is $x^2 + 4$.)

In Exercises 41–50, use the given zero of $f$ to find all the zeros of $f$.

**41.** $f(x) = 3x^3 - 5x^2 + 48x - 80$,    $4i$

**42.** $f(x) = 2x^3 + 3x^2 + 8x + 12$,    $2i$

**43.** $f(x) = x^4 - 2x^3 + 37x^2 - 72x + 36$,    $6i$

**44.** $f(x) = x^3 - 7x^2 - x + 87$,    $5 + 2i$

**45.** $f(x) = 4x^3 + 23x^2 + 34x - 10$,    $-3 + i$

**46.** $f(x) = 3x^3 - 4x^2 + 8x + 8$,    $1 - \sqrt{3}i$

**47.** $f(x) = x^4 + 3x^3 - 5x^2 - 21x + 22$,    $-3 + \sqrt{2}i$

**48.** $f(x) = x^3 + 4x^2 + 14x + 20$,    $-1 - 3i$

**49.** $f(x) = 8x^3 - 14x^2 + 18x - 9$,    $\frac{1}{2}(1 - \sqrt{5}i)$

**50.** $f(x) = 25x^3 - 55x^2 - 54x - 18$,    $\frac{1}{5}(-2 + \sqrt{2}i)$

**51.** *Graphical Analysis*    Find a fourth-degree polynomial that has four real zeros. Find a fourth-degree polynomial that has two real zeros. Find a fourth-degree polynomial that has no real zeros. Graph each polynomial and describe the similarities and differences.

**52.** *Graphical Analysis*    Find a sixth-degree polynomial that has six real zeros. Find a sixth-degree polynomial that has four real zeros. Find a sixth-degree polynomial that has two real zeros. Find a sixth-degree polynomial that has no real zeros. Graph each polynomial and describe the similarities and differences.

**53.** *Profit*    The demand and cost equations for a product are $p = 140 - 0.0001x$ and $C = 80x + 150,000$, where $p$ is the unit price, $C$ is the total cost, and $x$ is the number of units produced. The total profit obtained by producing and selling $x$ units is given by

$$P = R - C = xp - C.$$

Determine a price $p$ that would yield a profit of $9 million. Use a graphing utility to explain why this is not possible.

**54.** *Revenue*    The demand equation for a certain product is given by $p = 140 - 0.0001x$, where $p$ is the unit price (in dollars) of the product and $x$ is the number of units produced. The total revenue obtained by producing and selling $x$ units is given by

$$R = xp.$$

Determine a price $p$ that would yield a revenue of $50 million. Use a graphing utility to explain why this is not possible.

**55.** *Reasoning*    The imaginary number $2i$ is a zero of $f(x) = x^3 - 2ix^2 - 4x + 8i$, but the complex conjugate of $2i$ is not a zero of $f(x)$. Is this a contradiction of the conjugate pair statement on page 273? Explain.

**56.** *Reasoning*    The imaginary number $1 - 2i$ is a zero of $f(x) = x^3 - (1 - 2i)x^2 - 9x + 9(1 - 2i)$, but $1 + 2i$ is not a zero of $f(x)$. Is this a contradiction of the conjugate pair statement on page 273? Explain.

**57.** *Think About It*    A student claims that the polynomial $f(x) = x^4 - 7x^2 + 12$ may be factored over the rational numbers as

$$f(x) = \left(x - \sqrt{3}\right)\left(x + \sqrt{3}\right)(x - 2)(x + 2).$$

Do you agree with this claim? Explain your answer.

**58.** *Think About It*    A student claims that any polynomial with real coefficients can be factored over the real numbers. Give a counterexample to this claim.

## 4.7    Rational Functions

### Objectives

- Find the domain of a rational function.
- Find the vertical and horizontal asymptotes of the graph of a rational function.
- Sketch the graph of a rational function.
- Use a rational function model to solve an application problem.

## Introduction

A **rational function** is one that can be written in the form

$$f(x) = \frac{p(x)}{q(x)}$$

where $p(x)$ and $q(x)$ are polynomials and $q(x)$ is not the zero polynomial. In this section we assume that $p(x)$ and $q(x)$ have no common factors. Unlike polynomial functions, whose domains consist of all real numbers, rational functions often have restricted domains. In general, the *domain* of a rational function of $x$ includes all real numbers except $x$-values that make the denominator zero.

### Example 1    Finding the Domain of a Rational Function

Find the domain of $f$ and discuss its behavior near any excluded $x$-values.

$$f(x) = \frac{1}{x}$$

### Solution

The domain of $f$ is all real numbers except $x = 0$. To determine the behavior of $f$ near this $x$-value, evaluate $f(x)$ to the left and right of $x = 0$.

| $x$ approaches 0 from the left | | | | | |
|---|---|---|---|---|---|
| $x$ | $-1$ | $-0.5$ | $-0.1$ | $-0.01$ | $-0.001$ | $\longrightarrow 0$ |
| $f(x)$ | $-1$ | $-2$ | $-10$ | $-100$ | $-1000$ | $\longrightarrow -\infty$ |

| | | $x$ approaches 0 from the right | | | |
|---|---|---|---|---|---|
| $x$ | $0 \longleftarrow$ | $0.001$ | $0.01$ | $0.1$ | $0.5$ | $1$ |
| $f(x)$ | $\infty \longleftarrow$ | $1000$ | $100$ | $10$ | $2$ | $1$ |

Note that as $x$ approaches 0 *from the left*, $f(x)$ decreases without bound, whereas as $x$ approaches 0 *from the right*, $f(x)$ increases without bound. The graph of $f$ is shown in Figure 4.46.

Figure 4.46

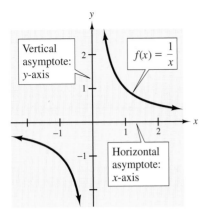

Vertical asymptote: $y$-axis

$f(x) = \dfrac{1}{x}$

Horizontal asymptote: $x$-axis

**Figure 4.47**

## Horizontal and Vertical Asymptotes

In Example 1, the behavior of $f(x) = 1/x$ near $x = 0$ is described by saying that the line $x = 0$ is a **vertical asymptote** of the graph of $f$, as shown in Figure 4.47. In this figure, note that the graph of $f$ also has a **horizontal asymptote**—the line $y = 0$. This means that the values of $f(x) = 1/x$ approach zero as $x$ increases or decreases without bound.

---

▶ **Definition of Vertical and Horizontal Asymptotes**

1. The line $x = a$ is a **vertical asymptote** of the graph of $f$ if

$$f(x) \to \infty \qquad \text{or} \qquad f(x) \to -\infty$$

as $x \to a$, either from the right or from the left.

2. The line $y = b$ is a **horizontal asymptote** of the graph of $f$ if

$$f(x) \to b$$

as $x \to \infty$ or $x \to -\infty$.

---

The graph of a rational function can never intersect its vertical asymptote. It may or may not intersect its horizontal asymptote. In either case, the distance between the horizontal asymptote and the points on the graph must approach zero (as $x \to \infty$ or $x \to -\infty$). Figure 4.48 shows the horizontal and vertical asymptotes of the graphs of three rational functions.

(a)

(b)

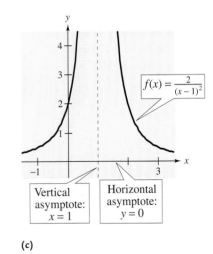

(c)

**Figure 4.48**

The graphs of $f(x) = 1/x$ (Figure 4.47) and $f(x) = (2x + 1)/(x + 1)$ (Figure 4.48a) are called **hyperbolas.**

Use a graphing utility to graph

$$g(x) = \frac{2x^2}{x^2 + 1}$$

in a standard viewing window. Graph $g(x)$ again with Xmin $= -100$ and Xmax $= 100$. Graph $g(x)$ a third time with Xmin $= -1000$ and Xmax $= 1000$. Describe the left and right behavior of the graphs. Complete the table for the function.

| $x$ | 10 | 100 | 1000 |
|---|---|---|---|
| $g(x)$ | | | |

| $x$ | $-10$ | $-100$ | $-1000$ |
|---|---|---|---|
| $g(x)$ | | | |

As the $x$-values become very large or very small, what do the values of $g(x)$ approach? What is the horizontal asymptote of $g(x)$?

▶ **Horizontal Asymptote of a Rational Function**

Let $f$ be the rational function given by

$$f(x) = \frac{a_n x^n + a_{n-1} x^{n-1} + \cdots + a_1 x + a_0}{b_m x^m + b_{m-1} x^{m-1} + \cdots + b_1 x + b_0}, \quad a_n \neq 0, b_n \neq 0.$$

**1.** If $n < m$, the $x$-axis is a horizontal asymptote.

**2.** If $n = m$, the line $y = a_n/b_m$ is a horizontal asymptote.

**3.** If $n > m$, there is no horizontal asymptote.

---

### Example 2    Horizontal Asymptotes of Rational Functions

**a.** $f(x) = \dfrac{2x}{3x^2 + 1}$      See Figure 4.49(a).

Because the degree of the numerator is *less than* the degree of the denominator, the graph of $f$ has the $x$-axis as a horizontal asymptote.

**b.** $g(x) = \dfrac{2x^2}{3x^2 + 1}$      See Figure 4.49(b).

Because the degree of the numerator is *equal to* the degree of the denominator, the graph of $g$ has the line $y = \frac{2}{3}$ as a horizontal asymptote.

**c.** $h(x) = \dfrac{2x^3}{3x^2 + 1}$      See Figure 4.49(c).

Because the degree of the numerator is *greater than* the degree of the denominator, the graph of $h$ has no horizontal asymptote.

(a)

(b)

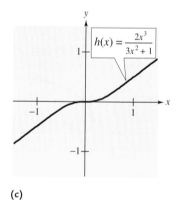

(c)

Figure 4.49

# Sketching the Graph of a Rational Function

▶ **Guidelines for Graphing Rational Functions**

Let $f(x) = p(x)/q(x)$, where $p(x)$ and $q(x)$ have no common factors.

1. Find and plot the $y$-intercept (if any) by evaluating $f(0)$.

2. Find the zeros of the numerator (if any) by solving the equation $p(x) = 0$. Then plot the corresponding $x$-intercepts.

3. Find the zeros of the denominator (if any) by solving the equation $q(x) = 0$. Then sketch the corresponding vertical asymptotes.

4. Find and sketch the horizontal asymptote (if any) by using the rule for finding the horizontal asymptote of a rational function.

5. Plot at least one point both *between and beyond* each $x$-intercept and vertical asymptote.

6. Use smooth curves to complete the graph between and beyond the vertical asymptotes.

Most graphing utilities do not produce good graphs of rational functions (the presence of vertical asymptotes is a problem). To obtain a reasonable graph, we suggest setting the calculator to Dot mode.

Testing for symmetry can be useful, especially for simple rational functions. For example, the graph of $f(x) = 1/x$ is symmetrical with respect to the origin, and the graph of $g(x) = 1/x^2$ is symmetrical with respect to the $y$-axis.

**Example 3**   Sketching the Graph of a Rational Function

Sketch the graph of $g(x) = \dfrac{3}{x - 2}$.

**Solution**

Begin by noting that the numerator and denominator have no common factors.

| | |
|---|---|
| *y-intercept:* | $\left(0, -\frac{3}{2}\right)$, because $g(0) = -\frac{3}{2}$ |
| *x-intercept:* | None, numerator has no zeros. |
| *Vertical asymptote:* | $x = 2$, zero of denominator |
| *Horizontal asymptote:* | $y = 0$, degree of $p(x) <$ degree of $q(x)$ |
| *Additional points:* | |

| $x$ | $-4$ | $1$ | $3$ | $5$ |
|---|---|---|---|---|
| $g(x)$ | $-0.5$ | $-3$ | $3$ | $1$ |

By plotting the intercepts, asymptotes, and a few additional points, you can obtain the graph shown in Figure 4.50. In the figure, note that the graph of $g$ is a vertical stretch and a right shift of the graph of $y = 1/x$.

Figure 4.50

**Figure 4.51**

**Figure 4.52**

| Example 4 | Sketching the Graph of a Rational Function |

Sketch the graph of

$$f(x) = \frac{2x - 1}{x}.$$

**Solution**

Begin by noting that the numerator and denominator have no common factors.

*y-intercept:* None, because $f(0)$ is not defined

*x-intercept:* $\left(\frac{1}{2}, 0\right)$, from $2x - 1 = 0$

*Vertical asymptote:* $x = 0$, zero of denominator

*Horizontal asymptote:* $y = 2$, degree of $p(x)$ = degree of $q(x)$

*Additional points:*

| $x$ | $-4$ | $-1$ | $\frac{1}{4}$ | $4$ |
|---|---|---|---|---|
| $f(x)$ | 2.25 | 3 | $-2$ | 1.75 |

By plotting the intercepts, asymptotes, and a few additional points, you can obtain the graph shown in Figure 4.51. Confirm the graph with your graphing utility.

| Example 5 | Sketching the Graph of a Rational Function |

Sketch the graph of $f(x) = \dfrac{x}{x^2 - x - 2}$.

**Solution**

By factoring the denominator, you have

$$f(x) = \frac{x}{x^2 - x - 2} = \frac{x}{(x + 1)(x - 2)}.$$

So, the numerator and denominator have no common factors.

*y-intercept:* $(0, 0)$

*x-intercept:* $(0, 0)$

*Vertical asymptotes:* $x = -1, x = 2$, zeros of denominator

*Horizontal asymptote:* $y = 0$, degree of $p(x)$ < degree of $q(x)$

*Additional points:*

| $x$ | $-3$ | $-0.5$ | $1$ | $3$ |
|---|---|---|---|---|
| $f(x)$ | $-0.3$ | 0.4 | $-0.5$ | 0.75 |

By plotting the intercepts, asymptotes, and a few additional points, you can obtain the graph shown in Figure 4.52. Confirm the graph with your graphing utility.

## Applications

There are many examples of asymptotic behavior in business and biology. For instance, the following two examples describe the asymptotic behavior related to the cost of removing smokestack emissions *and* the average cost of producing a product.

**Example 6**   Cost-Benefit Model

A utility company burns coal to generate electricity. The cost of removing a certain *percent* of the pollutants from the stack emission is typically not a linear function. That is, if it costs $C$ dollars to remove 25% of the pollutants, it would cost more than $2C$ dollars to remove 50% of the pollutants. As the percent of removed pollutants approaches 100%, the cost tends to become prohibitive. Suppose that the cost $C$ of removing $p$ percent of the smokestack pollutants is given by

$$C = \frac{80,000p}{100 - p}.$$

Suppose that you are a member of a state legislature that is considering a law that will require utility companies to remove 90% of the pollutants from their smokestack emissions. If the current law requires 85% removal, how much additional expense is the new law asking the utility company to incur?

### Solution

The graph of this function is shown in Figure 4.53. Note that the graph has a vertical asymptote at $p = 100$. Because the current law requires 85% removal, the current cost to the utility company is

$$C = \frac{80,000(85)}{100 - 85} \qquad \text{Substitute 85 for } p.$$

$$\approx \$453,333. \qquad \text{Use a calculator.}$$

If the new law increases the percent removal to 90%, the cost to the utility company will be

$$C = \frac{80,000(90)}{100 - 90} \qquad \text{Substitute 90 for } p.$$

$$\approx \$720,000. \qquad \text{Use a calculator.}$$

So, the new law would require the utility company to spend an additional

$$\$720,000 - \$453,333 = \$266,667.$$

How much would it cost to remove another 2% of the pollutants?

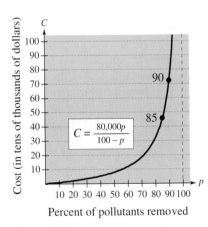

**Figure 4.53**

| Example 7 | Average Cost of Producing a Product |  |

A business has a cost of $C = 0.5x + 5000$, where $C$ is measured in dollars and $x$ is the number of units produced. The *average cost per unit* is given by

$$\overline{C} = \frac{C}{x} = \frac{0.5x + 5000}{x}.$$

Find the average cost per unit when $x = 1000$, $10{,}000$, and $100{,}000$. What is the horizontal asymptote for this function, and what does it represent?

**Solution**

When $x = 1000$, the average cost per unit is

$$\overline{C} = \frac{0.5(1000) + 5000}{1000} = \$5.50.$$

When $x = 10{,}000$, the average cost per unit is

$$\overline{C} = \frac{0.5(10{,}000) + 5000}{10{,}000} = \$1.00.$$

When $x = 100{,}000$, the average cost per unit is

$$\overline{C} = \frac{0.5(100{,}000) + 5000}{100{,}000} = \$0.55.$$

As shown in Figure 4.54, the horizontal asymptote is $C = 0.50$. This line represents the least possible unit cost for the product. This example points out one of the major problems of a small business—it is difficult to have competitively low prices with low levels of production.

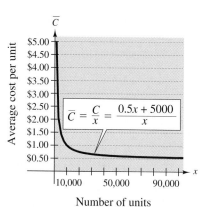

**Figure 4.54**    *As $x \to \infty$, the average cost per unit approaches $0.50.*

| Example 8 | Per Capita Land Area |  |

A model for the population $P$ (in millions) of the United States from 1960 to 1996 is $P = 2.284t + 182.002$, where $t$ represents the calendar year, with $t = 0$ corresponding to 1960. A model for the land area $A$ (in millions of acres) of the United States from 1960 to 1996 is $A = 2316.013$. Construct a rational function for per capita land area $L$. Sketch a graph of the rational function. If the model continues to be accurate for the years beyond 1996, what will be the per capita land area in 2007?    (Source: U.S. Bureau of the Census)

**Solution**

The rational function for the per capita land area $L$ is

$$L = \frac{A}{P} = \frac{2316.013}{2.284t + 182.002}.$$

The graph of the function is shown in Figure 4.55. To find the per capita land area in 2007, substitute $t = 47$ into $L$.

$$L = \frac{2316.013}{2.284t + 182.002} = \frac{2316.013}{2.284(47) + 182.002} = \frac{2316.013}{289.35} \approx 8.00$$

The per capita land area will be approximately 8 acres per person in 2007.

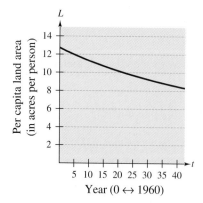

**Figure 4.55**

## Discussing the Concept    Analyzing a Slant Asymptote

By use of long division, the rational function given by

$$f(x) = \frac{x^2 + 2x + 3}{x - 1}$$

can be written as

$$f(x) = \frac{x^2 + 2x + 3}{x - 1}$$

$$= x + 3 + \frac{6}{x - 1}.$$

As $x$ approaches $-\infty$ or $\infty$, the value of the fraction $6/(x - 1)$ is insignificant and the graph of $f$ is close to the graph of $y = x + 3$ (see figure).

$$f(x) = \frac{x^2 + 2x + 3}{x - 1}$$

$$y = x + 3$$

The line

$$y = x + 3 \qquad \text{Slant asymptote}$$

is called a **slant asymptote** of the graph of $f$. Complete the table to show that the values of $f(x)$ and $y$ are nearly the same for large values of $x$.

| $x$ | $-10{,}000$ | $-1000$ | $-100$ | $100$ | $1000$ | $10{,}000$ |
|-----|------------|---------|--------|-------|--------|-----------|
| $f(x)$ | | | | | | |
| $y$ | | | | | | |

Use long division to show that

$$g(x) = \frac{2x^3 - 5}{x^2}$$

also has a slant asymptote. What is the equation of this slant asymptote? Graph $g$ and complete a table for $g$ similar to the one above for $f$. What can you conclude?

# Warm Up

The following warm-up exercises involve skills that were covered in earlier sections. You will use these skills in the exercise set for this section.

In Exercises 1–6, factor the polynomial.

**1.** $x^2 - 4x$              **2.** $2x^3 - 6x$

**3.** $x^2 - 3x - 10$       **4.** $x^2 - 7x + 10$

**5.** $x^3 + 4x^2 + 3x$      **6.** $x^3 - 4x^2 - 2x + 8$

In Exercises 7–10, sketch the graph of the equation.

**7.** $y = 2$            **8.** $x = -1$

**9.** $y = x - 2$       **10.** $y = -x + 1$

---

## 4.7    Exercises

LAB   8

In Exercises 1–8, find the domain of the function.

**1.** $f(x) = \dfrac{3}{(x + 2)^2}$      **2.** $f(x) = \dfrac{1}{(x - 4)^2}$

**3.** $f(x) = \dfrac{x - 7}{3 - x}$      **4.** $f(x) = \dfrac{1 - 5x}{1 + 2x}$

**5.** $f(x) = \dfrac{3x^2 + 1}{x^2 + 9}$      **6.** $f(x) = \dfrac{3x^2 + x - 5}{x^2 + 1}$

**7.** $f(x) = \dfrac{5x^4}{x^2 + 1}$      **8.** $f(x) = \dfrac{2x}{x + 1}$

In Exercises 9–14, match the function with its graph. [The graphs are labeled (a), (b), (c), (d), (e), and (f).]

(a)

(b)

(c)

(d)

(e)

(f)

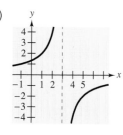

**9.** $f(x) = -\dfrac{4}{x - 3}$      **10.** $f(x) = \dfrac{2}{x + 5}$

**11.** $f(x) = \dfrac{x + 1}{x}$      **12.** $f(x) = \dfrac{3 - 4x}{x}$

**13.** $f(x) = \dfrac{x + 4}{x - 2}$      **14.** $f(x) = -\dfrac{x + 2}{x + 1}$

In Exercises 15–18, compare the graph of $f(x) = 1/x$ with the graph of $g$.

**15.** $g(x) = f(x) - 2 = \dfrac{1}{x} - 2$

**16.** $g(x) = f(x + 1) = \dfrac{1}{x + 1}$

**17.** $g(x) = -f(x) = -\dfrac{1}{x}$

**18.** $g(x) = -f(x + 1) = -\dfrac{1}{x + 1}$

In Exercises 19–22, compare the graph of $f(x) = 4/x^2$ with the graph of $g$.

**19.** $g(x) = f(x) + 3 = \dfrac{4}{x^2} + 3$

**20.** $g(x) = f(x - 1) = \dfrac{4}{(x - 1)^2}$

**21.** $g(x) = -f(x) = -\dfrac{4}{x^2}$     **22.** $g(x) = \dfrac{1}{8}f(x) = \dfrac{1}{2x^2}$

In Exercises 23–26, compare the graph of $f(x) = 8/x^3$ with the graph of $g$.

**23.** $g(x) = f(x) + 5 = \dfrac{8}{x^3} + 5$

**24.** $g(x) = f(x - 3) = \dfrac{8}{(x - 3)^3}$

**25.** $g(x) = -f(x) = -\dfrac{8}{x^3}$     **26.** $g(x) = \dfrac{1}{4}f(x) = \dfrac{2}{x^3}$

In Exercises 27–50, sketch the graph of the rational function. As sketching aids, check for intercepts, vertical asymptotes, and horizontal asymptotes.

**27.** $f(x) = \dfrac{1}{x - 4}$     **28.** $f(x) = \dfrac{1}{x + 6}$

**29.** $f(x) = \dfrac{-1}{x + 1}$     **30.** $f(x) = \dfrac{-1}{2 - x}$

**31.** $f(x) = \dfrac{x + 4}{x - 5}$     **32.** $f(x) = \dfrac{x - 2}{x - 3}$

**33.** $f(x) = \dfrac{2 + x}{1 - x}$     **34.** $f(x) = \dfrac{3 - x}{2 - x}$

**35.** $f(t) = \dfrac{3t + 1}{t}$     **36.** $f(t) = \dfrac{1 - 2t}{t}$

**37.** $g(x) = \dfrac{1}{x + 2} + 2$     **38.** $h(x) = \dfrac{1}{x - 3} + 1$

**39.** $C(x) = \dfrac{5 + 2x}{1 + x}$     **40.** $P(x) = \dfrac{1 - 3x}{1 - x}$

**41.** $f(x) = \dfrac{1}{x^2} + 2$     **42.** $f(x) = 2 - \dfrac{3}{x^2}$

**43.** $h(x) = \dfrac{x^2}{x^2 - 9}$     **44.** $g(x) = \dfrac{x}{x^2 - 9}$

**45.** $g(s) = \dfrac{s}{s^2 + 1}$     **46.** $h(t) = \dfrac{4}{t^2 + 1}$

**47.** $f(x) = \dfrac{x}{x^2 - 3x - 4}$     **48.** $f(x) = \dfrac{-x}{x^2 + x - 6}$

**49.** $f(x) = \dfrac{3x}{x^2 - x - 2}$     **50.** $f(x) = \dfrac{2x}{x^2 + x - 2}$

In Exercises 51–54, find all possible asymptotes (vertical, horizontal, and/or slant) of the given function. (Refer to Discussing the Concept on page 285.) Sketch the graph of $f(x)$.

**51.** $f(x) = \dfrac{x^2 - 7x + 12}{x - 3}$     **52.** $f(x) = \dfrac{x^2 - 9}{x + 3}$

**53.** $f(x) = \dfrac{x^2}{x + 1}$     **54.** $f(x) = \dfrac{x^3 + x}{x^2 - 1}$

In Exercises 55–57, find a counterexample to show that the statement is incorrect.

**55.** Every rational function has a vertical asymptote.

**56.** Every rational function has at least one asymptote.

**57.** A rational function can have only one vertical asymptote.

**58.** Is it possible for a rational function to have all three types of asymptotes (vertical, horizontal, and slant)? Why or why not?

**59.** *Seizure of Illegal Drugs*   The cost in millions of dollars for the federal government to seize $p$ percent of a certain illegal drug as it enters the country is

$$C = \frac{528p}{100 - p}, \quad 0 \le p < 100.$$

(a) Find the cost of seizing 25%.

(b) Find the cost of seizing 50%.

(c) Find the cost of seizing 75%.

(d) According to this model, would it be possible to seize 100% of the drug? Explain.

**60.** *Smokestack Emission*   The cost in dollars of removing $p$ percent of the air pollutants in the stack emission of a utility company that burns coal to generate electricity is

$$C = \frac{80,000p}{100 - p}, \quad 0 \le p < 100.$$

(a) Find the cost of removing 15%.

(b) Find the cost of removing 50%.

(c) Find the cost of removing 90%.

(d) According to the model, would it be possible to remove 100% of the pollutants? Explain.

**61.** *Population of Deer*    The Game Commission introduces 50 deer into newly acquired state game lands. The population of the herd is given by

$$N = \frac{10(5 + 3t)}{1 + 0.04t}, \quad t \geq 0$$

where $t$ is time in years.

(a) Find the population when $t$ is 5, 10, and 25.

(b) What is the limiting size of the herd as time progresses?

**62.** *Population of Elk*    The Game Commission introduces 40 elk into newly acquired state game lands. The population of the herd is given by

$$N = \frac{10(4 + 2t)}{1 + 0.03t}, \quad t \geq 0$$

where $t$ is time in years.

(a) Find the population when $t$ is 5, 10, and 25.

(b) What is the limiting size of the herd as time progresses?

**63.** *Average Cost*    The cost of producing $x$ units is

$$C = 150,000 + 0.25x$$

and so the average cost per unit is

$$\overline{C} = \frac{C}{x} = \frac{150,000 + 0.25x}{x}, \quad x > 0.$$

(a) Sketch the graph of this average cost function.

(b) Find the average cost of producing $x = 1000$, $x = 10,000$, and $x = 100,000$ units. What can you conclude?

**64.** *Average Cost*    The cost of producing $x$ units is

$$C = 250,000 + 3x$$

and so the average cost per unit is

$$\overline{C} = \frac{C}{x} = \frac{250,000 + 3x}{x}, \quad x > 0.$$

(a) Sketch the graph of this average cost function.

(b) Find the average cost of producing $x = 1000$, $x = 10,000$, and $x = 100,000$ units. What can you conclude?

**65.** *Human Memory Model*    Psychologists have developed mathematical models to predict memory performance as a function of the number of trials $n$ of a certain task. Consider the learning curve

$$P = \frac{0.5 + 0.9(n - 1)}{1 + 0.9(n - 1)}, \quad n > 0$$

where $P$ is the percent of correct responses after $n$ trials.

(a) Complete the following table.

| $n$ | 1 | 2 | 3 | 4 | 5 | 6 | 7 | 8 | 9 | 10 |
|-----|---|---|---|---|---|---|---|---|---|----|
| $P$ |   |   |   |   |   |   |   |   |   |    |

(b) According to this model, what is the limiting percent of correct responses as $n$ increases?

**66.** *Human Memory Model*    Consider the learning curve

$$P = \frac{0.6 + 0.85(n - 1)}{1 + 0.85(n - 1)}, \quad n > 0$$

where $P$ is the percent of correct responses after $n$ trials.

(a) Complete the following table.

| $n$ | 1 | 2 | 3 | 4 | 5 | 6 | 7 | 8 | 9 | 10 |
|-----|---|---|---|---|---|---|---|---|---|----|
| $P$ |   |   |   |   |   |   |   |   |   |    |

(b) According to this model, what is the limiting percent of correct responses as $n$ increases?

**67.** *Average Recycling Cost*    The cost of recycling a waste product is

$$C = 350,000 + 5x, \quad x > 0$$

where $x$ is the number of pounds of waste. The average recycling cost is

$$\overline{C} = \frac{C}{x} = \frac{350,000 + 5x}{x}, \quad x > 0.$$

(a) Use a graphing utility to sketch the graph of $\overline{C}$.

(b) Find the average cost of recycling for $x = 1000$, $x = 10,000$, and $x = 100,000$ pounds. What can you conclude?

**68.** *Average Recycling Cost* The cost of recycling a waste product is

$$C = 600,000 + 10x, \quad x > 0$$

where $x$ is the number of pounds of waste. The average recycling cost is

$$\overline{C} = \frac{C}{x} = \frac{600,000 + 10x}{x}, \quad x > 0.$$

(a) Use a graphing utility to sketch the graph of $\overline{C}$.

(b) Find the average cost of recycling for $x = 1000$, $x = 10,000$, and $x = 100,000$ pounds. What can you conclude?

**69.** *Sale of Long-Playing Albums* The sales of long-playing albums in the United States from 1975 to 1996 can be approximated by the model

$$L = \frac{136.69 - 5.72t}{1 - 0.137t + 0.006t^2}, \quad 5 \le t \le 26$$

where $L$ is the number of albums sold (in millions) and $t = 5$ represents 1975. From the figure, notice the drop in sales of albums after about 1980. How would you account for this drop in sales? (Source: Recording Industry Association of America)

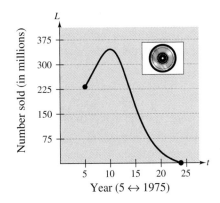

Year (5 ↔ 1975)

**70.** *1500-Meter Run* The winning times for the 1500-meter run at the Olympics from 1900 to 1996 are shown in the figure. A *quadratic* model that approximates these times is

$$y = 4.16 - 0.0118t + 0.00006t^2$$

where $y$ is the winning time and $t = 0$ represents 1900. How well did this model predict the winning time for 2000? Does this model have a horizontal asymptote? From the winning times, do you think that a good model should have a horizontal asymptote? (Source: ESPN 1998 *Information Please Almanac*)

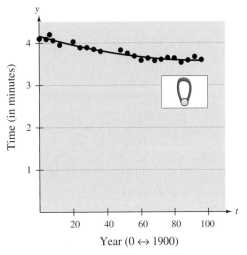

Year (0 ↔ 1900)

**Figure for 70**

**71.** *Drug Concentration* The concentration of a medication in the bloodstream $t$ hours after injection into muscle tissue is given by

$$C(t) = \frac{2t + 1}{t^2 + 4}, \quad t > 0.$$

(a) Use a graphing utility to graph the function. Estimate when the concentration is greatest.

(b) Does this function have a horizontal asymptote? If so, discuss the meaning of the asymptote in terms of the concentration of the medication.

**72.** *Drug Concentration* The concentration of a medication in the bloodstream $t$ minutes after sublingual (under the tongue) application is given by

$$C(t) = \frac{3t - 1}{2t^2 + 5}, \quad t > 0.$$

(a) Use a graphing utility to graph the function. Estimate when the concentration is greatest.

(b) Does this function have a horizontal asymptote? If so, discuss the meaning of the asymptote in terms of the concentration of the medication.

## Chapter Project

## Analyzing a Bouncing Ball

| Time (sec) | Height (ft) |
|------------|-------------|
| 1.5482 | 0.30946 |
| 1.5912 | 0.82161 |
| 1.6342 | 1.2761 |
| 1.6772 | 1.6716 |
| 1.7202 | 2.0157 |
| 1.7632 | 2.2950 |
| 1.8062 | 2.5162 |
| 1.8492 | 2.6837 |
| 1.8922 | 2.7932 |
| 1.9352 | 2.8468 |
| 1.9782 | 2.8428 |
| 2.0212 | 2.7797 |
| 2.0642 | 2.6563 |
| 2.1072 | 2.4756 |
| 2.1502 | 2.2315 |
| 2.1932 | 1.9297 |
| 2.2362 | 1.5635 |
| 2.2792 | 1.1437 |
| 2.3222 | 0.66756 |
| 2.3652 | 0.13378 |

The data in the table at the left was collected with a *Texas Instruments CBR* unit. The table shows values for the first full bounce of a soccer ball that was dropped from a height of 5 feet. Use this information to investigate the following questions.

1. *Inspecting the Data*   Study the table at the left. What pattern do you see in the data?

2. *Graphing the Data*   Enter the data given in the table in a graphing utility. Use the graphing utility to make a scatter plot of the data. How does the scatter plot compare to your answer to Question 1? Do the points appear to resemble a quadratic pattern?

3. *Fitting a Model to the Data*   Use the trace feature of your graphing utility to estimate the vertex of the parabola. The standard form of the quadratic equation is $y = a(x - h)^2 + k$, where $(h, k)$ is the vertex. Use your estimated vertex $(h, k)$ and any ordered pair $(x, y)$ from the data table to determine a value for $a$ in the quadratic equation. Write the quadratic equation in the general form $y = ax^2 + bx + c$.

4. *Fitting a Model to the Data*   Use the quadratic regression program of your graphing utility to develop another quadratic model for the data. Write the quadratic model in the general form $y = ax^2 + bx + c$.

5. *Comparing Actual Data with the Models*   Use the table feature of your graphing utility to evaluate heights of the ball that both models give for different times during the bounce. How do the two models' values compare to the actual values from the table?

6. *Interpreting the Models*   Earth's gravitational constant is $-16$ ft/sec$^2$. How do the values of $a$ in both quadratic models compare to Earth's gravitational constant? Does this make sense in the context of the data? Explain.

7. *Interpreting the Model*   Of the models that you found in Questions 3 and 4, select the one that appears to best fit the actual data. Use your graphing utility to graph the quadratic equation.

   (a) Use the trace feature to approximate the soccer ball's maximum height on its first bounce.

   (b) How does this value compare to the initial height from which the ball was dropped?

   (c) Imagine that this experiment was performed on the moon, using the same ball and dropping it onto the same surface from the same initial height. Do you think that the maximum height of the first bounce would be greater than, less than, or equal to the value you determined in part (a)? Could the maximum height exceed 5 feet? Explain your reasoning.

## CHAPTER SUMMARY

After studying this chapter, you should have acquired the following skills.
These skills are keyed to the Review Exercises that begin on page 292.
Answers to odd-numbered Review Exercises are given in the back of the book.

**4.1** · Sketch the graph of a quadratic function and identify its vertex        Review Exercises 1–4
and intercepts.

· Find a quadratic function given its vertex and a point on the graph.        Review Exercises 5, 6

· Construct and use a quadratic model to solve an application problem.    Review Exercises 7–12

**4.2** · Apply the Leading Coefficient Test to determine right and left        Review Exercises 13–16
behavior of the graph of a polynomial function.

· Find the real zeros of a polynomial function.        Review Exercises 17–20

**4.3** · Divide one polynomial by a second polynomial using long division.    Review Exercises 21, 22

· Simplify a rational expression using long division.        Review Exercises 23, 24

· Use synthetic division to divide two polynomials.        Review Exercises 25–28, 31, 32

· Use the Remainder Theorem and synthetic division to evaluate        Review Exercises 29, 30
a polynomial.

**4.4** · Find all possible rational zeros of a function using the Rational        Review Exercises 33, 34
Zero Test.

· Approximate the real zeros of a polynomial function using the        Review Exercises 35, 36
Intermediate Value Theorem.

· Find and approximate the real zeros of a polynomial function.        Review Exercises 37–44

· Apply techniques for approximating real zeros to solve an        Review Exercises 45, 46
application problem.

**4.5** · Find the complex conjugate of a complex number.        Review Exercises 47–50

· Perform operations with complex numbers and write the results        Review Exercises 51–66
in standard form.

· Solve a polynomial equation.        Review Exercises 67–70

· Plot a complex number in the complex plane.        Review Exercises 71, 72

**4.6** · Use the Fundamental Theorem of Algebra and the Linear        Review Exercises 73–78
Factorization Theorem to write a polynomial as the product of
linear factors.

· Find a polynomial with integer coefficients whose zeros are given.    Review Exercises 79, 80

· Factor a polynomial over the real and complex numbers.        Review Exercises 81, 82

· Find all real and complex zeros of a polynomial function.        Review Exercises 83–86

**4.7** · Find the domain of a rational function.        Review Exercises 87–90

· Find the vertical and horizontal asymptotes of the graph of a        Review Exercises 87–90
rational function.

· Sketch the graph of a rational function.        Review Exercises 91–94

· Use a rational function model to solve an application problem.        Review Exercises 95–100

## REVIEW EXERCISES

In Exercises 1–4, sketch the graph of the quadratic function. Identify the vertex and intercepts.

1. $f(x) = (x + 3)^2 - 2$

2. $g(x) = -x^2 + 6x + 8$

3. $h(x) = 3x^2 - 12x + 11$

4. $f(x) = \frac{1}{3}(x^2 + 5x - 4)$

In Exercises 5 and 6, find the quadratic function that has the indicated vertex and whose graph passes through the given point.

5. Vertex: $(-4, -2)$; point: $(-3, -5)$

6. Vertex: $(2, 5)$; point: $(4, 7)$

7. *Maximum Area* The perimeter of a rectangle is 200 feet. Let $x$ represent the width of the rectangle and write a quadratic function that expresses the area of the rectangle in terms of its width. Of all possible rectangles with perimeters of 200 feet, what are the measurements of the one that has the greatest area?

8. *Maximum Revenue* Find the number of units that produces a maximum revenue $R$ for

$R = 800x - 0.01x^2$

where $R$ is the total revenue in dollars and $x$ is the number of units produced.

9. *Minimum Cost* A manufacturer has daily production costs of

$C = 20{,}000 - 40x + 0.055x^2$

where $C$ is the total cost in dollars and $x$ is the number of units produced.

(a) Use a graphing utility to graph the cost function.

(b) Graphically estimate the number of units that should be produced to yield a minimum cost.

(c) Explain how to confirm the result of part (b) algebraically.

10. *Maximum Profit* The profit for a company is given by

$P = -0.0001x^2 + 130x - 300{,}000$

where $x$ is the number of units produced.

(a) Use a graphing utility to graph the profit function.

(b) Graphically estimate the number of units that should be produced to yield a maximum profit.

(c) Explain how to confirm the result of part (b) algebraically.

11. *Regression Problem* Let $x$ be the angle (in degrees) at which a baseball is hit with a 30-hertz backspin at an initial speed of 40 miles per hour and let $d(x)$ be the distance the ball travels. The following table gives the distances for the different angles at which the ball is hit. (Source: *The Physics of Sports*)

| $x$ | 10° | 15° | 30° | 36° | 42° | 43° |
|---|---|---|---|---|---|---|
| $d(x)$ | 61.2 | 83.0 | 130.4 | 139.4 | 143.2 | 143.3 |

| $x$ | 44° | 45° | 48° | 54° | 60° |
|---|---|---|---|---|---|
| $d(x)$ | 142.8 | 142.7 | 140.7 | 132.8 | 119.7 |

(a) Make a scatter plot using the data in the table.

(b) Use a quadratic regression program to find a mathematical model for $d(x)$.

(c) Use a graphing utility to graph your model for $d(x)$ and the scatter plot for the data.

(d) Use the model to approximate the angle that will result in a maximum distance.

12. *Regression Problem* Let $x$ be the crude protein intake (in kilograms per day) for pigs and let $s(x)$ be the shoulder weight. The following table gives the shoulder weight for pigs given the crude protein intake.

| $x$ | 0.195 | 0.238 | 0.297 | 0.341 | 0.401 | 0.427 |
|---|---|---|---|---|---|---|
| $s(x)$ | 8.13 | 8.74 | 9.68 | 9.69 | 9.81 | 8.99 |

(a) Make a scatter plot using the data in the table.

(b) Use a quadratic regression program to find a mathematical model for $s(x)$.

(c) Use a graphing utility to graph your model for $s(x)$ and the scatter plot for the data.

(d) Use the model to approximate the amount of crude protein that will result in a maximum shoulder weight. (Source: *Livestock Research for Rural Development*)

In Exercises 13–16, determine the right and left behavior of the graph of the function.

**13.** $f(x) = \frac{1}{2}x^3 + 2x$

**14.** $f(x) = 5 - 3x^2 + 4x^4 - x^6$

**15.** $f(x) = -x^5 + 3x^2 + 11$

**16.** $f(x) = \frac{3}{4}(x^4 + 3x^2 + 2)$

In Exercises 17–20, find all real zeros of the function.

**17.** $f(x) = 25 - x^2$

**18.** $f(x) = x^4 - 6x^2 + 8$

**19.** $f(x) = x^3 - 7x^2 + 10x$

**20.** $f(x) = x^3 - 6x^2 - 3x + 18$

In Exercises 21 and 22, divide by long division.

| Dividend | Divisor |
|---|---|
| **21.** $2x^3 - 5x^2 - x$ | $2x + 1$ |
| **22.** $x^4 - 5x^3 + 10x^2 - 12$ | $x^2 - 2x + 4$ |

In Exercises 23 and 24, simplify the expression.

**23.** $\dfrac{x^3 + 9x^2 + 2x - 48}{x - 2}$    **24.** $\dfrac{x^4 + 5x^3 - 20x - 16}{x^2 - 4}$

In Exercises 25 and 26, divide by synthetic division.

| Dividend | Divisor |
|---|---|
| **25.** $x^3 - 6x + 9$ | $x + 3$ |
| **26.** $x^5 - x^4 + x^3 - 13x^2 + x + 6$ | $x + 2$ |

In Exercises 27 and 28, use synthetic division to show that $x$ is a solution of the equation. Then factor the polynomial completely.

**27.** $x^3 - 4x^2 - 11x + 30 = 0, \quad x = 5$

**28.** $3x^3 + 23x^2 + 37x - 15 = 0, \quad x = \frac{1}{3}$

In Exercises 29 and 30, use synthetic division to evaluate the function.

**29.** $f(x) = 6 + 2x^2 - 3x^3$
   (a) $f(2)$    (b) $f(-1)$

**30.** $f(x) = 2x^4 + 3x^3 + 6$
   (a) $f(\frac{1}{2})$    (b) $f(-1)$

**31.** *Measurements of a Room*    A rectangular room has a volume of

$$x^3 + 13x^2 + 50x + 56$$

cubic feet. The height of the room is $x + 2$. Find the number of square feet of floor space in the room.

**32.** *Profit*    The profit for a product is given by

$$P = -130x^3 + 6500x^2 - 200,000, \quad 0 \le x \le 50$$

where $P$ is the profit (in dollars) and $x$ is the advertising expense (in tens of thousands of dollars). For this product, the advertising expense was \$400,000 ($x = 40$), and the profit was \$1,880,000. Use a graphing utility to graph the function and use the result to find another advertising expense that would have produced the same profit.

In Exercises 33 and 34, use the Rational Zero Test to list all possible rational zeros of $f$. Verify using a graphing utility.

**33.** $f(x) = -4x^3 + 8x^2 - 3x + 15$

**34.** $f(x) = 3x^4 + 4x^3 - 5x^2 + 10x - 8$

In Exercises 35 and 36, use the Intermediate Value Theorem to approximate the zero of $f(x)$ in the interval $[a, b]$. Give your approximation to the nearest tenth.

**35.** $f(x) = x^3 - 4x + 3, \quad [-3, -2]$

**36.** $f(x) = x^5 + 5x^2 + x - 1, \quad [0, 1]$

In Exercises 37–42, find and approximate all real zeros of the function.

**37.** $f(x) = x^3 + 2x^2 - 5x - 6$

**38.** $g(x) = 2x^3 - 15x^2 + 24x + 16$

**39.** $h(x) = 3x^4 - 27x^2 + 60$

**40.** $f(x) = x^5 - 4x^3 + 3x$

**41.** $C(x) = 3x^4 + 3x^3 - 7x^2 - x + 2$

**42.** $p(x) = x^4 - x^3 - 2x - 4$

In Exercises 43 and 44, approximate the real zeros of $f$ using a graphing utility. Use an accuracy of 0.001.

**43.** $f(x) = 5x^3 - 11x - 3$

**44.** $f(x) = 2x^4 - 9x^3 - 5x^2 + 10x + 12$

**45.** *Advertising Costs* A company that manufactures motorcycles estimates that the profit from selling the top-of-the-line model is given by

$$P = -35x^3 + 2000x^2 - 27,500, \quad 10 \le x \le 55$$

where $P$ is the profit (in dollars) and $x$ is the advertising expense (in tens of thousands of dollars). According to this model, how much money should the company spend on advertising to obtain a profit of $538,000?

**46.** *Age of the Groom* The average age of the groom in a marriage for a given age of the bride can be approximated by the model

$$y = -0.00428x^2 + 1.442x - 3.136, \quad 20 \le x \le 55$$

where $y$ is the groom's age and $x$ is the bride's age. For what age of the bride is the average age of the groom 30? (Source: U.S. National Center for Health Statistics)

In Exercises 47–50, write the complex number in standard form and find its complex conjugate.

**47.** $\sqrt{-32}$

**48.** $12$

**49.** $-3 + \sqrt{-16}$

**50.** $2 - \sqrt{-18}$

In Exercises 51–66, perform the indicated operation and write the result in standard form.

**51.** $(7 - 4i) + (-2 + 5i)$

**52.** $(14 + 6i) - (-1 - 2i)$

**53.** $\left(1 + \sqrt{-12}\right)\left(5 - \sqrt{-3}\right)$

**54.** $\left(3 - \sqrt{-4}\right)\left(4 - \sqrt{-49}\right)$

**55.** $(5 + 8i)(5 - 8i)$

**56.** $\left(\frac{1}{2} + \frac{3}{4}i\right)\left(\frac{1}{2} - \frac{3}{4}i\right)$

**57.** $-2i(4 - 5i)$

**58.** $-3(-2 + 4i)$

**59.** $(3 + 4i)^2$

**60.** $(2 - 5i)^2$

**61.** $\dfrac{8 - i}{2 + i}$

**62.** $\dfrac{3 - 4i}{1 - 5i}$

**63.** $\dfrac{4 - 3i}{i}$

**64.** $\dfrac{2}{(1 + i)^2}$

**65.** $(3 + 2i)^2 + (3 - 2i)^2$

**66.** $(1 + i)^2 - (1 - i)^2$

In Exercises 67–70, solve the equation.

**67.** $2x^2 - x + 3 = 0$

**68.** $3x^2 + 6x + 11 = 0$

**69.** $4x^2 + 11x + 3 = 0$

**70.** $9x^2 - 2x + 5 = 0$

In Exercises 71 and 72, plot the complex number.

**71.** $-3 + 2i$

**72.** $-1 - 4i$

In Exercises 73–78, write the polynomial as a product of linear factors.

**73.** $f(x) = x^4 - 81$

**74.** $h(x) = 2x^3 - 5x^2 + 4x - 10$

**75.** $f(t) = t^3 + 5t^2 + 3t + 15$

**76.** $h(x) = x^4 + 17x^2 + 16$

**77.** $g(x) = 4x^3 - 8x^2 + 9x - 18$

**78.** $f(x) = x^5 - 2x^4 + x^3 - x^2 + 2x - 1$

In Exercises 79 and 80, find a polynomial with integer coefficients that has the given zeros.

**79.** $3, 4i, -4i$

**80.** $2, -3, 1 - 2i, 1 + 2i$

In Exercises 81 and 82, write the polynomial (a) as the product of factors that are irreducible over the rationals, (b) as the product of linear and quadratic factors that are irreducible over the reals, and (c) in completely factored form.

**81.** $f(x) = x^4 + 5x^2 - 24$

**82.** $f(x) = x^4 - 3x^3 - 11x^2 + 15x + 30$

In Exercises 83–86, use the given zero of $f$ to find all the zeros of $f$.

**83.** $f(x) = 4x^3 - x^2 + 64x - 16, \quad -4i$

**84.** $h(x) = 50 - 75x + 2x^2 - 3x^3, \quad 5i$

**85.** $f(x) = x^4 + 7x^3 + 24x^2 + 58x + 40, \quad 1 + 3i$

**86.** $f(x) = x^4 + 4x^3 + 8x^2 + 4x + 7, \quad 2 - \sqrt{3}i$

In Exercises 87–90, find the domain of the function and identify any horizontal or vertical asymptotes.

**87.** $f(x) = \dfrac{5}{x - 6}$

**88.** $f(x) = \dfrac{2x^2 + 5x - 3}{x^2 + 2}$

**89.** $f(x) = \dfrac{x^2}{x^2 - 4}$

**90.** $f(x) = \dfrac{2x}{x^2 - 2x - 8}$

In Exercises 91–94, sketch the graph of the rational function. As sketching aids, check for intercepts, symmetry, vertical asymptotes, and horizontal asymptotes.

**91.** $P(x) = \dfrac{3 - x}{x + 2}$

**92.** $f(x) = \dfrac{4}{(x - 1)^2}$

**93.** $g(x) = \dfrac{1}{x + 3} + 2$

**94.** $h(x) = \dfrac{-3x}{2x^2 + 3x - 5}$

**95.** *Average Cost*   The cost of producing $x$ units is $C = 100,000 + 0.9x$, and so the average cost per unit is

$$\overline{C} = \frac{C}{x} = \frac{100,000 + 0.9x}{x}, \quad x > 0.$$

Sketch the graph of this average cost function and find the average cost of producing $x = 1000$, $x = 10,000$, and $x = 100,000$ units.

**96.** *Population of Fish*   The Parks and Wildlife Commission introduces 50,000 game fish into a large lake. The population of the fish (in thousands) is

$$N = \frac{20(4 + 3t)}{1 + 0.05t}, \quad t \geq 0$$

where $t$ is time in years.

(a) Find the population when $t = 5, 10,$ and $25$.

(b) What is the limiting number of this population of fish in the lake as time progresses?

**97.** *Human Memory Model*   Consider the learning curve

$$P = \frac{0.5 + 0.9(n - 1)}{1 + 0.9(n - 1)}, \quad n > 0$$

where $P$ is the percent of correct responses after $n$ trials. Complete the table for this model.

| $n$ | 1 | 2 | 3 | 4 | 5 | 6 | 7 | 8 | 9 | 10 |
|---|---|---|---|---|---|---|---|---|---|---|
| $P$ | | | | | | | | | | |

**98.** *Smokestack Emission*   The cost in dollars of removing $p$ percent of the air pollutants in the stack emission of a utility company that burns coal to generate electricity is

$$C = \frac{95,000p}{100 - p}, \quad 0 \leq p \leq 100.$$

(a) Find the cost of removing 25%.

(b) Find the cost of removing 60%.

(c) Find the cost of removing 99%.

(d) According to this model, would it be possible to remove 100% of the pollutants? Explain.

**99.** *Average Recycling Cost*   The cost of recycling a waste product is

$$C = 450,000 + 6x, \quad x > 0$$

where $x$ is the number of pounds of waste. The average recycling cost is

$$\overline{C} = \frac{C}{x} = \frac{450,000 + 6x}{x}, \quad x > 0.$$

(a) Sketch the graph of $\overline{C}$.

(b) Find the average cost of recycling for $x = 1000$ pounds, $x = 10,000$ pounds, and $x = 100,000$ pounds. What can you conclude?

**100.** *Cost Per Unit*   A manufacturer determines that the variable cost for a new product is $2.25 per unit and the fixed costs are $55,000. The product is to be sold for $3.95. Let $x$ be the number of units sold.

(a) Write the total cost function $C$ as a function of the number of units sold.

(b) Write the total revenue function $R$ as a function of the number of units sold.

(c) Write the total profit function $P$ as a function of the number of units sold. (*Hint:* $P = R - C$)

(d) Write the average cost per unit $\overline{C} = C/x$ as a function of $x$.

(e) Complete the table. What do you observe about $\overline{C}$ as $x$ increases?

| $x$ | 100 | 1000 | 10,000 | 100,000 | 1,000,000 |
|---|---|---|---|---|---|
| $\overline{C}$ | | | | | |

## Chapter Test

Take this test as you would take a test in class. After you are done, check your work against the answers given in the back of the book.

1. Graph $f(x) = -(x + 1)^2 + 2$. Indicate the $x$- and $y$-intercepts (if any), the vertex, and the line of symmetry.

2. Describe the left and right behavior of the graph of $f(x)$.
   (a) $f(x) = 12x^3 - 5x^2 - 49x + 15$
   (b) $f(x) = 5x^4 - 3x^3 + 2x^2 + 11x + 12$

3. List all the possible rational zeros of $f(x) = 2x^3 - 5x^2 + 2x - 5$. Use synthetic division to show that $x = \frac{5}{2}$ is a zero of $f(x)$. Using this result, completely factor the polynomial.

4. Simplify: $\dfrac{x^4 + 4x^3 - 19x^2 - 106x - 120}{x^2 - 3x - 10}$.

In Exercises 5–12, perform the indicated operation and write the result in standard form.

5. $(12 + 3i) + (4 - 6i)$

6. $(10 - 2i) - (3 + 7i)$

7. $\left(5 + \sqrt{-12}\right)\left(3 - \sqrt{-12}\right)$

8. $(4 + 3i)(2 - 5i)$

9. $(2 - 3i)^2$

10. $(5 + 2i)^2$

11. $\dfrac{1 + i}{1 - i}$

12. $\dfrac{5 - 2i}{i}$

In Exercises 13 and 14, solve the quadratic equation.

13. $x^2 + 5x + 7 = 0$

14. $2x^2 - 5x + 11 = 0$

15. Find all zeros of $f(x) = x^3 + 2x^2 + 5x + 10$, given that $\sqrt{5}i$ is a zero.

16. Find a polynomial with integer coefficients that has $2, 5, 3i$, and $-3i$ as zeros.

17. Graph $f(x) = \dfrac{2x}{x + 1}$. Indicate all $x$- and $y$-intercepts and any asymptotes. What is the domain of $f(x)$?

18. The profit for a company is $P(x) = 300 + 10x - 0.002x^2$, where $x$ is the number of units produced. What production level will yield a maximum profit?

19. The average cost of recycling a waste product $x$ (in pounds) is $\overline{C}(x) = \dfrac{450,000 + 5x}{x}, x > 0$. Find the average cost of recycling $x = 10,000$ pounds, $x = 100,000$ pounds, and $x = 1,000,000$ pounds. What can you conclude?

20. A company estimates that the profit from selling its product is
$$P = -11x^3 + 900x^2 - 50,000, \quad 0 \le x$$
where $P$ is the profit (in dollars) and $x$ is the advertising expense (in tens of thousands of dollars). How much should the company spend on advertising to obtain a profit of $222,000? Explain your reasoning.

# 5

# Exponential and Logarithmic Functions

Charles D. Winters/Photo Researchers, Inc.

When a warm object is placed in a cool room, its temperature will gradually change to the temperature of the room. (The same is true for a cool object placed in a warm room.)

According to Newton's Law of Cooling, the rate at which the object's temperature $T$ changes is proportional to the difference between its temperature and the temperature of the room $L$. This statement can be translated as

$$T = L + Ce^{kt}$$

where $k$ is a constant and $L + C$ is the original temperature of the object.

The table below and the scatter plot above show the temperature $T$ (in degrees Fahrenheit) for several times $t$ (in minutes) for 300 ml of hot water in a ceramic cup.

| $t$ | 0 | 2 | 8 | 10 | 21 | 24 | 28 |
|---|---|---|---|---|---|---|---|
| $T$ | 182 | 180 | 160 | 156 | 138 | 134 | 128 |

The chapter project related to this information is on page 351.

## 5.1 Exponential Functions

### Objectives

- Evaluate an exponential expression.
- Sketch the graph of an exponential function.
- Use the compound interest formula.
- Use an exponential model to solve an application problem.

## Exponential Functions

So far, this text has dealt only with **algebraic functions,** which include polynomial functions and rational functions. In this chapter you will study two types of nonalgebraic functions— *exponential* functions and *logarithmic* functions. These functions are examples of **transcendental functions.**

> ▶ Definition of Exponential Function
>
> The **exponential function** $f$ **with base** $a$ is denoted by
>
> $$f(x) = a^x$$
>
> where $a > 0$, $a \neq 1$, and $x$ is any real number.

The base $a = 1$ is excluded from the definition of exponential function because it yields $f(x) = 1^x = 1$, which is a constant function.

You already know how to evaluate $a^x$ for integer and rational values of $x$. For example, you know that $4^3 = 64$ and $4^{1/2} = 2$. However, to evaluate $4^x$ for any real number $x$, you need to interpret forms with *irrational* exponents. For the purposes of this text, it is sufficient to think of

$$a^{\sqrt{2}} \quad (\text{where } \sqrt{2} \approx 1.414214)$$

as that value having the successively closer approximations

$$a^{1.4}, \, a^{1.41}, \, a^{1.414}, \, a^{1.4142}, \, a^{1.41421}, \, a^{1.414214}, \, \ldots \, .$$

Example 1 shows how to use a calculator to evaluate an exponential function.

### Example 1   Evaluating Exponential Expressions

**Scientific Calculator**

| Number | Keystrokes | Display |
|--------|------------|---------|
| $2^{-\pi}$ | 2 $\boxed{y^x}$ $\pi$ $\boxed{+/-}$ $\boxed{=}$ | 0.1133147 |

**Graphing Calculator**

| Number | Keystrokes | Display |
|--------|------------|---------|
| $2^{-\pi}$ | 2 $\boxed{\wedge}$ $\boxed{(-)}$ $\pi$ $\boxed{\text{ENTER}}$ | 0.1133147 |

# Graphs of Exponential Functions

The graphs of all exponential functions have similar characteristics, as shown in Examples 2, 3, and 4.

### Example 2    Graphs of $y = a^x$

In the same coordinate plane, sketch the graphs of the following functions.

**a.** $f(x) = 2^x$    **b.** $g(x) = 4^x$

#### Solution

The table below lists some values for each function, and Figure 5.1 shows their graphs. Note that both graphs are increasing. Moreover, the graph of $g(x) = 4^x$ is increasing more rapidly than the graph of $f(x) = 2^x$.

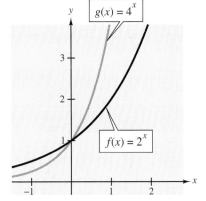

Figure 5.1

| $x$ | $-2$ | $-1$ | 0 | 1 | 2 | 3 |
|---|---|---|---|---|---|---|
| $2^x$ | $\frac{1}{4}$ | $\frac{1}{2}$ | 1 | 2 | 4 | 8 |
| $4^x$ | $\frac{1}{16}$ | $\frac{1}{4}$ | 1 | 4 | 16 | 64 |

### Example 3    Graphs of $y = a^{-x}$

In the same coordinate plane, sketch the graphs of the following functions.

**a.** $F(x) = 2^{-x}$    **b.** $G(x) = 4^{-x}$

#### Solution

The table below lists some values for each function, and Figure 5.2 shows their graphs. Note that both graphs are decreasing. Moreover, the graph of $G(x) = 4^{-x}$ is decreasing more rapidly than the graph of $F(x) = 2^{-x}$.

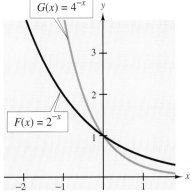

Figure 5.2

| $x$ | $-3$ | $-2$ | $-1$ | 0 | 1 | 2 |
|---|---|---|---|---|---|---|
| $2^{-x}$ | 8 | 4 | 2 | 1 | $\frac{1}{2}$ | $\frac{1}{4}$ |
| $4^{-x}$ | 64 | 16 | 4 | 1 | $\frac{1}{4}$ | $\frac{1}{16}$ |

The tables in Examples 2 and 3 were evaluated by hand. You could, of course, use the table feature of a graphing utility to construct tables with even more values.

Comparing the functions in Examples 2 and 3, observe that

$$F(x) = 2^{-x} = f(-x) \quad \text{and} \quad G(x) = 4^{-x} = g(-x).$$

Consequently, the graph of $F$ is a reflection (in the $y$-axis) of the graph of $f$. The graphs of $G$ and $g$ have the same relationship.

The graphs in Figures 5.1 and 5.2 are typical of the exponential functions $a^x$ and $a^{-x}$. They have one $y$-intercept and one horizontal asymptote (the $x$-axis), and they are continuous. The basic characteristics of these exponential functions are summarized in Figure 5.3.

Graph of $y = a^x, a > 1$

- Domain: $(-\infty, \infty)$
- Range: $(0, \infty)$
- Intercept: $(0, 1)$
- Increasing
- $x$-axis is a horizontal asymptote
  $(a^x \to 0 \text{ as } x \to -\infty)$
- Continuous

Graph of $y = a^{-x}, a > 1$

- Domain: $(-\infty, \infty)$
- Range: $(0, \infty)$
- Intercept: $(0, 1)$
- Decreasing
- $x$-axis is a horizontal asymptote
  $(a^{-x} \to 0 \text{ as } x \to \infty)$
- Continuous
- Reflection of graph of $y = a^x$
  about $y$-axis

Figure 5.3

In the following example, notice how the graph of $y = a^x$ can be used to sketch the graphs of functions of the form $f(x) = b \pm a^{x+c}$.

## Study Tip

Notice in Example 4(b) that shifting the graph downward two units also shifts the horizontal asymptote of $f(x) = 3^x$ from the $x$-axis ($y = 0$) to the line $y = -2$.

**Example 4**    Sketching Graphs of Exponential Functions

Each of the following graphs is a transformation of the graph of $f(x) = 3^x$, as shown in Figure 5.4.

**a.** Because $g(x) = 3^{x+1} = f(x + 1)$, the graph of $g$ can be obtained by shifting the graph of $f$ one unit to the left.

**b.** Because $h(x) = 3^x - 2 = f(x) - 2$, the graph of $h$ can be obtained by shifting the graph of $f$ down two units.

**c.** Because $k(x) = -3^x = -f(x)$, the graph of $k$ can be obtained by reflecting the graph of $f$ in the $x$-axis.

**d.** Because $j(x) = 3^{-x} = f(-x)$, the graph of $j$ can be obtained by reflecting the graph of $f$ in the $y$-axis.

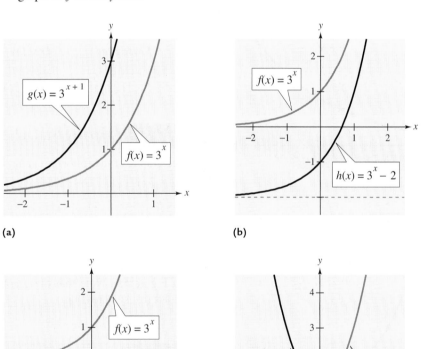

(a)                     (b)

(c)                     (d)
Figure 5.4

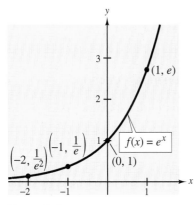

Figure 5.5

# The Natural Base $e$

In many applications, the most convenient choice for a base is the irrational number

$$e \approx 2.718281828 \ldots$$

called the **natural base.** The function $f(x) = e^x$ is called the **natural exponential function.** Its graph is shown in Figure 5.5. Be sure you see that for the exponential function $f(x) = e^x$, $e$ is the constant $2.718281828 \ldots$, whereas $x$ is the variable.

### Example 5     Evaluating the Natural Exponential Function

**Scientific Calculator**

| Number | Keystrokes | Display |
|---|---|---|
| $e^2$ | 2 $\boxed{e^x}$ $\boxed{=}$ | 7.3890561 |
| $e^{-1}$ | 1 $\boxed{+/-}$ $\boxed{e^x}$ $\boxed{=}$ | 0.3678794 |

**Graphing Calculator**

| Number | Keystrokes | Display |
|---|---|---|
| $e^2$ | $\boxed{e^x}$ 2 $\boxed{\text{ENTER}}$ | 7.3890561 |
| $e^{-1}$ | $\boxed{e^x}$ $\boxed{(-)}$ 1 $\boxed{\text{ENTER}}$ | 0.3678794 |

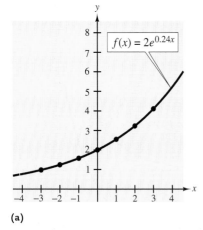

(a)

### Example 6     Graphing Natural Exponential Functions

Sketch the graphs of the following natural exponential functions.

**a.** $f(x) = 2e^{0.24x}$

**b.** $g(x) = \frac{1}{2}e^{-0.58x}$

### Solution

To sketch these two graphs, you can use a calculator to plot several points on each graph, as shown in the table. Then, connect the points with smooth curves, as shown in Figure 5.6. Note that the graph in part (a) is increasing whereas the graph in part (b) is decreasing.

(b)

Figure 5.6

| $x$ | $-3$ | $-2$ | $-1$ | $0$ | $1$ | $2$ | $3$ |
|---|---|---|---|---|---|---|---|
| $f(x) = 2e^{0.24x}$ | 0.974 | 1.238 | 1.573 | 2.000 | 2.543 | 3.232 | 4.109 |
| $g(x) = \frac{1}{2}e^{-0.58x}$ | 2.849 | 1.595 | 0.893 | 0.500 | 0.280 | 0.157 | 0.088 |

## Compound Interest

One of the most familiar examples of exponential growth is that of an investment earning *continuously compounded interest.* The formula for the balance in an account that is compounded $n$ times per year is $A = P(1 + r/n)^{nt}$, where $A$ is the balance in the account, $P$ is the initial deposit, $r$ is the annual interest rate (in decimal form), and $t$ is the number of years. Using exponential functions, you will *develop* this formula and show how it leads to continuous compounding.

Suppose a principal $P$ is invested at an annual percentage rate $r$, compounded once a year. If the interest is added to the principal at the end of the year, the balance is

$$P_1 = P + Pr = P(1 + r).$$

This pattern of multiplying the previous principal by $1 + r$ is then repeated each successive year, as shown below.

| *Year* | *Balance After Each Compounding* |
|--------|-----------------------------------|
| 0 | $P = P$ |
| 1 | $P_1 = P(1 + r)$ |
| 2 | $P_2 = P_1(1 + r) = P(1 + r)(1 + r) = P(1 + r)^2$ |
| 3 | $P_3 = P_2(1 + r) = P(1 + r)^2(1 + r) = P(1 + r)^3$ |
| | $\vdots$ |
| $t$ | $P_t = P(1 + r)^t$ |

To accommodate more frequent (quarterly, monthly, or daily) compounding of interest, let $n$ be the number of compoundings per year and $t$ the number of years. Then the rate per compounding is $r/n$ and the account balance after $t$ years is

$$A = P\left(1 + \frac{r}{n}\right)^{nt}. \qquad \text{Amount with } n \text{ compoundings}$$

If you let the number of compoundings $n$ increase without bound, the process approaches what is called **continuous compounding.** In the formula for $n$ compoundings per year, let $m = n/r$. This produces

$$A = P\left(1 + \frac{r}{n}\right)^{nt}$$

$$= P\left(1 + \frac{1}{m}\right)^{mrt}$$

$$= P\left[\left(1 + \frac{1}{m}\right)^m\right]^{rt}.$$

As $m$ increases without bound, it can be shown that $\left[1 + (1/m)\right]^m$ approaches $e$. From this, you can conclude that the formula for continuous compounding is $A = Pe^{rt}$.

·DISCOVERY·

Use a calculator and the formula $A = P(1 + r/n)^{nt}$ to calculate the amount in an account when $P = \$3000$, $r = 6\%$, $t$ is 10 years, and the number of compoundings is (1) by the day, (2) by the hour, (3) by the minute, and (4) by the second. Use these results to present an argument that increasing the number of compoundings does not mean unlimited growth of the amount in the account.

> ▶ Formulas for Compound Interest
>
> After $t$ years, the balance $A$ in an account with principal $P$ and annual percentage rate $r$ (in decimal form) is given by the following formulas.
>
> **1.** For $n$ compoundings per year: $A = P\left(1 + \dfrac{r}{n}\right)^{nt}$
>
> **2.** For continuous compounding: $A = Pe^{rt}$

Be sure that the annual percentage rate is expressed in decimal form. For instance, 6% should be expressed as 0.06 when using compound interest formulas.

**Example 7**   Compounding $n$ Times and Continuously

A total of \$12,000 is invested at an annual rate of 9%. Find the balance after 5 years if it is compounded

**a.** quarterly.    **b.** continuously.

**Solution**

**a.** For quarterly compoundings, you have $n = 4$. Thus, in 5 years at 9%, the balance is

$$A = P\left(1 + \frac{r}{n}\right)^{nt} \qquad \text{Formula for compound interest}$$

$$= 12{,}000\left(1 + \frac{0.09}{4}\right)^{4(5)} \qquad \text{Substitute for } P, r, n, \text{ and } t.$$

$$= \$18{,}726.11. \qquad \text{Use a calculator.}$$

**b.** Compounding continuously, the balance is

$$A = Pe^{rt} \qquad \text{Formula for continuous compounding}$$

$$= 12{,}000e^{0.09(5)} \qquad \text{Substitute for } P, r, \text{ and } t.$$

$$= \$18{,}819.75. \qquad \text{Use a calculator.}$$

Note that continuous compounding yields

$$\$18{,}819.75 - \$18{,}726.11 = \$93.64$$

more than quarterly compounding. This is typical of the two types of compounding. That is, for a given principal, interest rate, and time, continuous compounding will always yield a larger balance than compounding $n$ times a year.

## Another Application

**Example 8**    Radioactive Decay

In 1957, a nuclear reactor accident occurred in the Soviet Union. The explosion spread radioactive chemicals over thousands of square miles, and the government set aside a large region as a "permanent preserve." To see why this area was permanently declared off-limits to people, consider the following model.

$$P = 10e^{-0.00002845t}$$

This model represents the amount of plutonium that remains (from an initial amount of 10 pounds) after $t$ years. Sketch the graph of this function over the interval from $t = 0$ to $t = 100,000$. How much of the 10 pounds will remain after 100,000 years?

### Solution

The graph of this function is shown in Figure 5.7. Note from this graph that plutonium has a *half-life* of about 24,360 years. That is, after 24,360 years, *half* of the original amount will remain. After another 24,360 years, one-quarter of the original amount will remain, and so on. After 100,000 years, there will still be

$$P = 10e^{-0.00002845(100,000)} = 10e^{-2.845} \approx 0.58 \text{ pounds}$$

of the original amount of plutonium remaining.

**Figure 5.7**

---

| Discussing the Concept | Exponential Growth |
| --- | --- |

The pattern 3, 6, 9, 12, 15, . . . is given by $f(n) = 3n$ and is an example of **linear growth.** The pattern 3, 9, 27, 81, 243, . . . is given by $f(n) = 3^n$ and is an example of **exponential growth.** Explain the difference between these two types of growth. For each of the following patterns, indicate whether the pattern represents linear growth or exponential growth, and find a linear or exponential function that represents the pattern. Give several other examples of linear and exponential growth.

**a.** $\frac{1}{2}, \frac{1}{4}, \frac{1}{8}, \frac{1}{16}, \frac{1}{32}, \ldots$     **b.** 4, 8, 12, 16, 20, . . .

**c.** $\frac{2}{3}, \frac{4}{3}, 2, \frac{8}{3}, \frac{10}{3}, 4, \ldots$     **d.** 5, 25, 125, 625, . . .

# Warm Up

The following warm-up exercises involve skills that were covered in earlier sections. You will use these skills in the exercise set for this section.

In Exercises 1–10, use the properties of exponents to simplify the expression.

**1.** $5^{2x}(5^{-x})$

**2.** $3^{-x}(3^{3x})$

**3.** $\dfrac{4^{5x}}{4^{2x}}$

**4.** $\dfrac{10^{2x}}{10^{x}}$

**5.** $(4^x)^2$

**6.** $(4^{2x})^5$

**7.** $\left(\dfrac{2^x}{3^x}\right)^{-1}$

**8.** $(4^{6x})^{1/2}$

**9.** $(2^{3x})^{-1/3}$

**10.** $(16^x)^{1/4}$

## 5.1 Exercises

LAB 5

In Exercises 1–10, use a calculator to evaluate the expression. (Round the result to three decimal places.)

**1.** $(2.6)^{1.3}$

**2.** $(1.07)^{50}$

**3.** $100(1.03)^{-1.4}$

**4.** $1500(2^{-5/2})$

**5.** $6^{-\sqrt{2}}$

**6.** $1.3^{\sqrt{5}}$

**7.** $e^4$

**8.** $e^{-5}$

**9.** $e^{2/3}$

**10.** $e^{-2.7}$

In Exercises 11–18, match the function with its graph. [The graphs are labeled (a), (b), (c), (d), (e), (f), (g), and (h).]

(a)

(b)

(c)

(d)

(e)

(f)

(g)

(h)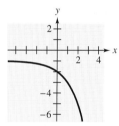

**11.** $f(x) = 2^x$

**12.** $f(x) = 2^{-x}$

**13.** $f(x) = -2^x$

**14.** $f(x) = -2^x - 1$

**15.** $f(x) = 2^x + 3$

**16.** $f(x) = 2^{-x} - 1$

**17.** $f(x) = 2^{x+1}$

**18.** $f(x) = 2^{x-3}$

 In Exercises 19–36, sketch the graph of the function. Use a graphing utility or make the sketch by hand.

**19.** $g(x) = 4^x$

**20.** $f(x) = \left(\frac{3}{2}\right)^x$

**21.** $f(x) = 4^x + 2$

**22.** $h(x) = \left(\frac{3}{2}\right)^{-x}$

**23.** $h(x) = 4^{x-3}$

**24.** $g(x) = \left(\frac{3}{2}\right)^{x+2}$

**25.** $g(x) = 4^{-x} - 2$

**26.** $f(x) = \left(\frac{3}{2}\right)^{-x} + 2$

**27.** $y = 2^{-x^2}$

**28.** $y = 3^{-x^2}$

**29.** $y = e^{-0.1x}$

**30.** $y = e^{0.2x}$

**31.** $f(x) = 2e^{0.12x}$

**32.** $f(x) = 3e^{-0.2x}$

**33.** $f(x) = e^{2x}$

**34.** $h(x) = e^{x-2}$

**35.** $g(x) = 1 + e^{-x}$

**36.** $N(t) = 1000e^{-0.2t}$

**37.** *Graphical Analysis*    Using a graphing utility and the standard viewing window, graph $f(x) = 20^x$, $f(x) = 50^x$, and $f(x) = 100^x$. How are the three graphs alike? How are the three graphs different? Modify the viewing window to observe the graphs when $x$ is between $-1$ and $-2$. How are the three graphs alike? How are the three graphs different?

**38.** *Graphical Analysis*    Using a graphing utility and the standard viewing window, graph $y = 0.1^x$, $y = 0.01^x$, and $y = 0.005^x$. How are the three graphs alike? How are the three graphs different? Modify the viewing window to observe the graphs when $x$ is between 1 and 2. How are the three graphs alike? How are the three graphs different?

*Compound Interest*    In Exercises 39–42, complete the table to find the balance $A$ for $P$ dollars invested at rate $r$ for $t$ years, compounded $n$ times per year.

| $n$ | 1 | 2 | 4 | 12 | 365 | Continuous |
|---|---|---|---|---|---|---|
| $A$ | | | | | | |

**39.** $P = \$5000$, $r = 8\%$, $t = 5$ years

**40.** $P = \$1000$, $r = 10\%$, $t = 10$ years

**41.** $P = \$2500$, $r = 12\%$, $t = 20$ years

**42.** $P = \$1000$, $r = 10\%$, $t = 40$ years

*Compound Interest*    In Exercises 43–46, complete the table to find the amount $P$ that must be invested at rate $r$ to obtain a balance of $A = \$100,000$ in $t$ years.

| $t$ | 1 | 10 | 20 | 30 | 40 | 50 |
|---|---|---|---|---|---|---|
| $P$ | | | | | | |

**43.** $r = 9\%$, compounded continuously

**44.** $r = 12\%$, compounded continuously

**45.** $r = 10\%$, compounded monthly

**46.** $r = 7\%$, compounded daily

**47.** *Trust Fund*    You deposit $25,000 in a trust fund that pays 8.75% interest, compounded continuously, on the day that your grandchild is born. What will the balance of this account be on your grandchild's 25th birthday?

**48.** *Trust Fund*    You deposit $5000 in a trust fund that pays 7.5% interest, compounded continuously. In the trust fund, you specify that the balance of the account will be given to your college after the money has earned interest for 50 years. How much will your college receive after 50 years?

**49.** *Demand Function*    The demand function for a certain product is given by

$$p = 500 - 0.5e^{0.004x}.$$

(a) Find the price $p$ for a demand of $x = 1000$ units.

(b) Find the price $p$ for a demand of $x = 1500$ units.

(c) Use a graphing utility to graph the demand function.

(d) Does the graph indicate what a good price would be? Explain your reasoning.

**50.** *Demand Function*    The demand function for a certain product is given by

$$p = 5000\left(1 - \frac{4}{4 + e^{-0.002x}}\right).$$

(a) Find the price $p$ for a demand of $x = 100$ units.

(b) Find the price $p$ for a demand of $x = 500$ units.

(c) Use a graphing utility to graph the demand function.

(d) Does the graph indicate what a good price would be? Explain your reasoning.

**51.** *Bacteria Growth*    A certain type of bacteria increases according to the model

$$P(t) = 100e^{0.2197t}$$

where $t$ is time in hours. Find (a) $P(0)$, (b) $P(5)$, and (c) $P(10)$.

**52.** *Population Growth* The population of a town increases according to the model

$$P(t) = 2500e^{0.0293t}$$

where $t$ is time in years, with $t = 0$ corresponding to 1990. Use the model to approximate the population in (a) 1995, (b) 2000, and (c) 2005.

**53.** *Radioactive Decay* Five pounds of the element plutonium ($Pu^{230}$) is released in a nuclear accident. The amount of plutonium that is present after $t$ years is given by

$$P = 5e^{-0.00002845t}.$$

(a) Use a graphing utility to graph this function over the interval from $t = 0$ to $t = 100,000$.

(b) How much will remain after 100,000 years?

(c) Use the graph to estimate the half-life of $Pu^{230}$. Explain your reasoning.

**54.** *Radioactive Decay* One hundred grams of radium ($Ra^{226}$) is stored in a container. The amount of radium present after $t$ years is given by

$$R = 100e^{-0.0004279t}.$$

(a) Use a graphing utility to graph this function over the interval from $t = 0$ to $t = 10,000$.

(b) How much of the 100 grams will remain after 10,000 years?

(c) Use the graph to estimate the half-life of $Ra^{226}$. Explain your reasoning.

**55.** *New York Stock Exchange* The total number of shares of stocks traded on the New York Stock Exchange between 1989 and 1995 can be approximated by the function

$$y = 1139.4e^{0.2672t} + 26,081.2, \quad 9 \le t \le 15$$

where $y$ represents the number of traded shares (in millions) and $t = 9$ represents 1989. (Source: New York Stock Exchange)

(a) Use the graph at the top of the next column to estimate *graphically* the total number of shares traded in 1990, 1992, and 1995.

(b) Use the model to confirm *algebraically* the estimates obtained in part (a).

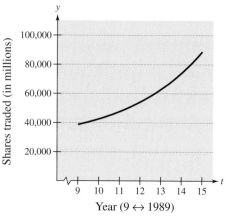

Figure for 55

**56.** *New York Stock Exchange* The total number of shares of stocks listed on the New York Stock Exchange between 1989 and 1995 can be approximated by the function

$$y = 106.37e^{0.0582t} - 98.56, \quad 9 \le t \le 15$$

where $y$ represents the number of shares listed (in billions) and $t = 9$ represents 1989. (Source: New York Stock Exchange)

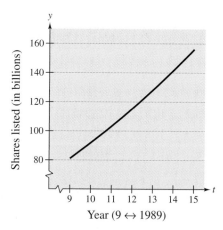

(a) Use the graph to estimate *graphically* the total number of shares listed in 1990, 1993, and 1995.

(b) Use the model to confirm *algebraically* the estimates obtained in part (a).

**57.** *Age at First Marriage* From 1980 to 1997, the average age of an American woman at her first marriage could be approximated by the model

$$A = -1.29 + \frac{28.21}{1 + e^{-0.0583t - 1.5724}}, \quad 0 \le t \le 17$$

where $A$ represents the average age and $t = 0$ represents 1980 (see figure). (Source: U.S. Bureau of the Census)

(a) Use the graph to estimate the average age of an American woman at her first marriage in 1980, 1990, and 1997.

(b) Algebraically confirm the results of part (a).

Figure for 58

**59.** *Writing* Compare the results of Exercises 57 and 58. What can you conclude about the differences in men's and women's ages at first marriages?

**60.** *Compound Interest* A bank offers two types of interest accounts. The first account receives 8% interest compounded quarterly. The second account receives 6% interest compounded continuously. Which account is better? Why?

**61.** *Compound Interest* The day you were born, money was deposited in a trust fund paying 7.5% interest compounded continuously. How much money should have been deposited so that your trust fund will be worth (a) $100,000, (b) $500,000, and (c) $1,000,000 on your next birthday? (*Note:* Answers will vary depending on your age.)

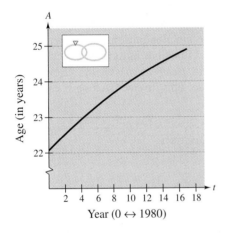

**58.** *Age at First Marriage* From 1980 to 1997, the average age of an American man at his first marriage could be approximated by the model

$$A = 10.37 + \frac{18.69}{1 + e^{-0.0517t - 1.2014}}, \quad 0 \le t \le 17$$

where $A$ represents the average age and $t = 0$ represents 1980 (see figure). (Source: U.S. Bureau of the Census)

(a) Use the graph to estimate the average age of an American man at his first marriage in 1980, 1990, and 1997.

(b) Algebraically confirm the results of part (a).

## 5.2    Logarithmic Functions

### Objectives

- Evaluate a logarithmic expression.
- Find the domain, vertical asymptote, and *x*-intercept of a logarithmic model.
- Sketch the graph of a logarithmic function.
- Use a logarithmic model to solve an application problem.

## Logarithmic Functions

In Section 3.5, you studied the concept of the inverse of a function. There, you learned that if a function has the property that no horizontal line intersects the graph of a function more than once, the function must have an inverse. By looking back at the graphs of the exponential functions introduced in Section 5.1, you will see that every function of the form $f(x) = a^x$ passes the "Horizontal Line Test" and therefore must have an inverse. This inverse function is called the **logarithmic function with base *a*.**

> ▶ **Definition of a Logarithmic Function**
>
> For $x > 0$ and $0 < a \neq 1$,
>
> $$y = \log_a x \text{ if and only if } x = a^y.$$
>
> The function given by
>
> $$f(x) = \log_a x$$
>
> is called the **logarithmic function with base *a*.**

The equations $y = \log_a x$ and $x = a^y$ are equivalent. The first equation is in logarithmic form and the second is in exponential form.

When evaluating logarithms, remember that *a logarithm is an exponent.* This means that $\log_a x$ is the exponent to which $a$ must be raised to obtain $x$. For instance, $\log_2 8 = 3$ because 2 must be raised to the third power to obtain 8.

### Study Tip

By the definition of logarithmic function,

$$\overset{a \quad y \qquad x}{\underset{3^4 = 81}{\diagdown \mid \diagup}}$$

can be written as

$$\overset{a \qquad x \quad y}{\underset{\log_3 81 = 4.}{\diagdown \mid \diagup}}$$

### Example 1    Evaluating Logarithmic Expressions

**a.** $\log_2 32 = 5$       because   $2^5 = 32$.

**b.** $\log_3 27 = 3$       because   $3^3 = 27$.

**c.** $\log_4 2 = \dfrac{1}{2}$       because   $4^{1/2} = \sqrt{4} = 2$.

**d.** $\log_{10} \dfrac{1}{100} = -2$   because   $10^{-2} = \dfrac{1}{10^2} = \dfrac{1}{100}$.

**e.** $\log_3 1 = 0$       because   $3^0 = 1$.

**f.** $\log_2 2 = 1$       because   $2^1 = 2$.

The logarithmic function with base 10 is called the **common logarithmic function.** On most calculators, this function is denoted by LOG.

| Example 2 | Evaluating Logarithmic Expressions on a Calculator |

**Scientific Calculator**

| *Number* | *Keystrokes* | *Display* |
|---|---|---|
| **a.** $\log_{10} 10$ | 10 LOG | 1 |
| **b.** $2 \log_{10} 2.5$ | 2.5 LOG $\times$ 2 $=$ | 0.7958800 |
| **c.** $\log_{10}(-2)$ | 2 $+/-$ LOG | ERROR |

**Graphing Calculator**

| *Number* | *Keystrokes* | *Display* |
|---|---|---|
| **a.** $\log_{10} 10$ | LOG 10 ENTER | 1 |
| **b.** $2 \log_{10} 2.5$ | 2 $\times$ LOG 2.5 ENTER | 0.7958800 |
| **c.** $\log_{10}(-2)$ | LOG $(-)$ 2 ENTER | ERROR |

Many calculators display an error message when you try to evaluate $\log_{10}(-2)$. The reason for this is that the domain of every logarithmic function is the set of *positive real numbers.*

### Study Tip

Because $\log_a x$ is the inverse function of $a^x$, it follows that the domain of $\log_a x$ is the range of $a^x$, $(0, \infty)$. In other words, $\log_a x$ is defined only if $x$ is positive.

The following properties follow directly from the definition of the logarithmic function with base $a$.

> ▶ Properties of Logarithms
>
> **1.** $\log_a 1 = 0$ because $a^0 = 1$.
>
> **2.** $\log_a a = 1$ because $a^1 = a$.
>
> **3.** $\log_a a^x = x$ because $a^x = a^x$.
>
> **4.** If $\log_a x = \log_a y$, then $x = y$.

| Example 3 | Using Properties of Logarithms |

**a.** Solve the equation $\log_2 x = \log_2 3$ for $x$.

**b.** Solve the equation $\log_5 x = 1$ for $x$.

### Solution

**a.** Using Property 4, you can conclude that $x = 3$.

**b.** Using Property 2, you can conclude that $x = 5$.

## Graphs of Logarithmic Functions

To sketch the graph of $y = \log_a x$, you can use the fact that the graphs of inverse functions are reflections of each other in the line $y = x$.

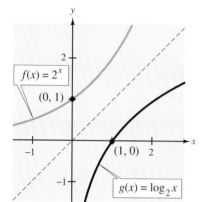

Figure 5.8    *Inverse Functions*

**Example 4**    Graphs of Exponential and Logarithmic Functions

In the same coordinate plane, sketch the graphs of the following functions.

**a.** $f(x) = 2^x$

**b.** $g(x) = \log_2 x$

### Solution

**a.** For $f(x) = 2^x$, construct a table of values, as follows.

| $x$ | $-2$ | $-1$ | 0 | 1 | 2 | 3 |
|---|---|---|---|---|---|---|
| $f(x) = 2^x$ | $\frac{1}{4}$ | $\frac{1}{2}$ | 1 | 2 | 4 | 8 |

By plotting these points and connecting them with a smooth curve, you obtain the graph shown in Figure 5.8.

**b.** Because $g(x) = \log_2 x$ is the inverse of $f(x) = 2^x$, the graph of $g$ is obtained by reflecting the graph of $f$ in the line $y = x$, as shown in Figure 5.8.

### Study Tip

In Example 4(a) you plotted the ordered pairs $(x, f(x))$ and connected the points with a smooth curve. To plot the logarithmic function in Example 4(b) you can plot the ordered pairs $(f(x), x)$ from the exponential function and connect the points with a smooth curve.

Before you can confirm the result of Example 4 with a graphing utility, you need to know how to enter $\log_2 x$. A procedure for doing this is discussed in Section 5.3.

**Example 5**    Sketching the Graph of a Logarithmic Function

Sketch the graph of the common logarithmic function $f(x) = \log_{10} x$.

### Solution

Begin by constructing a table of values. Note that some of the values can be obtained without a calculator, whereas others require a calculator. Next, plot the points and connect them with a smooth curve, as shown in Figure 5.9.

Figure 5.9

| | Without Calculator | | | | With Calculator | | |
|---|---|---|---|---|---|---|---|
| $x$ | $\frac{1}{100}$ | $\frac{1}{10}$ | 1 | 10 | 2 | 5 | 8 |
| $\log_{10} x$ | $-2$ | $-1$ | 0 | 1 | 0.301 | 0.699 | 0.903 |

The nature of the graph in Figure 5.9 is typical of functions of the form $f(x) = \log_a x$, $a > 1$. They have one $x$-intercept and one vertical asymptote. Notice how slowly the graph rises for $x > 1$. In Figure 5.9 you would need to move out to $x = 1000$ before the graph rose to $y = 3$. The basic characteristics of logarithmic graphs are summarized in Figure 5.10. Note that the vertical asymptote occurs at $x = 0$, where $\log_a x$ is *undefined*.

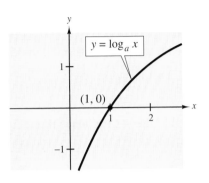

**Graph of $y = \log_a x$, $a > 1$**

- Domain: $(0, \infty)$
- Range: $(-\infty, \infty)$
- Intercept: $(1, 0)$
- Increasing
- $y$-axis is a vertical asymptote
  ($\log_a x \to -\infty$ as $x \to 0^+$)
- Continuous
- Reflection of graph of $y = a^x$ about the line $y = x$

Figure 5.10

In the following example, the graph of $\log_a x$ is used to sketch the graphs of functions of the form $y = b \pm \log_a(x + c)$.

## Study Tip

Notice in Example 6(a) that shifting the graph of $f(x)$ one unit to the right also shifts the vertical asymptote from the $y$-axis ($x = 0$) to the line $x = 1$.

**Example 6**   Sketching the Graphs of Logarithmic Functions

The graph of each of the following functions is similar to the graph of $f(x) = \log_{10} x$, as shown in Figure 5.11.

**a.** Because $g(x) = \log_{10}(x - 1) = f(x - 1)$, the graph of $g$ can be obtained by shifting the graph of $f$ one unit to the right.

**b.** Because $h(x) = 2 + \log_{10} x = 2 + f(x)$, the graph of $h$ can be obtained by shifting the graph of $f$ two units up.

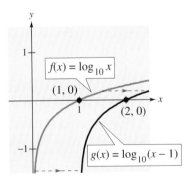

(a) Right shift of 1 unit

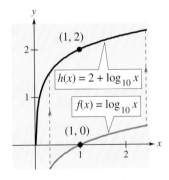

(b) Upward shift of 2 units

Figure 5.11

## The Natural Logarithmic Function

As with exponential functions, the most widely used base for logarithmic functions is the number $e$, where

$$e \approx 2.718281828\ldots.$$

The logarithmic function with base $e$ is the **natural logarithmic function** and is denoted by the special symbol $\ln x$, read as "el en of $x$."

---

▶ **The Natural Logarithmic Function**

The function defined by

$$f(x) = \log_e x = \ln x, \quad x > 0$$

is called the **natural logarithmic function.**

---

The four properties of logarithms listed on page 311 are also valid for natural logarithms.

---

▶ **Properties of Natural Logarithms**

**1.** $\ln 1 = 0$ because $e^0 = 1$.

**2.** $\ln e = 1$ because $e^1 = e$.

**3.** $\ln e^x = x$ because $e^x = e^x$.

**4.** If $\ln x = \ln y$, then $x = y$.

---

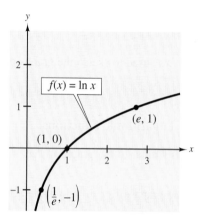

$f(x) = \ln x$

$(e, 1)$

$(1, 0)$

$\left(\frac{1}{e}, -1\right)$

**Figure 5.12**

The graph of the natural logarithmic function is shown in Figure 5.12. Try using a graphing utility to confirm this graph. What is the domain of the natural logarithmic function?

---

**Example 7**  Evaluating the Natural Logarithmic Function

**a.** $\ln \dfrac{1}{e} = \ln e^{-1} = -1$     Property 3

**b.** $\ln e^2 = 2$     Property 3

**c.** $\ln e^0 = 0$     Property 1

**d.** $2 \ln e = 2(1) = 2$     Property 2

On most calculators, the natural logarithm is denoted by $\boxed{\text{LN}}$, as illustrated in Example 8.

**Example 8**    Evaluating the Natural Logarithmic Function

**Scientific Calculator**

| | Number | Keystrokes | Display |
|---|---|---|---|
| **a.** | ln 2 | 2 $\boxed{\text{LN}}$ | 0.6931472 |
| **b.** | ln 0.3 | .3 $\boxed{\text{LN}}$ | −1.2039728 |
| **c.** | ln $e^2$ | 2 $\boxed{e^x}$ $\boxed{\text{LN}}$ | 2 |
| **d.** | ln(−1) | 1 $\boxed{+/-}$ $\boxed{\text{LN}}$ | ERROR |

**Graphing Calculator**

| | Number | Keystrokes | Display |
|---|---|---|---|
| **a.** | ln 2 | $\boxed{\text{LN}}$ 2 $\boxed{\text{ENTER}}$ | 0.6931472 |
| **b.** | ln 0.3 | $\boxed{\text{LN}}$ .3 $\boxed{\text{ENTER}}$ | −1.2039728 |
| **c.** | ln $e^2$ | $\boxed{\text{LN}}$ $\boxed{e^x}$ 2 $\boxed{\text{ENTER}}$ | 2 |
| **d.** | ln(−1) | $\boxed{\text{LN}}$ $\boxed{(-)}$ 1 $\boxed{\text{ENTER}}$ | ERROR |

In Example 8, be sure you see that ln(−1) gives an error message. This occurs because the domain of ln $x$ is the set of positive real numbers (see Figure 5.12). So, ln(−1) is undefined.

**Example 9**    Finding the Domains of Logarithmic Functions

Find the domains of the following functions.

**a.** $f(x) = \ln(x - 2)$    **b.** $g(x) = \ln(2 - x)$    **c.** $h(x) = \ln x^2$

**Solution**

**a.** Because $\ln(x - 2)$ is defined only if $x - 2 > 0$, it follows that the domain of $f$ is $(2, \infty)$.

**b.** Because $\ln(2 - x)$ is defined only if $2 - x > 0$, it follows that the domain of $g$ is $(-\infty, 2)$. The graph of $g$ is shown in Figure 5.13.

**c.** Because $\ln x^2$ is defined only if $x^2 > 0$, it follows that the domain of $h$ is all real numbers except $x = 0$.

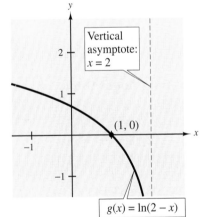

Figure 5.13

## Application

**Example 10**   Human Memory Model

Students participating in a psychological experiment attended several lectures on a subject. Every month for a year after that, the students were tested to see how much of the material they remembered. The average scores for the group were given by the *human memory model*

$$f(t) = 75 - 6\ln(t + 1), \quad 0 \le t \le 12$$

where $t$ is the time in months.

**a.** What was the average score on the original $(t = 0)$ exam?

**b.** What was the average score at the end of $t = 2$ months?

**c.** What was the average score at the end of $t = 6$ months?

Solution

**a.** The original average score was

$$f(0) = 75 - 6\ln(0 + 1) = 75 - 6(0) = 75.$$

**b.** After 2 months, the average score was

$$f(2) = 75 - 6\ln 3 \approx 75 - 6(1.0986) \approx 68.4.$$

**c.** After 6 months, the average score was

$$f(6) = 75 - 6\ln 7 \approx 75 - 6(1.9459) \approx 63.3.$$

The graph of $f$ is shown in Figure 5.14.

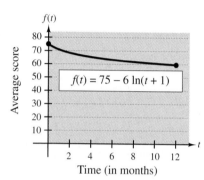

Figure 5.14

| Discussing the Concept | **Transforming Logarithmic Functions** |
| --- | --- |

Use a graphing utility to graph $f(x) = \ln x$. How will the graphs of $h(x) = \ln x + 5$, $j(x) = \ln(x + 5)$, $k(x) = \ln(x - 3)$, and $l(x) = \ln x - 4$ differ from the graph of $f$? Confirm your predictions with a graphing utility.

How will the basic graph of $f$ be affected when a constant $c$ is introduced: $g(x) = c \ln x$? Use a graphing utility to graph $g$ with several different positive values of $c$, and summarize the effect of $c$.

# Warm Up

The following warm-up exercises involve skills that were covered in earlier sections. You will use these skills in the exercise set for this section.

In Exercises 1–4, solve for $x$.

**1.** $2^x = 8$

**2.** $4^x = 1$

**3.** $10^x = 0.1$

**4.** $e^x = e$

In Exercises 5 and 6, evaluate the expression. (Round the result to three decimal places.)

**5.** $e^2$

**6.** $e^{-1}$

In Exercises 7–10, describe how the graph of $g$ is related to the graph of $f$.

**7.** $g(x) = f(x + 2)$

**8.** $g(x) = -f(x)$

**9.** $g(x) = -1 + f(x)$

**10.** $g(x) = f(-x)$

## 5.2    Exercises

LAB 6

In Exercises 1–16, evaluate the expression without a calculator.

**1.** $\log_3 9$

**2.** $\log_5 125$

**3.** $\log_2\left(\frac{1}{16}\right)$

**4.** $\log_6\left(\frac{1}{36}\right)$

**5.** $\log_8 2$

**6.** $\log_{64} 4$

**7.** $\log_7 7$

**8.** $\log_{12} 1$

**9.** $\log_{10} 0.0001$

**10.** $\log_{10} 100$

**11.** $\ln e$

**12.** $\ln e^{10}$

**13.** $\ln e^{-4}$

**14.** $\ln \dfrac{1}{e^3}$

**15.** $\log_a a^5$

**16.** $\log_a 1$

In Exercises 17–26, use the definition of a logarithm to write the equation in logarithmic form. For instance, the logarithmic form of $2^3 = 8$ is $\log_2 8 = 3$.

**17.** $4^4 = 256$

**18.** $7^3 = 343$

**19.** $81^{1/4} = 3$

**20.** $9^{3/2} = 27$

**21.** $6^{-2} = \frac{1}{36}$

**22.** $10^{-3} = 0.001$

**23.** $e^1 = e$

**24.** $e^4 = 54.5981\ldots$

**25.** $e^x = 4$

**26.** $e^{-x} = 2$

In Exercises 27–34, use a calculator to evaluate the logarithm. (Round to three decimal places.)

**27.** $\log_{10} 345$

**28.** $\log_{10}\left(\frac{4}{5}\right)$

**29.** $\log_{10} 145$

**30.** $\log_{10} 12.5$

**31.** $\ln 18.42$

**32.** $\ln 36.7$

**33.** $\ln\left(\sqrt{3}\right)$

**34.** $\ln\left(\sqrt{5}\right)$

**35.** *Population Growth*   The population of a town will double in $t = (10 \ln 2)/(\ln 67 - \ln 50)$ years. Find $t$.

**36.** *Work*   The work (in foot-pounds) done in compressing a volume of 9 cubic feet at a pressure of 15 pounds per square inch to a volume of 3 cubic feet is $W = 19{,}440(\ln 9 - \ln 3)$. Find $W$.

In Exercises 37–40, sketch the graphs of $f$ and $g$ in the same coordinate plane.

**37.** $f(x) = 3^x$, $g(x) = \log_3 x$

**38.** $f(x) = 5^x$, $g(x) = \log_5 x$

**39.** $f(x) = e^x$, $g(x) = \ln x$

**40.** $f(x) = 10^x$, $g(x) = \log_{10} x$

In Exercises 41–46, match the function with its graph. [The graphs are labeled (a), (b), (c), (d), (e), and (f).]

(a)

(b)

(c)

(d)

(e)

(f)

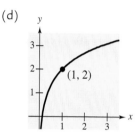

**41.** $f(x) = \ln x + 2$

**42.** $f(x) = -\ln x$

**43.** $f(x) = -3 \ln(x - 2)$

**44.** $f(x) = 4 \ln(-x)$

**45.** $f(x) = -3 \ln x + 2$

**46.** $f(x) = -3 \ln x - 2$

In Exercises 47–54, find the domain, vertical asymptote, and $x$-intercept of the logarithmic function. Then sketch its graph.

**47.** $f(x) = \log_2 x$

**48.** $g(x) = \log_8 x$

**49.** $h(x) = \log_2 (x + 4)$

**50.** $f(x) = \log_8(x - 2)$

**51.** $f(x) = -\log_2 x$

**52.** $h(x) = -\log_8(x + 1)$

**53.** $g(x) = \ln(-x)$

**54.** $f(x) = \ln(3 - x)$

**55.** *Human Memory Model*   Students in a mathematics class were given an exam and then retested monthly with an equivalent exam. The average score for the class was given by the human memory model

$$f(t) = 80 - 17 \log_{10}(t + 1), \quad 0 \le t \le 12$$

where $t$ is the time in months.

(a) What was the average score on the original exam?

(b) What was the average score after 4 months?

(c) What was the average score after 10 months?

**56.** *Human Memory Model*   Students in a seventh-grade class were given an exam. During the next 2 years, the same students were retested several times. The average score was given by the model

$$f(t) = 90 - 15 \log_{10}(t + 1), \quad 0 \le t \le 24$$

where $t$ is the time in months.

(a) What was the average score on the original exam?

(b) What was the average score after 6 months?

(c) What was the average score after 18 months?

**57.** *Investment Time*   A principal $P$, invested at $9\frac{1}{2}\%$ and compounded continuously, increases to an amount that is $K$ times the original principal after $t$ years, where $t$ is given by

$$t = \frac{\ln K}{0.095}.$$

(a) Complete the table.

| $K$ | 1 | 2 | 4 | 6 | 8 | 10 | 12 |
|-----|---|---|---|---|---|----|----|
| $t$ |   |   |   |   |   |    |    |

(b) Use the table in part (a) to graph the function.

**58.** *Investment Time*   A principal $P$, invested at 7.75% and compounded continuously, increases to an amount that is $K$ times the original principal after $t$ years, where $t$ is given by

$$t = \frac{\ln K}{0.0775}.$$

Use a graphing utility to graph this function.

*Skill Retention Model*    In Exercises 59 and 60, participants in an industrial psychology study were taught a simple mechanical task and tested monthly on this mechanical task for a period of 1 year. The average score for the class was given by the model

$$f(t) = 95 - 12 \log_{10}(t + 1), \quad 0 \le t \le 12$$

where $t$ is the time in months.

**59.** Use a graphing utility to graph this function. Sketch the graph and discuss the domain and range of this function.

**60.** Based on the graph of $f(t)$, do you think the study's participants practiced the simple mechanical task very often? Cite the behavior of the graph to justify your answer.

*Median Age of U.S. Population*    In Exercises 61–64, use the model

$$A = 18.10 - 0.016t + 5.121 \ln t, \quad 10 \le t \le 80$$

which approximates the median age of the United States population from 1980 to 2050. In the model, $A$ is the median age and $t = 10$ represents 1980 (see figure).    (Source: U.S. Bureau of the Census)

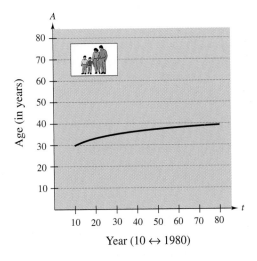

Year (10 $\leftrightarrow$ 1980)

**61.** Use the model to approximate the median age in the United States in 1980.

**62.** Use the model to approximate the median age in the United States in 1990.

**63.** Use the model to approximate the change in median age in the United States from 1980 to 2000.

**64.** Use the model to approximate the change in median age in the United States from 1980 to 2050.

*Monthly Payment*    In Exercises 65–68, use the model

$$t = \frac{5.315}{-6.7968 + \ln x}, \quad x > 1000$$

which approximates the length of a home mortgage (of $120,000 at 10%) in terms of the monthly payment. In the model, $t$ is the length of the mortgage in years and $x$ is the monthly payment in dollars (see figure).

Monthly payment (in dollars)

**65.** Use the model to approximate the length of a home mortgage (for $120,000 at 10%) that has a monthly payment of $1167.41.

**66.** Use the model to approximate the length of a home mortgage (for $120,000 at 10%) that has a monthly payment of $1068.45.

**67.** Approximate the total amount paid over the term of the mortgage for a monthly payment of $1167.41.

**68.** Approximate the total amount paid over the term of the mortgage for a monthly payment of $1068.45.

## 5.3    Properties of Logarithms

### Objectives

- Evaluate a logarithm using the change-of-base formula.
- Rewrite logarithmic expressions using the properties of logarithms.
- Construct a model using properties of exponents and properties of logarithms to relate two quantities.

**The Granger Collection**

### John Napier

**1550–1617**

John Napier, a Scottish mathematician, developed logarithms as a way to simplify some of the tedious calculations of his day. Beginning in 1594, Napier worked about 20 years on logarithms, but he was only partially successful in his quest. Nonetheless, the development of logarithms was a step forward and received immediate recognition.

### Study Tip

Notice in Examples 1 and 2 that the result is the same whether the common logarithm or the natural logarithm is used in the change-of-base formula.

## Change of Base

Most calculators have only two types of log keys, one for common logarithms (base 10) and one for natural logarithms (base $e$). Although common logs and natural logs are the most frequently used, you may occasionally need to evaluate logarithms to other bases. To do this, you can use the following *change-of-base formula*.

---

▶ **Change-of-Base Formula**

Let $a$, $b$, and $x$ be positive real numbers such that $a \neq 1$ and $b \neq 1$. Then $\log_a x$ is given by

$$\log_a x = \frac{\log_b x}{\log_b a}.$$

---

One way to look at the change-of-base formula is that logarithms to base $a$ are simply *constant multiples* of logarithms to base $b$. The constant multiplier is

$$\frac{1}{\log_b a}.$$

**Example 1**    Changing Bases Using Common Logarithms

**a.** $\log_4 30 = \dfrac{\log_{10} 30}{\log_{10} 4} \approx \dfrac{1.47712}{0.60206} \approx 2.4534$

**b.** $\log_2 14 = \dfrac{\log_{10} 14}{\log_{10} 2} \approx \dfrac{1.14613}{0.30103} \approx 3.8074$

**Example 2**    Changing Bases Using Natural Logarithms

**a.** $\log_4 30 = \dfrac{\ln 30}{\ln 4} \approx \dfrac{3.40120}{1.38629} \approx 2.4534$

**b.** $\log_2 14 = \dfrac{\ln 14}{\ln 2} \approx \dfrac{2.63906}{0.693147} \approx 3.8074$

# Properties of Logarithms

You know from the preceding section that the logarithmic function with base $a$ is the *inverse* of the exponential function with base $a$. So, it makes sense that the properties of exponents should have corresponding properties involving logarithms. For instance, the exponential property $a^0 = 1$ has the corresponding logarithmic property $\log_a 1 = 0$.

---

▶ **Properties of Logarithms**

Let $a$ be a positive number such that $a \neq 1$, and let $n$ be a real number. If $u$ and $v$ are positive real numbers, the following properties are true.

**1.** $\log_a(uv) = \log_a u + \log_a v$      **1.** $\ln(uv) = \ln u + \ln v$

**2.** $\log_a \dfrac{u}{v} = \log_a u - \log_a v$      **2.** $\ln \dfrac{u}{v} = \ln u - \ln v$

**3.** $\log_a u^n = n \log_a u$      **3.** $\ln u^n = n \ln u$

---

There is no general property that can be used to rewrite $\log_a(u \pm v)$. Specifically, $\log_a(x + y)$ is not equal to $\log_a x + \log_a y$.

**Example 3**    Using Properties of Logarithms

Write the logarithm in terms of $\ln 2$ and $\ln 3$.

**a.** $\ln 6$     **b.** $\ln \dfrac{2}{27}$

Solution

**a.** $\ln 6 = \ln(2 \cdot 3)$        Rewrite 6 as $2 \cdot 3$.

$\qquad = \ln 2 + \ln 3$        Property 1

**b.** $\ln \dfrac{2}{27} = \ln 2 - \ln 27$        Property 2

$\qquad = \ln 2 - \ln 3^3$        Rewrite 27 as $3^3$.

$\qquad = \ln 2 - 3 \ln 3$        Property 3

**Example 4**    Using Properties of Logarithms

Use the properties of logarithms to verify that $-\ln \frac{1}{2} = \ln 2$.

Solution

$$-\ln \frac{1}{2} = -\ln(2^{-1}) = -(-1)\ln 2 = \ln 2$$

Try checking this result on your calculator.

## Rewriting Logarithmic Expressions

The properties of logarithms are useful for rewriting logarithmic expressions in forms that simplify the operations of algebra. This is true because they convert complicated products, quotients, and exponential forms into simpler sums, differences, and products, respectively. Examples 5 and 6 illustrate some cases.

### ·DISCOVERY·

Use a calculator to find the approximate value of $\ln \sqrt[3]{e^2}$. Now find the exact value by rewriting $\ln \sqrt[3]{e^2}$ with a rational exponent using the properties of logarithms. How do the two values compare?

**Example 5** Rewriting the Logarithm of a Product

$$\log_{10} 5x^3y = \log_{10} 5 + \log_{10} x^3y$$

$$= \log_{10} 5 + \log_{10} x^3 + \log_{10} y$$

$$= \log_{10} 5 + 3 \log_{10} x + \log_{10} y$$

**Example 6** Rewriting the Logarithm of a Quotient

$$\ln \frac{\sqrt{3x - 5}}{7} = \ln(3x - 5)^{1/2} - \ln 7 = \frac{1}{2}\ln(3x - 5) - \ln 7$$

In Examples 5 and 6, the properties of logarithms were used to *expand* logarithmic expressions. In Examples 7 and 8, this procedure is reversed and the properties of logarithms are used to *condense* logarithmic expressions.

**Example 7** Condensing a Logarithmic Expression

$$\frac{1}{2}\log_{10} x - 3 \log_{10}(x + 1) = \log_{10} x^{1/2} - \log_{10}(x + 1)^3$$

$$= \log_{10} \frac{\sqrt{x}}{(x + 1)^3}$$

**Example 8** Condensing a Logarithmic Expression

$$2 \ln(x + 2) - \ln x = \ln(x + 2)^2 - \ln x = \ln \frac{(x + 2)^2}{x}$$

When applying the properties of logarithms to a logarithmic function, you should be careful to check the domain of the function. For example, the domain of $f(x) = \ln x^2$ is all real $x \neq 0$, whereas the domain of $g(x) = 2 \ln x$ is all real $x > 0$.

# Application

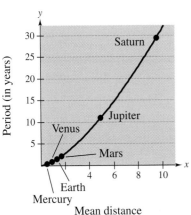

Figure 5.15

| Example 9 | Finding a Mathematical Model  |

The table gives the mean distance $x$ and the period $y$ of the six planets that are closest to the sun. In the table, the mean distance is given in terms of astronomical units (where the earth's mean distance is defined as 1.0), and the period is given in terms of years. Find an equation that expresses $y$ as a function of $x$.

| Planet | Mercury | Venus | Earth | Mars | Jupiter | Saturn |
|---|---|---|---|---|---|---|
| Period, $y$ | 0.241 | 0.615 | 1.0 | 1.881 | 11.861 | 29.457 |
| Mean distance, $x$ | 0.387 | 0.723 | 1.0 | 1.523 | 5.203 | 9.541 |

## Solution

The points in the table are plotted in Figure 5.15. From this figure it is not clear how to find an equation that relates $y$ and $x$. To solve this problem, take the natural log of each of the $x$- and $y$-values given in the table. This produces the following results.

| Planet | Mercury | Venus | Earth | Mars | Jupiter | Saturn |
|---|---|---|---|---|---|---|
| $\ln y$ | $-1.423$ | $-0.486$ | 0.0 | 0.632 | 2.473 | 3.383 |
| $\ln x$ | $-0.949$ | $-0.324$ | 0.0 | 0.421 | 1.649 | 2.256 |

Now, by plotting the points in the second table, you can see that all six of the points appear to lie on a line (see Figure 5.16). The slope of this line is $\frac{3}{2}$, and you can therefore conclude that

$$\ln y = \frac{3}{2} \ln x.$$

Figure 5.16

| Discussing the Concept | **Understanding Properties of Logarithms** |

Describe how you would demonstrate that

$$\log_a(x \pm y) \neq \log_a x \pm \log_a y.$$

Give an algebraic argument and an argument using a graphing utility.

# Warm Up

The following warm-up exercises involve skills that were covered in earlier sections. You will use these skills in the exercise set for this section.

In Exercises 1–4, evaluate the expression without using a calculator.

**1.** $\log_7 49$

**2.** $\log_2\left(\frac{1}{32}\right)$

**3.** $\ln \dfrac{1}{e^2}$

**4.** $\log_{10} 0.001$

In Exercises 5–8, simplify the expression.

**5.** $e^2 e^3$

**6.** $\dfrac{e^2}{e^3}$

**7.** $(e^2)^3$

**8.** $(e^2)^0$

In Exercises 9 and 10, rewrite the expression in exponential form.

**9.** $\dfrac{1}{x^2}$

**10.** $\sqrt{x}$

## 5.3 Exercises

In Exercises 1–4, write the logarithm in terms of common logarithms.

**1.** $\log_5 8$

**2.** $\log_7 12$

**3.** $\log_3 n$

**4.** $\ln 20$

In Exercises 5–8, write the logarithm in terms of natural logarithms.

**5.** $\log_5 8$

**6.** $\log_7 12$

**7.** $\log_3 n$

**8.** $\log_{10} 20$

In Exercises 9–16, evaluate the logarithm. (Round to three decimal places.)

**9.** $\log_2 6$

**10.** $\log_8 3$

**11.** $\log_5\left(\frac{1}{3}\right)$

**12.** $\log_9 0.4$

**13.** $\log_{15} 1250$

**14.** $\log_{20} 125$

**15.** $\log_{1/4} 10$

**16.** $\log_{1/3} 0.015$

In Exercises 17–36, use the properties of logarithms to write the expression as a sum, difference, and/or multiple of logarithms.

**17.** $\log_3 4n$

**18.** $\log_6 6x$

**19.** $\log_5 \dfrac{x}{25}$

**20.** $\log_{10} \dfrac{y}{2}$

**21.** $\log_2 x^4$

**22.** $\log_2 z^{-3}$

**23.** $\ln \sqrt{z}$

**24.** $\ln \sqrt[3]{t}$

**25.** $\ln xyz$

**26.** $\ln \dfrac{xy}{z}$

**27.** $\ln \sqrt{a - 1}$

**28.** $\ln\left(\dfrac{x^2 - 1}{x^3}\right)$

**29.** $\ln \dfrac{(z - 1)^2}{z}$

**30.** $\ln \sqrt{\dfrac{x^2}{y^3}}$

**31.** $\log_b \dfrac{x^2}{y^3}$

**32.** $\log_b \dfrac{x^4}{z^2}$

**33.** $\ln \sqrt[3]{\dfrac{x}{y}}$

**34.** $\ln \dfrac{x}{\sqrt{x^2 + 1}}$

**35.** $\ln \dfrac{\sqrt{y}}{z^5}$

**36.** $\ln \sqrt{x^2(x + 2)}$

In Exercises 37–52, write the expression as the logarithm of a single quantity.

**37.** $\log_3 x + \log_3 5$

**38.** $\log_5 y + \log_5 x$

**39.** $\ln 8 - \ln x$

**40.** $\log_{10} 4 - \log_{10} z$

**41.** $2 \log_{10}(x + 4)$

**42.** $-4 \log_{10} 2x$

**43.** $5 \ln x - 2 \ln 3$

**44.** $2 \ln 8 + 5 \ln z$

**45.** $\frac{1}{3} \ln 5x$

**46.** $\frac{3}{2} \ln(z - 2)$

**47.** $\ln(x - 2) - \ln(x + 2)$

**48.** $3 \ln x + 2 \ln y - 4 \ln z$

**49.** $2 \ln 3 - \frac{1}{2} \ln(x^2 + 1)$

**50.** $\frac{3}{2} \ln t^6 - \frac{3}{4} \ln t^4$

**51.** $\ln x - \ln(x + 2) - \ln(x - 2)$

**52.** $\ln(x + 1) + 2 \ln(x - 1) + 3 \ln x$

In Exercises 53–66, approximate the logarithm using the properties of logarithms, given $\log_b 2 \approx 0.3562$, $\log_b 3 \approx 0.5646$, and $\log_b 5 \approx 0.8271$.

**53.** $\log_b 10$

**54.** $\log_b 15$

**55.** $\log_b\left(\frac{2}{3}\right)$

**56.** $\log_b\left(\frac{5}{3}\right)$

**57.** $\log_b 8$

**58.** $\log_b 18 \ (18 = 2 \cdot 3^2)$

**59.** $\log_b \sqrt{2}$

**60.** $\log_b\left(\frac{9}{2}\right)$

**61.** $\log_b 40$

**62.** $\log_b \sqrt[3]{75}$

**63.** $\log_b(2b)^{-2}$

**64.** $\log_b(3b^2)$

**65.** $\log_b \sqrt{5b}$

**66.** $\log_b 1$

In Exercises 67–72, find the *exact* value of the logarithm.

**67.** $\log_4 64$

**68.** $\log_8 \sqrt[4]{8}$

**69.** $\log_4 16^{1.2}$

**70.** $\log_5\left(\frac{1}{125}\right)$

**71.** $\ln e^{4.5}$

**72.** $\ln \sqrt[4]{e^3}$

In Exercises 73–80, use the properties of logarithms to simplify the given logarithmic expression.

**73.** $\log_9 27$

**74.** $\log_5\left(\frac{1}{15}\right)$

**75.** $\log_7 \sqrt{70}$

**76.** $\log_2(4^2 \cdot 3^4)$

**77.** $\log_5\left(\frac{1}{250}\right)$

**78.** $\log_{10}\left(\frac{9}{300}\right)$

**79.** $\ln(5e^6)$

**80.** $\ln \dfrac{6}{e^2}$

*Curve Fitting*    In Exercises 81–84, find an equation that relates $x$ and $y$. Explain the steps used to find the equation.

**81.**

| $x$ | 1 | 2 | 3 | 4 | 5 | 6 |
|---|---|---|---|---|---|---|
| $y$ | 1 | 1.189 | 1.316 | 1.414 | 1.495 | 1.565 |

**82.**

| $x$ | 1 | 2 | 3 | 4 | 5 | 6 |
|---|---|---|---|---|---|---|
| $y$ | 1 | 1.587 | 2.080 | 2.520 | 2.924 | 3.302 |

**83.**

| $x$ | 1 | 2 | 3 | 4 | 5 | 6 |
|---|---|---|---|---|---|---|
| $y$ | 2.5 | 2.102 | 1.9 | 1.768 | 1.672 | 1.597 |

**84.**

| $x$ | 1 | 2 | 3 | 4 | 5 | 6 |
|---|---|---|---|---|---|---|
| $y$ | 0.5 | 2.828 | 7.794 | 16 | 27.951 | 44.091 |

**85.** *Graphical Analysis*    Use a graphing utility to graph $f(x) = \ln(x/3)$ and $g(x) = \ln x - \ln 3$ in the same viewing rectangle. What do you observe about the two graphs? What property of logarithms is being demonstrated graphically?

**86.** *Graphical Analysis*    Use a graphing utility to graph $f(x) = \ln 5x$ and $g(x) = \ln 5 + \ln x$ in the same viewing rectangle. What do you observe about the two graphs? What property is demonstrated?

**87.** *Reasoning*    You are helping another student learn the properties of logarithms. How would you use a graphing utility to demonstrate to this student the logarithm property $\log_a u^v = v \log_a u$ ($u$ and $v$ are positive real numbers and $a$ is a positive number such that $a \neq 1$)? What two functions could you use? Briefly describe your explanation of this property using these functions and their graphs.

**88.** *Reasoning*    An algebra student claims that the following is true:

$$\log_a \frac{x}{y} = \frac{\log_a x}{\log_a y} = \log_a x - \log_a y.$$

Discuss how you would use a graphing utility to demonstrate that this claim is not true. Describe how to demonstrate the actual property of logarithms that is hidden in this faulty claim.

**89.** *Nail Length* The approximate lengths and diameters (in inches) of common nails are given in the table. Find a mathematical model that relates the diameter $y$ of a common nail to its length $x$.

| Length, $x$ | 1.00 | 2.00 | 3.00 |
|---|---|---|---|
| Diameter, $y$ | 0.070 | 0.111 | 0.146 |

| Length, $x$ | 4.00 | 5.00 | 6.00 |
|---|---|---|---|
| Diameter, $y$ | 0.176 | 0.204 | 0.231 |

**90.** *Galloping Speeds of Animals* Four-legged animals run with two different types of motion: trotting and galloping. An animal that is trotting has at least one foot on the ground at all times, whereas an animal that is galloping sometimes has all four feet off the ground. The number of strides per minute at which an animal breaks from a trot to a gallop depends on the weight of the animal. Use the table at the top of the right-hand column to find a model that relates an animal's lowest galloping speed (in strides per minute) to its weight (in pounds).

| Weight, $x$ | 25 | 35 | 50 |
|---|---|---|---|
| Galloping speed, $y$ | 191.5 | 182.7 | 173.8 |

| Weight, $x$ | 75 | 500 | 1000 |
|---|---|---|---|
| Galloping speed, $y$ | 164.2 | 125.9 | 114.2 |

**91.** Complete the proof of the logarithmic property $\log_a(uv) = \log_a u + \log_a v$.

Let $\log_a u = x$ and $\log_a v = y$.

$a^x = \boxed{\phantom{xx}}$ and $a^y = \boxed{\phantom{xx}}$  Rewrite in exponential form.

$uv = \boxed{\phantom{xx}} \cdot \boxed{\phantom{xx}} = a^{\boxed{\phantom{x}}}$  Multiply and substitute for $u$ and $v$.

$\boxed{\phantom{xx}} = x + y$  Rewrite in logarithmic form.

$\log_a uv = \boxed{\phantom{xx}} + \boxed{\phantom{xx}}$  Substitute for $x$ and $y$.

**92.** Complete the proof of the logarithmic property

$$\log_a \frac{u}{v} = \log_a u - \log_a v.$$

Let $\log_a u = x$ and $\log_a v = y$.

$a^x = \boxed{\phantom{xx}}$ and $a^y = \boxed{\phantom{xx}}$  Rewrite in exponential form.

$\dfrac{u}{v} = \dfrac{\boxed{\phantom{xx}}}{\boxed{\phantom{xx}}} = a^{\boxed{\phantom{x}}}$  Divide and substitute for $u$ and $v$.

$\boxed{\phantom{xx}} = x - y$  Rewrite in logarithmic form.

$\log_a \dfrac{u}{v} = \boxed{\phantom{xx}} - \boxed{\phantom{xx}}$  Substitute for $x$ and $y$.

## Math Matters • Archimedes and the Gold Crown

Archimedes was a citizen of Syracuse, a city on the island of Sicily. He was born in 287 B.C. and is often regarded (with Isaac Newton and Carl Gauss) as one of the three greatest mathematicians who ever lived. The following story gives some idea of Archimedes's ingenuity.

The king of Syracuse ordered a crown to be made of pure gold. When the crown was delivered, the king suspected that some silver had been mixed with the gold. He asked Archimedes to find a way of checking whether the crown was pure gold (without destroying the crown). One day,

while Archimedes was at a public bath, he noticed that the level of the water rose when he stepped into the bath. He suddenly realized how he could solve the problem, and became so excited that he ran home, stark naked, shouting "Eureka!" (which means "I have found it!").

Archimedes's idea was that he could measure the volume of the crown by putting it into a bowl of water. If the crown contained any silver, the volume of the crown would be greater than the volume of an equal weight of pure gold, as shown in the accompanying figure.

1. Crown and block of pure gold balance.

2. Crown and block of pure silver balance.

3. Gold block makes water level rise.

4. Silver block makes water level rise higher.

5. Crown makes water level rise more than gold block.

## Mid-Chapter Quiz

Take this quiz as you would take a quiz in class. After you are done, check your work against the answers given in the back of the book.

**Figure for 1–4**

In Exercises 1–4, use the graph of $f(x) = 3^x$ to sketch the graph of the function.

**1.** $g(x) = 3^x - 2$

**2.** $h(x) = 3^{-x}$

**3.** $k(x) = \log_3 x$

**4.** $j(x) = \log_3 (x - 1)$

**5.** For $P = \$2000$, $r = 8\%$, and $t = 10$ years, find the balance of an account compounded monthly and continuously.

**6.** The average waiting time between incoming calls at a switchboard is 2 minutes. The probability of waiting less than $t$ minutes until the next incoming call is approximated by

$$F(t) = 1 - e^{-t/2}.$$

If a call has just come in, find the probability that the next call will be within

(a) 1 minute.     (b) 2 minutes.

**7.** *Bacteria Growth*   A bacteria population is modeled by $P(t) = 100e^{0.3012t}$, where $t$ is the time in hours. Find (a) $P(0)$, (b) $P(6)$, and (c) $P(12)$.

**8.** *Demand Function*   Use the demand function $p = 600 - 0.4e^{0.003x}$ to find the price for a demand of $x = 1000$ units.

In Exercises 9–12, evaluate the expression.

**9.** $\log_{10} 100$     **10.** $\ln e^4$     **11.** $\log_4 \frac{1}{16}$     **12.** $\ln 1$

**13.** Sketch the graphs of $f(x) = 3^x$ and $g(x) = \log_3 x$ in the same coordinate plane. Identify the domains of $f(x)$ and $g(x)$. Discuss the special relationship between $f(x)$ and $g(x)$ that is demonstrated by their graphs.

In Exercises 14 and 15, expand the logarithmic expression.

**14.** $\log \sqrt[3]{\dfrac{xy}{z}}$

**15.** $\ln\left(\dfrac{x^2 + 3}{x^3}\right)$

In Exercises 16 and 17, condense the logarithmic expression.

**16.** $\ln x + \ln y - \ln 3$

**17.** $-3 \log_{10} 4x$

In Exercises 18 and 19, find the *exact* value of the logarithm.

**18.** $\log_7 \sqrt{343}$

**19.** $\ln \sqrt[5]{e^6}$

**20.** Find an equation that relates $x$ and $y$.

| $x$ | 1 | 2 | 3 | 4 | 6 | 8 |
|---|---|---|---|---|---|---|
| $y$ | 1 | 1.260 | 1.442 | 1.587 | 1.817 | 2.000 |

## 5.4    Solving Exponential and Logarithmic Equations

### Objectives

- Solve an exponential equation.
- Solve a logarithmic equation.
- Use an exponential or logarithmic model to solve an application problem.

## Introduction

So far in this chapter, you have studied the definitions, graphs, and properties of exponential and logarithmic functions. In this section, you will study procedures for *solving equations* involving these exponential and logarithmic functions. As a simple example, consider the exponential equation

$$2^x = 32.$$

You can obtain the solution by rewriting the equation in the form

$$2^x = 2^5$$

which implies that $x = 5$. Although this method works in some cases, it does not work for an equation as simple as

$$e^x = 7.$$

To solve for $x$, you can take the natural logarithm of both sides to obtain

| | |
|---|---|
| $e^x = 7$ | Original equation |
| $\ln e^x = \ln 7$ | Take log of both sides. |
| $x = \ln 7$ | $\ln e^x = x$ because $e^x = e^x$. |

Here are some guidelines for solving exponential and logarithmic equations.

---

▶ **Solving Exponential and Logarithmic Equations**

**1.** *To solve an exponential equation,* first isolate the exponential expression, then take the logarithm of both sides and solve for the variable.

**2.** *To solve a logarithmic equation,* rewrite the equation in exponential form and solve for the variable.

---

These two guidelines are based on the following **inverse properties** of exponential and logarithmic functions.

| *Base a* | *Base e* |
|---|---|
| **1.** $\log_a a^x = x$ | $\ln e^x = x$ |
| **2.** $a^{\log_a x} = x$ | $e^{\ln x} = x$ |

# Solving Exponential Equations

### Example 1   Solving an Exponential Equation

Solve $e^x = 72$.

**Solution**

| | |
|---|---|
| $e^x = 72$ | Original equation |
| $\ln e^x = \ln 72$ | Take log of both sides. |
| $x = \ln 72$ | Inverse property of logs and exponents |
| $x \approx 4.277$ | Use a calculator. |

The solution is $\ln 72$. Check this in the original equation.

### Example 2   Solving an Exponential Equation

Solve $e^x + 5 = 60$.

**Solution**

| | |
|---|---|
| $e^x + 5 = 60$ | Original equation |
| $e^x = 55$ | Subtract 5 from both sides. |
| $\ln e^x = \ln 55$ | Take log of both sides. |
| $x = \ln 55$ | Inverse property of logs and exponents |
| $x \approx 4.007$ | Use a calculator. |

The solution is $\ln 55$. Check this in the original equation.

---

## •TECHNOLOGY•

When solving an exponential or logarithmic equation, remember that you can check your solution graphically by "graphing the left and right sides separately" and estimating the x-coordinate of the point of intersection. For instance, to check the solution of the equation in Example 1, you can sketch the graphs of

   $y = e^x$

and

   $y = 72$

on the same screen, as shown below. Notice that the graphs intersect when $x \approx 4.277$, which confirms the solution found in Example 1.

### Example 3   Solving an Exponential Equation

Solve $4e^{2x} = 5$.

**Solution**

| | |
|---|---|
| $4e^{2x} = 5$ | Original equation |
| $e^{2x} = \dfrac{5}{4}$ | Divide both sides by 4. |
| $\ln e^{2x} = \ln \dfrac{5}{4}$ | Take log of both sides. |
| $2x = \ln \dfrac{5}{4}$ | Inverse property of logs and exponents |
| $x = \dfrac{1}{2} \ln \dfrac{5}{4}$ | Divide both sides by 2. |
| $x \approx 0.112$ | Use a calculator. |

The solution is $\frac{1}{2} \ln \frac{5}{4}$. Check this in the original equation.

When an equation involves two or more exponential expressions, you can still use a procedure similar to that demonstrated in the first three examples. However, the algebra is a bit more complicated. Study the next example carefully.

### Example 4   Solving an Exponential Equation

Solve $e^{2x} - 3e^x + 2 = 0$.

**Solution**

| | | |
|---|---|---|
| $e^{2x} - 3e^x + 2 = 0$ | | Original equation |
| $(e^x)^2 - 3e^x + 2 = 0$ | | Quadratic form |
| $(e^x - 2)(e^x - 1) = 0$ | | Factor. |
| $e^x - 2 = 0$ $\implies$ $x = \ln 2$ | | Set 1st factor equal to 0. |
| $e^x - 1 = 0$ $\implies$ $x = 0$ | | Set 2nd factor equal to 0. |

The solutions are $\ln 2$ and $0$. Check these in the original equation.

Figure 5.17

To check the result of Example 4 graphically, you can sketch the graph of $y = e^{2x} - 3e^x + 2$ with a graphing utility. The graph should have two $x$-intercepts: $x = \ln 2$ and $x = 0$, as shown in Figure 5.17.

## Solving Logarithmic Equations

To solve a logarithmic equation such as

$$\ln x = 3 \qquad \text{Logarithmic form}$$

write the equation in exponential form as follows.

$$e^{\ln x} = e^3 \qquad \text{Exponentiate both sides.}$$

$$x = e^3 \qquad \text{Exponential form}$$

This procedure is called *exponentiating* both sides of an equation.

### Example 5  Solving a Logarithmic Equation

Solve $\ln x = 2$.

**Solution**

| | |
|---|---|
| $\ln x = 2$ | Original equation |
| $e^{\ln x} = e^2$ | Exponentiate both sides. |
| $x = e^2$ | Inverse property of exponents and logs |
| $x \approx 7.389$ | Use a calculator. |

The solution is $e^2$. Check this in the original equation.

### Example 6  Solving a Logarithmic Equation

Solve $5 + 2 \ln x = 4$.

**Solution**

| | |
|---|---|
| $5 + 2 \ln x = 4$ | Original equation |
| $2 \ln x = -1$ | Subtract 5 from both sides. |
| $\ln x = -\dfrac{1}{2}$ | Divide both sides by 2. |
| $e^{\ln x} = e^{-1/2}$ | Exponentiate both sides. |
| $x = e^{-1/2}$ | Inverse property of exponents and logs |
| $x \approx 0.607$ | Use a calculator. |

The solution is $e^{-1/2}$. Check this in the original equation.

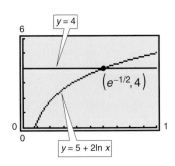

**Figure 5.18**

To check the result of Example 6 graphically, you can sketch the graphs of $y = 5 + 2 \ln x$ and $y = 4$ in the same coordinate plane, as shown in Figure 5.18.

### Example 7    Solving a Logarithmic Equation

Solve $2 \ln 3x = 4$.

**Solution**

$$2 \ln 3x = 4 \qquad \text{Original equation}$$

$$\ln 3x = 2 \qquad \text{Divide both sides by 2.}$$

$$e^{\ln 3x} = e^2 \qquad \text{Exponentiate both sides.}$$

$$3x = e^2 \qquad \text{Inverse property of exponents and logs}$$

$$x = \frac{1}{3}e^2 \qquad \text{Divide both sides by 3.}$$

$$x \approx 2.463 \qquad \text{Use a calculator.}$$

The solution is $\frac{1}{3}e^2$. Check this in the original equation.

### Example 8    Solving a Logarithmic Equation

Solve $\ln x - \ln(x - 1) = 1$.

**Solution**

$$\ln x - \ln(x - 1) = 1 \qquad \text{Original equation}$$

$$\ln \frac{x}{x - 1} = 1 \qquad \text{Property 2 of logarithms}$$

$$\frac{x}{x - 1} = e^1 \qquad \text{Exponentiate both sides.}$$

$$x = ex - e \qquad \text{Multiply both sides by } x - 1.$$

$$x - ex = -e \qquad \text{Subtract } ex \text{ from both sides.}$$

$$x(1 - e) = -e \qquad \text{Factor.}$$

$$x = \frac{-e}{1 - e} \qquad \text{Divide both sides by } 1 - e.$$

$$x = \frac{e}{e - 1} \qquad \text{Simplify.}$$

The solution is $e/(e - 1)$. Check this in the original equation.

In solving exponential or logarithmic equations, the following properties are useful. Can you see where these properties were used in the examples in this section?

**1.** $x = y$      if and only if      $\log_a x = \log_a y$.

**2.** $x = y$      if and only if      $a^x = a^y, \ a > 0, a \neq 1$.

# Applications

**Example 9** Doubling an Investment

You have deposited \$500 in an account that pays 6.75% interest, compounded continuously. (a) How long will it take your money to double? (b) How long will it take your money to triple?

**Solution**

Using the formula for compound interest with continuous compounding, you can find that the balance in the account is given by

$$A = Pe^{rt} = 500e^{0.0675t}.$$

**a.** To find the time required for the balance to double, let $A = 1000$, and solve the resulting equation for $t$.

| | |
|---|---|
| $500e^{0.0675t} = 1000$ | Original equation |
| $e^{0.0675t} = 2$ | Divide both sides by 500. |
| $\ln e^{0.0675t} = \ln 2$ | Take log of both sides. |
| $0.0675t = \ln 2$ | Inverse property of logs and exponents |
| $t = \dfrac{1}{0.0675} \ln 2$ | Divide both sides by 0.0675. |
| $t \approx 10.27$ | Use a calculator. |

The balance in the account will double after approximately 10.27 years.

**b.** To find the time required for the balance to triple, let $A = 1500$, and solve the resulting equation for $t$.

| | |
|---|---|
| $500e^{0.0675t} = 1500$ | Original equation |
| $e^{0.0675t} = 3$ | Divide both sides by 500. |
| $\ln e^{0.0675t} = \ln 3$ | Take log of both sides. |
| $0.0675t = \ln 3$ | Inverse property of logs and exponents |
| $t = \dfrac{1}{0.0675} \ln 3$ | Divide both sides by 0.0675. |
| $t \approx 16.28$ | Use a calculator. |

The balance in the account will triple after approximately 16.28 years.

Notice that it took 10.27 years to earn the first \$500 and only 6.01 years to earn the second \$500. This result is graphically demonstrated in Figure 5.19.

Figure 5.19

## Example 10   Shipments of Computers

From 1984 to 1998, the number of computers shipped can be approximated by

$$y = 13{,}516.186e^{0.00628t^2}$$

where $y$ is the number of computers (in thousands) shipped worldwide and $t = 4$ represents 1984 (see Figure 5.20). According to the model, during which year did the number of computers shipped reach 83,000,000?   (Source: *The Wall Street Journal Almanac 1998*)

### Solution

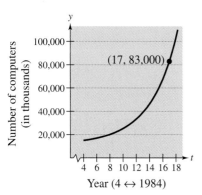

Figure 5.20

| | |
|---|---|
| $13{,}516.186e^{0.00628t^2} = y$ | Given model |
| $13{,}516.186e^{0.00628t^2} = 83{,}000$ | Let $y = 83{,}000$. |
| $e^{0.00628t^2} \approx 6.141$ | Divide both sides by 13,516.186. |
| $\ln e^{0.00628t^2} \approx \ln 6.141$ | Take ln of both sides. |
| $0.00628t^2 \approx 1.815$ | Inverse property of exponents and logs |
| $t^2 \approx 289.013$ | Divide both sides by 0.00628. |
| $t \approx 17$ | Extract positive square root. |

The solution is $t \approx 17$ years. Because $t = 4$ represents 1984, it follows that the number of computers shipped worldwide would have reached 83,000,000 in 1997.

---

| Discussing the Concept | Analyzing Relationships Numerically |
|---|---|

Use a calculator to fill in the following table row-by-row. Discuss the resulting pattern. What can you conclude? Find two equations that summarize the relationships you've discovered.

| $x$ | $\frac{1}{2}$ | 1 | 2 | 10 | 25 | 50 |
|---|---|---|---|---|---|---|
| $e^x$ | | | | | | |
| $\ln(e^x)$ | | | | | | |
| $\ln x$ | | | | | | |
| $e^{\ln x}$ | | | | | | |

# Warm Up

The following warm-up exercises involve skills that were covered in earlier sections. You will use these skills in the exercise set for this section.

In Exercises 1–6, solve for $x$.

1. $x \ln 2 = \ln 3$
2. $(x - 1) \ln 4 = 2$
3. $2xe^2 = e^3$
4. $4xe^{-1} = 8$
5. $x^2 - 4x + 5 = 0$
6. $2x^2 - 3x + 1 = 0$

In Exercises 7–10, simplify the expression.

7. $\log_{10} 10^x$
8. $\log_{10} 10^{2x}$
9. $\ln e^{2x}$
10. $\ln e^{-x^2}$

## 5.4 Exercises

In Exercises 1–10, solve for $x$.

1. $5^x = 125$
2. $2^x = 64$
3. $7^x = \frac{1}{49}$
4. $4^x = \frac{1}{256}$
5. $4^{2x-1} = 64$
6. $3^{x-1} = 27$
7. $\log_4 x = 3$
8. $\log_5 5x = 2$
9. $\log_{10} x = -1$
10. $\ln(2x - 1) = 0$

In Exercises 11–16, apply the inverse properties of $\ln x$ and $e^x$ to simplify the expression.

11. $\ln e^{x^2}$
12. $\ln e^{2x-1}$
13. $-8 + e^{\ln x^3}$
14. $-1 + \ln e^{2x}$
15. $e^{\ln x^2}$
16. $e^{\ln(5x+2)}$

In Exercises 17–44, solve the exponential equation.

17. $e^x = 20$
18. $e^x = 5$
19. $e^x = 14$
20. $e^x = 58$
21. $3e^x = 23$
22. $5e^x = 47$
23. $e^x - 5 = 10$
24. $e^x + 6 = 38$

25. $3e^x - 8 = 23$
26. $4e^x + 10 = 605$
27. $7 - 2e^x = 5$
28. $-14 + 3e^x = 11$
29. $e^{4x} = 20$
30. $e^{5x} = 40$
31. $5e^{-x} = 28$
32. $3e^{-2x} = 14$
33. $500e^{-x} = 300$
34. $8e^{-5x} = 112$
35. $500e^{0.06t} = 1500$
36. $750e^{0.07t} = 2000$
37. $e^{2x} + 3e^x - 10 = 0$
38. $e^{2x} - 6e^x + 8 = 0$
39. $25e^{2x+1} = 962$
40. $80e^{-t/2} + 20 = 70$
41. $\dfrac{1}{1 - 2e^{-t}} = 1.5$
42. $\dfrac{10,000}{1 + 19e^{-t/5}} = 2000$
43. $\dfrac{2}{1 - 4e^{-2t}} = 2.5$
44. $\dfrac{5000}{1 + 7e^{-t/3}} = 1000$

In Exercises 45–62, solve the logarithmic equation.

45. $\ln x = 7$
46. $\ln x = 3$
47. $\ln x = -8$
48. $\ln x = -5$
49. $\ln 2x = 1$
50. $\ln 4x = 2.5$
51. $2 \ln x = 7$
52. $3 \ln x = 12$

**53.** $2 \ln 4x = 0$

**54.** $3 \ln 5x = 10$

**55.** $5 + \ln 2x = 4$

**56.** $-6 + \ln 3x = 0$

**57.** $\ln 2x^2 = 5$

**58.** $\ln x^2 = 1$

**59.** $\ln x - \ln 2 = 5$

**60.** $\ln x + \ln(x + 2) = 3$

**61.** $\ln x + \ln(x - 2) = 1$

**62.** $\ln \sqrt{(x + 2)} = 1$

*Compound Interest*   In Exercises 63 and 64, find the time required for a $1000 investment to double at interest rate $r$, compounded continuously.

**63.** $r = 0.085$

**64.** $r = 0.12$

*Compound Interest*   In Exercises 65 and 66, find the time required for a $1000 investment to triple at interest rate $r$, compounded continuously.

**65.** $r = 0.0675$

**66.** $r = 0.075$

**67.** *Demand Function*   The demand function for a product is given by

$$p = 500 - 0.5(e^{0.004x}).$$

(a) Find the demand $x$ for a price of $p = \$350$.

(b) Find the demand $x$ for a price of $p = \$300$.

(c) Use a graphing utility to confirm graphically the results found in parts (a) and (b).

**68.** *Demand Function*   The demand function for a product is given by

$$p = 5000\left(1 - \frac{4}{4 + e^{-0.002x}}\right).$$

(a) Find the demand $x$ for a price of $p = \$600$.

(b) Find the demand $x$ for a price of $p = \$400$.

(c) Use a graphing utility to confirm graphically the results found in parts (a) and (b).

**69.** *Forest Yield*   The yield $V$ (in millions of cubic feet per acre) for a forest at age $t$ years is given by

$$V = 6.7e^{-48.1/t}, \quad t \geq 0.$$

(a) Use a graphing utility to find the time necessary to have a yield of 1.3 million cubic feet.

(b) Use a graphing utility to find the time necessary to have a yield of 2 million cubic feet.

**70.** *Human Memory Model*   In a group project on learning theory, a mathematical model for the percent $P$ of correct responses after $n$ trials was found to be

$$P = \frac{0.83}{1 + e^{-0.2n}}, \quad n \geq 0.$$

(a) After how many trials will 60% of the responses be correct? (That is, for what value of $n$ will $P = 0.6$?)

(b) Use a graphing utility to graph the memory model and confirm the result found in part (a).

(c) Write a paragraph describing the memory model.

In Exercises 71 and 72, use a graphing utility to graph the function.

**71.** $y = e^{-x^2}$

**72.** $y = 2e^{-x^2}$

**73.** *Average Heights*   The percent of American males (between 20 and 29 years old) who are less than $x$ inches tall is given by

$$p = -0.016 + \frac{1.037}{1 + e^{-0.5345(x - 69.41)}}, \quad 60 \leq x \leq 75$$

where $p$ is the percent (in decimal form) and $x$ is the height in inches (see figure). (Source: U.S. National Center for Health Statistics)

(a) What is the median height for American males between 20 and 29 years old? (In other words, for what value of $x$ is $p$ equal to 0.5?)

(b) Write a paragraph describing the height model.

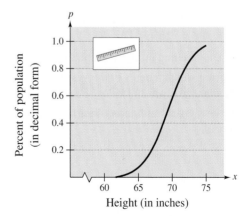

**74.** *Average Heights*    The percent of American females (between 20 and 29 years old) who are less than $x$ inches tall is given by

$$p = -0.013 + \frac{1.027}{1 + e^{-0.5868(x-64.14)}}, \quad 55 \le x \le 72$$

where $p$ is the percent (in decimal form) and $x$ is the height in inches (see figure). (Source: U.S. National Center for Health Statistics)

(a) What is the median height for American females between 20 and 29 years old? (In other words, for what value of $x$ is $p$ equal to 0.5?)

(b) Write a paragraph describing the height model.

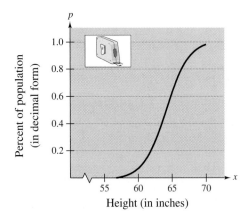

Height (in inches)

**75.** *Temperature in Baltimore*    The average monthly *low* temperature in Baltimore, Maryland can be approximated by the model

$$y = 21 + 46.3e^{-(t-7.2)^2/13}, \quad 1 \le t \le 12$$

where $y$ is the average monthly low temperature (in degrees Fahrenheit) and $t$ represents the month of the year, with $t = 1$ corresponding to mid-January. (Source: *PC U.S.A.*)

(a) Use a graphing utility to graph the model.

(b) During which months of the year is the low temperature likely to fall below freezing?

(c) Write a paragraph describing the temperature model.

**76.** *Temperature in Baltimore*    The average monthly *high* temperature in Baltimore, Maryland can be approximated by the model

$$y = 38.5 + 50.4e^{-(t-7.1)^2/13.3}, \quad 1 \le t \le 12$$

where $y$ is the average monthly high temperature (in degrees Fahrenheit) and $t$ represents the month of the year, with $t = 1$ corresponding to mid-January (see figure). During which months of the year is the high temperature likely to rise above 80°? (Source: *PC U.S.A.*)

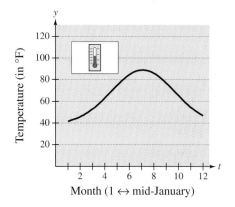

Month (1 ↔ mid-January)

**77.** *Native Prairie Grasses*    The number of native prairie grasses per acre $A$ is approximated by the model

$$A = 10.5 \cdot 10^{0.04x}, \quad 0 \le x \le 24$$

where $x$ is the number of months since the acre was plowed. Use this model to approximate the number of months since the field was plowed in a test plot if $A = 280$.

**78.** *Suburban Wildlife*    The variety of suburban non-domesticated wildlife is approximated by the model

$$V = 15 \cdot 10^{0.02x}, \quad 0 \le x \le 36$$

where $x$ is the number of months since the development was completed. Use this model to approximate the number of months since the development was completed if $V = 50$.

## 5.5   Exponential and Logarithmic Models

### Objectives

- Identify common mathematical models involving exponential and logarithmic functions.
- Construct and use a model for exponential growth or exponential decay.
- Use a Gaussian or logistics growth model to solve an application problem.
- Use a base 10 logarithmic model to solve an application problem.

## Introduction

The five most common types of mathematical models involving exponential functions and logarithmic functions are as follows.

1. Exponential growth:    $y = ae^{bx}, \quad b > 0$
2. Exponential decay:    $y = ae^{-bx}, \quad b > 0$
3. Gaussian model:    $y = ae^{-(x-b)^2/c}$
4. Logistics growth model:    $y = \dfrac{a}{1 + be^{-(x-c)/d}}$
5. Logarithmic models:    $y = \ln(ax + b), \quad y = \log_{10}(ax + b)$

The basic shape of each of these graphs is shown in Figure 5.21.

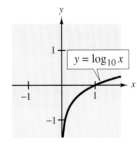

Figure 5.21

# Exponential Growth and Decay

**Example 1** Population Increase

The world population (in millions) from 1988 through 1996 is shown in the table.

| Year | 1988 | 1989 | 1990 | 1991 | 1992 | 1993 | 1994 | 1995 | 1996 |
|------|------|------|------|------|------|------|------|------|------|
| Population | 5109 | 5194 | 5282 | 5336 | 5448 | 5529 | 5610 | 5691 | 5771 |

(Source: U.S. Bureau of the Census)

An exponential growth model that approximates these data is given by

$$P = -14{,}577.979 + 19{,}044.426e^{0.00415t}, \quad 8 \leq t \leq 16$$

where $P$ is the population (in millions) and $t = 8$ represents 1988. Compare the values given in the model with the estimates given by the Census Bureau. According to the model, when will the world population reach 7 billion?

**Solution**

The following table compares the two sets of population figures. The model is shown in Figure 5.22.

| Year | 1988 | 1989 | 1990 | 1991 | 1992 | 1993 | 1994 | 1995 | 1996 |
|------|------|------|------|------|------|------|------|------|------|
| Population | 5109 | 5194 | 5282 | 5336 | 5448 | 5529 | 5610 | 5691 | 5771 |
| Model | 5109 | 5191 | 5273 | 5356 | 5439 | 5522 | 5606 | 5690 | 5774 |

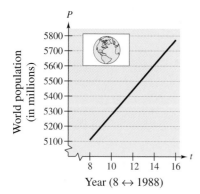

Figure 5.22

To find when the world population will reach 7 billion, let $P = 7000$ in the model and solve for $t$.

| | |
|---|---|
| $-14{,}577.979 + 19{,}044.426e^{0.00415t} = P$ | Given model |
| $-14{,}577.979 + 19{,}044.426e^{0.00415t} = 7000$ | Let $P = 7000$. |
| $19{,}044.426e^{0.00415t} = 21{,}577.979$ | Add 14,577.979 to both sides. |
| $e^{0.00415t} \approx 1.13303$ | Divide both sides by 19,044.426. |
| $\ln e^{0.00415t} \approx \ln 1.13303$ | Take ln of both sides. |
| $0.00415t \approx 0.1249$ | Inverse property of exponents and logs |
| $t \approx 30.1$ | Divide both sides by 0.00415. |

According to the model, the world population will reach 7 billion in 2010.

An exponential model increases (or decreases) by the same percent each year. What is the annual percent increase for the exponential model above?

In Example 1, you were given the exponential growth model. But suppose this model were not given; how could you find such a model? One technique for doing this is demonstrated in Example 2.

**Example 2**    Finding an Exponential Growth Model

Find an exponential growth model whose graph passes through the points $(0, 4453)$ and $(7, 5024)$, as shown in Figure 5.23(a).

Solution

The general form of the model is

$$y = ae^{bx}.$$

From the fact that the graph passes through the point $(0, 4453)$, you know that $y = 4453$ when $x = 0$. By substituting these values into the general model, you have

$$4453 = ae^0 \quad \Longrightarrow \quad a = 4453.$$

In a similar way, from the fact that the graph passes through the point $(7, 5024)$, you know that $y = 5024$ when $x = 7$. By substituting these values into the model, you obtain

$$5024 = 4453e^{7b} \quad \Longrightarrow \quad b = \frac{1}{7}\ln\frac{5024}{4453} \approx 0.01724.$$

So, the exponential growth model is

$$y = 4453e^{0.01724x}.$$

The graph of the model is shown in Figure 5.23(b).

(a)

(b)

Figure 5.23

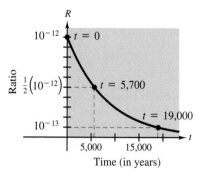

**Figure 5.24**

In living organic material, the ratio of radioactive carbon isotopes (carbon 14) to the number of nonradioactive carbon isotopes (carbon 12) is about 1 to $10^{12}$. When organic material dies, its carbon 12 content remains fixed, whereas its radioactive carbon 14 begins to decay with a half-life of about 5700 years. To estimate the age of dead organic material, scientists use the following formula, which denotes the ratio of carbon 14 to carbon 12 present at any time $t$ (in years).

$$R = \frac{1}{10^{12}} e^{-t/8223}$$

The graph of $R$ is shown in Figure 5.24. Note that $R$ decreases as the time $t$ increases.

**Example 3**    Carbon Dating

The ratio of carbon 14 to carbon 12 in a newly discovered fossil is

$$R = \frac{1}{10^{13}}.$$

Estimate the age of the fossil.

**Solution**

In the carbon dating model, substitute the given value of $R$ to obtain the following.

$$\frac{1}{10^{12}} e^{-t/8223} = R \qquad \text{Given model}$$

$$\frac{e^{-t/8223}}{10^{12}} = \frac{1}{10^{13}} \qquad \text{Let } R \text{ equal } \frac{1}{10^{13}}.$$

$$e^{-t/8223} = \frac{1}{10} \qquad \text{Multiply both sides by } 10^{12}.$$

$$\ln e^{-t/8223} = \ln \frac{1}{10} \qquad \text{Take log of both sides.}$$

$$-\frac{t}{8223} \approx -2.3026 \qquad \text{Inverse property of logs and exponents}$$

$$t \approx 18,934 \qquad \text{Multiply both sides by } -8223.$$

So, to the nearest thousand years, you can estimate the age of the fossil to be 19,000 years.

The carbon dating model in Example 3 assumed that the carbon 14/carbon 12 ratio was one part in 10,000,000,000,000. Suppose an error in measurement occurred and the actual ratio was only one part in 8,000,000,000,000. The fossil age corresponding to the actual ratio would then be approximately 17,000 years. Try checking this result.

## Gaussian Models

As mentioned at the beginning of this section, Gaussian models are of the form

$$y = ae^{-(x-b)^2/c}.$$

This type of model is commonly used in probability and statistics to represent populations that are **normally distributed.** For *standard* normal distributions, the model takes the form

$$y = \frac{1}{\sigma\sqrt{2\pi}}e^{-x^2/2\sigma^2}$$

where $\sigma$ is the standard deviation ($\sigma$ is the lowercase Greek letter sigma). The graph of a Gaussian model is called a **bell-shaped curve.** Try assigning a value to $\sigma$ and sketching a normal distribution curve with a graphing utility. Can you see why it is called a bell-shaped curve?

| Example 4 | SAT Scores |  |
|---|---|---|

In 1998, the Scholastic Aptitude Test (SAT) scores for males roughly followed a normal distribution given by

$$y = 0.0035e^{-(x-531)^2/26,000}, \quad 200 \le x \le 800$$

where $x$ is the SAT score for mathematics. Sketch the graph of this function. From the graph, estimate the average SAT score.   (Source: College Board)

### Solution

The graph of the function is given in Figure 5.25. From the graph, you can see that the average mathematics score for males in 1998 was 531.

Figure 5.25

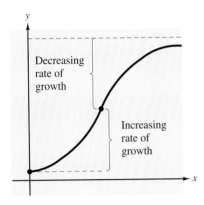

Figure 5.26    *Logistics Curve*

# Logistics Growth Models

Some populations initially have rapid growth, followed by a declining rate of growth, as indicated by the graph in Figure 5.26. One model for describing this type of growth pattern is the **logistics curve** given by the function

$$y = \frac{a}{1 + be^{-(x-c)/d}}$$

where $y$ is the population size and $x$ is the time. An example is a bacteria culture allowed to grow initially under ideal conditions, followed by less favorable conditions that inhibit growth. A logistics growth curve is also called a **sigmoidal curve.**

**Example 5**    Spread of a Virus

On a college campus of 5000 students, one student returned from vacation with a contagious flu virus. The spread of the virus through the student body is given by

$$y = \frac{5000}{1 + 4999e^{-0.8t}}, \quad t \geq 0$$

where $y$ is the total number infected after $t$ days. The college will cancel classes when 40% or more of the students are ill.

**a.** How many are infected after 5 days?

**b.** After how many days will the college cancel classes?

## Solution

**a.** After 5 days, the number infected is

$$y = \frac{5000}{1 + 4999e^{-0.8(5)}} = \frac{5000}{1 + 4999e^{-4}} \approx 54.$$

**b.** In this case, the number infected is $(0.40)(5000) = 2000$. Therefore, we solve for $t$ in the following equation.

$$2000 = \frac{5000}{1 + 4999e^{-0.8t}}$$

$$1 + 4999e^{-0.8t} = 2.5$$

$$e^{-0.8t} \approx 0.0003$$

$$\ln(e^{-0.8t}) \approx \ln 0.0003$$

$$-0.8t \approx -8.1115$$

$$t \approx 10.1$$

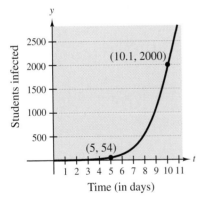

Figure 5.27

So, after 10 days, at least 40% of the students will be infected, and the college will cancel classes. The graph of the function is shown in Figure 5.27.

## Logarithmic Models

*This photo was taken in Anchorage, Alaska, in 1964 after a devastating earthquake that measured 9.2 on the Richter scale. This was the second largest quake registered in the world since 1900. (The largest took place in Chile in 1960, measuring 9.5 on the Richter scale.)* (Source: National Earthquake Information Center, U.S. Geological Survey)

**Example 6**    Magnitudes of Earthquakes

On the Richter scale, the magnitude $R$ of an earthquake of intensity $I$ is given by

$$R = \log_{10} \frac{I}{I_0}$$

where $I_0 = 1$ is the minimum intensity used for comparison. Find the intensity per unit of area for the following earthquakes. (Intensity is a measure of the wave energy of an earthquake.)

**a.** San Francisco in 1906, $R = 8.3$

**b.** San Francisco Bay Area in 1989, $R = 6.9$

**Solution**

**a.** Because $I_0 = 1$ and $R = 8.3$, you have

$$8.3 = \log_{10} I$$
$$I = 10^{8.3} \approx 199{,}526{,}000.$$

**b.** For $R = 6.9$, you have $6.9 = \log_{10} I$, and

$$I = 10^{6.9} \approx 7{,}943{,}000.$$

Note that an increase of 1.4 units on the Richter scale (from 6.9 to 8.3) represents an intensity change by a factor of

$$\frac{199{,}526{,}000}{7{,}943{,}000} \approx 25.$$

In other words, the "great San Francisco earthquake" of 1906 had a magnitude about 25 times greater than that of the 1989 earthquake.

---

| Discussing the Concept | Exponential and Logarithmic Models |
|---|---|

Use your school's library or some other reference source to find an application that fits one of the five models discussed in this section. After you have collected data for the model, use a graphing utility to make a scatter plot of the data. Then use a built-in regression program to find a model that describes the data you have found.

# Warm Up

The following warm-up exercises involve skills that were covered in earlier sections. You will use these skills in the exercise set for this section.

In Exercises 1–6, sketch the graph of the equation.

**1.** $y = e^{0.1x}$

**2.** $y = e^{-0.25x}$

**3.** $y = e^{-x^2/5}$

**4.** $y = \dfrac{2}{1 + e^{-x}}$

**5.** $y = \log_{10} 2x$

**6.** $y = \ln 4x$

In Exercises 7 and 8, solve the equation algebraically.

**7.** $3e^{2x} = 7$

**8.** $4 \ln 5x = 14$

In Exercises 9 and 10, solve the equation graphically.

**9.** $2e^{-0.2x} = 0.002$

**10.** $6 \ln 2x = 12$

## 5.5    Exercises

LAB 7

*Compound Interest*    In Exercises 1–10, complete the table for a savings account in which interest is compounded continuously.

| Initial Investment | Annual % rate | Time to Double | Amount after 10 Years |
|---|---|---|---|
| **1.** $5000 | 7% | | |
| **2.** $1000 | $9\frac{1}{4}\%$ | | |
| **3.** $500 | | 10 yr | |
| **4.** $10,000 | | 5 yr | |
| **5.** $1000 | | | $2281.88 |
| **6.** $2000 | | | $3000 |
| **7.** | 11% | | $19,205 |
| **8.** | 8% | | $20,000 |
| **9.** $5000 | | | $11,127.70 |
| **10.** $250 | | | $600 |

*Radioactive Decay*    In Exercises 11–16, complete the table for the radioactive isotopes.

| Isotope | Half-Life (Years) | Initial Quantity | Amount after 1000 Years |
|---|---|---|---|
| **11.** $Ra^{226}$ | 1620 | 4 g | |
| **12.** $Ra^{226}$ | 1620 | | 0.15 g |
| **13.** $C^{14}$ | 5730 | | 3.5 g |
| **14.** $C^{14}$ | 5730 | 8 g | |
| **15.** $Pu^{230}$ | 24,360 | | 1.6 g |
| **16.** $Pu^{230}$ | 24,360 | | 0.38 g |

In Exercises 17–20, classify the model as exponential growth or exponential decay.

**17.** $y = 3e^{0.5t}$

**18.** $y = 2e^{-0.6t}$

**19.** $y = 20e^{-1.5t}$

**20.** $y = 4e^{0.07t}$

In Exercises 21–24, find the constant $k$ such that the exponential function $y = Ce^{kt}$ passes through the points on the graph.

**21.**

**22.**

**23.**

**24.**

**25.** *Population*   The population $P$ of a city is given by

$$P = 105,300e^{0.015t}$$

where $t$ is the time in years, with $t = 0$ corresponding to 1990. Sketch the graph of this equation. According to this model, in what year will the city have a population of 150,000?

**26.** *Population*   The population $P$ of a city is given by

$$P = 240,360e^{0.012t}$$

where $t$ is the time in years, with $t = 0$ corresponding to 1990. Sketch the graph of this equation. According to this model, in what year will the city have a population of 300,000?

**27.** *Population*   The population $P$ of a city is given by

$$P = 25,000e^{kt}$$

where $t$ is the time in years, with $t = 0$ corresponding to 1990. In 1945, the population was 3350. Find the value of $k$ and use this result to predict the population in the year 2010.

**28.** *Population*   The population $P$ of a city is given by

$$P = 140,500e^{kt}$$

where $t = 0$ represents 1990. In 1960, the population was 100,250. Find the value of $k$ and use this result to predict the population in the year 2005.

**29.** *Bacteria Growth*   The number of bacteria $N$ in a culture is given by the model

$$N = 100e^{kt}$$

where $t$ is the time in hours, with $t = 0$ corresponding to the time when $N = 100$. When $t = 4$, there are 250 bacteria. How long does it take the population to double in size? To triple in size?

**30.** *Bacteria Growth*   The number of bacteria $N$ in a culture is given by the model

$$N = 250e^{kt}$$

where $t$ is the time in hours, with $t = 0$ corresponding to the time when $N = 250$. When $t = 10$, there are 280 bacteria. How long does it take the population to double in size? To triple in size?

**31.** *Radioactive Decay*   What percent of a present amount of radioactive radium ($Ra^{226}$) will remain after 100 years? Use the fact that radioactive radium has a half-life of 1620 years.

**32.** *Radioactive Decay*   Find the half-life of a radioactive material if, after 1 year, 99.57% of the initial amount remains.

**33.** *Learning Curve*   The management at a factory has found that the maximum number of units a worker can produce in a day is 30. The learning curve for the number of units $N$ produced per day after a new employee has worked $t$ days is given by

$$N = 30(1 - e^{kt}).$$

After 20 days on the job, a particular worker produced 19 units in 1 day.

(a) Find the learning curve for this worker (that is, find the value of $k$).

(b) How many days should pass before this worker is producing 25 units per day?

**34.** *Sales*   The sales $S$ (in thousands of units) of a new product after it has been on the market for $t$ years are given by $S = 100(1 - e^{kt})$.

(a) Find $S$ as a function of $t$ if 15,000 units have been sold after 1 year.

(b) How many units will have been sold after 5 years?

**35.** *Snowmobile Sales* From 1988 to 1996, the sales of snowmobiles can be approximated by the function

$$y = \begin{cases} -4.3 + 33.0t + 42{,}630.9e^{-t}, & 8 \le t \le 12 \\ 334.3 + 644.4e^{-0.173(t-15.7)^2}, & 12 < t \le 16 \end{cases}$$

where $y$ is the sales (in millions of dollars) and $t = 8$ represents 1988. (Source: National Sporting Goods Association)

(a) Use a graphing utility to graph this model.

(b) Describe the change in the growth pattern that occurred in 1992.

**36.** *Textbook Sales* The sales of a college textbook that was published in 1991 and revised in 1995 can be approximated by the function

$$y = \begin{cases} e^{9.84 + 0.30t - 0.22t^2}, & 1 \le t \le 4 \\ e^{5.52 + 2.05t - 0.22t^2}, & 5 \le t \le 8 \end{cases}$$

where $y$ is the number of copies sold and $t = 1$ represents 1991.

(a) Use a graphing utility to graph this model.

(b) The graph is typical of college textbook sales. How would you explain this sales pattern?

**37.** *Women's Heights* The distribution for the heights of American women (between 30 and 39 years of age) can be approximated by the function

$$p = 0.148e^{-(x-64.9)^2/14.45},$$

where $x$ is the height in inches and $p$ is the percent (in decimal form). (Source: U.S. National Center for Health Statistics)

(a) Use a graphing utility to graph this model.

(b) Use the graph to determine the average height of women in this age bracket.

**38.** *Men's Heights* The distribution for the heights of American men (between 30 and 39 years of age) can be approximated by the function

$$p = 0.140e^{-(x-70)^2/16.27}, \quad 62 \le x \le 75$$

where $x$ is the height in inches and $p$ is the percent (in decimal form). (Source: U.S. National Center for Health Statistics)

(a) Use a graphing utility to graph this model.

(b) Use the graph to determine the average height of men in this age bracket.

**39.** *Stocking a Lake with Fish* A certain lake is stocked with 500 fish, and the fish population increases according to the logistics curve

$$P = \frac{10{,}000}{1 + 19e^{-t/5}}, \quad 0 \le t$$

where $t$ is measured in months.

(a) Use a graphing utility to graph this model.

(b) Find the fish population after 5 months.

(c) After how many months will the fish population be 2000?

**40.** *High School Graduates* The percent of the American population (who are 25 years old or older) who have *not* completed high school has been decreasing according to the model

$$P = 7.56 + \frac{124.71}{1 + e^{(t+25.2)/17.36}}, \quad 0 \le t \le 17$$

where $P$ is the percent of the population and $t$ is the calendar year, with $t = 0$ corresponding to 1980. (Source: U.S. Bureau of the Census)

(a) Use a graphing utility to graph this model.

(b) According to this model, when did the percent of the population who have not completed high school fall below 25%?

(c) Use the given equation to construct a model that represents the percent of the population who *have* completed high school.

**41.** *Endangered Species* A conservation organization releases 100 animals of an endangered species into a game preserve. The organization believes that the preserve has a carrying capacity of 1000 animals and that the growth of the herd will be modeled by the logistics curve

$$p = \frac{1000}{1 + 9e^{-kt}}, \quad t \ge 0$$

where $t$ is measured in years.

(a) Find $k$ if the herd size is 134 after 2 years.

(b) Find the population after 5 years.

**42.** *Sales and Advertising*    After discontinuing all advertising for a certain product (in 1995), the manufacturer of the product found that the sales began to drop according to the model

$$S = \frac{500,000}{1 + 0.6e^{kt}}, \quad 0 \le t \le 6$$

where $S$ represents the number of units sold and $t$ represents the calendar year, with $t = 0$ corresponding to 1995.

(a) Find $k$ if the company sold 300,000 units in 1997.

(b) According to this model, what were the sales in 2000?

**43.** *Value of Cassettes*    The dollar value $D$ (in millions) of prerecorded music cassettes from 1990 to 1996 is given by

$$D = 1585.80 + \frac{1907.60}{1 + e^{(t-4.686)/0.7852}}, \quad 0 \le t \le 6$$

where $t = 0$ represents 1990 (see figure). Why do you think the dollar value of cassettes was falling during this period?    (Source: Recording Industry Association of America, Inc.)

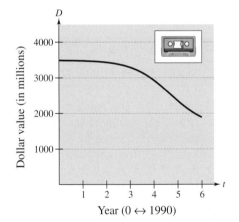

Year (0 ↔ 1990)

**44.** *Sound Recording Sales*    From 1987 to 1995, the sales $S$ (in millions of dollars) of sound recordings in the United States were given by

$$S = 5184 + \frac{10,403}{1 + e^{-(t-6)/2.16}}, \quad 0 \le t \le 8$$

where $t = 0$ represents 1987 (see figure). In what year will sales reach $14,000? Why do you think sales of sound recordings were increasing during this time period?    (Source: Recording Industry Association of America, Inc.)

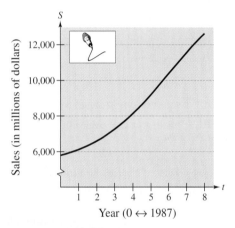

Year (0 ↔ 1987)

*Earthquake Magnitudes*    In Exercises 45 and 46, use the Richter scale (see Example 6) for measuring the magnitude of earthquakes.

**45.** Find the magnitude $R$ (on the Richter scale) of an earthquake of intensity $I$ (let $I_0 = 1$).

(a) $I = 80,500,000$

(b) $I = 48,275,000$

(c) $I = 40,000,000$

(d) $I = 20,000,000$

**46.** Find the intensity $I$ of an earthquake measuring $R$ on the Richter scale (let $I_0 = 1$).

(a) Colombia in 1906, $R = 8.6$

(b) Los Angeles in 1971, $R = 6.7$

*Intensity of Sound* In Exercises 47 and 48, use the information to determine the level of sound (in decibels) for the sound intensity. The level of sound $\beta$, in decibels, of a sound with an intensity of $I$ is given by

$$\beta(I) = 10 \log_{10} \frac{I}{I_0}$$

where $I_0$ is an intensity of $10^{-16}$ watts per square centimeter, corresponding roughly to the faintest sound that can be heard.

**47.** (a) $I = 10^{-14}$ watts per cm² (faint whisper)

(b) $I = 10^{-9}$ watts per cm² (busy street corner)

(c) $I = 10^{-6.5}$ watts per cm² (air hammer)

(d) $I = 10^{-4}$ watts per cm² (threshold of pain)

**48.** (a) $I = 10^{-13}$ watts per cm² (whisper)

(b) $I = 10^{-7.5}$ watts per cm² (DC-8 4 miles away)

(c) $I = 10^{-7}$ watts per cm² (diesel truck at 25 feet)

(d) $I = 10^{-4.5}$ watts per cm² (automobile horn at 3 feet)

**49.** *Noise Level* Due to the installation of acoustical tiling, the noise level in an auditorium was reduced from 78 to 72 decibels. Find the percent decrease in the intensity level of the noise as a result of the installation of the tiling.

**50.** *Noise Level* Due to the installation of a muffler, the noise level of an automobile engine was reduced from 88 to 76 decibels. Find the percent decrease in the intensity level of the noise as a result of the installation of the muffler.

**51.** *Estimating the Time of Death* At 8:30 A.M., a coroner was called to the home of a person who had died during the night. In order to estimate the time of death, the coroner took the person's temperature twice. At 9:00 A.M. the temperature was 85.7°, and at 9:30 A.M. the temperature was 82.8°. From these two temperature readings, the coroner was able to determine that the time elapsed since death and the body temperature were related by the formula

$$t = -2.5 \ln \frac{T - 70}{98.6 - 70}$$

where $t$ is the time in hours that have elapsed since the person died and $T$ is the temperature (in degrees Fahrenheit) of the person's body. Assume the person had a normal body temperature of 98.6° at death, and the room temperature was a constant 70°. (This formula is derived from a general cooling principle called Newton's Law of Cooling.) Use this formula to estimate the time of death of the person.

**52.** *Thawing a Package of Steaks* Suppose you take a 3-pound package of steaks out of the freezer at noon. Will the steaks be thawed in time to be grilled on the barbecue at 6 P.M.? To answer this question, assume that the room temperature is 70° Fahrenheit and that the freezer temperature is 24°. Use the formula

$$t = -3.8 \ln \frac{T - 70}{24 - 70}$$

where $t$ is the time (with $t = 0$ corresponding to noon) and $T$ is the temperature of the package of steaks.

**53.** *Population Growth* The population of a city is given in the table, with $t = 0$ corresponding to 1990. Make a scatter plot of these data. Use the exponential regression program of your graphing utility to find a model for the data. Graph the model with the data points. Use the model to predict the city's population in the years 2000 and 2002.

| $t$ | 0 | 2 | 4 | 6 | 8 |
|---|---|---|---|---|---|
| Population | 35,478 | 52,369 | 65,923 | 71,589 | 79,963 |

**54.** *Worker's Productivity* The numbers of units per day $n$ that a new worker can produce after $t$ days on the job are listed in the table. Make a scatter plot of these data. Do the data fit an exponential model or a logarithmic model? Use the regression program of your graphing utility to find the model. Graph the model with the data points. Is the model a good fit?

| $t$ | 5 | 10 | 12 | 15 | 21 |
|---|---|---|---|---|---|
| $n$ | 6 | 13 | 17 | 22 | 31 |

# Chapter Project

## Newton's Law of Cooling

| $t$ | $T_1$ | $T_2$ |
|-----|-------|-------|
| 0   | 182   | 176   |
| 2   | 180   | 170   |
| 8   | 160   | 158   |
| 10  | 156   | 156   |
| 21  | 138   | 140   |
| 24  | 134   | 136   |
| 28  | 128   | 132   |

When a warm object is placed in a cool room, its temperature will gradually change to the temperature of the room. (The same is true for a cool object placed in a warm room.)

According to Newton's Law of Cooling, the rate at which the object's temperature $T$ changes is proportional to the difference between its temperature and the temperature of the room $L$. This statement can be translated as

$$T = L + Ce^{kt}$$

where $k$ is a constant and $L + C$ is the original temperature of the object.

To confirm Newton's Law of Cooling experimentally, you can place a warm object in a cool room and monitor its temperature. The table at the left shows the temperatures (in degrees Fahrenheit) of 300 ml of water in a ceramic cup ($T_1$) and a Styrofoam cup ($T_2$) for several times $t$ (in minutes). The room temperature at the time of data collection was 70°F. The data was collected using a cooking thermometer and a digital watch.

1. *Make a Scatter Plot* Enter the time $t$ and the temperature of the water in the ceramic cup $T_1$ in your graphing utility and make a scatter plot. Enter the time $t$ and the temperature of the water in the Styrofoam cup $T_2$ in your graphing utility and make a scatter plot. If possible, plot both sets of data in the same viewing window. How are the scatter plots alike? How are the scatter plots different?

2. *Fit a Model to the Data* Use your graphing utility's exponential regression feature to find an exponential model for the temperature of the water in the ceramic cup. How well does the model fit the data? Create a table that shows the time $t$, the temperature of the water $T_1$, and the difference $d_1$ between $T_1$ and the room temperature. Now run the graphing utility's exponential regression program on the temperature difference data ($d_1$). For the model

$$y = a \cdot b^x$$

what are the values of $a$ and $b$?

3. *Change a Model's Form and Compare* Use the model from Question 2 and the fact that $b = e^{\ln b}$ to rewrite the model in the form $T = Ce^{kt}$. Now write the model for the temperature of a cooling object in the form

$$T = 70 + Ce^{kt}.$$

Construct a table that compares the values of this model to the actual data.

4. *Fit a Model to the Data* Repeat Questions 2 and 3 using the data for the Styrofoam cup ($T_2$).

5. *Compare the Models* Use your graphing utility to graph the models from Question 3 for both types of cups. What observations can you make? Which cup seems to keep the water warm for a longer period of time? Use the models to predict when the water in each cup will reach room temperature.

## CHAPTER SUMMARY

After studying this chapter, you should have acquired the following skills.
These skills are keyed to the Review Exercises that begin on page 353.
Answers to odd-numbered Review Exercises are given in the back of the book.

**5.1** · Match an exponential function with its graph.                                   Review Exercises 1–4

· Sketch the graph of an exponential function.                                          Review Exercises 9–16

· Use the compound interest formula.                                                    Review Exercises 17–20

· Use an exponential model to solve an application problem.                              Review Exercises 21, 22

**5.2** · Match a logarithmic function with its graph.                                   Review Exercises 5–8

· Evaluate a logarithmic expression.                                                    Review Exercises 23–28

· Write an exponential equation in logarithmic form.                                    Review Exercises 29, 30

· Find the domain, vertical asymptote, and *x*-intercept of a logarithmic model.        Review Exercises 33, 34

· Sketch the graph of a logarithmic function.                                           Review Exercises 31–34

· Use a logarithmic model to solve an application problem.                              Review Exercises 35–40

**5.3** · Evaluate a logarithm using the change-of-base formula.                        Review Exercises 41–44

· Rewrite logarithmic expressions using the properties of logarithms.                   Review Exercises 45–56

· Use properties of logarithms to evaluate a logarithmic expression.                    Review Exercises 57–64

· Construct a model using properties of exponents and properties of logarithms to relate two quantities.   Review Exercise 81

**5.4** · Solve an exponential equation.                                                Review Exercises 65–68, 75, 76

· Solve a logarithmic equation.                                                         Review Exercises 69–74

· Use an exponential or logarithmic model to solve an application problem.              Review Exercises 79, 80, 82, 83, 87, 89

**5.5** · Construct and use a model for exponential growth or exponential decay.        Review Exercises 77, 78

· Use a Gaussian or logistics growth model to solve an application problem.             Review Exercises 84, 85, 88

· Use a base 10 logarithmic model to solve an application problem.                      Review Exercise 86

## REVIEW EXERCISES

In Exercises 1–8, match the function with its graph. [The graphs are labeled (a), (b), (c), (d), (e), (f), (g), and (h).]

(a)

(b)

(c)

(d)

(e)

(f)

(g)

(h)
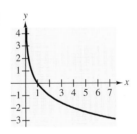

**1.** $f(x) = -3^x$

**2.** $f(x) = 3^{-x}$

**3.** $f(x) = -3^{-x}$

**4.** $f(x) = 2^x + 1$

**5.** $f(x) = \log_2 x$

**6.** $f(x) = \log_2(x - 1)$

**7.** $f(x) = -\log_2 x$

**8.** $f(x) = \log_2(x + 1)$

In Exercises 9–16, sketch the graph of the function.

**9.** $y = 4^x$

**10.** $y = e^{2x}$

**11.** $y = \left(\frac{1}{2}\right)^x$

**12.** $f(x) = \left(\frac{1}{2}\right)^{x+1}$

**13.** $f(x) = 3e^{0.2x}$

**14.** $f(x) = 4^{x-2}$

**15.** $y = 4^{-x^2}$

**16.** $f(x) = 100e^{-0.2x}$

In Exercises 17 and 18, complete the table to determine the balance $A$ for $P$ dollars invested at a rate $r$ for $t$ years and compounded $n$ times per year.

| $n$ | 1 | 2 | 4 | 12 | 365 | Continuous |
|---|---|---|---|---|---|---|
| $A$ | | | | | | |

**17.** $P = \$3500$,   $r = 10.5\%$,   $t = 10$ years

**18.** $P = \$2000$,   $r = 12\%$,   $t = 30$ years

In Exercises 19 and 20, complete the table to determine the amount $P$ that should be invested at a rate $r$ to produce a final balance of \$200,000 in $t$ years.

| $t$ | 1 | 10 | 20 | 30 | 40 | 50 |
|---|---|---|---|---|---|---|
| $P$ | | | | | | |

**19.** $r = 8\%$, compounded continuously

**20.** $r = 10\%$, compounded monthly

**21.** *Trust Fund*   Suppose the day your child was born you deposited \$50,000 in a trust fund that paid 8.75% interest, compounded continuously. Between what birthdays would your child become a millionaire?

**22.** *Population*   Suppose the population of a town is increasing at the rate of 5% per year. The population of the town in 1998 is 50,000.

(a) During which year will the population of the town reach 100,000?

(b) During which year will the population of the town reach 200,000?

In Exercises 23–28, evaluate the expression without using a calculator.

**23.** $\log_2 32$

**24.** $\log_9 3$

**25.** $\ln e^7$

**26.** $\log_a \dfrac{1}{a}$

**27.** $\ln e^{-1/2}$

**28.** $\ln 1$

In Exercises 29 and 30, use the definition of a logarithm to write the equation in logarithmic form.

**29.** $4^3 = 64$                **30.** $25^{3/2} = 125$

In Exercises 31 and 32, use the fact that $f$ and $g$ are inverses of each other to sketch their graphs in the same coordinate plane.

**31.** $f(x) = 10^x, \ g(x) = \log_{10} x$
**32.** $f(x) = e^x, \ g(x) = \ln x$

In Exercises 33 and 34, find the domain, vertical asymptote, and $x$-intercept of the logarithmic function and sketch its graph.

**33.** $f(x) = \log_3(x - 4)$      **34.** $g(x) = \ln(3 - x)$

**35.** *Human Memory Model*   Students in a sociology class were given an exam and then were retested monthly with an equivalent exam. The average score for the class was given by the human memory model

$$f(t) = 85 - 14 \log_{10}(t + 1), \quad 0 \le t \le 4$$

where $t$ is the time in months. How did the average score change over the 4-month period?

**36.** *Investment Time*   A principal $P$, invested at 8.75% and compounded continuously, increases to an amount that is $K$ times the original principal after $t$ years, where $t$ is given by

$$t = \frac{\ln K}{0.0875}.$$

Complete the following table and describe the result.

| $K$ | 1 | 2 | 3 | 4 | 6 | 8 | 10 |
|-----|---|---|---|---|---|---|----|
| $t$ |   |   |   |   |   |   |    |

*Human Memory Model*   In Exercises 37 and 38, use the following information. Students in a learning theory study were given an exam and were retested monthly for 6 months with an equivalent exam. The average scores for the class are given in the table, with $t = 1$ representing 1 month after the initial exam.

| $t$ | 1 | 2 | 3 | 4 | 5 | 6 |
|-----|---|---|---|---|---|---|
| Average score | 80.2 | 78.4 | 72.1 | 68.5 | 67.1 | 65.3 |

**37.** Make a scatter plot for the data. Find a mathematical model that represents the data.

**38.** Predict the average score after 8 months ($t = 8$) based on your model.

*Median Age of the United States Population*   In Exercises 39 and 40, use the model

$$A = 18.10 - 0.016t + 5.121 \ln t, \quad 10 \le t \le 80$$

which approximates the median age of the United States population from 1980 to 2050. In the model, assume $A$ is the median age and $t$ is the calendar year, with $t = 10$ corresponding to 1980.   (Source: U.S. Bureau of the Census)

**39.** Use the model to approximate the median ages in the United States in 1995 and 2010.

**40.** Based on your answers to Exercise 39, discuss the change in the median age in the United States from 1995 to 2010.

In Exercises 41–44, evaluate the logarithm using the change-of-base formula. Do each problem twice, once with common logarithms and once with natural logarithms. (Round your answer to three decimal places.)

**41.** $\log_3 10$                **42.** $\log_{1/4} 7$
**43.** $\log_{12} 200$             **44.** $\log_3 0.28$

In Exercises 45–50, use the properties of logarithms to write the expression as a sum, difference, and/or multiple of logarithms.

**45.** $\log_{10} \dfrac{x}{y}$        **46.** $\ln \dfrac{xy^2}{z}$

**47.** $\ln\left(x\sqrt{x - 3}\right)$      **48.** $\ln \sqrt[3]{\dfrac{x^3}{y^2}}$

**49.** $\log_{10}(y - 5)^3$           **50.** $\ln 2xyz$

In Exercises 51–56, write the expression as the logarithm of a single quantity.

**51.** $\log_4 2 + \log_4 3$
**52.** $\ln y + 2 \ln z$
**53.** $\frac{1}{2} \ln x$
**54.** $4 \log_3 x + \log_3 y - 2 \log_3 z$
**55.** $\ln x - \ln(x - 3) - \ln(x + 1)$
**56.** $\ln(x + 2) + 2 \ln x - 3 \ln(x + 4)$

In Exercises 57-60, approximate the logarithm using the properties of logarithms given that $\log_b 2 \approx 0.3562$, $\log_b 3 \approx 0.5646$, and $\log_b 5 \approx 0.8271$.

**57.** $\log_b 6$

**58.** $\log_b\left(\frac{4}{25}\right)$

**59.** $\log_b \sqrt{3}$

**60.** $\log_b 30$

In Exercises 61-64, find the *exact* value of the logarithm without using a calculator.

**61.** $\log_7 49$

**62.** $\log_6 \frac{1}{36}$

**63.** $\ln e^{3.2}$

**64.** $\ln \sqrt[5]{e^3}$

In Exercises 65-74, solve the equation.

**65.** $e^x = 8$

**66.** $e^{2x} = 17$

**67.** $4e^{-3x} = 146$

**68.** $e^{2x} - 7e^x + 10 = 0$

**69.** $\ln 3x = 8.2$

**70.** $2 \ln 4x = 15$

**71.** $-2 + \ln 5x = 0$

**72.** $\ln 4x^2 = 21$

**73.** $\ln x - \ln 3 = 2$

**74.** $\ln \sqrt{x + 1} = 2$

**75. *Demand Function*** The demand function for a certain product is given by

$$p = 600 - 0.3(e^{0.005x}).$$

Find the demand $x$ for a price of (a) $p = \$500$ and (b) $p = \$400$.

**76. *Demand Function*** The demand function for a certain product is given by

$$p = 4000\left(1 - \frac{3}{3 + e^{-0.004x}}\right).$$

Find the demand $x$ for a price of (a) $p = \$700$ and (b) $p = \$400$.

***Radioactive Decay*** In Exercises 77 and 78, complete the table for the radioactive isotope.

| Isotope | Half–Life | Initial Quantity | Amount After 1000 Years |
|---|---|---|---|
| **77.** C$^{14}$ | 5730 | 12 g | |
| **78.** Pu$^{230}$ | 24,360 | | 3.1 g |

**79. *Population*** The population $P$ of a city is given by

$$P = 270{,}000e^{0.019t}$$

where $t$ is the time in years, with $t = 0$ corresponding to 1990.

(a) Use a graphing utility to graph this equation.

(b) According to this model, in what year will the city have a population of 300,000?

**80. *Population*** The population $P$ of a city is given by

$$P = 50{,}000e^{kt}$$

where $t$ is the time in years, with $t = 0$ corresponding to 1990. In 1980, the population was 34,500.

(a) Find the value of $k$ and use this result to predict the population in the year 2020.

(b) Use a graphing utility to confirm the result of part (a).

**81. *Curve Fitting*** Use the properties of logarithms and exponentiation to find an equation that relates $x$ and $y$. (*Hint:* See Example 9 in Section 5.3.)

| $x$ | 1 | 2 | 3 | 4 | 5 | 6 |
|---|---|---|---|---|---|---|
| $y$ | 1 | 2.520 | 4.327 | 6.350 | 8.550 | 10.903 |

**82. *Bacteria Growth*** The number of bacteria $N$ in a culture is given by the model

$$N = 200e^{kt}$$

where $t$ is the time in hours, with $t = 0$ corresponding to the time when $N = 200$. When $t = 6$, $N = 500$. How long does it take the population to double in size?

**83.** *Learning Curve* The management at a factory has found that the maximum number of units a worker can produce in a day is 50. The learning curve for the number of units $N$ produced per day after a new employee has worked $t$ days is given by

$$N = 50(1 - e^{kt}).$$

After 20 days on the job, a particular worker produced 31 units in 1 day.

(a) Find the learning curve for this worker.

(b) How many days should pass before this worker is producing 45 units per day?

**84.** *Test Scores* The scores on a general aptitude test roughly follow a normal distribution given by

$$y = 0.0028e^{-[(x-300)^2]/20,000}, \quad 100 \le x \le 500.$$

Sketch the graph of this function. Estimate the average score on this test.

**85.** *Wildlife Management* A state parks and wildlife department releases 100 deer into a wilderness area. The department believes that the carrying capacity of the area is 500 deer and that the growth of the herd will be modeled by the logistics curve

$$P = \frac{500}{1 + 4e^{-kt}}, \quad 0 \le t$$

where $t$ is measured in years.

(a) Find $k$ if the herd size is 170 after 2 years.

(b) Find the population after 5 years.

**86.** *Earthquake Magnitudes* On the Richter scale, the magnitude $R$ of an earthquake of intensity $I$ is given by

$$R = \log_{10} \frac{I}{I_0}$$

where $I_0 = 1$ is the minimum intensity used for comparison. Find the intensity per unit of area for the following values of $R$.

(a) $R = 8.4$

(b) $R = 6.85$

(c) $R = 9.1$

**87.** *Thawing Time* Suppose that you take a 5-pound package of steaks out of a freezer at 11 A.M. Will the steaks be thawed in time to be grilled at 6 P.M.? Assume that the room temperature is 70° Fahrenheit and the freezer temperature is 20°. Use the formula

$$t = -3.8 \ln \frac{T - 70}{20 - 70}$$

where $t$ is the time (with $t = 0$ corresponding to 11 A.M.) and $T$ is the temperature of the package of steaks.

**88.** *Sales and Advertising* In 1997, after discontinuing all advertising for a product, the manufacturer of the product found that the sales began to drop according to the model

$$S = \frac{600,000}{1 + 0.4e^{kt}}$$

where $S$ represents the number of units sold and $t$ represents the calendar year, with $t = 0$ corresponding to 1997.

(a) Find $k$ if the company sold 400,000 units in 1999.

(b) According to this model, what will sales be in 2001? In 2002?

**89.** *Bacteria Growth* The number of bacteria $N$ in a culture is given by the model

$$N = 200e^{kt}$$

where $t$ is the time in hours, with $t = 0$ corresponding to the time when $N = 200$. When $t = 5$, there are 350 bacteria. How long does it take for the population to triple in size?

# Chapter Test

Take this test as you would take a test in class. After you are done, check your work against the answers given in the back of the book.

In Exercises 1–4, sketch the graph of the function.

**1.** $y = 2^x$

**2.** $y = 3^{-x}$

**3.** $y = \ln x$

**4.** $y = \log_3(x - 1)$

**5.** You deposit $30,000 into a fund that pays 7.25% interest, compounded continuously. When will the balance be greater than $100,000?

In Exercises 6–8, expand the logarithmic expression.

**6.** $\ln \dfrac{x^2 y^3}{z}$

**7.** $\log_{10} 3xyz^2$

**8.** $\ln\left(x \sqrt[3]{x - 2}\right)$

In Exercises 9 and 10, condense the expression.

**9.** $\ln y + 2 \ln z - 3 \ln x$

**10.** $\frac{2}{3}(\log_{10} x + \log_{10} y)$

In Exercises 11–14, solve the equation.

**11.** $e^{4x} = 21$

**12.** $e^{2x} - 8e^x + 12 = 0$

**13.** $-3 + \ln 4x = 0$

**14.** $\ln \sqrt{x + 2} = 3$

In Exercises 15–17, students in a psychology class were given an exam and then were retested monthly with an equivalent exam. The average score for the class is given by the human memory model $f(t) = 87 - 15 \log_{10}(t + 1)$, $0 \le t \le 4$, where $t$ is the time in months.

**15.** What was the average score on the original exam?

**16.** What was the average score after 2 months? After 4 months?

**17.** Suppose the students in this psychology class participated in a study that required that they continue taking an equivalent exam every 6 months for 2 years. If the model remained valid, what would the average score be after 12 months? After 18 months? What could this indicate about human memory?

**18.** The population $P$ of a city is given by $P = 70,000e^{0.023t}$, where $t = 0$ represents 2000. When will the city have a population of 100,000? Explain.

**19.** The number of bacteria $N$ in a culture is given by $N = 100e^{kt}$, where $t$ is the time in hours, with $t = 0$ corresponding to the time when $N = 100$. When $t = 8$, $N = 500$. How long does it take the population to double?

**20.** Carbon 14 has a half-life of 5730 years. You have an initial quantity of 10 grams. How many grams will remain after 10,000 years? After 20,000 years?

## Cumulative Test: Chapters 3–5

Take this test as you would take a test in class. After you are done, check your work against the answers given in the back of the book.

In Exercises 1–6, given $f(x) = x^2 + 1$ and $g(x) = 3x - 5$, write the indicated function.

**1.** $(f + g)(x)$        **2.** $(f - g)(x)$        **3.** $(fg)(x)$

**4.** $\left(\dfrac{f}{g}\right)(x)$       **5.** $(f \circ g)(x)$       **6.** $(g \circ f)(x)$

In Exercises 7–11, sketch the graph of the function. Describe the domain and range of the function.

**7.** $f(x) = (x - 2)^2 + 3$     **8.** $g(x) = \dfrac{2}{x - 3}$     **9.** $h(x) = 2^{-x}$

**10.** $f(x) = \log_4 (x - 1)$     **11.** $g(x) = \begin{cases} x + 5, & x < 0 \\ 5, & x = 0 \\ x^2 + 5, & x > 0 \end{cases}$

**12.** The profit for a company is given by

$$P = 200 + 10x - 0.001x^2$$

where $x$ is the number of units produced. What production level will yield a maximum profit?

In Exercises 13–15, perform the indicated operation and write the result in standard form.

**13.** $(10 + 2i)(3 - 4i)$     **14.** $(4 + 5i)^2$     **15.** $\dfrac{1 + 2i}{3 - i}$

**16.** Use the Quadratic Formula to solve $3x^2 - 5x + 7 = 0$.

**17.** Find all the zeros of $f(x) = x^4 + 10x^2 + 9$ given that $3i$ is a zero. Explain your reasoning.

In Exercises 18 and 19, solve the equation.

**18.** $e^{2x} - 11e^x + 24 = 0$     **19.** $\frac{1}{2} \ln(x - 3) = 4$

**20.** The population $P$ of a city is given by

$$P = 80{,}000e^{0.035t}$$

where $t$ is the time in years, with $t = 6$ corresponding to 1996. According to this model, in what year will the city have a population of 140,000?

# 6

# Systems of Equations and Inequalities

Consumer credit outstanding (in billions of dollars)

Year (0 ↔ 1990)

Steve Smith/FPG International

The amounts of U.S. consumer credit outstanding (in billions of dollars) held by finance companies $y_1$ and held by credit unions $y_2$ from 1988 to 1998 are shown in the table. Models for the data are given by

$$y_1 = -0.1123t^4 + 1.233t^3 - 0.97t^2 - 11.5t + 135.0,$$
$$-2 \le t \le 8 \qquad \text{and}$$

$$y_2 = -0.0685t^4 + 0.724t^3 - 0.32t^2 - 1.2t + 88.9,$$
$$-2 \le t \le 8 \qquad \text{where } t = 0 \text{ represents 1990.}$$

The graphs of these two models are shown above. If the graphs of the two models intersect, the amount of consumer credit outstanding for both would be the same. If the graphs do not intersect, what possible conclusion could you draw?   (Source: Federal Reserve Statistical Release, Consumer Credit Table G.19, Oct. 17, 1998 [Internet])

|       | 1988  | 1989  | 1990  | 1991  | 1992  | 1993  | 1994  | 1995  | 1996  | 1997  | 1998  |
|-------|-------|-------|-------|-------|-------|-------|-------|-------|-------|-------|-------|
| $y_1$ | 143.1 | 142.1 | 134.8 | 127.5 | 117.0 | 110.0 | 124.0 | 141.3 | 153.9 | 156.7 | 154.3 |
| $y_2$ | 83.1  | 88.3  | 90.4  | 88.5  | 88.3  | 95.4  | 108.2 | 125.7 | 136.1 | 147.6 | 149.4 |

The chapter project related to this information is on page 417.

359

### 6.1    Systems of Equations

**Objectives**

- Solve a system of equations by the method of substitution.
- Solve a system of equations graphically.
- Construct and use a system of equations to solve an application problem.

## The Method of Substitution

Up to this point in the text, most problems have involved either a function of one variable or a single equation in two variables. However, many problems in science, business, and engineering involve two or more equations in two or more variables. To solve such problems, you need to find solutions of a **system of equations.** Here is an example of a system of two equations in $x$ and $y$.

$$2x + \ y = 5 \qquad \text{Equation 1}$$

$$3x - 2y = 4 \qquad \text{Equation 2}$$

A **solution** of this system is an ordered pair that satisfies each equation in the system. Finding the set of all solutions is called **solving the system of equations.** For instance, the ordered pair $(2, 1)$ is a solution of this system. To check this, you can substitute 2 for $x$ and 1 for $y$ in *each* equation.

$$2x + y = 5 \qquad \text{Equation 1}$$

$$2(2) + 1 \overset{?}{=} 5 \qquad \text{Substitute 2 for } x \text{ and 1 for } y.$$

$$4 + 1 = 5 \qquad \text{Solution checks in Equation 1. } \checkmark$$

$$3x - 2y = 4 \qquad \text{Equation 2}$$

$$3(2) - 2(1) \overset{?}{=} 4 \qquad \text{Substitute 2 for } x \text{ and 1 for } y.$$

$$6 - 2 = 4 \qquad \text{Solution checks in Equation 2. } \checkmark$$

There are several different ways to solve systems of equations. In this chapter you will study three of the most common techniques, beginning with the **method of substitution.**

---

▶ Method of Substitution

1. *Solve* one of the equations for one variable in terms of the other.

2. *Substitute* the expression found in Step 1 into the other equation to obtain an equation of one variable.

3. *Solve* the equation obtained in Step 2.

4. *Back-substitute* the solution in Step 3 into the expression obtained in Step 1 to find the value of the other variable.

5. *Check* the solution to see that it satisfies *each* of the original equations.

Use a graphing utility to graph $y = -x + 4$ and $y = x - 2$ in the same viewing window. Use the trace feature to find the coordinates of the point of intersection. Are the coordinates the same as the solution found in Example 1? Explain.

## Study Tip

Because many steps are required to solve a system of equations, it is easy to make errors in arithmetic. We *strongly* suggest that you always *check your solution by substituting it into each equation in the original system.*

**Example 1**    Solving a System of Two Equations

Solve the following system of equations.

$$x + y = 4 \qquad \text{Equation 1}$$
$$x - y = 2 \qquad \text{Equation 2}$$

**Solution**

Begin by solving for $y$ in Equation 1.

$$y = 4 - x \qquad \text{Solve for } y \text{ in Equation 1.}$$

Next, substitute this expression for $y$ in Equation 2 and solve the resulting single-variable equation for $x$.

$$x - y = 2 \qquad \text{Equation 2}$$
$$x - (4 - x) = 2 \qquad \text{Substitute } 4 - x \text{ for } y.$$
$$x - 4 + x = 2 \qquad \text{Simplify.}$$
$$2x = 6 \qquad \text{Combine like terms.}$$
$$x = 3 \qquad \text{Divide both sides by 2.}$$

Finally, you can solve for $y$ by *back-substituting* $x = 3$ into the equation $y = 4 - x$, to obtain

$$y = 4 - x \qquad \text{Revised Equation 1}$$
$$y = 4 - 3 \qquad \text{Substitute 3 for } x.$$
$$y = 1 \qquad \text{Solve for } y.$$

The solution is the ordered pair $(3, 1)$. You can check this as follows.

*Check*

$$x + y = 4 \qquad \text{Equation 1}$$
$$3 + 1 \stackrel{?}{=} 4 \qquad \text{Substitute for } x \text{ and } y.$$
$$4 = 4 \qquad \text{Solution checks in Equation 1.} ✓$$
$$x - y = 2 \qquad \text{Equation 2}$$
$$3 - 1 \stackrel{?}{=} 2 \qquad \text{Substitute for } x \text{ and } y.$$
$$2 = 2 \qquad \text{Solution checks in Equation 2.} ✓$$

The term *back-substitution* implies that you work *backwards*. First you solve for one of the variables, and then you substitute that value *back* into one of the equations in the system to find the value of the other variable.

> **Example 2**    Solving a System by Substitution

A total of \$12,000 is invested in two funds paying 9% and 11% simple interest. If the yearly interest is \$1180, how much of the \$12,000 is invested at each rate?

**Solution**

*Verbal Model:*    | 9% fund | $+$ | 11% fund | $=$ | Total investment |

| 9% interest | $+$ | 11% interest | $=$ | Total interest |

*Labels:*    Amount in 9% fund $= x$                 (dollars)
Interest for 9% fund $= 0.09x$           (dollars)
Amount in 11% fund $= y$                 (dollars)
Interest for 11% fund $= 0.11y$          (dollars)
Total investment $= \$12,000$            (dollars)
Total interest $= \$1180$                (dollars)

*System:*    $x + y = 12{,}000$                 Equation 1

$0.09x + 0.11y = 1180$              Equation 2

To begin, it is convenient to multiply both sides of the second equation by 100 to obtain $9x + 11y = 118{,}000$. This eliminates the need to work with decimals.

$9x + 11y = 118{,}000$             Revised Equation 2

To solve this system, you can solve for $x$ in Equation 1.

$x = 12{,}000 - y$                 Solve for $x$ in Equation 1.

Then, substitute this expression for $x$ into Equation 2, and solve the resulting equation for $y$.

$$9x + 11y = 118{,}000$$             Revised Equation 2

$$9(12{,}000 - y) + 11y = 118{,}000$$    Substitute $12{,}000 - y$ for $x$.

$$108{,}000 - 9y + 11y = 118{,}000$$     Distributive Property

$$2y = 10{,}000$$                    Combine like terms.

$$y = 5000$$                         Divide both sides by 2.

Next, back-substitute the value $y = 5000$ to solve for $x$.

$x = 12{,}000 - y$                 Revised Equation 1

$x = 12{,}000 - 5000$              Substitute 5000 for $y$.

$x = 7000$                         Solve for $x$.

The solution is $(7000, 5000)$. Check this in the original statement of the problem.

The equations in Examples 1 and 2 are linear. The method of substitution can also be used to solve systems in which one or both of the equations are nonlinear.

### Example 3    Solving a System by Substitution: Two-Solution Case

Solve the following system of equations.

$$x^2 - x - y = \phantom{-}1 \qquad\qquad \text{Equation 1}$$

$$-x + y = -1 \qquad\qquad \text{Equation 2}$$

#### Solution

Begin by solving for $y$ in Equation 2 to obtain $y = x - 1$. Next, substitute this expression for $y$ into Equation 1 and solve for $x$.

$$x^2 - x - y = 1 \qquad\qquad \text{Equation 1}$$

$$x^2 - x - (x - 1) = 1 \qquad\qquad \text{Substitute } x - 1 \text{ for } y.$$

$$x^2 - 2x + 1 = 1 \qquad\qquad \text{Simplify.}$$

$$x^2 - 2x = 0 \qquad\qquad \text{Standard form}$$

$$x(x - 2) = 0 \qquad\qquad \text{Factor.}$$

$$x = 0, 2 \qquad\qquad \text{Solve for } x.$$

Back-substituting these values of $x$ to solve for the corresponding values of $y$ produces the two solutions $(0, -1)$ and $(2, 1)$. Check these solutions in the original system.

### Example 4    Solving a System by Substitution: No-Solution Case

Solve the following system of equations.

$$-x + y = 4 \qquad\qquad \text{Equation 1}$$

$$x^2 + y = 3 \qquad\qquad \text{Equation 2}$$

#### Solution

Begin by solving Equation 1 for $y$ to obtain $y = x + 4$. Next, substitute this expression for $y$ into Equation 2, and solve for $x$.

$$x^2 + y = 3 \qquad\qquad \text{Equation 2}$$

$$x^2 + (x + 4) = 3 \qquad\qquad \text{Substitute } x + 4 \text{ for } y.$$

$$x^2 + x + 1 = 0 \qquad\qquad \text{Simplify.}$$

$$x = \frac{-1 \pm \sqrt{1^2 - 4(1)(1)}}{2} \qquad\qquad \text{Quadratic Formula}$$

Because the discriminant is negative, the equation $x^2 + x + 1 = 0$ has no (real) solution. Hence, this system has no (real) solution.

## Graphical Approach to Finding Solutions

From Examples 2, 3, and 4, you can see that a system of two equations in two unknowns can have exactly one solution, more than one solution, or no solution. In practice, you can gain insight about the location and number of solutions of a system of equations by graphing each of the equations in the same coordinate plane. The solutions of the system correspond to the **points of intersection** of the graphs. For instance, in Figure 6.1(a) the two equations graph as two lines with a *single point* of intersection. The two equations in Example 3 graph as a parabola and a line with *two points* of intersection, as shown in Figure 6.1(b). Moreover, the two equations in Example 4 graph as a line and a parabola that happen to have *no points* of intersection, as shown in Figure 6.1(c).

One point of intersection          Two points of intersection          No points of intersection

(a) One solution          (b) Two solutions          (c) No solution

Figure 6.1

---

**Example 5**   Solving a System of Equations

Solve the following system of equations.

$$y = \ln x \qquad \text{Equation 1}$$

$$x + y = 1 \qquad \text{Equation 2}$$

### Solution

The graph of each equation is shown in Figure 6.2. From this, it is clear that there is only one point of intersection. Also, it appears that $(1, 0)$ is the solution point. You can confirm this by substituting in *both* equations.

*Check*

$$0 = \ln 1 \qquad \text{Equation 1 checks. } \checkmark$$

$$1 + 0 = 1 \qquad \text{Equation 2 checks. } \checkmark$$

Figure 6.2

## Applications

The total cost $C$ of producing $x$ units of a product typically has two components—the initial cost and the cost per unit. When enough units have been sold so that the total revenue $R$ equals the total cost, the sales are said to have reached the **break-even point.** You will find that the break-even point corresponds to the point of intersection of the cost and revenue curves.

> ### Example 6    An Application:  Break-Even Analysis
>
> A small business invests $10,000 in equipment to produce a product. Each unit of the product costs $0.65 to produce and is sold for $1.20. How many items must be sold before the business breaks even?

### Solution

The total cost of producing $x$ units is

| Total cost | = | Cost per unit | · | Number of units | + | Initial cost |
|---|---|---|---|---|---|---|

$$C = 0.65x + 10{,}000. \qquad \text{Equation 1}$$

The revenue obtained by selling $x$ units is

| Total revenue | = | Price per unit | · | Number of units |
|---|---|---|---|---|

$$R = 1.2x. \qquad \text{Equation 2}$$

Because the break-even point occurs when $R = C$, you have

$$1.2x = 0.65x + 10{,}000 \qquad \text{Equate } R \text{ and } C.$$

$$0.55x = 10{,}000 \qquad \text{Subtract } 0.65x \text{ from both sides.}$$

$$x = \frac{10{,}000}{0.55} \qquad \text{Divide both sides by 0.55.}$$

$$x \approx 18{,}182 \text{ units.} \qquad \text{Use a calculator.}$$

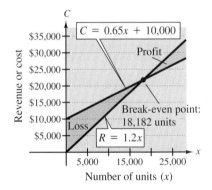

**Figure 6.3**

Note in Figure 6.3 that sales less than the break-even point correspond to an overall loss, whereas sales greater than the break-even point correspond to a profit.

---

Another way to view the solution in Example 6 is to consider the profit function

$$P = R - C.$$

The break-even point occurs when the profit is 0, which is the same as saying that $R = C$.

**Example 7**    CD Singles and Cassette Singles

From 1993 to 1996, the value of CD singles shipped each year in the United States was *increasing* and the value of cassette singles shipped each year was *decreasing.* Two models that approximate the values are

$$V = -112 + 47t \qquad \text{CD singles}$$

$$V = 415 - 37t \qquad \text{Cassette singles}$$

where $V$ is the value (in millions of dollars) shipped and $t = 3$ represents 1993 (see Figure 6.4). According to these two models, when would you expect the value of CD singles shipped to exceed the value of cassette singles shipped? (Source: Recording Industry Association of America)

**Solution**

Because the first equation has already been solved for $V$ in terms of $t$, you substitute this value into the second equation and solve for $t$, as follows.

$$-112 + 47t = 415 - 37t$$

$$47t + 37t = 415 + 112$$

$$84t = 527$$

$$t \approx 6.3$$

So, from the given models, you would expect that the value of CD singles shipped exceeded the value of cassette singles shipped sometime in 1996.

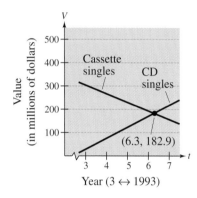

Figure 6.4

---

| Discussing the Concept | Interpreting Points of Intersection |
|---|---|

You plan to rent a 14-foot truck for a 2-day local move. At truck rental agency A, you can rent a truck for $29.95 per day plus $0.49 per mile. At agency B, you can rent a truck for $50 per day plus $0.25 per mile. The total cost y (in dollars) for the truck from agency A is

$$y = (\$29.95 \text{ per day})(2 \text{ days}) + 0.49x = 59.90 + 0.49x,$$

where x is the total number of miles the truck is driven. Write a total cost equation in terms of x and y for the total cost for the truck from agency B. Use a graphing utility to graph the two equations and find the point of intersection. Interpret the meaning of the point of intersection in the context of the problem. Which agency should you choose if you plan to travel a total of 100 miles over the 2-day move? Why? How does the situation change if you plan to drive 200 miles per day instead?

# Warm Up

The following warm-up exercises involve skills that were covered in earlier sections. You will use these skills in the exercise set for this section.

In Exercises 1–4, sketch the graph of the equation.

**1.** $y = -\frac{1}{3}x + 6$

**2.** $y = 2(x - 3)$

**3.** $x^2 + y^2 = 4$

**4.** $y = 5 - (x - 3)^2$

In Exercises 5–8, perform the indicated operations and simplify.

**5.** $(3x + 2y) - 2(x + y)$

**6.** $(-10u + 3v) + 5(2u - 8v)$

**7.** $x^2 + (x - 3)^2 + 6x$

**8.** $y^2 - (y + 1)^2 + 2y$

In Exercises 9 and 10, solve the equation.

**9.** $3x + (x - 5) = 15 + 4$

**10.** $y^2 + (y - 2)^2 = 2$

## 6.1    Exercises

In Exercises 1 and 2, decide whether each ordered pair is a solution of the system.

**1.** $x + 4y = -3$
$5x - y = 6$
(a) $(-1, -1)$
(b) $(1, -1)$

**2.** $2x - y = 2$
$x + 3y = 8$
(a) $(2, 1)$
(b) $(2, 2)$

In Exercises 3–12, use substitution to solve the system. Then use the graph to confirm your solution.

**3.** $x + y = -1$
$-2x + y = -7$

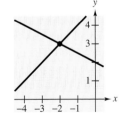

**4.** $x - y = -5$
$x + 2y = 4$

**5.** $x - y = -3$
$x^2 - y = -1$

**6.** $x^2 - y = 0$
$x^2 - 4x + y = 0$

**7.** $x - y = 0$
$x^3 - 5x + y = 0$

**8.** $y = x^3 - 3x^2 + 3$
$2x + y = 3$

**9.** $3x + y = 4$
$x^2 + y^2 = 16$

**10.** $3x - 4y = 18$
$x^2 + y^2 = 36$

**11.** $y = -x^2 + 1$
$y = x^2 - 1$

**12.** $y = x^2 - 3x - 4$
$y = -x^2 + 3x + 4$

In Exercises 13–32, use substitution to solve the system.

**13.**    $2x - y = -3$
$-3x - 4y = -1$

**14.**    $x + 2y = 1$
$5x - 4y = -23$

**15.** $2x - y + 2 = 0$
$4x + y - 5 = 0$

**16.** $6x - 3y - 4 = 0$
$x + 2y - 4 = 0$

**17.** $0.3x - 0.4y - 0.33 = 0$
$0.1x + 0.2y - 0.21 = 0$

**18.** $1.5x + 0.8y = 2.3$
$0.3x - 0.2y = 0.1$

**19.** $\frac{1}{5}x + \frac{1}{2}y = 8$
$x + y = 20$

**20.** $\frac{1}{2}x + \frac{3}{4}y = 10$
$\frac{3}{2}x - y = 4$

**21.** $x - y = 7$
$2x + y = 23$

**22.** $x - 2y = -2$
$3x - y = 6$

**23.** $y = 2x$
$y = x^2 + 1$

**24.** $x + y = 4$
$x^2 - y = 2$

**25.** $3x - 7y + 6 = 0$
$x^2 - y^2 = 4$

**26.** $x^2 + y^2 = 25$
$2x + y = 10$

**27.** $x^2 + y^2 = 5$
$x - y = 1$

**28.** $x^2 + y = 4$
$2x - y = 1$

**29.** $y = x^4 - 2x^2 + 1$
$y = 1 - x^2$

**30.** $y = x^3 - 2x^2 + x - 1$
$y = -x^2 + 3x - 1$

**31.** $xy - 2 = 0$
$y = \sqrt{x - 1}$

**32.** $xy = 3$
$y = \sqrt{x - 2}$

In Exercises 33–40, use a graphing utility to find graphically the point of intersection of the graphs. Then confirm your solution algebraically.

**33.** $y = x^2 + 3x - 1$
$y = -x^2 - 2x + 2$

**34.** $y = -2x^2 + x - 1$
$y = x^2 - 2x - 1$

**35.** $x - y + 3 = 0$
$x^2 - 4x + 7 = y$

**36.** $x - y = 3$
$x - y^2 = 1$

**37.** $y = e^x$
$x - y + 1 = 0$

**38.** $y = \sqrt{x}$
$y = x$

**39.** $2x - y + 3 = 0$
$x^2 + y^2 - 4x = 0$

**40.** $x^2 + y^2 = 8$
$y = x^2 + 4$

*Break-Even Analysis*    In Exercises 41–44, find the sales necessary to break even $(R = C)$ for the given cost $C$ of producing $x$ units and the given revenue $R$ obtained by selling $x$ units. (Round your answer to the nearest whole unit.)

**41.** $C = 8650x + 250{,}000;\ R = 9950x$

**42.** $C = 5.5\sqrt{x} + 10{,}000;\ R = 3.29x$

**43.** $C = 2.65x + 350{,}000;\ R = 4.15x$

**44.** $C = 0.08x + 50{,}000;\ R = 0.25x$

**45.** *Break-Even Point*    Suppose you are setting up a small business and have invested \$16,000 to produce an item that will sell for \$5.95. If each unit can be produced for \$3.45, how many units must you sell to break even?

**46.** *Break-Even Point*    Suppose you are setting up a small business and have made an initial investment of \$5000. The unit cost of the product is \$21.60, and the selling price is \$34.10. How many units must you sell to break even?

**47.** *Comparing Populations*    From 1990 to 1996, the population of the midwestern part of the United States was growing more slowly than that of the western part. Two models that represent the populations of the two regions are

$$P = 59.7 + 0.5t - 0.01t^2 \qquad \text{Midwest}$$

$$P = 53.1 + 0.93t \qquad \text{West}$$

where $P$ is the population in millions and $t = 0$ represents 1990. Use a graphing utility to determine when the population of the West will overtake the population of the Midwest.    (Source: U.S. Bureau of the Census)

**48.** *Comparing Populations*    In 1994, Detroit was the tenth most populated city in the United States and San Jose was the eleventh. The populations of Detroit and San Jose from 1990 to 1994 can be approximated by the models

$$P = 1027.9 - 8.4t - 0.143t^2 \qquad \text{Detroit}$$

$$P = 781.7 + 9.2t - 0.071t^2 \qquad \text{San Jose}$$

where $P$ is the population (within the city limits) in thousands and $t = 0$ represents 1990. Use a graphing utility to determine if the population of San Jose will overtake the population of Detroit.    (Source: U.S. Bureau of the Census)

**49.** *Gross National Product*    The annual gross national products $G$ (in billions of dollars) of the United States and Japan from 1985 to 1995 can be approximated by the models

$$G = 2692.1 + 291.26t + 0.74t^2 \qquad \text{United States}$$

$$G = 732.6 + 430.61t - 8.84t^2 \qquad \text{Japan}$$

where $t = 5$ represents 1985. Use a graphing utility to determine if the annual gross national product of Japan will exceed that of the United States. (Source: U.S. Arms Control and Disarmament Agency)

**50.** *Advertising Expenditures*    The amount of money $y$ (in millions of dollars) spent on advertising in newspapers and on television from 1991 to 1996 can be modeled by

$$y = 29{,}813 + 159.1t + 222.2t^2 \qquad \text{Newspapers}$$

$$y = 26{,}481 + 787.5t + 249.6t^2 \qquad \text{Television}$$

where $t = 1$ represents 1991. The amount of money spent for television advertising is growing faster than the amount of money spent for newspaper advertising. Use a graphing utility to determine when television advertising expenditures overtook newspaper advertising expenditures. (Source: McCann-Erickson, Inc.)

**51.** *Investment Portfolio*    A total of $25,000 is invested in two funds paying 8% and 8.5% simple interest. If the yearly interest is $2060, how much of the $25,000 is invested at each rate?

**52.** *Investment Portfolio*    A total of $18,000 is invested in two funds paying 7.75% and 8.25% simple interest. If the yearly interest is $1455, how much of the $18,000 is invested at each rate?

**53.** *Choice of Two Jobs*    You are offered two different sales jobs. One company offers a straight commission of 6% of the sales. The other company offers a salary of $250 per week *plus* 3% of the sales. How much would you have to sell in a week in order to make the straight commission offer better?

**54.** *Choice of Two Jobs*    You are offered two different sales jobs. One company offers an annual salary of $20,000 *plus* a year-end bonus of 1% of your total sales. The other company offers a salary of $15,000 *plus* a year-end bonus of 2% of your total sales. How much would you have to sell in a year in order to make the second offer better?

**55.** *Record Store or Other Stores?*    The purchases $y$ (in percent) of sound recordings from record stores and other stores by consumers from 1991 to 1996 can be approximated by the models

$$y = 66.0 - 3.76t + 0.179t^2 \qquad \text{Record stores}$$

$$y = 34.3 + 3.14t - 0.125t^2 \qquad \text{Other stores}$$

where $t = 1$ represents 1991. According to these models, when will the purchases from other stores overtake record store purchases? (Source: Recording Industry Association of America, Inc.)

## 6.2     Linear Systems in Two Variables

### Objectives

- Solve a linear system by the method of elimination.
- Interpret the solution of a linear system graphically.
- Construct and use a linear system to solve an application problem.

# The Method of Elimination

In Section 6.1, you studied two methods for solving a system of equations: substitution and graphing. In this section, you will study a third method called the **method of elimination.** The key step in the method of elimination is to obtain, for one of the variables, coefficients that differ only in sign so that *adding* the two equations eliminates this variable. The following system provides an example.

$$3x + 5y = \phantom{-}7 \qquad \text{Equation 1}$$
$$\underline{-3x - 2y = -1} \qquad \text{Equation 2}$$
$$3y = \phantom{-}6 \qquad \text{Add equations.}$$

Note that by adding the two equations, you eliminate the variable $x$ and obtain a single equation in $y$. Solving this equation for $y$ produces $y = 2$, which you can then back-substitute into one of the original equations to solve for $x$.

### Example 1     The Method of Elimination

Solve the following system of linear equations.

$$3x + 2y = 4 \qquad \text{Equation 1}$$
$$5x - 2y = 8 \qquad \text{Equation 2}$$

### Solution

Because the coefficients for $y$ differ only in sign, you can eliminate $y$ by adding the two equations.

$$3x + 2y = \phantom{-}4 \qquad \text{Equation 1}$$
$$\underline{5x - 2y = \phantom{-}8} \qquad \text{Equation 2}$$
$$8x \phantom{+ 2y} = 12 \qquad \text{Add equations.}$$

So, $x = \frac{3}{2}$. By back-substituting this value into the first equation, you can solve for $y$, as follows.

$$3x + 2y = 4 \qquad \text{Equation 1}$$
$$3\left(\tfrac{3}{2}\right) + 2y = 4 \qquad \text{Substitute } \tfrac{3}{2} \text{ for } x.$$
$$y = -\tfrac{1}{4} \qquad \text{Solve for } y.$$

The solution is $\left(\frac{3}{2}, -\frac{1}{4}\right)$. Check this in the original system.

### Study Tip

The method of substitution can also be used to solve the system in Example 1. Use substitution to solve the system. Which method do you think is easier? Many people find that the method of elimination is more efficient.

| Example 2 | The Method of Elimination |

Solve the following system of linear equations.

$$2x - 3y = -7 \qquad \text{Equation 1}$$

$$3x + y = -5 \qquad \text{Equation 2}$$

## Study Tip

To obtain coefficients (for one of the variables) that differ only in sign, you often need to multiply one or both of the equations by a suitable constant.

### Solution

For this system, you can obtain coefficients that differ only in sign by multiplying the second equation by 3.

$$2x - 3y = -7 \quad \Longrightarrow \quad 2x - 3y = -7 \qquad \text{Equation 1}$$

$$\underline{3x + y = -5} \quad \Longrightarrow \quad \underline{9x + 3y = -15} \qquad \text{Multiply Equation 2 by 3.}$$

$$11x \qquad = -22 \qquad \text{Add equations.}$$

By dividing both sides by 11, you can see that $x = -2$. By back-substituting this value of $x$ into Equation 1, you can solve for $y$.

$$2x - 3y = -7 \qquad \text{Equation 1}$$

$$2(-2) - 3y = -7 \qquad \text{Substitute } -2 \text{ for } x.$$

$$-3y = -3 \qquad \text{Add 4 to both sides.}$$

$$y = 1 \qquad \text{Solve for } y.$$

The solution is $(-2, 1)$. Check this in the original system, as follows.

*Check*

$$2(-2) - 3(1) \overset{?}{=} -7 \qquad \text{Substitute into Equation 1.}$$

$$-4 - 3 = -7 \qquad \text{Equation 1 checks. } \checkmark$$

$$3(-2) + 1 \overset{?}{=} -5 \qquad \text{Substitute into Equation 2.}$$

$$-6 + 1 = -5 \qquad \text{Equation 2 checks. } \checkmark$$

In Example 2, the two systems of linear equations

$$2x - 3y = -7 \qquad \text{and} \qquad 2x - 3y = -7$$

$$3x + y = -5 \qquad\qquad 9x + 3y = -15$$

are called **equivalent** because they have precisely the same solution set. The operations that can be performed on a system of linear equations to produce an equivalent system are (1) interchanging any two equations, (2) multiplying an equation by a nonzero constant, and (3) adding a multiple of one equation to any other equation in the system.

▶ The Method of Elimination

To use the **method of elimination** to solve a system of two linear equations in $x$ and $y$, use the following steps.

1. Obtain coefficients for $x$ (or $y$) that differ only in sign by multiplying all terms of one or both equations by suitably chosen constants.

2. Add the equations to eliminate one variable and solve the resulting equation.

3. Back-substitute the value obtained in Step 2 into either of the original equations and solve for the other variable.

4. Check your solution in both of the original equations.

---

**Example 3**     The Method of Elimination

Solve the following system of linear equations.

$$5x + 3y = \phantom{0}9 \qquad \text{Equation 1}$$

$$2x - 4y = 14 \qquad \text{Equation 2}$$

**Solution**

You can obtain coefficients that differ only in sign by multiplying Equation 1 by 4 and multiplying Equation 2 by 3.

$$5x + 3y = \phantom{0}9 \quad\Longrightarrow\quad 20x + 12y = 36 \qquad \text{Multiply Equation 1 by 4.}$$

$$2x - 4y = 14 \quad\Longrightarrow\quad \phantom{0}6x - 12y = 42 \qquad \text{Multiply Equation 2 by 3.}$$

$$26x \phantom{+ 12yyy} = 78 \qquad \text{Add equations.}$$

From this equation, you can see that $x = 3$. By back-substituting this value of $x$ into the second equation, you can solve for $y$, as follows.

$$2x - 4y = 14 \qquad \text{Equation 2}$$

$$2(3) - 4y = 14 \qquad \text{Substitute 3 for } x.$$

$$-4y = 8$$

$$y = -2 \qquad \text{Solve for } y.$$

The solution is $(3, -2)$. Check this in the original system.

Remember that you can check the solution of a system of equations graphically. For instance, to check the solution found in Example 3, sketch the graphs of both equations on the same screen, as shown in Figure 6.5. Notice that the two lines intersect at $(3, -2)$.

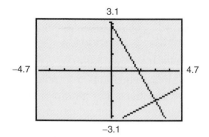

Figure 6.5

# Graphical Interpretation of Solutions

It is possible for a *general* system of equations to have exactly one solution, two or more solutions, or no solution. If a system of *linear* equations has two different solutions, it must have an *infinite* number of solutions. To see why this is true, consider the following graphical interpretations of systems of two linear equations in two variables. (Remember that the graph of a linear equation in two variables is a straight line.)

> ► **Graphical Interpretation of Solutions**
>
> For a system of two linear equations in two variables, the number of solutions is given by one of the following.
>
> | *Number of Solutions* | *Graphical Interpretation* |
> |---|---|
> | **1.** Exactly one solution | The two lines intersect at one point. |
> | **2.** Infinitely many solutions | The two lines are identical. |
> | **3.** No solution | The two lines are parallel. |
>
> These graphical interpretations are shown in Figure 6.6.

*Consistent*
*Two lines that intersect*
*Single point of intersection*
Figure 6.6

*Consistent*
*Two lines that coincide*
*Infinitely many points*

*Inconsistent*
*Two parallel lines*
*No point of intersection*

A system of linear equations is called **consistent** if it has at least one solution. A consistent system with exactly one solution is **independent.** A consistent system with infinitely many solutions is **dependent.** A system of linear equations is called **inconsistent** if it has no solution. In Examples 4 and 5, note how you can use the method of elimination to determine that a system of linear equations has no solution or infinitely many solutions.

**Figure 6.7**    *No Solution*

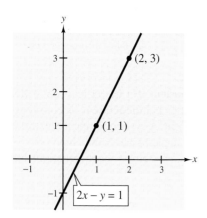

**Figure 6.8**    *Infinite Number of Solutions*

| Example 4 | The Method of Elimination: No-Solution Case |

Solve the following system of linear equations.

$$x - 2y = 3 \qquad \text{Equation 1}$$
$$-2x + 4y = 1 \qquad \text{Equation 2}$$

**Solution**

To obtain coefficients that differ only in sign, multiply Equation 1 by 2.

| | | |
|---|---|---|
| $x - 2y = 3$ | ⟹ | $2x - 4y = 6$ |
| $-2x + 4y = 1$ | ⟹ | $-2x + 4y = 1$ |
| | | $0 = 7$ |

Multiply Equation 1 by 2.

Equation 2

False statement

Because there are no values of $x$ and $y$ for which $0 = 7$, you can conclude that the system is inconsistent and has no solution. The lines corresponding to the two equations given in this system are shown in Figure 6.7. Note that the two lines are parallel, and therefore have no point of intersection.

In Example 4, note that the occurrence of a false statement, such as $0 = 7$, indicates that the system has no solution. In the next example, note that the occurrence of a statement that is true for all values of the variables, such as $0 = 0$, indicates that the system has infinitely many solutions.

| Example 5 | The Method of Elimination: Many-Solutions Case |

Solve the following system of linear equations.

$$2x - y = 1 \qquad \text{Equation 1}$$
$$4x - 2y = 2 \qquad \text{Equation 2}$$

**Solution**

To obtain coefficients that differ only in sign, multiply the second equation by $-\frac{1}{2}$.

| | | |
|---|---|---|
| $2x - y = 1$ | ⟹ | $2x - y = 1$ |
| $4x - 2y = 2$ | ⟹ | $-2x + y = -1$ |
| | | $0 = 0$ |

Equation 1

Multiply Equation 2 by $-\frac{1}{2}$.

Add equations.

Because the two equations turn out to be equivalent (have the same solution set), you can conclude that the system has infinitely many solutions. The solution set consists of all points $(x, y)$ lying on the line $2x - y = 1$ as shown in Figure 6.8. To represent the solution set as an ordered pair, let $x = a$, where $a$ is any real number. Then $y = 2a - 1$ and the solution set can be written as $(a, 2a - 1)$.

Example 6 illustrates a strategy for solving a system of linear equations that has decimal coefficients.

### Example 6    A Linear System Having Decimal Coefficients

Solve the following system of linear equations.

| | |
|---|---|
| $0.02x - 0.05y = -0.38$ | Equation 1 |
| $0.03x + 0.04y = \phantom{-}1.04$ | Equation 2 |

**Solution**

Because the coefficients in this system have two decimal places, you can begin by multiplying each equation by 100. (This produces a system in which the coefficients are all integers.)

| | |
|---|---|
| $2x - 5y = -38$ | Revised Equation 1 |
| $3x + 4y = \phantom{-}104$ | Revised Equation 2 |

Now, to obtain coefficients that differ only in sign, multiply the first equation by 3 and multiply the second equation by $-2$.

| | | | |
|---|---|---|---|
| $2x - 5y = -38$ | ⟹ | $6x - 15y = -114$ | Multiply Equation 1 by 3. |
| $3x + 4y = \phantom{-}104$ | ⟹ | $-6x - \phantom{1}8y = -208$ | Multiply Equation 2 by $-2$. |
| | | $-23y = -322$ | Add equations. |

So, you can conclude that

$$y = \frac{-322}{-23} = 14.$$

Back-substituting this value into Equation 2 produces the following.

| | |
|---|---|
| $3x + 4y = 104$ | Equation 2 |
| $3x + 4(14) = 104$ | Substitute 14 for $y$. |
| $3x = 48$ | |
| $x = 16$ | Solve for $x$. |

The solution is $(16, 14)$. Check this in the original system, as follows.

*Check*

| | |
|---|---|
| $0.02(16) - 0.05(14) \overset{?}{=} -0.38$ | Substitute into Equation 1. |
| $0.32 - 0.70 = -0.38$ | Equation 1 checks. ✓ |
| $0.03(16) + 0.04(14) \overset{?}{=} 1.04$ | Substitute into Equation 2. |
| $0.48 + 0.56 = 1.04$ | Equation 2 checks. ✓ |

## Applications

At this point, you may be asking the question "How can I tell which application problems can be solved using a system of linear equations?" The answer comes from the following considerations.

**1.** Does the problem involve more than one unknown quantity?

**2.** Are there two (or more) equations or conditions to be satisfied?

If one or both of these conditions occur, the appropriate mathematical model for the problem may be a system of linear equations. Example 7 shows how to construct such a model.

**Example 7**   An Application of a Linear System

An airplane flying into a headwind travels the 2000-mile flying distance between two cities in 4 hours and 24 minutes. On the return flight, the same distance is traveled in 4 hours. Find the ground speed of the plane and the speed of the wind, assuming that both remain constant.

### Solution

The two unknown quantities are the speeds of the wind and the plane. If $r_1$ is the speed of the plane and $r_2$ is the speed of the wind, then

$$r_1 - r_2 = \text{speed of the plane } against \text{ the wind}$$

$$r_1 + r_2 = \text{speed of the plane } with \text{ the wind}$$

as shown in Figure 6.9. Using the formula

$$\text{Distance} = (\text{rate})(\text{time})$$

for these two speeds, you obtain the following equations.

$$2000 = (r_1 - r_2)\left(4 + \frac{24}{60}\right)$$

$$2000 = (r_1 + r_2)(4)$$

These two equations simplify as follows.

$$5000 = 11r_1 - 11r_2 \qquad \text{Equation 1}$$

$$500 = r_1 + r_2 \qquad \text{Equation 2}$$

By elimination, the solution is

$$r_1 = \frac{5250}{11} \approx 477.27 \text{ miles per hour}$$

$$r_2 = \frac{250}{11} \approx 22.73 \text{ miles per hour.}$$

Original flight

$$r_1 - r_2$$

Return flight

$$r_1 + r_2$$

**Figure 6.9**

Example 8    Finding the Point of Equilibrium

The demand and supply functions for a certain type of calculator are given by

$$p = 150 - 0.00001x \qquad \text{Demand equation}$$

$$p = 60 + 0.00002x \qquad \text{Supply equation}$$

where $p$ is the price in dollars and $x$ represents the number of units. Find the point of equilibrium for this market. The point of equilibrium is the price $p$ and the number of units $x$ that satisfy both the demand and supply equations.

### Solution

Begin by substituting the value of $p$ given in Equation 2 into Equation 1.

$$p = 150 - 0.00001x \qquad \text{Equation 1}$$

$$60 + 0.00002x = 150 - 0.00001x \qquad \text{Substitute } 60 + 0.00002x \text{ for } p.$$

$$0.00003x = 90 \qquad \text{Add } 0.00001x \text{ and subtract } 60.$$

$$x = 3{,}000{,}000 \qquad \text{Divide both sides by } 0.00003.$$

So, the point of equilibrium occurs when the demand and supply are each 3 million units. (See Figure 6.10.) The price that corresponds to this $x$-value is obtained by back-substituting $x = 3{,}000{,}000$ into either of the original equations. For instance, back-substituting into the first equation produces

$$p = 150 - 0.00001(3{,}000{,}000) = 150 - 30 = \$120.$$

Try back-substituting $x = 3{,}000{,}000$ into the second equation to see that you obtain the same price.

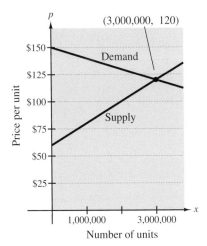

Figure 6.10

---

| Discussing the Concept | Creating Consistent and Inconsistent Systems |

For each of the following systems, find the value of $k$ such that the system has infinitely many solutions.

**a.** $4x + 3y = -8$
    $x + ky = -2$

**b.** $3x - 12y = 9$
    $x - 4y = k$

Is it possible to find values of $k$ such that the systems have no solutions? Explain why or why not for each system. Is it possible to find values of $k$ such that the systems have unique solutions? Explain why or why not for each system. Summarize the differences between these two systems of equations.

# Warm Up

The following warm-up exercises involve skills that were covered in earlier sections. You will use these skills in the exercise set for this section.

In Exercises 1 and 2, sketch the graph of the equation.

**1.** $2x + y = 4$          **2.** $5x - 2y = 3$

In Exercises 3 and 4, find an equation of the line passing through the two points.

**3.** $(-1, 3), (4, 8)$      **4.** $(2, 6), (5, 1)$

In Exercises 5 and 6, determine the slope of the line.

**5.** $3x + 6y = 4$      **6.** $7x - 4y = 10$

In Exercises 7–10, determine whether the lines represented by the pair of equations are parallel, perpendicular, or neither.

**7.** $2x - 3y = -10$
$\phantom{xx}3x + 2y = \phantom{-}11$

**8.** $\phantom{xx}4x - 12y = 5$
$\phantom{xxx}-2x + \phantom{1}6y = 3$

**9.** $5x + \phantom{x}y = 2$
$\phantom{xx}3x + 2y = 1$

**10.** $\phantom{xx}x - 3y = 2$
$\phantom{xxx}6x + 2y = 4$

## 6.2    Exercises

In Exercises 1–10, solve the linear system by elimination. Then use the graph to confirm your solution. Copy the graphs and label each line with the appropriate equation.

**1.** $3x - 2y = 2$
$\phantom{x}x + 2y = 6$

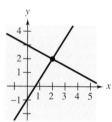

**2.** $-x + 3y = \phantom{-}2$
$\phantom{xx}x - 4y = -4$

**3.** $\phantom{x}x - 4y = 2$
$2x + \phantom{x}y = 4$

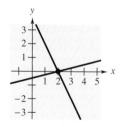

**4.** $2x - \phantom{x}y = \phantom{2}2$
$\phantom{x}4x + 3y = 24$

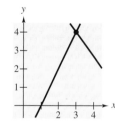

**5.** $\phantom{xx}x - \phantom{x}y = 1$
$\phantom{xx}-2x + 2y = 5$

**6.** $3x + 2y = \phantom{1}2$
$\phantom{x}6x + 4y = 14$

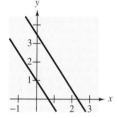

**7.** $\phantom{xx}3x - 2y = \phantom{-1}6$
$\phantom{xx}-6x + 4y = -12$

**8.** $2x + \phantom{1}4y = \phantom{2}8$
$\phantom{x}6x + 12y = 24$

**9.** $9x - 3y = -1$
$3x + 6y = -5$

**10.** $5x + 3y = 18$
$2x - 7y = -1$

In Exercises 11–30, solve the system by elimination.

**11.** $4x - 3y = 11$
$-6x + 3y = 3$

**12.** $3x - 5y = 2$
$2x + 5y = 13$

**13.** $3x - y = 17$
$5x + 5y = -5$

**14.** $x + 7y = 12$
$3x - 5y = 10$

**15.** $3x + 2y = 10$
$2x + 5y = 3$

**16.** $8r + 16s = 20$
$16r + 50s = 55$

**17.** $2u + v = 120$
$u + 2v = 120$

**18.** $5u + 6v = 24$
$3u + 5v = 18$

**19.** $6r - 5s = 3$
$-1.2r + s = 0.5$

**20.** $1.8x + 1.2y = 4$
$9x + 6y = 3$

**21.** $\dfrac{x}{4} + \dfrac{y}{6} = 1$
$x - y = 3$

**22.** $\dfrac{2}{3}x + \dfrac{1}{6}y = \dfrac{2}{3}$
$4x + y = 4$

**23.** $\dfrac{x + 3}{4} + \dfrac{y - 1}{3} = 1$
$x - y = 3$

**24.** $\dfrac{x - 1}{2} + \dfrac{y + 2}{3} = 4$
$x - 2y = 5$

**25.** $2.5x - 3y = 1.5$
$10x - 12y = 6$

**26.** $3.5x - 2y = 2.6$
$7x - 2.5y = 4$

**27.** $0.05x - 0.03y = 0.21$
$0.07x + 0.02y = 0.16$

**28.** $0.02x - 0.05y = -0.19$
$0.03x + 0.04y = 0.52$

**29.** $4b + 3m = 3$
$3b + 11m = 13$

**30.** $3b + 3m = 7$
$3b + 5m = 3$

In Exercises 31 and 32, the graphs of the two equations appear to be parallel. Are they? Explain your reasoning.

**31.** $200y - x = 200$
$199y - x = -198$

**32.** $25x - 24y = 0$
$13x - 12y = 120$

**33.** *Airplane Speed*    An airplane flying into a headwind travels the 1800-mile flying distance between two cities in 3 hours and 36 minutes. On the return flight, the distance is traveled in 3 hours. Find the ground speed of the plane and the speed of the wind, assuming that both remain constant.

**34.** *Airplane Speed*    Two planes start from the same airport and fly in opposite directions. The second plane starts one-half hour after the first plane, but its speed is 50 miles per hour faster. Find the ground speed of each plane if, 2 hours after the first plane starts its flight, the planes are 2000 miles apart.

**35.** *Acid Mixture*    Ten gallons of a 30% acid solution is obtained by mixing a 20% solution with a 50% solution. How much of each must be used?

**36.** *Fuel Mixture*    Five hundred gallons of 89-octane gasoline is obtained by mixing 87-octane gasoline with 90-octane gasoline. How much of each must be used? (Octane ratings can be interpreted as percents. A high octane rating indicates that the gasoline is knock resistant. A low octane rating indicates that the gasoline is knock prone.)

**37.** *Investment Portfolio*    A total of $12,000 is invested in two corporate bonds that pay 10.5% and 12% simple interest. The annual interest is $1380. How much is invested in each bond?

**38.** *Investment Portfolio*    A total of $32,000 is invested in two municipal bonds that pay 5.75% and 6.25% simple interest. The annual interest is $1930. How much is invested in each bond?

**39.** *Ticket Sales* You are the manager of a theater. On Saturday morning you are going over the ticket sales for Friday evening. A total of 750 tickets were sold. The tickets for adults and children sold for $7.50 and $4.00, respectively, and the receipts for the performance were $3808.50. However, your assistant manager did not record how many of each type of ticket were sold. From the information you have, can you determine how many of each type were sold?

**40.** *Shoe Sales* Suppose you are the manager of a shoe store. On Sunday morning you are going over the receipts for the previous week's sales. A total of 240 pairs of tennis shoes were sold. One style sold for $66.95 and the other sold for $84.95. The total receipts were $17,652. The cash register that was supposed to keep track of the number of each type of shoe sold malfunctioned. Can you recover the information? If so, how many of each type were sold?

*Supply and Demand* In Exercises 41–44, find the point of equilibrium for each pair of supply and demand equations.

| *Demand* | *Supply* |
|---|---|
| **41.** $p = 56 - 0.0001x$ | $p = 22 + 0.00001x$ |
| **42.** $p = 60 - 0.00001x$ | $p = 15 + 0.00004x$ |
| **43.** $p = 140 - 0.00002x$ | $p = 80 + 0.00001x$ |
| **44.** $p = 400 - 0.0002x$ | $p = 225 + 0.0005x$ |

**45.** *U.S. Exports and General Imports* The amounts of U.S. exports and general imports from 1993 to 1996 are shown in the table. The numbers in the table represent billions of dollars. (Source: U.S. Bureau of the Census)

| Year | 1993 | 1994 | 1995 | 1996 |
|---|---|---|---|---|
| Exports | 465.1 | 512.6 | 584.7 | 624.8 |
| Imports | 580.7 | 663.3 | 743.4 | 791.4 |

(a) Find a linear equation that represents the exports and imports.

(b) Assuming that the amounts for the given 4 years are representative of future years, will the exports ever equal the imports?

**46.** *Students Per Computer* From 1993 to 1997, the number of students per computer in elementary schools in the United States can be approximated by the models

$$y = 13.9 - 2.01t + 0.143t^2 \qquad \text{Public schools}$$
$$y = 14.8 - 1.52t + 0.021t^2 \qquad \text{Private schools}$$

where $y$ represents the average number of students per computer and $t = 0$ represents 1993. (Source: Market Data Retrieval)

(a) Use a graphing utility to graph this system.

(b) Use the graphs to predict when the number of students per computer for public elementary schools will equal the number of students per computer for private elementary schools.

**47.** *Delivery Service* A shipping department hires two delivery services to deliver 2400 packages. One service will deliver three times as many packages as the other. Find the number delivered by each.

**48.** *Reasoning* Design a system of two linear equations containing two variables that has (a) exactly one solution, (b) infinitely many solutions, and (c) no solutions. Solve each system algebraically and graph each system.

*Fitting a Line to Data* In Exercises 49–52, find the least squares regression line $y = ax + b$ for the points $(x_1, y_1), (x_2, y_2), \ldots, (x_n, y_n)$. To find the line, solve the system for $a$ and $b$. (If you are unfamiliar with summation notation, look at the discussion in Section 8.1.)

$$nb + \left(\sum_{i=1}^{n} x_i\right)a = \sum_{i=1}^{n} y_i$$
$$\left(\sum_{i=1}^{n} x_i\right)b + \left(\sum_{i=1}^{n} x_i^2\right)a = \sum_{i=1}^{n} x_i y_i$$

**49.**  $5b + 10a = 20.2$      **50.**  $5b + 10a = 11.7$
   $10b + 30a = 50.1$          $10b + 30a = 25.6$

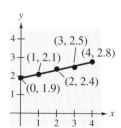

**51.** $7b + 21a = 35.1$
$21b + 91a = 114.2$

**52.** $6b + 15a = 23.6$
$15b + 55a = 48.8$

**53.** *Restaurant Sales*  The total sales (in billions of dollars) for full-service restaurants in the United States from 1992 to 1996 are shown in the table. (Source: National Restaurant Association)

(a) Use the technique demonstrated in Exercises 49–52 to find the line that best fits the data.

(b) Graph the line on a graphing utility and use the trace feature to predict the total restaurant sales in 1998.

(c) Use a linear regression program to find a model. Compare this model with the one you found in part (a).

| Year | $t$ | Total Sales (in billions of dollars) |
|------|-----|--------------------------------------|
| 1992 | 2 | 83.6 |
| 1993 | 3 | 87.0 |
| 1994 | 4 | 85.3 |
| 1995 | 5 | 95.7 |
| 1996 | 6 | 100.0 |

**54.** *SAT Scores*  The average math scores on the SAT for college-bound seniors from 1992 to 1996 are shown in the table. (Source: College Entrance Examination Board)

(a) Use the technique demonstrated in Exercises 49–52 to find the line that best fits the data.

(b) Graph the line on a graphing utility and use the trace feature to predict the average math score in 2000.

(c) Use a linear regression program to find a model. Compare this model with the one you found in part (a).

| Year | $t$ | Average SAT Score in Math |
|------|-----|---------------------------|
| 1992 | 2 | 501 |
| 1993 | 3 | 503 |
| 1994 | 4 | 504 |
| 1995 | 5 | 506 |
| 1996 | 6 | 508 |

## 6.3    Linear Systems in Three or More Variables

### Objectives

- Solve a linear system in row-echelon form using back-substitution.
- Use Gaussian elimination to solve a linear system.
- Solve a nonsquare linear system.
- Construct and use a linear system in three or more variables to solve an application problem.
- Find the equation of a circle or a parabola using a linear system in three or more variables.

**Chui-chang suan-shu**

(250 B.C.)

One of the most influential Chinese mathematics books was the *Chui-chang suan-shu* or *Nine Chapters on the Mathematical Art* (written in approximately 250 B.C.). Chapter Eight of the *Nine Chapters* contained solutions of systems of linear equations using positive and negative numbers. One such system was

$$3x + 2y + z = 39$$
$$2x + 3y + z = 34$$
$$x + 2y + 3z = 26.$$

This system was solved using column operations on a matrix.

## Row-Echelon Form and Back-Substitution

The method of elimination can be applied to a system of linear equations in more than two variables. In fact, this method easily adapts to computer use for solving linear systems with dozens of variables.

When elimination is used to solve a system of linear equations, the goal is to rewrite the system in a form to which back-substitution can be applied. To see how this works, consider the following two systems of linear equations.

$$
\begin{aligned}
x - 2y + 3z &= 9 \\
-x + 3y &= -4 \\
2x - 5y + 5z &= 17
\end{aligned}
\qquad
\begin{aligned}
x - 2y + 3z &= 9 \\
y + 3z &= 5 \\
z &= 2
\end{aligned}
$$

The system on the right is said to be in **row-echelon form,** which means that it has a "stair-step" pattern with leading coefficients of 1. After comparing the two systems, it should be clear that it is easier to solve the system on the right.

### Example 1    Using Back-Substitution

Solve the following system of linear equations.

$$
\begin{aligned}
x - 2y + 3z &= 9 \qquad &\text{Equation 1} \\
y + 3z &= 5 \qquad &\text{Equation 2} \\
z &= 2 \qquad &\text{Equation 3}
\end{aligned}
$$

**Solution**

From Equation 3, you know the value of $z$. To solve for $y$, substitute $z = 2$ into Equation 2 to obtain

$$
\begin{aligned}
y + 3(2) &= 5 \qquad &\text{Substitute 2 for } z. \\
y &= -1. \qquad &\text{Solve for } y.
\end{aligned}
$$

Finally, substitute $y = -1$ and $z = 2$ into Equation 1 to obtain

$$
\begin{aligned}
x - 2(-1) + 3(2) &= 9 \qquad &\text{Substitute } -1 \text{ for } y \text{ and 2 for } z. \\
x &= 1. \qquad &\text{Solve for } x.
\end{aligned}
$$

The solution is $x = 1$, $y = -1$, and $z = 2$, which can be written as the **ordered triple** $(1, -1, 2)$. Check this in the original system of equations.

# Gaussian Elimination

Two systems of equations are **equivalent** if they have the same solution set. To solve a system that is not in row-echelon form, first convert it to an *equivalent* system that is in row-echelon form. To see how this is done, let's take another look at the method of elimination, as applied to a system of two linear equations.

| **Example 2** | The Method of Elimination |

Solve the following system of linear equations.

$$3x - 2y = -1$$
$$x - y = 0$$

### Solution

$$x - y = 0 \qquad \text{You can interchange two equations in the system.}$$
$$3x - 2y = -1$$

$$-3x + 3y = 0 \qquad \text{Multiply the first equation by } -3.$$

$$-3x + 3y = 0 \qquad \text{You can add the multiple of the first equation to the}$$
$$\underline{3x - 2y = -1} \qquad \text{second equation to obtain a new equation.}$$
$$y = -1$$

$$x - y = 0 \qquad \text{New system in row-echelon form}$$
$$y = -1$$

Now, using back-substitution, you can determine that the solution is $y = -1$ and $x = -1$, which can be written as the ordered pair $(-1, -1)$. Check this in the original system of equations.

---

> ▶ **Operations That Produce Equivalent Systems**
>
> Each of the following **row operations** on a system of linear equations produces an *equivalent* system of linear equations.
>
> **1.** Interchange two equations.
>
> **2.** Multiply one of the equations by a nonzero constant.
>
> **3.** Add a multiple of one of the equations to another equation to replace the latter equation.

**Carl Friedrich Gauss**

(1777–1855)

The process of rewriting a system of equations in row-echelon form by using the three basic row operations is called **Gaussian elimination,** after the German mathematician Carl Friedrich Gauss. Example 2 shows the chain of equivalent systems used to solve a linear system in two variables.

## ·DISCOVERY·

The total numbers of sides and diagonals of regular polygons with three, four, and five sides are three, six, and ten, as shown in the figure.

3

6

10

15

Find a quadratic function $y = ax^2 + bx + c$, where $y$ represents the total number of sides and diagonals and $x$ represents the number of sides, that fits these data. Check to see if the quadratic function gives the correct answers for a polygon with six sides.

**Example 3**   Using Elimination to Solve a System

Solve the following system of linear equations.

$$\begin{aligned} x - 2y + 3z &= 9 & \text{Equation 1} \\ -x + 3y \phantom{{}+3z} &= -4 & \text{Equation 2} \\ 2x - 5y + 5z &= 17 & \text{Equation 3} \end{aligned}$$

### Solution

There are many ways to begin, but we suggest working from the upper left corner, saving the $x$ in the upper left position and eliminating the other $x$'s from the first column.

$$\begin{aligned} x - 2y + 3z &= 9 \\ y + 3z &= 5 \\ 2x - 5y + 5z &= 17 \end{aligned}$$

> Adding the first equation to the second equation produces a new second equation.

$$\begin{aligned} x - 2y + 3z &= 9 \\ y + 3z &= 5 \\ -y - z &= -1 \end{aligned}$$

> Adding $-2$ times the first equation to the third equation produces a new third equation.

Now that all but the first $x$ have been eliminated from the first column, go to work on the second column. (You need to eliminate $y$ from the third equation.)

$$\begin{aligned} x - 2y + 3z &= 9 \\ y + 3z &= 5 \\ 2z &= 4 \end{aligned}$$

> Adding the second equation to the third equation produces a new third equation.

Finally, you need a coefficient of 1 for $z$ in the third equation.

$$\begin{aligned} x - 2y + 3z &= 9 \\ y + 3z &= 5 \\ z &= 2 \end{aligned}$$

> Multiplying the third equation by $\frac{1}{2}$ produces a new third equation.

This is the same system that was solved in Example 1, and, as in that example, you can conclude that the solution is

$$x = 1, \quad y = -1, \quad \text{and} \quad z = 2.$$

In Example 3, you can check the solution by substituting $x = 1$, $y = -1$, and $z = 2$ into each original equation, as follows.

Equation 1:   $(1) - 2(-1) + 3(2) = 9$ ✓

Equation 2:   $-(1) + 3(-1) = -4$ ✓

Equation 3:   $2(1) - 5(-1) + 5(2) = 17$ ✓

The next example involves an inconsistent system—one that has no solution. The key to recognizing an inconsistent system is that at some stage in the elimination process you obtain an absurdity such as $0 = -2$.

---

**Example 4**    An Inconsistent System

Solve the following system of linear equations.

$$\begin{array}{rl} x - 3y + z = 1 & \text{Equation 1} \\ 2x - y - 2z = 2 & \text{Equation 2} \\ x + 2y - 3z = -1 & \text{Equation 3} \end{array}$$

**Solution**

$$\begin{array}{r} x - 3y + z = 1 \\ 5y - 4z = 0 \\ x + 2y - 3z = -1 \end{array}$$

 Adding $-2$ times the first equation to the second equation produces a new second equation.

$$\begin{array}{r} x - 3y + z = 1 \\ 5y - 4z = 0 \\ 5y - 4z = -2 \end{array}$$

Adding $-1$ times the first equation to the third equation produces a new third equation.

$$\begin{array}{r} x - 3y + z = 1 \\ 5y - 4z = 0 \\ 0 = -2 \end{array}$$

Adding $-1$ times the second equation to the third equation produces a new third equation.

Because the third "equation" is impossible, you can conclude that this system is inconsistent and therefore has no solution. Moreover, because this system is equivalent to the original system, you can conclude that the original system also has no solution.

---

As with a system of linear equations in two variables, the solution(s) of a system of linear equations in more than two variables must fall into one of three categories. Because an equation in three variables represents a plane in space, the possible solutions can be shown graphically, as in Figure 6.11.

---

▶ **The Number of Solutions of a Linear System**

For a system of linear equations, exactly one of the following is true.

**1.** There is exactly one solution. [See Figure 6.11(a).]

**2.** There are infinitely many solutions. [See Figures 6.11(b) and (c).]

**3.** There is no solution. [See Figures 6.11(d) and (e).]

---

**(a)** Solution: one point

**(b)** Solution: one line

**(c)** Solution: one plane

**(d)** Solution: none

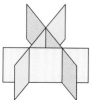

**(e)** Solution: none

Figure 6.11

**Example 5** A System with Infinitely Many Solutions

Solve the following system of linear equations.

$$\begin{array}{rrrcl} x + & y & - 3z & = & -1 \qquad \text{Equation 1} \\ & y & - z & = & 0 \qquad \text{Equation 2} \\ -x + & 2y & & = & 1 \qquad \text{Equation 3} \end{array}$$

**Solution**

$$\begin{array}{rcl} x + y - 3z & = & -1 \\ y - z & = & 0 \\ 3y - 3z & = & 0 \end{array}$$

Adding the first equation to the third equation produces a new third equation.

$$\begin{array}{rcl} x + y - 3z & = & -1 \\ y - z & = & 0 \\ 0 & = & 0 \end{array}$$

Adding $-3$ times the second equation to the third equation produces a new third equation.

This means that Equation 3 depends on Equations 1 and 2 in the sense that it gives us no additional information about the variables. So, the original system is equivalent to the system

$$\begin{array}{rcl} x + y - 3z & = & -1 \\ y - z & = & 0. \end{array}$$

In this last equation, solve for $y$ in terms of $z$ to obtain $y = z$. Back-substituting for $y$ into the previous equation produces $x = 2z - 1$. Finally, letting $z = a$, the solutions to the given system are all of the form

$$x = 2a - 1, \quad y = a, \quad \text{and} \quad z = a$$

where $a$ is a real number. So, every ordered triple of the form

$$(2a - 1, a, a), \qquad a \text{ is a real number}$$

is a solution of the system.

---

In Example 5, there are other ways to write the same infinite set of solutions. For instance, the solutions could have been written as

$$\left(b, \tfrac{1}{2}(b + 1), \tfrac{1}{2}(b + 1)\right), \qquad b \text{ is a real number.}$$

Try convincing yourself of this by substituting $a = 0$, $a = 1$, $a = 2$, and $a = 3$ into the solution given in Example 5. Then substitute $b = -1$, $b = 1$, $b = 3$, and $b = 5$ into the solution given above. In both cases, you should obtain the same ordered triples. So, when comparing descriptions of an infinite solution set, keep in mind that there is more than one way to describe the set.

## Nonsquare Systems

So far, each system of linear equations has been **square,** which means that the number of equations is equal to the number of variables. In a **nonsquare** system, the number of equations differs from the number of variables. A system of linear equations cannot have a unique solution unless there are at least as many equations as there are variables in the system.

### Example 6    A System with Fewer Equations than Variables

Solve the following system of linear equations.

$$x - 2y + z = 2 \qquad \text{Equation 1}$$
$$2x - y - z = 1 \qquad \text{Equation 2}$$

**Solution**

Begin by rewriting the system in row-echelon form, as follows.

$$x - 2y + z = 2$$
$$3y - 3z = -3$$

> Adding $-2$ times the first equation to the second equation produces a new second equation.

$$x - 2y + z = 2$$
$$y - z = -1$$

> Multiplying the second equation by $\frac{1}{3}$ produces a new second equation.

Solving for $y$ in terms of $z$, you get $y = z - 1$, and back-substitution into Equation 1 yields

$$x - 2(z - 1) + z = 2$$
$$x - 2z + 2 + z = 2$$
$$x = z.$$

Finally, by letting $z = a$, you have the solution

$$x = a, \qquad y = a - 1, \qquad \text{and} \qquad z = a$$

where $a$ is a real number. So, every ordered triple of the form

$$(a, a - 1, a), \qquad a \text{ is a real number}$$

is a solution of the system.

In Example 6, try choosing some values of $a$ to obtain different solutions of the system, such as $(1, 0, 1)$, $(2, 1, 2)$, and $(3, 2, 3)$. Then check each of the solutions in the original system.

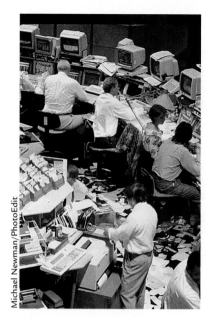

*In 1996, the amount Americans saved or invested increased by about 140.0 billion dollars—an average of about $529 per person.* (Source: U.S. Bureau of Economic Analysis)

# Applications

### Example 7  An Investment Portfolio

You have a portfolio totaling $450,000 and want to invest in (1) certificates of deposit, (2) municipal bonds, (3) blue-chip stocks, and (4) growth or speculative stocks. The certificates pay 9% annually, and the municipal bonds pay 6% annually. Over a 5-year period, you expect the blue-chip stocks to return 10% annually and the growth stocks to return 15% annually. You want a combined annual return of 8%, and you also want to have only one-third of the portfolio invested in stocks. How much should be allocated to each type of investment?

#### Solution

To solve this problem, let $C$, $M$, $B$, and $G$ represent the amounts in the four types of investments. Because the total investment is $450,000, you can write the following equation.

$$C + M + B + G = 450{,}000$$

A second equation can be derived from the fact that the combined annual return should be 8%.

$$0.09C + 0.06M + 0.10B + 0.15G = 0.08(450{,}000)$$

Finally, because only one-third of the total investment should be allocated to stocks, you can write

$$B + G = \tfrac{1}{3}(450{,}000).$$

These three equations make up the following system.

$$
\begin{array}{rcr}
C + M + B + G &=& 450{,}000 \\
0.09C + 0.06M + 0.10B + 0.15G &=& 36{,}000 \\
B + G &=& 150{,}000
\end{array}
$$

Using elimination, you find that the system has infinitely many solutions, which can be written as follows.

$$C = -\tfrac{5}{3}a + 100{,}000, \qquad M = \tfrac{5}{3}a + 200{,}000, \qquad B = -a + 150{,}000,$$
$$G = a$$

So, you have many different options. One possible solution is to choose $a = 30{,}000$, which yields the following portfolio.

1. Certificates of deposit:            $50,000
2. Municipal bonds:                    $250,000
3. Blue-chip stocks:                   $120,000
4. Growth or speculative stocks:       $30,000

| Example 8 | Data Analysis: Curve-Fitting |

Find a quadratic equation, $y = ax^2 + bx + c$, whose graph passes through the points $(-1, 3)$, $(1, 1)$, and $(2, 6)$.

### Solution

Because the graph of $y = ax^2 + bx + c$ passes through the points $(-1, 3)$, $(1, 1)$, and $(2, 6)$, you can write the following.

When $x = -1$, $y = 3$:    $a(-1)^2 + b(-1) + c = 3$

When $x = 1$, $y = 1$:    $a(1)^2 + b(1) + c = 1$

When $x = 2$, $y = 6$:    $a(2)^2 + b(2) + c = 6$

This produces the following system of linear equations.

$$a - b + c = 3 \qquad \text{Equation 1}$$
$$a + b + c = 1 \qquad \text{Equation 2}$$
$$4a + 2b + c = 6 \qquad \text{Equation 3}$$

The solution of this system is $a = 2$, $b = -1$, and $c = 0$. So, the equation of the parabola is $y = 2x^2 - x$, as shown in Figure 6.12.

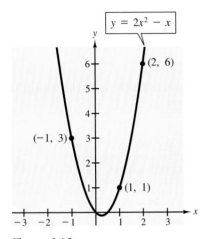

$y = 2x^2 - x$

**Figure 6.12**

---

## Discussing the Concept    Mathematical Modeling

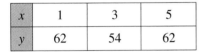

| $x$ | 1 | 3 | 5 |
|---|---|---|---|
| $y$ | 62 | 54 | 62 |

Suppose you work for an outerwear manufacturer, and the marketing department is concerned about sales trends in Arizona. Your manager has asked you to investigate climate data in hopes of explaining sales patterns and gives you the table at the left, which represents the average monthly temperature $y$ in degrees Fahrenheit for Phoenix, Arizona, for the month $x$, where $x = 1$ corresponds to November.    (Source: National Climatic Data Center)

Construct a scatter plot of the data; decide what type of mathematical model might be appropriate for the data; and use the methods you have learned thus far to fit an appropriate model. Your manager would like to know the average monthly temperatures for December and February. Explain to your manager how you found your model, what it represents, and how it can be used to find the December and February average temperatures. Investigate the usefulness of this model for the rest of the year. Would you recommend using the model to predict monthly average temperatures for the whole year or for just part of the year? Explain your reasoning.

# Warm Up

The following warm-up exercises involve skills that were covered in earlier sections. You will use these skills in the exercise set for this section.

In Exercises 1–4, solve the system of linear equations.

**1.** $x + y = 25$
$\qquad y = 10$

**2.** $2x - 3y = \quad 4$
$\qquad 6x \quad = -12$

**3.** $x + y = 32$
$\qquad x - y = 24$

**4.** $2r - \quad s = \quad 5$
$\qquad r + 2s = 10$

In Exercises 5–8, determine whether the ordered triple is a solution of the equation.

**5.** $5x - 3y + 4z = 2$
$\qquad (-1, -2, 1)$

**6.** $x - 2y + 12z = 9$
$\qquad (6, 3, 2)$

**7.** $2x - 5y + 3z = -9$
$\qquad (a - 2, a + 1, a)$

**8.** $-5x + y + z = 21$
$\qquad (a - 4, 4a + 1, a)$

In Exercises 9 and 10, solve for $x$ in terms of $a$.

**9.** $x + 2y - 3z = 4$
$\qquad y = 1 - a, z = a$

**10.** $x - 3y + 5z = 4$
$\qquad y = 2a + 3, z = a$

## 6.3  Exercises

In Exercises 1–26, solve the system of linear equations.

**1.** $x - \ y + z = 4$
$\quad x + 2y - z = 2$
$\quad 2x + 3y \quad = 2$

**2.** $x + y + \ z = 3$
$\quad 4x - y + 3z = 7$
$\quad \ \ 5y + 2z = 0$

**3.** $4x + \ y - 3z = \ 11$
$\quad 2x - 3y + 2z = \quad 9$
$\quad x + \ y + \ z = -3$

**4.** $\qquad 6y + 4z = -12$
$\quad 3x + 3y \qquad = \quad 9$
$\quad 2x \qquad - 3z = \quad 10$

**5.** $3x \qquad + 2z = \ 13$
$\quad x + 2y \qquad = -5$
$\quad \ \ -3y - \ z = \ 10$

**6.** $2x + 3y + \ z = -4$
$\quad 2x - 4y + 3z = \ 18$
$\quad 3x - 2y + 2z = \quad 9$

**7.** $3x - 2y + 4z = 1$
$\quad x + \ y - 2z = 3$
$\quad 2x - 3y + 6z = 8$

**8.** $5x - \ 3y + 2z = 3$
$\quad 2x + \ 4y - \ z = 7$
$\quad x - 11y + 4z = 3$

**9.** $3x + 3y + \ 5z = 1$
$\quad 3x + 5y + \ 9z = 0$
$\quad 5x + 9y + 17z = 0$

**10.** $2x + \ y - \ z = \ 13$
$\quad x + 2y + \ z = \quad 2$
$\quad 8x - 3y + 4z = -2$

**11.**
$$x + 2y - 7z = -4$$
$$2x + y + z = 13$$
$$3x + 9y - 36z = -33$$

**12.**
$$2x + y - 3z = 4$$
$$4x + 2z = 10$$
$$-2x + 3y - 13z = -8$$

**13.**
$$x + 4z = 13$$
$$4x - 2y + z = 7$$
$$2x - 2y - 7z = -19$$

**14.**
$$4x - y + 5z = 11$$
$$x + 2y - z = 5$$
$$5x - 8y + 13z = 7$$

**15.**
$$x - 2y + 5z = 2$$
$$3x + 2y - z = -2$$

**16.**
$$x - 3y + 2z = 18$$
$$5x - 13y + 12z = 80$$

**17.**
$$2x - 3y + z = -2$$
$$-4x + 9y = 7$$

**18.**
$$2x + 3y + 3z = 7$$
$$4x + 18y + 15z = 44$$

**19.**
$$x + 3w = 4$$
$$2y - z - w = 0$$
$$3y - 2w = 1$$
$$2x - y + 4z = 5$$

**20.**
$$x + y + z + w = 6$$
$$2x + 3y - w = 0$$
$$-3x + 4y + z + 2w = 4$$
$$x + 2y - z + w = 0$$

**21.**
$$x + 4z = 1$$
$$x + y + 10z = 10$$
$$2x - y + 2z = -5$$

**22.**
$$3x - 2y - 6z = -4$$
$$-3x + 2y + 6z = 1$$
$$x - y - 5z = -3$$

**23.**
$$4x + 3y + 5z = 10$$
$$5x + 2y + 10z = 13$$
$$3x + y - 2z = -9$$

**24.**
$$2x + 5y = 25$$
$$3x - 2y + 4z = 1$$
$$4x - 3y + z = 9$$

**25.**
$$5x + 5y - z = 0$$
$$10x + 5y + 2z = 0$$
$$5x + 15y - 9z = 0$$

**26.**
$$12x + 5y + z = 0$$
$$12x + 4y - z = 0$$

**27.** One solution for Exercise 15 is $(-a, 2a - 1, a)$. A student gives $(b, -2b - 1, -b)$ as a solution to the same exercise. Explain how both solutions are correct.

In Exercises 28–30, write three ordered triples of the given form.

**28.** $\left(a, a - 5, \frac{2}{3}a + 1\right)$

**29.** $(3a, 5 - a, a)$

**30.** $\left(\frac{1}{2}a, 3a, 5\right)$

In Exercises 31 and 32, find the equation of the parabola that passes through the points.

**31.**

**32.**

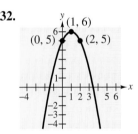

In Exercises 33 and 34, find the equation of the circle that passes through the points.

**33.**

**34.**

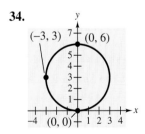

**35.** *Investment* An inheritance of $16,000 was divided among three investments yielding a total of $990 in interest per year. The interest rates for the three investments were 5%, 6%, and 7%. Find the amount placed in each investment if the 5% and 6% investments were $3000 and $2000 less than the 7% investment, respectively.

**36.** *Investment* Suppose you receive a total of $1520 a year in interest from three investments. The interest rates for the three investments are 5%, 7%, and 8%. The 5% investment is half of the 7% investment, and the 7% investment is $1500 less than the 8% investment. What is the amount of each investment?

**37.** *Investment* A company borrows $775,000. Some is borrowed at 8%, some at 9%, and some at 10%. How much is borrowed at each rate if the annual interest is $67,500 and the amount borrowed at 8% is four times the amount borrowed at 10%?

**38.** *Investment* A company borrows $800,000. Some is borrowed at 8%, some at 9%, and some at 10%. How much is borrowed at each rate if the annual interest is $67,000 and the amount borrowed at 8% is five times the amount borrowed at 10%?

**39.** *Grades of Paper* A manufacturer sells a 50-pound package of paper that consists of three grades of computer paper. Grade C costs $3.50 per pound, grade B costs $4.50 per pound, and grade A costs $6.00 per pound. Half of the 50-pound package consists of the two cheaper grades. The cost of the 50-pound package is $252.50. How many pounds of each grade of paper are there in the 50-pound package?

**40.** *Product Types* A company sells three types of products for $20, $10, and $5 per unit. In 1 year, the total revenue for the three products was $350,000, which corresponded to the sale of 32,500 units. The company sold half as many units of the $20 product as units of the $10 product. How many units of each product were sold?

**41.** *Crop Spraying* A mixture of 6 gallons of chemical A, 8 gallons of chemical B, and 13 gallons of chemical C is required to kill a certain destructive crop insect. Commercial spray X contains 1, 2, and 2 parts, respectively, of these chemicals. Commercial spray Y contains only chemical C. Commercial spray Z contains chemicals A, B, and C in equal amounts. How much of each type of commercial spray is needed to get the desired mixture?

**42.** *Chemistry* A chemist needs 10 liters of a 25% acid solution. The solution is to be mixed from three solutions whose concentrations are 10%, 20%, and 50%. How many liters of each solution should the chemist use to satisfy the following?

(a) Use as little as possible of the 50% solution.

(b) Use as much as possible of the 50% solution.

(c) Use 2 liters of the 50% solution.

*Investment Portfolio* In Exercises 43 and 44, you have a total of $500,000 that is to be invested in (1) certificates of deposit, (2) municipal bonds, (3) blue-chip stocks, and (4) growth or speculative stocks. How much should be put in each type of investment?

**43.** The certificates of deposit pay 10% annually, and the municipal bonds pay 8% annually. Over a 5-year period, you expect the blue-chip stocks to return 12% annually and the growth stocks to return 13% annually. You want a combined annual return of 10% and you also want to have only one-fourth of the portfolio in stocks.

**44.** The certificates of deposit pay 9% annually, and the municipal bonds pay 5% annually. Over a 5-year period, you expect the blue-chip stocks to return 12% annually and the growth stocks to return 14% annually. You want a combined annual return of 10% and you also want to have only one-fourth of the portfolio in stocks.

*Fitting a Parabola to Data* In Exercises 45–48, find the least squares regression parabola $y = ax^2 + bx + c$ for the points $(x_1, y_1)$, $(x_2, y_2)$, ..., $(x_n, y_n)$. To find the parabola, solve the system of linear equations for $a$, $b$, and $c$.

$$nc + \left(\sum_{i=1}^{n} x_i\right)b + \left(\sum_{i=1}^{n} x_i^2\right)a = \sum_{i=1}^{n} y_i$$

$$\left(\sum_{i=1}^{n} x_i\right)c + \left(\sum_{i=1}^{n} x_i^2\right)b + \left(\sum_{i=1}^{n} x_i^3\right)a = \sum_{i=1}^{n} x_i y_i$$

$$\left(\sum_{i=1}^{n} x_i^2\right)c + \left(\sum_{i=1}^{n} x_i^3\right)b + \left(\sum_{i=1}^{n} x_i^4\right)a = \sum_{i=1}^{n} x_i^2 y_i$$

**45.**

$$5c + 10a = 15.5$$
$$10b = 6.3$$
$$10c + 34a = 32.1$$

**46.**
$$5c \quad\ + 10a = 15.0$$
$$10b \quad\quad = 17.3$$
$$10c \quad\ + 34a = 34.5$$

**47.**
$$6c + 3b + 19a = 23.9$$
$$3c + 19b + 27a = -7.2$$
$$19c + 27b + 115a = 48.8$$

**48.**
$$6c + 3b + 19a = 13.1$$
$$3c + 19b + 27a = -2.6$$
$$19c + 27b + 115a = 29.0$$

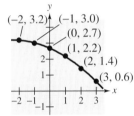

**49.** *Computer Use in Schools*  The total numbers of personal computers used in grades K through 12 in the United States from 1993 to 1997 are shown in the accompanying table. In the table, $y$ represents the number of personal computers in millions and $x = 0$ represents 1995. Find the least squares regression parabola $y = ax^2 + bx + c$ that best fits these data by solving the following system. Then use the model to predict the number of personal computers in use in 2000.  (Source: Market Data Retrieval)

$$5c \quad\quad + \quad 10a = 26.1$$
$$\quad 10b \quad\quad\quad = 6.86$$
$$10c \quad\quad + \quad 34a = 53.2$$

| $x$ | $-2$ | $-1$ | 0 | 1 | 2 |
|---|---|---|---|---|---|
| $y$ | 3.95 | 4.47 | 5.30 | 5.53 | 6.85 |

**50.** *Prices of Mobile Homes*  The average prices of new mobile homes sold in the United States from 1992 to 1996 are shown in the accompanying table. In the table, $y$ represents the average price (in thousands of dollars) and $x = 0$ represents 1993. Find the least squares regression parabola

$$y = ax^2 + bx + c$$

that best fits these data by solving the system. Then use the model to predict the average price of a new mobile home in 1999.  (Source: U.S. Bureau of the Census)

$$5c + 5b + 15a = 167.1$$
$$5c + 15b + 35a = 192.9$$
$$15c + 35b + 99a = 552.7$$

| $x$ | $-1$ | 0 | 1 | 2 | 3 |
|---|---|---|---|---|---|
| $y$ | 28.4 | 30.5 | 33.5 | 36.3 | 38.4 |

**51.** *Stopping Distance*  In testing of the new braking system in an automobile, the speed in miles per hour and the stopping distance in feet were recorded in the table below.

| Speed ($x$) | 30 | 40 | 50 |
|---|---|---|---|
| Stopping distance ($y$) | 56 | 110 | 193 |

| Speed ($x$) | 60 | 70 |
|---|---|---|
| Stopping distance ($y$) | 305 | 425 |

(a) Make a scatter plot of the data.

(b) Find the least squares regression parabola for the data.

(c) Use the quadratic regression program in your graphing utility to find a second model.

(d) How do your results from parts (b) and (c) compare? Use the better model to estimate the stopping distance if the speed is 75 miles per hour.

In Exercises 52 and 53, find the position equation $s = \frac{1}{2}gt^2 + v_0t + s_0$, where $g = -32$ ft/sec$^2$.

**52.** At $t = 1$ second, $s = 134$ feet
At $t = 2$ seconds, $s = 86$ feet
At $t = 3$ seconds, $s = 6$ feet

**53.** At $t = 1$ second, $s = 184$ feet
At $t = 2$ seconds, $s = 116$ feet
At $t = 3$ seconds, $s = 16$ feet

## Math Matters • Regular Polygons and Regular Polyhedra

A regular polygon is a polygon that has $n$ sides of equal length and $n$ equal angles. For instance, a regular three-sided polygon is called an equilateral triangle, a regular four-sided polygon is called a square, and so on. There are infinitely many different types of regular polygons. For instance, it is possible to construct a regular polygon with 100, 1000, or even 1,000,000 sides (they would look very much like circles, but it is still possible to do).

For solid figures, the story is quite different. A regular polyhedron is a solid figure each of whose sides is a regular polygon (of the same size) and each of whose angles is formed by the same number of sides. At first, one might think that there are infinitely many different types of regular polyhedra, but in fact it can be shown that there are only five. The five different types are tetrahedron (4 triangular sides), cube (6 square sides), octahedron (8 triangular sides), dodecahedron (12 pentagonal sides), and icosahedron (20 triangular sides), as shown in the figures.

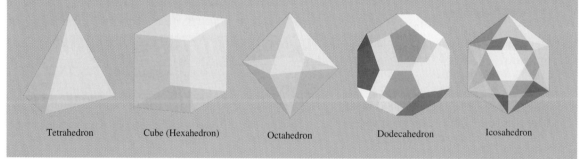

Tetrahedron    Cube (Hexahedron)    Octahedron    Dodecahedron    Icosahedron

## Mid-Chapter Quiz

Take this quiz as you would take a quiz in class. After you are done, check your work against the answers given in the back of the book.

In Exercises 1 and 2, write a system that has the given solution.

**1.** $(3, 2)$

**2.** $(5, -2, 3)$

In Exercises 3 and 4, solve algebraically. Use a graphing utility to verify the solution.

**3.** $y = 2\sqrt{x} + 1, y = -x + 4$

**4.** $x^2 + y^2 = 9, y = 2x + 1$

In Exercises 5 and 6, find the number of sales necessary to break even.

**5.** $C = 12.50x + 10{,}000, R = 19.95x$

**6.** $C = 3.79x + 400{,}000, R = 4.59x$

In Exercises 7 and 8, solve the system by substitution or elimination. Verify the solution graphically.

**7.** $2.5x - y = 6$
$\quad\ 3x + 4y = 2$

**8.** $\frac{1}{2}x + \frac{1}{3}y = 1$
$\quad\ x - 2y = -2$

In Exercises 9 and 10, find the point of equilibrium. Verify the solution graphically.

**9.** Demand: $p = 45 - 0.001x$;    supply: $p = 23 + 0.0002x$

**10.** Demand: $p = 95 - 0.0002x$;    supply: $p = 80 + 0.00001x$

In Exercises 11–13, solve the system of equations.

**11.** $2x + 3y - z = -7$
$\quad\ x \quad\ + 3z = 10$
$\quad\ 2y + z = -1$

**12.** $x + y - 2z = 12$
$\quad\ 2x - y - z = 6$
$\quad\ y - z = 6$

**13.** $3x + 2y + z = 17$
$\quad\ -x + y + z = 4$
$\quad\ x - y - z = 3$

In Exercises 14 and 15, write three ordered triples of the given form.

**14.** $(a, a - 2, 3a)$

**15.** $(2a, a + 5, a)$

**16.** Find the position equation $s = \frac{1}{2}gt^2 + v_0t + s_0$ given
$\quad t = 1$ second, $s = 144$ feet
$\quad t = 2$ seconds, $s = 106$ feet
$\quad t = 3$ seconds, $s = 36$ feet

**17.** How many solutions does the system $2x - 3y + z = 4, y - z = 7$ have?

**18.** *Acid Mixture*   Ten gallons of a 35% acid solution is obtained by mixing a 25% solution with a 40% solution. How much of each must be used?

**19.** *True or False?*   A system of equations may have infinitely many solutions.

**20.** *True or False?*   A square linear system must have exactly one solution.

## 6.4 Systems of Inequalities

### Objectives

- Sketch the graph of an inequality in two variables.
- Solve a system of inequalities.
- Construct and use a system of inequalities to solve an application problem.

## The Graph of an Inequality

The following statements are inequalities in two variables:

$$3x - 2y < 6 \quad \text{and} \quad 2x^2 + 3y^2 \geq 6.$$

An ordered pair $(a, b)$ is a **solution of an inequality** in $x$ and $y$ if the inequality is true when $a$ and $b$ are substituted for $x$ and $y$, respectively. The **graph** of an inequality is the collection of all solutions of the inequality. To sketch the graph of an inequality, begin by sketching the graph of the *corresponding equation*. The graph of the equation will normally separate the plane into two or more regions. In each such region, one of the following must be true.

1. *All* points in the region are solutions of the inequality.

2. *No* point in the region is a solution of the inequality.

So, you can determine whether the points in an entire region satisfy the inequality by simply testing *one* point in the region.

> ▶ Sketching the Graph of an Inequality in Two Variables
>
> 1. Replace the inequality sign by an equal sign, and sketch the graph of the resulting equation. (Use a dashed line for < or > and a solid line for ≤ or ≥.)
>
> 2. Test one point in each of the regions formed by the graph in Step 1. If the point satisfies the inequality, shade the entire region to denote that every point in the region satisfies the inequality.

### Example 1    Sketching the Graph of an Inequality

Sketch the graph of the inequality $y \geq x^2 - 1$.

### Solution

The graph of the corresponding *equation* $y = x^2 - 1$ is a parabola, as shown in Figure 6.13. By testing a point *above* the parabola $(0, 0)$ and a point *below* the parabola $(0, -2)$, you can see that the points that satisfy the inequality are those lying above (or on) the parabola.

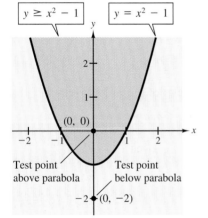

Figure 6.13

The inequality given in Example 1 is a nonlinear inequality in two variables. Most of the following examples involve **linear inequalities** such as $ax + by < c$. The graph of a linear inequality is a half-plane lying on one side of the line $ax + by = c$. The simplest linear inequalities are those corresponding to horizontal or vertical lines, as shown in Example 2.

## Study Tip

To graph a linear inequality, it can help to write the inequality in slope-intercept form. For instance, by writing $x - y < 2$ in the form

$$y > x - 2$$

you can see that the solution points lie *above* the line $x - y = 2$ (or $y = x - 2$), as shown in Figure 6.16.

---

| **Example 2** | Sketching the Graph of a Linear Inequality |

Sketch the graphs of the following linear inequalities.

**a.** $x > -2$    **b.** $y \leq 3$

### Solution

**a.** The graph of the corresponding equation $x = -2$ is a vertical line. The points that satisfy the inequality $x > -2$ are those lying to the right of this line, as shown in Figure 6.14.

**b.** The graph of the corresponding equation $y = 3$ is a horizontal line. The points that satisfy the inequality $y \leq 3$ are those lying below (or on) this line, as shown in Figure 6.15.

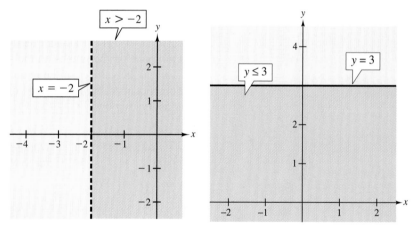

Figure 6.14                         Figure 6.15

| **Example 3** | Sketching the Graph of a Linear Inequality |

Sketch the graph of $x - y < 2$.

### Solution

The graph of the corresponding equation $x - y = 2$ is a line, as shown in Figure 6.16. Because the origin $(0, 0)$ satisfies the inequality, the graph consists of the half-plane lying above the line. (Try checking a point below the line. Regardless of which point you choose, you will see that it does not satisfy the inequality.)

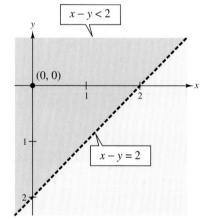

Figure 6.16

## Systems of Inequalities

Many practical problems in business, science, and engineering involve systems of linear inequalities. A **solution** of a system of inequalities in $x$ and $y$ is a point $(x, y)$ that satisfies each inequality in the system.

To sketch the graph of a system of inequalities in two variables, first sketch the graph of each individual inequality (on the same coordinate system) and then find the region that is *common* to every graph in the system. For systems of *linear* inequalities, it is helpful to find the vertices of the solution region.

### Example 4    Solving a System of Inequalities

Sketch the graph (and label the vertices) of the solution set of the following system.

$$x - y < 2$$
$$x > -2$$
$$y \leq 3$$

**Solution**

The graphs of these inequalities are shown in Figures 6.16, 6.14, and 6.15 on page 397, respectively. The triangular region common to all three graphs can be found by superimposing the graphs on the same coordinate plane, as shown in Figure 6.17. To find the vertices of the region, solve the three systems of corresponding equations obtained by taking *pairs* of equations representing the boundaries of the individual regions.

| *Vertex A:* $(-2, -4)$ | *Vertex B:* $(5, 3)$ | *Vertex C:* $(-2, 3)$ |
|---|---|---|
| *Obtained by solving the system* | *Obtained by solving the system* | *Obtained by solving the system* |

$$\begin{aligned} x - y &= 2 \\ x &= -2 \end{aligned} \qquad \begin{aligned} x - y &= 2 \\ y &= 3 \end{aligned} \qquad \begin{aligned} x &= -2 \\ y &= 3 \end{aligned}$$

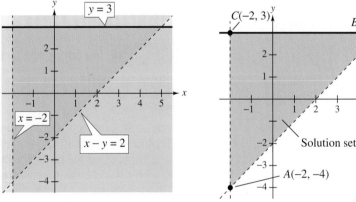

Figure 6.17

For the triangular region shown in Figure 6.17, each point of intersection of a pair of boundary lines corresponds to a vertex. With more complicated regions, two border lines can sometimes intersect at a point that is not a vertex of the region, as shown in Figure 6.18. In order to keep track of which points of intersection are actually vertices of the region, we suggest that you make a careful sketch of the region and refer to your sketch as you find each point of intersection.

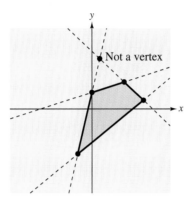

Figure 6.18    *Boundary lines can intersect at a point that is not a vertex.*

### Example 5    Solving a System of Inequalities

Sketch the region containing all points that satisfy the following system.

$$x^2 - y \leq 1$$

$$-x + y \leq 1$$

Solution

As shown in Figure 6.19, the points that satisfy the inequality $x^2 - y \leq 1$ are the points lying above (or on) the parabola given by

$$y = x^2 - 1. \qquad \text{Parabola}$$

The points satisfying the inequality $-x + y \leq 1$ are the points lying on or below the line given by

$$y = x + 1. \qquad \text{Line}$$

To find the points of intersection of the parabola and the line, solve the system of corresponding equations.

$$x^2 - y = 1$$

$$-x + y = 1$$

Using the method of substitution, you can find the solutions to be $(-1, 0)$ and $(2, 3)$, as shown in Figure 6.19.

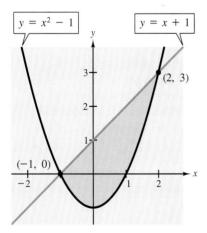

Figure 6.19

When solving a system of inequalities, you should be aware that the system might have no solution. For instance, the system

$x + y > 3$

$x + y < -1$

has no solution points, because the quantity $(x + y)$ cannot be both less than $-1$ and greater than 3, as shown in Figure 6.20.

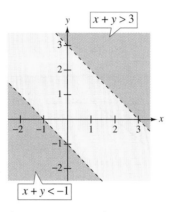

Figure 6.20 *No Solution*

Figure 6.21 *Unbounded Region*

Another possibility is that the solution set of a system of inequalities can be unbounded. For instance, the solution set of

$x + \phantom{2}y < 3$

$x + 2y > 3$

forms an *infinite wedge*, as shown in Figure 6.21.

## •TECHNOLOGY•

**Inequalities and Graphing Utilities**   A graphing utility can be used to sketch the graph of an inequality. A sketch of the graph of $y \geq x^2 - 2$ is shown below. Consult the user's guide to determine how to graph an inequality in two variables (or a system of inequalities) on your graphing utility. Then use your graphing utility to graph the following inequalities.

**a.** $y \leq 2x + 2$    **b.** $y \geq \frac{1}{2}x^2 - 4$    **c.** $y \leq x^3 - 4x^2 + 4$

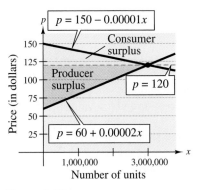

Figure 6.22

Figure 6.23

# Applications

Example 8 in Section 6.2 discussed the *point of equilibrium* for a demand and supply function. The next example discusses two related concepts that economists call **consumer surplus** and **producer surplus**. As shown in Figure 6.22, the consumer surplus is defined as the area of the region that lies *below* the demand curve, *above* the horizontal line passing through the equilibrium point, and to the right of the y-axis. Similarly, the producer surplus is defined as the area of the region that lies *above* the supply curve, *below* the horizontal line passing through the equilibrium point, and to the right of the y-axis. The consumer surplus is a measure of the amount that consumers would have been willing to pay *above what they actually paid*, whereas the producer surplus is a measure of the amount that producers would have been willing to receive *below what they actually received.*

**Example 6    Consumer and Producer Surplus**

The demand and supply functions for a certain type of calculator are given by

$$p = 150 - 0.00001x \qquad \text{Demand equation}$$

$$p = 60 + 0.00002x \qquad \text{Supply equation}$$

where $p$ is the price in dollars and $x$ represents the number of units. Find the consumer surplus and producer surplus for these two equations.

## Solution

Begin by finding the point of equilibrium by solving the equation

$$60 + 0.00002x = 150 - 0.00001x.$$

In Example 8 in Section 6.2, you saw that the solution is $x = 3,000,000$, which corresponds to an equilibrium price of $p = \$120$. Thus, the consumer surplus and producer surplus are the areas of the triangular regions given by the system of inequalities.

| Consumer Surplus | Producer Surplus |
|---|---|
| $p \leq 150 - 0.00001x$ | $p \geq 60 + 0.00002x$ |
| $p \geq 120$ | $p \leq 120$ |
| $x \geq 0$ | $x \geq 0$ |

In Figure 6.23, you can see that the consumer surplus is

Consumer surplus $= \frac{1}{2}(\text{base})(\text{height}) = \frac{1}{2}(30)(3,000,000) = \$45,000,000$

and the producer surplus is

Producer surplus $= \frac{1}{2}(\text{base})(\text{height}) = \frac{1}{2}(60)(3,000,000) = \$90,000,000.$

**Example 7**    Nutrition

The liquid portion of a diet is to provide at least 300 calories, 36 units of vitamin A, and 90 units of vitamin C daily. A cup of dietary drink X provides 60 calories, 12 units of vitamin A, and 10 units of vitamin C. A cup of dietary drink Y provides 60 calories, 6 units of vitamin A, and 30 units of vitamin C. Set up a system of linear inequalities that describes the minimum daily requirements for calories and vitamins.

**Solution**

Begin by letting $x$ and $y$ represent the following.

$$x = \text{number of cups of dietary drink X}$$

$$y = \text{number of cups of dietary drink Y}$$

To meet the minimum daily requirements, the following inequalities must be satisfied.

| | |
|---|---|
| For calories: | $60x + 60y \geq 300$ |
| For vitamin A: | $12x + 6y \geq 36$ |
| For vitamin C: | $10x + 30y \geq 90$ |
| | $x \geq 0$ |
| | $y \geq 0$ |

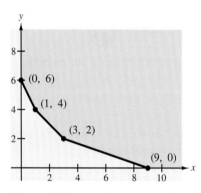

**Figure 6.24**

The last two inequalities are included because $x$ and $y$ cannot be negative. The graph of this system of inequalities is shown in Figure 6.24. (More is said about this application in Example 7 in Section 6.5.)

---

**Discussing the Concept**    **Writing a System of Inequalities**

Write a system of inequalities that includes $x \geq 0$ and $y \geq 0$. Find the solution set by sketching a graph of the system and identifying the vertices. Exchange your system of inequalities for the system of inequalities written by another student. Find the solution set and identify the vertices of the system that you received. Does your answer agree with the answer of the student who wrote the problem?

# Warm Up

The following warm-up exercises involve skills that were covered in earlier sections. You will use these skills in the exercise set for this section.

In Exercises 1–6, identify the graph of the given equation.

**1.** $x + y = 3$

**2.** $4x - y = 8$

**3.** $y = x^2 - 4$

**4.** $y = -x^2 + 1$

**5.** $x^2 + y^2 = 9$

**6.** $\dfrac{x^2}{4} + \dfrac{y^2}{9} = 1$

In Exercises 7–10, solve the system of equations.

**7.** $\begin{aligned} x + 2y &= 3 \\ 4x - 7y &= -3 \end{aligned}$

**8.** $\begin{aligned} 2x - 3y &= 4 \\ x + 5y &= 2 \end{aligned}$

**9.** $\begin{aligned} x^2 + y &= 5 \\ 2x - 4y &= 0 \end{aligned}$

**10.** $\begin{aligned} x^2 + y^2 &= 13 \\ x + y &= 5 \end{aligned}$

## 6.4   Exercises

In Exercises 1–6, match the inequality with its graph. [The graphs are labeled (a), (b), (c), (d), (e), and (f).]

(a)

(b)

(c)

(d)

(e)

(f)
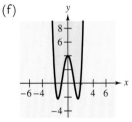

**1.** $2x + 3y \le 6$

**2.** $2x - y \ge -2$

**3.** $x^2 - y \le 2$

**4.** $y \le 4 - x^2$

**5.** $y > x^4 - 5x^2 + 4$

**6.** $3x^4 + y \le 6x^2$

In Exercises 7–20, graph the inequality.

**7.** $x \ge 2$

**8.** $x < 4$

**9.** $y > -1$

**10.** $y \le 3$

**11.** $y < 2 - x$

**12.** $y > 2x - 4$

**13.** $2y - x \ge 4$

**14.** $5x + 3y \ge -15$

**15.** $y^2 + 2x > 0$

**16.** $y^2 - x < 0$

**17.** $y \le \dfrac{1}{1 + x^2}$

**18.** $y < \ln x$

**19.** $x^2 + y^2 \le 4$

**20.** $x^2 + y^2 > 4$

In Exercises 21–40, graph the solution set of the system of inequalities.

**21.** $\begin{aligned} x + y &\le 2 \\ -x + y &\le 2 \\ y &\ge 0 \end{aligned}$

**22.** $\begin{aligned} 3x + 2y &< 6 \\ x &> 1 \\ y &> 0 \end{aligned}$

**23.** $\begin{aligned} x + y &\le 5 \\ x &\ge 2 \\ y &\ge 0 \end{aligned}$

**24.** $\begin{aligned} 2x + y &\ge 2 \\ x &\le 2 \\ y &\le 1 \end{aligned}$

**25.** $\begin{aligned} -3x + 2y &< 6 \\ x + 4y &> -2 \\ 2x + y &< 3 \end{aligned}$

**26.** $\begin{aligned} x - 7y &> -36 \\ 5x + 2y &> 5 \\ 6x - 5y &> 6 \end{aligned}$

**27.** $\begin{aligned} 2x + y &< 2 \\ 6x + 3y &> 2 \end{aligned}$

**28.** $\begin{aligned} 5x - 3y &< -6 \\ 5x - 3y &> -9 \end{aligned}$

**29.** $\begin{aligned} y &\ge -3 \\ y &\le 1 - x^2 \end{aligned}$

**30.** $\begin{aligned} x - y^2 &> 0 \\ y &> (x - 3)^2 - 4 \end{aligned}$

**31.** $\begin{aligned} x^2 + y^2 &\le 16 \\ x^2 + y^2 &< 1 \end{aligned}$

**32.** $\begin{aligned} x^2 + y^2 &\le 25 \\ x^2 + y^2 &\ge 9 \end{aligned}$

**33.** $\begin{aligned} x &> y^2 \\ x &< y + 2 \end{aligned}$

**34.** $\begin{aligned} x &< 2y - y^2 \\ 0 &< x + y \end{aligned}$

**35.** $\begin{aligned} y &\le \sqrt{3x} + 1 \\ y &\ge x + 1 \end{aligned}$

**36.** $\begin{aligned} y &< \sqrt{2x} + 3 \\ y &> x + 3 \end{aligned}$

**37.** $\begin{aligned} y &< x^3 - 2x + 1 \\ y &> -2x \\ x &\le 1 \end{aligned}$

**38.** $\begin{aligned} x &\ge 1 \\ x - 2y &\le 3 \\ 3x + 2y &\ge 9 \\ x + y &\le 6 \end{aligned}$

**39.** $\begin{aligned} y &\le e^x \\ y &\ge \ln x \\ x &\ge \tfrac{1}{2} \\ x &\le 2 \end{aligned}$

**40.** $\begin{aligned} y &\le e^{-x^2/2} \\ y &\ge 0 \\ -1 &\le x \le 1 \end{aligned}$

In Exercises 41–46, write a system of inequalities that describes the region.

**41.** Rectangular region with vertices at $(1, 1)$, $(5, 1)$, $(5, 5)$, and $(1, 5)$

**42.** Parallelogram region with vertices at $(0, 0)$, $(5, 0)$, $(1, 6)$, and $(6, 6)$

**43.** Triangular region with vertices at $(0, 0)$, $(7, 0)$, and $(3, 5)$

**44.** Triangular region with vertices at $(-2, 0)$, $(6, 0)$, and $(0, 5)$

**45.** Sector of a circle

**46.** Sector of a circle

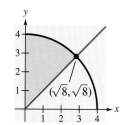

**47.** *Furniture Production* A furniture company can sell all the tables and chairs it produces. Each table requires 1 hour in the assembly center and $1\frac{1}{3}$ hours in the finishing center. Each chair requires $1\frac{1}{2}$ hours in the assembly center and $1\frac{1}{2}$ hours in the finishing center. The company's assembly center is available 12 hours per day, and its finishing center is available 15 hours per day. If $x$ is the number of tables produced per day and $y$ is the number of chairs, find a system of inequalities describing all possible production levels. Sketch the graph of the system.

**48.** *Computer Inventory* A store sells two models of a certain brand of computer. Because of the demand, it is necessary to stock three times as many units of model A as units of model B. The costs to the store for the two models are $900 and $1100, respectively. The management does not want more than $40,000 in computer inventory at any one time, and it wants at least six model A computers and two model B computers in inventory at all times. Devise a system of inequalities describing all possible inventory levels, and sketch the graph of the system.

**49.** *Investment*   A person plans to invest $20,000 in two different interest-bearing accounts. Each account is to contain at least $5000. Moreover, one account should have at least twice the amount that is in the other account.

(a) Find a system of inequalities that describes the amounts that can be deposited in each account.

(b) Sketch the graph of the system.

**50.** *Concert Ticket Sales*   Two types of tickets are to be sold for a concert. One type costs $15 and the other type costs $25. The promoter of the concert must sell at least 15,000 tickets, including at least 8000 of the $15 tickets and at least 4000 of the $25 tickets. Moreover, the gross receipts must total at least $275,000 in order for the concert to be held.

(a) Find a system of inequalities describing the different numbers of tickets that can be sold.

(b) Sketch the graph of the system.

**51.** *Diet Supplement*   A dietitian is asked to design a special diet supplement using two different foods. Each ounce of food X contains 20 units of calcium, 15 units of iron, and 10 units of vitamin B. Each ounce of food Y contains 10 units of calcium, 10 units of iron, and 20 units of vitamin B. The minimum daily requirements in the diet are 300 units of calcium, 150 units of iron, and 200 units of vitamin B.

(a) Find a system of inequalities describing the different amounts of food X and food Y that can be used in the diet.

(b) Sketch the graph of the system.

**52.** *Diet Supplement*   A dietitian is asked to design a special diet supplement using two different foods. Each ounce of food X contains 20 units of calcium, 15 units of iron, and 10 units of vitamin B. Each ounce of food Y contains 10 units of calcium, 10 units of iron, and 20 units of vitamin B. The minimum daily requirements in the diet are 280 units of calcium, 160 units of iron, and 180 units of vitamin B.

(a) Find a system of inequalities describing the different amounts of food X and food Y that can be used in the diet.

(b) Sketch the graph of the system.

*Consumer and Producer Surpluses*   In Exercises 53–56, find the consumer surplus and producer surplus for the pair of supply and demand equations.

| | *Demand* | *Supply* |
|---|---|---|
| **53.** | $p = 56 - 0.0001x$ | $p = 22 + 0.00001x$ |
| **54.** | $p = 60 - 0.00001x$ | $p = 15 + 0.00004x$ |
| **55.** | $p = 140 - 0.00002x$ | $p = 80 + 0.00001x$ |
| **56.** | $p = 400 - 0.0002x$ | $p = 225 + 0.0005x$ |

**57.** *Births in District of Columbia*   The number of babies $y$ (in thousands) born to D.C. residents each year from 1992 to 1996 can be approximated by the linear model

$$y = 12.6 - 0.70t$$

where $t = 2$ represents 1992. The *total* number of babies born to D.C. residents during this 5-year period can be approximated by finding the area of the trapezoid represented by the system

$$y \le 12.6 - 0.70t, \ y \ge 0, t \ge 1.5, t \le 6.5.$$

(Source: U.S. National Center for Health Statistics)

(a) Sketch this region using a graphing utility.

(b) Use the formula for the area of a trapezoid to approximate the total number of births.

**58.** *Spectator Amusements*   Annual consumer spending for admissions to spectator amusements in the United States from 1991 to 1995 can be approximated by the linear model

$$y = 1.13t + 14.6$$

where $y$ is consumer spending (in billions of dollars) and $t = 1$ represents 1991. The *total* amount of consumer spending during this 5-year period can be approximated by finding the area of the trapezoid represented by the system

$$y \le 1.13t + 14.6, y \ge 0, t \ge 0.5, t \le 5.5.$$

(Source: U.S. Bureau of Economic Analysis)

(a) Sketch this region using a graphing utility.

(b) Use the formula for the area of a trapezoid to approximate the total consumer spending.

## 6.5     Linear Programming

### Objectives

- Solve a linear programming problem.
- Use linear programming to minimize or maximize an objective function.
- Use linear programming to optimize an application.

## Linear Programming: A Graphical Approach

Many applications in business and economics involve a process called **optimization,** in which you are asked to find the minimum cost, the maximum profit, or the minimum use of resources. In this section you will study an optimization strategy called **linear programming.**

A two-dimensional linear programming problem consists of a linear **objective function** and a system of linear inequalities called **constraints.** The objective function gives the quantity that is to be maximized (or minimized), and the constraints determine the set of **feasible solutions.** For example, consider a linear programming problem in which you are asked to maximize the value of

$$z = ax + by \qquad \text{Objective function}$$

subject to a set of constraints that determines the region in Figure 6.25. Because every point in the region satisfies each constraint, it is not clear how you should go about finding the point that yields a maximum value of $z$. Fortunately, it can be shown that if there is an optimal solution, it must occur at one of the vertices. This means that *you can find the maximum value by testing $z$ at each of the vertices.*

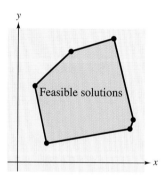

Feasible solutions

Figure 6.25

### ▶ Optimal Solution of a Linear Programming Problem

If a linear programming problem has a solution, it must occur at a vertex of the set of feasible solutions. If the problem has more than one solution, then at least one of them must occur at a vertex of the set of feasible solutions. In either case, the value of the objective function is unique.

| Example 1 | Solving a Linear Programming Problem |

Find the maximum value of

$$z = 3x + 2y \qquad \text{Objective function}$$

subject to the following constraints.

$$\left.\begin{array}{r} x \geq 0 \\ y \geq 0 \\ x + 2y \leq 4 \\ x - y \leq 1 \end{array}\right\} \qquad \text{Constraints}$$

**Solution**

The constraints form the region shown in Figure 6.26. At the four vertices of this region, the objective function has the following values.

At $(0, 0)$:  $\qquad z = 3(0) + 2(0) = 0$

At $(1, 0)$:  $\qquad z = 3(1) + 2(0) = 3$

At $(2, 1)$:  $\qquad z = 3(2) + 2(1) = 8 \qquad$ Maximum value of $z$

At $(0, 2)$:  $\qquad z = 3(0) + 2(2) = 4$

So, the maximum value of $z$ is 8, and this occurs when $x = 2$ and $y = 1$.

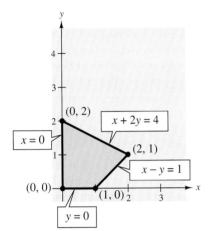

**Figure 6.26**

In Example 1, try testing some of the *interior* points in the region. You will see that the corresponding values of $z$ are less than 8. Here are some examples.

At $(1, 1)$:  $\qquad z = 3(1) + 2(1) = 5$

At $\left(1, \frac{1}{2}\right)$:  $\qquad z = 3(1) + 2\left(\frac{1}{2}\right) = 4$

At $\left(\frac{1}{2}, \frac{3}{2}\right)$:  $\qquad z = 3\left(\frac{1}{2}\right) + 2\left(\frac{3}{2}\right) = \frac{9}{2}$

To see why the maximum value of the objective function in Example 1 must occur at a vertex, consider writing the objective function in the form

$$y = -\frac{3}{2}x + \frac{z}{2} \qquad \text{Family of lines}$$

where $z/2$ is the $y$-intercept of the objective function. This equation represents a family of lines, each of slope $-\frac{3}{2}$. Of these infinitely many lines, you want the one that has the largest $z$-value while still intersecting the region determined by the constraints. In other words, of all the lines whose slope is $-\frac{3}{2}$, you want the one that has the largest $y$-intercept *and* intersects the given region, as shown in Figure 6.27. It should be clear that such a line will pass through one (or more) of the vertices of the region.

**Figure 6.27**

▶ Solving a Linear Programming Problem

To solve a linear programming problem involving two variables by the graphical method, use the following steps.

1. Sketch the region corresponding to the system of constraints. (The points inside or on the boundary of the region are called *feasible solutions.*)

2. Find the vertices of the region.

3. Test the objective function at each of the vertices and select the values of the variables that optimize the objective function. For a bounded region, both a minimum and a maximum value will exist. (For an unbounded region, *if* an optimal solution exists, it will occur at a vertex.)

## Study Tip

Remember that a vertex of the region can be found using a system of linear equations. The system will consist of the equations of the lines passing through the vertex.

These guidelines will work whether the objective function is to be maximized or minimized. For instance, the same test used in Example 1 to find the maximum value of $z$ can be used to conclude that the minimum value of $z$ is 0 and that this value occurs at the vertex $(0, 0)$.

### Example 2    Solving a Linear Programming Problem

Find the maximum value of the objective function

$$z = 4x + 6y \qquad \text{Objective function}$$

where $x \geq 0$ and $y \geq 0$, subject to the following constraints.

$$\left.\begin{array}{r} -x + \phantom{2}y \leq 11 \\ x + \phantom{2}y \leq 27 \\ 2x + 5y \leq 90 \end{array}\right\} \qquad \text{Constraints}$$

### Solution

The region bounded by the constraints is shown in Figure 6.28. By testing the objective function at each vertex, you obtain the following.

At $(0, 0)$:    $z = 4(0) \phantom{0} + 6(0) \phantom{00} = \phantom{000} 0$

At $(0, 11)$:   $z = 4(0) \phantom{0} + 6(11) = \phantom{00} 66$

At $(5, 16)$:   $z = 4(5) \phantom{0} + 6(16) = 116$

At $(15, 12)$:  $z = 4(15) + 6(12) = 132$    Maximum value of $z$

At $(27, 0)$:   $z = 4(27) + 6(0) \phantom{0} = 108$

So, the maximum value of $z$ is 132, and this occurs when $x = 15$ and $y = 12$.

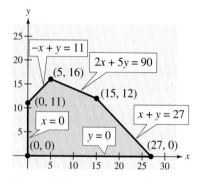

Figure 6.28

The next example shows that the same basic procedure can be used to solve a linear programming problem in which the objective function is to be *minimized*.

### Example 3    Minimizing an Objective Function

Find the minimum value of the objective function

$$z = 5x + 7y \qquad \text{Objection function}$$

where $x \geq 0$ and $y \geq 0$, subject to the following constraints.

$$\left.\begin{array}{r} 2x + 3y \geq 6 \\ 3x - y \leq 15 \\ -x + y \leq 4 \\ 2x + 5y \leq 27 \end{array}\right\} \qquad \text{Constraints}$$

#### Solution

The region bounded by the constraints is shown in Figure 6.29. By testing the objective function at each vertex, you obtain the following.

At $(0, 2)$:   $z = 5(0) + 7(2) = 14$    Minimum value of $z$

At $(0, 4)$:   $z = 5(0) + 7(4) = 28$

At $(1, 5)$:   $z = 5(1) + 7(5) = 40$

At $(6, 3)$:   $z = 5(6) + 7(3) = 51$

At $(5, 0)$:   $z = 5(5) + 7(0) = 25$

At $(3, 0)$:   $z = 5(3) + 7(0) = 15$

So, the minimum value of $z$ is 14, and this occurs when $x = 0$ and $y = 2$.

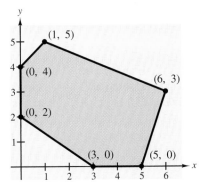

**Figure 6.29**

### Example 4    Maximizing an Objective Function

Find the maximum value of the objective function

$$z = 5x + 7y \qquad \text{Objective function}$$

where $x \geq 0$ and $y \geq 0$, subject to the following constraints.

$$\left.\begin{array}{r} 2x + 3y \geq 6 \\ 3x - y \leq 15 \\ -x + y \leq 4 \\ 2x + 5y \leq 27 \end{array}\right\} \qquad \text{Constraints}$$

#### Solution

This linear programming problem is identical to that given in Example 3 above, *except* that the objective function is maximized instead of minimized. Using the values of $z$ at the vertices shown above, you can conclude that the maximum value of $z$ is 51, and that this value occurs when $x = 6$ and $y = 3$.

### Study Tip

The steps used to find the minimum and maximum values in Examples 3 and 4 are precisely the same. In other words, once you have evaluated the objective function at the vertices of the feasible region, you simply choose the largest value as the maximum and the smallest value as the minimum.

It is possible for the maximum (or minimum) value in a linear programming problem to occur at *two* different vertices. For instance, at the vertices of the region shown in Figure 6.30, the objective function

$$z = 2x + 2y \qquad \text{Objective function}$$

has the following values.

At $(0, 0)$: $z = 2(0) + 2(0) = 0$

At $(0, 4)$: $z = 2(0) + 2(4) = 8$

At $(2, 4)$: $z = 2(2) + 2(4) = 12$     Maximum value of $z$

At $(5, 1)$: $z = 2(5) + 2(1) = 12$     Maximum value of $z$

At $(5, 0)$: $z = 2(5) + 2(0) = 10$

In this case, you can conclude that the objective function has a maximum value not only at the vertices $(2, 4)$ and $(5, 1)$; it also has a maximum value (of 12) at *any point on the line segment connecting these two vertices.* Note that the objective function

$$y = -x + \tfrac{1}{2}z$$

has the same slope as the line through the vertices $(2, 4)$ and $(5, 1)$.

Some linear programming problems have no optimal solution. This can occur if the region determined by the constraints is *unbounded.* Example 5 illustrates such a problem.

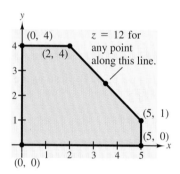

Figure 6.30

### Example 5   An Unbounded Region

Find the maximum value of

$$z = 4x + 2y \qquad \text{Objective function}$$

where $x \geq 0$ and $y \geq 0$, subject to the following constraints.

$$\left. \begin{array}{r} x + 2y \geq 4 \\ 3x + y \geq 7 \\ -x + 2y \leq 7 \end{array} \right\} \quad \text{Constraints}$$

### Solution

The region determined by the constraints is shown in Figure 6.31. For this unbounded region, there is no maximum value of $z$. To see this, note that the point $(x, 0)$ lies in the region for all values of $x \geq 4$. By choosing $x$ to be large, you can obtain values of

$$z = 4(x) + 2(0) = 4x$$

that are as large as you want. So, there is no maximum value of $z$. For this problem, there *is* a minimum value of $z = 10$, which occurs at the vertex $(2, 1)$.

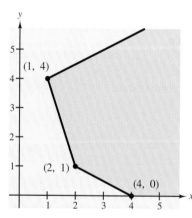

Figure 6.31

## Applications

Example 6 shows how linear programming can be used to find the maximum profit in a business application.

### Example 6    Maximum Profit

A manufacturer wants to maximize the profit for two products. The first product yields a profit of $1.50 per unit, and the second product yields a profit of $2.00 per unit. Market tests and available resources have indicated the following constraints.

**1.** The combined production level should not exceed 1200 units per month.

**2.** The demand for product II is no more than half the demand for product I.

**3.** The production level of product I is less than or equal to 600 units plus three times the production level of product II.

#### Solution

If you let $x$ be the number of units of product I and $y$ be the number of units of product II, the objective function (for the combined profit) is given by

$$P = 1.5x + 2y.$$    Objective function

The three constraints translate into the following linear inequalities.

**1.** $x + y \le 1200$  ⟹  $x + y \le 1200$

**2.** $y \le \frac{1}{2}x$  ⟹  $-x + 2y \le 0$

**3.** $x \le 3y + 600$  ⟹  $x - 3y \le 600$

Because neither $x$ nor $y$ can be negative, you also have the two additional constraints of $x \ge 0$ and $y \ge 0$. Figure 6.32 shows the region determined by the constraints. To find the maximum profit, test the value of $P$ at the vertices of the region.

At $(0, 0)$:        $P = 1.5(0) + 2(0) = 0$

At $(800, 400)$:    $P = 1.5(800) + 2(400) = 2000$    Maximum profit

At $(1050, 150)$:   $P = 1.5(1050) + 2(150) = 1875$

At $(600, 0)$:      $P = 1.5(600) + 2(0) = 900$

So, the maximum profit is $2000, and it occurs when the monthly production consists of 800 units of product I and 400 units of product II.

**Figure 6.32**

*y* axis labeled "Units: Product II"; *x* axis labeled "Units: Product I". Vertices: $(0, 0)$, $(800, 400)$, $(1050, 150)$, $(600, 0)$.

In Example 6, suppose the manufacturer improved the production of product I so that it yielded a profit of $2.50 per unit. How would this affect the number of units the manufacturer should sell to obtain a maximum profit?

**Example 7** Minimum Cost

The liquid part of a diet is to provide at least 300 calories, 36 units of vitamin A, and 90 units of vitamin C daily. A cup of dietary drink X costs $0.12 and provides 60 calories, 12 units of vitamin A, and 10 units of vitamin C. A cup of dietary drink Y costs $0.15 and provides 60 calories, 6 units of vitamin A, and 30 units of vitamin C. How many cups of each drink should be consumed each day to minimize the cost and still meet the daily requirements?

**Solution**

As in Example 7 on page 402, let $x$ be the number of cups of dietary drink X and let $y$ be the number of cups of dietary drink Y.

$$\left.\begin{array}{ll} \text{For calories:} & 60x + 60y \geq 300 \\ \text{For vitamin A:} & 12x + 6y \geq 36 \\ \text{For vitamin C:} & 10x + 30y \geq 90 \\ & x \geq 0 \\ & y \geq 0 \end{array}\right\} \quad \text{Constraints}$$

The cost $C$ is given by

$$C = 0.12x + 0.15y. \qquad \text{Objective function}$$

The graph of the region corresponding to the constraints is shown in Figure 6.33. To determine the minimum cost, test $C$ at each vertex of the region, as follows.

At $(0, 6)$: $C = 0.12(0) + 0.15(6) = 0.90$

At $(1, 4)$: $C = 0.12(1) + 0.15(4) = 0.72$

At $(3, 2)$: $C = 0.12(3) + 0.15(2) = 0.66$ $\qquad$ Minimum value of $C$

At $(9, 0)$: $C = 0.12(9) + 0.15(0) = 1.08$

So, the minimum cost is $0.66 per day, and this occurs when three cups of drink X and two cups of drink Y are consumed each day.

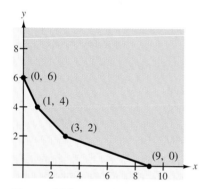

**Figure 6.33**

---

**Discussing the Concept** **Analysis of Constraints**

Explain what difficulties you might encounter with the following sets of linear programming constraints.

**a.**
$$\begin{array}{rl} x - y < & 0 \\ 3x + y > & 9 \\ -4x + y > & -2 \end{array}$$

**b.**
$$\begin{array}{rl} 2x + y > & 11 \\ x - y > & 0 \\ x \quad < & 4 \\ y < & 0 \end{array}$$

# Warm Up

The following warm-up exercises involve skills that were covered in earlier sections. You will use these skills in the exercise set for this section.

In Exercises 1–4, sketch the graph of the linear equation.

**1.** $y + x = 3$

**2.** $y - x = 12$

**3.** $x = 0$

**4.** $y = 4$

In Exercises 5–8, find the point of intersection of the two lines.

**5.** $x + y = 4, x = 0$

**6.** $x + 2y = 12, y = 0$

**7.** $x + y = 4, 2x + 3y = 9$

**8.** $x + 2y = 12, 2x + y = 9$

In Exercises 9 and 10, sketch the graph of the inequality.

**9.** $2x + 3y \geq 18$

**10.** $4x + 3y \geq 12$

## 6.5   Exercises

In Exercises 1–8, find the minimum and maximum values of the objective function, subject to the indicated constraints. (For each exercise, the graph of the region determined by the constraints is provided.)

**1.** *Objective function:*

$z = 6x + 5y$

*Constraints:*

$$x \geq 0$$
$$y \geq 0$$
$$x + y \leq 6$$

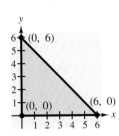

**2.** *Objective function:*

$z = 2x + 8y$

*Constraints:*

$$x \geq 0$$
$$y \geq 0$$
$$2x + y \leq 4$$

**3.** *Objective function:*

$z = 8x + 7y$

*Constraints:*

(See Exercise 1.)

**4.** *Objective function:*

$z = 7x + 3y$

*Constraints:*

(See Exercise 2.)

**5.** *Objective function:*

$z = 3x + 2y$

*Constraints:*

$$x \geq 0$$
$$y \geq 0$$
$$x + 3y \leq 15$$
$$4x + y \leq 16$$

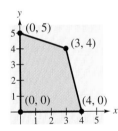

**6.** *Objective function:*

$z = 5x + 4y$

*Constraints:*

$$x \geq 0$$
$$2x + 3y \geq 6$$
$$3x - 2y \leq 9$$
$$x + 5y \leq 20$$

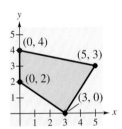

**7.** *Objective function:*

$z = 5x + 0.5y$

*Constraints:*

(See Exercise 5.)

**8.** *Objective function:*

$z = x + 6y$

*Constraints:*

(See Exercise 6.)

In Exercises 9–20, sketch the region determined by the constraints. Then find the minimum and maximum values of the objective function, subject to the indicated constraints.

**9.** *Objective function:*

$z = 6x + 10y$

*Constraints:*

$x \geq 0$

$y \geq 0$

$3x + 5y \leq 15$

**11.** *Objective function:*

$z = 9x + 4y$

*Constraints:*

$x \geq 0$

$y \geq 0$

$3x + 5y \leq 15$

**13.** *Objective function:*

$z = 4x + 5y$

*Constraints:*

$x \geq 0$

$y \geq 0$

$x + y \geq 8$

$3x + 5y \geq 30$

**15.** *Objective function:*

$z = 2x + 7y$

*Constraints:*

(See Exercise 13.)

**17.** *Objective function:*

$z = 5x + 2y$

*Constraints:*

$x \geq 0$

$y \geq 0$

$x + 2y \leq 40$

$x + y \geq 30$

$2x + 3y \geq 72$

**19.** *Objective function:*

$z = 2x + 5y$

*Constraints:*

(See Exercise 17.)

**10.** *Objective function:*

$z = 7x + 8y$

*Constraints:*

$x \geq 0$

$y \geq 0$

$x + 2y \leq 8$

**12.** *Objective function:*

$z = 7x + 2y$

*Constraints:*

$x \geq 0$

$y \geq 0$

$x + 2y \leq 8$

**14.** *Objective function:*

$z = 4x + 5y$

*Constraints:*

$x \geq 0$

$y \geq 0$

$2x + 2y \leq 10$

$x + 2y \leq 6$

**16.** *Objective function:*

$z = 2x - y$

*Constraints:*

(See Exercise 14.)

**18.** *Objective function:*

$z = x$

*Constraints:*

$x \geq 0$

$y \geq 0$

$2x + 3y \leq 60$

$2x + y \leq 28$

$4x + y \leq 48$

**20.** *Objective function:*

$z = y$

*Constraints:*

(See Exercise 18.)

In Exercises 21–24, maximize the objective function subject to the constraints

$3x + y \leq 15$   and   $4x + 3y \leq 30$

where $x \geq 0$ and $y \geq 0$.

**21.** $z = 2x + y$

**22.** $z = 5x + y$

**23.** $z = x + y$

**24.** $z = 3x + y$

In Exercises 25–28, maximize the objective function subject to the constraints

$x + 4y \leq 20$,   $x + y \leq 8$,   and   $3x + 2y \leq 21$

where $x \geq 0$ and $y \geq 0$.

**25.** $z = 2x + 5y$

**26.** $z = 3x + 5y$

**27.** $z = 10x + 7y$

**28.** $z = 10x + 4y$

In Exercises 29–32, find an objective function that has a maximum or a minimum value at the indicated vertex of the constraint region shown. (There are many correct answers.)

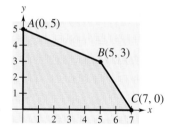

**29.** The maximum occurs at vertex $A$.

**30.** The maximum occurs at vertex $B$.

**31.** The maximum occurs at vertex $C$.

**32.** The minimum occurs at vertex $A$.

**33.** *Maximum Profit*   A store plans to sell two models of computers at costs of \$1100 and \$1500. The \$1100 model yields a profit of \$100 and the \$1500 model yields a profit of \$125. How many units of each model should be stocked to maximize profit, subject to the following constraints?

• The merchant estimates that the total monthly demand will not exceed 250 units.

• The merchant does not want to invest more than \$295,000 in computer inventory.

**34.** *Maximum Profit* A fruit grower has 150 acres of land available to raise two crops, A and B. The profit is $140 per acre for crop A and $235 per acre for crop B. Find the number of acres of each fruit that should be planted to maximize profit, subject to the following constraints.

- It takes 1 day to trim an acre of crop A and 2 days to trim an acre of crop B, and there are 240 days per year available for trimming.

- It takes 0.3 day to pick an acre of crop A and 0.1 day to pick an acre of crop B, and there are 30 days per year available for picking.

**35.** *Minimum Cost* A farming cooperative mixes two brands of cattle feed. Brand X costs $25 per bag, and brand Y costs $20 per bag. Find the number of bags of each brand that should be mixed to produce a mixture having a minimum cost per bag, subject to the following constraints.

- Brand X contains 2 units of nutritional element A, 2 units of element B, and 2 units of element C.

- Brand Y contains 1 unit of nutritional element A, 9 units of element B, and 3 units of element C.

- The minimum requirements of nutrients A, B, and C are 12 units, 36 units, and 24 units, respectively.

**36.** *Minimum Cost* Two gasolines, type A and type B, have octane ratings of 89 and 93, respectively. Type A costs $0.99 per gallon and type B costs $1.25 per gallon. Determine the blend of minimum cost with an octane rating of at least 91. (*Hint:* Let $x$ be the fraction of each gallon that is type A and $y$ be the fraction that is type B.)

**37.** *Maximum Profit* A manufacturer produces two models of bicycles. The times (in hours) required for assembling, painting, and packaging each model are as follows.

| Process | Model A | Model B |
|---|---|---|
| Assembling | 2 | 2.5 |
| Painting | 4 | 1 |
| Packaging | 1 | 0.75 |

The total times available for assembling, painting, and packaging are 4000 hours, 4800 hours, and 1500 hours, respectively. The profits per unit are $45 for model A and $50 for model B. How many of each should be made to obtain a maximum profit?

**38.** *Maximum Profit* A company makes two models of a product. The times (in hours) required for assembling, painting, and packaging are as follows.

| Process | Model A | Model B |
|---|---|---|
| Assembling | 2.5 | 3 |
| Painting | 2 | 1 |
| Packaging | 0.75 | 1.25 |

The total times available for assembling, painting, and packaging are 4000 hours, 2500 hours, and 1500 hours, respectively. The profits per unit are $50 for model A and $52 for model B. How many of each should be made to obtain a maximum profit?

**39.** *Maximum Revenue* An accounting firm has 900 hours of staff time and 100 hours of review time available each week. The firm charges $2000 for an audit and $300 for a tax return. What numbers of audits and tax returns will bring in a maximum revenue, subject to the following constraints?

- Each audit requires 100 hours of staff time and 10 hours of review time.

- Each tax return requires 12.5 hours of staff time and 2.5 hours of review time.

**40.** *Maximum Revenue* The accounting firm in Exercise 39 lowers its charge for an audit to $1800. What numbers of audits and tax returns will bring in a maximum revenue?

**41.** *Media Selection* A company has budgeted a maximum of $600,000 for advertising a product nationally. Each minute of television time costs $60,000 and each one-page newspaper ad costs $15,000. Each television ad is expected to be viewed by 15 million viewers, and each newspaper ad is expected to be seen by 3 million readers. The company's market research department recommends that at most 90% of the advertising budget be spent on television ads. How should the advertising budget be spent to maximize the total audience?

**42.** *Maximum Profit* A fruit juice company makes two drinks by blending apple and pineapple juices. The first uses 30% apple juice and 70% pineapple, and the second uses 60% apple and 40% pineapple. There are 1000 liters of apple juice and 1500 liters of pineapple juice available. The profit for the first drink is $0.60 per liter and the profit for the second drink is $0.50. Find the number of liters of each drink that should be produced in order to maximize profit.

**43.** *Investments* An investor has up to $250,000 to invest in two types of investments. Type A pays 8% annually and type B pays 10% annually. To have a well-balanced portfolio, the investor imposes the following conditions. At least one-fourth of the total portfolio is to be allocated to type A investments and at least one-fourth of the portfolio is to be allocated to type B investments. How much should be allocated to each type of investment to obtain a maximum return?

**44.** *Investments* An investor has up to $450,000 to invest in two types of investments. Type A pays 6% annually and type B pays 10% annually. To have a well-balanced portfolio, the investor imposes the following conditions. At least one-half of the total portfolio is to be allocated to type A investments and at least one-fourth of the portfolio is to be allocated to type B investments. How much should be allocated to each type of investment to obtain a maximum return?

In Exercises 45–50, the given linear programming problem has an unusual characteristic. Sketch a graph of the solution region for the problem and describe the unusual characteristic. (In each problem, the objective function is to be maximized.)

**45.** *Objective function:*

$$z = 2.5x + y$$

*Constraints:*

$$x \geq 0$$
$$y \geq 0$$
$$3x + 5y \leq 15$$
$$5x + 2y \leq 10$$

**46.** *Objective function:*

$$z = x + y$$

*Constraints:*

$$x \geq 0$$
$$y \geq 0$$
$$-x + y \leq 1$$
$$-x + 2y \leq 4$$

**47.** *Objective function:*

$$z = -x + 2y$$

*Constraints:*

$$x \geq 0$$
$$y \geq 0$$
$$x \leq 10$$
$$x + y \leq 7$$

**48.** *Objective function:*

$$z = x + y$$

*Constraints:*

$$x \geq 0$$
$$y \geq 0$$
$$-x + y \leq 1$$
$$-3x + y \geq 3$$

**49.** *Objective function:*

$$z = 3x + 4y$$

*Constraints:*

$$x \geq 0$$
$$y \geq 0$$
$$x + y \leq 1$$
$$2x + y \leq 4$$

**50.** *Objective function:*

$$z = x + 2y$$

*Constraints:*

$$x \geq 0$$
$$y \geq 0$$
$$x + 2y \leq 4$$
$$2x + y \leq 4$$

**51.** *Maximum Profit* A company makes two models of a patio furniture set. The times for assembling, finishing, and packaging model A are 3.0 hours, 2.5 hours, and 0.60 hour, respectively. The times for model B are 2.75 hours, 1.0 hour, and 1.25 hours. The total times available for assembling, finishing, and packaging are 3000 hours, 2400 hours, and 1200 hours, respectively. The profit per unit for model A is $80 and the profit per unit for model B is $65. How many of each model should be produced to obtain a maximum profit?

## Chapter Project

## Consumer Credit Outstanding

The following table gives the amounts $y_1$ of consumer credit outstanding (in billions of dollars) held by finance companies from 1988 to 1998. In the table, $t = 0$ represents 1990. From the scatter plot of the data in Figure A, what type of equation would you guess models the data?

| $t$ | $-2$ | $-1$ | 0 | 1 | 2 | 3 | 4 | 5 | 6 | 7 | 8 |
|---|---|---|---|---|---|---|---|---|---|---|---|
| $y_1$ | 143.1 | 142.1 | 134.8 | 127.5 | 117.0 | 110.0 | 124.0 | 141.3 | 153.9 | 156.7 | 154.3 |

The amounts $y_2$ of U.S. consumer credit outstanding (in billions of dollars) held by credit unions during the same time are listed in the table below. A scatter plot for this data is shown in Figure B. What is your suggested model for this data?

| $t$ | $-2$ | $-1$ | 0 | 1 | 2 | 3 | 4 | 5 | 6 | 7 | 8 |
|---|---|---|---|---|---|---|---|---|---|---|---|
| $y_2$ | 83.1 | 88.3 | 90.4 | 88.5 | 88.3 | 95.4 | 108.2 | 125.7 | 136.1 | 147.6 | 149.4 |

Figure A

Figure B

Use this information to investigate the following questions. (Source: Federal Reserve Statistical Release, Consumer Credit Table G.19, Oct. 17, 1998. [Internet])

1. *Fitting a Model to Data*   Use a graphing utility to find a quadratic regression model and a quartic regression model for the data in the first table. Graph the two models along with the scatter plot of the data. Which model is a better fit?

2. *Comparing a Model to Actual Data*   Make a numerical comparison by constructing a table of the values given by the best-fitting model found in Question 1 and the actual data. Does the numerical comparison confirm that the model you found is valid?

3. *Fitting a Model to Data*   Use a graphing utility to find a quadratic regression model and a quartic regression model for the data in the second table. Graph the two models along with the scatter plot of the data. Which model is a better fit?

4. *Comparing a Model to Actual Data*   Make a numerical comparison by constructing a table of the values given by the best-fitting model found in Question 3 and the actual data. Does the numerical comparison confirm that the model you found is valid?

5. *Comparing Two Models*   Use a graphing utility to graph both models in the same viewing window, where $-2 \le t \le 8$. Do the graphs intersect? Does that agree with your data? Modify your viewing window so that $-2 \le t \le 10$. Do the graphs intersect? Are the models valid for $8 < t \le 10$? Explain.

## CHAPTER SUMMARY

After studying this chapter, you should have acquired the following skills.
These skills are keyed to the Review Exercises that begin on page 419.
Answers to odd-numbered Review Exercises are given in the back of the book.

**6.1** · Solve a system of equations by the method of substitution.        Review Exercises 1–6

· Solve a system of equations graphically.        Review Exercises 7, 8, 10

· Construct and use a system of equations to solve an application problem.        Review Exercises 9, 11, 12, 21, 22, 52, 53

**6.2** · Solve a linear system by the method of elimination.        Review Exercises 13–18

· Interpret the solution of a linear system graphically.        Review Exercises 19, 20

· Construct and use a linear system to solve an application problem.        Review Exercises 23, 24, 40

**6.3** · Solve a linear system in row-echelon form using back-substitution.        Review Exercises 25, 26

· Use Gaussian elimination to solve a linear system.        Review Exercises 27–32

· Construct and use a linear system in three or more variables to solve an application problem.        Review Exercises 37–39

· Find the equation of a circle or a parabola using a linear system in three or more variables.        Review Exercises 33–36, 41

**6.4** · Sketch the graph of an inequality in two variables.        Review Exercises 42–47

· Solve a system of inequalities.        Review Exercises 50, 51

· Use a system of inequalities to model and solve an application problem.        Review Exercises 48–51, 54, 55

**6.5** · Use linear programming to minimize or maximize an objective function.        Review Exercises 56–63

· Use linear programming to optimize an application.        Review Exercises 64–71

## REVIEW EXERCISES

In Exercises 1–6, solve the system by the method of substitution.

**1.** $x + 3y = 10$
  $4x - 5y = -28$

**2.** $3x - y - 13 = 0$
  $4x + 3y - 26 = 0$

**3.** $\frac{1}{2}x + \frac{3}{5}y = -2$
  $2x + y = 6$

**4.** $1.3x + 0.9y = 7.5$
  $0.4x - 0.5y = -0.8$

**5.** $x^2 + y^2 = 100$
  $x + 2y = 20$

**6.** $y = x^3 - 2x^2 - 2x - 3$
  $y = -x^2 + 4x - 3$

In Exercises 7 and 8, use a graphing utility to find all points of intersection of the graphs of the equations.

**7.** $y = x^2 - 3x + 11$
  $y = -x^2 + 2x + 8$

**8.** $y = \sqrt{9 - x^2}$
  $y = e^x + 1$

**9.** *Break-Even Point* You are setting up a business and have made an initial investment of $20,000. The unit cost of the product is $3.25 and the selling price is $6.95. How many units must you sell to break even? (Round to the nearest whole unit.)

**10.** *Comparing Product Sales* Your company produces three types of CD players. The research and development department has designed a new CD player that is predicted to sell better than the three players combined. The sales equations for the CD players are

$S = 2150.78 - 56.8t + 0.45t^2$  Three players combined

$S = 121.27 + 52.12t + 3.98t^2$  New player

where $S$ is the sales in thousands and $t = 8$ represents 1998. Use a graphing utility to determine when the sales of the new CD player will overtake the combined sales of the other players.

**11.** *Investment Portfolio* A total of $50,000 is invested in two funds paying 6.75% and 7.5% simple interest. If the yearly interest is $3637.50, how much of the $50,000 is invested at each rate?

**12.** *Choice of Two Jobs* You are offered two different jobs selling computers. One company offers an annual salary of $28,000 plus a year-end bonus of 1.5% of your total sales. The other company offers a salary of $22,000 plus a year-end bonus of 3.5% of your total sales. How much would you have to sell in order to make the second offer better?

In Exercises 13–18, solve the system by elimination.

**13.** $2x - 3y = 21$
  $3x + y = 4$

**14.** $3u + 5v = 9$
  $12u + 10v = 22$

**15.** $1.25x - 2y = 3.5$
  $5x - 8y = 14$

**16.** $\dfrac{x - 2}{3} + \dfrac{y + 3}{4} = 5$
  $2x - y = 7$

**17.** $1.5x + 2.5y = 8.5$
  $6x + 10y = 24$

**18.** $\frac{3}{5}x + \frac{2}{7}y = 10$
  $x + 2y = 38$

In Exercises 19 and 20, describe the graph of the solution of the linear system.

**19.** $2x + y = -1$
  $3x - 2y = -5$

**20.** $x - 2y = -1$
  $-2x + 4y = 2$

**21.** *Acid Mixture* Twelve gallons of a 25% acid solution are obtained by mixing a 10% solution with a 50% solution.

(a) Write a system of equations that represents the problem and use a graphing utility to graph both equations.

(b) How much of each solution must be used?

**22.** *Compact Disc Sales* Suppose you are the manager of a music store. At the end of the week you are going over receipts for the previous week's sales. Eight hundred and fifty compact discs were sold. One type of CD sold for $11.95 and a second type sold for $15.95. The total CD receipts were $11,257.50. The cash register that was supposed to keep track of the number of each type of CD sold malfunctioned. Can you recover the information? If so, how many of each type of CD were sold?

*Supply and Demand* In Exercises 23 and 24, find the point of equilibrium for the given pair of demand and supply equations.

| Demand | Supply |
|--------|--------|
| **23.** $p = 37 - 0.0002x$ | $p = 22 + 0.00001x$ |
| **24.** $p = 120 - 0.0001x$ | $p = 45 + 0.0002x$ |

In Exercises 25–32, solve the system of equations.

**25.** $4x - 3y + 2z = 1$
$\quad\quad\; 2y - 4z = 2$
$\quad\quad\quad\quad\; 3z = 6$

**26.** $2x + y - 4z = 6$
$\quad\quad\; 3y + z = 2$
$\quad\quad\quad\quad\; 2z = -8$

**27.** $2x + y + z = 6$
$\quad x - 4y - z = 3$
$\quad x + y + z = 4$

**28.** $\quad x + 3y - z = 13$
$\quad 2x \quad\quad - 5z = 23$
$\quad 4x - y - 2z = 14$

**29.** $\quad x + y + z = 10$
$-2x + 3y + 4z = 22$

**30.** $x + y + z = 1$
$2x - 4y + 3z = -20$
$2x + 3y + 2z = 5$

**31.** $2x + 6y - z = 1$
$\quad x - 3y + z = 2$
$\quad \frac{3}{2}x + \frac{3}{2}y = 6$

**32.** $x + y + z + w = 8$
$\quad\quad 4y + 5z - 2w = 3$
$2x + 3y - z = -2$
$3x + 2y \quad - 4w = -20$

In Exercises 33 and 34, find the equation of the parabola $y = ax^2 + bx + c$ that passes through the points.

**33.** $(0, -6), (1, -3), (2, 4)$
**34.** $(-5, 0), (1, -6), (2, 14)$

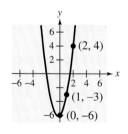

Figure for 33          Figure for 34

In Exercises 35 and 36, find the equation of the circle $x^2 + y^2 + Dx + Ey + F = 0$ that passes through the points.

**35.** $(2, 2), (5, -1), (-1, -1)$
**36.** $(4, 2), (1, 3), (-2, -6)$

 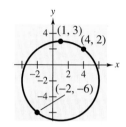

Figure for 35          Figure for 36

**37.** *Investment* Suppose you receive $1520 a year in interest from three investments. The interest rates for the three investments are 6%, 7%, and 8%. The 7% investment is six times the 6% investment, and the 8% investment is $1000 more than the 6% investment. What is the amount of each investment?

**38.** *Product Types* A company sells three types of products for $20, $15, and $10 per unit. In 1 year, the total revenue for the three products was $1,500,000, which corresponded to the sale of 110,000 units. The company sold half as many units of the $20 product as units of the $15 product. How many units of each product were sold?

**39.** *Investment Portfolio* Suppose an investor has a portfolio totaling $500,000 that is to be allocated among the following types of investments: (1) certificates of deposit, (2) municipal bonds, (3) blue-chip stocks, and (4) growth or speculative stocks. The certificates of deposit pay 8% annually, and the municipal bonds pay 6% annually. Over a 5-year period, the investor expects the blue-chip stocks to return 10% annually and the growth stocks to return 15% annually. The investor wishes a combined return of $8\frac{1}{2}\%$ and also wants to have only two-fifths of the portfolio invested in stocks. How much should be allocated to each type of investment if the amount invested in certificates of deposit is twice that invested in municipal bonds?

**40.** *Fitting a Line to Data*   Find the least squares regression line $y = ax + b$ for the points

$(0, 1.6), (1, 2.4), (2, 3.6), (3, 4.7), (4, 5.5)$.

To find the equation of the line, solve the system of linear equations for $a$ and $b$.

$$5b + 10a = 17.8$$
$$10b + 30a = 45.7$$

**41.** *Fitting a Parabola to Data*   Find the least squares regression parabola $y = ax^2 + bx + c$ for the points

$(-2, 0.4), (-1, 0.9), (0, 1.9), (1, 2.1), (2, 3.8)$.

To find the parabola, solve the system of linear equations for $a$, $b$, and $c$.

$$5c \quad\; + 10a = \;\, 9.1$$
$$\quad 10b \qquad = \;\, 8.0$$
$$10c \quad\; + 34a = 19.8$$

In Exercises 42–47, sketch the graph of the solution set of the system of inequalities.

**42.** $2x + 3y < 9$
$\quad\;\; x > 0$
$\quad\;\; y > 0$

**43.** $2x - y > 6$
$\quad\;\; x < 5$
$\quad\;\; y \le 8$

**44.** $3x - y > -4$
$\quad 2x + y > -1$
$\quad 7x + y < \;\; 4$

**45.** $x + y > \;\; 4$
$\quad 3x + y < 10$

**46.** $\quad\; x^2 + y^2 \le 9$
$\quad x^2 - x - 2 \le y$

**47.** $\ln x < \; y$
$\quad\;\; y > -1$
$\quad\;\; x < 4$

In Exercises 48 and 49, derive a set of inequalities that describes the region.

**48.** Parallelogram

**49.** Triangle

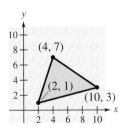

**50.** *Video Cassette Recorder Inventory*   A store sells two models of a certain brand of VCR. Because of the demand, it is necessary to stock twice as many units of model A as units of model B. The costs to the store for the two models are $200 and $300, respectively. The management does not want more than $4000 in VCR inventory at any one time, and it wants at least four model A VCRs and two model B VCRs in inventory at all times. Derive a system of inequalities that describes all possible inventory levels, and sketch the graph of the system.

**51.** *Concert Ticket Sales*   Two types of tickets are to be sold for a concert. One type costs $20 per ticket and the other type costs $30 per ticket. The promoter of the concert must sell at least 16,000 tickets, including at least 9000 of the $20 tickets and at least 4000 of the $30 tickets. Moreover, the gross receipts must total at least $400,000 in order for the concert to be held. Find a system of inequalities describing the different numbers of tickets that can be sold, and sketch the graph of the system.

**52.** *Morning or Evening Paper?*   The total circulations of morning and evening newspapers in the United States from 1980 to 1996 can be approximated by the models

$$y = 31.22 + 0.93t \qquad \text{Morning}$$
$$y = 32.40 - 1.22t \qquad \text{Evening}$$

where $y$ is the circulation (in millions) and $t$ represents the year, with $t = 0$ corresponding to 1980. During which year did the circulation of morning papers overtake the circulation of evening papers? (Source: Editor and Publisher Company)

**53.** *News Programs*   A local television station tracked the numbers of viewers of its two prime news shows for 1 year. The numbers of viewers for the two news shows can be approximated by the models

$$y = 9500 + 350t \qquad \text{6 P.M. news}$$
$$y = 15,000 - 275t \qquad \text{10 P.M. news}$$

where $y$ is the number of viewers and $t = 1$ corresponds to the month of January. During what month did the number of viewers of the 6 P.M. news program surpass the number of viewers of the 10 P.M. news program?

In Exercises 54 and 55, find the consumer surplus and producer surplus for the given pair of demand and supply equations.

| Demand | Supply |
|--------|--------|
| **54.** $p = 160 - 0.0001x$ | $p = 70 + 0.0002x$ |
| **55.** $p = 130 - 0.0002x$ | $p = 30 + 0.0003x$ |

In Exercises 56–59, find the minimum and maximum values of the objective function, subject to the indicated constraints. (For each exercise, the graph of the constraints is provided.)

**56.** *Objective function:*

$z = 5x + 6y$

*Constraints:*

$x \geq 0$

$y \geq 0$

$x + y \leq 8$

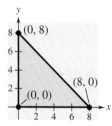

**57.** *Objective function:*

$z = 15x + 12y$

*Constraints:*

$x \geq 0$

$y \geq 0$

$x + 3y \leq 12$

$3x + 2y \leq 15$

**58.** *Objective function:*

$z = 8x + 3y$

*Constraints:*

$0 \leq x \leq 50$

$0 \leq y \leq 35$

$4x + 5y \leq 275$

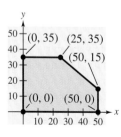

**59.** *Objective function:*

$z = 50x + 60y$

*Constraints:*

$x \geq 0$

$y \geq 0$

$3x + 4y \geq 1200$

$5x + 6y \leq 3000$

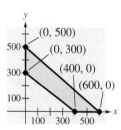

In Exercises 60–63, sketch the region determined by the constraints. Find the minimum and maximum values of the objective function, subject to the indicated constraints.

**60.** *Objective function:*

$z = 6x + 8y$

*Constraints:*

$x \geq 0$

$y \geq 0$

$x + 4y \leq 16$

$3x + 2y \leq 18$

**61.** *Objective function:*

$z = 5x + 8y$

*Constraints:*

$0 \leq x \leq 5$

$y \geq 0$

$x + 2y \leq 12$

$2x + 3y \leq 19$

**62.** *Objective function:*

$z = 8x + 3y$

*Constraints:*

$0 \leq x \leq 5$

$0 \leq y \leq 7$

$x + y \leq 9$

$3x + y \leq 17$

**63.** *Objective function:*

$z = 10x + 11y$

*Constraints:*

$x \geq 0$

$y \geq 0$

$2x + 5y \leq 30$

$x + y \geq 3$

$2x + y \leq 14$

**64.** *Maximum Profit*  A merchant plans to sell two models of camcorders at prices of $800 and $1000. The $800 model yields a profit of $75 and the $1000 model yields a profit of $125. The merchant estimates that the total monthly demand will not exceed 300 units. Find the number of units of each model that should be stocked in order to maximize profit. Assume that the merchant does not want to invest more than $280,000 in camcorder inventory.

**65.** *Maximum Profit*  A merchant plans to sell two models of color televisions at prices of $300 and $500. The $300 model yields a profit of $30 and the $500 model yields a profit of $45. The merchant estimates that the total monthly demand will not exceed 100 units. Find the number of units of each model that should be stocked in order to maximize profit. Assume that the merchant does not want to invest more than $40,000 in color television inventory.

**66.** *Maximum Profit* A factory manufactures two color television set models: a basic model that yields $100 profit and a deluxe model that yields a profit of $200. The times (in hours) required for assembling, finishing, and packaging are as follows.

| Process | Basic Model | Deluxe Model |
|---|---|---|
| Assembling | 2 | 5 |
| Finishing | 1 | 2 |
| Packaging | 1 | 1 |

The total times available for assembling, finishing, and packaging are 3000 hours, 1400 hours, and 1000 hours, respectively. How many of each model should be produced for maximum profit?

**67.** *Maximum Profit* A company makes two models of a product. The times (in hours) required for assembling, finishing, and packaging each model are as follows.

| Process | Model A | Model B |
|---|---|---|
| Assembling | 3.5 | 8 |
| Finishing | 2.5 | 2 |
| Packaging | 1.3 | 0.7 |

The total times available for assembling, finishing, and packaging are 5600 hours, 2000 hours, and 910 hours, respectively. The profits per unit are $100 for model A and $150 for model B. How many of each model should be produced to obtain a maximum profit?

**68.** *Minimum Cost* A pet supply company mixes two brands of dry dog food. Brand X costs $15 per bag and contains 8 units of nutritional element A, 1 unit of nutritional element B, and 2 units of nutritional element C. Brand Y costs $30 per bag and contains 2 units of nutritional element A, 1 unit of nutritional element B, and 7 units of nutritional element C. Each bag of mixed dog food must contain at least 16 units of element A, 5 units of element B, and 20 units of element C. Find the number of bags of each brand that should be mixed to produce a mixture having a minimum cost per bag.

**69.** *Minimum Cost* Two gasolines, type A and type B, have octane ratings of 85 and 89, respectively. Type A costs $1.01 per gallon and type B costs $1.13 per gallon. Determine the blend of minimum cost that has an octane rating of at least 87. (Hint: Let $x$ be the fraction of each gallon that is type A and let $y$ be the fraction of each gallon that is type B.)

**70.** *Maximum Revenue* An accounting firm has 800 hours of staff time and 90 hours of review time available each week. The firm charges $2400 for an audit and $300 for a tax return. Each audit requires 100 hours of staff time and 10 hours of review time. Each tax return requires 10 hours of staff time and 2 hours of review time. What number of audits and tax returns will bring in maximum revenue?

**71.** *Maximum Profit* Suppose the accounting firm in Exercise 70 realizes a profit of $800 for each audit and $120 for each tax return. What combination of audits and tax returns will yield a maximum profit?

## Chapter Test

Take this test as you would take a test in class. After you are done, check your work against the answers given in the back of the book.

In Exercises 1–8, solve the system of equations using the indicated method.

**1.** *Substitution*

$$3x - 2y = -2$$
$$4x + 3y = 20$$

**2.** *Substitution*

$$x + y = 3$$
$$x^2 + y = 9$$

**3.** *Substitution*

$$2x - y = 13$$
$$5x + 4y = 13$$

**4.** *Graphing*

$$5x - y = 6$$
$$2x^2 + y = 8$$

**5.** *Graphing*

$$1.5x - 2.25y = 8$$
$$2.5x + 2y = 5.75$$

**6.** *Elimination*

$$2x - 4y + z = 11$$
$$x + 2y + 3z = 9$$
$$3y + 5z = 12$$

**7.** *Elimination*

$$3x - 2y + z = 16$$
$$5x - z = 6$$
$$2x - y - z = 3$$

**8.** *Elimination*

$$-x + y - 2z = 3$$
$$x - 4y - 2z = 1$$
$$x + 2y + 6z = 5$$

**9.** Find the point of equilibrium for a system that has a demand function of $p = 45 - 0.0003x$ and a supply function of $p = 29 + 0.00002x$.

**10.** A system of linear equations reduces to $-23 = 0$. What can you conclude?

**11.** A system of linear equations reduces to $0 = 0$. What can you conclude?

**12.** Find an equation of the parabola passing through $(1, -5), (3, 1)$, and $(0, -2)$.

**13.** A total of \$50,000 is invested in two funds paying 8% and 8.5% simple interest. The annual interest is \$4150. How much is invested in each fund?

In Exercises 14–17, sketch the inequality.

**14.** $x \geq 0$     **15.** $y \geq 0$     **16.** $x + 3y \leq 12$     **17.** $3x + 2y \leq 15$

**18.** Sketch the solution of the system of inequalities comprised of the inequalities in Exercises 14–17.

**19.** Find the minimum and maximum values of the objective function $z = 6x + 7y$, subject to the constraints given in Exercises 14–17.

**20.** A manufacturer produces two models of stair climbers. The times (in hours) required for assembling, painting, and packaging each model are as follows.

• Assembling: 3.5 hours for model A; 8 hours for model B

• Painting: 2.5 hours for model A; 2 hours for model B

• Packaging: 1.3 hours for model A; 0.9 hour for model B

The total times available for assembling, painting, and packaging are 5600 hours, 2000 hours, and 900 hours, respectively. The profits per unit are \$200 for model A and \$275 for model B. How many of each type should be produced to obtain a maximum profit? Explain your reasoning.

# 7

# Matrices and Determinants

Month

Your advertising group has been hired by Company C, a cable/internet service provider, to develop an advertising campaign to increase its market share. Currently your client competes with two other service providers, Companies A and B. With this new campaign, you predict that each month (1) 92% of A's subscribers will remain with A, 3% will switch to B, and 5% will switch to C; (2) 91% of B's subscribers will remain with B, 7% will switch to A, and 1% will change to C; and (3) 93% of C's subscribers will remain with C, 6% will change to B, and 1% will change to A. The current percents of the market are given in the matrix

$$x_0 = \begin{bmatrix} 0.2 \\ 0.7 \\ 0.1 \end{bmatrix} \begin{matrix} \text{Company A} \\ \text{Company B} \\ \text{Company C} \end{matrix}$$

The market shares for Company C for the first six months of the campaign are shown in the table and bar graph.

| Month | 0   | 1     | 2     | 3     | 4     | 5     | 6     |
|-------|-----|-------|-------|-------|-------|-------|-------|
| Share | 0.1 | 0.137 | 0.169 | 0.196 | 0.219 | 0.239 | 0.255 |

The chapter project related to this information is on page 484.

Will and Deni McIntyre/Tony Stone Images

425

## 7.1     Matrices and Linear Systems

### Objectives

- Determine the order of a matrix.
- Perform elementary row operations on a matrix.
- Write a matrix in row-echelon form or reduced row-echelon form.
- Solve a system of linear equations using Gaussian elimination or Gauss-Jordan elimination.

## Matrices

In this section you will study a streamlined technique for solving systems of linear equations. This technique involves the use of a rectangular array of real numbers called a **matrix.** The plural of matrix is *matrices.*

---

▶ **Definition of a Matrix**

If $m$ and $n$ are positive integers, an $m \times n$ **matrix** (read "$m$ by $n$") is a rectangular array

$$
\begin{bmatrix}
a_{11} & a_{12} & a_{13} & \cdots & a_{1n} \\
a_{21} & a_{22} & a_{23} & \cdots & a_{2n} \\
a_{31} & a_{32} & a_{33} & \cdots & a_{3n} \\
\vdots & \vdots & \vdots & & \vdots \\
a_{m1} & a_{m2} & a_{m3} & \cdots & a_{mn}
\end{bmatrix}
\Bigg\} \; m \text{ rows}
$$

$n$ columns

in which each **entry,** $a_{ij}$, of the matrix is a number. An $m \times n$ matrix has $m$ **rows** (horizontal lines) and $n$ **columns** (vertical lines). A matrix that has only one row is called a **row matrix,** and a matrix that has only one column is called a **column matrix.**

---

The entry in the $i$th row and $j$th column is denoted by the *double subscript* notation $a_{ij}$. That is, $a_{21}$ refers to the entry in row 2, column 1. A matrix having $m$ rows and $n$ columns is said to be of **order** $m \times n$. If $m = n$, the matrix is **square** of order $n$. For a square matrix, the entries $a_{11}, a_{22}, a_{33}, \ldots$ are the **main diagonal** entries.

### Example 1     Examples of Matrices

The following matrices have the indicated orders.

**a.** *Order:* $1 \times 1$

$$[2]$$

**b.** *Order:* $1 \times 4$

$$\begin{bmatrix} 1 & -3 & 0 & \frac{1}{2} \end{bmatrix}$$

**c.** *Order:* $2 \times 2$

$$\begin{bmatrix} 0 & 0 \\ 0 & 0 \end{bmatrix}$$

**d.** *Order:* $3 \times 2$

$$\begin{bmatrix} 5 & 0 \\ 2 & -2 \\ -7 & 4 \end{bmatrix}$$

A matrix derived from a system of linear equations (each written in standard form with the constant term on the right) is the **augmented matrix** of the system. Moreover, the matrix derived from the coefficients of the system (but that does not include the constant terms) is the **coefficient matrix** of the system. Note in the matrices the use of 0 for the missing $y$-variable in the third equation, and also note the fourth column of constant terms in the augmented matrix.

$$
\begin{array}{ccc}
\textit{System} & \textit{Augmented Matrix} & \textit{Coefficient Matrix}
\end{array}
$$

$$
\begin{array}{rrrrr}
x & - & 4y & + & 3z & = & 5 \\
-x & + & 3y & - & z & = & -3 \\
2x & & & - & 4z & = & 6
\end{array}
\qquad
\left[
\begin{array}{rrr:r}
1 & -4 & 3 & 5 \\
-1 & 3 & -1 & -3 \\
2 & 0 & -4 & 6
\end{array}
\right]
\qquad
\left[
\begin{array}{rrr}
1 & -4 & 3 \\
-1 & 3 & -1 \\
2 & 0 & -4
\end{array}
\right]
$$

When forming either the coefficient matrix or the augmented matrix of a system, you should begin by vertically aligning the variables in the equations and using 0's for the missing variables.

$$
\begin{array}{ccc}
\textit{Given System} & \textit{Line Up Variables} & \textit{Form Augmented Matrix}
\end{array}
$$

$$
\begin{array}{rrr}
x + 3y & = & 9 \\
-y + 4z & = & -2 \\
x - 5z & = & 0
\end{array}
\qquad
\begin{array}{rrrr}
x & + & 3y & & = 9 \\
& & -y & + 4z & = -2 \\
x & & & - 5z & = 0
\end{array}
\qquad
\left[
\begin{array}{rrr:r}
1 & 3 & 0 & 9 \\
0 & -1 & 4 & -2 \\
1 & 0 & -5 & 0
\end{array}
\right]
$$

## Elementary Row Operations

In Section 6.3, you studied three operations that can be used on a system of linear equations to produce an equivalent system.

**1.** Interchange two equations.

**2.** Multiply an equation by a nonzero constant.

**3.** Add a multiple of an equation to another equation.

In matrix terminology, these three operations correspond to **elementary row operations.** An elementary row operation on an augmented matrix of a given system of linear equations produces a new augmented matrix corresponding to a new (but equivalent) system of linear equations. Two matrices are **row-equivalent** if one can be obtained from the other by a sequence of elementary row operations.

---

▶ **Elementary Row Operations**

**1.** Interchange two rows.

**2.** Multiply a row by a nonzero constant.

**3.** Add a multiple of a row to another row.

## Study Tip

Although elementary row operations are simple to perform, they involve a lot of arithmetic. Because it is easy to make a mistake, we suggest that you get in the habit of noting, next to the row you are changing, the elementary row operation performed in each step so that you can go back and check your work.

The next example demonstrates each of the elementary row operations that can be performed on a matrix to produce an equivalent matrix.

**Example 2**  Elementary Row Operations

**a.** Interchange the first and second rows.

*Original Matrix*

$$\begin{bmatrix} 0 & 1 & 3 & 4 \\ -1 & 2 & 0 & 3 \\ 2 & -3 & 4 & 1 \end{bmatrix}$$

*New Row-Equivalent Matrix*

$$\begin{matrix} R_2 \\ R_1 \end{matrix} \begin{bmatrix} -1 & 2 & 0 & 3 \\ 0 & 1 & 3 & 4 \\ 2 & -3 & 4 & 1 \end{bmatrix}$$

**b.** Multiply the first row by $\frac{1}{2}$.

*Original Matrix*

$$\begin{bmatrix} 2 & -4 & 6 & -2 \\ 1 & 3 & -3 & 0 \\ 5 & -2 & 1 & 2 \end{bmatrix}$$

*New Row-Equivalent Matrix*

$$\frac{1}{2}R_1 \rightarrow \begin{bmatrix} 1 & -2 & 3 & -1 \\ 1 & 3 & -3 & 0 \\ 5 & -2 & 1 & 2 \end{bmatrix}$$

**c.** Add $-2$ times the first row to the third row.

*Original Matrix*

$$\begin{bmatrix} 1 & 2 & -4 & 3 \\ 0 & 3 & -2 & -1 \\ 2 & 1 & 5 & -2 \end{bmatrix}$$

*New Row-Equivalent Matrix*

$$-2R_1 + R_3 \rightarrow \begin{bmatrix} 1 & 2 & -4 & 3 \\ 0 & 3 & -2 & -1 \\ 0 & -3 & 13 & -8 \end{bmatrix}$$

Note that we write the elementary row operation beside the row that we are changing.

---

**·TECHNOLOGY·**

Most graphing utilities can perform elementary row operations on matrices. Consult your user's guide to determine how to enter a matrix and perform elementary row operations. The screen on the left shows how one graphing utility displays the original matrix in Example 2(a). The screen on the right shows the new row-equivalent matrix in Example 2(a).

```
[A]
[ 0   1   3   4]
[ -1  2   0   3]
[ 2  -3   4   1]
```

```
[A]
[ -1  2   0   3]
[ 0   1   3   4]
[ 2  -3   4   1]
```

In Example 3 of Section 6.3, you used Gaussian elimination with back-substitution to solve a system of linear equations. The next example demonstrates the matrix version of Gaussian elimination. The two methods are essentially the same. The basic difference is that with matrices you do not need to keep writing the variables.

### Example 3    Using Elementary Row Operations

*Linear System*

$$x - 2y + 3z = 9$$
$$-x + 3y \quad\quad = -4$$
$$2x - 5y + 5z = 17$$

*Associated Augmented Matrix*

$$\begin{bmatrix} 1 & -2 & 3 & \vdots & 9 \\ -1 & 3 & 0 & \vdots & -4 \\ 2 & -5 & 5 & \vdots & 17 \end{bmatrix}$$

Add the first equation to the second equation.

Add the first row to the second row $(R_1 + R_2)$.

$$x - 2y + 3z = 9$$
$$y + 3z = 5$$
$$2x - 5y + 5z = 17$$

$$R_1 + R_2 \longrightarrow \begin{bmatrix} 1 & -2 & 3 & \vdots & 9 \\ 0 & 1 & 3 & \vdots & 5 \\ 2 & -5 & 5 & \vdots & 17 \end{bmatrix}$$

Add $-2$ times the first equation to the third equation.

Add $-2$ times the first row to the third row $(-2R_1 + R_3)$.

$$x - 2y + 3z = 9$$
$$y + 3z = 5$$
$$-y - z = -1$$

$$-2R_1 + R_3 \longrightarrow \begin{bmatrix} 1 & -2 & 3 & \vdots & 9 \\ 0 & 1 & 3 & \vdots & 5 \\ 0 & -1 & -1 & \vdots & -1 \end{bmatrix}$$

Add the second equation to the third equation.

Add the second row to the third row $(R_2 + R_3)$.

$$x - 2y + 3z = 9$$
$$y + 3z = 5$$
$$2z = 4$$

$$R_2 + R_3 \longrightarrow \begin{bmatrix} 1 & -2 & 3 & \vdots & 9 \\ 0 & 1 & 3 & \vdots & 5 \\ 0 & 0 & 2 & \vdots & 4 \end{bmatrix}$$

Multiply the third equation by $\frac{1}{2}$.

Multiply the third row by $\frac{1}{2}$.

$$x - 2y + 3z = 9$$
$$y + 3z = 5$$
$$z = 2$$

$$\tfrac{1}{2}R_3 \longrightarrow \begin{bmatrix} 1 & -2 & 3 & \vdots & 9 \\ 0 & 1 & 3 & \vdots & 5 \\ 0 & 0 & 1 & \vdots & 2 \end{bmatrix}$$

At this point, you can use back-substitution to find that the solution is $x = 1$, $y = -1$, and $z = 2$, as was done in Example 3 of Section 6.3.

Remember that you can check a solution by substituting the values of $x$, $y$, and $z$ into each equation in the original system.

The last matrix in Example 3 is said to be in **row-echelon form.** The term *echelon* refers to the stair-step pattern formed by the nonzero elements of the matrix. To be in this form, a matrix must have the following properties.

---

▶ **Row-Echelon Form and Reduced Row-Echelon Form**

A matrix in **row-echelon form** has the following properties.

1. All rows consisting entirely of zeros occur at the bottom of the matrix.

2. For each row that does not consist entirely of zeros, the first nonzero entry is 1 (called a **leading 1**).

3. For two successive (nonzero) rows, the leading 1 in the higher row is farther to the left than the leading 1 in the lower row.

A matrix in *row-echelon form* is in **reduced row-echelon form** if every column that has a leading 1 has zeros in every position above and below its leading 1.

---

**Example 4**  Row-Echelon Form

The following matrices are in row-echelon form.

a. $\begin{bmatrix} 1 & 2 & -1 & 4 \\ 0 & 1 & 0 & 3 \\ 0 & 0 & 1 & -2 \end{bmatrix}$
b. $\begin{bmatrix} 0 & 1 & 0 & 5 \\ 0 & 0 & 1 & 3 \\ 0 & 0 & 0 & 0 \end{bmatrix}$

c. $\begin{bmatrix} 1 & -5 & 2 & -1 & 3 \\ 0 & 0 & 1 & 3 & -2 \\ 0 & 0 & 0 & 1 & 4 \\ 0 & 0 & 0 & 0 & 1 \end{bmatrix}$
d. $\begin{bmatrix} 1 & 0 & 0 & -1 \\ 0 & 1 & 0 & 2 \\ 0 & 0 & 1 & 3 \\ 0 & 0 & 0 & 0 \end{bmatrix}$

The matrices in (b) and (d) also happen to be in *reduced* row-echelon form. The following matrices are not in row-echelon form.

e. $\begin{bmatrix} 1 & 2 & -3 & 4 \\ 0 & 2 & 1 & -1 \\ 0 & 0 & 1 & -3 \end{bmatrix}$
f. $\begin{bmatrix} 1 & 2 & -1 & 2 \\ 0 & 0 & 0 & 0 \\ 0 & 1 & 2 & -4 \end{bmatrix}$

---

Every matrix is row-equivalent to a matrix in row-echelon form. For instance, in Example 4, you can change the matrix in part (e) to row-echelon form by multiplying its second row by $\frac{1}{2}$. What elementary row operation could you perform on the matrix in part (f) so that it would be in row-echelon form?

# Gaussian Elimination with Back-Substitution

Example 5    Gaussian Elimination with Back-Substitution

Solve the following system.

$$\begin{aligned} y + z - 2w &= -3 \\ x + 2y - z \phantom{ - 2w} &= 2 \\ 2x + 4y + z - 3w &= -2 \\ x - 4y - 7z - w &= -19 \end{aligned}$$

## Study Tip

Gaussian elimination with back-substitution works well for solving systems of linear equations by hand or with a computer. For this algorithm, the order in which the elementary row operations are performed is important. We suggest operating from *left to right by columns,* using elementary row operations to obtain zeros in all entries directly below the leading 1's.

**Solution**

$$\begin{array}{c} R_2 \\ R_1 \end{array} \left[ \begin{array}{cccc:c} 1 & 2 & -1 & 0 & 2 \\ 0 & 1 & 1 & -2 & -3 \\ 2 & 4 & 1 & -3 & -2 \\ 1 & -4 & -7 & -1 & -19 \end{array} \right]$$

First column has leading 1 in upper left corner.

$$\begin{array}{c} \\ \\ -2R_1 + R_3 \rightarrow \\ -R_1 + R_4 \rightarrow \end{array} \left[ \begin{array}{cccc:c} 1 & 2 & -1 & 0 & 2 \\ 0 & 1 & 1 & -2 & -3 \\ 0 & 0 & 3 & -3 & -6 \\ 0 & -6 & -6 & -1 & -21 \end{array} \right]$$

First column has zeros below its leading 1.

$$\begin{array}{c} \\ \\ \\ 6R_2 + R_4 \rightarrow \end{array} \left[ \begin{array}{cccc:c} 1 & 2 & -1 & 0 & 2 \\ 0 & 1 & 1 & -2 & -3 \\ 0 & 0 & 3 & -3 & -6 \\ 0 & 0 & 0 & -13 & -39 \end{array} \right]$$

Second column has zeros below its leading 1.

$$\begin{array}{c} \\ \\ \frac{1}{3}R_3 \rightarrow \\ \\ \end{array} \left[ \begin{array}{cccc:c} 1 & 2 & -1 & 0 & 2 \\ 0 & 1 & 1 & -2 & -3 \\ 0 & 0 & 1 & -1 & -2 \\ 0 & 0 & 0 & -13 & -39 \end{array} \right]$$

Third column has zeros below its leading 1.

$$\begin{array}{c} \\ \\ \\ -\frac{1}{13}R_4 \rightarrow \end{array} \left[ \begin{array}{cccc:c} 1 & 2 & -1 & 0 & 2 \\ 0 & 1 & 1 & -2 & -3 \\ 0 & 0 & 1 & -1 & -2 \\ 0 & 0 & 0 & 1 & 3 \end{array} \right]$$

Fourth column has a leading 1.

The matrix is now in row-echelon form, and the corresponding system is

$$\begin{aligned} x + 2y - z \phantom{ - 2w} &= 2 \\ y + z - 2w &= -3 \\ z - w &= -2 \\ w &= 3 . \end{aligned}$$

Using back-substitution, you can determine that the solution is $x = -1$, $y = 2$, $z = 1$, and $w = 3$. Check this in the original system of equations.

> ▶ Gaussian Elimination with Back-Substitution
>
> 1. Write the augmented matrix of the system of linear equations.
>
> 2. Use elementary row operations to rewrite the augmented matrix in row-echelon form.
>
> 3. Write the system of linear equations corresponding to the matrix in row-echelon form, and use back-substitution to find the solution.

When solving a system of linear equations, remember that it is possible for the system to have no solution. If, in the elimination process, you obtain a row with zeros except for the last entry, it is unnecessary to continue the elimination process. You can simply conclude that the system is inconsistent.

**Example 6**   A System with No Solution

Solve the following system.

$$x - y + 2z = 4$$
$$x \qquad + z = 6$$
$$2x - 3y + 5z = 4$$
$$3x + 2y - z = 1$$

**Solution**

$$\begin{bmatrix} 1 & -1 & 2 & \vdots & 4 \\ 1 & 0 & 1 & \vdots & 6 \\ 2 & -3 & 5 & \vdots & 4 \\ 3 & 2 & -1 & \vdots & 1 \end{bmatrix} \quad \begin{matrix} \\ -R_1 + R_2 \rightarrow \\ -2R_1 + R_3 \rightarrow \\ -3R_1 + R_4 \rightarrow \end{matrix} \begin{bmatrix} 1 & -1 & 2 & \vdots & 4 \\ 0 & 1 & -1 & \vdots & 2 \\ 0 & -1 & 1 & \vdots & -4 \\ 0 & 5 & -7 & \vdots & -11 \end{bmatrix}$$

$$R_2 + R_3 \rightarrow \begin{bmatrix} 1 & -1 & 2 & \vdots & 4 \\ 0 & 1 & -1 & \vdots & 2 \\ 0 & 0 & 0 & \vdots & -2 \\ 0 & 5 & -7 & \vdots & -11 \end{bmatrix}$$

Note that the third row of this matrix consists of zeros except for the last entry. This means that the original system of linear equations is *inconsistent*. You can see why this is true by converting back to a system of linear equations.

$$x - y + 2z = 4$$
$$y - z = 2$$
$$0 = -2$$
$$5y - 7z = -11$$

Because the third equation is not possible, the system has no solution.

## Gauss-Jordan Elimination

With Gaussian elimination, we apply elementary row operations to a matrix to obtain a (row-equivalent) row-echelon form. A second method of elimination called **Gauss-Jordan elimination,** after Carl Friedrich Gauss and Wilhelm Jordan (1842–1899), continues the reduction process until a *reduced* row-echelon form is obtained. We demonstrate this procedure in the following example.

### Study Tip

Either Gaussian elimination or Gauss-Jordan elimination can be used to solve a system of equations. The method you use depends on your preference.

**Example 7**    Gauss-Jordan Elimination

Use Gauss-Jordan elimination to solve the following system.

$$x - 2y + 3z = \phantom{-}9$$
$$-x + 3y \phantom{+ 5z} = -4$$
$$2x - 5y + 5z = \phantom{-}17$$

### Solution

In Example 3 we used Gaussian elimination to obtain the row-echelon form

$$\begin{bmatrix} 1 & -2 & 3 & \vdots & 9 \\ 0 & 1 & 3 & \vdots & 5 \\ 0 & 0 & 1 & \vdots & 2 \end{bmatrix}.$$

Now, rather than using back-substitution, apply elementary row operations until you obtain a matrix in reduced row-echelon form. To do this, you must produce zeros above each of the leading 1's, as follows.

$$2R_2 + R_1 \rightarrow \begin{bmatrix} 1 & 0 & 9 & \vdots & 19 \\ 0 & 1 & 3 & \vdots & 5 \\ 0 & 0 & 1 & \vdots & 2 \end{bmatrix}$$

Second column has zeros above its leading 1.

$$\begin{matrix} -9R_3 + R_1 \rightarrow \\ -3R_3 + R_2 \rightarrow \end{matrix} \begin{bmatrix} 1 & 0 & 0 & \vdots & 1 \\ 0 & 1 & 0 & \vdots & -1 \\ 0 & 0 & 1 & \vdots & 2 \end{bmatrix}$$

Third column has zeros above its leading 1.

Now, converting back to a system of linear equations, you have

$$x = \phantom{-}1$$
$$y = -1$$
$$z = \phantom{-}2.$$

The beauty of Gauss-Jordan elimination is that, from the reduced row-echelon form, you can simply read the solution.

The elimination procedures described in this section employ an algorithmic approach that is easily adapted to computer use. However, the procedure makes no effort to avoid fractional coefficients. For instance, if the system given in Example 7 had been listed as

$$2x - 5y + 5z = 17$$
$$x - 2y + 3z = 9$$
$$-x + 3y = -4$$

the procedure would have required multiplication of the first row by $\frac{1}{2}$, which would have introduced fractions in the first row. For hand computations, fractions can sometimes be avoided by judiciously choosing the order in which the elementary row operations are applied.

## Study Tip

Remember that the solution set for a system with an infinite number of solutions can be written in several ways. For example, the solution set in Example 8 could have been written as

$$\left( \frac{1 - 5b}{3}, b, \frac{b + 1}{3} \right)$$

where $b$ is a real number.

### Example 8  A System with an Infinite Number of Solutions

Solve the following system.

$$2x + 4y - 2z = 0$$
$$3x + 5y = 1$$

**Solution**

$$\begin{bmatrix} 2 & 4 & -2 & \vdots & 0 \\ 3 & 5 & 0 & \vdots & 1 \end{bmatrix} \qquad \frac{1}{2}R_1 \rightarrow \begin{bmatrix} 1 & 2 & -1 & \vdots & 0 \\ 3 & 5 & 0 & \vdots & 1 \end{bmatrix}$$

$$-3R_1 + R_2 \rightarrow \begin{bmatrix} 1 & 2 & -1 & \vdots & 0 \\ 0 & -1 & 3 & \vdots & 1 \end{bmatrix}$$

$$-R_2 \rightarrow \begin{bmatrix} 1 & 2 & -1 & \vdots & 0 \\ 0 & 1 & -3 & \vdots & -1 \end{bmatrix}$$

$$-2R_2 + R_1 \rightarrow \begin{bmatrix} 1 & 0 & 5 & \vdots & 2 \\ 0 & 1 & -3 & \vdots & -1 \end{bmatrix}$$

The corresponding system of equations is

$$x + 5z = 2$$
$$y - 3z = -1.$$

Solving for $x$ and $y$ in terms of $z$, you have $x = -5z + 2$ and $y = 3z - 1$. Then, letting $z = a$, the solution set has the form

$$(-5a + 2, 3a - 1, a)$$

where $a$ is a real number. Try substituting values for $a$ to obtain a few solutions. Then check each solution in the original system of equations.

It is worth noting that the row-echelon form of a matrix is not unique. That is, two different sequences of elementary row operations may yield different row-echelon forms. For instance, the following sequence of elementary row operations on the matrix in Example 3 produces a slightly different row-echelon form.

$$\begin{bmatrix} 1 & -2 & 3 & \vdots & 9 \\ -1 & 3 & 0 & \vdots & -4 \\ 2 & -5 & 5 & \vdots & 17 \end{bmatrix} \qquad \begin{matrix} R_2 \\ R_1 \end{matrix} \begin{bmatrix} -1 & 3 & 0 & \vdots & -4 \\ 1 & -2 & 3 & \vdots & 9 \\ 2 & -5 & 5 & \vdots & 17 \end{bmatrix}$$

$$-R_1 \longrightarrow \begin{bmatrix} 1 & -3 & 0 & \vdots & 4 \\ 1 & -2 & 3 & \vdots & 9 \\ 2 & -5 & 5 & \vdots & 17 \end{bmatrix}$$

$$\begin{matrix} -R_1 + R_2 \longrightarrow \\ -2R_1 + R_3 \longrightarrow \end{matrix} \begin{bmatrix} 1 & -3 & 0 & \vdots & 4 \\ 0 & 1 & 3 & \vdots & 5 \\ 0 & 1 & 5 & \vdots & 9 \end{bmatrix}$$

$$-R_2 + R_3 \longrightarrow \begin{bmatrix} 1 & -3 & 0 & \vdots & 4 \\ 0 & 1 & 3 & \vdots & 5 \\ 0 & 0 & 2 & \vdots & 4 \end{bmatrix}$$

$$\tfrac{1}{2}R_3 \longrightarrow \begin{bmatrix} 1 & -3 & 0 & \vdots & 4 \\ 0 & 1 & 3 & \vdots & 5 \\ 0 & 0 & 1 & \vdots & 2 \end{bmatrix}$$

The corresponding system of linear equations is

$$\begin{aligned} x - 3y &= 4 \\ y + 3z &= 5 \\ z &= 2. \end{aligned}$$

Try using back-substitution on this system to see that you obtain the same solution that was obtained in Example 3.

| Discussing the Concept | Unique Solutions Versus Infinite Solutions |
|---|---|

Construct one linear system of three variables that has a unique solution and another linear system of three variables that has an infinite number of solutions. Exchange systems with another student. Solve the systems you receive using the methods of this section. Compare and discuss your solutions with each other. Did you both write the infinite solution in the same form? If not, verify that your answers represent the same solution set.

# Warm Up

The following warm-up exercises involve skills that were covered in earlier sections. You will use these skills in the exercise set for this section.

In Exercises 1–4, evaluate the expression.

**1.** $2(-1) - 3(5) + 7(2)$

**2.** $-4(-3) + 6(7) + 8(-3)$

**3.** $11\left(\frac{1}{2}\right) - 7\left(-\frac{3}{2}\right) - 5(2)$

**4.** $\frac{2}{3}\left(\frac{1}{2}\right) + \frac{4}{3}\left(-\frac{1}{3}\right)$

In Exercises 5 and 6, decide whether $x = 1$, $y = 3$, and $z = -1$ is a solution of the system.

**5.**
$$\begin{aligned} 4x - 2y + 3z &= -5 \\ x + 3y - z &= 11 \\ -x + 2y &= 5 \end{aligned}$$

**6.**
$$\begin{aligned} -x + 2y + z &= 4 \\ 2x - 3z &= 5 \\ 3x + 5y - 2z &= 21 \end{aligned}$$

In Exercises 7–10, use back-substitution to solve the system of linear equations.

**7.**
$$\begin{aligned} 2x - 3y &= 4 \\ y &= 2 \end{aligned}$$

**8.**
$$\begin{aligned} 5x + 4y &= 0 \\ y &= -3 \end{aligned}$$

**9.**
$$\begin{aligned} x - 3y + z &= 0 \\ y - 3z &= 8 \\ z &= 2 \end{aligned}$$

**10.**
$$\begin{aligned} 2x - 5y + 3z &= -2 \\ y - 4z &= 0 \\ z &= 1 \end{aligned}$$

## 7.1    Exercises

In Exercises 1–6, determine the order of the matrix.

**1.** $\begin{bmatrix} 5 & -1 & 6 \\ 3 & 1 & -2 \end{bmatrix}$

**2.** $\begin{bmatrix} 4 & -1 \end{bmatrix}$

**3.** $\begin{bmatrix} 6 & 4 & 1 \\ 8 & 3 & 0 \\ -1 & 2 & 1 \\ 1 & 5 & 4 \end{bmatrix}$

**4.** $\begin{bmatrix} 1 \\ 0 \\ 3 \\ 5 \\ 6 \end{bmatrix}$

**5.** $\begin{bmatrix} 33 & 45 \\ -9 & 20 \\ 12 & 15 \\ 16 & -2 \end{bmatrix}$

**6.** $\begin{bmatrix} 12 & -2 & 4 \\ -3 & 4 & 0 \\ -8 & 12 & 2 \end{bmatrix}$

In Exercises 7–10, determine whether the matrix is in row-echelon form. If it is, determine if it is also in reduced row-echelon form.

**7.** $\begin{bmatrix} 1 & 0 & 0 & 0 \\ 0 & 1 & 1 & 5 \\ 0 & 0 & 0 & 0 \end{bmatrix}$

**8.** $\begin{bmatrix} 1 & 0 & 2 & 1 \\ 0 & 1 & -3 & 10 \\ 0 & 0 & 1 & 0 \end{bmatrix}$

**9.** $\begin{bmatrix} 2 & 0 & 4 & 0 \\ 0 & -1 & 3 & 6 \\ 0 & 0 & 1 & 5 \end{bmatrix}$

**10.** $\begin{bmatrix} 1 & 3 & 0 & 0 & 0 & 0 \\ 0 & 0 & 1 & 8 & 1 & 0 \\ 0 & 0 & 0 & 0 & 1 & 1 \\ 0 & 0 & 0 & 0 & 1 & 1 \end{bmatrix}$

**11.** Use a graphing utility to perform the row operations to reduce the matrix to row-echelon form.

$$\begin{bmatrix} 1 & 1 & 2 \\ 3 & 4 & -3 \\ 2 & -1 & 1 \end{bmatrix}$$

(a) Add $-3$ times Row 1 to Row 2.

(b) Add $-2$ times Row 1 to Row 3.

(c) Add 3 times Row 2 to Row 3.

(d) Multiply Row 3 by $-\frac{1}{30}$.

**12.** Use a graphing utility to perform the operations to reduce the matrix to *reduced* row-echelon form.

$$\begin{bmatrix} 7 & 1 \\ 0 & 2 \\ -3 & 4 \\ 4 & 1 \end{bmatrix}$$

(a) Add Row 3 to Row 4.

(b) Interchange Rows 1 and 4.

(c) Add 3 times Row 1 to Row 3.

(d) Add $-7$ times Row 1 to Row 4.

(e) Multiply Row 2 by $\frac{1}{2}$.

(f) Add the appropriate multiple of Row 2 to Rows 1, 3, and 4.

In Exercises 13–16, write the matrix in row-echelon form. (*Note:* Row-echelon forms are not unique.)

**13.** $\begin{bmatrix} 1 & 2 & -1 & 5 \\ 3 & 2 & 1 & 11 \\ 4 & 8 & 1 & 10 \end{bmatrix}$

**14.** $\begin{bmatrix} 1 & 2 & -1 & 3 \\ 3 & 7 & -5 & 14 \\ -2 & -1 & -3 & 8 \end{bmatrix}$

**15.** $\begin{bmatrix} 1 & -1 & -1 & 1 \\ 5 & -4 & 1 & 8 \\ -6 & 8 & 18 & 0 \end{bmatrix}$

**16.** $\begin{bmatrix} 1 & -3 & 0 & -7 \\ -3 & 10 & 1 & 23 \\ 1 & 0 & 1 & 12 \\ 4 & -10 & 2 & -24 \end{bmatrix}$

In Exercises 17–20, write the matrix in *reduced* row-echelon form.

**17.** $\begin{bmatrix} 4 & 4 & 8 \\ 1 & 2 & 2 \\ -3 & 6 & -9 \end{bmatrix}$   **18.** $\begin{bmatrix} 1 & 3 & 2 \\ 5 & 15 & 9 \\ 2 & 6 & 10 \end{bmatrix}$

**19.** $\begin{bmatrix} 1 & 2 & 3 & -5 \\ 1 & 2 & 4 & -9 \\ -2 & -4 & -4 & 3 \\ 4 & 8 & 11 & -14 \end{bmatrix}$

**20.** $\begin{bmatrix} 1 & -3 \\ -1 & 8 \\ 0 & 4 \\ -2 & 10 \end{bmatrix}$

In Exercises 21–24, write the system of linear equations represented by the augmented matrix. (Use variables $w$, $x$, $y$, and $z$.)

**21.** $\begin{bmatrix} 2 & 4 & \vdots & 6 \\ -1 & 3 & \vdots & -8 \end{bmatrix}$   **22.** $\begin{bmatrix} 7 & -2 & \vdots & 7 \\ -8 & 3 & \vdots & -3 \end{bmatrix}$

**23.** $\begin{bmatrix} 1 & 0 & 2 & \vdots & -10 \\ 0 & 3 & -1 & \vdots & 5 \\ 4 & 2 & 0 & \vdots & 3 \end{bmatrix}$

**24.** $\begin{bmatrix} 5 & 8 & 2 & 0 & \vdots & -1 \\ -2 & 15 & 5 & 1 & \vdots & 9 \\ 1 & 6 & -7 & 0 & \vdots & -3 \end{bmatrix}$

In Exercises 25–28, write the system of equations represented by the augmented matrix. Use back-substitution to find the solution. (Use $w$, $x$, $y$, and $z$.)

**25.** $\begin{bmatrix} 1 & -5 & \vdots & 6 \\ 0 & 1 & \vdots & -2 \end{bmatrix}$

**26.** $\begin{bmatrix} 1 & 2 & -1 & \vdots & 3 \\ 0 & 1 & -2 & \vdots & -3 \\ 0 & 0 & 1 & \vdots & 4 \end{bmatrix}$

**27.** $\begin{bmatrix} 1 & 3 & -1 & \vdots & 15 \\ 0 & 1 & 4 & \vdots & -12 \\ 0 & 0 & 1 & \vdots & -5 \end{bmatrix}$

**28.** $\begin{bmatrix} 1 & 2 & -2 & 0 & \vdots & -1 \\ 0 & 1 & 1 & 2 & \vdots & 9 \\ 0 & 0 & 1 & 0 & \vdots & 2 \\ 0 & 0 & 0 & 1 & \vdots & -3 \end{bmatrix}$

In Exercises 29–32, an augmented matrix that represents a system of linear equations (in variables $x, y$, and $z$) has been reduced using Gauss-Jordan elimination. Write the solution represented by the augmented matrix.

**29.** $\begin{bmatrix} 1 & 0 & \vdots & -4 \\ 0 & 1 & \vdots & 6 \end{bmatrix}$     **30.** $\begin{bmatrix} 1 & 0 & \vdots & 9 \\ 0 & 1 & \vdots & -3 \end{bmatrix}$

**31.** $\begin{bmatrix} 1 & 0 & 0 & \vdots & -4 \\ 0 & 1 & 0 & \vdots & -8 \\ 0 & 0 & 1 & \vdots & 2 \end{bmatrix}$

**32.** $\begin{bmatrix} 1 & 0 & 0 & \vdots & 3 \\ 0 & 1 & 0 & \vdots & -1 \\ 0 & 0 & 1 & \vdots & 0 \end{bmatrix}$

In Exercises 33–36, write the augmented matrix for the system of linear equations.

**33.** $\begin{aligned} 2x - y &= 3 \\ 5x + 7y &= 12 \end{aligned}$     **34.** $\begin{aligned} 8x + 3y &= 25 \\ 3x - 9y &= 12 \end{aligned}$

**35.** $\begin{aligned} x + 10y - 3z &= 2 \\ 5x - 3y + 4z &= 0 \\ 2x + 4y &= 6 \end{aligned}$

**36.** $\begin{aligned} 9w - 3x + 20y + z &= 13 \\ 12w - 8y &= 5 \\ w + 2x + 3y - 4z &= -2 \\ -w - x + y + z &= 1 \end{aligned}$

In Exercises 37–58, solve the system of equations. Use Gaussian elimination with back-substitution or Gauss-Jordan elimination.

**37.** $\begin{aligned} x + 2y &= 7 \\ 2x + y &= 8 \end{aligned}$     **38.** $\begin{aligned} 2x + 6y &= 16 \\ 2x + 3y &= 7 \end{aligned}$

**39.** $\begin{aligned} -3x + 5y &= -22 \\ 3x + 4y &= 4 \\ 4x - 8y &= 32 \end{aligned}$     **40.** $\begin{aligned} x + 2y &= 0 \\ x + y &= 6 \\ 3x - 2y &= 8 \end{aligned}$

**41.** $\begin{aligned} 8x - 4y &= 7 \\ 5x + 2y &= 1 \end{aligned}$     **42.** $\begin{aligned} 2x - y &= -0.1 \\ 3x + 2y &= 1.6 \end{aligned}$

**43.** $\begin{aligned} -x + 2y &= 1.5 \\ 2x - 4y &= 3 \end{aligned}$     **44.** $\begin{aligned} x - 3y &= 5 \\ -2x + 6y &= -10 \end{aligned}$

**45.** $\begin{aligned} 2x + 3z &= 3 \\ 4x - 3y + 7z &= 5 \\ 8x - 9y + 15z &= 9 \end{aligned}$     **46.** $\begin{aligned} 2x - y + 3z &= 24 \\ 2y - z &= 14 \\ 7x - 5y &= 6 \end{aligned}$

**47.** $\begin{aligned} x + y - 5z &= 3 \\ x - 2z &= 1 \\ 2x - y - z &= 0 \end{aligned}$     **48.** $\begin{aligned} x - 3z &= -2 \\ 3x + y - 2z &= 5 \\ 2x + 2y + z &= 4 \end{aligned}$

**49.** $\begin{aligned} x + 2y + z &= 8 \\ 3x + 7y + 6z &= 26 \end{aligned}$

**50.** $\begin{aligned} 4x + 12y - 7z - 20w &= 22 \\ 3x + 9y - 5z - 28w &= 30 \end{aligned}$

**51.** $\begin{aligned} 3x + 3y + 12z &= 6 \\ x + y + 4z &= 2 \\ 2x + 5y + 20z &= 10 \\ -x + 2y + 8z &= 4 \end{aligned}$

**52.** $\begin{aligned} 2x + 10y + 2z &= 6 \\ x + 5y + 2z &= 6 \\ x + 5y + z &= 3 \\ -3x - 15y - 3z &= -9 \end{aligned}$

**53.** $\begin{aligned} 2x + y - z + 2w &= -16 \\ 3x + 4y + w &= 1 \\ x + 5y + 2z + 6w &= -3 \\ 5x + 2y - z - w &= 3 \end{aligned}$

**54.** $\begin{aligned} x + 2y + 2z + 4w &= 11 \\ 3x + 6y + 5z + 12w &= 30 \end{aligned}$

**55.** $\begin{aligned} x + 2y &= 0 \\ -x - y &= 0 \end{aligned}$     **56.** $\begin{aligned} x + 2y &= 0 \\ 2x + 4y &= 0 \end{aligned}$

**57.** $\begin{aligned} x + y + z &= 0 \\ 2x + 3y + z &= 0 \\ 3x + 5y + z &= 0 \end{aligned}$

**58.** $\begin{aligned} x + 2y + z + 3w &= 0 \\ x - y + w &= 0 \\ y - z + 2w &= 0 \end{aligned}$

**59.** *Borrowing Money*   A small corporation borrowed $500,000 to expand its product line. Some of the money was borrowed at 9%, some at 10%, and some at 12%. How much was borrowed at each rate if the annual interest was $52,000 and the amount borrowed at 10% was two and one-half times the amount borrowed at 9%?

**60.** *Borrowing Money*   A company borrowed $800,000 to expand its product line. Some of the money was borrowed at 10%, some at 10.5%, and some at 11.5%. How much was borrowed at each rate if the annual interest was $83,500 and the amount borrowed at 10% was four times the amount borrowed at 11.5%?

**61.** You and a friend solve the following system of equations independently.

$$2x - 4y - 3z = 3$$
$$x + 3y + z = -1$$
$$5x + y - 2z = 2$$

You write your solution set as $(a, -a, 2a - 1)$, where $a$ is any real number. Your friend's solution set is $\left(\frac{1}{2}b + \frac{1}{2}, -\frac{1}{2}b - \frac{1}{2}, b\right)$, where $b$ is any real number. Are you both correct? Explain. If you let $a = 3$, what value of $b$ should be selected so that you have the same ordered triple?

**62.** Describe how you would explain to another student that the augmented matrix below represents a dependent system of equations. Describe a way to write the infinitely many solutions to this system.

$$\begin{bmatrix} 1 & -2 & 3 & \vdots & -6 \\ 0 & 1 & 2 & \vdots & 5 \\ 0 & 0 & 0 & \vdots & 0 \end{bmatrix}$$

**63.** *Average Salary*   From 1987 to 1996, the average annual salary for professors at public colleges and universities in the United States increased in a pattern that was approximately linear (see figure). Find the least squares regression line $y = a + bt$ for the data shown in the figure by solving the following system using a graphing utility. ($y$ represents the average salary in thousands of dollars and $t = 0$ represents 1991.)

$$10a + 5b = 553.1$$
$$5a + 85b = 444.9$$

Use the result to predict the average salary for a professor at a public college in 1999. (Source: American Association of University Professors)

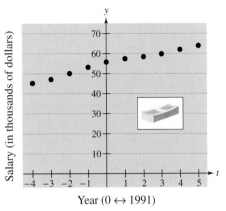

Year (0 ↔ 1991)

Figure for 63

**64.** *Average Salary*   From 1987 to 1996, the average annual salary for professors at private (non-church-related) colleges and universities in the United States increased in a pattern that was approximately linear (see figure). Find the least squares regression line $y = a + bt$ for the data shown in the figure by solving the following system using a graphing utility. ($y$ represents the average salary in thousands of dollars and $t = 0$ represents 1991.)

$$10a + 5b = 634.4$$
$$5a + 85b = 558.7$$

Use the result to predict the average salary for a professor at a private college in 1999. (Source: American Association of University Professors)

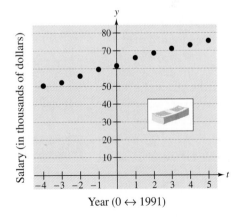

Year (0 ↔ 1991)

## 7.2    Operations with Matrices

### Objectives

- Determine if two matrices are equal.
- Add or subtract two matrices.
- Multiply a matrix by a scalar.
- Find the product of two matrices.
- Solve a matrix equation.

The Granger Collection

**Arthur Cayley**

**1821–1895**

A British mathematician, Arthur Cayley, invented matrices around 1858. Cayley was a Cambridge University graduate and a lawyer by profession. His groundbreaking work on matrices was begun as he studied the theory of transformations. Cayley also was instrumental in the development of determinants. Cayley and two American mathematicians, Benjamin Peirce (1809–1880) and his son Charles S. Peirce (1839–1914), are credited with developing "matrix algebra."

## Equality of Matrices

In Section 7.1, you used matrices to solve systems of linear equations. Matrices, however, can do much more than this. There is a rich mathematical theory of matrices, and its applications are numerous. This section and the next introduce some fundamentals of matrix theory. It is standard mathematical convention to represent matrices in any of the following three ways.

**1.** A matrix can be denoted by an uppercase letter such as $A$, $B$, or $C$.

**2.** A matrix can be denoted by a representative element enclosed in brackets, such as $[a_{ij}]$, $[b_{ij}]$, or $[c_{ij}]$.

**3.** A matrix can be denoted by a rectangular array of numbers such as

$$A = [a_{ij}] = \begin{bmatrix} a_{11} & a_{12} & a_{13} & \cdots & a_{1n} \\ a_{21} & a_{22} & a_{23} & \cdots & a_{2n} \\ a_{31} & a_{32} & a_{33} & \cdots & a_{3n} \\ \vdots & \vdots & \vdots & & \vdots \\ a_{m1} & a_{m2} & a_{m3} & \cdots & a_{mn} \end{bmatrix}.$$

Two matrices $A = [a_{ij}]$ and $B = [b_{ij}]$ are **equal** if they have the same order $(m \times n)$ and

$$a_{ij} = b_{ij}$$

for $1 \le i \le m$ and $1 \le j \le n$. In other words, two matrices are equal if their corresponding entries are equal.

### Example 1    Equality of Matrices

Solve for $a_{11}, a_{12}, a_{21},$ and $a_{22}$ in the following matrix equation.

$$\begin{bmatrix} a_{11} & a_{12} \\ a_{21} & a_{22} \end{bmatrix} = \begin{bmatrix} 2 & -1 \\ -3 & 0 \end{bmatrix}$$

### Solution

Because two matrices are equal only if their corresponding entries are equal, you can conclude that

$$a_{11} = 2, \quad a_{12} = -1, \quad a_{21} = -3, \quad \text{and} \quad a_{22} = 0.$$

# Matrix Addition and Scalar Multiplication

You can **add** two matrices (of the same order) by adding their corresponding entries.

---

▶ **Definition of Matrix Addition**

If $A = [a_{ij}]$ and $B = [b_{ij}]$ are matrices of order $m \times n$, their **sum** is the $m \times n$ matrix given by

$$A + B = [a_{ij} + b_{ij}].$$

The sum of two matrices of different orders is undefined.

---

**Example 2**    Addition of Matrices

**a.** $\begin{bmatrix} -1 & 2 \\ 0 & 1 \end{bmatrix} + \begin{bmatrix} 1 & 3 \\ -1 & 2 \end{bmatrix} = \begin{bmatrix} -1+1 & 2+3 \\ 0-1 & 1+2 \end{bmatrix} = \begin{bmatrix} 0 & 5 \\ -1 & 3 \end{bmatrix}$

**b.** $\begin{bmatrix} 0 & 1 & -2 \\ 1 & 2 & 3 \end{bmatrix} + \begin{bmatrix} 0 & 0 & 0 \\ 0 & 0 & 0 \end{bmatrix} = \begin{bmatrix} 0 & 1 & -2 \\ 1 & 2 & 3 \end{bmatrix}$

**c.** $\begin{bmatrix} 1 \\ -3 \\ -2 \end{bmatrix} + \begin{bmatrix} -1 \\ 3 \\ 2 \end{bmatrix} = \begin{bmatrix} 0 \\ 0 \\ 0 \end{bmatrix}$

**d.** The sum of

$$A = \begin{bmatrix} 2 & 1 & 0 \\ 4 & 0 & -1 \\ 3 & -2 & 2 \end{bmatrix} \quad \text{and} \quad B = \begin{bmatrix} 0 & 1 \\ -1 & 3 \\ 2 & 4 \end{bmatrix}$$

is undefined.

---

When working with matrices, numbers are usually referred to as **scalars.** In this text, scalars will always be real numbers. You can multiply a matrix $A$ by a scalar $c$ by multiplying each entry in $A$ by $c$.

---

▶ **Definition of Scalar Multiplication**

If $A = [a_{ij}]$ is an $m \times n$ matrix and $c$ is a scalar, the **scalar multiple** of $A$ by $c$ is the $m \times n$ matrix given by

$$cA = [ca_{ij}].$$

---

The symbol $-A$ represents the scalar product $(-1)A$. Moreover, if $A$ and $B$ are of the same order, then $A - B$ represents the sum of $A$ and $(-1)B$. That is,

$$A - B = A + (-1)B.$$    Subtraction of matrices

**Example 3**    Scalar Multiplication and Matrix Subtraction

For the following matrices, find (a) $3A$, (b) $-B$, and (c) $3A - B$.

$$A = \begin{bmatrix} 2 & 2 & 4 \\ -3 & 0 & -1 \\ 2 & 1 & 2 \end{bmatrix} \quad \text{and} \quad B = \begin{bmatrix} 2 & 0 & 0 \\ 1 & -4 & 3 \\ -1 & 3 & 2 \end{bmatrix}$$

**Solution**

**a.** $3A = 3 \begin{bmatrix} 2 & 2 & 4 \\ -3 & 0 & -1 \\ 2 & 1 & 2 \end{bmatrix}$    Scalar multiplication

$$= \begin{bmatrix} 3(2) & 3(2) & 3(4) \\ 3(-3) & 3(0) & 3(-1) \\ 3(2) & 3(1) & 3(2) \end{bmatrix}$$    Multiply each entry by 3.

$$= \begin{bmatrix} 6 & 6 & 12 \\ -9 & 0 & -3 \\ 6 & 3 & 6 \end{bmatrix}$$    Simplify.

**b.** $-B = (-1) \begin{bmatrix} 2 & 0 & 0 \\ 1 & -4 & 3 \\ -1 & 3 & 2 \end{bmatrix}$    Definition of negation

$$= \begin{bmatrix} -2 & 0 & 0 \\ -1 & 4 & -3 \\ 1 & -3 & -2 \end{bmatrix}$$    Multiply each entry by $-1$.

**c.** $3A - B = \begin{bmatrix} 6 & 6 & 12 \\ -9 & 0 & -3 \\ 6 & 3 & 6 \end{bmatrix} - \begin{bmatrix} 2 & 0 & 0 \\ 1 & -4 & 3 \\ -1 & 3 & 2 \end{bmatrix}$    Matrix subtraction

$$= \begin{bmatrix} 4 & 6 & 12 \\ -10 & 4 & -6 \\ 7 & 0 & 4 \end{bmatrix}$$    Subtract corresponding entries.

It is often convenient to rewrite the scalar multiple $cA$ by factoring $c$ out of every entry in the matrix. For instance, in the following example, the scalar $\frac{1}{2}$ has been factored out of the matrix.

$$\begin{bmatrix} \frac{1}{2} & -\frac{3}{2} \\ \frac{5}{2} & \frac{1}{2} \end{bmatrix} = \frac{1}{2} \begin{bmatrix} 1 & -3 \\ 5 & 1 \end{bmatrix}$$

The properties of matrix addition and scalar multiplication are similar to those of addition and multiplication of real numbers.

---

▶ **Properties of Matrix Addition and Scalar Multiplication**

If $A$, $B$, and $C$ are $m \times n$ matrices and $c$ and $d$ are scalars, the following properties are true.

1. $A + B = B + A$                Commutative Property of Matrix Addition

2. $A + (B + C) = (A + B) + C$    Associative Property of Matrix Addition

3. $(cd)A = c(dA)$                Associative Property of Scalar Multiplication

4. $1A = A$                       Scalar Identity

5. $c(A + B) = cA + cB$           Distributive Property

6. $(c + d)A = cA + dA$           Distributive Property

---

Note that the Associative Property of Matrix Addition allows you to write expressions such as $A + B + C$ without ambiguity because the same sum occurs no matter how the matrices are grouped. In other words, you obtain the same sum whether you group $A + B + C$ as $(A + B) + C$ or as $A + (B + C)$. This same reasoning applies to sums of four or more matrices.

**Example 4**    Addition of More than Two Matrices

By adding corresponding entries, you obtain the following sum of four matrices.

$$\begin{bmatrix} 1 \\ 2 \\ -3 \end{bmatrix} + \begin{bmatrix} -1 \\ -1 \\ 2 \end{bmatrix} + \begin{bmatrix} 0 \\ 1 \\ 4 \end{bmatrix} + \begin{bmatrix} 2 \\ -3 \\ -2 \end{bmatrix} = \begin{bmatrix} 2 \\ -1 \\ 1 \end{bmatrix}$$

---

**•TECHNOLOGY•**

Most graphing utilities can add and subtract matrices and multiply matrices by scalars. Consult your user's guide for your graphing utility's capabilities and keystrokes. Use your graphing utility to find (a) $A + B$, (b) $A - B$, (c) $4A$, and (d) $4A + B$.

$$A = \begin{bmatrix} 2 & -3 \\ -1 & 0 \end{bmatrix} \quad \text{and} \quad B = \begin{bmatrix} -1 & 4 \\ 2 & -5 \end{bmatrix}$$

One important property of addition of real numbers is that the number $0$ is the additive identity. That is, $c + 0 = c$ for any real number $c$. For matrices, a similar property holds. That is, if $A$ is an $m \times n$ matrix and $O$ is the $m \times n$ **zero matrix** consisting entirely of zeros, then $A + O = A$.

In other words, $O$ is the **additive identity** for the set of all $m \times n$ matrices. For example, the following matrices are the additive identities for the set of all $2 \times 3$ and $2 \times 2$ matrices.

$$O = \begin{bmatrix} 0 & 0 & 0 \\ 0 & 0 & 0 \end{bmatrix} \quad \text{and} \quad O = \begin{bmatrix} 0 & 0 \\ 0 & 0 \end{bmatrix}.$$

$\underbrace{\qquad}$ Zero $2 \times 3$ matrix $\qquad\qquad$ $\underbrace{\qquad}$ Zero $2 \times 2$ matrix

The algebra of real numbers and the algebra of matrices have many similarities. For example, compare the following solutions.

| Real Numbers (Solve for x.) | $m \times n$ Matrices (Solve for X.) |
|---|---|
| $x + a = b$ | $X + A = B$ |
| $x + a + (-a) = b + (-a)$ | $X + A + (-A) = B + (-A)$ |
| $x + 0 = b - a$ | $X + O = B - A$ |
| $x = b - a$ | $X = B - A$ |

### Example 5 Solving a Matrix Equation

Solve for $X$ in the equation $3X + A = B$, where

$$A = \begin{bmatrix} 1 & -2 \\ 0 & 3 \end{bmatrix} \quad \text{and} \quad B = \begin{bmatrix} -3 & 4 \\ 2 & 1 \end{bmatrix}.$$

**Solution**

Begin by solving the equation for $X$ to obtain

$$3X = B - A \implies X = \frac{1}{3}(B - A).$$

Now, using the matrices $A$ and $B$, you have

$$X = \frac{1}{3}\left( \begin{bmatrix} -3 & 4 \\ 2 & 1 \end{bmatrix} - \begin{bmatrix} 1 & -2 \\ 0 & 3 \end{bmatrix} \right)$$

$$= \frac{1}{3}\begin{bmatrix} -4 & 6 \\ 2 & -2 \end{bmatrix}$$

$$= \begin{bmatrix} -\frac{4}{3} & 2 \\ \frac{2}{3} & -\frac{2}{3} \end{bmatrix}.$$

## Matrix Multiplication

The third basic matrix operation is **matrix multiplication.** At first glance the definition may seem unusual. You will see later, however, that this definition of the product of two matrices has many practical applications.

> ▶ **Definition of Matrix Multiplication**
>
> If $A = [a_{ij}]$ is an $m \times n$ matrix and $B = [b_{ij}]$ is an $n \times p$ matrix, the **product** $AB$ is an $m \times p$ matrix
>
> $$AB = [c_{ij}]$$
>
> where $c_{ij} = a_{i1}b_{1j} + a_{i2}b_{2j} + a_{i3}b_{3j} + \cdots + a_{in}b_{nj}.$

The definition of matrix multiplication indicates a *row-by-column* multiplication, where the entry in the $i$th row and $j$th column of the product $AB$ is obtained by multiplying the entries in the $i$th row of $A$ by the corresponding entries in the $j$th column of $B$ and then adding the results. Example 6 illustrates this process.

**Example 6    Finding the Product of Two Matrices**

Find the product $AB$ where

$$A = \begin{bmatrix} -1 & 3 \\ 4 & -2 \\ 5 & 0 \end{bmatrix} \quad \text{and} \quad B = \begin{bmatrix} -3 & 2 \\ -4 & 1 \end{bmatrix}.$$

**Solution**

First, note that the product $AB$ is defined because the number of columns of $A$ is equal to the number of rows of $B$. Moreover, the product $AB$ has order $3 \times 2$, and is of the form

$$\begin{bmatrix} -1 & 3 \\ 4 & -2 \\ 5 & 0 \end{bmatrix} \begin{bmatrix} -3 & 2 \\ -4 & 1 \end{bmatrix} = \begin{bmatrix} c_{11} & c_{12} \\ c_{21} & c_{22} \\ c_{31} & c_{32} \end{bmatrix}.$$

To find the entries of the product, multiply each row of $A$ by each column of $B$, as follows. Use a graphing utility to check this result.

$$AB = \begin{bmatrix} -1 & 3 \\ 4 & -2 \\ 5 & 0 \end{bmatrix} \begin{bmatrix} -3 & 2 \\ -4 & 1 \end{bmatrix}$$

$$= \begin{bmatrix} (-1)(-3) + (3)(-4) & (-1)(2) + (3)(1) \\ (4)(-3) + (-2)(-4) & (4)(2) + (-2)(1) \\ (5)(-3) + (0)(-4) & (5)(2) + (0)(1) \end{bmatrix}$$

$$= \begin{bmatrix} -9 & 1 \\ -4 & 6 \\ -15 & 10 \end{bmatrix}$$

Be sure you understand that for the product of two matrices to be defined, the number of columns of the first matrix must equal the number of rows of the second matrix. That is, the middle two indices must be the same and the outside two indices give the order of the product, as shown in the following diagram.

$$\underset{m \times n}{A} \qquad \underset{n \times p}{B} \quad = \quad \underset{m \times p}{AB}$$

equal

order of $AB$

**Example 7**  Matrix Multiplication

a. $\begin{bmatrix} 1 & 0 & 3 \\ 2 & -1 & -2 \end{bmatrix} \begin{bmatrix} -2 & 4 & 2 \\ 1 & 0 & 0 \\ -1 & 1 & -1 \end{bmatrix} = \begin{bmatrix} -5 & 7 & -1 \\ -3 & 6 & 6 \end{bmatrix}$

$2 \times 3$ $\qquad\qquad$ $3 \times 3$ $\qquad\qquad$ $2 \times 3$

b. $\begin{bmatrix} 3 & 4 \\ -2 & 5 \end{bmatrix} \begin{bmatrix} 1 & 0 \\ 0 & 1 \end{bmatrix} = \begin{bmatrix} 3 & 4 \\ -2 & 5 \end{bmatrix}$

$2 \times 2$ $\qquad$ $2 \times 2$ $\qquad$ $2 \times 2$

c. $\begin{bmatrix} 1 & 2 \\ 1 & 1 \end{bmatrix} \begin{bmatrix} -1 & 2 \\ 1 & -1 \end{bmatrix} = \begin{bmatrix} 1 & 0 \\ 0 & 1 \end{bmatrix}$

$2 \times 2$ $\qquad$ $2 \times 2$ $\qquad$ $2 \times 2$

d. $\begin{bmatrix} 1 & -2 & -3 \end{bmatrix} \begin{bmatrix} 2 \\ -1 \\ 1 \end{bmatrix} = \begin{bmatrix} 1 \end{bmatrix}$

$1 \times 3$ $\qquad$ $3 \times 1$ $\qquad$ $1 \times 1$

e. $\begin{bmatrix} 2 \\ -1 \\ 1 \end{bmatrix} \begin{bmatrix} 1 & -2 & -3 \end{bmatrix} = \begin{bmatrix} 2 & -4 & -6 \\ -1 & 2 & 3 \\ 1 & -2 & -3 \end{bmatrix}$

$3 \times 1$ $\qquad$ $1 \times 3$ $\qquad\qquad$ $3 \times 3$

f. The product $AB$ for the following matrices is not defined.

$$A = \begin{bmatrix} -2 & 1 \\ 1 & -3 \\ 1 & 4 \end{bmatrix} \quad \text{and} \quad B = \begin{bmatrix} -2 & 3 & 1 & 4 \\ 0 & 1 & -1 & 2 \\ 2 & -1 & 0 & 1 \end{bmatrix}$$

$3 \times 2$ $\qquad\qquad\qquad$ $3 \times 4$

In parts (d) and (e) of Example 7, note that the two products are different. Matrix multiplication is not, in general, commutative. That is, for most matrices, $AB \neq BA$.

The general pattern for matrix multiplication is as follows. To obtain the entry in the $i$th row and the $j$th column of the product $AB$, use the $i$th row of $A$ and the $j$th column of $B$.

$$\begin{bmatrix} a_{11} & a_{12} & a_{13} & \cdots & a_{1n} \\ a_{21} & a_{22} & a_{23} & \cdots & a_{2n} \\ a_{31} & a_{32} & a_{33} & \cdots & a_{3n} \\ & \vdots & & & \vdots \\ a_{i1} & a_{i2} & a_{i3} & \cdots & a_{in} \\ & \vdots & & & \vdots \\ a_{m1} & a_{m2} & a_{m3} & \cdots & a_{mn} \end{bmatrix} \begin{bmatrix} b_{12} & b_{11} & \cdots & b_{1j} & \cdots & b_{1p} \\ b_{22} & b_{21} & \cdots & b_{2j} & \cdots & b_{2p} \\ b_{32} & b_{31} & \cdots & b_{3j} & \cdots & b_{3p} \\ & \vdots & & \vdots & & \vdots \\ b_{n2} & b_{n1} & \cdots & b_{nj} & \cdots & b_{np} \end{bmatrix} = \begin{bmatrix} c_{11} & c_{12} & \cdots & c_{1j} & \cdots & c_{1p} \\ c_{21} & c_{22} & \cdots & c_{2j} & \cdots & c_{2p} \\ & \vdots & & \vdots & & \vdots \\ c_{i1} & c_{i2} & \cdots & c_{ij} & \cdots & c_{ip} \\ & \vdots & & \vdots & & \vdots \\ c_{m1} & c_{m2} & \cdots & c_{mj} & \cdots & c_{mp} \end{bmatrix}$$

$$a_{i1}b_{1j} + a_{i2}b_{2j} + a_{i3}b_{3j} + \cdots + a_{in}b_{nj} = c_{ij}$$

▶ **Properties of Matrix Multiplication**

Let $A$, $B$, and $C$ be matrices and let $c$ be a scalar.

1. $A(BC) = (AB)C$ — Associative Property of Multiplication

2. $A(B + C) = AB + AC$ — Left Distributive Property

3. $(A + B)C = AC + BC$ — Right Distributive Property

4. $c(AB) = (cA)B = A(cB)$

The $n \times n$ matrix that consists of 1's on its main diagonal and 0's elsewhere is called the **identity matrix of order $n$** and is denoted by

$$I_n = \begin{bmatrix} 1 & 0 & 0 & \cdots & 0 \\ 0 & 1 & 0 & \cdots & 0 \\ 0 & 0 & 1 & \cdots & 0 \\ \vdots & \vdots & \vdots & & \vdots \\ 0 & 0 & 0 & \cdots & 1 \end{bmatrix}.$$    Identity matrix

Note that an identity matrix must be *square*. When the order is understood to be $n$, you can denote $I_n$ simply by $I$. If $A$ is an $n \times n$ matrix, the identity matrix has the property that $AI_n = A$ and $I_nA = A$. For example,

$$\begin{bmatrix} 3 & -2 & 5 \\ 1 & 0 & 4 \\ -1 & 2 & -3 \end{bmatrix} \begin{bmatrix} 1 & 0 & 0 \\ 0 & 1 & 0 \\ 0 & 0 & 1 \end{bmatrix} = \begin{bmatrix} 3 & -2 & 5 \\ 1 & 0 & 4 \\ -1 & 2 & -3 \end{bmatrix}$$

and

$$\begin{bmatrix} 1 & 0 & 0 \\ 0 & 1 & 0 \\ 0 & 0 & 1 \end{bmatrix} \begin{bmatrix} 3 & -2 & 5 \\ 1 & 0 & 4 \\ -1 & 2 & -3 \end{bmatrix} = \begin{bmatrix} 3 & -2 & 5 \\ 1 & 0 & 4 \\ -1 & 2 & -3 \end{bmatrix}.$$

## Applications

One application of matrix multiplication is representation of a system of linear equations. Note how the system

$$a_{11}x_1 + a_{12}x_2 + a_{13}x_3 = b_1$$

$$a_{21}x_1 + a_{22}x_2 + a_{23}x_3 = b_2$$

$$a_{31}x_1 + a_{32}x_2 + a_{33}x_3 = b_3$$

can be written as the matrix equation $AX = B$, where $A$ is the *coefficient matrix* of the system, and $X$ and $B$ are column matrices.

$$\begin{bmatrix} a_{11} & a_{12} & a_{13} \\ a_{21} & a_{22} & a_{23} \\ a_{31} & a_{32} & a_{33} \end{bmatrix} \begin{bmatrix} x_1 \\ x_2 \\ x_3 \end{bmatrix} = \begin{bmatrix} b_1 \\ b_2 \\ b_3 \end{bmatrix}$$

$$A \qquad \times X \ = \ B$$

### Example 8  Solving a System of Linear Equations

Solve the matrix equation $AX = B$ for $X$, where

Coefficient matrix    Constant matrix

$$A = \begin{bmatrix} 1 & -2 & 1 \\ 0 & 1 & 2 \\ 2 & 3 & -2 \end{bmatrix} \quad \text{and} \quad B = \begin{bmatrix} -4 \\ 4 \\ 2 \end{bmatrix}.$$

### Solution

As a system of linear equations, $AX = B$ is as follows.

$$x_1 - 2x_2 + x_3 = -4$$

$$x_2 + 2x_3 = 4$$

$$2x_1 + 3x_2 - 2x_3 = 2$$

Using Gauss-Jordan elimination on the augmented matrix of this system, you obtain the following reduced row-echelon matrix.

$$\begin{bmatrix} 1 & 0 & 0 & \vdots & -1 \\ 0 & 1 & 0 & \vdots & 2 \\ 0 & 0 & 1 & \vdots & 1 \end{bmatrix}$$

So, the solution of the system of linear equations is $x_1 = -1$, $x_2 = 2$, and $x_3 = 1$, and the solution of the matrix equation is

$$X = \begin{bmatrix} x_1 \\ x_2 \\ x_3 \end{bmatrix} = \begin{bmatrix} -1 \\ 2 \\ 1 \end{bmatrix}.$$

Check this solution in the original system of equations.

**Example 9**   Softball Team Expenses

Two softball teams submit equipment lists to their sponsors.

|         | Women's Team | Men's Team |
|---------|--------------|------------|
| Bats    | 12           | 15         |
| Balls   | 45           | 38         |
| Gloves  | 15           | 17         |

Each bat costs $55, each ball costs $4, and each glove costs $80. Use matrices to find the total cost of equipment for each team.

Solution

The equipment lists and the costs per item can be written in matrix form as

$$E = \begin{bmatrix} 12 & 15 \\ 45 & 38 \\ 15 & 17 \end{bmatrix} \quad \text{and} \quad C = \begin{bmatrix} 55 & 4 & 80 \end{bmatrix}.$$

The total cost of equipment for each team is given by the product

$$CE = \begin{bmatrix} 55 & 4 & 80 \end{bmatrix} \begin{bmatrix} 12 & 15 \\ 45 & 38 \\ 15 & 17 \end{bmatrix} = \begin{bmatrix} 2040 & 2337 \end{bmatrix}.$$

So, the total cost of equipment for the women's team is $2040, and the total cost of equipment for the men's team is $2337.

---

### Discussing the Concept     Matrix Multiplication

Discuss the requirements for matrix order $m \times n$ for multiplication of two matrices. Determine which of the following matrix multiplications $AB$ is (are) defined. For each case in which $AB$ is defined, what is the order of the resulting matrix?

**a.** $A$ is of order $1 \times 3$
   $B$ is of order $2 \times 1$

**b.** $A$ is of order $2 \times 3$
   $B$ is of order $2 \times 3$

**c.** $A$ is of order $3 \times 4$
   $B$ is of order $4 \times 2$

**d.** $A$ is of order $3 \times 1$
   $B$ is of order $3 \times 3$

Discuss why matrix multiplication is not, in general, commutative. Give an example of two $2 \times 2$ matrices such that $AB \neq BA$. Find an example of two $2 \times 2$ matrices such that $AB = BA$.

# Warm Up

The following warm-up exercises involve skills that were covered in earlier sections. You will use these skills in the exercise set for this section.

In Exercises 1 and 2, evaluate the expression.

**1.** $-3\left(-\frac{5}{6}\right) + 10\left(-\frac{3}{4}\right)$

**2.** $-22\left(\frac{5}{2}\right) + 6(8)$

In Exercises 3 and 4, determine whether the matrix is in *reduced* row-echelon form.

**3.** $\begin{bmatrix} 0 & 1 & 0 & -5 \\ 1 & 0 & 3 & 2 \\ 0 & 0 & 1 & 0 \end{bmatrix}$

**4.** $\begin{bmatrix} 1 & 0 & 0 & 2 & 3 \\ 0 & 0 & 0 & 0 & 0 \\ 0 & 1 & 1 & 3 & 10 \end{bmatrix}$

In Exercises 5 and 6, write the augmented matrix for the system of linear equations.

**5.** $\begin{aligned} -5x + 10y &= 12 \\ 7x - 3y &= 0 \\ -x + 7y &= 25 \end{aligned}$

**6.** $\begin{aligned} 10x + 15y - 9z &= 42 \\ 6x - 5y &= 0 \end{aligned}$

In Exercises 7–10, solve the system of linear equations represented by the augmented matrix.

**7.** $\left[\begin{array}{cc:c} 1 & 0 & 0 \\ 0 & 1 & 2 \end{array}\right]$

**8.** $\left[\begin{array}{ccc:c} 1 & 0 & -1 & 2 \\ 0 & 1 & 1 & 3 \end{array}\right]$

**9.** $\left[\begin{array}{ccc:c} 1 & 2 & 1 & 0 \\ 0 & 0 & 1 & -1 \\ 0 & 0 & 0 & 0 \end{array}\right]$

**10.** $\left[\begin{array}{ccc:c} 1 & -1 & 0 & 3 \\ 0 & 1 & -2 & 1 \\ 0 & 0 & 1 & -1 \end{array}\right]$

## 7.2 Exercises

In Exercises 1–4, find x and y.

**1.** $\begin{bmatrix} 4 & x \\ -1 & y \end{bmatrix} = \begin{bmatrix} 4 & -3 \\ -1 & 2 \end{bmatrix}$

**2.** $\begin{bmatrix} x & -7 \\ 9 & y \end{bmatrix} = \begin{bmatrix} 5 & -7 \\ 9 & -8 \end{bmatrix}$

**3.** $\begin{bmatrix} -4 & 3 \\ 6 & -1 \\ 8 & 2 \\ 5 & 9 \end{bmatrix} = \begin{bmatrix} x-2 & 3 \\ 6 & -1 \\ 8 & -x \\ 5 & 2y-1 \end{bmatrix}$

**4.** $\begin{bmatrix} x+2 & 8 & -3 \\ 1 & 2y & 2x \\ 7 & -2 & y+2 \end{bmatrix} = \begin{bmatrix} 2x+6 & 8 & -3 \\ 1 & 18 & -8 \\ 7 & -2 & 11 \end{bmatrix}$

In Exercises 5–10, find (a) $A + B$, (b) $A - B$, (c) $3A$, and (d) $3A - 2B$.

**5.** $A = \begin{bmatrix} 5 & -2 \\ 3 & 1 \end{bmatrix}$, $B = \begin{bmatrix} 3 & 1 \\ -2 & 6 \end{bmatrix}$

**6.** $A = \begin{bmatrix} 7 & 4 \\ -4 & 5 \end{bmatrix}$, $B = \begin{bmatrix} -3 & 1 \\ 8 & -4 \end{bmatrix}$

**7.** $A = \begin{bmatrix} 6 & -1 \\ 2 & 4 \\ -3 & 5 \end{bmatrix}$, $B = \begin{bmatrix} 1 & 4 \\ -1 & 5 \\ 1 & 10 \end{bmatrix}$

**8.** $A = \begin{bmatrix} 6 & 8 & -3 & 2 & 1 \\ -4 & 2 & 1 & 5 & -2 \end{bmatrix}$,

$B = \begin{bmatrix} 6 & 0 & 4 & -1 & 3 \\ 4 & 5 & -2 & 1 & 2 \end{bmatrix}$

**9.** $A = \begin{bmatrix} 2 & 2 & -1 & 0 \\ 1 & 1 & -2 & 0 \\ 1 & -1 & 3 & 2 \end{bmatrix}$,

$B = \begin{bmatrix} 1 & 1 & -1 & 1 \\ -3 & 4 & 9 & -6 \\ 0 & -7 & 8 & 2 \end{bmatrix}$

**10.** $A = \begin{bmatrix} 3 \\ 2 \\ -1 \end{bmatrix}$, $B = \begin{bmatrix} -4 \\ 6 \\ 2 \end{bmatrix}$

In Exercises 11–16, find (a) $AB$, (b) $BA$, and, if possible, (c) $A^2$. (*Note:* $A^2 = AA$.)

**11.** $A = \begin{bmatrix} 1 & 2 \\ 4 & 2 \end{bmatrix}$, $B = \begin{bmatrix} 2 & -1 \\ -1 & 8 \end{bmatrix}$

**12.** $A = \begin{bmatrix} 2 & -1 \\ 1 & 4 \end{bmatrix}$, $B = \begin{bmatrix} 0 & 0 \\ 3 & -3 \end{bmatrix}$

**13.** $A = \begin{bmatrix} -1 & 2 & 3 \\ 4 & 1 & -1 \end{bmatrix}$, $B = \begin{bmatrix} 1 & 3 \\ -1 & -2 \\ 2 & 4 \end{bmatrix}$

**14.** $A = \begin{bmatrix} 1 & -1 & 7 \\ 2 & -1 & 8 \\ 3 & 1 & -1 \end{bmatrix}$, $B = \begin{bmatrix} 1 & 1 & 2 \\ 2 & 1 & 1 \\ 1 & -3 & 2 \end{bmatrix}$

**15.** $A = \begin{bmatrix} -4 & 2 & 3 \end{bmatrix}$, $B = \begin{bmatrix} 1 \\ 0 \\ 5 \end{bmatrix}$

**16.** $A = \begin{bmatrix} 3 & 2 & 1 & 0 \end{bmatrix}$, $B = \begin{bmatrix} 2 \\ 3 \\ 1 \\ 0 \end{bmatrix}$

In Exercises 17–24, find $AB$, if possible.

**17.** $A = \begin{bmatrix} 3 & -2 \\ 4 & 5 \\ 1 & -1 \end{bmatrix}$, $B = \begin{bmatrix} -1 & 4 & -2 & 5 \\ 2 & 1 & 3 & -1 \end{bmatrix}$

**18.** $A = \begin{bmatrix} 0 & -1 & 0 \\ 4 & 0 & 2 \\ 8 & -1 & 7 \end{bmatrix}$, $B = \begin{bmatrix} 2 & 1 \\ -3 & 4 \\ 1 & 6 \end{bmatrix}$

**19.** $A = \begin{bmatrix} -1 & 3 \\ 4 & -5 \\ 0 & 2 \end{bmatrix}$, $B = \begin{bmatrix} 1 & 2 \\ 0 & 7 \end{bmatrix}$

**20.** $A = \begin{bmatrix} 1 & 0 & 0 \\ 0 & 4 & 0 \\ 0 & 0 & -2 \end{bmatrix}$, $B = \begin{bmatrix} 3 & 0 & 0 \\ 0 & -1 & 0 \\ 0 & 0 & 5 \end{bmatrix}$

**21.** $A = \begin{bmatrix} 5 & 0 & 0 \\ 0 & -8 & 0 \\ 0 & 0 & 7 \end{bmatrix}$, $B = \begin{bmatrix} \frac{1}{5} & 0 & 0 \\ 0 & -\frac{1}{8} & 0 \\ 0 & 0 & \frac{1}{2} \end{bmatrix}$

**22.** $A = \begin{bmatrix} 0 & 1 & 0 \\ 3 & 0 & 2 \\ 5 & 0 & 0 \end{bmatrix}$, $B = \begin{bmatrix} 4 \\ -2 \\ 0 \\ 1 \end{bmatrix}$

**23.** $A = \begin{bmatrix} 6 \\ -2 \\ 1 \\ 6 \end{bmatrix}$, $B = \begin{bmatrix} 10 & 12 \end{bmatrix}$

**24.** $A = \begin{bmatrix} 1 & 0 & 3 & -2 & 4 \\ 6 & 13 & 8 & -17 & 10 \end{bmatrix}$, $B = \begin{bmatrix} 1 & 6 \\ 4 & 2 \end{bmatrix}$

In Exercises 25–28, solve for $X$ given

$$A = \begin{bmatrix} -2 & -1 \\ 1 & 0 \\ 3 & -4 \end{bmatrix} \quad \text{and} \quad B = \begin{bmatrix} 0 & 3 \\ 2 & 0 \\ -4 & -1 \end{bmatrix}.$$

**25.** $X = 3A - 2B$        **26.** $2X = 2A - B$

**27.** $2X + 3A = B$        **28.** $2A + 4B = -2X$

In Exercises 29–34, find matrices $A$, $X$, and $B$ such that the system of linear equations can be written as the matrix equation $AX = B$. Solve the system of equations.

**29.**  $\begin{aligned} -x + y &= 4 \\ -2x + y &= 0 \end{aligned}$        **30.** $\begin{aligned} 2x + 3y &= 5 \\ x + 4y &= 10 \end{aligned}$

**31.**  $\begin{aligned} x + 2y &= 3 \\ 3x - y &= 2 \end{aligned}$        **32.** $\begin{aligned} 2x - 4y + z &= 0 \\ -x + 3y + z &= 1 \\ x + y &= 3 \end{aligned}$

**33.**  $\begin{aligned} x - 2y + 3z &= 9 \\ -x + 3y - z &= -6 \\ 2x - 5y + 5z &= 17 \end{aligned}$

**34.**  $\begin{aligned} x + y - 3z &= -1 \\ -x + 2y &= 1 \\ -y + z &= 0 \end{aligned}$

**35.** If $a, b,$ and $c$ are real numbers such that $c \neq 0$ and $ac = bc$, then $a = b$. However, if $A, B,$ and $C$ are matrices such that $AC = BC$, then $A$ is *not* necessarily equal to $B$. Illustrate this using the following matrices.

$$A = \begin{bmatrix} 1 & 2 & 3 \\ 0 & 5 & 4 \\ 3 & -2 & 1 \end{bmatrix}, \quad B = \begin{bmatrix} 4 & -6 & 3 \\ 5 & 4 & 4 \\ -1 & 0 & 1 \end{bmatrix},$$

and $C = \begin{bmatrix} 0 & 0 & 0 \\ 0 & 0 & 0 \\ 4 & -2 & 3 \end{bmatrix}$

**36.** If $a$ and $b$ are real numbers such that $ab = 0$, then $a = 0$ or $b = 0$. However, if $A$ and $B$ are matrices such that $AB = O$, then it is *not* necessarily true that $A = O$ or $B = O$. Illustrate this using the following matrices.

$$A = \begin{bmatrix} 3 & 3 \\ 4 & 4 \end{bmatrix} \quad \text{and} \quad B = \begin{bmatrix} 1 & -1 \\ -1 & 1 \end{bmatrix}$$

Find another example of two nonzero matrices whose product is the zero matrix.

**37.** *Factory Production* A certain corporation has four factories, each of which manufactures two products. The number of units of product $i$ produced at factory $j$ in one day is represented by $a_{ij}$ in the matrix

$$A = \begin{bmatrix} 100 & 90 & 70 & 30 \\ 40 & 20 & 60 & 60 \end{bmatrix}.$$

Find the output if production is increased by 10%. (*Hint:* Because an increase of 10% corresponds to 100% + 10%, multiply the matrix by 1.10.)

**38.** *Factory Production* A certain corporation has three factories, each of which manufactures two products. The number of units of product $i$ produced at factory $j$ in one day is represented by $a_{ij}$ in the matrix

$$A = \begin{bmatrix} 60 & 40 & 20 \\ 30 & 90 & 60 \end{bmatrix}.$$

Find the output if production is decreased by 20%. (*Hint:* Because a decrease of 20% corresponds to 100% − 20%, multiply the matrix by 0.80.)

**39.** *Hotel Pricing* A convention planning service has identified three suitable hotels for a convention. The quoted room rates are for single, double, triple, and quadruple occupancy. The current cost for each type of room by hotel is represented by the matrix $A$.

|  | Hotel $x$ | Hotel $y$ | Hotel $z$ |  |
|---|---|---|---|---|
| | 79 | 85 | 109 | Single |
| $A =$ | 87 | 93 | 120 | Double |
| | 95 | 110 | 129 | Triple |
| | 95 | 125 | 135 | Quadruple |

(Occupancy)

If room rates were guaranteed not to increase more than 10% by the time of the convention, what would be the maximum rate per room per hotel?

**40.** *Vacation Packages* A vacation service has identified four resort hotels with a special all-inclusive package (room and meals included) at a popular travel destination. The quoted room rates are for a double or family (maximum of four people) occupancy for 5 days and 4 nights. The current cost for each type of room by hotel is represented by the matrix $A$.

|  | Hotel $w$ | Hotel $x$ | Hotel $y$ | Hotel $z$ |  |
|---|---|---|---|---|---|
| $A =$ | 595 | 650 | 727 | 983 | Double |
| | 985 | 1020 | 1105 | 1097 | Family |

(Occupancy)

If room rates were guaranteed not to increase more than 15% by next season, what would be the maximum rate per package per hotel?

**41.** *Inventory Levels* A company sells five different models of computers through three retail outlets. The inventories of the three models at the three outlets are given by the matrix $S$.

|  | Model A | B | C | D | E |  |
|---|---|---|---|---|---|---|
| | 3 | 2 | 2 | 3 | 0 | 1 |
| $S =$ | 0 | 2 | 3 | 4 | 3 | 2 |
| | 4 | 2 | 1 | 3 | 2 | 3 |

(Outlet)

The wholesale and retail prices for each model are given by the matrix $T$.

$$T = \begin{array}{c} \\ \overbrace{\hspace{3cm}}^{\text{Price}} \\ \end{array}$$

| | Wholesale | Retail | |
|---|---|---|---|
| | $840 | $1100 | A |
| | $1200 | $1350 | B |
| $T =$ | $1450 | $1650 | C | Model
| | $2650 | $3000 | D |
| | $3050 | $3200 | E |

(a) What is the total retail price of the inventory at Outlet 1?

(b) What is the total wholesale price of the inventory at Outlet 3?

(c) Compute the product $ST$ and use the context of the problem to interpret the result.

**42.** *Labor/Wage Requirements* A company that manufactures boats has the following labor-hour and wage requirements.

*Labor-Hour Requirements (per boat)*

Department

| | Cutting | Assembly | Packaging | |
|---|---|---|---|---|
| | 1.0 hour | 0.5 hour | 0.2 hour | Small |
| $S =$ | 1.6 hours | 1.0 hour | 0.2 hour | Medium |
| | 2.5 hours | 2.0 hours | 0.4 hour | Large |

Boat size

*Wage Requirements (per hour)*

Plant

| | A | B | |
|---|---|---|---|
| | $12 | $10 | Cutting |
| $T =$ | $9 | $8 | Assembly |
| | $6 | $5 | Packaging |

Department

(a) What is the labor cost for a medium-sized boat at Plant B?

(b) What is the labor cost for a large-sized boat at Plant A?

(c) Compute $ST$ and interpret the result.

*Think About It* In Exercises 43 and 44, find a matrix $B$ such that $AB$ is the identity matrix. Is there more than one correct result?

**43.** $A = \begin{bmatrix} 1 & 3 \\ 1 & 2 \end{bmatrix}$   **44.** $A = \begin{bmatrix} 2 & 1 \\ 5 & 2 \end{bmatrix}$

**45.** *Voting Preference* The matrix

From

$$P = \begin{bmatrix} & R & D & I \\ 0.6 & 0.1 & 0.1 & R \\ 0.2 & 0.7 & 0.1 & D \\ 0.2 & 0.2 & 0.8 & I \end{bmatrix} \text{ To}$$

is called a stochastic matrix. Each entry $p_{ij}$ $(i \neq j)$ represents the proportion of the voting population that changes from party $i$ to party $j$, and $p_{ii}$ represents the proportion that remains loyal to the party from one election to the next. Use a graphing utility to find $P^2$. (This matrix gives the transition probabilities from the first election to the third.)

**46.** *Voting Preference* Use a graphing utility to find $P^3$, $P^4$, $P^5$, $P^6$, $P^7$, and $P^8$ for the matrix given in Exercise 45. Can you detect a pattern as $P$ is raised to higher and higher powers?

**47.** *Contract Bonuses* Professional athletes frequently have bonus or incentive clauses in their contracts. For example, a defensive football player might receive a bonus for a sack, interception, and/or key tackle. Suppose that in a contract a sack is worth $2000, an interception is worth $1000, and a key tackle is worth $700. Use matrices to calculate the bonuses for defensive players A, B, and C if the following matrix describes the numbers of sacks, interceptions, and key tackles in a game.

| Player | Sacks | Interceptions | Key Tackles |
|---|---|---|---|
| Player A | 2 | 0 | 4 |
| Player B | 0 | 1 | 5 |
| Player C | 1 | 3 | 2 |

## 7.3    The Inverse of a Square Matrix

### Objectives

- Verify that a matrix is the inverse of a given matrix.
- Find the inverse of a matrix.
- Use an inverse matrix to solve a system of linear equations.

## The Inverse of a Matrix

This section further develops the algebra of matrices. To begin, consider the real number equation $ax = b$. To solve this equation for $x$, multiply both sides of the equation by $a^{-1}$ (provided $a \neq 0$).

$$ax = b$$
$$(a^{-1}a)x = a^{-1}b$$
$$(1)x = a^{-1}b$$
$$x = a^{-1}b$$

The number $a^{-1}$ is called the *multiplicative inverse of a* because it has the property that $a^{-1}a = 1$. The definition of a multiplicative inverse of a matrix is similar.

---

▶ **Definition of an Inverse of a Square Matrix**

Let $A$ be an $n \times n$ matrix. If there exists a matrix $A^{-1}$ such that

$$AA^{-1} = I_n = A^{-1}A$$

$A^{-1}$ is called the **inverse** of $A$. (The symbol $A^{-1}$ is read "$A$ inverse.")

---

### Study Tip

Recall that it is not always true that $AB = BA$, even if both products are defined. However, if $A$ and $B$ are both square matrices and $AB = I_n$, it can be shown that $BA = I_n$. So, in Example 1, you need only to check that $AB = I_2$.

### Example 1    The Inverse of a Matrix

Show that $B$ is the inverse of $A$, where

$$A = \begin{bmatrix} -1 & 2 \\ -1 & 1 \end{bmatrix} \quad \text{and} \quad B = \begin{bmatrix} 1 & -2 \\ 1 & -1 \end{bmatrix}.$$

**Solution**

To show that $B$ is the inverse of $A$, show that $AB = I = BA$, as follows.

$$AB = \begin{bmatrix} -1 & 2 \\ -1 & 1 \end{bmatrix}\begin{bmatrix} 1 & -2 \\ 1 & -1 \end{bmatrix} = \begin{bmatrix} -1+2 & 2-2 \\ -1+1 & 2-1 \end{bmatrix} = \begin{bmatrix} 1 & 0 \\ 0 & 1 \end{bmatrix}$$

$$BA = \begin{bmatrix} 1 & -2 \\ 1 & -1 \end{bmatrix}\begin{bmatrix} -1 & 2 \\ -1 & 1 \end{bmatrix} = \begin{bmatrix} -1+2 & 2-2 \\ -1+1 & 2-1 \end{bmatrix} = \begin{bmatrix} 1 & 0 \\ 0 & 1 \end{bmatrix}$$

If a matrix $A$ has an inverse, $A$ is called **invertible** (or **nonsingular**); otherwise, $A$ is called **singular.** A nonsquare matrix cannot have an inverse. To see this, note that if $A$ is of order $m \times n$ and $B$ is of order $n \times m$ (where $m \neq n$), the products $AB$ and $BA$ are of different orders and therefore cannot be equal to each other. Not all square matrices have inverses (see the matrix at the bottom of page 457). If, however, a matrix does have an inverse, that inverse is unique. The following example shows how to use a system of equations to find the inverse.

**Example 2**    Finding the Inverse of a Matrix

Find the inverse of the matrix

$$A = \begin{bmatrix} 1 & 4 \\ -1 & -3 \end{bmatrix}.$$

Solution

To find the inverse of $A$, try to solve the matrix equation $AX = I$ for $X$.

$$\overset{A}{\begin{bmatrix} 1 & 4 \\ -1 & -3 \end{bmatrix}} \overset{X}{\begin{bmatrix} x_{11} & x_{12} \\ x_{21} & x_{22} \end{bmatrix}} = \overset{I}{\begin{bmatrix} 1 & 0 \\ 0 & 1 \end{bmatrix}}$$

$$\begin{bmatrix} x_{11} + 4x_{21} & x_{12} + 4x_{22} \\ -x_{11} - 3x_{21} & -x_{12} - 3x_{22} \end{bmatrix} = \begin{bmatrix} 1 & 0 \\ 0 & 1 \end{bmatrix}$$

Equating corresponding entries, you obtain the following two systems of linear equations.

$$\begin{aligned} x_{11} + 4x_{21} &= 1 \\ -x_{11} - 3x_{21} &= 0 \end{aligned} \qquad \begin{aligned} x_{12} + 4x_{22} &= 0 \\ -x_{12} - 3x_{22} &= 1 \end{aligned}$$

From the first system you can determine that $x_{11} = -3$ and $x_{21} = 1$, and from the second system you can determine that $x_{12} = -4$ and $x_{22} = 1$. Therefore, the inverse of $A$ is

$$X = A^{-1} = \begin{bmatrix} -3 & -4 \\ 1 & 1 \end{bmatrix}.$$

You can use matrix multiplication to check this result.

Check

$$AA^{-1} = \begin{bmatrix} 1 & 4 \\ -1 & -3 \end{bmatrix} \begin{bmatrix} -3 & -4 \\ 1 & 1 \end{bmatrix} = \begin{bmatrix} 1 & 0 \\ 0 & 1 \end{bmatrix} \checkmark$$

$$A^{-1}A = \begin{bmatrix} -3 & -4 \\ 1 & 1 \end{bmatrix} \begin{bmatrix} 1 & 4 \\ -1 & -3 \end{bmatrix} = \begin{bmatrix} 1 & 0 \\ 0 & 1 \end{bmatrix} \checkmark$$

## Finding Inverse Matrices

In Example 2, note that the two systems of linear equations have the *same coefficient matrix A*. Rather than solve the two systems represented by

$$\begin{bmatrix} 1 & 4 & \vdots & 1 \\ -1 & -3 & \vdots & 0 \end{bmatrix} \quad \text{and} \quad \begin{bmatrix} 1 & 4 & \vdots & 0 \\ -1 & -3 & \vdots & 1 \end{bmatrix}$$

separately, you can solve them *simultaneously* by **adjoining** the identity matrix to the coefficient matrix to obtain

$$\begin{array}{cc} A & I \end{array}$$
$$\begin{bmatrix} 1 & 4 & \vdots & 1 & 0 \\ -1 & -3 & \vdots & 0 & 1 \end{bmatrix}.$$

Then, applying Gauss-Jordan elimination to this matrix, you can solve *both* systems with a single elimination process, as follows.

$$\begin{bmatrix} 1 & 4 & \vdots & 1 & 0 \\ -1 & -3 & \vdots & 0 & 1 \end{bmatrix}$$

$$R_1 + R_2 \rightarrow \begin{bmatrix} 1 & 4 & \vdots & 1 & 0 \\ 0 & 1 & \vdots & 1 & 1 \end{bmatrix}$$

$$-4R_2 + R_1 \rightarrow \begin{bmatrix} 1 & 0 & \vdots & -3 & -4 \\ 0 & 1 & \vdots & 1 & 1 \end{bmatrix}$$

Thus, from the "doubly augmented" matrix $[A : I]$, you obtained the matrix $[I : A^{-1}]$.

$$\begin{array}{cccc} A & & I & \end{array} \qquad \begin{array}{cccc} I & & A^{-1} & \end{array}$$
$$\begin{bmatrix} 1 & 4 & \vdots & 1 & 0 \\ -1 & -3 & \vdots & 0 & 1 \end{bmatrix} \implies \begin{bmatrix} 1 & 0 & \vdots & -3 & -4 \\ 0 & 1 & \vdots & 1 & 1 \end{bmatrix}$$

This procedure (or algorithm) works for an arbitrary square matrix that has an inverse.

---

▶ **Finding an Inverse Matrix**

Let $A$ be a square matrix of order $n$.

1. Write the $n \times 2n$ matrix that consists of the given matrix $A$ on the left and the $n \times n$ identity matrix $I$ on the right to obtain $[A : I]$. Note that we separate the matrices $A$ and $I$ by a dotted line. We call this process **adjoining** the matrices $A$ and $I$.

2. If possible, row reduce $A$ to $I$ using elementary row operations on the *entire* matrix $[A : I]$. The result will be the matrix $[I : A^{-1}]$. If this is not possible, $A$ is not invertible.

3. Check your work by multiplying to see that $AA^{-1} = I = A^{-1}A$.

**Example 3**    Finding the Inverse of a Matrix

Find the inverse of the matrix

$$A = \begin{bmatrix} 1 & -1 & 0 \\ 1 & 0 & -1 \\ 6 & -2 & -3 \end{bmatrix}.$$

**Solution**

Begin by adjoining the identity matrix to $A$ to form the matrix

$$[A \;\vdots\; I] = \begin{bmatrix} 1 & -1 & 0 & \vdots & 1 & 0 & 0 \\ 1 & 0 & -1 & \vdots & 0 & 1 & 0 \\ 6 & -2 & -3 & \vdots & 0 & 0 & 1 \end{bmatrix}.$$

Use elementary row operations to obtain the form $[I \;\vdots\; A^{-1}]$, as follows.

$$\begin{bmatrix} 1 & -1 & 0 & \vdots & 1 & 0 & 0 \\ 1 & 0 & -1 & \vdots & 0 & 1 & 0 \\ 6 & -2 & -3 & \vdots & 0 & 0 & 1 \end{bmatrix}$$

$$\begin{matrix} \\ -R_1 + R_2 \rightarrow \\ -6R_1 + R_3 \rightarrow \end{matrix} \begin{bmatrix} 1 & -1 & 0 & \vdots & 1 & 0 & 0 \\ 0 & 1 & -1 & \vdots & -1 & 1 & 0 \\ 0 & 4 & -3 & \vdots & -6 & 0 & 1 \end{bmatrix}$$

$$\begin{matrix} R_2 + R_1 \rightarrow \\ \\ -4R_2 + R_3 \rightarrow \end{matrix} \begin{bmatrix} 1 & 0 & -1 & \vdots & 0 & 1 & 0 \\ 0 & 1 & -1 & \vdots & -1 & 1 & 0 \\ 0 & 0 & 1 & \vdots & -2 & -4 & 1 \end{bmatrix}$$

$$\begin{matrix} R_3 + R_1 \rightarrow \\ R_3 + R_2 \rightarrow \\ \\ \end{matrix} \begin{bmatrix} 1 & 0 & 0 & \vdots & -2 & -3 & 1 \\ 0 & 1 & 0 & \vdots & -3 & -3 & 1 \\ 0 & 0 & 1 & \vdots & -2 & -4 & 1 \end{bmatrix}$$

So, the matrix $A$ is invertible and its inverse is

$$A^{-1} = \begin{bmatrix} -2 & -3 & 1 \\ -3 & -3 & 1 \\ -2 & -4 & 1 \end{bmatrix}.$$

Try confirming this result by multiplying $A$ and $A^{-1}$ to obtain $I$.

---

The process shown in Example 3 applies to any $n \times n$ matrix $A$. If $A$ has an inverse, this process will find it. On the other hand, if $A$ does not have an inverse, this process will tell us so. For instance, the following matrix has no inverse.

$$A = \begin{bmatrix} 1 & 2 & 0 \\ 3 & -1 & 2 \\ -2 & 3 & -2 \end{bmatrix}$$

Explain how the elimination process shows that this matrix is singular.

# The Inverse of a $2 \times 2$ Matrix (Quick Method)

Using Gauss-Jordan elimination to find the inverse of a matrix works well (even as a computer technique) for matrices of order $3 \times 3$ or greater. For $2 \times 2$ matrices, however, many people prefer to use a formula for the inverse rather than Gauss-Jordan elimination. This simple formula, which works *only* for $2 \times 2$ matrices, is explained as follows. If $A$ is a $2 \times 2$ matrix given by

$$A = \begin{bmatrix} a & b \\ c & d \end{bmatrix}$$

then $A$ is invertible if and only if $ad - bc \neq 0$. Moreover, if $ad - bc \neq 0$, the inverse is given by

$$A^{-1} = \frac{1}{ad - bc} \begin{bmatrix} d & -b \\ -c & a \end{bmatrix}.$$

Try verifying this inverse by multiplication.

The denominator $ad - bc$ is called the **determinant** of the $2 \times 2$ matrix $A$. You will study determinants in the next section.

### Example 4    Finding the Inverse of a $2 \times 2$ Matrix

If possible, find the inverses of the following matrices.

**a.** $A = \begin{bmatrix} 3 & -1 \\ -2 & 2 \end{bmatrix}$

**b.** $B = \begin{bmatrix} 3 & -1 \\ -6 & 2 \end{bmatrix}$

**Solution**

**a.** For the matrix $A$, apply the formula for the inverse of a $2 \times 2$ matrix to obtain

$$ad - bc = (3)(2) - (-1)(-2) = 4.$$

Because this quantity is not zero, the inverse is formed by interchanging the entries on the main diagonal and changing the signs of the other two entries, as follows.

$$A^{-1} = \frac{1}{4} \begin{bmatrix} 2 & 1 \\ 2 & 3 \end{bmatrix} = \begin{bmatrix} \frac{1}{2} & \frac{1}{4} \\ \frac{1}{2} & \frac{3}{4} \end{bmatrix}$$

**b.** For the matrix $B$, you have

$$ad - bc = (3)(2) - (-1)(-6) = 0$$

which means that $B$ is not invertible.

## Systems of Linear Equations

We know that a system of linear equations can have exactly one solution, infinitely many solutions, or no solution. If the coefficient matrix $A$ of a *square* system (a system that has the same number of equations as variables) is invertible, the system has a unique solution, which is given as follows.

**·TECHNOLOGY·**

The formula $X = A^{-1}B$ is used in most graphing utilities to solve linear systems that have invertible coefficient matrices. That is, you enter the $n \times n$ coefficient matrix $[A]$ and the $n \times 1$ column matrix $[B]$. The solution $X$ is given by $[A]^{-1}[B]$. Use your graphing utility to verify the solution in Example 5.

▶ **A System of Equations with a Unique Solution**

If $A$ is an invertible matrix, the system of linear equations represented by $AX = B$ has a unique solution given by

$$X = A^{-1}B.$$

**Example 5**    Solving a System of Equations Using an Inverse

Use an inverse matrix to solve the following system.

$$2x + 3y + z = -1$$
$$3x + 3y + z = 1$$
$$2x + 4y + z = -2$$

**Solution**

$$X = A^{-1}B = \begin{bmatrix} -1 & 1 & 0 \\ -1 & 0 & 1 \\ 6 & -2 & -3 \end{bmatrix} \begin{bmatrix} -1 \\ 1 \\ -2 \end{bmatrix} = \begin{bmatrix} 2 \\ -1 \\ -2 \end{bmatrix}$$

So, the solution is $x = 2$, $y = -1$, and $z = -2$.

**Discussing the Concept    Methods of Problem Solving**

Describe how the method of solving Example 5 is similar to the method used to solve a simple equation such as

$$2x = 10.$$

How are the two methods different?

# Warm Up

The following warm-up exercises involve skills that were covered in earlier sections. You will use these skills in the exercise set for this section.

In Exercises 1–8, perform the indicated matrix operations.

1. $4\begin{bmatrix} 1 & 6 \\ 0 & -4 \\ 12 & 2 \end{bmatrix}$

2. $\dfrac{1}{2}\begin{bmatrix} 11 & 10 & 48 \\ 1 & 0 & 16 \\ 0 & 2 & 8 \end{bmatrix}$

3. $\begin{bmatrix} 1 & -10 & 3 \\ 4 & 1 & 0 \end{bmatrix} - 2\begin{bmatrix} 3 & -4 & 8 \\ 0 & 7 & 1 \end{bmatrix}$

4. $\begin{bmatrix} 5 & 20 \\ -7 & 15 \end{bmatrix} - 3\begin{bmatrix} 6 & 3 \\ 4 & -2 \end{bmatrix}$

5. $\begin{bmatrix} 1 & -2 \\ -1 & 3 \end{bmatrix}\begin{bmatrix} 3 & 2 \\ 1 & 1 \end{bmatrix}$

6. $\begin{bmatrix} 1 & 0 \\ 0 & 1 \end{bmatrix}\begin{bmatrix} 6 & 5 \\ 3 & -2 \end{bmatrix}$

7. $\begin{bmatrix} 2 & 0 & 0 \\ 0 & -1 & 0 \\ 0 & 0 & 3 \end{bmatrix}\begin{bmatrix} \frac{1}{2} & 0 & 0 \\ 0 & -1 & 0 \\ 0 & 0 & \frac{1}{3} \end{bmatrix}$

8. $\begin{bmatrix} 1 & -1 & 0 \\ 1 & 0 & -1 \\ 6 & -2 & -3 \end{bmatrix}\begin{bmatrix} -2 & -3 & 1 \\ -3 & -3 & 1 \\ -2 & -4 & 1 \end{bmatrix}$

In Exercises 9 and 10, rewrite the matrix in reduced row-echelon form.

9. $\begin{bmatrix} 3 & -2 & 1 & 0 \\ 4 & -3 & 0 & 1 \end{bmatrix}$

10. $\begin{bmatrix} 1 & 1 & 2 & 1 & 0 & 0 \\ -1 & 0 & 3 & 0 & 1 & 0 \\ 1 & 2 & 8 & 0 & 0 & 1 \end{bmatrix}$

## 7.3 Exercises

LAB **II**

In Exercises 1–10, show that $B$ is the inverse of $A$.

1. $A = \begin{bmatrix} 7 & 4 \\ 5 & 3 \end{bmatrix},\ B = \begin{bmatrix} 3 & -4 \\ -5 & 7 \end{bmatrix}$

2. $A = \begin{bmatrix} -4 & 1 \\ -9 & 2 \end{bmatrix},\ B = \begin{bmatrix} 2 & -1 \\ 9 & -4 \end{bmatrix}$

3. $A = \begin{bmatrix} 2 & -1 \\ 5 & -4 \end{bmatrix},\ B = \begin{bmatrix} \frac{4}{3} & -\frac{1}{3} \\ \frac{5}{3} & -\frac{2}{3} \end{bmatrix}$

4. $A = \begin{bmatrix} 1 & -2 \\ 3 & -10 \end{bmatrix},\ B = \begin{bmatrix} \frac{5}{2} & -\frac{1}{2} \\ \frac{3}{4} & -\frac{1}{4} \end{bmatrix}$

5. $A = \begin{bmatrix} -2 & 2 & 3 \\ 1 & -1 & 0 \\ 0 & 1 & 4 \end{bmatrix},\ B = \dfrac{1}{3}\begin{bmatrix} -4 & -5 & 3 \\ -4 & -8 & 3 \\ 1 & 2 & 0 \end{bmatrix}$

6. $A = \begin{bmatrix} 2 & -17 & 11 \\ -1 & 11 & -7 \\ 0 & 3 & -2 \end{bmatrix},\ B = \begin{bmatrix} 1 & 1 & 2 \\ 2 & 4 & -3 \\ 3 & 6 & -5 \end{bmatrix}$

7. $A = \begin{bmatrix} -1 & 0 & 2 \\ 1 & -2 & 0 \\ 1 & 0 & 3 \end{bmatrix},\ B = \dfrac{1}{10}\begin{bmatrix} -6 & 0 & 4 \\ -3 & -5 & 2 \\ 2 & 0 & 2 \end{bmatrix}$

8. $A = \begin{bmatrix} -1 & 1 & -3 \\ 2 & -1 & 4 \\ -1 & 1 & -2 \end{bmatrix},\ B = \begin{bmatrix} 2 & 1 & -1 \\ 0 & 1 & 2 \\ -1 & 0 & 1 \end{bmatrix}$

9. $A = \begin{bmatrix} 2 & 0 & 1 & 1 \\ 3 & 0 & 0 & 1 \\ -1 & 1 & -2 & 1 \\ 4 & -1 & 1 & 0 \end{bmatrix},$

$B = \begin{bmatrix} -1 & 2 & -1 & -1 \\ -4 & 9 & -5 & -6 \\ 0 & 1 & -1 & -1 \\ 3 & -5 & 3 & 3 \end{bmatrix}$

**10.** $A = \begin{bmatrix} -1 & 1 & 0 & -1 \\ 1 & -1 & 2 & 0 \\ -1 & 1 & 2 & 0 \\ 0 & -1 & 1 & 1 \end{bmatrix}$,

$B = \dfrac{1}{4} \begin{bmatrix} -4 & 1 & 1 & -4 \\ -4 & -1 & 3 & -4 \\ 0 & 1 & 1 & 0 \\ -4 & -2 & 2 & 0 \end{bmatrix}$

In Exercises 11–36, find the inverse of the matrix (if it exists).

**11.** $\begin{bmatrix} 8 & 4 \\ -2 & -2 \end{bmatrix}$

**12.** $\begin{bmatrix} 1 & 2 \\ 3 & 7 \end{bmatrix}$

**13.** $\begin{bmatrix} 2 & 3 \\ 6 & 9 \end{bmatrix}$

**14.** $\begin{bmatrix} -7 & 33 \\ 4 & -19 \end{bmatrix}$

**15.** $\begin{bmatrix} -1 & 1 \\ -2 & 1 \end{bmatrix}$

**16.** $\begin{bmatrix} 2 & 3 \\ 1 & 4 \end{bmatrix}$

**17.** $\begin{bmatrix} 0 & 4 \\ -3 & 6 \end{bmatrix}$

**18.** $\begin{bmatrix} 11 & 1 \\ -1 & 0 \end{bmatrix}$

**19.** $\begin{bmatrix} 2 & 7 & 1 \\ -3 & -9 & 2 \end{bmatrix}$

**20.** $\begin{bmatrix} -2 & 5 \\ 6 & -15 \\ 0 & 1 \end{bmatrix}$

**21.** $\begin{bmatrix} 1 & 1 & 1 \\ 3 & 5 & 4 \\ 3 & 6 & 5 \end{bmatrix}$

**22.** $\begin{bmatrix} 1 & 2 & 2 \\ 3 & 7 & 9 \\ -1 & -4 & -7 \end{bmatrix}$

**23.** $\begin{bmatrix} 1 & 2 & -1 \\ 3 & 7 & -10 \\ -5 & -7 & -15 \end{bmatrix}$

**24.** $\begin{bmatrix} 10 & 5 & -7 \\ -5 & 1 & 4 \\ 3 & 2 & -2 \end{bmatrix}$

**25.** $\begin{bmatrix} 1 & 1 & 2 \\ 3 & 1 & 0 \\ -2 & 0 & 3 \end{bmatrix}$

**26.** $\begin{bmatrix} 3 & 2 & 2 \\ 2 & 2 & 2 \\ -4 & 4 & 3 \end{bmatrix}$

**27.** $\begin{bmatrix} 3 & 0 & 0 \\ 0 & -2 & 0 \\ 0 & 0 & 4 \end{bmatrix}$

**28.** $\begin{bmatrix} 2 & 0 & 0 \\ 0 & 3 & 0 \\ 0 & 0 & 5 \end{bmatrix}$

**29.** $\begin{bmatrix} 1 & 0 & 0 \\ 3 & 4 & 0 \\ 2 & 5 & 5 \end{bmatrix}$

**30.** $\begin{bmatrix} 1 & 0 & 0 \\ 3 & 0 & 0 \\ 2 & 5 & 5 \end{bmatrix}$

**31.** $\begin{bmatrix} 1 & 0 & 3 & 0 \\ 0 & 2 & 0 & 4 \\ 1 & 0 & 3 & 0 \\ 0 & 2 & 0 & 4 \end{bmatrix}$

**32.** $\begin{bmatrix} 1 & 3 & -2 & 0 \\ 0 & 2 & 4 & 6 \\ 0 & 0 & -2 & 1 \\ 0 & 0 & 0 & 5 \end{bmatrix}$

**33.** $\begin{bmatrix} -8 & 0 & 0 & 0 \\ 0 & 1 & 0 & 0 \\ 0 & 0 & 4 & 0 \\ 0 & 0 & 0 & -5 \end{bmatrix}$

**34.** $\begin{bmatrix} -1 & 0 & 1 & 0 \\ 0 & 2 & 0 & -1 \\ 2 & 0 & -1 & 0 \\ 0 & -1 & 0 & 1 \end{bmatrix}$

**35.** $\begin{bmatrix} 1 & -2 & -1 & -2 \\ 3 & -5 & -2 & -3 \\ 2 & -5 & -2 & -5 \\ -1 & 4 & 4 & 11 \end{bmatrix}$

**36.** $\begin{bmatrix} 4 & 8 & -7 & 14 \\ 2 & 5 & -4 & 6 \\ 0 & 2 & 1 & -7 \\ 3 & 6 & -5 & 10 \end{bmatrix}$

In Exercises 37–40, use an inverse matrix to solve the system of linear equations. (Use the inverse matrix found in Exercise 15.)

**37.** $-x + y = 4$
$\quad -2x + y = 0$

**38.** $-x + y = -3$
$\quad -2x + y = 5$

**39.** $-x + y = 20$
$\quad -2x + y = 10$

**40.** $-x + y = 0$
$\quad -2x + y = 7$

In Exercises 41–44, use an inverse matrix to solve the system of linear equations. (Use the inverse matrix found in Exercise 16.)

**41.** $2x + 3y = 5$
$\quad x + 4y = 10$

**42.** $2x + 3y = 0$
$\quad x + 4y = 3$

**43.** $2x + 3y = 4$
$\quad x + 4y = 2$

**44.** $2x + 3y = 1$
$\quad x + 4y = -2$

In Exercises 45 and 46, use an inverse matrix to solve the system of linear equations. (Use the inverse matrix found in Exercise 26.)

**45.** $\quad 3x + 2y + 2z = 0$
$\quad 2x + 2y + 2z = 5$
$\quad -4x + 4y + 3z = 2$

**46.** $\quad 3x + 2y + 2z = -1$
$\quad 2x + 2y + 2z = 2$
$\quad -4x + 4y + 3z = 0$

In Exercises 47 and 48, use an inverse matrix to solve the system of linear equations. (Use the inverse matrix found in Exercise 35.)

**47.**
$$x_1 - 2x_2 - x_3 - 2x_4 = 0$$
$$3x_1 - 5x_2 - 2x_3 - 3x_4 = 1$$
$$2x_1 - 5x_2 - 2x_3 - 5x_4 = -1$$
$$-x_1 + 4x_2 + 4x_3 + 11x_4 = 2$$

**48.**
$$x_1 - 2x_2 - x_3 - 2x_4 = 1$$
$$3x_1 - 5x_2 - 2x_3 - 3x_4 = -2$$
$$2x_1 - 5x_2 - 2x_3 - 5x_4 = 0$$
$$-x_1 + 4x_2 + 4x_3 + 11x_4 = -3$$

**49.** *Research and Development Expenditures*    The amounts of their own funds that universities and colleges spent on research and development for the years 1982 to 1995 are shown in the figure. The least squares regression line $y = a + bt$ for these data is found by solving the system

$$14a + 91b = 35.98$$
$$91a + 819b = 288.47$$

where $y$ is the expenditures (in billions of dollars) and $t = 0$ represents 1982.    (Source: U.S. National Science Foundation)

(a) Use a graphing utility to find an inverse matrix to solve this system and find the equation of the least squares regression line.

(b) Use the result of part (a) to approximate the expenditures in 1999.

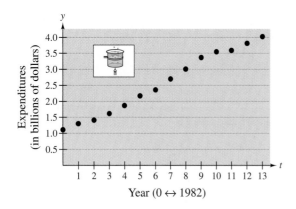

Year (0 ↔ 1982)

**50.** *Service Industry in the United States*    The total numbers of people employed in the "service industry" (excluding self-employed people) from 1992 to 1996 are shown in the figure. The least squares regression parabola $y = at^2 + bt + c$ for these data is found by solving the system

$$5c + 10b + 30a = 215.1$$
$$10c + 30b + 100a = 440.1$$
$$30c + 100b + 353a = 1330.1$$

where $y$ is the number of workers (in millions) and $t = 0$ represents 1992.    (Source: U.S. Bureau of Labor Statistics)

(a) Use a graphing utility to find an inverse matrix to solve this system and write the equation of the least squares regression line.

(b) Use the result of part (a) to approximate the number of workers in the service industry in 2000.

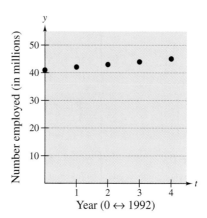

Year (0 ↔ 1992)

In Exercises 51 and 52, develop for the given matrix a system of equations that will have the given solution. Use an inverse matrix to verify that the system of equations gives the desired solution.

**51.**
$$\begin{bmatrix} 2 & 1 & 3 \\ 4 & 0 & -2 \\ 0 & 3 & 2 \end{bmatrix} \qquad \begin{array}{l} x = 2 \\ y = -3 \\ z = 5 \end{array}$$

**52.**
$$\begin{bmatrix} 1 & 0 & 2 \\ 1 & 1 & 1 \\ 2 & -1 & 0 \end{bmatrix} \qquad \begin{array}{l} x = 5 \\ y = -2 \\ z = 1 \end{array}$$

*Bond Investment*    In Exercises 53–56, you are investing in AAA-rated bonds, A-rated bonds, and B-rated bonds. Your average yield is 8% on AAA bonds, 6% on A bonds, and 7% on B bonds. Twice as much is invested in B bonds as in A bonds. The desired system of linear equations (where $x, y$, and $z$ represent the amounts invested in AAA, A, and B bonds) is as follows.

$$x + \quad y + \quad z = \text{(total investment)}$$
$$0.08x + 0.06y + 0.07z = \text{(annual return)}$$
$$2y - \quad z = 0$$

Use the inverse of the coefficient matrix of this system to find the amount invested in each type of bond, with the given total investment.

**53.** Total investment $=$ \$37,500; annual return $=$ \$2620

**54.** Total investment $=$ \$40,000; annual return $=$ \$2900

**55.** Total investment $=$ \$38,000; annual return $=$ \$2850

**56.** Total investment $=$ \$42,000; annual return $=$ \$3200

*Acquisition of Raw Materials*    In Exercises 57–60, consider a company that produces computer chips, resistors, and transistors. Each computer chip requires 2 units of copper, 2 units of zinc, and 1 unit of glass. Each resistor requires 1 unit of copper, 3 units of zinc, and 2 units of glass. Each transistor requires 3 units of copper, 2 units of zinc, and 2 units of glass. The desired system of linear equations (where $x, y$, and $z$ represent the numbers of computer chips, resistors, and transistors) is as follows.

$$2x + \quad y + 3z = \text{(units of copper)}$$
$$2x + 3y + 2z = \text{(units of zinc)}$$
$$x + 2y + 2z = \text{(units of glass)}$$

Use the inverse of the coefficient matrix of this system to find the numbers of computer chips, resistors, and transistors that the company can produce with the given amounts of raw materials.

**57.** 70 units of copper
80 units of zinc
55 units of glass

**58.** 85 units of copper
105 units of zinc
70 units of glass

**59.** 200 units of copper
250 units of zinc
150 units of glass

**60.** 300 units of copper
400 units of zinc
250 units of glass

*Raw Materials*    In Exercises 61 and 62, consider a company that specializes in gourmet chocolate baked goods—chocolate muffins, cookies, and brownies. Each chocolate muffin requires 2 units of chocolate, 3 units of flour, and 2 units of sugar. Each chocolate cookie requires 1 unit of chocolate, 1 unit of flour, and 1 unit of sugar. Each chocolate brownie requires 2 units of chocolate, 1 unit of flour, and 1.5 units of sugar. Find the numbers of muffins, cookies, and brownies that the company can produce with the given amounts of raw materials.

**61.** 550 units of chocolate
525 units of flour
500 units of sugar

**62.** 800 units of chocolate
750 units of flour
700 units of sugar

In Exercises 63 and 64, use the following matrices.

$$A = \begin{bmatrix} 4 & 3 \\ -2 & 1 \end{bmatrix}, \quad B = \begin{bmatrix} 1 & -2 \\ 3 & 4 \end{bmatrix}, \quad C = \begin{bmatrix} 13 & 4 \\ 1 & 8 \end{bmatrix}$$

**63.** Find $AB$ and $BA$. What do you observe about the two products?

**64.** Find $C^{-1}$, $A^{-1} \cdot B^{-1}$, and $B^{-1} \cdot A^{-1}$. What do you observe about the three matrices?

In Exercises 65 and 66, find a value for $k$ (values for $k$ will vary) that will make the matrix invertible and then find a value for $k$ that will make the matrix singular.

**65.** $\begin{bmatrix} 4 & 3 \\ -2 & k \end{bmatrix}$

**66.** $\begin{bmatrix} 2k + 1 & 3 \\ -7 & 1 \end{bmatrix}$

## Mid-Chapter Quiz

Take this quiz as you would take a quiz in class. After you are done, check your work against the answers given in the back of the book.

In Exercises 1 and 2, write a matrix of the given order.

**1.** $3 \times 4$

**2.** $1 \times 3$

In Exercises 3 and 4, write the augmented matrix for the system of equations.

**3.** $3x + 2y = -2$
$5x - y = 19$

**4.** $x \qquad + 3z = -5$
$x + 2y - z = 3$
$3x \qquad + 4z = 0$

**5.** Use Gaussian elimination with back substitution to solve the augmented matrix found in Exercise 3.

**6.** Use Gauss-Jordan elimination to solve the augmented matrix found in Exercise 4.

In Exercises 7–12, use the following matrices to find the indicated matrix.

$$A = \begin{bmatrix} 1 & -2 \\ 3 & 4 \end{bmatrix}, \quad B = \begin{bmatrix} -1 & 2 & -3 \\ 2 & 0 & 5 \end{bmatrix}, \quad C = \begin{bmatrix} 0 & -2 \\ 3 & 1 \end{bmatrix}$$

**7.** $2A + 3C$

**8.** $AB$

**9.** $A - 2C$

**10.** $C^2$

**11.** $A^{-1}$

**12.** $B^{-1}$

In Exercises 13 and 14, solve for $X$ using matrices $A$ and $C$ from Exercises 7–12.

**13.** $X = 2A - 3C$

**14.** $2X = 3A - C$

In Exercises 15 and 16, find matrices $A, X,$ and $B$ such that the system can be written as $AX = B$. Then solve for $X$.

**15.** $x - 3y = 10$
$-2x + y = -10$

**16.** $2x - y + z = 3$
$3x \qquad - z = 15$
$4y + 3z = -1$

$$L = \begin{bmatrix} 1.0 & 0.6 & 0.2 \\ 2.4 & 1.0 & 0.2 \\ 2.8 & 2.0 & 0.5 \end{bmatrix} \begin{matrix} \text{Model A} \\ \text{Model B} \\ \text{Model C} \end{matrix}$$

Assembly · Finishing · Packaging

*Labor Requirements*
*(in hours)*

$$W = \begin{bmatrix} 15 & 12 \\ 7 & 9 \\ 6 & 8 \end{bmatrix} \begin{matrix} \text{Assembly} \\ \text{Finishing} \\ \text{Packaging} \end{matrix}$$

Plant 1 · Plant 2

*Wage Requirements*
*(in dollars per hour)*

*Labor/Wage Requirements* In Exercises 17–20, a hang-glider manufacturer has the labor-hour and wage requirements indicated at the left.

**17.** What is the labor cost for model A?

**18.** What is the labor cost for model B?

**19.** What is the labor cost for model C?

**20.** Find $LW$ and interpret the result.

## 7.4　The Determinant of a Square Matrix

### Objectives

- Evaluate the determinant of a $2 \times 2$ matrix.
- Find the minors and cofactors of a matrix.
- Find the determinant of a square matrix.
- Find the determinant of a triangular matrix.

## The Determinant of a $2 \times 2$ Matrix

Every *square* matrix can be associated with a real number called its **determinant.** Determinants have many uses, and several will be discussed in this and the next section. Historically, the use of determinants arose from special number patterns that occur when systems of linear equations are solved. For instance, the system

$$a_1 x + b_1 y = c_1$$

$$a_2 x + b_2 y = c_2$$

has a solution given by

$$x = \frac{c_1 b_2 - c_2 b_1}{a_1 b_2 - a_2 b_1} \qquad \text{and} \qquad y = \frac{a_1 c_2 - a_2 c_1}{a_1 b_2 - a_2 b_1}$$

provided that $a_1 b_2 - a_2 b_1 \neq 0$. Note that the denominator of each fraction is the same. This denominator is called the **determinant** of the coefficient matrix of the system.

| *Coefficient Matrix* | *Determinant* |
|---|---|
| $A = \begin{bmatrix} a_1 & b_1 \\ a_2 & b_2 \end{bmatrix}$ | $\det(A) = a_1 b_2 - a_2 b_1$ |

The determinant of the matrix $A$ can also be denoted by vertical bars on both sides of the matrix, as indicated in the following definition.

---

▶ **Definition of the Determinant of a $2 \times 2$ Matrix**

The **determinant** of the matrix

$$A = \begin{bmatrix} a_1 & b_1 \\ a_2 & b_2 \end{bmatrix}$$

is given by

$$\det(A) = |A| = \begin{vmatrix} a_1 & b_1 \\ a_2 & b_2 \end{vmatrix} = a_1 b_2 - a_2 b_1.$$

---

In this text, $\det(A)$ and $|A|$ are used interchangeably to represent the determinant of $A$. Although vertical bars are also used to denote the absolute value of a real number, the context will show which use is intended.

A convenient method for remembering the formula for the determinant of a $2 \times 2$ matrix is shown in the following diagram.

$$\det(A) = \begin{vmatrix} a_1 & b_1 \\ a_2 & b_2 \end{vmatrix} = a_1 b_2 - a_2 b_1$$

Note that the determinant is given by the difference of the products of the two diagonals of the matrix. In Example 1 you will see that the determinant of a matrix can be positive, zero, or negative.

**Example 1**   The Determinant of a $2 \times 2$ Matrix

Find the determinant of each matrix.

**a.** $A = \begin{bmatrix} 2 & -3 \\ 1 & 2 \end{bmatrix}$   **b.** $B = \begin{bmatrix} 2 & 1 \\ 4 & 2 \end{bmatrix}$   **c.** $C = \begin{bmatrix} 0 & 3 \\ 2 & 4 \end{bmatrix}$

**Solution**

**a.** $\det(A) = \begin{vmatrix} 2 & -3 \\ 1 & 2 \end{vmatrix} = 2(2) - 1(-3) = 4 + 3 = 7$

**b.** $\det(B) = \begin{vmatrix} 2 & 1 \\ 4 & 2 \end{vmatrix} = 2(2) - 4(1) = 4 - 4 = 0$

**c.** $\det(C) = \begin{vmatrix} 0 & 3 \\ 2 & 4 \end{vmatrix} = 0(4) - 2(3) = 0 - 6 = -6$

The determinant of a matrix of order $1 \times 1$ is defined simply as the entry of the matrix. For instance, if $A = [-2]$, then $\det(A) = -2$.

**·TECHNOLOGY·**

Most graphing utilities can evaluate the determinant of a matrix. Consult your user's guide to find how to evaluate a determinant. Use your graphing utility to find the determinant of A. The result should be 7, as in Example 1(a). Try evaluating the determinant of B with your graphing utility. What happens when you try to evaluate the determinant of a nonsquare matrix?

$$A = \begin{bmatrix} 2 & -3 \\ 1 & 2 \end{bmatrix} \qquad B = \begin{bmatrix} 3 & -1 & 1 \\ 0 & 2 & 1 \end{bmatrix}$$

## Minors and Cofactors

To define the determinant of a square matrix of order $3 \times 3$ or higher, it is convenient to introduce the concepts of **minors** and **cofactors.**

*Sign Pattern for Cofactors*

$$\begin{bmatrix} + & - & + \\ - & + & - \\ + & - & + \end{bmatrix}$$

$3 \times 3$ matrix

$$\begin{bmatrix} + & - & + & - \\ - & + & - & + \\ + & - & + & - \\ - & + & - & + \end{bmatrix}$$

$4 \times 4$ matrix

$$\begin{bmatrix} + & - & + & - & + & \cdots \\ - & + & - & + & - & \cdots \\ + & - & + & - & + & \cdots \\ - & + & - & + & - & \cdots \\ + & - & + & - & + & \cdots \\ \vdots & \vdots & \vdots & \vdots & \vdots & \end{bmatrix}$$

$n \times n$ matrix

▶ **Minors and Cofactors of a Square Matrix**

If $A$ is a square matrix, the **minor** $M_{ij}$ of the entry $a_{ij}$ is the determinant of the matrix obtained by deleting the $i$th row and $j$th column of $A$. The **cofactor** $C_{ij}$ of the entry $a_{ij}$ is given by

$$C_{ij} = (-1)^{i+j}M_{ij}.$$

**Example 2**    Finding the Minors and Cofactors of a Matrix

Find all the minors and cofactors of $A = \begin{bmatrix} 0 & 2 & 1 \\ 3 & -1 & 2 \\ 4 & 0 & 1 \end{bmatrix}$.

**Solution**

To find the minor $M_{11}$, delete the first row and first column of $A$ and evaluate the determinant of the resulting matrix.

$$\begin{bmatrix} \boxed{0} & 2 & 1 \\ 3 & -1 & 2 \\ 4 & 0 & 1 \end{bmatrix}, \quad M_{11} = \begin{vmatrix} -1 & 2 \\ 0 & 1 \end{vmatrix} = -1(1) - 0(2) = -1$$

Similarly, to find $M_{12}$, delete the first row and second column.

$$\begin{bmatrix} 0 & \boxed{2} & 1 \\ 3 & -1 & 2 \\ 4 & 0 & 1 \end{bmatrix}, \quad M_{12} = \begin{vmatrix} 3 & 2 \\ 4 & 1 \end{vmatrix} = 3(1) - 4(2) = -5$$

Continuing this pattern, you obtain the following minors.

$$M_{11} = -1 \qquad M_{12} = -5 \qquad M_{13} = 4$$
$$M_{21} = 2 \qquad M_{22} = -4 \qquad M_{23} = -8$$
$$M_{31} = 5 \qquad M_{32} = -3 \qquad M_{33} = -6$$

Now, to find the cofactors, combine the checkerboard pattern of signs with these minors to obtain the following.

$$C_{11} = -1 \qquad C_{12} = 5 \qquad C_{13} = 4$$
$$C_{21} = -2 \qquad C_{22} = -4 \qquad C_{23} = 8$$
$$C_{31} = 5 \qquad C_{32} = 3 \qquad C_{33} = -6$$

## The Determinant of a Square Matrix

The definition given below is called **inductive** because it uses determinants of matrices of order $n - 1$ to define the determinant of a matrix of order $n$.

---

▶ **Determinant of a Square Matrix**

If $A$ is a square matrix (of order $2 \times 2$ or greater), the determinant of $A$ is the sum of the entries in any row (or column) of $A$ multiplied by their respective cofactors. For instance, expanding along the first row yields

$$|A| = a_{11}C_{11} + a_{12}C_{12} + \cdots + a_{1n}C_{1n}.$$

Applying this definition to find a determinant is called **expanding by cofactors.**

---

Try checking that for a $2 \times 2$ matrix this definition yields $|A| = a_{11}a_{22} - a_{12}a_{21}$, as previously defined.

**Example 3**   The Determinant of a Matrix of Order $3 \times 3$

Find the determinant of

$$A = \begin{bmatrix} 0 & 2 & 1 \\ 3 & -1 & 2 \\ 4 & 0 & 1 \end{bmatrix}.$$

**Solution**

Note that this is the same matrix that was given in Example 2. There we found the cofactors of the entries in the first row to be

$$C_{11} = -1, \quad C_{12} = 5, \quad \text{and} \quad C_{13} = 4.$$

Therefore, by the definition of a determinant, you have the following.

$$|A| = a_{11}C_{11} + a_{12}C_{12} + a_{13}C_{13} \qquad \text{First-row expansion}$$
$$= 0(-1) + 2(5) + 1(4)$$
$$= 14$$

In Example 3 the determinant was found by expanding by the cofactors in the first row. You could have used any row or column. For instance, you could have expanded along the second row to obtain

$$|A| = a_{21}C_{21} + a_{22}C_{22} + a_{23}C_{23} \qquad \text{Second-row expansion}$$
$$= 3(-2) + (-1)(-4) + 2(8)$$
$$= 14.$$

When expanding by cofactors, you do not need to find cofactors of zero entries, because zero times its cofactor is zero.

$$a_{ij}C_{ij} = (0)C_{ij} = 0$$

So, the row (or column) containing the most zeros is usually the best choice for expansion by cofactors. This is demonstrated in the next example.

### Example 4    The Determinant of a Matrix of Order $4 \times 4$

Find the determinant of

$$A = \begin{bmatrix} 1 & -2 & 3 & 0 \\ -1 & 1 & 0 & 2 \\ 0 & 2 & 0 & 3 \\ 3 & 4 & 0 & 2 \end{bmatrix}.$$

**Solution**

After inspecting this matrix, you can see that three of the entries in the third column are zeros. Thus, you can eliminate some of the work in the expansion by using the third column.

$$|A| = 3(C_{13}) + 0(C_{23}) + 0(C_{33}) + 0(C_{43})$$

Because $C_{23}$, $C_{33}$, and $C_{43}$ have zero coefficients, you need only find the cofactor $C_{13}$. To do this, delete the first row and third column of $A$ and evaluate the determinant of the resulting matrix.

$$C_{13} = (-1)^{1+3} \begin{vmatrix} -1 & 1 & 2 \\ 0 & 2 & 3 \\ 3 & 4 & 2 \end{vmatrix}$$

$$= \begin{vmatrix} -1 & 1 & 2 \\ 0 & 2 & 3 \\ 3 & 4 & 2 \end{vmatrix}$$

Expanding by cofactors in the second row yields the following.

$$C_{13} = 0(-1)^3 \begin{vmatrix} 1 & 2 \\ 4 & 2 \end{vmatrix} + 2(-1)^4 \begin{vmatrix} -1 & 2 \\ 3 & 2 \end{vmatrix} + 3(-1)^5 \begin{vmatrix} -1 & 1 \\ 3 & 4 \end{vmatrix}$$

$$= 0 + 2(1)(-8) + 3(-1)(-7)$$

$$= 5$$

So, you obtain

$$|A| = 3C_{13} = 3(5) = 15.$$

Try using a graphing utility to confirm the result of Example 4.

There is an alternative method that is commonly used for evaluating the determinant of a $3 \times 3$ matrix $A$. (This method works *only* for $3 \times 3$ matrices.) To apply this method, copy the first and second columns of $A$ to form fourth and fifth columns. The determinant of $A$ is then obtained by adding the products of the three "downward diagonals" and subtracting the products of the three "upward diagonals," as shown in the following diagram.

Subtract these three products

$$
\begin{array}{ccccc}
a_{11} & a_{12} & a_{13} & a_{11} & a_{12} \\
a_{21} & a_{22} & a_{23} & a_{21} & a_{22} \\
a_{31} & a_{32} & a_{33} & a_{31} & a_{32}
\end{array}
$$

Add these three products

So, the determinant of the $3 \times 3$ matrix $A$ is given by

$$|A| = a_{11}a_{22}a_{33} + a_{12}a_{23}a_{31} + a_{13}a_{21}a_{32}$$
$$- a_{31}a_{22}a_{13} - a_{32}a_{23}a_{11} - a_{33}a_{21}a_{12}.$$

### Example 5  The Determinant of a 3 × 3 Matrix

Find the determinant of

$$A = \begin{bmatrix} 0 & 2 & 1 \\ 3 & -1 & 2 \\ 4 & -4 & 1 \end{bmatrix}.$$

### Solution

Because $A$ is a $3 \times 3$ matrix, you can use the alternative procedure for finding $|A|$. Begin by recopying the first two columns and then computing the six diagonal products, as follows.

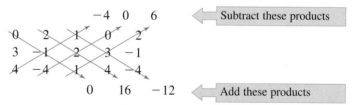

Now, by adding the lower three products and subtracting the upper three products, you find the determinant of $A$ to be

$$|A| = 0 + 16 - 12 - (-4) - 0 - 6 = 2.$$

### Study Tip

Be sure you understand that the diagonal process illustrated in Example 5 is valid *only* for matrices of order $3 \times 3$. For matrices of higher orders, another method must be used.

# Triangular Matrices

Evaluating determinants of matrices of order 4 or higher can be tedious. There is, however, an important exception: the determinant of a **triangular** matrix. A square matrix is **upper triangular** if it has all zero entries below its main diagonal and **lower triangular** if it has all zero entries above its main diagonal. A matrix that is both upper and lower triangular is called **diagonal.** That is, a diagonal matrix is one in which all entries above and below the main diagonal are zero.

$$\text{Upper Triangular Matrix} \qquad\qquad \text{Lower Triangular Matrix}$$

$$\begin{bmatrix} a_{11} & a_{12} & a_{13} & \cdots & a_{1n} \\ 0 & a_{22} & a_{23} & \cdots & a_{2n} \\ 0 & 0 & a_{33} & \cdots & a_{3n} \\ \vdots & \vdots & \vdots & & \vdots \\ 0 & 0 & 0 & \cdots & a_{nn} \end{bmatrix} \qquad \begin{bmatrix} a_{11} & 0 & 0 & \cdots & 0 \\ a_{21} & a_{22} & 0 & \cdots & 0 \\ a_{31} & a_{32} & a_{33} & \cdots & 0 \\ \vdots & \vdots & \vdots & & \vdots \\ a_{n1} & a_{n2} & a_{n3} & \cdots & a_{nn} \end{bmatrix}$$

To find the determinant of a triangular matrix of any order, simply form the product of the entries on the main diagonal.

### Example 6    The Determinant of a Triangular Matrix

a. $\begin{vmatrix} 2 & 0 & 0 & 0 \\ 4 & -2 & 0 & 0 \\ -5 & 6 & 1 & 0 \\ 1 & 5 & 3 & 3 \end{vmatrix} = (2)(-2)(1)(3) = -12$

b. $\begin{vmatrix} -1 & 0 & 0 & 0 & 0 \\ 0 & 3 & 0 & 0 & 0 \\ 0 & 0 & 2 & 0 & 0 \\ 0 & 0 & 0 & 4 & 0 \\ 0 & 0 & 0 & 0 & -2 \end{vmatrix} = (-1)(3)(2)(4)(-2) = 48$

## Discussing the Concept    The Determinant of a Triangular Matrix

Explain why the determinant of a $3 \times 3$ triangular matrix is the product of its main-diagonal entries.

$$\begin{vmatrix} a_{11} & a_{12} & a_{13} \\ 0 & a_{22} & a_{23} \\ 0 & 0 & a_{33} \end{vmatrix} = a_{11}a_{22}a_{33}$$

Is this true of higher-order triangular matrices? Explain.

## Warm Up

The following warm-up exercises involve skills that were covered in earlier sections. You will use these skills in the exercise set for this section.

In Exercises 1–4, perform the indicated matrix operations.

**1.** $\begin{bmatrix} 1 & -2 \\ 0 & 3 \end{bmatrix} + \begin{bmatrix} 2 & 7 \\ 4 & -3 \end{bmatrix}$

**2.** $\begin{bmatrix} -2 & 5 \\ 3 & -2 \end{bmatrix} - \begin{bmatrix} 0 & -3 \\ 1 & 2 \end{bmatrix}$

**3.** $3\begin{bmatrix} 3 & -4 & 2 \\ 1 & 0 & -1 \\ 0 & 1 & -2 \end{bmatrix}$

**4.** $4\begin{bmatrix} 0 & 2 & 3 \\ -1 & 2 & 3 \\ -2 & 1 & -2 \end{bmatrix}$

In Exercises 5–10, perform the indicated arithmetic operations.

**5.** $[(1)(3) + (-3)(2)] - [(1)(4) + (3)(5)]$

**6.** $[(4)(4) + (-1)(-3)] - [(-1)(2) + (-2)(7)]$

**7.** $\dfrac{4(7) - 1(-2)}{(-5)(-2) - 3(4)}$

**8.** $\dfrac{3(6) - 2(7)}{6(-5) - 2(1)}$

**9.** $-5(-1)^2[6(-2) - 7(-3)]$

**10.** $4(-1)^3[3(6) - 2(7)]$

## 7.4    Exercises

In Exercises 1–24, find the determinant of the matrix.

**1.** $[-2]$

**2.** $[6]$

**3.** $\begin{bmatrix} 1 & 3 \\ 2 & 7 \end{bmatrix}$

**4.** $\begin{bmatrix} -3 & 4 \\ -2 & 1 \end{bmatrix}$

**5.** $\begin{bmatrix} 5 & 6 \\ 2 & 3 \end{bmatrix}$

**6.** $\begin{bmatrix} -7 & -4 \\ 8 & 7 \end{bmatrix}$

**7.** $\begin{bmatrix} 9 & 3 \\ 12 & 4 \end{bmatrix}$

**8.** $\begin{bmatrix} -5 & -2 \\ 10 & 4 \end{bmatrix}$

**9.** $\begin{bmatrix} \frac{2}{3} & 0 \\ -1 & 6 \end{bmatrix}$

**10.** $\begin{bmatrix} 9 & -\frac{1}{4} \\ 8 & 0 \end{bmatrix}$

**11.** $\begin{bmatrix} 2 & -1 & 0 \\ 4 & 2 & 1 \\ 4 & 2 & 1 \end{bmatrix}$

**12.** $\begin{bmatrix} -2 & 2 & 3 \\ 1 & -1 & 0 \\ 0 & 1 & 4 \end{bmatrix}$

**13.** $\begin{bmatrix} 0.3 & 0.2 & 0.2 \\ 0.2 & 0.2 & 0.2 \\ -0.4 & 0.4 & 0.3 \end{bmatrix}$

**14.** $\begin{bmatrix} 0.1 & 0.2 & 0.3 \\ -0.3 & 0.2 & 0.2 \\ 0.5 & 0.4 & 0.4 \end{bmatrix}$

**15.** $\begin{bmatrix} 1 & 4 & -2 \\ 3 & 6 & -6 \\ -2 & 1 & 4 \end{bmatrix}$

**16.** $\begin{bmatrix} -1 & 3 & 1 \\ 4 & 2 & 5 \\ -2 & 1 & 6 \end{bmatrix}$

**17.** $\begin{bmatrix} 6 & 3 & -7 \\ 0 & 0 & 0 \\ 4 & -6 & 3 \end{bmatrix}$

**18.** $\begin{bmatrix} 5 & 0 & 3 \\ -4 & 0 & 8 \\ 3 & 0 & -6 \end{bmatrix}$

**19.** $\begin{bmatrix} 2 & 0 & 0 \\ 4 & -3 & 0 \\ 6 & 5 & 1 \end{bmatrix}$

**20.** $\begin{bmatrix} 1 & 0 & 0 \\ -4 & -1 & 0 \\ 5 & 1 & 5 \end{bmatrix}$

**21.** $\begin{bmatrix} 2 & 3 & -1 & -1 \\ 0 & -1 & -3 & 5 \\ 0 & 0 & -2 & 7 \\ 0 & 0 & 0 & -4 \end{bmatrix}$

**22.** $\begin{bmatrix} 4 & 0 & 0 & 0 \\ 1 & -4 & 0 & 0 \\ 2 & 1 & -1 & 0 \\ 6 & -2 & 3 & -1 \end{bmatrix}$

**23.** $\begin{bmatrix} 1 & 0 & 0 & 0 & 0 \\ 0 & 2 & 0 & 0 & 0 \\ 0 & 0 & 3 & 0 & 0 \\ 0 & 0 & 0 & 4 & 0 \\ 0 & 0 & 0 & 0 & 5 \end{bmatrix}$

**24.** $\begin{bmatrix} -2 & 0 & 0 & 0 & 0 \\ 0 & 3 & 0 & 0 & 0 \\ 0 & 0 & -1 & 0 & 0 \\ 0 & 0 & 0 & 2 & 0 \\ 0 & 0 & 0 & 0 & -4 \end{bmatrix}$

In Exercises 25–28, find (a) all minors and (b) all cofactors of the given matrix.

**25.** $\begin{bmatrix} 3 & 4 \\ 2 & -5 \end{bmatrix}$  **26.** $\begin{bmatrix} 11 & 0 \\ -3 & 2 \end{bmatrix}$

**27.** $\begin{bmatrix} 3 & -2 & 8 \\ 3 & 2 & -6 \\ -1 & 3 & 6 \end{bmatrix}$  **28.** $\begin{bmatrix} -2 & 9 & 4 \\ 7 & -6 & 0 \\ 6 & 7 & -6 \end{bmatrix}$

In Exercises 29–34, find the determinant of the matrix by the method of expansion by cofactors. Expand using the indicated row or column.

**29.** $\begin{bmatrix} 4 & 1 & -3 \\ 6 & 5 & -2 \\ -1 & 3 & -4 \end{bmatrix}$  **30.** $\begin{bmatrix} -3 & 4 & 2 \\ 6 & 3 & 1 \\ 4 & -7 & -8 \end{bmatrix}$

(a) Row 3   (a) Row 2

(b) Column 2   (b) Column 3

**31.** $\begin{bmatrix} 7 & 0 & -4 \\ 2 & -3 & 0 \\ 5 & 8 & 1 \end{bmatrix}$  **32.** $\begin{bmatrix} 10 & -5 & 5 \\ 30 & 0 & 10 \\ 0 & 10 & 1 \end{bmatrix}$

(a) Row 1   (a) Row 3

(b) Column 3   (b) Column 1

**33.** $\begin{bmatrix} 6 & 0 & -3 & 5 \\ 4 & 13 & 6 & -8 \\ -1 & 0 & 7 & 4 \\ 8 & 6 & 0 & 2 \end{bmatrix}$

(a) Row 2

(b) Column 2

**34.** $\begin{bmatrix} 10 & 8 & 3 & -7 \\ 4 & 0 & 5 & -6 \\ 0 & 3 & 2 & 7 \\ 1 & 0 & -3 & 2 \end{bmatrix}$

(a) Row 3

(b) Column 1

In Exercises 35–44, find the determinant of the matrix. Use a graphing utility to confirm your result.

**35.** $\begin{bmatrix} 1 & 4 & -2 \\ 3 & 2 & 0 \\ -1 & 4 & 3 \end{bmatrix}$  **36.** $\begin{bmatrix} 2 & -1 & 3 \\ 1 & 4 & 4 \\ 1 & 0 & 2 \end{bmatrix}$

**37.** $\begin{bmatrix} 2 & 4 & 6 \\ 0 & 3 & 1 \\ 0 & 0 & -5 \end{bmatrix}$  **38.** $\begin{bmatrix} -3 & 0 & 0 \\ 7 & 11 & 0 \\ 1 & 2 & 2 \end{bmatrix}$

**39.** $\begin{bmatrix} 3 & 6 & -5 & 4 \\ -2 & 0 & 6 & 0 \\ 1 & 1 & 2 & 2 \\ 0 & 3 & -1 & -1 \end{bmatrix}$

**40.** $\begin{bmatrix} 2 & 6 & 6 & 2 \\ 2 & 7 & 3 & 6 \\ 1 & 5 & 0 & 1 \\ 3 & 7 & 0 & 7 \end{bmatrix}$

**41.** $\begin{bmatrix} 5 & 3 & 0 & 6 \\ 4 & 6 & 4 & 12 \\ 0 & 2 & -3 & 4 \\ 0 & 1 & -2 & 2 \end{bmatrix}$  **42.** $\begin{bmatrix} 1 & 4 & 3 & 2 \\ -5 & 6 & 2 & 1 \\ 0 & 0 & 0 & 0 \\ 3 & -2 & 1 & 5 \end{bmatrix}$

**43.** $\begin{bmatrix} 3 & 2 & 4 & -1 & 5 \\ -2 & 0 & 1 & 3 & 2 \\ 1 & 0 & 0 & 4 & 0 \\ 6 & 0 & 2 & -1 & 0 \\ 3 & 0 & 5 & 1 & 0 \end{bmatrix}$

**44.** $\begin{bmatrix} 5 & 2 & 0 & 0 & -2 \\ 0 & 1 & 4 & 3 & 2 \\ 0 & 0 & 2 & 6 & 3 \\ 0 & 0 & 3 & 4 & 1 \\ 0 & 0 & 0 & 0 & 2 \end{bmatrix}$

In Exercises 45 and 46, find a $4 \times 4$ *upper* triangular matrix whose determinant is equal to the given value and a $4 \times 4$ *lower* triangular matrix whose determinant is equal to the given value. Use a graphing utility to confirm your results.

**45.** $-18$  **46.** 28

In Exercises 47 and 48, explain why the determinant of the matrix is equal to zero.

**47.** $\begin{bmatrix} 3 & 4 & -2 & 7 \\ 1 & 3 & -1 & 2 \\ 0 & 5 & 7 & 1 \\ 1 & 3 & -1 & 2 \end{bmatrix}$  **48.** $\begin{bmatrix} 3 & 2 & -1 \\ -6 & -4 & 2 \\ 5 & -7 & 9 \end{bmatrix}$

## Math Matters • Guess the Number

Here is a guessing game that can be made using the four cards shown in the figure (note that the fourth card has numbers on the front *and* back). To play the game, ask someone to think of a number between 1 and 15. Ask the person if the number is on the first card. If it is, place the card face up with the "YES" on top. If it isn't, place the card face up with the "NO" on top. Repeat this with each of the four cards (using the same number), stacking the cards one on top of another. Be sure that the fourth card is used last. After all four cards are in a stack, turn the stack over, The person's number will be the number that is showing through a window in the cards. Can you explain why this card game works?

## 7.5   Applications of Matrices and Determinants

### Objectives

- Find the area of a triangle using a determinant.
- Determine whether three points are collinear using a determinant.
- Use a determinant to find the equation of a line.
- Encode and decode a cryptogram using a matrix.

## Area of a Triangle

In this section, you will study some additional applications of matrices and determinants. The first involves a formula for finding the area of a triangle whose vertices are given by three points on a rectangular coordinate system.

> ▶ **Area of a Triangle**
>
> The area of a triangle with vertices $(x_1, y_1)$, $(x_2, y_2)$, and $(x_3, y_3)$ is given by
>
> $$\text{Area} = \pm\frac{1}{2} \begin{vmatrix} x_1 & y_1 & 1 \\ x_2 & y_2 & 1 \\ x_3 & y_3 & 1 \end{vmatrix}$$
>
> where the symbol $(\pm)$ indicates that the appropriate sign should be chosen to yield a positive area.

**Example 1**   Finding the Area of a Triangle

Find the area of the triangle whose vertices are $(1, 0)$, $(2, 2)$, and $(4, 3)$, as shown in Figure 7.1.

**Solution**

Let $(x_1, y_1) = (1, 0)$, $(x_2, y_2) = (2, 2)$, and $(x_3, y_3) = (4, 3)$. Then, to find the area of the triangle, evaluate the determinant

$$\begin{vmatrix} x_1 & y_1 & 1 \\ x_2 & y_2 & 1 \\ x_3 & y_3 & 1 \end{vmatrix} = \begin{vmatrix} 1 & 0 & 1 \\ 2 & 2 & 1 \\ 4 & 3 & 1 \end{vmatrix} = 1\begin{vmatrix} 2 & 1 \\ 3 & 1 \end{vmatrix} - 0\begin{vmatrix} 2 & 1 \\ 4 & 1 \end{vmatrix} + 1\begin{vmatrix} 2 & 2 \\ 4 & 3 \end{vmatrix}$$

$$= 1(-1) - 0(-2) + 1(-2)$$

$$= -3.$$

Using this value, you can conclude that the area of the triangle is

$$\text{Area} = -\frac{1}{2}\begin{vmatrix} 1 & 0 & 1 \\ 2 & 2 & 1 \\ 4 & 3 & 1 \end{vmatrix} = -\frac{1}{2}(-3) = \frac{3}{2}.$$

Figure 7.1

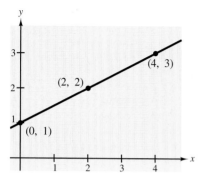

**Figure 7.2**

## Lines in the Plane

Suppose the three points in Example 1 had been on the same line. What would have happened had the area formula been applied to three such points? The answer is that the determinant would have been zero. Consider, for instance, the three collinear points $(0, 1)$, $(2, 2)$, and $(4, 3)$, as shown in Figure 7.2. The area of the "triangle" that has these three points as vertices is

$$\frac{1}{2}\begin{vmatrix} 0 & 1 & 1 \\ 2 & 2 & 1 \\ 4 & 3 & 1 \end{vmatrix} = \frac{1}{2}\left(0\begin{vmatrix} 2 & 1 \\ 3 & 1 \end{vmatrix} - 1\begin{vmatrix} 2 & 1 \\ 4 & 1 \end{vmatrix} + 1\begin{vmatrix} 2 & 2 \\ 4 & 3 \end{vmatrix}\right)$$

$$= \frac{1}{2}[0(-1) - 1(-2) + 1(-2)] = 0.$$

This result is generalized as follows.

> ▶ **Test for Collinear Points**
>
> Three points $(x_1, y_1)$, $(x_2, y_2)$, and $(x_3, y_3)$ are collinear (lie on the same line) if and only if
>
> $$\begin{vmatrix} x_1 & y_1 & 1 \\ x_2 & y_2 & 1 \\ x_3 & y_3 & 1 \end{vmatrix} = 0.$$

**Example 2** **Testing for Collinear Points**

Determine whether the points $(-2, -2)$, $(1, 1)$, and $(7, 5)$ lie on the same line. (See Figure 7.3.)

**Solution**

Letting $(x_1, y_1) = (-2, -2)$, $(x_2, y_2) = (1, 1)$, and $(x_3, y_3) = (7, 5)$, you have

$$\begin{vmatrix} x_1 & y_1 & 1 \\ x_2 & y_2 & 1 \\ x_3 & y_3 & 1 \end{vmatrix} = \begin{vmatrix} -2 & -2 & 1 \\ 1 & 1 & 1 \\ 7 & 5 & 1 \end{vmatrix}$$

$$= -2\begin{vmatrix} 1 & 1 \\ 5 & 1 \end{vmatrix} - (-2)\begin{vmatrix} 1 & 1 \\ 7 & 1 \end{vmatrix} + 1\begin{vmatrix} 1 & 1 \\ 7 & 5 \end{vmatrix}$$

$$= -2(-4) - (-2)(-6) + 1(-2)$$

$$= -6.$$

Because the value of this determinant is *not* zero, you can conclude that the three points do not lie on the same line.

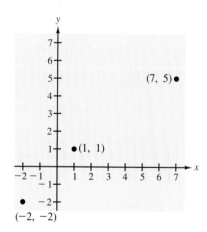

**Figure 7.3**

The test for collinear points can be adapted to another use. That is, if you are given two points on a rectangular coordinate system, you can find an equation of the line passing through the two points, as follows.

---

▶ **Two-Point Form of the Equation of a Line**

An equation of the line passing through the distinct points $(x_1, y_1)$ and $(x_2, y_2)$ is given by

$$\begin{vmatrix} x & y & 1 \\ x_1 & y_1 & 1 \\ x_2 & y_2 & 1 \end{vmatrix} = 0.$$

---

**Example 3**    Finding an Equation of a Line

Find an equation of the line passing through the two points $(2, 4)$ and $(-1, 3)$, as shown in Figure 7.4.

**Solution**

Applying the determinant formula for the equation of a line produces

$$\begin{vmatrix} x & y & 1 \\ 2 & 4 & 1 \\ -1 & 3 & 1 \end{vmatrix} = 0.$$

To evaluate this determinant, you can expand by cofactors along the first row to obtain the following.

$$x \begin{vmatrix} 4 & 1 \\ 3 & 1 \end{vmatrix} - y \begin{vmatrix} 2 & 1 \\ -1 & 1 \end{vmatrix} + 1 \begin{vmatrix} 2 & 4 \\ -1 & 3 \end{vmatrix} = x - 3y + 10 = 0$$

So, an equation of the line is

$$x - 3y + 10 = 0.$$

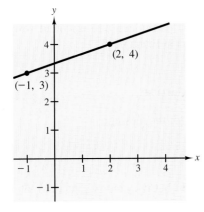

Figure 7.4

Note that this method of finding the equation of a line works for all lines, including horizontal and vertical lines. For instance, the equation of the vertical line through $(2, 0)$ and $(2, 2)$ is

$$\begin{vmatrix} x & y & 1 \\ 2 & 0 & 1 \\ 2 & 2 & 1 \end{vmatrix} = 0$$

$$4 - 2x = 0$$

$$x = 2.$$

*Both governments and private industry use codes to transmit messages. Many different systems are used. The first mechanical device for coding and decoding was invented in 1917 by an American engineer, Gilbert Vernam.*

# Cryptography

A **cryptogram** is a message written according to a secret code. (The Greek word *kryptos* means "hidden.") Matrix multiplication can be used to **encode** and **decode** messages. To begin, you need to assign a number to each letter in the alphabet (with 0 assigned to a blank space), as follows.

| | | |
|---|---|---|
| 0 = __ | 9 = I | 18 = R |
| 1 = A | 10 = J | 19 = S |
| 2 = B | 11 = K | 20 = T |
| 3 = C | 12 = L | 21 = U |
| 4 = D | 13 = M | 22 = V |
| 5 = E | 14 = N | 23 = W |
| 6 = F | 15 = O | 24 = X |
| 7 = G | 16 = P | 25 = Y |
| 8 = H | 17 = Q | 26 = Z |

Then the message is converted to numbers and partitioned into **uncoded row matrices,** each having $n$ entries, as demonstrated in Example 4.

### Example 4  Forming Uncoded Row Matrices

Write the uncoded row matrices of order $1 \times 3$ for the message

MEET ME MONDAY.

#### Solution

Partitioning the message (including blank spaces, but ignoring punctuation) into groups of three produces the following uncoded row matrices.

$$[13 \quad 5 \quad 5] \quad [20 \quad 0 \quad 13] \quad [5 \quad 0 \quad 13] \quad [15 \quad 14 \quad 4] \quad [1 \quad 25 \quad 0]$$
$$\text{M} \quad \text{E} \quad \text{E} \quad \text{T} \qquad \text{M} \quad \text{E} \qquad \text{M} \quad \text{O} \quad \text{N} \quad \text{D} \quad \text{A} \quad \text{Y}$$

Note that a blank space is used to fill out the last uncoded row matrix.

To **encode** a message, choose an $n \times n$ invertible matrix $A$ and multiply the uncoded row matrices by $A$ to obtain **coded row matrices.** The uncoded matrix should be on the left whereas the encoding matrix $A$ should be on the right. Here is an example.

*Uncoded Matrix*   *Encoding Matrix A*   *Coded Matrix*

$$[13 \quad 5 \quad 5] \begin{bmatrix} 1 & -2 & 2 \\ -1 & 1 & 3 \\ 1 & -1 & -4 \end{bmatrix} = [13 \quad -26 \quad 21]$$

This technique is further illustrated in Example 5.

Example 5    Encoding a Message

Use the following matrix to encode the message MEET ME MONDAY.

$$A = \begin{bmatrix} 1 & -2 & 2 \\ -1 & 1 & 3 \\ 1 & -1 & -4 \end{bmatrix}$$

### Solution

The coded row matrices are obtained by multiplying each of the uncoded row matrices found in Example 4 by the matrix $A$, as follows.

Uncoded Matrix    Encoding Matrix A    Coded Matrix

$$[13 \quad 5 \quad 5] \begin{bmatrix} 1 & -2 & 2 \\ -1 & 1 & 3 \\ 1 & -1 & -4 \end{bmatrix} = [13 \quad -26 \quad 21]$$

$$[20 \quad 0 \quad 13] \begin{bmatrix} 1 & -2 & 2 \\ -1 & 1 & 3 \\ 1 & -1 & -4 \end{bmatrix} = [33 \quad -53 \quad -12]$$

$$[5 \quad 0 \quad 13] \begin{bmatrix} 1 & -2 & 2 \\ -1 & 1 & 3 \\ 1 & -1 & -4 \end{bmatrix} = [18 \quad -23 \quad -42]$$

$$[15 \quad 14 \quad 4] \begin{bmatrix} 1 & -2 & 2 \\ -1 & 1 & 3 \\ 1 & -1 & -4 \end{bmatrix} = [5 \quad -20 \quad 56]$$

$$[1 \quad 25 \quad 0] \begin{bmatrix} 1 & -2 & 2 \\ -1 & 1 & 3 \\ 1 & -1 & -4 \end{bmatrix} = [-24 \quad 23 \quad 77]$$

So, the sequence of coded row matrices is

$$[13 \; -26 \; 21][33 \; -53 \; -12][18 \; -23 \; -42][5 \; -20 \; 56][-24 \; 23 \; 77].$$

Finally, removing the matrix notation produces the following cryptogram.

$$13 \; -26 \; 21 \; 33 \; -53 \; -12 \; 18 \; -23 \; -42 \; 5 \; -20 \; 56 \; -24 \; 23 \; 77$$

For those who do not know the matrix $A$, decoding the cryptogram found in Example 5 is difficult. But for an authorized receiver who knows the matrix $A$, decoding is simple. The receiver need only multiply the coded row matrices by $A^{-1}$ (on the right) to retrieve the uncoded row matrices. Here is an example.

$$\underbrace{[13 \quad -26 \quad 21]}_{\text{Coded}} A^{-1} = \underbrace{[13 \quad 5 \quad 5]}_{\text{Uncoded}}$$

Example 6 Decoding a Message

Use the inverse of the matrix $A = \begin{bmatrix} 1 & -2 & 2 \\ -1 & 1 & 3 \\ 1 & -1 & -4 \end{bmatrix}$ to decode the cryptogram

13 −26 21 33 −53 −12 18 −23 −42 5 −20 56 −24 23 77.

Solution

Partition the message into groups of three to form the coded row matrices. Multiply each coded row matrix on the right by $A^{-1}$ to obtain the decoded row matrices.

*Coded Matrix*  *Decoding Matrix* $A^{-1}$  *Decoded Matrix*

$\begin{bmatrix} 13 & -26 & 21 \end{bmatrix} \begin{bmatrix} -1 & -10 & -8 \\ -1 & -6 & -5 \\ 0 & -1 & -1 \end{bmatrix} = \begin{bmatrix} 13 & 5 & 5 \end{bmatrix}$

$\begin{bmatrix} 33 & -53 & -12 \end{bmatrix} \begin{bmatrix} -1 & -10 & -8 \\ -1 & -6 & -5 \\ 0 & -1 & -1 \end{bmatrix} = \begin{bmatrix} 20 & 0 & 13 \end{bmatrix}$

$\begin{bmatrix} 18 & -23 & -42 \end{bmatrix} \begin{bmatrix} -1 & -10 & -8 \\ -1 & -6 & -5 \\ 0 & -1 & -1 \end{bmatrix} = \begin{bmatrix} 5 & 0 & 13 \end{bmatrix}$

$\begin{bmatrix} 5 & -20 & 56 \end{bmatrix} \begin{bmatrix} -1 & -10 & -8 \\ -1 & -6 & -5 \\ 0 & -1 & -1 \end{bmatrix} = \begin{bmatrix} 15 & 14 & 4 \end{bmatrix}$

$\begin{bmatrix} -24 & 23 & 77 \end{bmatrix} \begin{bmatrix} -1 & -10 & -8 \\ -1 & -6 & -5 \\ 0 & -1 & -1 \end{bmatrix} = \begin{bmatrix} 1 & 25 & 0 \end{bmatrix}$

So, the message is as follows.

$\begin{bmatrix} 13 & 5 & 5 \end{bmatrix}$ $\begin{bmatrix} 20 & 0 & 13 \end{bmatrix}$ $\begin{bmatrix} 5 & 0 & 13 \end{bmatrix}$ $\begin{bmatrix} 15 & 14 & 4 \end{bmatrix}$ $\begin{bmatrix} 1 & 25 & 0 \end{bmatrix}$

M E E T  M E  M O N D A Y

Discussing the Concept  **Decoding**

Show how to use a graphing utility with matrix operations to decode the following cryptogram. (Use the matrix in Example 6.)

12 −25 15  28 −32 −89  10 −10 −49  12

−12 −51 17 −31  10  10 −28  55  4 −8 8

# Warm Up

The following warm-up exercises involve skills that were covered in earlier sections. You will use these skills in the exercise set for this section.

In Exercises 1–6, evaluate the determinant.

**1.** $\begin{vmatrix} 4 & 3 \\ -3 & -2 \end{vmatrix}$

**2.** $\begin{vmatrix} 10 & -20 \\ -1 & 2 \end{vmatrix}$

**3.** $\begin{vmatrix} 4 & 0 \\ -3 & -2 \end{vmatrix}$

**4.** $\begin{vmatrix} x & x^2 \\ 1 & 2x \end{vmatrix}$

**5.** $\begin{vmatrix} 4 & 0 & -2 \\ 3 & 1 & 2 \\ -8 & 0 & 6 \end{vmatrix}$

**6.** $\begin{vmatrix} 3 & 2 & 5 \\ 0 & 0 & -4 \\ -6 & 1 & 1 \end{vmatrix}$

In Exercises 7 and 8, find the inverse of the matrix.

**7.** $A = \begin{bmatrix} 1 & 3 \\ 2 & 7 \end{bmatrix}$

**8.** $A = \begin{bmatrix} 10 & 5 & -2 \\ -4 & -2 & 1 \\ 1 & 1 & 0 \end{bmatrix}$

In Exercises 9 and 10, perform the indicated matrix multiplication.

**9.** $\begin{bmatrix} 0.1 & 0.2 & 0.2 \\ 0.4 & 0.3 & 0.5 \\ 0.5 & 0.5 & 0.3 \end{bmatrix} \begin{bmatrix} 0.4 \\ 0.5 \\ 0.1 \end{bmatrix}$

**10.** $\begin{bmatrix} 2 & 5 & 8 \end{bmatrix} \begin{bmatrix} 1 & 2 & -1 \\ 1 & 2 & 2 \\ 2 & 5 & 0 \end{bmatrix}$

## 7.5    Exercises

In Exercises 1–10, use a determinant to find the area of the triangle with the given vertices.

**1.**

**2.**

**3.**

**4.**

**5.**

**6.**
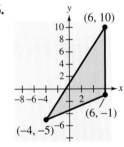

**7.** $(-2, 4), (2, 3), (-1, 5)$

**8.** $(0, -2), (-1, 4), (3, 5)$

**9.** $(-3, 5), (2, 6), (3, -5)$

**10.** $(-2, 4), (1, 5), (3, -2)$

In Exercises 11 and 12, find a value of $x$ such that the triangle has an area of 4.

**11.** $(-5, 1), (0, 2), (-2, x)$

**12.** $(-4, 2), (-3, 5), (-1, x)$

**13.** *Area of a Region*   A large region of forest has been infected with gypsy moths. The region is roughly triangular, as shown in the figure. From the northernmost vertex $A$ of the region, the distances to the other vertices are 25 miles south and 10 miles east (for vertex $B$), and 20 miles south and 28 miles east (for vertex $C$). Use a graphing utility to approximate the number of square miles in this region.

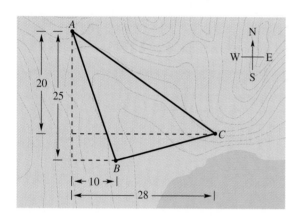

**14.** *Area of a Region*   You own a triangular tract of land, as shown in the figure. To estimate the number of square feet in the tract, you start at one vertex, walk 65 feet east and 50 feet north to the second vertex, and then walk 85 feet west and 30 feet north to the third vertex. Use a graphing utility to determine how many square feet there are in the tract of land.

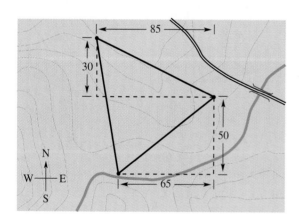

In Exercises 15–20, use a determinant to determine whether the points are collinear.

**15.** $(-4, -7), (0, -4), (4, -1)$

**16.** $(2, 4), (4, 5), (-2, 2)$

**17.** $(-1, -7), (0, -3), (1, 2)$

**18.** $(-2, -11), (4, 13), (2, 5)$

**19.** $(1, 7), (0, 4), (-1, 2)$

**20.** $(4, 3), (3, 1), (2, -1)$

In Exercises 21–26, use a determinant to find an equation of the line through the points.

**21.** $(0, 0), (5, 3)$

**22.** $(0, 0), (-2, 2)$

**23.** $(-4, 3), (2, 1)$

**24.** $(10, 7), (-2, -7)$

**25.** $\left(-\frac{1}{2}, 3\right), \left(\frac{5}{2}, 1\right)$

**26.** $\left(\frac{2}{3}, 4\right), (6, 12)$

In Exercises 27 and 28, find $x$ such that the points are collinear.

**27.** $(2, -5), (4, x), (5, -2)$

**28.** $(-6, 2), (-5, x), (-3, 5)$

In Exercises 29 and 30, find the uncoded $1 \times 2$ row matrices for the message. Then encode the message using the matrix.

| Message | Matrix |
|---|---|
| **29.** COME HOME SOON | $\begin{bmatrix} 1 & 2 \\ 3 & 5 \end{bmatrix}$ |
| **30.** HELP IS ON THE WAY | $\begin{bmatrix} -2 & 3 \\ -1 & 1 \end{bmatrix}$ |

In Exercises 31 and 32, find the uncoded $1 \times 3$ row matrices for the message. Then encode the message using the matrix.

| Message | Matrix |
|---|---|
| **31.** CALL ME TOMORROW | $\begin{bmatrix} 1 & -1 & 0 \\ 1 & 0 & -1 \\ -6 & 2 & 3 \end{bmatrix}$ |
| **32.** PLEASE SEND MONEY | $\begin{bmatrix} 4 & 2 & 1 \\ -3 & -3 & -1 \\ 3 & 2 & 1 \end{bmatrix}$ |

In Exercises 33–36, write a cryptogram for the message using the matrix

$$A = \begin{bmatrix} 1 & 2 & 2 \\ 3 & 7 & 9 \\ -1 & -4 & -7 \end{bmatrix}.$$

**33.** LANDING SUCCESSFUL

**34.** BEAM ME UP SCOTTY

**35.** HAPPY BIRTHDAY

**36.** OPERATION OVERLORD

In Exercises 37–40, use $A^{-1}$ to decode the cryptogram.

**37.** $A = \begin{bmatrix} 1 & 2 \\ 3 & 5 \end{bmatrix}$

11, 21, 64, 112, 25, 50, 29, 53, 23, 46, 40, 75, 55, 92

**38.** $A = \begin{bmatrix} 2 & 3 \\ 3 & 4 \end{bmatrix}$

85, 120, 6, 8, 10, 15, 84, 117, 42, 56, 90, 125, 60, 80, 30, 45, 19, 26

**39.** $A = \begin{bmatrix} 4 & 2 & 1 \\ -3 & -3 & -1 \\ 3 & 2 & 1 \end{bmatrix}$

110, 59, 32, 79, 40, 20, 95, 50, 25, 99, 53, 29, −22, −32, −9

**40.** $A = \begin{bmatrix} 1 & -1 & 0 \\ 1 & 0 & -1 \\ -6 & 2 & 3 \end{bmatrix}$

9, −1, −9, 38, −19, −19, 28, −9, −19, −80, 25, 41, −64, 21, 31, −7, −4, 7

In Exercises 41 and 42, decode the cryptogram by using the inverse of the matrix

$$A = \begin{bmatrix} 1 & 2 & 2 \\ 3 & 7 & 9 \\ -1 & -4 & -7 \end{bmatrix}.$$

**41.** 20, 17, −15, −9, −44, −83, 64, 136, 157, 24, 31, 12, 4, −37, −102

**42.** 13, −9, −59, 61, 112, 106, −17, −73, −131, 11, 24, 29, 65, 144, 172

**43.** The following cryptogram was encoded with a 2 × 2 matrix.

8, 21, −15, −10, −13, −13, 5, 10, 5, 25, 5, 19, −1, 6, 20, 40, −18, −18, 1, 16

The last word of the message is __RON. What is the message?

**44.** The following cryptogram was encoded with a 2 × 2 matrix.

5, 2, 25, 11, −2, −7, −15, −15, 32, 14, −8, −13, 38, 19, −19, −19, 37, 16

The last word of the message is __SUE. What is the message?

## Chapter Project

## Market Share

Your advertising group has been hired by Company C, a cable/internet service provider, to develop an advertising campaign that will increase its market share from 10% to at least 30% in one year. Currently, your client competes with two other cable/internet service providers, Company A and Company B. Your group has designed a campaign that you believe will meet the 30% objective. With this new campaign, you predict that each month

(1) 92% of A's subscribers will remain with A, 3% will switch to B, and 5% will switch to C;

(2) 91% of B's subscribers will remain with B, 7% will switch to A, and 1% will change to C; and

(3) 93% of C's subscribers will remain with C, 6% will change to B, and 1% will change to A.

$$x_0 = \begin{bmatrix} 0.2 \\ 0.7 \\ 0.1 \end{bmatrix} \begin{matrix} \text{Company A} \\ \text{Company B} \\ \text{Company C} \end{matrix}$$

The current percents of the market are given in the matrix at the left. After one month, the shares of the three companies will be given by

$$x_1 = px_0 = \begin{bmatrix} 0.92 & 0.03 & 0.05 \\ 0.07 & 0.91 & 0.02 \\ 0.01 & 0.06 & 0.93 \end{bmatrix} \begin{bmatrix} 0.2 \\ 0.7 \\ 0.1 \end{bmatrix} = \begin{bmatrix} 0.210 \\ 0.653 \\ 0.137 \end{bmatrix} \begin{matrix} \text{Company A} \\ \text{Company B} \\ \text{Company C} \end{matrix}.$$

Use this information to investigate the following questions.

**1.** *Making a Table* Construct a table that shows the percent market share for each of the three companies during each of the first 12 months. The matrix operations on your graphing utility can help you with the calculations.

**2.** *Fitting a Model to the Data* Enter the results for Company C from Question 1 into the statistical package of your graphing utility.

(a) Sketch a scatter plot of the data.

(b) What type of model do you think best fits the data? Select a curve-fitting program on your graphing utility to model the data.

(c) Graph the model you found in part (b) for the data for Company C.

(d) Repeat (a), (b), and (c) using the data for Company A and Company B from Question 1.

**3.** *Comparing Models* Graph the models for Companies A, B, and C in the same viewing window. Will Company C's market share reach 30% within 12 months? Based on the model for Company C's market share, will Company C maintain a 30% or better market share?

**4.** *Making Predictions* Are the models you found good predictors over long periods of time? Would you expect this type of market to fluctuate or be static? As Company C begins to take more of the market share, what action do you predict Companies A and B will take? What possible effect on this system could their actions have?

## CHAPTER SUMMARY

After studying this chapter, you should have acquired the following skills.
These skills are keyed to the Review Exercises that begin on page 486.
Answers to odd-numbered Review Exercises are given in the back of the book.

**7.1** · Determine the order of a matrix. — Review Exercises 1, 2

· Perform elementary row operations on a matrix. — Review Exercises 3–6

· Write a matrix in row-echelon form or reduced row-echelon form. — Review Exercises 3–6

· Solve a system of linear equations using Gaussian elimination or Gauss-Jordan elimination. — Review Exercises 7–18

**7.2** · Add or subtract two matrices. — Review Exercises 19–22

· Multiply a matrix by a scalar. — Review Exercises 19–22, 37, 38

· Find the product of two matrices. — Review Exercises 23–34, 39, 40

· Solve a matrix equation. — Review Exercises 35, 36

**7.3** · Verify that a matrix is the inverse of a given matrix. — Review Exercises 41, 42

· Find the inverse of a matrix. — Review Exercises 43–46

· Use an inverse matrix to solve a system of linear equations. — Review Exercises 47–54

**7.4** · Evaluate the determinant of a 2 × 2 matrix. — Review Exercises 55–58

· Find the determinant of a square matrix. — Review Exercises 59, 60, 65, 66

· Find the determinant of a triangular matrix. — Review Exercises 61–64

**7.5** · Find the area of a triangle using a determinant. — Review Exercises 67, 68

· Determine whether three points are collinear using a determinant. — Review Exercises 69–72

· Use a determinant to find the equation of a line. — Review Exercises 73, 74

· Encode and decode a cryptogram using a matrix. — Review Exercises 75–78

In Exercises 1 and 2, determine the order of the matrix.

**1.** $\begin{bmatrix} 3 & 7 & 4 & -2 & -3 \\ -1 & 8 & 6 & 1 & 2 \end{bmatrix}$

**2.** $\begin{bmatrix} 5 \\ -1 \\ 2 \\ 4 \end{bmatrix}$

In Exercises 3 and 4, write the matrix in row-echelon form.

**3.** $\begin{bmatrix} 1 & 3 & 0 & 2 \\ 3 & 10 & 1 & 8 \\ 2 & 3 & 3 & 10 \end{bmatrix}$

**4.** $\begin{bmatrix} 1 & 2 & -1 & 0 \\ -2 & -3 & 3 & 4 \\ 4 & 0 & 1 & 3 \end{bmatrix}$

In Exercises 5 and 6, write the matrix in reduced row-echelon form.

**5.** $\begin{bmatrix} 1 & 2 & 3 \\ -2 & 0 & 2 \\ 2 & 1 & 2 \end{bmatrix}$

**6.** $\begin{bmatrix} 2 & 3 & 1 & -5 \\ 1 & 0 & 5 & 2 \\ -1 & 4 & 3 & 6 \\ 0 & -2 & 6 & -8 \end{bmatrix}$

In Exercises 7-14, use matrices to solve the system.

**7.** $4x - 3y = 18$
$\quad x + y = 1$

**8.** $2x + 4y = 16$
$\quad -x + 3y = 17$

**9.** $2x + 3y - z = 13$
$\quad 3x \quad\quad + z = 8$
$\quad x - 2y + 3z = -4$

**10.** $3x + 4y + 2z = 5$
$\quad 2x + 3y \quad\quad = 7$
$\quad\quad 2y - 3z = 12$

**11.** $x + 2y + 2z = 10$
$\quad 2x + 3y + 5z = 20$

**12.** $3x + 10y + 4z = 20$
$\quad x + 3y - 2z = 8$

**13.** $2x + y - 3z = 4$
$\quad x + 2y + 2z = 10$
$\quad x \quad\quad - 2z = 12$
$\quad x + y + z = 6$

**14.** $2x + 4y + 2z = 10$
$\quad x \quad\quad + 3z = 9$
$\quad 3x - 2y \quad\quad = 4$
$\quad x + y + z = 8$

**15.** *Reasoning* Discuss how you know that the systems of equations in Exercises 11 and 12 will have infinitely many solutions.

**16.** *Reasoning* Discuss the three possible outcomes for the solution of a system of linear equations.

**17.** *Borrowing Money* A company borrowed $600,000 to expand its product line. Some of the money was borrowed at 9%, some at 10%, and some at 12%. How much was borrowed at each rate if the annual interest was $63,000 and the amount borrowed at 10% was three times the amount borrowed at 9%?

**18.** *Borrowing Money* A company borrowed $700,000 to expand its product line. Some of the money was borrowed at 9.5%, some at 10.5%, and some at 11%. How much was borrowed at each rate if the annual interest was $70,000 and the amount borrowed at 9.5% was four times the amount borrowed at 11%?

In Exercises 19-22, find (a) $A + B$, (b) $A - B$, (c) $4A$, and (d) $4A - 3B$.

**19.** $A = \begin{bmatrix} -1 & 5 \\ 2 & 1 \end{bmatrix}$, $B = \begin{bmatrix} 4 & 2 \\ -6 & 3 \end{bmatrix}$

**20.** $A = \begin{bmatrix} 1 & 0 & 2 \\ -1 & 3 & 5 \\ 2 & -2 & 3 \end{bmatrix}$, $B = \begin{bmatrix} 2 & 0 & 1 \\ 3 & -4 & 6 \\ 1 & 2 & -3 \end{bmatrix}$

**21.** $A = \begin{bmatrix} 1 & 3 & -2 & 6 \\ 0 & 1 & 3 & 2 \end{bmatrix}$,
$B = \begin{bmatrix} 2 & 1 & 4 & -5 \\ 3 & -6 & 3 & -2 \end{bmatrix}$

**22.** $A = \begin{bmatrix} 3 \\ -2 \\ 3 \end{bmatrix}$, $B = \begin{bmatrix} -1 \\ 4 \\ 5 \end{bmatrix}$

In Exercises 23–26, find (a) $AB$, (b) $BA$, and, if possible, (c) $A^2$. (*Note:* $A^2 = AA$.)

**23.** $A = \begin{bmatrix} 1 & -3 & 4 \end{bmatrix}$, $B = \begin{bmatrix} 2 \\ -2 \\ -1 \end{bmatrix}$

**24.** $A = \begin{bmatrix} 2 & -3 \\ 4 & 5 \end{bmatrix}$, $B = \begin{bmatrix} -1 & 0 \\ 2 & -1 \end{bmatrix}$

**25.** $A = \begin{bmatrix} 1 & 0 & 2 \\ 3 & 1 & -2 \\ 1 & 1 & 1 \end{bmatrix}$, $B = \begin{bmatrix} 2 & 0 & 0 \\ 1 & -2 & 1 \\ 5 & 4 & -2 \end{bmatrix}$

**26.** $A = \begin{bmatrix} 0 & 2 & 3 \\ 1 & -1 & 0 \\ 1 & 2 & 1 \end{bmatrix}$, $B = \begin{bmatrix} 1 & 1 & 4 \\ 2 & 3 & 0 \\ 4 & 3 & 2 \end{bmatrix}$

In Exercises 27–32, find $AB$, if possible.

**27.** $A = \begin{bmatrix} 1 & 4 \\ -2 & -1 \\ 3 & 2 \end{bmatrix}$, $B = \begin{bmatrix} -4 \\ 3 \end{bmatrix}$

**28.** $A = \begin{bmatrix} 3 \\ 2 \\ 4 \\ 6 \end{bmatrix}$, $B = \begin{bmatrix} 2 & 0 & -1 \end{bmatrix}$

**29.** $A = \begin{bmatrix} 4 & 0 & 0 \\ 0 & 3 & 0 \\ 0 & 0 & -2 \end{bmatrix}$, $B = \begin{bmatrix} \frac{1}{4} & 0 & 0 \\ 0 & \frac{1}{3} & 0 \\ 0 & 0 & -\frac{1}{2} \end{bmatrix}$

**30.** $A = \begin{bmatrix} 3 & 1 \\ 4 & 7 \\ 1 & 1 \end{bmatrix}$, $B = \begin{bmatrix} 1 & 2 & -2 \\ 3 & 4 & 0 \\ 0 & 1 & 0 \end{bmatrix}$

**31.** $A = \begin{bmatrix} 1 & 2 & 3 & 6 & -1 \\ 2 & 8 & 0 & 0 & 2 \end{bmatrix}$,

$B = \begin{bmatrix} 3 & 2 \\ 4 & -1 \end{bmatrix}$

**32.** $A = \begin{bmatrix} 0 & 0 & 2 \\ 1 & 0 & 6 \\ 0 & 2 & 2 \end{bmatrix}$, $B = \begin{bmatrix} 3 & 4 & 0 & 1 \\ 2 & 1 & 0 & 0 \\ 0 & 0 & 1 & 1 \end{bmatrix}$

**33.** *Think About It*   Give an example of two matrices $A$ and $B$ such that the products of $AB$ and $BA$ exist.

**34.** *Think About It*   Give an example of two matrices $A$ and $B$ such that the product of $AB$ exists but $BA$ does not exist.

In Exercises 35 and 36, solve for $X$ given

$$A = \begin{bmatrix} 1 & -2 \\ 0 & 1 \\ 2 & 3 \end{bmatrix} \quad \text{and} \quad B = \begin{bmatrix} 0 & 1 \\ 1 & 1 \\ 3 & 5 \end{bmatrix}.$$

**35.** $X = 5A - 3B$       **36.** $3X = 2A + 3B$

**37.** *Factory Production*   A corporation has four factories, each of which manufactures three products. The number of units of product $i$ produced at factory $j$ in one day is represented by $a_{ij}$ in the matrix

$$A = \begin{bmatrix} 80 & 70 & 90 & 40 \\ 50 & 30 & 80 & 20 \\ 90 & 60 & 100 & 50 \end{bmatrix}.$$

Find the production levels if production is increased by 20%.

**38.** *Factory Production*   A corporation has three factories, each of which manufactures four products. The number of units of product $j$ produced at factory $i$ in one day is represented by $a_{ij}$ in the matrix

$$A = \begin{bmatrix} 30 & 80 & 70 & 20 \\ 50 & 100 & 90 & 90 \\ 60 & 70 & 80 & 100 \end{bmatrix}.$$

Find the production levels if production is decreased by 10%. (*Hint:* Because a 10% decrease corresponds to $100\% - 10\%$, multiply the given matrix by 0.90.)

**39.** *Inventory Levels*   A company sells four different models of car sound systems through three retail outlets. The inventories of the four models at the three outlets are given by matrix $S$.

$$S = \begin{array}{c} \\ \\ \end{array} \overset{\text{Model}}{\overset{\text{A} \quad \text{B} \quad \text{C} \quad \text{D}}{\begin{bmatrix} 3 & 2 & 1 & 4 \\ 1 & 3 & 4 & 3 \\ 5 & 3 & 2 & 2 \end{bmatrix}}} \begin{array}{l} 1 \\ 2 \\ 3 \end{array} \Big\} \text{Outlet}$$

The wholesale and retail prices of the four models are given by matrix $T$.

$$T = \overset{\text{Price}}{\overset{\text{Wholesale} \quad \text{Retail}}{\begin{bmatrix} 300 & 500 \\ 400 & 650 \\ 200 & 350 \\ 800 & 1200 \end{bmatrix}}} \begin{array}{l} \text{A} \\ \text{B} \\ \text{C} \\ \text{D} \end{array} \Big\} \text{Outlet}$$

Use a graphing utility to compute $ST$ and interpret the result.

**40.** *Labor/Wage Requirements* A company that manufactures racing bicycles has the following labor-hour and wage requirements.

*Labor-Hour Requirements (per bicycle)*

Department

$$S = \begin{bmatrix} 0.9 \text{ hour} & 0.8 \text{ hour} & 0.2 \text{ hour} \\ 1.5 \text{ hours} & 1.0 \text{ hour} & 0.4 \text{ hour} \\ 3.5 \text{ hours} & 3.0 \text{ hours} & 0.5 \text{ hour} \end{bmatrix} \begin{matrix} \text{Basic} \\ \text{Light} \\ \text{Ultra-} \\ \text{Light} \end{matrix}$$ Models

Cutting    Assembly    Packaging

*Wage Requirements (per hour)*

Plant

A   B

$$T = \begin{bmatrix} \$14 & \$12 \\ \$8 & \$9 \\ \$7 & \$6 \end{bmatrix} \begin{matrix} \text{Cutting} \\ \text{Assembly} \\ \text{Packaging} \end{matrix}$$ Department

(a) What is the labor cost for a light racing bicycle at Plant A?

(b) What is the labor cost for an ultralight racing bicycle at Plant B?

(c) Use a graphing utility to compute $ST$ and interpret the result.

In Exercises 41 and 42, show that $B$ is the inverse of $A$.

**41.** $A = \begin{bmatrix} 1 & 2 & 1 \\ 3 & 6 & 4 \\ 0 & 1 & 3 \end{bmatrix}$, $B = \begin{bmatrix} -14 & 5 & -2 \\ 9 & -3 & 1 \\ -3 & 1 & 0 \end{bmatrix}$

**42.** $A = \begin{bmatrix} 2 & 0 & 1 & 2 \\ 3 & 0 & 0 & 1 \\ -1 & 1 & 2 & 0 \\ 0 & -1 & 2 & 2 \end{bmatrix}$,

$B = \dfrac{1}{9}\begin{bmatrix} -4 & 6 & 1 & 1 \\ 10 & -6 & 2 & -7 \\ -7 & 6 & 4 & 4 \\ 12 & -9 & -3 & -3 \end{bmatrix}$

In Exercises 43–46, find the inverse of the matrix (if it exists).

**43.** $\begin{bmatrix} 1 & 3 \\ 2 & 5 \end{bmatrix}$      **44.** $\begin{bmatrix} -2 & 1 \\ 4 & 3 \end{bmatrix}$

**45.** $\begin{bmatrix} -1 & 0 & 0 & 0 \\ 0 & 2 & 0 & 0 \\ 0 & 0 & 4 & 0 \\ 0 & 0 & 0 & 6 \end{bmatrix}$    **46.** $\begin{bmatrix} 3 & 2 & 2 \\ 0 & 2 & 1 \\ 1 & 0 & 1 \end{bmatrix}$

In Exercises 47 and 48, use an inverse matrix to solve the system of linear equations. (Use the inverse matrix found in Exercise 43.)

**47.** $x + 3y = 15$      **48.** $x + 3y = 7$
$\phantom{x}2x + 5y = 26$        $\phantom{x}2x + 5y = 11$

In Exercises 49 and 50, use an inverse matrix to solve the system of linear equations. (Use the inverse matrix found in Exercise 46.)

**49.** $3x + 2y + 2z = 13$    **50.** $3x + 2y + 2z = 12$
$\phantom{3x + }2y + \phantom{2}z = 4$         $\phantom{3x + }2y + \phantom{2}z = 13$
$\phantom{3}x \phantom{ + 2y} + \phantom{2}z = 5$         $\phantom{3}x \phantom{ + 2y} + \phantom{2}z = 3$

*Computer Models* In Exercises 51 and 52, consider a company that produces three computer models. Model A requires 2 units of plastic, 2 units of computer chips, and 1 unit of computer "cards." Model B requires 1 unit of plastic, 3 units of computer chips, and 2 units of computer "cards." Model C requires 3 units of plastic, 2 units of computer chips, and 2 units of computer "cards." A system of linear equations (where $x$, $y$, and $z$ represent models A, B, and C) is as follows.

$2x + \phantom{3}y + 3z =$ (units of plastic)

$2x + 3y + 2z =$ (units of computer chips)

$\phantom{2}x + 2y + 2z =$ (units of computer "cards")

Use the inverse of the coefficient matrix of this system to find the numbers of models A, B, and C that the company can produce with the given amounts of components.

**51.** 700 units of plastic
     900 units of computer chips
     600 units of computer "cards"

**52.** 600 units of plastic
     800 units of computer chips
     500 units of computer "cards"

*Investment Portfolio* In Exercises 53 and 54, consider a person who invests in blue-chip stocks, common stocks, and municipal bonds. The average yields are 8% on blue-chip stocks, 6% on common stocks, and 7% on municipal bonds. Twice as much is invested in municipal bonds as in common stocks. A system of linear equations (where x, y, and z represent the amounts invested in blue-chip stocks, common stocks, and municipal bonds) is as follows.

$$x + \quad y + \quad z = \text{(total investment)}$$

$$0.08x + 0.06y + 0.07z = \text{(annual return)}$$

$$2y - \quad z = 0$$

Use the inverse of the coefficient matrix of this system to find the amount invested in each type of stock or bond for the indicated total investment.

**53.** Total investment = \$50,000; annual return = \$3720

**54.** Total investment = \$45,000; annual return = \$3380

In Exercises 55–66, find the determinant of the matrix.

**55.** $\begin{bmatrix} 8 & 4 \\ 3 & 2 \end{bmatrix}$

**56.** $\begin{bmatrix} 7 & 2 \\ 9 & -3 \end{bmatrix}$

**57.** $\begin{bmatrix} 5 & 2 \\ 0 & 0 \end{bmatrix}$

**58.** $\begin{bmatrix} 3 & 0 \\ 0 & -7 \end{bmatrix}$

**59.** $\begin{bmatrix} 1 & 2 & 3 \\ 8 & 6 & 7 \\ 0 & 2 & -1 \end{bmatrix}$

**60.** $\begin{bmatrix} -2 & 3 & 3 \\ -1 & 0 & 5 \\ 1 & 2 & -1 \end{bmatrix}$

**61.** $\begin{bmatrix} 1 & 3 & 2 & 4 \\ 0 & -1 & 2 & 2 \\ 0 & 0 & 3 & 0 \\ 0 & 0 & 0 & 4 \end{bmatrix}$

**62.** $\begin{bmatrix} 2 & 0 & 0 & 0 \\ 3 & 4 & 0 & 0 \\ 5 & 1 & -2 & 0 \\ 6 & 3 & 1 & 1 \end{bmatrix}$

**63.** $\begin{bmatrix} 3 & 0 & 0 & 0 \\ 0 & 2 & 0 & 0 \\ 0 & 0 & -1 & 0 \\ 0 & 0 & 0 & -10 \end{bmatrix}$

**64.** $\begin{bmatrix} 2 & 0 & 0 & 0 \\ 0 & 5 & 0 & 0 \\ 0 & 0 & 4 & 0 \\ 0 & 0 & 0 & 3 \end{bmatrix}$

**65.** $\begin{bmatrix} 1 & 3 & 0 & 7 \\ 5 & 2 & 4 & 2 \\ 0 & 1 & 0 & 1 \\ 2 & 1 & 0 & 0 \end{bmatrix}$

**66.** $\begin{bmatrix} 1 & 0 & 0 & 0 & 0 \\ 0 & 2 & 0 & 0 & 0 \\ 0 & 0 & 0 & 8 & 0 \\ 0 & 0 & 0 & 0 & \frac{1}{2} \end{bmatrix}$

In Exercises 67 and 68, use a determinant to find the area of the triangle with the given vertices.

**67.** $(3, 4), (2, -3), (-1, -4)$

**68.** $(1, -4), (-2, 3), (0, 6)$

In Exercises 69–72, use a determinant to ascertain whether the points are collinear.

**69.** $(0, 3), (1, 5), (2, 8)$

**70.** $(2, 6), (-2, 3), (0, 5)$

**71.** $(-4, 1), (6, 6), (0, 3)$

**72.** $(-3, -1), (0, 5), (-4, -3)$

In Exercises 73 and 74, use a determinant to find an equation of the line through the points $(x_1, y_1)$ and $(x_2, y_2)$.

**73.** $(-7, 3), (8, 2)$      **74.** $(5, -4), (-3, 2)$

In Exercises 75 and 76, use the matrix to encode the message.

| *Message* | *Matrix* |
|---|---|
| **75.** TRANSMIT NOW | $\begin{bmatrix} 2 & 3 \\ 3 & 4 \end{bmatrix}$ |
| **76.** CALL AT MIDNIGHT | $\begin{bmatrix} 1 & 2 & 2 \\ 3 & 7 & 9 \\ -1 & -4 & -7 \end{bmatrix}$ |

In Exercises 77 and 78, use $A^{-1}$ to decode the cryptogram.

**77.** $A = \begin{bmatrix} 1 & 2 \\ -1 & 3 \end{bmatrix}$

14, 53, −17, 96, 5, 10, 12, 64, 5, 10, 3, 11, 25, 50

**78.** $A = \begin{bmatrix} 1 & -1 & 0 \\ 1 & 0 & -1 \\ -6 & 2 & 3 \end{bmatrix}$

−14, −1, 10, −38, 2, 27, −94, 18, 57, 7, −11, −1, −96, 20, 57, −74, 23, 35, 17, −12, −5

## Chapter Test

Take this test as you would take a test in class. After you are done, check your work against the answers given in the back of the book.

In Exercises 1 and 2, write an augmented matrix that represents the system.

**1.** $3x + y = 1$
$\phantom{3}x - y = 7$

**2.** $3x + 4y + 2z = \phantom{-}4$
$\phantom{3}2x + 3y \phantom{+ 2z} = -2$
$\phantom{3x + }2y - 3z = -13$

In Exercises 3-5, use matrices to solve the system.

**3.** $x + 2y + 3z = 16$
$5x + 4y - \phantom{2}z = 22$

**4.** $x - 2y + \phantom{3}z = 14$
$\phantom{x - 2}y - 3z = \phantom{1}2$
$\phantom{x - 2y + }z = -6$

**5.** $2x - 3y + z = 14$
$\phantom{2}x + 2y \phantom{+ z} = -4$
$\phantom{2x + }y - z = -4$

In Exercises 6-9, use the matrices to find the indicated matrix.

$$A = \begin{bmatrix} 1 & 3 \\ 2 & 4 \end{bmatrix}, \quad B = \begin{bmatrix} 2 & -1 & 3 \\ 4 & 0 & 1 \end{bmatrix}, \quad C = \begin{bmatrix} 0 & -2 \\ 3 & 5 \end{bmatrix}, \quad D = \begin{bmatrix} 3 \\ 2 \\ -1 \end{bmatrix}$$

**6.** $2A + C$    **7.** $AB$    **8.** $BD$    **9.** $A^2$

In Exercises 10-12, find the inverse of the matrix.

**10.** $A = \begin{bmatrix} 2 & -1 \\ -3 & 4 \end{bmatrix}$    **11.** $A = \begin{bmatrix} 1 & 0 \\ 0 & 1 \end{bmatrix}$    **12.** $A = \begin{bmatrix} 3 & 4 & 2 \\ 2 & 3 & 0 \\ 0 & 2 & -3 \end{bmatrix}$

In Exercises 13-15, find the determinant of the matrix.

**13.** $\begin{bmatrix} 3 & -1 \\ 4 & 7 \end{bmatrix}$    **14.** $\begin{bmatrix} 3 & 2 & -1 \\ 1 & 0 & 2 \\ 4 & 5 & 2 \end{bmatrix}$    **15.** $\begin{bmatrix} 2 & 0 & 0 \\ 0 & 5 & 0 \\ 0 & 0 & -2 \end{bmatrix}$

**16.** Use the inverse found in Exercise 12 to solve the system in Exercise 2.

**17.** Find two nonzero matrices whose product is a zero matrix.

**18.** Use a determinant to decide whether $(2, 9)$, $(-2, 1)$, and $(3, 11)$ are collinear.

**19.** Find the area of the triangle whose vertices are $(-2, 4)$, $(0, 5)$, and $(3, -1)$.

**20.** A manufacturer produces three models of a product, which are shipped to two warehouses. The number of units $i$ that are shipped to warehouse $j$ is represented by $a_{ij}$ in matrix $A$ at the left. The price per unit is represented by matrix $B$. Find the product $BA$ and interpret the result.

$$A = \begin{bmatrix} 1000 & 3000 \\ 2000 & 4000 \\ 5000 & 8000 \end{bmatrix}$$

$$B = [\$50 \quad \$35 \quad \$28]$$

Matrices for 20

# Sequences, Series, and Probability

Jamie Squire/Allsport

In an economy, one person's expenditures become another person's income. If individuals in an economy spend 75% of their income, then for an initial one dollar the first person spends $0.75, the second person spends $0.75 × 75% = $0.56, the third person spends $0.56 × 75% = $0.42, and so on.

The table and graph show the cycle of income in a community that has hosted a large event such as the Super Bowl. It was predicted that Super Bowl XXXIII (held in Miami, Florida, in January of 1999) would have a total economic impact of $380 million on southern Florida. (Source: *Miami Herald*, January 19, 1999)

| Cycle | 1 | 2 | 3 | 4 | 5 | 6 |
|---|---|---|---|---|---|---|
| Total spending (in millions of dollars) | 92.000 | 73.460 | 57.598 | 45.122 | 37.020 | 29.491 |

The chapter project related to this information is on page 564.

### Objectives

- Find the terms of a sequence.
- Evaluate a factorial expression.
- Find the sum of a finite sequence.
- Use summation notation to write the sum of a sequence.
- Use a sequence to solve an application problem.

## Sequences

In mathematics, the word *sequence* is used in much the same way as in ordinary English. Saying that a collection is listed *in sequence* means that it is ordered so that it has a first member, a second member, a third member, and so on.

Mathematically, you can think of a sequence as a *function* whose domain is the set of positive integers. Rather than using function notation, however, sequences are usually written using subscript notation, as indicated in the definition below.

---

▶ **Definition of Sequence**

An **infinite sequence** is a function whose domain is the set of positive integers. The function values

$$a_1, a_2, a_3, a_4, \ldots, a_n, \ldots$$

are the **terms** of the sequence. If the domain of the function consists of the first $n$ positive integers only, the sequence is called a **finite sequence.**

---

On occasion it is convenient to begin subscripting a sequence with 0 instead of 1 so that the terms of the sequence become $a_0, a_1, a_2, a_3, \ldots$.

**Example 1**     Finding Terms of a Sequence

**a.** The first four terms of the sequence given by $a_n = 3n - 2$ are

$$a_1 = 3(1) - 2 = 1 \qquad \text{1st term}$$

$$a_2 = 3(2) - 2 = 4 \qquad \text{2nd term}$$

$$a_3 = 3(3) - 2 = 7 \qquad \text{3rd term}$$

$$a_4 = 3(4) - 2 = 10. \qquad \text{4th term}$$

**b.** The first four terms of the sequence given by $a_n = 3 + (-1)^n$ are

$$a_1 = 3 + (-1)^1 = 3 - 1 = 2 \qquad \text{1st term}$$

$$a_2 = 3 + (-1)^2 = 3 + 1 = 4 \qquad \text{2nd term}$$

$$a_3 = 3 + (-1)^3 = 3 - 1 = 2 \qquad \text{3rd term}$$

$$a_4 = 3 + (-1)^4 = 3 + 1 = 4. \qquad \text{4th term}$$

The terms of a sequence need not all be positive, as shown in Example 2.

### Example 2    Finding Terms of a Sequence

The first four terms of the sequence given by $a_n = (-1)^n/(2n - 1)$ are

$$a_1 = \frac{(-1)^1}{2(1) - 1} = \frac{-1}{2 - 1} = -1 \qquad \text{1st term}$$

$$a_2 = \frac{(-1)^2}{2(2) - 1} = \frac{1}{4 - 1} = \frac{1}{3} \qquad \text{2nd term}$$

$$a_3 = \frac{(-1)^3}{2(3) - 1} = \frac{-1}{6 - 1} = -\frac{1}{5} \qquad \text{3rd term}$$

$$a_4 = \frac{(-1)^4}{2(4) - 1} = \frac{1}{8 - 1} = \frac{1}{7}. \qquad \text{4th term}$$

Try finding the first four terms of the sequence whose $n$th term is

$$a_n = \frac{(-1)^{n+1}}{2n - 1}.$$

How do they differ from the first four terms of the sequence in Example 2?

It is important to realize that simply listing the first few terms is not sufficient to define a sequence—the $n$th term *must be given.* To see this, consider the following sequences, both of which have the same first three terms.

$$\frac{1}{2}, \frac{1}{4}, \frac{1}{8}, \frac{1}{16}, \cdots \cdots \frac{1}{2^n}, \cdots$$

$$\frac{1}{2}, \frac{1}{4}, \frac{1}{8}, \frac{1}{15}, \cdots \cdots \frac{6}{(n + 1)(n^2 - n + 6)}, \cdots$$

### •TECHNOLOGY•

Use the table feature of your graphing utility to find the terms of the sequence. Be sure to start the table with $n = 1$ and set the table to increment $n$ by 1.

**a.** $a_n = 2^n + 1$

**b.** $a_n = (-1)^n(2^n + 1)$

**c.** $a_n = (-1)^{n+1}(2^n + 1)$

Describe the differences in the terms of the sequences.

## Factorial Notation

Some very important sequences in mathematics involve terms that are defined with special types of products called **factorials.**

### •TECHNOLOGY•

Most scientific and graphing calculators have a factorial key, ☐! , built in. Use the ☐! key to verify the factorials at the right. Try finding such values as 8!, 15!, and 20!. How does the calculator's display change for increasingly large values of $n$?

> ▶ Definition of Factorial
>
> If $n$ is a positive integer, $n$ **factorial** is defined by
>
> $$n! = 1 \cdot 2 \cdot 3 \cdot 4 \cdots (n-1) \cdot n.$$
>
> As a special case, zero factorial is defined as $0! = 1$.

Here are some values of $n!$ for the first several nonnegative integers. Notice that 0! is 1 by definition.

$$0! = 1$$

$$1! = 1$$

$$2! = 1 \cdot 2 = 2$$

$$3! = 1 \cdot 2 \cdot 3 = 6$$

$$4! = 1 \cdot 2 \cdot 3 \cdot 4 = 24$$

$$5! = 1 \cdot 2 \cdot 3 \cdot 4 \cdot 5 = 120$$

The value of $n$ does not have to be very large before the value of $n!$ becomes huge. For instance, $10! = 3,628,800$.

### Example 3 Finding Terms of a Sequence Involving Factorials

List the first five terms of the sequence whose $n$th term is $a_n = 2/n!$. Begin with $n = 0$.

**Solution**

$$a_0 = \frac{2}{0!} = \frac{2}{1} = 2 \qquad \text{0th term}$$

$$a_1 = \frac{2}{1!} = \frac{2}{1} = 2 \qquad \text{1st term}$$

$$a_2 = \frac{2}{2!} = \frac{2}{2} = 1 \qquad \text{2nd term}$$

$$a_3 = \frac{2}{3!} = \frac{2}{6} = \frac{1}{3} \qquad \text{3rd term}$$

$$a_4 = \frac{2}{4!} = \frac{2}{24} = \frac{1}{12} \qquad \text{4th term}$$

Factorials follow the same conventions for order of operations as do exponents. For instance,

$$2n! = 2(n!) = 2(1 \cdot 2 \cdot 3 \cdot 4 \cdots n)$$

whereas $(2n)! = 1 \cdot 2 \cdot 3 \cdot 4 \cdots 2n.$

### Example 4    Finding Terms of a Sequence Involving Factorials

List the first five terms of the sequence whose $n$th term is $a_n = 2^n/n!$. Begin with $n = 0$.

**Solution**

$$a_0 = \frac{2^0}{0!} = \frac{1}{1} = 1 \qquad \text{0th term}$$

$$a_1 = \frac{2^1}{1!} = \frac{2}{1} = 2 \qquad \text{1st term}$$

$$a_2 = \frac{2^2}{2!} = \frac{4}{2} = 2 \qquad \text{2nd term}$$

$$a_3 = \frac{2^3}{3!} = \frac{8}{6} = \frac{4}{3} \qquad \text{3rd term}$$

$$a_4 = \frac{2^4}{4!} = \frac{16}{24} = \frac{2}{3} \qquad \text{4th term}$$

When working with fractions involving factorials, you will often find that the fractions can be reduced.

### Example 5    Evaluating Factorial Expressions

Evaluate the following factorial expressions.

**a.** $\dfrac{n!}{(n-1)!}$    **b.** $\dfrac{8!}{2! \cdot 6!}$    **c.** $\dfrac{2! \cdot 6!}{3! \cdot 5!}$

**Solution**

**a.** $\dfrac{n!}{(n-1)!} = \dfrac{1 \cdot 2 \cdot 3 \cdots (n-1) \cdot n}{1 \cdot 2 \cdot 3 \cdots (n-1)} = n$

**b.** $\dfrac{8!}{2! \cdot 6!} = \dfrac{1 \cdot 2 \cdot 3 \cdot 4 \cdot 5 \cdot 6 \cdot 7 \cdot 8}{1 \cdot 2 \cdot 1 \cdot 2 \cdot 3 \cdot 4 \cdot 5 \cdot 6} = \dfrac{7 \cdot 8}{2} = 28$

**c.** $\dfrac{2! \cdot 6!}{3! \cdot 5!} = \dfrac{1 \cdot 2 \cdot 1 \cdot 2 \cdot 3 \cdot 4 \cdot 5 \cdot 6}{1 \cdot 2 \cdot 3 \cdot 1 \cdot 2 \cdot 3 \cdot 4 \cdot 5} = \dfrac{6}{3} = 2$

## Series and Summation Notation

The sum of the terms of an infinite sequence is called an **infinite series.** The sum of the first $n$ terms of a sequence is called a **finite series** or the **$n$th partial sum.** There is a convenient notation for the sum of the terms of a finite sequence. It is called **summation notation** or **sigma notation** because it involves the use of the uppercase Greek letter sigma, written as $\Sigma$.

---

▶ **Definition of Summation Notation**

The sum of the first $n$ terms of a sequence is represented by

$$\sum_{i=1}^{n} a_i = a_1 + a_2 + a_3 + a_4 + \cdots + a_n$$

where $i$ is called the **index of summation,** $n$ is the **upper limit of summation,** and 1 is the **lower limit of summation.**

---

| Example 6 | Summation Notation for Sums |
|---|---|

**a.** $\displaystyle\sum_{i=1}^{5} 3i = 3(1) + 3(2) + 3(3) + 3(4) + 3(5)$

$$= 3(1 + 2 + 3 + 4 + 5)$$

$$= 3(15) = 45$$

**b.** $\displaystyle\sum_{k=3}^{6} (1 + k^2) = (1 + 3^2) + (1 + 4^2) + (1 + 5^2) + (1 + 6^2)$

$$= 10 + 17 + 26 + 37 = 90$$

**c.** $\displaystyle\sum_{i=0}^{8} \frac{1}{i!} = \frac{1}{0!} + \frac{1}{1!} + \frac{1}{2!} + \frac{1}{3!} + \frac{1}{4!} + \frac{1}{5!} + \frac{1}{6!} + \frac{1}{7!} + \frac{1}{8!}$

$$= 1 + 1 + \frac{1}{2} + \frac{1}{6} + \frac{1}{24} + \frac{1}{120} + \frac{1}{720} + \frac{1}{5040} + \frac{1}{40,320}$$

$$\approx 2.71828$$

Note that the sum in part (c) is very close to the irrational number $e \approx 2.718281828$. It can be shown that as more terms of the sequence whose $n$th term is $1/n!$ are added, the sum becomes closer and closer to $e$.

---

In Example 6, note that the lower index of a summation does not have to be 1. Also note that the index does not have to be the letter $i$. For instance, in part (b), the letter $k$ is the index.

| Example 7 | Writing a Sum in Summation Notation |

Write the partial sum in summation notation.

**a.** $\dfrac{3}{1+2} + \dfrac{3}{2+2} + \dfrac{3}{3+2} + \dfrac{3}{4+2} + \dfrac{3}{5+2}$

**b.** $2 - 4 + 8 - 16$

**Solution**

**a.** Begin by looking for similarities and differences in each of the terms. Each term has 3 in the numerator. In the denominator, 2 is added to a number that is increasing by 1 for each term. Let $i$ be the index of summation, let 1 be the lower limit of summation, and let 5 be the upper limit of summation. Then, using summation notation, you can write

$$\frac{3}{1+2} + \frac{3}{2+2} + \frac{3}{3+2} + \frac{3}{4+2} + \frac{3}{5+2} = \sum_{i=1}^{5} \frac{3}{i+2}.$$

**b.** This series has terms with alternating signs, and each term can be written as a power of 2. Let $n$ be the index of summation, let 1 be the lower limit of summation, and let 4 be the upper limit of summation. Then, using summation notation, you can write

$$2 - 4 + 8 - 16 = (-1)^{1+1}(2) + (-1)^{2+1}(4) + (-1)^{3+1}(8) + (-1)^{4+1}(16)$$

$$= (-1)^{1+1}(2)^1 + (-1)^{2+1}(2)^2 + (-1)^{3+1}(2)^3 + (-1)^{4+1}(2)^4$$

$$= \sum_{n=1}^{4} (-1)^{n+1}(2)^n.$$

---

▶ **Properties of Sums**

**1.** $\displaystyle\sum_{i=1}^{n} ca_i = c\sum_{i=1}^{n} a_i,$   $c$ is any constant

**2.** $\displaystyle\sum_{i=1}^{n} (a_i + b_i) = \sum_{i=1}^{n} a_i + \sum_{i=1}^{n} b_i$

**3.** $\displaystyle\sum_{i=1}^{n} (a_i - b_i) = \sum_{i=1}^{n} a_i - \sum_{i=1}^{n} b_i$

The sum of the terms of any *finite* sequence must be a finite number. Variations in the upper and lower limits of summation can produce quite different-looking summation notations for *the same sum*. For example, consider the following two sums.

$$\sum_{i=1}^{5} 3(2^i) = 3\sum_{i=1}^{5} 2^i = 3(2^1 + 2^2 + 2^3 + 2^4 + 2^5)$$

$$\sum_{i=0}^{4} 3(2^{i+1}) = 3\sum_{i=0}^{4} 2^{i+1} = 3(2^1 + 2^2 + 2^3 + 2^4 + 2^5)$$

## Application

Sequences have many applications in business and science. One is illustrated in Example 8.

### Example 8     Population of the United States

From 1953 to 1996, the resident population of the United States can be approximated by the model

$$a_n = \sqrt{26{,}176 + 882n + 2.89n^2}, \qquad n = 0, 1, \ldots, 43$$

where $a_n$ is the population in millions and $n$ represents the calendar year, with $n = 0$ corresponding to 1953. Find the last five terms of this finite sequence. (Source: U.S. Bureau of the Census)

### Solution

The last five terms of this finite sequence are as follows.

$$a_{39} = \sqrt{26{,}176 + 882(39) + 2.89(39)^2} \approx 254.9 \qquad \text{1992 population}$$

$$a_{40} = \sqrt{26{,}176 + 882(40) + 2.89(40)^2} \approx 257.1 \qquad \text{1993 population}$$

$$a_{41} = \sqrt{26{,}176 + 882(41) + 2.89(41)^2} \approx 259.2 \qquad \text{1994 population}$$

$$a_{42} = \sqrt{26{,}176 + 882(42) + 2.89(42)^2} \approx 261.4 \qquad \text{1995 population}$$

$$a_{43} = \sqrt{26{,}176 + 882(43) + 2.89(43)^2} \approx 263.5 \qquad \text{1996 population}$$

The bar graph in Figure 8.1 graphically represents the population given by this sequence for the entire 44-year period from 1953 to 1996.

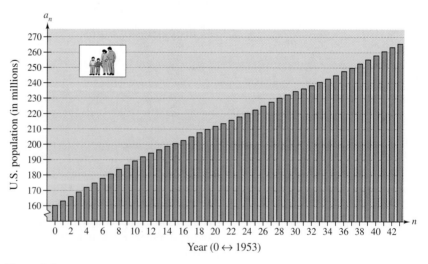

Figure 8.1

| Discussing the Concept | A Summation Program |
|---|---|

A graphing calculator can be programmed to calculate the sum of the first $n$ terms of a sequence. Consult your user's guide for specific instructions on programming your calculator. Use the algorithm below to write the summation program, or access our website at *www.hmco.com* for programs written for selected graphing calculators.

### Summation Program

To find the sum of a finite sequence, write the program for your graphing calculator from the algorithm given below and enter it into your calculator. Then enter the $n$th term of the sequence in $Y_1$ and run the program. For instance, to find the sum

$$\sum_{i=1}^{10} 2^i$$

enter 2^$X$ into $Y_1$. Then run the program, entering 1 as the lower limit and 10 as the upper limit. The last sum displayed is the final sum. If you wish to view each partial sum, add a pause command after displaying the value of $S$. For the example above, 10 numbers will be displayed, with the final sum being 2046.

### Summation Algorithm

1. Enter $M$ as the lower limit of summation.
2. Enter $N$ as the upper limit of summation.
3. Set $S$ equal to 0 so that $S$ can be used to store the sums.
4. Use a FOR loop to sum from the lower limit $M$ to the upper limit $N$ incrementing by $X$.
5. Add the sum of $Y_1$ and $S$ and store the result in $S$.
6. Display the value of $S$.
7. End the FOR loop when $Y_1$ has been calculated and summed for the upper limit.

Try using this program to find the following sums.

**a.** $\displaystyle\sum_{i=1}^{15} (i - 2)^2$    **b.** $\displaystyle\sum_{i=1}^{20} 3^{(i-1)}$    **c.** $\displaystyle\sum_{i=1}^{10} \sqrt{e^i}$

# Warm Up

The following warm-up exercises involve skills that were covered in earlier sections. You will use these skills in the exercise set for this section.

**1.** Find $f(2)$ for $f(n) = \dfrac{2n}{n^2 + 1}$.

**2.** Find $f(3)$ for $f(n) = \dfrac{4}{3(n + 1)}$.

In Exercises 3–6, factor the expression.

**3.** $4n^2 - 1$

**4.** $4n^2 - 8n + 3$

**5.** $n^2 - 3n + 2$

**6.** $n^2 + 3n + 2$

In Exercises 7–10, perform the indicated operations and/or simplify.

**7.** $\left(\dfrac{2}{3}\right)\left(\dfrac{3}{4}\right)\left(\dfrac{4}{5}\right)\left(\dfrac{5}{6}\right)$

**8.** $\dfrac{2 \cdot 4 \cdot 6 \cdot 8}{2^4}$

**9.** $\dfrac{1}{2 \cdot 2} + \dfrac{1}{2 \cdot 3} + \dfrac{1}{2 \cdot 4}$

**10.** $\dfrac{1}{1 \cdot 2} + \dfrac{1}{2 \cdot 3} + \dfrac{1}{3 \cdot 4}$

## 8.1 Exercises

In Exercises 1–22, write the first five terms of the sequence. (Assume that $n$ begins with 1.)

**1.** $a_n = 3n - 2$

**2.** $a_n = 4 - 2n$

**3.** $a_n = 2^n$

**4.** $a_n = (-2)^n$

**5.** $a_n = \left(\frac{1}{2}\right)^n$

**6.** $a_n = \left(-\frac{1}{2}\right)^n$

**7.** $a_n = \dfrac{2}{n + 1}$

**8.** $a_n = \dfrac{n + 1}{3}$

**9.** $a_n = \dfrac{(-1)^n}{n}$

**10.** $a_n = \dfrac{(-1)^{n+1}}{n}$

**11.** $a_n = 3 - \dfrac{1}{2^n}$

**12.** $a_n = \dfrac{3^n}{4^n}$

**13.** $a_n = \dfrac{3^n}{n!}$

**14.** $a_n = \dfrac{n!}{n}$

**15.** $a_n = \dfrac{n^2 - 1}{n + 1}$

**16.** $a_n = \dfrac{n^2 - 1}{n^2 + 2}$

**17.** $a_n = \dfrac{(-1)^n}{n^2}$

**18.** $a_n = \dfrac{(-1)^n n}{n + 1}$

**19.** $a_n = \dfrac{3n!}{(n - 1)!}$

**20.** $a_n = \dfrac{(n + 1)!}{n!}$

**21.** $a_n = \dfrac{5n^2 - n + 1}{n^2}$

**22.** $a_n = \dfrac{3n^2 - n + 4}{2n^2 + 1}$

In Exercises 23–26, evaluate the expression.

**23.** $\dfrac{8!}{6!}$

**24.** $\dfrac{10!}{9!}$

**25.** $\dfrac{28!}{25!}$

**26.** $\dfrac{49!}{50!}$

In Exercises 27–30, simplify the expression. Then evaluate the expression when $n = 0$ and $n = 1$.

**27.** $\dfrac{(n + 1)!}{n!}$

**28.** $\dfrac{(n + 2)!}{n!}$

**29.** $\dfrac{(2n - 1)!}{(2n + 1)!}$

**30.** $\dfrac{(2n + 2)!}{(2n)!}$

In Exercises 31–44, write an expression for the most apparent $n$th term of the sequence. (Assume that $n$ begins with 1.)

**31.** $1, 3, 5, 7, 9, \ldots$

**32.** $5, 8, 11, 14, 17, \ldots$

**33.** $2, \frac{1}{2}, \frac{2}{9}, \frac{1}{8}, \frac{2}{25}, \ldots$

**34.** $1, \frac{1}{4}, \frac{1}{9}, \frac{1}{16}, \frac{1}{25}, \ldots$

**35.** $\frac{2}{3}, \frac{3}{4}, \frac{4}{5}, \frac{5}{6}, \frac{6}{7}, \ldots$

**36.** $\frac{2}{1}, \frac{3}{3}, \frac{4}{5}, \frac{5}{7}, \frac{6}{9}, \ldots$

**37.** $\frac{1}{2}, -\frac{1}{4}, \frac{1}{8}, -\frac{1}{16}, \frac{1}{32}, \ldots$

**38.** $\frac{1}{3}, \frac{2}{9}, \frac{4}{27}, \frac{8}{81}, \frac{16}{243}, \ldots$

**39.** $1 + \frac{1}{1}, 1 + \frac{1}{2}, 1 + \frac{1}{3}, 1 + \frac{1}{4}, 1 + \frac{1}{5}, \ldots$

**40.** $1 + \frac{1}{2}, 1 + \frac{3}{4}, 1 + \frac{7}{8}, 1 + \frac{15}{16}, 1 + \frac{31}{32}, \ldots$

**41.** $1, \frac{1}{2}, \frac{1}{6}, \frac{1}{24}, \frac{1}{120}, \ldots$

**42.** $2, -4, 6, -8, 10, \ldots$

**43.** $1, -1, 1, -1, 1, \ldots$

**44.** $1, 2, \dfrac{2^2}{2}, \dfrac{2^3}{6}, \dfrac{2^4}{24}, \dfrac{2^5}{120}, \ldots$

In Exercises 45–58, find the sum.

**45.** $\displaystyle\sum_{i=1}^{4} (6i + 3)$

**46.** $\displaystyle\sum_{i=1}^{6} (4i - 2)$

**47.** $\displaystyle\sum_{i=1}^{7} -3i$

**48.** $\displaystyle\sum_{i=1}^{5} -3(i - 2)$

**49.** $\displaystyle\sum_{k=1}^{4} 10$

**50.** $\displaystyle\sum_{k=1}^{5} 4$

**51.** $\displaystyle\sum_{i=0}^{4} i^2$

**52.** $\displaystyle\sum_{i=0}^{5} 3i^2$

**53.** $\displaystyle\sum_{k=2}^{6} \frac{1}{2k}$

**54.** $\displaystyle\sum_{j=3}^{5} \frac{1}{j}$

**55.** $\displaystyle\sum_{i=1}^{4} (i - 1)^2$

**56.** $\displaystyle\sum_{k=2}^{5} (k + 1)(k - 3)$

**57.** $\displaystyle\sum_{i=1}^{4} (-1)^i(2i + 4)$

**58.** $\displaystyle\sum_{i=1}^{4} (-2)^i$

In Exercises 59–68, use summation notation to write the sum.

**59.** $\dfrac{1}{3(1)} + \dfrac{1}{3(2)} + \dfrac{1}{3(3)} + \cdots + \dfrac{1}{3(9)}$

**60.** $\dfrac{5}{1 + 1} + \dfrac{5}{1 + 2} + \dfrac{5}{1 + 3} + \cdots + \dfrac{5}{1 + 15}$

**61.** $\left[2\left(\frac{1}{8}\right) + 3\right] + \left[2\left(\frac{2}{8}\right) + 3\right] + \cdots + \left[2\left(\frac{8}{8}\right) + 3\right]$

**62.** $\left[1 - \left(\frac{1}{6}\right)^2\right] + \left[1 - \left(\frac{2}{6}\right)^2\right] + \cdots + \left[1 - \left(\frac{6}{6}\right)^2\right]$

**63.** $3 - 9 + 27 - 81 + 243 - 729$

**64.** $1 - \frac{1}{2} + \frac{1}{4} - \frac{1}{8} + \cdots - \frac{1}{128}$

**65.** $\dfrac{1}{1^2} - \dfrac{1}{2^2} + \dfrac{1}{3^2} - \dfrac{1}{4^2} + \cdots - \dfrac{1}{20^2}$

**66.** $\dfrac{1}{1 \cdot 3} + \dfrac{1}{2 \cdot 4} + \dfrac{1}{3 \cdot 5} + \cdots + \dfrac{1}{10 \cdot 12}$

**67.** $\frac{1}{4} + \frac{3}{8} + \frac{7}{16} + \frac{15}{32} + \frac{31}{64}$

**68.** $\frac{1}{2} + \frac{2}{4} + \frac{6}{8} + \frac{24}{16} + \frac{120}{32} + \frac{720}{64}$

**69.** *Compound Interest*   A deposit of $5000 is made in an account that earns 8% interest compounded quarterly. The balance in the account after $n$ quarters is given by

$$A_n = 5000\left(1 + \frac{0.08}{4}\right)^n, \quad n = 1, 2, 3, \ldots.$$

(a) Compute the first eight terms of this sequence.

(b) Find the balance in this account after 10 years by computing the 40th term of the sequence.

(c) Is the balance after 20 years twice the balance after 10 years? Explain.

**70.** *Compound Interest*   A deposit of $100 is made *each* month in an account that earns 12% interest compounded monthly. The balance in the account after $n$ months is given by

$$A_n = 100(101)[(1.01)^n - 1], \quad n = 1, 2, 3, \ldots.$$

(a) Compute the first six terms of this sequence.

(b) Find the balance after 5 years by computing the 60th term of the sequence.

(c) Find the balance after 20 years by computing the 240th term of the sequence.

**71.** *Ratio of Men to Women* In 1940, the population of the United States had about the same number of men as women. After that, there were more women than men. The ratio of men to women is approximately given by the model

$$a_n = 1.04 - 0.034n + 0.0032n^2,$$

$$n = 1, 2, \ldots, 7$$

where $a_n$ is the ratio of men to women and $n$ is the year, with $n = 1, 2, 3, \ldots, 7$ corresponding to 1940, 1950, 1960, . . . , 2000. (Source: U.S. Bureau of the Census)

(a) Use a graphing utility to find the terms of this finite sequence.

(b) Construct a bar graph that represents the sequence.

(c) In 1990, the population of the United States was 250 million. How many of these were women? How many were men?

**72.** *Total Sales* The total number of new trucks sold annually from 1991 to 1996 can be approximated by the model

$$a_n = 4292.3 + 819.0n - 59.23n^2,$$

$$n = 0, 1, 2, 3, 4, 5$$

where $a_n$ is the number of trucks sold (in thousands) and $n = 0$ represents 1991. Find the total number sold from 1991 to 1996 by evaluating the sum

$$\sum_{n=0}^{5} (4292.3 + 819.0n - 59.23n^2).$$

(Source: American Automobile Manufacturers Association)

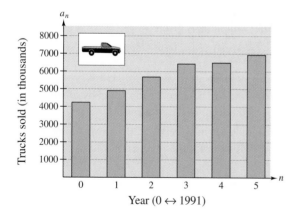

**73.** *Hospital Costs* The average cost of a day in a hospital from 1988 to 1995 is given by the model

$$a_n = 578.1 + 61.99n - 0.702n^2,$$

$$n = 0, 1, \ldots, 7$$

where $a_n$ is the average cost in dollars and $n = 0$ represents 1988. Use a graphing utility to find the terms of this finite sequence and construct a bar graph that represents the sequence. Predict the average cost of a day in a hospital in 2000. (Source: American Hospital Association)

**74.** *Total Revenue* The total annual sales for MCI Communications from 1987 to 1996 can be approximated by the model

$$a_n = 4222.3 + 812.25n + 78.83n^2,$$

$$n = 0, 1, \ldots, 9$$

where $a_n$ is the annual sales (in millions of dollars) and $n = 0$ represents 1987. Find the total revenue from 1987 to 1996 by evaluating the sum

$$\sum_{n=0}^{9} (4222.3 + 812.25n + 78.83n^2).$$

(Source: MCI Telecommunications Corporation)

**75.** *Dividends* A company declares dividends per share of common stock that can be approximated by the model $a_n = 0.35n + 1.25$, $n = 0, 1, 2, 3, 4$, where $a_n$ is the dividend in dollars and $n = 0$ represents 1995. Approximate the sum of the dividends per share of common stock for the years 1995 through 1999 by evaluating

$$\sum_{n=0}^{4} (0.35n + 1.25).$$

## 8.2    Arithmetic Sequences and Series

### Objectives

- Determine whether a sequence is arithmetic.
- Find the $n$th term of an arithmetic sequence.
- Find a formula for an arithmetic sequence.
- Find the sum of an arithmetic sequence.
- Use an arithmetic sequence to solve an application problem.

## Arithmetic Sequences

A sequence whose consecutive terms have a common difference is called an **arithmetic sequence.**

·DISCOVERY·

Determine the arithmetic rule used to generate the terms of each sequence.

**a.** $1, 4, 7, 10, 13, \ldots$

**b.** $6, 12, 18, 24, \ldots$

**c.** $100, 90, 80, 70, \ldots$

▶ **Definition of Arithmetic Sequence**

A sequence is **arithmetic** if the differences between consecutive terms are the same. Thus, the sequence

$$a_1, a_2, a_3, a_4, \ldots, a_n, \ldots$$

is arithmetic if there is a number $d$ such that

$$a_2 - a_1 = d, \quad a_3 - a_2 = d, \quad a_4 - a_3 = d,$$

and so on. The number $d$ is the **common difference** of the arithmetic sequence.

**Example 1**    Examples of Arithmetic Sequences

**a.** The sequence whose $n$th term is $4n + 3$ is arithmetic. For this sequence, the common difference between consecutive terms is 4.

$$7, 11, 15, 19, \ldots, 4n + 3, \ldots$$

$$11 - 7 = 4$$

**b.** The sequence whose $n$th term is $7 - 5n$ is arithmetic. For this sequence, the common difference between consecutive terms is $-5$.

$$2, -3, -8, -13, \ldots, 7 - 5n, \ldots$$

$$-3 - 2 = -5$$

**c.** The sequence whose $n$th term is $\frac{1}{4}(n + 3)$ is arithmetic. For this sequence, the common difference between consecutive terms is $\frac{1}{4}$.

$$1, \frac{5}{4}, \frac{3}{2}, \frac{7}{4}, \ldots, \frac{n + 3}{4}, \ldots$$

$$\frac{5}{4} - 1 = \frac{1}{4}$$

In Example 1, notice that each of the arithmetic sequences has an $n$th term that is of the form $dn + c$, where the common difference of the sequence is $d$. This result is summarized as follows.

---

▶ **The $n$th Term of an Arithmetic Sequence**

The $n$th term of an arithmetic sequence has the form

$$a_n = dn + c$$

where $d$ is the common difference between consecutive terms of the sequence and $c = a_1 - d$.

---

**Example 2**    Finding the $n$th Term of an Arithmetic Sequence

Find a formula for the $n$th term of the arithmetic sequence whose common difference is 3 and whose first term is 2.

**Solution**

Because the sequence is arithmetic, you know that the formula for the $n$th term is of the form $a_n = dn + c$. Moreover, because the common difference is $d = 3$, the formula must have the form

$$a_n = 3n + c.$$

Because $a_1 = 2$, it follows that

$$c = a_1 - d = 2 - 3 = -1.$$

So, the formula for the $n$th term is

$$a_n = 3n - 1.$$

The sequence therefore has the following form.

$$2, 5, 8, 11, 14, \ldots, 3n - 1, \ldots$$

---

Another way to find a formula for the $n$th term of the sequence in Example 2 is to begin by writing the terms of the sequence.

| $a_1$ | $a_2$ | $a_3$ | $a_4$ | $a_5$ | $a_6$ | $a_7$ | |
|-------|-------|-------|-------|-------|-------|-------|-----|
| 2 | $2 + 3$ | $5 + 3$ | $8 + 3$ | $11 + 3$ | $14 + 3$ | $17 + 3$ | $\ldots$ |
| 2 | 5 | 8 | 11 | 14 | 17 | 20 | $\ldots$ |

From these terms, you can reason that the $n$th term is of the form

$$a_n = dn + c = 3n - 1.$$

| Example 3 | Finding the $n$th Term of an Arithmetic Sequence |

The fourth term of an arithmetic sequence is 20, and the 13th term is 65. Write the first several terms of this sequence.

### Solution

The fourth and 13th terms of the sequence are related by

$$a_{13} = a_4 + 9d.$$

Using $a_4 = 20$ and $a_{13} = 65$, you can conclude that $d = 5$, which implies that the sequence is as follows.

$a_1$   $a_2$   $a_3$   $a_4$   $a_5$   $a_6$   $a_7$   $a_8$   $a_9$   $a_{10}$   $a_{11}$

5,   10,   15,   20,   25,   30,   35,   40,   45,   50,   55,   . . .

## Study Tip

If you substitute $a_1 - d$ for $c$ in the formula $a_n = dn + c$, then the $n$th term of an arithmetic sequence has the alternative recursion formula

$$a_n = a_1 + (n - 1)d.$$

Use this formula to solve Example 4 and Example 9.

If you know the $n$th term of an arithmetic sequence *and* you know the common difference of the sequence, you can find the $(n + 1)$th term by using the **recursion formula**

$$a_{n+1} = a_n + d.$$

With this formula, you can find any term of an arithmetic sequence, *provided* that you know the previous term. For instance, if you know the first term, you can find the second term. Then, knowing the second term, you can find the third term, and so on.

| Example 4 | Using a Recursion Formula |

Find the ninth term of the arithmetic sequence whose first two terms are 2 and 9.

### Solution

For this sequence, the common difference is $d = 9 - 2 = 7$. There are two ways to find the ninth term. One way is simply to write out the first nine terms (by repeatedly adding 7).

2, 9, 16, 23, 30, 37, 44, 51, 58

Another way to find the ninth term is first to find a formula for the $n$th term. Because the first term is 2, it follows that

$$c = a_1 - d = 2 - 7 = -5.$$

So, a formula for the $n$th term is $a_n = 7n - 5$, which implies that the ninth term is $a_9 = 7(9) - 5 = 58$.

## The Sum of an Arithmetic Series

The *sum* of the terms of an arithmetic sequence is called an **arithmetic series.** There is a simple formula for the sum of a finite arithmetic series. Be sure you see that this formula works only for *arithmetic* series.

> ▶ **The Sum of a Finite Arithmetic Series**
>
> The sum of a finite arithmetic series with $n$ terms is given by
>
> $$S = \frac{n}{2}(a_1 + a_n).$$

·**DISCOVERY**·

To *develop* the formula for the sum of a finite arithmetic series, consider writing the sum in two different ways. One way is to write the sum as

$$S = a_1 + (a_1 + d)$$
$$+ (a_1 + 2d) + \cdots$$
$$+ [a_1 + (n - 1)d].$$

In the second way, you repeatedly subtract $d$ from the $n$th term to obtain

$$S = a_n + (a_n - d)$$
$$+ (a_n - 2d) + \cdots$$
$$+ [a_n - (n - 1)d].$$

Can you discover a way to combine these two versions of $S$ to obtain the following formula?

$$S = \frac{n}{2}(a_1 + a_n)$$

**Example 5**  Finding the Sum of an Arithmetic Series

Find the sum: $1 + 3 + 5 + 7 + 9 + 11 + 13 + 15 + 17 + 19$.

**Solution**

To begin, notice that the series is arithmetic (with a common difference of 2). Moreover, the series has 10 terms. So, the sum of the series is

$$S = 1 + 3 + 5 + 7 + 9 + 11 + 13 + 15 + 17 + 19$$

$$= \frac{n}{2}(a_1 + a_n)$$

$$= \frac{10}{2}(1 + 19)$$

$$= 5(20)$$

$$= 100.$$

**Example 6**  Finding the Sum of an Arithmetic Series

Find the sum of the integers from 1 to 100.

**Solution**

The integers from 1 to 100 form an arithmetic series that has 100 terms. So, you can use the formula for the sum of an arithmetic series, as follows.

$$S = 1 + 2 + 3 + 4 + 5 + 6 + \cdots + 99 + 100$$

$$= \frac{n}{2}(a_1 + a_n)$$

$$= \frac{100}{2}(1 + 100)$$

$$= 50(101)$$

$$= 5050$$

Example 7    Finding the Sum of an Arithmetic Series

Find the sum of the first 150 terms of the arithmetic sequence

$$5, 16, 27, 38, 49, \ldots .$$

### Solution

For this arithmetic series, you have $a_1 = 5$ and $d = 16 - 5 = 11$. Thus, $c = a_1 - d = 5 - 11 = -6$, and the $n$th term is

$$a_n = 11n - 6.$$

So, $a_{150} = 11(150) - 6 = 1644$, and the sum of the first 150 terms is as follows.

$$S = \frac{n}{2}(a_1 + a_n)$$

$$= \frac{150}{2}(5 + 1644)$$

$$= 75(1649)$$

$$= 123,675$$

## Applications

Example 8    Seating Capacity

An auditorium has 20 rows of seats. There are 20 seats in the first row, 22 seats in the second row, 24 seats in the third row, and so on (see Figure 8.2). How many seats are there in all 20 rows?

### Solution

The number of seats in the rows forms an arithmetic sequence in which the common difference is $d = 2$. Because $c = a_1 - d = 20 - 2 = 18$, you can determine that the formula for the $n$th term in the sequence is $a_n = 2n + 18$. So, the 20th term in the sequence is $a_{20} = 2(20) + 18 = 58$, and the total number of seats is

$$S = 20 + 22 + 24 + \cdots + 58$$

$$= \frac{n}{2}(a_1 + a_{20})$$

$$= \frac{20}{2}(20 + 58)$$

$$= 10(78)$$

$$= 780.$$

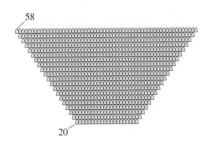

58

20

**Figure 8.2**

| **Example 9** | Total Sales |  |

A small business sells $10,000 worth of products during its first year. The owner of the business has set a goal of increasing annual sales by $7500 each year for 9 years. Assuming that this goal is met, find the total sales during the first 10 years this business is in operation.

### Solution

The annual sales form an arithmetic sequence in which $a_1 = 10,000$ and $d = 7500$. So, $c = a_1 - d = 10,000 - 7500 = 2500$, and the $n$th term of the sequence is

$$a_n = 7500n + 2500.$$

This implies that the 10th term of the sequence is $a_{10} = 77,500$. So, the total sales for the first 10 years are as follows.

$$S = \frac{n}{2}(a_1 + a_{10})$$

$$= \frac{10}{2}(10,000 + 77,500)$$

$$= 5(87,500)$$

$$= \$437,500$$

---

| **Discussing the Concept** | **Numerical Relationships** |

Decide whether it is possible to fill in the blanks in each of the following such that the resulting sequence is arithmetic. If so, find a recursion formula for the sequence.

a. $-7,$ ___, ___, ___, ___, ___, $11$

b. $17,$ ___, ___, ___, ___, ___, ___, ___, ___, $71$

c. $2, 6,$ ___, ___, $162$

d. $4, 7.5,$ ___, ___, ___, ___, ___, ___, ___, ___, $39$

e. $8, 12,$ ___, ___, ___, $60.75$

# Warm Up

The following warm-up exercises involve skills that were covered in earlier sections. You will use these skills in the exercise set for this section.

In Exercises 1 and 2, find the sum.

**1.** $\sum_{i=1}^{6} (2i - 1)$

**2.** $\sum_{i=1}^{10} (4i + 2)$

In Exercises 3 and 4, find the distance between the two real numbers.

**3.** $\frac{5}{2}, 8$

**4.** $\frac{4}{3}, \frac{14}{3}$

In Exercises 5 and 6, evaluate the function as indicated.

**5.** Find $f(3)$ for $f(n) = 10 + (n - 1)4$.

**6.** Find $f(10)$ for $f(n) = 1 + (n - 1)\frac{1}{3}$.

In Exercises 7–10, evaluate the expression.

**7.** $\frac{11}{2}(1 + 25)$

**8.** $\frac{16}{2}(4 + 16)$

**9.** $\frac{20}{2}[2(5) + (12 - 1)3]$

**10.** $\frac{8}{2}[2(-3) + (15 - 1)5]$

## 8.2    Exercises

In Exercises 1–10, determine whether the sequence is arithmetic. If it is, find the common difference.

**1.** $3, 7, 11, 15, 19, \ldots$

**2.** $20, 17, 14, 11, 8, \ldots$

**3.** $-3, -2, 0, 3, 7, \ldots$

**4.** $-12, -8, -4, 0, 4, 8, \ldots$

**5.** $3, \frac{5}{2}, 2, \frac{3}{2}, 1, \frac{1}{2}, \ldots$

**6.** $\frac{9}{4}, 2, \frac{7}{4}, \frac{3}{2}, \frac{5}{4}, 1, \ldots$

**7.** $\frac{1}{3}, \frac{2}{3}, \frac{4}{3}, \frac{8}{3}, \frac{16}{3}, \frac{32}{3}, \ldots$

**8.** $\ln 1, \ln 2, \ln 3, \ln 4, \ln 5, \ldots$

**9.** $5.3, 5.7, 6.1, 6.5, 6.9, \ldots$

**10.** $1^2, 2^2, 3^2, 4^2, 5^2, \ldots$

In Exercises 11–18, write the first five terms of the specified sequence. Determine whether the sequence is arithmetic. If it is, find the common difference.

**11.** $a_n = 5 + 3n$

**12.** $a_n = 100 - 3n$

**13.** $a_n = (2 + n) - (1 + n)$

**14.** $a_n = 1 + (n - 1)4$

**15.** $a_n = (2^n)n$

**16.** $a_n = 2^{n-1}$

**17.** $a_n = \dfrac{1}{n + 1}$

**18.** $a_n = (-1)^n$

In Exercises 19–28, find a formula for $a_n$ for the arithmetic sequence.

**19.** $a_1 = 4, d = 5$

**20.** $a_1 = -2, d = 3$

**21.** $a_1 = 50, d = -12$

**22.** $a_1 = 120, d = -15$

**23.** $a_1 = 7, a_5 = 27$

**24.** $a_1 = -5, a_7 = -17$

**25.** $a_3 = 8, a_9 = 32$

**26.** $a_6 = 29, a_{18} = 95$

**27.** 2, 8, 14, 20, 26, . . .

**28.** 3, 13, 23, 33, 43, . . .

In Exercises 29–36, write the first five terms of the arithmetic sequence.

**29.** $a_1 = 10, d = 4$

**30.** $a_1 = 2, d = \frac{1}{3}$

**31.** $a_1 = 5, a_{n+1} = a_n + 9$

**32.** $a_1 = 7, a_{n+1} = a_n - 2$

**33.** $a_1 = 2, a_{12} = 46$

**34.** $a_5 = 28, a_{10} = 53$

**35.** $a_8 = 26, a_{12} = 42$

**36.** $a_4 = 16, a_{10} = 46$

In Exercises 37–44, find the sum of the first $n$ terms of the arithmetic sequence.

**37.** 4, 8, 12, 16, . . . , $n = 6$

**38.** 3, 11, 19, 27, . . . , $n = 10$

**39.** $-10, -8, -6, -4, \ldots, n = 12$

**40.** $-23, -16, -9, -2, \ldots, n = 10$

**41.** 0.5, 0.9, 1.3, 1.7, . . . , $n = 10$

**42.** 1.50, 1.45, 1.40, 1.35, . . . , $n = 20$

**43.** $a_1 = 100, a_{25} = 220, n = 25$

**44.** $a_1 = 15, a_{100} = 307, n = 100$

In Exercises 45–54, find the indicated sum.

**45.** $\displaystyle\sum_{n=1}^{50} n$

**46.** $\displaystyle\sum_{n=1}^{100} 2n$

**47.** $\displaystyle\sum_{n=1}^{100} 5n$

**48.** $\displaystyle\sum_{n=51}^{100} 7n$

**49.** $\displaystyle\sum_{n=0}^{50} (1000 - 5n)$

**50.** $\displaystyle\sum_{n=1}^{250} (1000 - n)$

**51.** $\displaystyle\sum_{n=1}^{20} (2n + 5)$

**52.** $\displaystyle\sum_{n=1}^{500} (n + 3)$

**53.** $\displaystyle\sum_{n=1}^{100} \frac{n + 4}{2}$

**54.** $\displaystyle\sum_{n=0}^{100} \frac{8 - 3n}{16}$

**55.** Find the sum of the first 100 odd integers.

**56.** Find the sum of the integers from $-10$ to 50.

**57.** *Job Offer* A person accepts a position with a company and will receive a salary of $27,500 for the first year. The person is guaranteed a raise of $1500 per year for the first 5 years.

(a) What will the salary be during the sixth year of employment?

(b) Use a graphing utility to find how much the company will have paid the person by the end of the sixth year.

**58.** *Job Offer* A person accepts a position with a company and will receive a salary of $26,800 for the first year. The person is guaranteed a raise of $1750 per year for the first 5 years.

(a) What will the salary be during the sixth year of employment?

(b) Use a graphing utility to find how much the company will have paid the person by the end of the sixth year.

**59.** *Seating Capacity* Determine the seating capacity of an auditorium with 30 rows of seats if there are 15 seats in the first row, 20 seats in the second row, 25 seats in the third row, and so on.

**60.** *Baling Hay* As a farmer bales a field of hay, each trip around the field gets shorter. Suppose that on the first round there are 267 bales and on the second round there are 253 bales. Assume that the decrease will be the same on each round and that there will be 11 more trips. How many bales of hay will the farmer get from the field?

**61.** *Total Sales* The annual sales for Campbell Soup Company from 1992 to 1997 can be approximated by the model

$$a_n = 354.1n + 4410.1, \quad n = 5, 6, 7, 8, 9, 10$$

where $a_n$ is the total annual sales (in millions of dollars) and $n = 5$ represents 1992. (Source: Campbell Soup Company)

(a) Sketch a bar graph showing the annual sales for Campbell Soup Company from 1992 to 1997.

(b) Find the total sales from 1992 to 1997.

**62.** *Total Revenue*   The annual revenue for Paychex, Inc. from 1993 to 1998 can be modeled by

$$a_n = 193 - 23.3n + 7.58n^2, \quad n = 3, 4, 5, 6, 7, 8$$

where $a_n$ is the annual revenue (in millions of dollars) and $n = 3$ represents 1993 (see figure). Find the total revenue from 1993 to 1998. (Source: Paychex, Inc.)

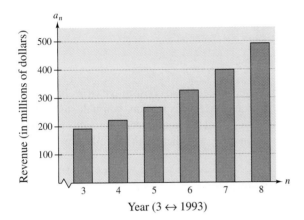

Year (3 ↔ 1993)

**63.** *Falling Object*   A heavy object (with negligible air resistance) is dropped from a plane. During the first second of fall, the object falls 16 feet; during the second second, it falls 48 feet; during the third second, it falls 80 feet; during the fourth second, it falls 112 feet. If this pattern continues, how many feet will the object fall in 8 seconds?

**64.** *Falling Object*   A heavy object (with negligible air resistance) is dropped from a plane. During the first second of fall, the object falls 4.9 meters; during the second second, it falls 14.7 meters; during the third second, it falls 24.5 meters; during the fourth second, it falls 34.3 meters. How many meters will the object fall in 10 seconds?

**65.** *Number of Logs*   Logs are stacked in a pile, as shown in the figure. The top row has 15 logs and the bottom row has 21 logs. How many logs are in the stack?

—15

—21

**66.** *Brick Pattern*   A brick patio is roughly the shape of a trapezoid (see figure). The patio has 20 rows of bricks. The first row has 14 bricks and the 20th row has 33 bricks. How many bricks are in the patio?

33—

14—

**67.** *Total Profit*   A small company makes a profit of $20,000 during its first year. The company president sets a goal of increasing profit by $5000 each year for 4 years. Assuming that this goal is met, find the total profit during the first 5 years of business.

**68.** *Total Sales*   An entrepreneur sells $16,000 worth of gumballs in the first year of operation. The entrepreneur has set a goal of increasing annual sales by $4000 each year for 9 years. Assuming that this goal is met, find the total sales during the first 10 years of this gumball business.

## 8.3    Geometric Sequences and Series

### Objectives

- Determine whether a sequence is geometric.
- Find the $n$th term of a geometric sequence.
- Find the sum of a finite or infinite geometric sequence.
- Use a geometric sequence to solve an application problem.

## Geometric Sequences

In Section 8.2, you learned that a sequence whose consecutive terms have a common *difference* is an arithmetic sequence. In this section, you will study another important type of sequence called a **geometric sequence.** Consecutive terms of a geometric sequence have a common *ratio*, as indicated in the following definition.

> ▶ **Definition of Geometric Sequence**
>
> A sequence is **geometric** if the ratios of consecutive terms are the same. Thus, a sequence is geometric if there is a number $r$, $r \neq 0$, such that
>
> $$\frac{a_2}{a_1} = r, \quad \frac{a_3}{a_2} = r, \quad \frac{a_4}{a_3} = r,$$
>
> and so on. The number $r$ is the **common ratio** of the geometric sequence.

**Example 1**    Examples of Geometric Sequences

**a.** The sequence whose $n$th term is $2^n$ is geometric. For this sequence, the common ratio between consecutive terms is 2.

$$2, 4, 8, 16, \ldots, 2^n, \ldots$$

$\frac{4}{2} = 2$

**b.** The sequence whose $n$th term is $4(3^n)$ is geometric. For this sequence, the common ratio between consecutive terms is 3.

$$12, 36, 108, 324, \ldots, 4(3^n), \ldots$$

$\frac{36}{12} = 3$

**c.** The sequence whose $n$th term is $\left(-\frac{1}{3}\right)^n$ is geometric. For this sequence, the common ratio between consecutive terms is $-\frac{1}{3}$.

$$-\frac{1}{3}, \frac{1}{9}, -\frac{1}{27}, \frac{1}{81}, \ldots, \left(-\frac{1}{3}\right)^n, \ldots$$

$\frac{1/9}{-1/3} = -\frac{1}{3}$

In Example 1, notice that each of the geometric sequences has an $n$th term that is of the form $ar^n$, where the common ratio of the sequence is $r$.

> ▶ **The $n$th Term of a Geometric Sequence**
>
> The $n$th term of a geometric sequence has the form
>
> $$a_n = a_1 r^{n-1}$$
>
> where $r$ is the common ratio of consecutive terms of the sequence. Thus, every geometric sequence can be written in the following form.
>
> $$a_1, \quad a_2, \quad a_3, \quad a_4, \quad a_5, \quad \ldots\ldots, \quad a_n, \ldots\ldots$$
> $$\downarrow \quad \downarrow \quad \downarrow \quad \downarrow \quad \downarrow \qquad\qquad \downarrow$$
> $$a_1, \ a_1 r, \ a_1 r^2, \ a_1 r^3, \ a_1 r^4, \ \ldots, \ a_1 r^{n-1}, \ \ldots$$

If you know the $n$th term of a geometric sequence, you can find the $(n + 1)$th term by multiplying by $r$. That is, $a_{n+1} = r a_n$.

**· TECHNOLOGY ·**

You can generate the geometric sequence in Example 2 with a graphing utility using the following steps.

3 [ENTER]

2 [X] [ANS]

Now press [ENTER] repeatedly to generate the terms of the sequence.

**Example 2**   Finding the Terms of a Geometric Sequence

Write the first five terms of the geometric sequence whose first term is $a_1 = 3$ and whose common ratio is $r = 2$.

**Solution**

Starting with 3, repeatedly multiply by 2 to obtain the following.

$$a_1 = 3 \qquad\qquad \text{1st term}$$
$$a_2 = 3(2^1) = 6 \qquad\qquad \text{2nd term}$$
$$a_3 = 3(2^2) = 12 \qquad\qquad \text{3rd term}$$
$$a_4 = 3(2^3) = 24 \qquad\qquad \text{4th term}$$
$$a_5 = 3(2^4) = 48 \qquad\qquad \text{5th term}$$

**Example 3**   Finding a Term of a Geometric Sequence

Find the 15th term of the geometric sequence whose first term is 20 and whose common ratio is 1.05.

**Solution**

$$a_{15} = a_1 r^{n-1} \qquad\qquad \text{Formula for geometric sequence}$$
$$= 20(1.05)^{15-1} \qquad\qquad \text{Substitute for } a_1, r, \text{ and } n.$$
$$\approx 39.599 \qquad\qquad \text{Use a calculator.}$$

| Example 4 | Finding a Term of a Geometric Sequence |

Find the 12th term of the geometric sequence

$$5, 15, 45, \ldots .$$

**Solution**

The common ratio of this sequence is $r = \frac{15}{5} = 3$. Because the first term is $a_1 = 5$, you can determine the 12th term $(n = 12)$ to be

$$
\begin{aligned}
a_{12} &= a_1 r^{n-1} &&\text{Formula for geometric sequence}\\
&= 5(3)^{12-1} &&\text{Substitute 12 for } n.\\
&= 5(177,147) &&\text{Use a calculator.}\\
&= 885,735. &&\text{Simplify.}
\end{aligned}
$$

## Study Tip

Remember that $r$ is the common ratio of consecutive terms of a sequence. So in Example 5,

$$
\begin{aligned}
a_{10} &= a_1 r^9\\
&= a_1 \cdot r \cdot r \cdot r \cdot r^6\\
&= a_1 \cdot \frac{a_2}{a_1} \cdot \frac{a_3}{a_2} \cdot \frac{a_4}{a_3} \cdot r^6\\
&= a_4 r^6.
\end{aligned}
$$

If you know any two terms of a geometric sequence, you can use that information to find a formula for the $n$th term of the sequence.

| Example 5 | Finding a Term of a Geometric Sequence |

The fourth term of a geometric sequence is 125, and the 10th term is $\frac{125}{64}$. Find the 14th term. (Assume that the terms of the sequence are positive.)

**Solution**

The 10th term is related to the fourth term by the equation

$$a_{10} = a_4 r^6.$$

Because $a_{10} = 125/64$ and $a_4 = 125$, you can solve for $r$ as follows.

$$\frac{125}{64} = 125 r^6$$

$$\frac{1}{64} = r^6$$

$$\frac{1}{2} = r$$

You can obtain the 14th term by multiplying the 10th term by $r^4$.

$$
\begin{aligned}
a_{14} &= a_{10} r^4\\
&= \frac{125}{64}\left(\frac{1}{2}\right)^4 = \frac{125}{1024}
\end{aligned}
$$

## The Sum of a Geometric Series

The sum of the terms of a geometric sequence is called a **geometric series.** The formula for the sum of a *finite* geometric series is as follows.

> ▶ **The Sum of a Finite Geometric Series**
>
> The sum of the geometric series
>
> $$a_1 + a_1 r + a_1 r^2 + a_1 r^3 + a_1 r^4 + \cdots + a_1 r^{n-1}$$
>
> with common ratio $r \neq 1$ is given by
>
> $$S = a_1 \left( \frac{1 - r^n}{1 - r} \right).$$

·DISCOVERY·

To *develop* the formula for the sum of a finite geometric series, consider the following two equations.

$$S = a_1 + a_1 r + a_1 r^2$$
$$+ \cdots + a_1 r^{n-2}$$
$$+ a_1 r^{n-1}$$
$$rS = a_1 r + a_1 r^2 + a_1 r^3$$
$$+ \cdots + a_1 r^{n-1}$$
$$+ a_1 r^n$$

Can you discover a way to simplify the difference between these two equations to obtain the formula for the sum of a finite geometric series?

**Example 6**    Finding the Sum of a Finite Geometric Series

Find the sum $\displaystyle\sum_{n=1}^{12} 4(0.3)^n$.

**Solution**

By writing out a few terms, you have

$$\sum_{n=1}^{12} 4(0.3)^n = 4(0.3) + 4(0.3)^2 + 4(0.3)^3 + \cdots + 4(0.3)^{12}.$$

Now, because $a_1 = 4(0.3)$, $r = 0.3$, and $n = 12$, you can apply the formula for the sum of a finite geometric series to obtain

$$\sum_{n=1}^{12} 4(0.3)^n = a_1 \left( \frac{1 - r^n}{1 - r} \right)$$
$$= 4(0.3) \left[ \frac{1 - (0.3)^{12}}{1 - 0.3} \right]$$
$$\approx 1.714.$$

When using the formula for the sum of a finite geometric series, be careful to check that the index begins at $i = 1$. If the index begins at $i = 0$, you must adjust the formula for the $n$th partial sum. For instance, if the index in Example 6 had begun with $n = 0$, the sum would have been

$$\sum_{n=0}^{12} 4(0.3)^n = 4 + \sum_{n=1}^{12} 4(0.3)^n \approx 4 + 1.714 = 5.714.$$

The formula for the sum of a *finite* geometric series can, depending on the value of $r$, be extended to produce a formula for the sum of an *infinite* geometric series. Specifically, if the common ratio $r$ has the property that $|r| < 1$, it can be shown that $r^n$ becomes arbitrarily close to zero as $n$ increases without bound. Consequently,

$$a_1\left(\frac{1 - r^n}{1 - r}\right) \to a_1\left(\frac{1 - 0}{1 - r}\right) \quad \text{as} \quad n \to \infty.$$

This result is summarized as follows.

---

▶ **The Sum of an Infinite Geometric Series**

If $|r| < 1$, the infinite geometric series

$$a_1 + a_1 r + a_1 r^2 + a_1 r^3 + \cdots + a_1 r^{n-1} + \cdots$$

has the sum

$$S = \frac{a_1}{1 - r}.$$

---

**Example 7**   Finding the Sum of an Infinite Geometric Series

Find the sum of the infinite geometric series.

**a.** $\displaystyle\sum_{n=1}^{\infty} 4(0.6)^{n-1}$   **b.** $\displaystyle\sum_{n=1}^{\infty} 3(0.1)^{n-1}$

**Solution**

**a.** $\displaystyle\sum_{n=1}^{\infty} 4(0.6)^{n-1} = 4 + 4(0.6) + 4(0.6)^2 + 4(0.6)^3 + \cdots + 4(0.6)^{n-1} + \cdots$

$$= \frac{4}{1 - (0.6)} \qquad \frac{a_1}{1 - r}$$

$$= 10$$

**b.** $\displaystyle\sum_{n=1}^{\infty} 3(0.1)^{n-1} = 3 + 3(0.1) + 3(0.1)^2 + 3(0.1)^3 + \cdots + 3(0.1)^{n-1} + \cdots$

$$= \frac{3}{1 - (0.1)} \qquad \frac{a_1}{1 - r}$$

$$= \frac{10}{3}$$

$$\approx 3.33$$

## Application

**Example 8**   Compound Interest

A deposit of \$50 is made on the first day of each month in a savings account that pays 6% compounded monthly. What is the balance at the end of 2 years?

Solution

The first deposit will gain interest for 24 months, and its balance will be

$$A_{24} = 50\left(1 + \frac{0.06}{12}\right)^{24} = 50(1.005)^{24}.$$

The second deposit will gain interest for 23 months, and its balance will be

$$A_{23} = 50\left(1 + \frac{0.06}{12}\right)^{23} = 50(1.005)^{23}.$$

The last deposit will gain interest for only 1 month, and its balance will be

$$A_1 = 50\left(1 + \frac{0.06}{12}\right)^{1} = 50(1.005).$$

The total balance in the account will be the sum of the balances of the 24 deposits. Using the formula for the sum of a finite geometric series, with $A_1 = 50(1.005)$ and $r = 1.005$, you have

$$S = 50(1.005)\left[\frac{1 - (1.005)^{24}}{1 - 1.005}\right] = \$1277.96.$$

| Discussing the Concept | An Experiment |
|---|---|

You will need a piece of string or yarn, a pair of scissors, and a tape measure. Measure out any length of string at least 5 feet long. Double over the string and cut it in half. Take one of the resulting halves, double it over, and cut it in half. Continue this process until you are no longer able to cut a length of string in half. How many cuts were you able to make? Construct a sequence of the resulting string lengths after each cut, starting with the original length of the string. Find a formula for the $n$th term of this sequence. How many cuts could you theoretically make? Discuss why you were not able to make that many cuts.

# Warm Up

The following warm-up exercises involve skills that were covered in earlier sections. You will use these skills in the exercise set for this section.

In Exercises 1–4, evaluate the expression.

**1.** $\left(\frac{4}{5}\right)^3$

**2.** $\left(\frac{3}{4}\right)^2$

**3.** $2^{-4}$

**4.** $4(3^4)$

In Exercises 5–8, simplify the expression.

**5.** $(2n)(3n^2)$

**6.** $n(3n)^3$

**7.** $\dfrac{4n^5}{n^2}$

**8.** $\dfrac{(2n)^3}{8n}$

In Exercises 9 and 10, use sigma notation to write the sum.

**9.** $2 + 2(3) + 2(3^2) + 2(3^3)$

**10.** $3 + 3(2) + 3(2^2) + 3(2^3)$

## 8.3 Exercises

LAB 9

In Exercises 1–10, determine whether the sequence is geometric. If it is, find its common ratio and write a formula for $a_n$.

**1.** $1, 4, 16, 64, \ldots$

**2.** $2, 10, 50, 250, \ldots$

**3.** $4, -12, 36, -108, \ldots$

**4.** $1, -2, 4, -8, \ldots$

**5.** $1, -\frac{1}{2}, \frac{1}{4}, -\frac{1}{8}, \ldots$

**6.** $\frac{3}{2}, 1, \frac{2}{3}, \frac{4}{9}, \ldots$

**7.** $\frac{1}{2}, \frac{2}{3}, \frac{3}{4}, \frac{4}{5}, \ldots$

**8.** $9, -6, 4, -\frac{8}{3}, \ldots$

**9.** $1, \frac{1}{2}, \frac{1}{3}, \frac{1}{4}, \ldots$

**10.** $\frac{1}{5}, \frac{2}{3}, \frac{3}{9}, \frac{4}{11}, \ldots$

In Exercises 11–18, write the first five terms of the geometric sequence.

**11.** $a_1 = 4, r = 2$

**12.** $a_1 = 6, r = 3$

**13.** $a_1 = 1, r = \frac{1}{3}$

**14.** $a_1 = 1, r = \frac{2}{3}$

**15.** $a_1 = 7, r = -\frac{1}{5}$

**16.** $a_1 = \frac{1}{2}, r = -\frac{1}{3}$

**17.** $a_1 = 1, r = e$

**18.** $a_1 = 2, r = e^{0.1}$

In Exercises 19–30, find the indicated term of the geometric sequence.

**19.** $a_1 = 16, r = \frac{1}{4}$, 5th term

**20.** $a_1 = 9, r = \frac{2}{3}$, 7th term

**21.** $a_1 = 6, r = -\frac{1}{3}$, 12th term

**22.** $a_1 = 1, r = -\frac{2}{3}$, 10th term

**23.** $a_1 = 100, r = e$, 9th term

**24.** $a_1 = 6, r = e^{0.1}$, 9th term

**25.** $a_1 = 500, r = 1.02$, 40th term

**26.** $a_1 = 1000, r = 1.005$, 60th term

**27.** $a_1 = 16, a_4 = \frac{27}{4}$, 3rd term

**28.** $a_2 = 3, a_5 = \frac{3}{64}$, 1st term

**29.** $a_2 = -18, a_5 = \frac{2}{3}$, 6th term

**30.** $a_3 = \frac{16}{3}, a_5 = \frac{64}{27}$, 7th term

In Exercises 31–38, find the indicated sum.

**31.** $\displaystyle\sum_{n=1}^{10} 8(2^n)$

**32.** $\displaystyle\sum_{n=1}^{6} 3(4^n)$

**33.** $\displaystyle\sum_{i=1}^{10} 8\left(\tfrac{1}{4}\right)^{i-1}$

**34.** $\displaystyle\sum_{i=1}^{10} 5\left(\tfrac{1}{3}\right)^{i-1}$

**35.** $\displaystyle\sum_{n=0}^{8} 2^n$

**36.** $\displaystyle\sum_{n=0}^{8} (-2)^n$

**37.** $\displaystyle\sum_{n=0}^{20} 3\left(\tfrac{3}{2}\right)^n$

**38.** $\displaystyle\sum_{n=0}^{15} 2\left(\tfrac{4}{3}\right)^n$

In Exercises 39–46, find the sum, if it exists.

**39.** $\displaystyle\sum_{n=0}^{\infty} \left(\tfrac{1}{2}\right)^n = 1 + \tfrac{1}{2} + \tfrac{1}{4} + \tfrac{1}{8} + \cdots$

**40.** $\displaystyle\sum_{n=0}^{\infty} 2\left(\tfrac{2}{3}\right)^n = 2 + \tfrac{4}{3} + \tfrac{8}{9} + \tfrac{16}{27} + \cdots$

**41.** $\displaystyle\sum_{n=1}^{\infty} \left(-\tfrac{1}{2}\right)^{n-1} = 1 - \tfrac{1}{2} + \tfrac{1}{4} - \tfrac{1}{8} + \cdots$

**42.** $\displaystyle\sum_{n=1}^{\infty} 2\left(-\tfrac{2}{3}\right)^{n-1} = 2 - \tfrac{4}{3} + \tfrac{8}{9} - \tfrac{16}{27} + \cdots$

**43.** $\displaystyle\sum_{n=0}^{\infty} 4\left(\tfrac{1}{4}\right)^n = 4 + 1 + \tfrac{1}{4} + \tfrac{1}{16} + \cdots$

**44.** $\displaystyle\sum_{n=0}^{\infty} 3\left(\tfrac{1}{10}\right)^n = 3 + 0.3 + 0.03 + 0.003 + \cdots$

**45.** $\displaystyle\sum_{n=0}^{\infty} \left(\tfrac{3}{2}\right)^n = 1 + \tfrac{3}{2} + \tfrac{9}{4} + \tfrac{27}{8} + \cdots$

**46.** $\displaystyle\sum_{n=0}^{\infty} 5\left(\tfrac{4}{3}\right)^n = 5 + \tfrac{20}{3} + \tfrac{80}{9} + \tfrac{320}{27} + \cdots$

**47.** *Compound Interest*   A deposit of $100 is made at the beginning of each month for 5 years in an account that pays 6% compounded monthly. The balance $A$ in the account at the end of 5 years is

$$A = 100\left(1 + \frac{0.06}{12}\right)^1 + \cdots + 100\left(1 + \frac{0.06}{12}\right)^{60}.$$

(a) Find the balance after 5 years.

(b) How much would the balance increase if the interest rate were raised to 7%?

(c) At 7% interest, find the balance after 10 years.

**48.** *Compound Interest*   A deposit of $50 is made at the beginning of each month for 5 years in an account that pays 5% compounded monthly. Use a graphing utility to find the balance $A$ in the account at the end of 5 years.

$$A = 50\left(1 + \frac{0.05}{12}\right)^1 + \cdots + 50\left(1 + \frac{0.05}{12}\right)^{60}$$

**49.** *Compound Interest*   Suppose you deposit $75 in an account at the beginning of each month for 10 years. The account pays 8% compounded monthly. Use a graphing utility to find your balance at the end of 10 years. If the interest were compounded continuously, what would the balance be?

**50.** *Compound Interest*   Suppose you deposit $200 in an account at the beginning of each month for 40 years. The account pays 6% compounded monthly. What would your balance be at the end of 40 years? If the interest were compounded continuously, what would the balance be?

**51.** *Profit*   The annual profit for the Hershey Foods Corp. from 1988 to 1997 can be approximated by the model

$$a_n = 165.7e^{0.079n}, \quad n = 0, 1, 2, 3, \ldots, 9$$

where $a_n$ is the annual profit (in millions of dollars) and $n$ represents the year, with $n = 0$ corresponding to 1988 (see figure). Use the formula for the sum of a finite geometric series to approximate the total profit earned during this 10-year period. (Source: Hershey Foods Corp.)

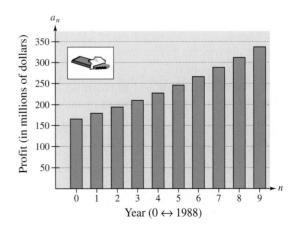

Year (0 ↔ 1988)

**52.** *Profit*    The annual profit for the Bristol-Myers Squibb Company from 1991 to 1997 can be approximated by the model

$$a_n = 1949.91e^{0.078n}, \quad n = 0, 1, 2, 3, \ldots, 6$$

where $a_n$ is the annual profit (in millions of dollars) and $n$ represents the year, with $n = 0$ corresponding to 1991 (see figure). Use the formula for the sum of a finite geometric series to approximate the total profit earned during this 7-year period.    (Source: Bristol-Myers Squibb Company)

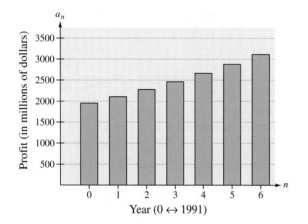

Year (0 ↔ 1991)

**53.** *Sales*    The annual sales for the La-Z-Boy Chair Company from 1992 to 1997 can be modeled by

$$a_n = 637e^{0.095n}, \quad n = 0, 1, 2, 3, 4, 5$$

where $a_n$ is the annual sales (in millions of dollars) and $n$ represents the year, with $n = 0$ corresponding to 1992. Use the formula for the sum of a finite geometric series to approximate the total sales earned during this 6-year period.    (Source: La-Z-Boy Chair Company)

**54.** *Sales*    The annual sales for Rubbermaid, Inc. from 1987 to 1997 can be approximated by the model

$$a_n = 1.24e^{0.073n}, \quad n = 0, 1, 2, 3, \ldots, 10$$

where $a_n$ is the annual sales (in billions of dollars) and $n$ represents the year, with $n = 0$ corresponding to 1987. Use the formula for the sum of a finite geometric series to approximate the total sales earned during this 11-year period.    (Source: Rubbermaid, Inc.)

**55.** *Would You Take This Job?*    Suppose you go to work at a company that pays $0.01 for the first day, $0.02 for the second day, $0.04 for the third day, and so on. If the daily wage keeps doubling, what will your total income be for working 29 days? 30 days? 31 days?

**56.** *Gambler's Ruin*    Suppose you were planning to visit a gambling casino and a friend suggested the following gambling strategy. Find a game that pays $2 for each dollar bet (such as the pass line at craps). Start by betting $2, and every time you win, return the $2 bet. If you lose the $2 bet, then bet $4. If you lose the $4 bet, then bet $8. For instance, if you lose four times in a row and then win, you will have bet $2 + 4 + 8 + 16 + 32$ and you will have won $64 on the last bet (which puts you $2 ahead). What is wrong with this gambling strategy? How much money would you have to take to cover the possibility of losing 12 times in a row?

**57.** *Area*    The sides of a square are 16 inches in length. A new square is formed by connecting the midpoints of the sides of the original square, and two of the triangles are shaded (see figure). If this process is repeated five more times, what will the total area of the shaded region be?

**58.** *Area*    The sides of a square are 16 inches in length. The square is divided into nine smaller squares and the center square is shaded (see figure). Each of the eight unshaded squares is then divided into nine smaller squares and the center square of each is shaded. If this process is repeated four more times, what will the total area of the shaded region be?

## Math Matters • The Snowflake Curve

To create a **snowflake curve,** begin with an equilateral triangle (called Stage 1). To form Stage 2, trisect each side of the triangle and at each of the middle sections construct an equilateral triangle pointing outward. To form each additional stage, repeat this process. The following diagram shows the first six stages of the snowflake curve, which is the curve obtained by continuing this process infinitely many times.

Stage 1 (3 sides)          Stage 2 (12 sides)          Stage 3 (48 sides)

Stage 4 (192 sides)        Stage 5 (768 sides)         Stage 6 (3072 sides)

In this figure, note that the $n$th stage of the snowflake curve has $3(4^{n-1})$ sides. If each side in Stage 1 has a length of 1, then each side in Stage 2 will have a length of $\frac{1}{3}$, each side in Stage 3 will have a length of $\left(\frac{1}{3}\right)^2 = \frac{1}{9}$, and so on. Thus, the total perimeter of the $n$th stage is

$$3(4^{n-1}) \cdot \left(\frac{1}{3}\right)^{n-1} = 3\left(\frac{4}{3}\right)^{n-1}.$$

Because the value of this expression is unbounded as $n$ approaches infinity, you can conclude that the perimeter of the snowflake curve is infinite. This seems to be a paradox because it is difficult to conceive of a curve that is infinitely long being drawn on a piece of paper that has a finite area.

## Mid-Chapter Quiz

Take this quiz as you would take a quiz in class. After you are done, check your work against the answers given in the back of the book.

In Exercises 1 and 2, write the first five terms of the sequence. (Begin with $n = 1$.)

**1.** $a_n = 2n + 1$

**2.** $a_n = \dfrac{n}{(n + 1)^2}$

In Exercises 3 and 4, evaluate the expression.

**3.** $\dfrac{5!}{3!}$

**4.** $\dfrac{30!}{27!}$

In Exercises 5–8, write an expression for $a_n$. (Begin with $n = 1$.)

**5.** $2, \frac{1}{2}, \frac{2}{9}, \frac{1}{8}, \frac{2}{25}, \ldots$

**6.** $\frac{1}{2}, \frac{1}{3}, \frac{1}{4}, \frac{1}{5}, \frac{1}{6}, \ldots$

**7.** *Arithmetic:* $a_1 = 2, d = 5$

**8.** *Arithmetic:* $a_1 = 7, d = 3$

In Exercises 9–12, decide whether the sequence is arithmetic, geometric, or neither. Explain your reasoning.

**9.** $a_n = 2 + 3(n - 1)$

**10.** $3, -6, 18, -72, \ldots$

**11.** $1, \frac{1}{2}, \frac{1}{4}, \frac{1}{8}, \ldots$

**12.** $a_n = 3(2)^n$

In Exercises 13–18, evaluate the sum.

**13.** $\displaystyle\sum_{i=1}^{5} (2i - 1)$

**14.** $\displaystyle\sum_{i=1}^{4} 2i^2$

**15.** $\displaystyle\sum_{n=1}^{25} 2n$

**16.** $\displaystyle\sum_{n=1}^{6} 2^n$

**17.** $\displaystyle\sum_{n=0}^{\infty} \left(\tfrac{1}{3}\right)^n$

**18.** $\displaystyle\sum_{n=0}^{\infty} 5\left(\tfrac{1}{5}\right)^n$

**19.** *Job Offer*   A person accepts a position with a company and will receive a salary of $28,500 for the first year. The person is guaranteed a raise of $2000 per year for the first 5 years.

   (a) What will the salary be during the sixth year of employment?

   (b) How much will the company have paid the person by the end of the sixth year?

**20.** *Compound Interest*   Suppose that you deposit $75 in an account at the beginning of each month for 20 years. The account pays 7% compounded monthly. What will your balance be at the end of the 20 years?

## Objectives

- Find a binomial coefficient.
- Use Pascal's Triangle to determine a binomial coefficient.
- Expand a binomial using binomial coefficients.
- Find a specific term in a binomial expansion.

## Binomial Coefficients

Recall that a **binomial** is a polynomial that has two terms. In this section, you will study a formula that gives a quick method of raising a binomial to a power. To begin, let's look at the expansion of $(x + y)^n$ for several values of $n$.

$$(x + y)^0 = 1$$

$$(x + y)^1 = x + y$$

$$(x + y)^2 = x^2 + 2xy + y^2$$

$$(x + y)^3 = x^3 + 3x^2y + 3xy^2 + y^3$$

$$(x + y)^4 = x^4 + 4x^3y + 6x^2y^2 + 4xy^3 + y^4$$

$$(x + y)^5 = x^5 + 5x^4y + 10x^3y^2 + 10x^2y^3 + 5xy^4 + y^5$$

There are several observations you can make about these expansions.

1. In each expansion, there are $n + 1$ terms.
2. In each expansion, $x$ and $y$ have symmetrical roles. The powers of $x$ decrease by 1 in successive terms, whereas the powers of $y$ increase by 1.
3. The sum of the powers of each term is $n$. For instance, in the expansion of $(x + y)^5$, the sum of the powers of each term is 5.

$$\overset{4 + 1 = 5 \qquad 3 + 2 = 5}{(x + y)^5 = x^5 + 5x^4y^1 + 10x^3y^2 + 10x^2y^3 + 5xy^4 + y^5}$$

4. The coefficients increase and then decrease in a symmetric pattern.

The coefficients of a binomial expansion are called **binomial coefficients.** To find them, you can use the following theorem.

---

▶ **The Binomial Theorem**

In the expansion of $(x + y)^n$

$$(x + y)^n = x^n + nx^{n-1}y + \cdots + {}_nC_r x^{n-r}y^r + \cdots + nxy^{n-1} + y^n$$

the coefficient of $x^{n-r}y^r$ is given by

$${}_nC_r = \frac{n!}{(n - r)!r!}.$$

---

**Example 1**    Finding Binomial Coefficients

Find the binomial coefficients.

**a.** $_8C_2$    **b.** $_{10}C_3$    **c.** $_7C_0$    **d.** $_8C_8$

Solution

**a.** $_8C_2 = \dfrac{8!}{6! \cdot 2!} = \dfrac{(8 \cdot 7) \cdot 6!}{6! \cdot 2!} = \dfrac{8 \cdot 7}{2 \cdot 1} = 28$

**b.** $_{10}C_3 = \dfrac{10!}{7! \cdot 3!} = \dfrac{(10 \cdot 9 \cdot 8) \cdot 7!}{7! \cdot 3!} = \dfrac{10 \cdot 9 \cdot 8}{3 \cdot 2 \cdot 1} = 120$

**c.** $_7C_0 = \dfrac{7!}{7! \cdot 0!} = 1$

**d.** $_8C_8 = \dfrac{8!}{0! \cdot 8!} = 1$

---

When $r \neq 0$ and $r \neq n$, as in parts (a) and (b) above, there is a simple pattern for evaluating binomial coefficients.

$$\underbrace{\overbrace{8C_2 = \dfrac{8 \cdot 7}{2 \cdot 1}}^{2 \text{ factors}}}_{2 \text{ factorial}} \quad \text{and} \quad \underbrace{\overbrace{_{10}C_3 = \dfrac{10 \cdot 9 \cdot 8}{3 \cdot 2 \cdot 1}}^{3 \text{ factors}}}_{3 \text{ factorial}}$$

**Example 2**    Finding Binomial Coefficients

Find the binomial coefficients.

**a.** $_7C_3$    **b.** $_7C_4$    **c.** $_{12}C_1$    **d.** $_{12}C_{11}$

Solution

**a.** $_7C_3 = \dfrac{7 \cdot 6 \cdot 5}{3 \cdot 2 \cdot 1} = 35$

**b.** $_7C_4 = \dfrac{7 \cdot 6 \cdot 5 \cdot 4}{4 \cdot 3 \cdot 2 \cdot 1} = 35$

**c.** $_{12}C_1 = \dfrac{12}{1} = 12$

**d.** $_{12}C_{11} = \dfrac{12!}{1! \cdot 11!} = \dfrac{(12) \cdot 11!}{1! \cdot 11!} = \dfrac{12}{1} = 12$

---

It is not a coincidence that the results in parts (a) and (b) of Example 2 are the same and that the results in parts (c) and (d) are the same. In general, it is true that

$$_nC_r = {_nC_{n-r}}.$$

This shows the symmetric property of binomial coefficients that was identified earlier.

## Pascal's Triangle

There is a convenient way to remember a pattern for binomial coefficients. By arranging the coefficients in a triangular pattern, you obtain the following array, which is called **Pascal's Triangle.** This triangle is named after the famous French mathematician Blaise Pascal (1623–1662).

$$
\begin{array}{ccccccccccccccc}
 & & & & & & & 1 \\
 & & & & & & 1 & & 1 \\
 & & & & & 1 & & 2 & & 1 \\
 & & & & 1 & & 3 & & 3 & & 1 \\
 & & & 1 & & 4 & & 6 & & 4 & & 1 \\
 & & 1 & & 5 & & 10 & & 10 & & 5 & & 1 \\
 & 1 & & 6 & & 15 & & 20 & & 15 & & 6 & & 1 \\
1 & & 7 & & 21 & & 35 & & 35 & & 21 & & 7 & & 1
\end{array}
$$

The first and last number in each row of Pascal's Triangle is 1. Every other number in each row is formed by adding the two numbers immediately above the number. Pascal noticed that numbers in this triangle are precisely the same numbers that are the coefficients of binomial expansions, as follows.

$$(x + y)^0 = 1$$

$$(x + y)^1 = 1x + 1y$$

$$(x + y)^2 = 1x^2 + 2xy + 1y^2$$

$$(x + y)^3 = 1x^3 + 3x^2y + 3xy^2 + 1y^3$$

$$(x + y)^4 = 1x^4 + 4x^3y + 6x^2y^2 + 4xy^3 + 1y^4$$

$$(x + y)^5 = 1x^5 + 5x^4y + 10x^3y^2 + 10x^2y^3 + 5xy^4 + 1y^5$$

$$(x + y)^6 = 1x^6 + 6x^5y + 15x^4y^2 + 20x^3y^3 + 15x^2y^4 + 6xy^5 + 1y^6$$

$$(x + y)^7 = 1x^7 + 7x^6y + 21x^5y^2 + 35x^4y^3 + 35x^3y^4 + 21x^2y^5 + 7xy^6 + 1y^7$$

The top row in Pascal's Triangle is called the *zero row* because it corresponds to the binomial expansion $(x + y)^0 = 1$.

Similarly, the next row is called the *first row* because it corresponds to the binomial expansion $(x + y)^1 = 1x + 1y$. In general, the *nth row* in Pascal's Triangle gives the coefficients of $(x + y)^n$.

**Example 3**   Using Pascal's Triangle

Use the seventh row of Pascal's Triangle to find the binomial coefficients.

$$_8C_0,\ _8C_1,\ _8C_2,\ _8C_3,\ _8C_4,\ _8C_5,\ _8C_6,\ _8C_7,\ _8C_8$$

**Solution**

## Binomial Expansions

As mentioned at the beginning of this section, when you write out the coefficients for a binomial that is raised to a power, you are **expanding a binomial.** The formulas for binomial coefficients give you an easy way to expand binomials, as demonstrated in the next four examples.

圖 方 蔡 七 法 古

**Pascal's Triangle**

"Pascal's" Triangle and forms of the Binomial Theorem were known in Eastern cultures prior to the Western "discovery" of the theorem. A Chinese text *Precious Mirror* contains a triangle of binomial expansions through the eighth power.

**Example 4**    Expanding a Binomial

Write the expansion for the expression

$(x + 1)^3$.

### Solution

The binomial coefficients from the third row of Pascal's Triangle are

1, 3, 3, 1.

So, the expansion is as follows.

$$(x + 1)^3 = (1)x^3 + (3)x^2(1) + (3)x(1^2) + (1)(1^3)$$
$$= x^3 + 3x^2 + 3x + 1$$

To expand binomials representing *differences,* rather than sums, you alternate signs. Here are two examples.

$$(x - 1)^3 = x^3 - 3x^2 + 3x - 1$$
$$(x - 1)^4 = x^4 - 4x^3 + 6x^2 - 4x + 1$$

**Example 5**    Expanding a Binomial

Write the expansion for the expression

$(x - 3)^4$.

### Solution

The binomial coefficients from the fourth row of Pascal's Triangle are

1, 4, 6, 4, 1.

So, the expansion is as follows.

$$(x - 3)^4 = (1)x^4 - (4)x^3(3) + (6)x^2(3^2) - (4)x(3^3) + (1)(3^4)$$
$$= x^4 - 12x^3 + 54x^2 - 108x + 81$$

### Example 6    Expanding a Binomial

Write the expansion for $(x - 2y)^4$.

**Solution**

Use the fourth row of Pascal's Triangle, as follows.

$$(x - 2y)^4 = (1)x^4 - (4)x^3(2y) + (6)x^2(2y)^2 - (4)x(2y)^3 + (1)(2y)^4$$

$$= x^4 - 8x^3y + 24x^2y^2 - 32xy^3 + 16y^4$$

### Example 7    Finding a Term in a Binomial Expansion

Find the sixth term of $(a + 2b)^8$.

**Solution**

From the Binomial Theorem, you can see that the $(r + 1)$th term is $_nC_r x^{n-r}y^r$. So in this case, $6 = r + 1$ means that $r = 5$. Because $n = 8$, $x = a$, and $y = 2b$, the sixth term in the binomial expansion is

$$_8C_5 a^{8-5}(2b)^5 = 56 \cdot a^3 \cdot (2b)^5$$

$$= 56(2^5)a^3b^5$$

$$= 1792a^3b^5$$

### Discussing the Concept    Finding a Pattern

By adding the terms in each of the rows of Pascal's Triangle, you obtain the following.

Row 0:   $1 = 1$

Row 1:   $1 + 1 = 2$

Row 2:   $1 + 2 + 1 = 4$

Row 3:   $1 + 3 + 3 + 1 = 8$

Row 4:   $1 + 4 + 6 + 4 + 1 = 16$

Find a pattern for this sequence. Then use the pattern to find the sum of the terms in the 10th row of Pascal's Triangle. Check your answer by actually adding the terms of the 10th row.

# Warm Up

The following warm-up exercises involve skills that were covered in earlier sections. You will use these skills in the exercise set for this section.

In Exercises 1–6, perform the indicated operations and/or simplify.

**1.** $5x^2(x^3 + 3)$  **2.** $(x + 5)(x^2 - 3)$

**3.** $(x + 4)^2$  **4.** $(2x - 3)^2$

**5.** $x^2y(3xy^{-2})$  **6.** $(-2z)^5$

In Exercises 7–10, evaluate the expression.

**7.** $5!$  **8.** $\dfrac{8!}{5!}$

**9.** $\dfrac{10!}{7!}$  **10.** $\dfrac{6!}{3!3!}$

## 8.4 Exercises

In Exercises 1–10, evaluate $_nC_r$.

**1.** $_6C_2$  **2.** $_9C_4$

**3.** $_{12}C_0$  **4.** $_8C_8$

**5.** $_{20}C_{15}$  **6.** $_{20}C_5$

**7.** $_{100}C_{98}$  **8.** $_{10}C_4$

**9.** $_{100}C_2$  **10.** $_{10}C_6$

In Exercises 11–30, use the Binomial Theorem to expand the expression. Simplify your answer.

**11.** $(x + 1)^4$  **12.** $(x + 1)^6$

**13.** $(x + 2)^3$  **14.** $(x + 2)^4$

**15.** $(x - 2)^4$  **16.** $(x - 2)^5$

**17.** $(x + y)^5$  **18.** $(x + y)^6$

**19.** $(s + 2)^4$  **20.** $(s + 3)^5$

**21.** $(2 + 3s)^6$  **22.** $(2 + 3s)^5$

**23.** $(x - y)^5$  **24.** $(2x - y)^5$

**25.** $(1 - 2x)^3$  **26.** $(1 - 3x)^4$

**27.** $(x^2 - y^2)^5$  **28.** $(x^2 + y^2)^6$

**29.** $\left(\dfrac{1}{x} + y\right)^5$  **30.** $\left(\dfrac{x}{2} - 3y\right)^3$

**31.** How many terms are in the expanded form of $(x + y)^5$? $(x + y)^6$? $(x + y)^7$? $(x + y)^8$? How many terms are in $(x + y)^n$?

**32.** How are the expansions of the binomials $(x + y)^n$ and $(x - y)^n$ alike? How are they different?

In Exercises 33–38, use the Binomial Theorem to expand the complex number. Simplify your answer by using the fact that $i^2 = -1$.

**33.** $(1 + i)^4$  **34.** $(2 - i)^5$

**35.** $\left(4 - \sqrt{-4}\right)^6$  **36.** $\left(5 + \sqrt{-9}\right)^3$

**37.** $\left(\dfrac{-1}{2} + \dfrac{\sqrt{3}}{2}i\right)^3$  **38.** $\left(\dfrac{3}{4} - \dfrac{\sqrt{2}}{4}i\right)^4$

In Exercises 39–42, expand the expression using Pascal's Triangle to determine the coefficients.

**39.** $(2t - 1)^5$  **40.** $(x - 4)^5$

**41.** $(3 + 2z)^4$  **42.** $(3y + 2)^5$

In Exercises 43–50, find the indicated term in the expansion of the binomial expression.

**43.** $(x + 2)^{10}$  $ax^6$

**44.** $(x^2 + 3)^{12}$  $ax^8$

**45.** $(x - 2y)^{10}$  $ax^8y^2$

**46.** $(4x - y)^{10}$  $ax^3y^7$

**47.** $(3x - 2y)^9$  $ax^4y^5$

**48.** $(2x - 3y)^{11}$  $ax^3y^8$

**49.** $(x^2 - 1)^8$  $ax^8$

**50.** $(x^2 + 2)^{12}$  $ax^{10}$

In Exercises 51–56, use the Binomial Theorem to expand the expression. [In the study of probability, it is sometimes necessary to use the expansion of $(p + q)^n$, where $p + q = 1$.]

**51.** $\left(\frac{1}{2} + \frac{1}{2}\right)^7$

**52.** $\left(\frac{1}{4} + \frac{3}{4}\right)^{10}$

**53.** $\left(\frac{1}{3} + \frac{2}{3}\right)^8$

**54.** $(0.3 + 0.7)^{12}$

**55.** $(0.6 + 0.4)^5$

**56.** $(0.35 + 0.65)^6$

**57.** *Finding a Pattern*   Describe the pattern formed by the sums of the numbers along the diagonal segments of Pascal's Triangle (see figure).

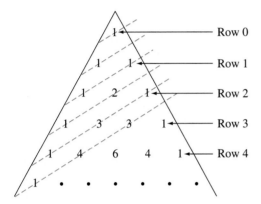

**58.** *Finding a Pattern*   Use each of the encircled groups of numbers in the figure to form a $2 \times 2$ matrix. Find the determinant of the matrix. Then describe the pattern.

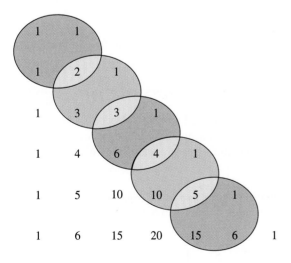

**59.** *Finding a Pattern*   Each of the "flowers" in the figure is taken from Pascal's Triangle. Copy the flowers and fill in the missing numbers. Which rows did the flowers come from?

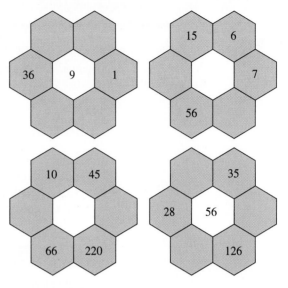

**60.** *Life Insurance*   The average amount of life insurance per household $f(t)$ (in thousands of dollars) from 1970 through 1996 can be approximated by

$$f(t) = 0.123t^2 + 1.25t + 19.8, \quad 0 \le t \le 26$$

where $t = 0$ represents 1970 (see figure). You want to adjust this model so that $t = 0$ corresponds to 1980 rather than 1970. To do this, you shift the graph of $f$ 10 units *to the left* and obtain

$$g(t) = f(t + 10).$$

(Source: American Council of Life Insurance)

(a) Write $g(t)$ in standard form.

(b) Use a graphing utility to graph $f$ and $g$ in the same viewing window.

## 8.5 Counting Principles

### Objectives

- Solve simple counting problems.
- Use the Fundamental Counting Principle.
- Determine the number of permutations of *n* elements taken *r* at a time.
- Determine the number of combinations of *n* elements taken *r* at a time.

## Simple Counting Problems

The next two sections of this chapter contain a brief introduction to some of the basic counting principles and their application to probability. In the next section, you will see that much of probability has to do with counting the number of ways an event can occur.

**Example 1** Selecting Pairs of Numbers at Random

Eight pieces of paper are numbered from 1 to 8 and placed in a box. One piece of paper is drawn from the box, its number is written down, and the piece of paper is replaced in the box. Then, a second piece of paper is drawn from the box, and its number is written down. Finally, the two numbers are added together. How many different ways can a total of 12 be obtained?

**Solution**

To solve this problem, count the different ways that a total of 12 can be obtained using two numbers from 1 to 8.

| First number | 4 | 5 | 6 | 7 | 8 |
|---|---|---|---|---|---|
| Second number | 8 | 7 | 6 | 5 | 4 |

From this list, you can see that a total of 12 can occur in five different ways.

**Example 2** Selecting Pairs of Numbers at Random

Eight pieces of paper are numbered from 1 to 8 and placed in a box. Two pieces of paper are drawn from the box, and the numbers on the pieces of paper are written down and totaled. How many different ways can a total of 12 be obtained?

**Solution**

To solve this problem, you can count the different ways that a total of 12 can be obtained *using two different numbers* from 1 to 8.

| First number | 4 | 5 | 7 | 8 |
|---|---|---|---|---|
| Second number | 8 | 7 | 5 | 4 |

So, a total of 12 can be obtained in four different ways.

## Counting Principles

Examples 1 and 2 describe simple counting problems in which you can *list* each possible way that an event can occur. When it is possible, this is always the best way to solve a counting problem. However, some events can occur in so many different ways that it is not feasible to write out the entire list. In such cases, you must rely on formulas and counting principles. The most important of these is the **Fundamental Counting Principle.**

---

▶ **Fundamental Counting Principle**

Let $E_1$ and $E_2$ be two events. The first event $E_1$ can occur in $m_1$ different ways. After $E_1$ has occurred, $E_2$ can occur in $m_2$ different ways. The number of ways that the two events can occur is $m_1 \cdot m_2$.

---

The Fundamental Counting Principle can be extended to three or more events. For instance, the number of ways that three events $E_1, E_2,$ and $E_3$ can occur is $m_1 \cdot m_2 \cdot m_3$.

**Example 3**   Using the Fundamental Counting Principle

How many different pairs of letters from the English alphabet are possible?

**Solution**

This experiment has two events. The first event is the choice of the first letter, and the second event is the choice of the second letter. Because the English alphabet contains 26 letters, it follows that the number of pairs of letters is $26 \cdot 26 = 676.$

**Example 4**   Using the Fundamental Counting Principle

Telephone numbers in the United States have 10 digits. The first three are the *area code* and the next seven are the *local telephone number.* How many different telephone numbers are possible within each area code? (Note that a local telephone number cannot begin with 0 or 1.)

**Solution**

Because the first digit cannot be 0 or 1, there are only eight choices for the first digit. For each of the other six digits, there are 10 choices.

So, the number of local telephone numbers that are possible within each area code is $8 \cdot 10 \cdot 10 \cdot 10 \cdot 10 \cdot 10 \cdot 10 = 8,000,000.$

## Permutations

One important application of the Fundamental Counting Principle is in determining the number of ways that $n$ elements can be arranged (in order). An ordering of $n$ elements is called a **permutation** of the elements.

---

▶ **Definition of Permutation**

A **permutation** of $n$ different elements is an ordering of the elements such that one element is first, one is second, one is third, and so on.

---

Example 5 **Finding the Number of Permutations of $n$ Elements**

How many permutations are possible for the letters A, B, C, D, E, and F?

**Solution**

Consider the following reasoning.

| | |
|---|---|
| First position: | Any of the *six* letters. |
| Second position: | Any of the remaining *five* letters. |
| Third position: | Any of the remaining *four* letters. |
| Fourth position: | Any of the remaining *three* letters. |
| Fifth position: | Any of the remaining *two* letters. |
| Sixth position: | The *one* remaining letter. |

So, the numbers of choices for the six positions are as follows.

Permutations of six letters

6   5   4   3   2   1

The total number of permutations of the six letters is $6! = 720$.

---

▶ **Number of Permutations of $n$ Elements**

The number of permutations of $n$ elements is given by

$$n \cdot (n - 1) \cdots 4 \cdot 3 \cdot 2 \cdot 1 = n!.$$

In other words, there are $n!$ different ways that $n$ elements can be ordered.

---

Occasionally, you are interested in ordering a *subset* of a collection of elements rather than the entire collection. For example, you might want to choose (and order) *r* elements out of a collection of *n* elements. Such an ordering is called a **permutation of *n* elements taken *r* at a time.**

| Example 6 | Counting Horse Race Finishes |

Eight horses are running in a race. In how many different ways can these horses come in first, second, and third? (Assume that there are no ties.)

Solution

Here are the different possibilities.

| | |
|---|---|
| Win (first position): | *Eight* choices |
| Place (second position): | *Seven* choices |
| Show (third position): | *Six* choices |

Using the Fundamental Counting Principle, multiply these three numbers together to obtain the following.

Different orders of horses

8      7      6

So, there are $8 \cdot 7 \cdot 6 = 336$ different orders.

---

### ▶ Permutations of *n* Elements Taken *r* at a Time

The number of permutations of *n* elements taken *r* at a time is

$$_nP_r = \frac{n!}{(n-r)!} = n(n-1)(n-2) \cdots (n-r+1).$$

Using this formula, you can rework Example 6 to find that the number of permutations of eight horses taken three at a time is

$$_8P_3 = \frac{8!}{5!}$$

$$= \frac{8 \cdot 7 \cdot 6 \cdot 5!}{5!}$$

$$= 336$$

which is the same answer obtained in Example 6.

Remember that for permutations, order is important. So, if you are looking at the possible permutations of the letters $A$, $B$, $C$, and $D$ taken three at a time, the permutations $(A, B, D)$ and $(B, A, D)$ would be different because the *order* of the elements is different.

Suppose, however, that you are asked to find the possible permutations of the letters $A$, $A$, $B$, and $C$. The total number of permutations of the four letters would be $_4P_4 = 4!$. However, not all of these arrangements would be *distinguishable* because there are two $A$'s in the list. To find the number of distinguishable permutations, you can use the following formula.

---

▶ **Distinguishable Permutations**

Suppose a set of $n$ objects has $n_1$ of one kind of object, $n_2$ of a second kind, $n_3$ of a third kind, and so on, with $n = n_1 + n_2 + n_3 + \cdots + n_k$. Then the number of **distinguishable permutations** of the $n$ objects is

$$\frac{n!}{n_1! \cdot n_2! \cdot n_3! \cdot \cdots \cdot n_k!}.$$

---

**Example 7** Distinguishable Permutations

In how many distinguishable ways can the letters in BANANA be written?

**Solution**

This word has six letters, of which three are A's, two are N's, and one is a B. Thus, the number of distinguishable ways the letters can be written is

$$\frac{6!}{3! \cdot 2! \cdot 1!} = \frac{6 \cdot 5 \cdot 4 \cdot 3!}{3! \cdot 2!} = 60.$$

The 60 different "words" are as follows.

| | | | | | |
|---|---|---|---|---|---|
| AAABNN | AAANBN | AAANNB | AABANN | AABNAN | AABNNA |
| AANABN | AANANB | AANBAN | AANBNA | AANNAB | AANNBA |
| ABAANN | ABANAN | ABANNA | ABNAAN | ABNANA | ABNNAA |
| ANAABN | ANAANB | ANABAN | ANABNA | ANANAB | ANANBA |
| ANBAAN | ANBANA | ANBNAA | ANNAAB | ANNABA | ANNBAA |
| BAAANN | BAANAN | BAANNA | BANAAN | BANANA | BANNAA |
| BNAAAN | BNAANA | BNANAA | BNNAAA | NAAABN | NAAANB |
| NAABAN | NAABNA | NAANAB | NAANBA | NABAAN | NABANA |
| NABNAA | NANAAB | NANABA | NANBAA | NBAAAN | NBAANA |
| NBANAA | NBNAAA | NNAAAB | NNAABA | NNABAA | NNBAAA |

## Combinations

When one counts the number of possible permutations of a set of elements, *order* is important. As a final topic in this section, we look at a method of selecting subsets of a larger set in which order is *not important.* Such subsets are called **combinations of _n_ elements taken _r_ at a time.** For instance, the combinations

$$\{A, B, C\} \quad \text{and} \quad \{B, A, C\}$$

are equivalent because both sets contain the same three elements, and the order in which the elements are listed is *not important.* Hence, you would count only one of the two sets. A common example of how a combination occurs is a card game in which the player is free to reorder the cards after they have been dealt.

| **Example 8** | Combinations of *n* Elements Taken *r* at a Time |

In how many different ways can three letters be chosen from the letters A, B, C, D, and E? (The order of the three letters is not important.)

### Solution

The following subsets represent the different combinations of three letters that can be chosen from five letters.

$$\{A, B, C\} \quad \{A, B, D\}$$
$$\{A, B, E\} \quad \{A, C, D\}$$
$$\{A, C, E\} \quad \{A, D, E\}$$
$$\{B, C, D\} \quad \{B, C, E\}$$
$$\{B, D, E\} \quad \{C, D, E\}$$

From this list, you can conclude that there are 10 different ways that three letters can be chosen from five letters.

---

### •TECHNOLOGY•

Most graphing utilities have keys that will evaluate the formulas for the number of permutations or combinations of *n* elements taken *r* at a time. Consult the user's guide for directions. Evaluate $_8P_5$ (the number of permutations of eight elements taken five at a time). The display should be 6720. Now evaluate $_8C_5$ (the number of combinations of eight elements taken five at a time). The display should be 56. Use your graphing utility to evaluate the following.

**a.** $_{36}P_4$    **b.** $_{36}C_4$    **c.** $_{25}P_3$    **d.** $_{25}C_3$

> ▶ **Combinations of $n$ Elements Taken $r$ at a Time**
>
> The number of combinations of $n$ elements taken $r$ at a time is
>
> $$_nC_r = \frac{n!}{(n-r)!r!}.$$

Note that the formula for $_nC_r$ is the same one given for binomial coefficients. To see how this formula is used, let's solve the counting problem in Example 8. In that problem, you are asked to find the number of combinations of five elements taken three at a time. Thus, $n = 5$, $r = 3$, and the number of combinations is

$$_5C_3 = \frac{5!}{2!3!} = \frac{5 \cdot 4 \cdot \overset{2}{\cancel{3}}}{3 \cdot 2 \cdot 1} = 10$$

which is the same answer obtained in Example 8.

---

**Example 9**    **Counting Card Hands**

A standard poker hand consists of five cards dealt from a deck of 52. How many different poker hands are possible? (After the cards are dealt, the player may reorder them, and therefore order is not important.)

**Solution**

You can find the number of different poker hands by using the formula for the number of combinations of 52 elements taken five at a time, as follows.

$$_{52}C_5 = \frac{52!}{47!5!} = \frac{52 \cdot 51 \cdot 50 \cdot 49 \cdot 48}{5 \cdot 4 \cdot 3 \cdot 2 \cdot 1} = 2{,}598{,}960 \text{ different hands}$$

---

**Discussing the Concept**    **Problem Posing**

According to NASA, each space shuttle astronaut consumes an average of 2100 calories per day. An evening meal normally consists of a main dish, a vegetable dish, and a dessert. The space shuttle food and beverage list contains 21 items classified as main dishes, 15 vegetable dishes, and 20 desserts. How many different evening meal menus are possible? Create and solve two other problems that could be asked about the evening meal menus. (*Hint:* You may want to think of variety versus repetition and length of mission.)    (Source: NASA)

# Warm Up

The following warm-up exercises involve skills that were covered in earlier sections. You will use these skills in the exercise set for this section.

In Exercises 1–4, evaluate the expression.

**1.** $13 \cdot 8^2 \cdot 2^3$

**2.** $10^2 \cdot 9^3 \cdot 4$

**3.** $\dfrac{12!}{2!(7!)(3!)}$

**4.** $\dfrac{25!}{22!}$

In Exercises 5 and 6, find the binomial coefficient.

**5.** $_{12}C_7$

**6.** $_{25}C_{22}$

In Exercises 7–10, simplify the expression.

**7.** $\dfrac{n!}{(n-4)!}$

**8.** $\dfrac{(2n)!}{4(2n-3)!}$

**9.** $\dfrac{2 \cdot 4 \cdot 6 \cdot 8 \cdots (2n)}{2^n}$

**10.** $\dfrac{3 \cdot 6 \cdot 9 \cdot 12 \cdots (3n)}{3^n}$

## 8.5 Exercises

1. *Job Applicants* A small college needs two additional faculty members: a chemist and a statistician. In how many ways can these positions be filled if there are four applicants for the chemistry position and five for the position in statistics?

2. *Computer Systems* A customer in a computer store can choose one of five monitors, one of four keyboards, and one of six computers. If all the choices are compatible, how many different systems could be chosen?

3. *Toboggan Ride* Six people are lining up for a ride on a toboggan, but only two of the six are willing to take the first position. With that constraint, in how many ways can the six people be seated on the toboggan? Draw a diagram to illustrate the number of ways that people can sit.

4. *Course Schedule* A college student is preparing her course schedule for the next semester. She may select one of five mathematics courses, one of three science courses, and one of eight courses from the social sciences and humanities. In how many ways can she select her schedule?

5. *License Plate Numbers* In a certain state, each automobile license plate number consists of two letters followed by a four-digit number. How many distinct license plate numbers can be formed?

6. *License Plate Numbers* In a certain state, each automobile license plate number consists of two letters following by a four-digit number. To avoid confusion between "O" and "Zero" and "I" and "One," the letters "O" and "I" are not used. How many distinct license plate numbers can be formed?

7. *True-False Exam* In how many ways can a six-question true-false exam be answered? (Assume that no questions are omitted.)

8. *True-False Exam* In how many ways can a 10-question true-false exam be answered? (Assume that no questions are omitted.)

9. *Three-Digit Numbers* How many three-digit numbers can be formed under the following conditions?

   (a) Leading digits cannot be zero.

   (b) Leading digits cannot be zero and no repetition of digits is allowed.

   (c) Leading digits cannot be zero and the number must be a multiple of 5.

**10.** *Four-Digit Numbers* How many four-digit numbers can be formed under the following conditions?

(a) Leading digits cannot be zero.

(b) Leading digits cannot be zero and no repetition of digits is allowed.

(c) Leading digits cannot be zero and the number must be a multiple of 5.

**11.** *Combination Lock* A combination lock will open when the right choice of three numbers (from 1 to 40, inclusive) is selected. How many different lock combinations are possible?

**12.** *Combination Lock* A combination lock will open when the right choice of three numbers (from 1 to 50, inclusive) is selected. How many different lock combinations are possible?

**13.** *Concert Seats* Four couples have reserved seats in a given row for a concert. In how many different ways can they be seated, given the following conditions?

(a) There are no restrictions.

(b) The two members of each couple wish to sit together.

**14.** *Single File* In how many orders can five girls and three boys walk through a doorway single-file, given the following conditions?

(a) There are no restrictions.

(b) The boys go before the girls.

(c) The girls go before the boys.

In Exercises 15–24, evaluate $_nP_r$.

**15.** $_4P_3$

**16.** $_5P_4$

**17.** $_8P_3$

**18.** $_{10}P_5$

**19.** $_{12}P_4$

**20.** $_{20}P_2$

**21.** $_{100}P_2$

**22.** $_{48}P_3$

**23.** $_6P_6$

**24.** $_7P_7$

**25.** *Posing for a Photograph* In how many ways can five children line up in one row to have their picture taken?

**26.** *Riding in a Car* In how many ways can eight people sit in an eight-passenger van?

**27.** *Choosing Officers* From a pool of 12 candidates, the offices of president, vice-president, secretary, and treasurer will be filled. In how many different ways can the offices be filled if each of the 12 candidates can hold any office?

**28.** *Assembly Line Production* There are four processes involved in assembling a certain product, and these processes can be performed in any order. The management wants to test each order to determine which is the least time consuming. How many different orders will have to be tested?

**29.** *Forming an Experimental Group* In order to conduct a certain experiment, four students are randomly selected from a class of 26. How many different groups of four students are possible?

**30.** *Test Questions* A student may answer any 12 questions from a total of 15 questions on an exam. How many different ways can the student select the questions?

**31.** *Lottery Choices* There are 40 numbers in a particular state lottery. In how many ways can a player select six of the numbers? (The order of selection is not important.)

**32.** *Lottery Choices* There are 50 numbers in a particular state lottery. In how many ways can a player select eight of the numbers? (The order of selection is not important.)

**33.** *Number of Subsets* How many subsets of four elements can be formed from a set of 100 elements?

**34.** *Number of Subsets* How many subsets of five elements can be formed from a set of 80 elements?

**35.** *Forming a Committee* A committee composed of three graduate students and two undergraduate students is to be selected from a group of 10 graduates and six undergraduates. How many different committees can be formed?

**36.** *Defective Units* A shipment of 12 microwave ovens contains three defective units. In how many ways can a vending company purchase four of these units and receive (a) all good units, (b) two good units, and (c) at least two good units?

**37.** *Job Applicants*   An employer interviews 10 people for four openings in the company. Four of the 10 people are women. If all 10 are qualified, in how many ways can the employer fill the four positions if (a) the selection is random and (b) exactly two are women?

**38.** *Forming a Committee*   Four people are randomly chosen from a group of four couples. In how many ways can this happen, given the following conditions?

(a) There are no restrictions.

(b) The selection must include one member from each couple.

**39.** *Poker Hand*   Five cards are selected from an ordinary deck of 52 playing cards. In how many ways can you get a full house? (A full house consists of three of one kind and two of another. For example, A-A-A-5-5 and K-K-K-10-10 are full houses.)

**40.** *Poker Hand*   Five cards are selected from an ordinary deck of 52 playing cards. In how many ways can you get a straight flush? (A straight flush consists of five cards that are in order and of the same suit. For example, A♥, 2♥, 3♥, 4♥, 5♥ and 10♠, J♠, Q♠, K♠, A♠ are straight flushes.)

*Diagonals of a Polygon*   In Exercises 41–44, find the number of diagonals of the polygon. (A line segment connecting any two nonadjacent vertices is called a *diagonal* of the polygon.)

**41.** Pentagon             **42.** Hexagon

**43.** Octagon             **44.** Decagon (10 sides)

In Exercises 45–50, find the number of distinguishable permutations of the group of letters.

**45.** A, S, S, E, S, S

**46.** P, A, P, A, Y, A

**47.** R, E, F, E, R, E, E

**48.** P, A, R, A, L, L, E, L

**49.** A, L, G, E, B, R, A

**50.** M, I, S, S, I, S, S, I, P, P, I

In Exercises 51–53, use the example of the tree diagram to determine the possible arrangements needed to answer the question.

A college theater company is planning a tour with performances in Phoenix, Dallas, and New Orleans. If there are no restrictions on the order of the performances, how many ways can the tour be arranged?

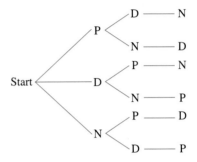

To determine the possible arrangements, we count the paths. A path moves from the start to the end of the last branch. In this case, there are six paths, so there are six possible ways to arrange the itinerary of the tour. The tree diagram verifies visually what we know by using counting principles. That is, the number of possible arrangements would be $3 \cdot 2 \cdot 1 = 6$.

**51.** Use a tree diagram to show the possible arrangements for the college theater company if the city of Houston is added to its tour.

**52.** Use a tree diagram to describe visually the possible arrangements for the genders of three children within a family.

**53.** Use a tree diagram to describe visually the possible arrangements for the genders of four children within a family.

**54.** Complete the following table.

| $n$ | $_nP_n$ | $n!$ |
|---|---|---|
| 4 | | |
| 5 | | |
| 6 | | |
| 7 | | |

What do you observe about the values in the second and third columns? Explain.

## 8.6  Probability

### Objectives

- Determine the sample space of an experiment.
- Find the probability of an event.
- Find the probability of mutually exclusive events and independent events.
- Find the probability of the complement of an event.

## The Probability of an Event

In measuring the uncertainties in everyday life, we often use ambiguous terminology, such as "fairly certain" and "highly unlikely." In mathematics we attempt to assign a number to the likelihood of the occurrence of an event. We call this measurement the **probability** that the event will occur. For example, if we toss a fair coin, we say that the probability that it will land face up is $\frac{1}{2}$.

Any happening whose result is uncertain is called an **experiment.** The possible results of the experiment are **outcomes,** the set of all possible outcomes of an experiment is the **sample space** of the experiment, and any subcollection of a sample space is an **event.**

For instance, when a six-sided die is tossed, the sample space can be represented by the numbers from 1 through 6, each having *equally likely* outcomes.

To describe sample spaces in such a way that each outcome is equally likely, you must sometimes distinguish between various outcomes in ways that appear artificial. Example 1 illustrates such a situation.

**Example 1**  Finding the Sample Space

Find the sample spaces for the following.

**a.** One coin is tossed.    **b.** Two coins are tossed.    **c.** Three coins are tossed.

**Solution**

**a.** Because the coin will land either heads up (denoted by $H$) or tails up (denoted by $T$), the sample space is $S = \{H, T\}$.

**b.** Because either coin can land heads up or tails up, the possible outcomes are

$HH$ = heads up on both coins

$HT$ = heads up on first coin and tails up on second coin

$TH$ = tails up on first coin and heads up on second coin

$TT$ = tails up on both coins.

The sample space is $S = \{HH, HT, TH, TT\}$. This list distinguishes between the two cases $HT$ and $TH$, even though these two outcomes appear to be similar.

**c.** Following the notation of part (b), the sample space is

$S = \{HHH, HHT, HTH, HTT, THH, THT, TTH, TTT\}$.

To calculate the probability of an event, count the number of outcomes in the event and in the sample space. The *number of outcomes* in event $E$ is denoted by $n(E)$, and the number of outcomes in the sample space $S$ is denoted by $n(S)$. The probability that event $E$ will occur is given by $n(E)/n(S)$.

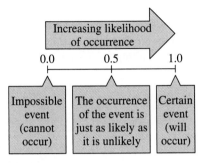

Figure 8.3

> ▶ **The Probability of an Event**
>
> If an event $E$ has $n(E)$ equally likely outcomes and its sample space $S$ has $n(S)$ equally likely outcomes, the **probability** of event $E$ is
>
> $$P(E) = \frac{n(E)}{n(S)}.$$

Because the number of outcomes in an event must be less than or equal to the number of outcomes in the sample space, the probability of an event must be a number between 0 and 1. That is, for any event $E$, it must be true that $0 \le P(E) \le 1$, as indicated in Figure 8.3.

> ▶ **Properties of the Probability of an Event**
>
> Let $E$ be an event that is a subset of a finite sample space $S$.
>
> **1.** $0 \le P(E) \le 1$
>
> **2.** If $P(E) = 0$, $E$ *cannot occur* and is called an **impossible event.**
>
> **3.** If $P(E) = 1$, $E$ *must occur* and is called a **certain event.**

**Example 2**   Finding the Probability of an Event

**a.** Two coins are tossed. What is the probability that both land heads up?

**b.** A card is drawn from a standard deck of playing cards. What is the probability that it is an ace?

**Solution**

**a.** Following the procedure in Example 1(b), let $E = \{HH\}$ and $S = \{HH, HT, TH, TT\}$. The probability of getting two heads is

$$P(E) = \frac{n(E)}{n(S)} = \frac{1}{4}.$$

**b.** Because there are 52 cards in a standard deck of playing cards and there are four aces (one in each suit), the probability of drawing an ace is

$$P(E) = \frac{n(E)}{n(S)} = \frac{4}{52} = \frac{1}{13}.$$

**Figure 8.4**

**Example 3**      Finding the Probability of an Event

Two six-sided dice are tossed. What is the probability that the total of the two dice is 7? (See Figure 8.4.)

**Solution**

Because there are six possible outcomes on each die, you can use the Fundamental Counting Principle to conclude that there are $6 \cdot 6$ or 36 different outcomes when two dice are tossed. To find the probability of rolling a total of 7, you must first count the number of ways this can occur.

| First Die | 1 | 2 | 3 | 4 | 5 | 6 |
|-----------|---|---|---|---|---|---|
| Second Die | 6 | 5 | 4 | 3 | 2 | 1 |

So, a total of 7 can be rolled in six ways, which means that the probability of rolling a 7 is

$$P(E) = \frac{n(E)}{n(S)} = \frac{6}{36} = \frac{1}{6}.$$

You could have written out each sample space in Examples 2 and 3 and simply counted the outcomes in the desired events. For larger sample spaces, however, you must make more use of the counting principles discussed in Section 8.5.

**Example 4**      Finding the Probability of an Event

Twelve-sided dice can be constructed (in the shape of regular dodecahedrons) so that each of the numbers from 1 to 6 appears twice on each die, as shown in Figure 8.5. Prove that these dice can be used in any game requiring ordinary six-sided dice without changing the probabilities of different outcomes.

**Solution**

For an ordinary six-sided die, each of the numbers 1, 2, 3, 4, 5, and 6 occurs only once, so the probability of any particular number coming up is

$$P(E) = \frac{n(E)}{n(S)} = \frac{1}{6}.$$

For one of the twelve-sided dice, each number occurs twice, so the probability of any particular number coming up is

$$P(E) = \frac{n(E)}{n(S)} = \frac{2}{12} = \frac{1}{6}.$$

**Figure 8.5**

*Although popular in the early 1800s, lotteries were banned in more and more states until, by 1894, no state allowed lotteries. In 1964, New Hampshire became the first state to reinstitute a state lottery. Today, lotteries are conducted by almost all states.*

**Example 5**   The Probability of Winning a Lottery

In a state lottery, a player chooses six numbers from 1 to 40. If these six numbers match the six numbers drawn by the lottery commission, the player wins (or shares) the top prize. What is the probability of winning the top prize?

### Solution

To find the number of elements in the sample space, use the formula for the number of combinations of 40 elements taken six at a time.

$$n(S) = {}_{40}C_6 = \frac{40 \cdot 39 \cdot 38 \cdot 37 \cdot 36 \cdot 35}{6 \cdot 5 \cdot 4 \cdot 3 \cdot 2 \cdot 1} = 3{,}838{,}380.$$

If a person buys only one ticket, the probability of winning is

$$P(E) = \frac{n(E)}{n(S)} = \frac{1}{3{,}838{,}380}.$$

**Example 6**   Random Selection

The total number of colleges and universities in the United States in 1996 is shown in Figure 8.6. One institution is selected at random. What is the probability that the institution is in one of the three southern regions?   (Source: U.S. National Center for Education Statistics)

### Solution

From the figure, the total number of colleges and universities is 3696. Because there are $607 + 265 + 298 = 1170$ colleges and universities in the three southern regions, the probability that the institution is in one of these regions is

$$P(E) = \frac{n(E)}{n(S)} = \frac{1170}{3696} \approx 0.317.$$

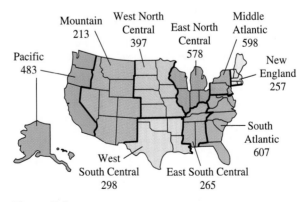

**Figure 8.6**

## Mutually Exclusive Events

Two events $A$ and $B$ (from the same sample space) are **mutually exclusive** if $A$ and $B$ have no outcomes in common. In the terminology of sets, the **intersection of $A$ and $B$** is the empty set and

$$P(A \cap B) = 0.$$

For instance, if two dice are tossed, the event $A$ of rolling a total of 6 and the event $B$ of rolling a total of 9 are mutually exclusive. To find the probability that one or the other of two mutually exclusive events will occur, you can *add* their individual probabilities.

---

### ▶ Probability of the Union of Two Events

If $A$ and $B$ are events in the same sample space, the probability of $A$ *or* $B$ occurring is given by

$$P(A \cup B) = P(A) + P(B) - P(A \cap B).$$

If $A$ and $B$ are mutually exclusive, then

$$P(A \cup B) = P(A) + P(B).$$

---

**Example 7** The Probability of a Union

One card is selected from a standard deck of 52 playing cards. What is the probability that the card is either a heart or a face card?

### Solution

Because the deck has 13 hearts, the probability of selecting a heart (event $A$) is $P(A) = \frac{13}{52}$. Similarly, because the deck has 12 face cards, the probability of selecting a face card (event $B$) is $P(B) = \frac{12}{52}$. Because three of the cards are hearts and face cards (see Figure 8.7), it follows that $P(A \cap B) = \frac{3}{52}$. Finally, applying the formula for the probability of the union of two events, you can conclude that the probability of selecting a heart or a face card is

$$P(A \cup B) = P(A) + P(B) - P(A \cap B)$$

$$= \frac{13}{52} + \frac{12}{52} - \frac{3}{52}$$

$$= \frac{22}{52}$$

$$\approx 0.423.$$

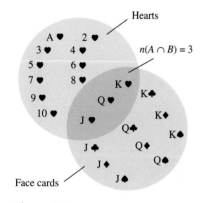

Hearts

$n(A \cap B) = 3$

Face cards

**Figure 8.7**

| Example 8 | Probability of Mutually Exclusive Events |

The personnel department of a company has compiled data showing the number of years of service for each employee. The results are shown in the table.

| Years of Service | Number of Employees |
|---|---|
| 0–4 | 157 |
| 5–9 | 89 |
| 10–14 | 74 |
| 15–19 | 63 |
| 20–24 | 42 |
| 25–29 | 38 |
| 30–34 | 37 |
| 35–39 | 21 |
| 40–44 | 8 |

If an employee is chosen at random, what is the probability that the employee has had nine or fewer years of service?

Solution

To begin, add the number of employees and find that the total is 529. Next, let event $A$ represent choosing an employee with 0 to 4 years of service and let event $B$ represent choosing an employee with 5 to 9 years of service. Then

$$P(A) = \frac{157}{529} \quad \text{and} \quad P(B) = \frac{89}{529}.$$

Because $A$ and $B$ have no outcomes in common, you can conclude that these two events are mutually exclusive and that

$$P(A \cup B) = P(A) + P(B) = \frac{157}{529} + \frac{89}{529}$$

$$= \frac{246}{529}$$

$$\approx 0.465.$$

So, the probability of choosing an employee who has nine or fewer years of service is about 0.465.

## Independent Events

Two events are **independent** if the occurrence of one has no effect on the occurrence of the other. To find the probability that two independent events will occur, *multiply* the probabilities of each. For instance, rolling a total of 12 with two six-sided dice has no effect on the outcome of future rolls of the dice.

---

▶ **Probability of Independent Events**

If $A$ and $B$ are independent events, the probability that both $A$ and $B$ will occur is

$$P(A \text{ and } B) = P(A) \cdot P(B).$$

---

| **Example 9** | Probability of Independent Events |  |

A random number generator on a computer selects three integers from 1 to 20. What is the probability that all three numbers are less than or equal to 5?

**Solution**

The probability of selecting a number from 1 to 5 is

$$P(A) = \frac{5}{20} = \frac{1}{4}.$$

So, the probability that all three numbers are less than or equal to 5 is

$$P(A) \cdot P(A) \cdot P(A) = \left(\frac{1}{4}\right)\left(\frac{1}{4}\right)\left(\frac{1}{4}\right)$$

$$= \frac{1}{64}.$$

| **Example 10** | Probability of Independent Events |  |

In 1996, 57.4% of the population of the United States were 30 years old or older. Suppose that in a survey, 10 people were chosen at random from the population. What is the probability that all 10 were 30 years old or older?

**Solution**

Let $A$ represent choosing a person who was 30 years old or older. Because the probability of choosing a person who was 30 years old or older was 0.574, you can conclude that the probability that all 10 people were 30 years old or older is

$$[P(A)]^{10} = (0.574)^{10}$$

$$\approx 0.0039.$$

## The Complement of an Event

The **complement of an event** $A$ is the collection of all outcomes in the sample space that are *not* in $A$. The complement of event $A$ is denoted by $A'$. Because $P(A \text{ or } A') = 1$ and because $A$ and $A'$ are mutually exclusive, it follows that $P(A) + P(A') = 1$. So, the probability of $A'$ is given by

$$P(A') = 1 - P(A).$$

For instance, if the probability of *winning* a certain game is

$$P(A) = \frac{1}{4}$$

the probability of *losing* the game is

$$P(A') = 1 - \frac{1}{4} = \frac{3}{4}.$$

> ### ▶ Probability of a Complement
>
> Let $A$ be an event and let $A'$ be its complement. If the probability of $A$ is $P(A)$, then the probability of the complement is
>
> $$P(A') = 1 - P(A).$$

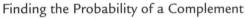

**Example 11**    Finding the Probability of a Complement

A manufacturer has determined that a certain machine averages one faulty unit for every 1000 it produces. What is the probability that an order of 200 units will have one or more faulty units?

### Solution

To solve this problem as stated, you would need to find the probability of having exactly one faulty unit, exactly two faulty units, exactly three faulty units, and so on. However, using complements, you can simply find the probability that all units are perfect and then subtract this value from 1. Because the probability that any given unit is perfect is $999/1000$, the probability that all 200 units are perfect is

$$P(A) = \left(\frac{999}{1000}\right)^{200} \approx 0.8186.$$

So, the probability that at least one unit is faulty is

$$P(A') = 1 - P(A) \approx 0.1814.$$

## Discussing the Concept   An Experiment in Probability

In this section you have been finding probabilities from a *theoretical* point of view. Another way to find probabilities is from an *experimental* point of view. For instance, suppose you wanted to find the probability of obtaining a given total when two six-sided dice are tossed. The following BASIC program simulates the tossing of a pair of dice 5000 times.

*BASIC Program:*

```
10      RANDOMIZE
20      DIM TALLY (12)
30      FOR I=1 TO 5000
40      ROLLONE=INT (6*RND)+1
50      ROLLTWO=INT (6*RND)+1
60      DICETOTAL=ROLLONE+ROLLTWO
70      TALLY(DICETOTAL)=TALLY(DICETOTAL)+1
80      NEXT
90      FOR I=2 TO 12
100     PRINT "TOTAL OF", I,"OCCURRED", TALLY(I), "TIMES"
110     NEXT
120     END
```

When we ran this program, the printout was as follows.

```
TOTAL OF    2   OCCURRED    139   TIMES
TOTAL OF    3   OCCURRED    264   TIMES
TOTAL OF    4   OCCURRED    443   TIMES
TOTAL OF    5   OCCURRED    553   TIMES
TOTAL OF    6   OCCURRED    691   TIMES
TOTAL OF    7   OCCURRED    810   TIMES
TOTAL OF    8   OCCURRED    715   TIMES
TOTAL OF    9   OCCURRED    557   TIMES
TOTAL OF   10   OCCURRED    398   TIMES
TOTAL OF   11   OCCURRED    270   TIMES
TOTAL OF   12   OCCURRED    160   TIMES
```

In Example 3 you found that the theoretical probability of tossing a total of 7 on a pair of dice is $\frac{1}{6} \approx 0.167$. From this experiment, calculate the experimental probability of tossing a total of 7 on a pair of dice. If you increase the number of trials from 5000 to 10,000, do you expect the experimental result to get closer to the theoretical result? Explain.

# Warm Up

The following warm-up exercises involve skills that were covered in earlier sections. You will use these skills in the exercise set for this section.

In Exercises 1–8, evaluate the expression.

1. $\frac{1}{4} + \frac{5}{8} - \frac{5}{16}$

2. $\frac{4}{15} + \frac{3}{5} - \frac{1}{3}$

3. $\frac{5 \cdot 4}{5!}$

4. $\frac{5!22!}{27!}$

5. $\frac{4!}{8!12!}$

6. $\frac{9 \cdot 8 \cdot 7 \cdot 6 \cdot 5}{9!}$

7. $\frac{{}_5C_3}{{}_{10}C_3}$

8. $\frac{{}_{10}C_2 \cdot {}_{10}C_2}{{}_{20}C_4}$

In Exercises 9 and 10, evaluate the expression. (Round your answer to three decimal places.)

9. $\left(\frac{99}{100}\right)^{100}$

10. $1 - \left(\frac{89}{100}\right)^{50}$

## 8.6    Exercises

*Heads or Tails?*    In Exercises 1–4, a coin is tossed three times. Find the probability of the event.

1. Getting exactly one head

2. Getting a tail on the second toss

3. Getting at least one tail

4. Getting at least two heads

*Tossing a Die*    In Exercises 5–10, two six-sided dice are tossed. Find the probability of the event.

5. The sum is 5.

6. The sum is less than 10.

7. The sum is at least 7.

8. The total is 2, 3, or 8.

9. The sum is odd and no more than 7.

10. The sum is odd or a prime number.

*Drawing a Card*    In Exercises 11–14, a card is selected from a standard deck of 52 cards. Find the probability of the event.

11. Getting a face card

12. Not getting a face card

13. Getting a black card that is not a face card

14. Getting a card that is a 6 or less (aces are low)

*Drawing Marbles*    In Exercises 15–18, two marbles are drawn (without replacement) from a bag containing one green, two yellow, and three red marbles. Find the probability of the event.

15. Drawing exactly one red marble

16. Drawing two green marbles

17. Drawing neither of the yellow marbles

18. Drawing marbles of different colors

In Exercises 19 and 20, you are given the probability that an event *will* happen. Find the probability that the event *will not* happen.

19. $p = 0.4$

20. $p = 0.72$

In Exercises 21 and 22, you are given the probability that an event *will not* happen. Find the probability that the event *will* happen.

21. $p = 0.2$

22. $p = 0.68$

23. Use information from news sources, the Internet, or an original experiment to write an example of two mutually exclusive events.

24. Use information from news sources, the Internet, or an original experiment to write an example of two independent events.

25. *Winning an Election* Three people have been nominated for president of a college class. From a small poll, it is estimated that the probability of Jane winning the election is 0.46, and the probability of Larry winning the election is 0.32. What is the probability of the third candidate winning the election?

26. *Winning an Election* Taylor, Moore, and Jenkins are candidates for a public office. It is estimated that Moore and Jenkins have about the same probability of winning, and Taylor is believed to be twice as likely to win as either of the others. Find each candidate's probability of winning the election.

27. *College Bound* In a high school graduating class of 198 students, 43 are on the honor roll. Of these, 37 are going on to college, and of the other 155 students, 102 are going on to college. If a student is selected at random from the class, what are the probabilities that the person chosen is

    (a) going to college?

    (b) not going to college?

    (c) on the honor roll, but not going to college?

28. *Alumni Association* The alumni office of a college is sending a survey to selected members of the class of 1995. Of the 1254 people who graduated that year, 672 are women, and of those, 124 went to graduate school. Of the 582 male graduates, 198 went to graduate school. If an alumnus is selected at random, what are the probabilities that the person is

    (a) female?    (b) male?

    (c) female and did not attend graduate school?

29. *Random Number Generator* Two integers (from 1 to 30, inclusive) are chosen by a random number generator on a computer. What are the probabilities that (a) the numbers are both even, (b) one number is even and one is odd, (c) both numbers are less than 10, and (d) the same number is chosen twice?

30. *Random Number Generator* Two integers (from 1 to 40, inclusive) are chosen by a random number generator on a computer. What are the probabilities that (a) the numbers are both even, (b) one number is even and one is odd, (c) both numbers are less than 30, and (d) the same number is chosen twice?

31. *Preparing for a Test* An instructor gives her class a list of eight study problems, from which she will select five to be answered on an exam. If a given student knows how to solve six of the problems, find the probability that the student will be able to answer all five questions on the exam.

32. *Preparing for a Test* An instructor gives his class a list of 20 study problems, from which he will select 10 to be answered on an exam. If a given student knows how to solve 15 of the problems, find the probability that the student will be able to answer all 10 questions on the exam.

33. *Drawing Cards from a Deck* Two cards are selected at random from an ordinary deck of 52 playing cards. Find the probabilities that two aces are selected under the following conditions.

    (a) The cards are drawn in sequence, with the first card being replaced and the deck reshuffled prior to the second drawing.

    (b) The two cards are drawn consecutively, without replacement.

34. *Drawing Cards from a Deck* Two cards are selected at random from an ordinary deck of 52 playing cards. Find the probabilities that two hearts are selected under the following conditions.

    (a) The first card being replaced and the deck reshuffled prior to the second drawing.

    (b) The two cards are drawn, without replacement.

35. *Game Show* On a game show, you are given five digits to arrange in the proper order to represent the price of a car. If you are correct, you win the car. Find the probabilities of winning if

    (a) You must guess the position of each digit.

    (b) You know the first digit, but must guess the remaining four.

    (c) You know the first and last digits, but must guess the remaining three.

36. *Game Show* On a game show, you are given six digits to arrange in the proper order to represent the price of a house. If you are correct, you win the house. Find the probabilities of winning if

    (a) You guess the position of each digit.

    (b) You know the first digit, but must guess the remaining five.

    (c) You know the first and last digits, but must guess the remaining four.

37. *Letter Mix-Up* Five letters and envelopes are addressed to five different people. If the letters are randomly inserted into the envelopes, what are the probabilities that (a) exactly one will be inserted in the correct envelope and (b) at least one will be inserted in the correct envelope?

**38.** *Payroll Mix-Up*    Three paychecks and envelopes are addressed to three different people. If the paychecks get mixed up and are randomly inserted into the envelopes, what are the probabilities that (a) exactly one will be inserted in the correct envelope and (b) at least one will be inserted in the correct envelope?

**39.** *Poker Hand*    Five cards are drawn from a standard deck of 52 cards. What is the probability of getting a full house? (See Exercise 39 in Section 8.5.)

**40.** *Poker Hand*    Five cards are drawn from a standard deck of 52 cards. What is the probability of getting a straight flush? (See Exercise 40 in Section 8.5.)

**41.** *Defective Units*    A shipment of 1000 compact disc players contains four defective units. A retail outlet has ordered 20 units. (a) What is the probability that all 20 units are good? (b) What is the probability that at least one unit is defective?

**42.** *Defective Units*    A shipment of 12 stereos contains three defective units. Four of the units are shipped to a retail store. What are the probabilities that (a) all four units are good, (b) exactly two units are good, and (c) at least two units are good?

**43.** *Backup System*    A space vehicle has an independent backup system for one of its communication networks. The probability that either system will function satisfactorily for the duration of a flight is 0.985. What are the probabilities that during a given flight (a) both systems function satisfactorily, (b) at least one system functions satisfactorily, and (c) both systems fail?

**44.** *Backup Vehicle*    A fire company keeps two rescue vehicles to serve the community. Because of the demand on the company's time and the chance of mechanical failure, the probability that a specific vehicle is available when needed is 90%. If the availability of one vehicle is *independent* of the other, find the probabilities that (a) both vehicles are available at a given time, (b) neither vehicle is available at a given time, and (c) at least one vehicle is available at a given time.

**45.** *Making a Sale*    A sales representative makes a sale at approximately one-third of the offices she calls on. If, on a given day, she goes to four offices, what are the probabilities that she will make a sale at (a) all four offices, (b) none of the offices, and (c) at least one office?

**46.** *Making a Sale*    A sales representative makes a sale at approximately one-fourth of the businesses he calls on. If, on a given day, he goes to five businesses, what

are the probabilities that he will make a sale at (a) all five businesses, (b) none of the businesses, and (c) at least one of the businesses?

**47.** *A Boy or a Girl?*    Assume that the probability of the birth of a child of a particular gender is 50%. In a family with six children, what are the probabilities that

(a) all six children are girls?

(b) all six children are of the same gender?

(c) there is at least one girl?

**48.** *A Boy or a Girl?*    Assume that the probability of the birth of a child of a particular gender is 50%. In a family with four children, what are the probabilities that

(a) all four children are boys?

(b) all four children are of the same gender?

(c) there is at least one boy?

*Probability*    In Exercises 49–52, consider $n$ independent trials of an experiment in which each trial has two possible outcomes, called success and failure. The probability of a success on each trial is $p$, and the probability of a failure is $q = 1 - p$. In this context, the term $_nC_k\, p^k\, q^{n-k}$ in the expansion of $(p + q)^n$ gives the probability of $k$ successes in the $n$ trials of the experiment.

**49.** A fair coin is tossed eight times. To find the probability of obtaining five heads, evaluate the term

$$_8C_5\left(\frac{1}{2}\right)^5\left(\frac{1}{2}\right)^3$$

in the expansion of $\left(\frac{1}{2} + \frac{1}{2}\right)^8$.

**50.** The probability of a baseball player getting a hit during any given time at bat is $\frac{1}{5}$. To find the probability that the player gets four hits during the next 10 times at bat, evaluate the term

$$_{10}C_4\left(\frac{1}{5}\right)^4\left(\frac{4}{5}\right)^6$$

in the expansion of $\left(\frac{1}{5} + \frac{4}{5}\right)^{10}$.

**51.** The probability of a sales representative making a sale with any one customer is $\frac{1}{4}$. The sales representative makes 10 contacts a day. To find the probability of making four sales, evaluate the term

$$_{10}C_4\left(\frac{1}{4}\right)^4\left(\frac{3}{4}\right)^6$$

in the expansion of $\left(\frac{1}{4} + \frac{3}{4}\right)^{10}$.

**52.** To find the probability that the baseball player in Exercise 50 makes five hits during the next 10 times at bat if the probability of a hit is $\frac{1}{3}$, evaluate the term

$$_{10}C_5\left(\frac{1}{3}\right)^5\left(\frac{2}{3}\right)^5$$

in the expansion of $\left(\frac{1}{3} + \frac{2}{3}\right)^{10}$.

**53.** *Is That Cash or Charge?*   According to a survey by *USA Today*, the methods used by Christmas shoppers to pay for gifts are as shown in the pie graph. Suppose two Christmas shoppers are chosen at random. What is the probability that both shoppers paid for their gifts in cash only?

How Shoppers Pay for Gifts

**54.** *Watching the Ads*   According to a survey by Video Storyboard Tests, Inc., television viewers pay varying amounts of attention to television advertisements as compared with the television programs (see figure). Suppose two television viewers are chosen at random. What is the probability that both viewers pay no attention to the television ads they see?

Watching the Ads

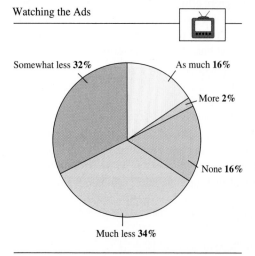

**55.** *Number of Telephones*   According to a survey by Maritz Ameripoll, the numbers of telephones in American households are as shown in the figure. Suppose three households are chosen at random. What is the probability that all three households have at least four telephones?

Telephones in the Home

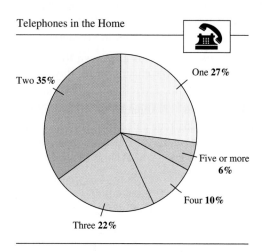

**56.** *Flexible Work Hours*   In a survey by Robert Hall International, people were asked if they would prefer to work flexible hours—even if it meant slower career advancement—so that they could spend more time with their families. The results of the survey are shown in the figure. Suppose three people are chosen at random. What is the probability that all three people prefer flexible work hours?

Flexible Work Hours

## 8.7 Mathematical Induction

### Objectives

- Use mathematical induction to prove a formula.
- Find the sums of powers of integers.
- Find a formula for a finite sum.
- Use finite differences to find a quadratic model.

## Introduction

In this section you will study a form of mathematical proof called **mathematical induction.** It is important that you clearly see the logical need for it, so let's look at the following problem.

$$S_1 = 1 = 1^2$$

$$S_2 = 1 + 3 = 2^2$$

$$S_3 = 1 + 3 + 5 = 3^2$$

$$S_4 = 1 + 3 + 5 + 7 = 4^2$$

$$S_5 = 1 + 3 + 5 + 7 + 9 = 5^2$$

Judging from the pattern formed by these first five sums, it appears that the sum of the first $n$ integers is

$$S_n = 1 + 3 + 5 + 7 + 9 + \cdots + (2n - 1) = n^2.$$

Although this particular formula *is* valid, it is important for you to see that recognizing a pattern and then simply *jumping to the conclusion* that the pattern must be true for all values of $n$ is *not* a logically valid method of proof. There are many examples in which a pattern appears to be developing for small values of $n$ and then at some point the pattern fails. One of the most famous cases of this was the conjecture by the French mathematician Pierre de Fermat (1601–1665), who speculated that all numbers of the form

$$F_n = 2^{2^n} + 1, \quad n = 0, 1, 2, \ldots$$

are prime. For $n = 0, 1, 2, 3,$ and 4, the conjecture is true.

$$F_0 = 3, F_1 = 5, F_2 = 17, F_3 = 257, F_4 = 65{,}537$$

The size of the next Fermat number ($F_5 = 4{,}294{,}967{,}297$) is so great that it was difficult for Fermat to determine whether it was prime or not. However, another well-known mathematician, Leonhard Euler (1707–1783), later found the factorization

$$F_5 = 4{,}294{,}967{,}297 = 641(6{,}700{,}417)$$

which proved that $F_5$ is not prime and therefore Fermat's conjecture was false.

Just because a rule, pattern, or formula seems to work for several values of $n$, you cannot simply decide that it is valid for all values of $n$ without going through a *legitimate proof.*

> ▶ **The Principle of Mathematical Induction**
>
> Let $P_n$ be a statement involving the positive integer $n$. If
>
> **1.** $P_1$ is true, and
>
> **2.** the truth of $P_k$ implies the truth of $P_{k+1}$ for every positive $k$,
>
> then $P_n$ must be true for all positive integers $n$.

It is important to recognize that both parts of the Principle of Mathematical Induction are necessary. To apply the Principle of Mathematical Induction, you need to be able to determine the statement $P_{k+1}$ for a given statement $P_k$.

**Example 1**    A Preliminary Example

Find $P_{k+1}$ for the following.

**a.** $P_k: S_k = \dfrac{k^2(k+1)^2}{4}$

**b.** $P_k: S_k = 1 + 5 + 9 + \cdots + [4(k-1) - 3] + (4k - 3)$

**c.** $P_k: 3^k \geq 2k + 1$

Solution

**a.** $P_{k+1}: S_{k+1} = \dfrac{(k+1)^2(k+1+1)^2}{4}$        Replace $k$ by $k + 1$.

$\qquad\qquad = \dfrac{(k+1)^2(k+2)^2}{4}.$        Simplify.

**b.** $P_{k+1}: S_{k+1} = 1 + 5 + 9 + \cdots + \{4[(k+1) - 1] - 3\} + [4(k+1) - 3]$

$\qquad\qquad = 1 + 5 + 9 + \cdots + (4k - 3) + (4k + 1).$

**c.** $P_{k+1}: 3^{k+1} \geq 2(k+1) + 1$

$\qquad\qquad 3^{k+1} \geq 2k + 3.$

**Figure 8.8**

A well-known illustration used to explain why the Principle of Mathematical Induction works is the unending line of dominoes shown in Figure 8.8. If the line actually contains infinitely many dominoes, it is clear that you could not knock the entire line down by knocking down only *one domino* at a time. However, suppose it were true that each domino would knock down the next one as it fell. Then you could knock them all down simply by pushing the first one and starting a chain reaction. Mathematical induction works in the same way. If the truth of $P_k$ implies the truth of $P_{k+1}$ and if $P_1$ is true, the chain reaction proceeds as follows: $P_1$ implies $P_2$, $P_2$ implies $P_3$, $P_3$ implies $P_4$, and so on.

**Example 2**    Using Mathematical Induction

Use mathematical induction to prove the following formula.

$$S_n = 1 + 3 + 5 + 7 + \cdots + (2n - 1)$$
$$= n^2$$

**Solution**

Mathematical induction consists of two distinct parts. First, you must show that the formula is true when $n = 1$.

**1.** When $n = 1$, the formula is valid, because

$$S_1 = 1 = 1^2.$$

The second part of mathematical induction has two steps. The first step is to assume that the formula is valid for *some* integer $k$. The second step is to use this assumption to prove that the formula is valid for the next integer, $k + 1$.

**2.** Assuming that the formula

$$S_k = 1 + 3 + 5 + 7 + \cdots + (2k - 1)$$
$$= k^2$$

is true, you must show that the formula $S_{k+1} = (k + 1)^2$ is true.

$$\begin{aligned} S_{k+1} &= 1 + 3 + 5 + 7 + \cdots + (2k - 1) + [2(k + 1) - 1] \\ &= [1 + 3 + 5 + 7 + \cdots + (2k - 1)] + (2k + 2 - 1) \\ &= S_k + (2k + 1) \qquad \text{\small Group terms to form } S_k. \\ &= k^2 + 2k + 1 \qquad \text{\small Replace } S_k \text{ by } k^2. \\ &= (k + 1)^2 \end{aligned}$$

Combining the results of parts (1) and (2), you can conclude by mathematical induction that the formula is valid for *all* positive integer values of $n$.

---

## Study Tip

When using mathematical induction to prove a *summation* formula (such as the one in Example 2), it is helpful to think of $S_{k+1}$ as $S_{k+1} = S_k + a_{k+1}$, where $a_{k+1}$ is the $(k + 1)$th term of the original sum.

---

It occasionally happens that a statement involving natural numbers is not true for the first $k - 1$ positive integers but is true for all values of $n \geq k$. In these instances, you use a slight variation of the Principle of Mathematical Induction in which you verify $P_k$ rather than $P_1$. This variation is called the **extended principle of mathematical induction.** To see the validity of this variation, note from Figure 8.8 that all but the first $k - 1$ dominoes can be knocked down by knocking over the $k$th domino. This suggests that you can prove a statement $P_n$ to be true for $n \geq k$ by showing that $P_k$ is true and that $P_k$ implies $P_{k+1}$. In Exercises 35–38 of this section you are asked to apply this extension of mathematical induction.

> **Example 3** Using Mathematical Induction

Use mathematical induction to prove the following formula.

$$S_n = 1^2 + 2^2 + 3^2 + 4^2 + \cdots + n^2$$

$$= \frac{n(n + 1)(2n + 1)}{6}$$

**Solution**

1. When $n = 1$, the formula is valid, because

$$S_1 = 1^2 = \frac{1(2)(3)}{6}.$$

2. Assuming that

$$S_k = 1^2 + 2^2 + 3^2 + 4^2 + \cdots + k^2$$

$$= \frac{k(k + 1)(2k + 1)}{6}$$

you must show that

$$S_{k+1} = \frac{(k + 1)(k + 2)(2k + 3)}{6}.$$

To do this, write the following.

$$S_{k+1} = S_k + a_{k+1}$$

$$= (1^2 + 2^2 + 3^2 + 4^2 + \cdots + k^2) + (k + 1)^2$$

$$= \frac{k(k + 1)(2k + 1)}{6} + (k + 1)^2$$

$$= \frac{k(k + 1)(2k + 1) + 6(k + 1)^2}{6}$$

$$= \frac{(k + 1)[k(2k + 1) + 6(k + 1)]}{6}$$

$$= \frac{(k + 1)(2k^2 + 7k + 6)}{6}$$

$$= \frac{(k + 1)(k + 2)(2k + 3)}{6}$$

Combining the results of parts (1) and (2), you can conclude by mathematical induction that the formula is valid for *all* $n \geq 1$.

When proving a formula with mathematical induction, the only statement that you *need* to verify is $P_1$. As a check, however, it is good to try verifying other statements. For instance, in Example 3, try verifying $S_2$ and $S_3$.

# Sums of Powers of Integers

The formula in Example 3 is one of a collection of useful summation formulas. We summarize this and other formulas dealing with the sums of various powers of the first $n$ positive integers as follows.

---

▶ **Sums of Powers of Integers**

**1.** $1 + 2 + 3 + 4 + \cdots + n = \dfrac{n(n + 1)}{2}$

**2.** $1^2 + 2^2 + 3^2 + 4^2 + \cdots + n^2 = \dfrac{n(n + 1)(2n + 1)}{6}$

**3.** $1^3 + 2^3 + 3^3 + 4^3 + \cdots + n^3 = \dfrac{n^2(n + 1)^2}{4}$

**4.** $1^4 + 2^4 + 3^4 + 4^4 + \cdots + n^4 = \dfrac{n(n + 1)(2n + 1)(3n^2 + 3n - 1)}{30}$

**5.** $1^5 + 2^5 + 3^5 + 4^5 + \cdots + n^5 = \dfrac{n^2(n + 1)^2(2n^2 + 2n - 1)}{12}$

---

Each of these formulas for sums can be proven by mathematical induction.

**Example 4**    Finding a Sum of Powers of Integers

Find $\displaystyle\sum_{n=1}^{7} n^3 = 1^3 + 2^3 + 3^3 + 4^3 + 5^3 + 6^3 + 7^3$.

## Solution

Using the formula for the sum of the cubes of the first $n$ positive integers, you obtain the following.

$$\sum_{n=1}^{7} n^3 = 1^3 + 2^3 + 3^3 + 4^3 + 5^3 + 6^3 + 7^3$$

$$= \frac{7^2(7 + 1)^2}{4}$$

$$= \frac{49(64)}{4}$$

$$= 784$$

Check this sum by adding the numbers 1, 8, 27, 64, 125, 216, and 343.

| Example 5 | Proving an Inequality by Mathematical Induction |

Prove that $n < 2^n$ for all positive integers $n$.

**Solution**

**1.** For $n = 1$, the formula is true, because

$$1 < 2^1.$$

**2.** Assuming that

$$k < 2^k$$

you need to show that $k + 1 < 2^{k+1}$. For $n = k$, you have

$$2^{k+1} = 2(2^k) > 2(k) = 2k. \qquad \text{By assumption}$$

Because $2k = k + k > k + 1$ for all $k > 1$, it follows that

$$2^{k+1} > 2k > k + 1$$

or

$$k + 1 < 2^{k+1}.$$

So, $n < 2^n$ for all integers $n \geq 1$.

---

## Pattern Recognition

Although choosing a formula on the basis of a few observations does *not* guarantee the validity of the formula, pattern recognition *is* important. Once you have a pattern that you think works, you can try using mathematical induction to prove your formula.

---

▶ **Finding a Formula for the *n*th Term of a Sequence**

To find a formula for the *n*th term of a sequence, consider the following guidelines.

**1.** Calculate the first several terms of the sequence. It is often a good idea to write the terms in both simplified and factored forms.

**2.** Try to find a recognizable pattern for the terms and write a formula for the *n*th term of the sequence. This is your *hypothesis* or *conjecture*. You might try computing one or two more terms in the sequence to test your hypothesis.

**3.** Use mathematical induction to prove your hypothesis.

Example 6    Finding a Formula for a Finite Sum

Find a formula for the following finite sum.

$$\frac{1}{1 \cdot 2} + \frac{1}{2 \cdot 3} + \frac{1}{3 \cdot 4} + \frac{1}{4 \cdot 5} + \cdots + \frac{1}{n(n + 1)}$$

Solution

Begin by writing out the first few sums.

$$S_1 = \frac{1}{1 \cdot 2} = \frac{1}{2} = \frac{1}{1 + 1}$$

$$S_2 = \frac{1}{1 \cdot 2} + \frac{1}{2 \cdot 3} = \frac{4}{6} = \frac{2}{3} = \frac{2}{2 + 1}$$

$$S_3 = \frac{1}{1 \cdot 2} + \frac{1}{2 \cdot 3} + \frac{1}{3 \cdot 4} = \frac{9}{12} = \frac{3}{4} = \frac{3}{3 + 1}$$

$$S_4 = \frac{1}{1 \cdot 2} + \frac{1}{2 \cdot 3} + \frac{1}{3 \cdot 4} + \frac{1}{4 \cdot 5} = \frac{48}{60} = \frac{4}{5} = \frac{4}{4 + 1}$$

From this sequence, it appears that the formula for the $k$th sum is

$$S_k = \frac{1}{1 \cdot 2} + \frac{1}{2 \cdot 3} + \frac{1}{3 \cdot 4} + \frac{1}{4 \cdot 5} + \cdots + \frac{1}{k(k + 1)} = \frac{k}{k + 1}.$$

To prove the validity of this hypothesis, use mathematical induction, as follows. Note that you have already verified the formula for $n = 1$, so you can begin by assuming that the formula is valid for $n = k$ and trying to show that it is valid for $n = k + 1$.

$$S_{k+1} = \left[ \frac{1}{1 \cdot 2} + \frac{1}{2 \cdot 3} + \frac{1}{3 \cdot 4} + \frac{1}{4 \cdot 5} + \cdots + \frac{1}{k(k + 1)} \right] + \frac{1}{(k + 1)(k + 2)}$$

$$= \frac{k}{k + 1} + \frac{1}{(k + 1)(k + 2)}$$

$$= \frac{k(k + 2) + 1}{(k + 1)(k + 2)}$$

$$= \frac{k^2 + 2k + 1}{(k + 1)(k + 2)}$$

$$= \frac{(k + 1)^2}{(k + 1)(k + 2)}$$

$$= \frac{k + 1}{k + 2}$$

So, the hypothesis is valid.

## Finite Differences

The **first differences** of a sequence are found by subtracting consecutive terms. The **second differences** are found by subtracting consecutive first differences. The first and second differences of the sequence 3, 5, 8, 12, 17, 23, . . . are as follows.

For this sequence, the second differences are all the same. When this happens, the sequence has a perfect quadratic model. If the first differences are all the same, the sequence has a linear model. That is, it is arithmetic.

### Example 7  Finding a Quadratic Model

Find the quadratic model for the sequence

$$3, 5, 8, 12, 17, 23, \ldots .$$

**Solution**

You know that the model has the form

$$a_n = an^2 + bn + c.$$

By substituting 1, 2, and 3 for $n$, you can obtain a system of three linear equations in three variables.

$a_1 = a(1)^2 + b(1) + c = 3$      Substitute 1 for $n$.

$a_2 = a(2)^2 + b(2) + c = 5$      Substitute 2 for $n$.

$a_3 = a(3)^2 + b(3) + c = 8$      Substitute 3 for $n$.

You now have a system of three equations in $a$, $b$, and $c$.

$a + \ b + c = 3$      Equation 1

$4a + 2b + c = 5$      Equation 2

$9a + 3b + c = 8$      Equation 3

Using the techniques discussed in Chapter 6, you can find the solution to be $a = \frac{1}{2}$, $b = \frac{1}{2}$, and $c = 2$. Thus, the quadratic model is

$$a_n = \frac{1}{2}n^2 + \frac{1}{2}n + 2.$$

Try checking the values of $a_1$, $a_2$, and $a_3$.

| Discussing the Concept | Mathematical Modeling |
| --- | --- |

Use finite differences to determine whether each sequence in the table can be represented by either a linear or quadratic model.

| $n$ | 1 | 2 | 3 | 4 | 5 | 6 |
| --- | --- | --- | --- | --- | --- | --- |
| $a_n$ | 12 | 14 | 22 | 36 | 56 | 82 |
| $b_n$ | $-23.5$ | $-20.0$ | $-16.5$ | $-13.0$ | $-9.5$ | $-6.0$ |
| $c_n$ | 7 | 13 | 20 | 26 | 33 | 39 |
| $d_n$ | 0.8 | 4.2 | 9.2 | 15.8 | 24.0 | 33.8 |

If the sequence can be represented by a linear or quadratic model, find the appropriate model, and predict the value of the tenth term. Discuss how the finite differences technique complements other modeling techniques you have learned thus far.

# Warm Up

The following warm-up exercises involve skills that were covered in previous courses and earlier sections. You will use these skills in the exercise set for this section.

In Exercises 1–4, find the sum.

**1.** $\displaystyle\sum_{k=3}^{6} (2k - 3)$

**2.** $\displaystyle\sum_{j=1}^{5} (j^2 - j)$

**3.** $\displaystyle\sum_{k=2}^{5} \frac{1}{k}$

**4.** $\displaystyle\sum_{i=1}^{2} \left(1 + \frac{1}{i}\right)$

In Exercises 5–10, simplify the expression.

**5.** $\dfrac{2(k + 1) + 3}{5}$

**6.** $\dfrac{3(k + 1) - 2}{6}$

**7.** $2 \cdot 2^{2(k+1)}$

**8.** $\dfrac{3^{2k}}{3^{2(k+1)}}$

**9.** $\dfrac{k + 1}{k^2 + k}$

**10.** $\dfrac{\sqrt{32}}{\sqrt{50}}$

## 8.7   Exercises

In Exercises 1–4, find $P_{k+1}$ for the given $P_k$.

**1.** $P_k = \dfrac{5}{k(k + 1)}$

**2.** $P_k = \dfrac{1}{(k + 1)(k + 3)}$

**3.** $P_k = \dfrac{k^2(k + 1)^2}{4}$

**4.** $P_k = \dfrac{k}{2}(3k - 1)$

In Exercises 5–18, use mathematical induction to prove the formula for every positive integer $n$.

**5.** $2 + 4 + 6 + 8 + \cdots + 2n = n(n + 1)$

**6.** $3 + 7 + 11 + 15 + \cdots + (4n - 1) = n(2n + 1)$

**7.** $2 + 7 + 12 + 17 + \cdots + (5n - 3) = \dfrac{n}{2}(5n - 1)$

**8.** $1 + 4 + 7 + 10 + \cdots + (3n - 2) = \dfrac{n}{2}(3n - 1)$

**9.** $1 + 2 + 2^2 + 2^3 + \cdots + 2^{n-1} = 2^n - 1$

**10.** $2(1 + 3 + 3^2 + 3^3 + \cdots + 3^{n-1}) = 3^n - 1$

**11.** $1 + 2 + 3 + 4 + \cdots + n = \dfrac{n(n + 1)}{2}$

**12.** $1^2 + 2^2 + 3^2 + 4^2 + \cdots + n^2 = \dfrac{n(n + 1)(2n + 1)}{6}$

**13.** $1^3 + 2^3 + 3^3 + 4^3 + \cdots + n^3 = \dfrac{n^2(n + 1)^2}{4}$

**14.** $\left(1 + \dfrac{1}{1}\right)\left(1 + \dfrac{1}{2}\right)\left(1 + \dfrac{1}{3}\right) \cdots \left(1 + \dfrac{1}{n}\right) = n + 1$

**15.** $\displaystyle\sum_{i=1}^{n} i^5 = \dfrac{n^2(n + 1)^2(2n^2 + 2n - 1)}{12}$

**16.** $\displaystyle\sum_{i=1}^{n} i^4 = \dfrac{n(n + 1)(2n + 1)(3n^2 + 3n - 1)}{30}$

**17.** $\displaystyle\sum_{i=1}^{n} i(i + 1) = \dfrac{n(n + 1)(n + 2)}{3}$

**18.** $\displaystyle\sum_{i=1}^{n} \dfrac{1}{(2i - 1)(2i + 1)} = \dfrac{n}{2n + 1}$

In Exercises 19–28, find the sum using the formulas for the sums of powers of integers.

**19.** $\displaystyle\sum_{n=1}^{20} n$

**20.** $\displaystyle\sum_{n=1}^{50} n$

**21.** $\displaystyle\sum_{n=1}^{6} n^2$

**22.** $\displaystyle\sum_{n=1}^{10} n^3$

**23.** $\displaystyle\sum_{n=1}^{5} n^4$

**24.** $\displaystyle\sum_{n=1}^{8} n^5$

**25.** $\displaystyle\sum_{n=1}^{6} (n^2 - n)$

**26.** $\displaystyle\sum_{n=1}^{10} (n^3 - n^2)$

**27.** $\displaystyle\sum_{i=1}^{6} (6i - 8i^3)$

**28.** $\displaystyle\sum_{j=1}^{4} \left(2 + \tfrac{5}{2}j - \tfrac{3}{2}j^2\right)$

In Exercises 29–34, find a formula for the sum of the first $n$ terms of the sequence.

**29.** $1, 5, 9, 13, \ldots$

**30.** $25, 22, 19, 16, \ldots$

**31.** $1, \frac{9}{10}, \frac{81}{100}, \frac{729}{1000}, \ldots$

**32.** $3, -\frac{9}{2}, \frac{27}{4}, -\frac{81}{8}, \ldots$

**33.** $\dfrac{1}{4}, \dfrac{1}{12}, \dfrac{1}{24}, \dfrac{1}{40}, \cdots, \dfrac{1}{2n(n+1)}, \cdots$

**34.** $\dfrac{1}{2 \cdot 3}, \dfrac{1}{3 \cdot 4}, \dfrac{1}{4 \cdot 5}, \dfrac{1}{5 \cdot 6}, \cdots, \dfrac{1}{(n+1)(n+2)}, \cdots$

In Exercises 35–38, prove the inequality for the indicated integer values of $n$.

**35.** $n! > 2^n, \quad n \geq 4$

**36.** $\left(\frac{4}{3}\right)^n > n, \quad n \geq 7$

**37.** $\dfrac{1}{\sqrt{1}} + \dfrac{1}{\sqrt{2}} + \dfrac{1}{\sqrt{3}} + \cdots + \dfrac{1}{\sqrt{n}} > \sqrt{n}, \quad n \geq 2$

**38.** $\left(\dfrac{x}{y}\right)^{n+1} < \left(\dfrac{x}{y}\right)^n$, if $n \geq 1$ and $0 < x < y$.

In Exercises 39–46, use mathematical induction to prove the given property for all positive integers $n$.

**39.** $(ab)^n = a^n b^n$

**40.** $\left(\dfrac{a}{b}\right)^n = \dfrac{a^n}{b^n}$

**41.** If $x_1 \neq 0, x_2 \neq 0, \ldots, x_n \neq 0$, then
$(x_1 x_2 x_3 \cdots x_n)^{-1} = x_1^{-1} x_2^{-1} x_3^{-1} \cdots x_n^{-1}$.

**42.** If $x_1 > 0, x_2 > 0, \ldots, x_n > 0$, then
$\ln(x_1 x_2 x_3 \cdots x_n) = \ln x_1 + \ln x_2 + \ln x_3 + \cdots + \ln x_n$.

**43.** Generalized Distributive Law:
$x(y_1 + y_2 + \cdots + y_n) = xy_1 + xy_2 + \cdots + xy_n$

**44.** $(a + bi)^n$ and $(a - bi)^n$ are complex conjugates for all $n \geq 1$.

**45.** A factor of $(n^3 + 3n^2 + 2n)$ is 3.

**46.** A factor of $(2^{2n-1} + 3^{2n-1})$ is 5.

**47.** In your own words, explain what is meant by a proof by mathematical induction.

**48.** *Think About It*   What conclusion can be drawn from the given information about the sequence of statements $P_n$?

  (a) $P_3$ is true and $P_k$ implies $P_{k+1}$.

  (b) $P_1, P_2, P_3, \ldots, P_{50}$ are all true.

  (c) $P_1, P_2,$ and $P_3$ are all true, but the truth of $P_k$ does not imply that $P_{k+1}$ is true.

  (d) $P_2$ is true and $P_{2k}$ implies $P_{2k+2}$.

In Exercises 49–52, write the first five terms of the sequence.

**49.** $a_0 = 1$
$a_n = a_{n-1} + 2$

**50.** $a_0 = 10$
$a_n = 4a_{n-1}$

**51.** $a_0 = 4$
$a_1 = 2$
$a_n = a_{n-1} - a_{n-2}$

**52.** $a_0 = 0$
$a_1 = 2$
$a_n = a_{n-1} + 2a_{n-2}$

In Exercises 53–62, write the first five terms of the sequence where $a_1 = f(1)$. Then calculate the first and second differences of the sequence. Does the sequence have a linear model, a quadratic model, or neither?

**53.** $f(1) = 0$
$a_n = a_{n-1} + 3$

**54.** $f(1) = 2$
$a_n = n - a_{n-1}$

**55.** $f(1) = 3$
$a_n = a_{n-1} - n$

**56.** $f(2) = -3$
$a_n = -2a_{n-1}$

**57.** $a_0 = 0$
$a_n = a_{n-1} + n$

**58.** $a_0 = 2$
$a_n = (a_{n-1})^2$

**59.** $f(1) = 2$
$a_n = a_{n-1} + 2$

**60.** $f(1) = 0$
$a_n = a_{n-1} + 2n$

**61.** $a_0 = 1$
$a_n = a_{n-1} + n^2$

**62.** $a_0 = 0$
$a_n = a_{n-1} - 1$

In Exercises 63–66, find a quadratic model for the sequence with the indicated terms.

**63.** $a_0 = 3, a_1 = 3, a_4 = 15$

**64.** $a_0 = 7, a_1 = 6, a_3 = 10$

**65.** $a_0 = -3, a_2 = 1, a_4 = 9$

**66.** $a_0 = 3, a_2 = 0, a_6 = 36$

## Chapter Project

## The Multiplier Effect

| Cycle, $x$ | Total spending (in dollars), $y$ |
|---|---|
| 1 | 92,000,000 |
| 2 | 73,460,000 |
| 3 | 57,598,000 |
| 4 | 45,122,000 |
| 5 | 37,020,000 |
| 6 | 29,491,000 |
| 7 | 23,592,960 |
| 8 | 18,854,365 |
| 9 | 15,099,500 |
| 10 | 12,080,600 |
| 11 | 9,578,675 |
| 12 | 7,740,950 |
| 13 | 6,185,752 |
| 14 | 4,950,800 |
| 15 | 3,980,200 |
| 16 | 3,150,600 |
| 17 | 2,540,300 |
| 18 | 2,025,700 |
| 19 | 1,600,300 |
| 20 | 1,300,050 |

The ideas of economist John Maynard Keynes have been influential in the development of economic policy at the national, state, and local levels. Keynesian economics brought the term "multiplier effect" into popular use. The "multiplier effect" is based on the idea that the expenditures of one person become the income of another. As the second person spends his or her income, then his/her expenditures become a third person's income. This pattern continues as one person's expenditures become another person's income. If individuals in an economy spend 75% of their income (some amount will go to income taxes, forced retirement contributions, voluntary savings/investments, and so forth), then a person spends $0.75 of every dollar earned. For one dollar earned the first person would spend $0.75. The next person spends only $0.56 of the original dollar ($0.75 × 75%). Each subsequent person would spend progressively smaller and smaller amounts of the original dollar, so the income "leaks" out of the spending system. The greater the leakage, the smaller the "multiplier effect" will be.

The table shows the cycle of income in a community that has hosted a large event such as the Super Bowl. In the table, $x$ corresponds to the cycle number of a dollar and $y$ is the total spending in dollars for that cycle. Use this information to investigate the following questions.

**1.** *Fitting the Model to Data*   Use a graphing utility to make a scatter plot of the data in the table. What type or types of graphs does the scatter plot suggest might fit the data? Use the statistics package of a graphing utility to find a quadratic model ($y = ax^2 + bx + c$), a power series model ($y = ax^b$), and an exponential model ($y = a \cdot b^x$).

**2.** *Comparing a Model to Actual Data*   Numerically compare the values given by the models found in Question 1 with the actual data in the table by constructing a table. Graphically compare the models with the data by graphing each model over the scatter plot.

**3.** *Discussing the Models*   Which model fits the data best? Explain. Find the total economic impact of this event on a community by summing the total dollars spent after 20 cycles and the total according to the model you selected. Did the model fit the data well? Explain.

**4.** *Further Explorations*   Use a reference to further investigate the "multiplier effect." Can you find examples of situations where the "multiplier effect" could apply?

# CHAPTER SUMMARY

After studying this chapter, you should have acquired the following skills.
These skills are keyed to the Review Exercises that begin on page 566.
Answers to odd-numbered Review Exercises are given in the back of the book.

| | | |
|---|---|---|
| **8.1** | · Find the terms of a sequence. | Review Exercises 1–6 |
| | · Evaluate a factorial expression. | Review Exercises 7, 8 |
| | · Write an expression for the $n$th term of a sequence. | Review Exercises 9–14 |
| | · Find the sum of a finite sequence. | Review Exercises 15–18 |
| | · Use summation notation to write the sum of a sequence. | Review Exercises 19–22 |
| | · Use a sequence to solve an application problem. | Review Exercises 23, 24 |
| **8.2** | · Determine whether a sequence is arithmetic. | Review Exercises 25–30 |
| | · Find the $n$th term of an arithmetic sequence. | Review Exercises 31, 32 |
| | · Write the first five terms of an arithmetic sequence. | Review Exercises 33, 34 |
| | · Find the sum of an arithmetic sequence. | Review Exercises 35–38 |
| | · Use an arithmetic sequence to solve an application problem. | Review Exercises 39, 40 |
| **8.3** | · Determine whether a sequence is geometric. | Review Exercises 41–44 |
| | · Write the first five terms of a geometric sequence. | Review Exercises 45–48 |
| | · Find the $n$th term of a geometric sequence. | Review Exercises 49–52 |
| | · Find the sum of a finite or infinite geometric sequence. | Review Exercises 53–58 |
| | · Use a geometric sequence to solve an application problem. | Review Exercises 59–62 |
| **8.4** | · Find a binomial coefficient. | Review Exercises 63–66 |
| | · Use Pascal's Triangle to determine a binomial coefficient. | Review Exercises 67, 68 |
| | · Expand a binomial using binomial coefficients. | Review Exercises 69–74, 77, 78 |
| | · Find a specific term in a binomial expansion. | Review Exercises 75, 76 |
| **8.5** | · Use the Fundamental Counting Principle. | Review Exercises 79–85 |
| | · Determine the number of permutations of $n$ elements taken $r$ at a time. | Review Exercises 86–89 |
| | · Determine the number of combinations of $n$ elements taken $r$ at a time. | Review Exercises 90–93 |
| **8.6** | · Find the probability of an event. | Review Exercises 94–97 |
| | · Find the probability of mutually exclusive events and independent events. | Review Exercises 100–106 |
| | · Find the probability of the complement of an event. | Review Exercises 98, 99, 102, 104–106 |
| **8.7** | · Use mathematical induction to prove a formula. | Review Exercises 107–110 |
| | · Find the sum of powers of integers. | Review Exercises 111–114 |
| | · Find a formula for a finite sum. | Review Exercises 115, 116 |
| | · Use finite differences to find a quadratic model. | Review Exercises 117, 118 |

## REVIEW EXERCISES

In Exercises 1–6, write the first five terms of the sequence. (Begin with $n = 1$.)

**1.** $a_n = 4n + 1$

**2.** $a_n = 2n - 3$

**3.** $a_n = \dfrac{2n}{2n + 1}$

**4.** $a_n = \dfrac{5n - 2}{n}$

**5.** $a_n = (-1)^{n+1} \dfrac{2^n}{n + 1}$

**6.** $a_n = (-1)^n \dfrac{3^n - 1}{3^n + 1}$

In Exercises 7 and 8, evaluate the fraction.

**7.** $\dfrac{16!}{12!}$

**8.** $\dfrac{24!}{22!}$

In Exercises 9–14, write an expression for the apparent $n$th term of the sequence. (Begin with $n = 1$.)

**9.** $1, 2, 4, 8, \ldots$

**10.** $4, -4, 4, -4, \ldots$

**11.** $1, \frac{1}{2}, \frac{1}{3}, \frac{1}{4}, \frac{1}{5}, \ldots$

**12.** $-\frac{2}{3}, \frac{3}{4}, -\frac{4}{5}, \frac{5}{6}, -\frac{6}{7}, \ldots$

**13.** $3, -9, 27, -81, 243, \ldots$

**14.** $-2, 4, -8, 16, -32, \ldots$

In Exercises 15–18, find the sum.

**15.** $\displaystyle\sum_{i=1}^{6} (2i - 3)$

**16.** $\displaystyle\sum_{i=1}^{6} (3i + 2)$

**17.** $\displaystyle\sum_{k=1}^{6} (k - 1)(k + 2)$

**18.** $\displaystyle\sum_{k=1}^{8} k^2 + 1$

In Exercises 19–22, use summation notation to write the sum.

**19.** $\dfrac{3}{1 + 1} + \dfrac{3}{1 + 2} + \dfrac{3}{1 + 3} + \dfrac{3}{1 + 4} + \cdots + \dfrac{3}{1 + 12}$

**20.** $1 + \frac{1}{2} + \frac{1}{3} + \frac{1}{4} + \frac{1}{5} + \cdots + \frac{1}{60}$

**21.** $2 - 4 + 8 - 16 + 32 - 64 + \cdots + 8192$

**22.** $2 + 8 + 18 + 32 + \cdots + 98$

**23.** *Compound Interest* Suppose on your next birthday you deposit $2000 in an account that earns 8% compounded quarterly. The balance in the account after $n$ quarters is given by

$$A_n = 2000\left(1 + \frac{0.08}{4}\right)^n, \quad n = 1, 2, 3, \ldots$$

(a) Compute the first eight terms of the sequence.

(b) Find the balance in this account 10 years from the date of deposit by computing the 40th term of the sequence.

(c) Assuming you do not withdraw money from the account, find the balance 30 years from the date of deposit by computing the 120th term of the sequence.

**24.** *Compound Deposit* A deposit of $20,000 is made in an account that earns 6% compounded monthly. The balance in the account after $n$ months is given by

$$A_n = 20{,}000\left(1 + \frac{0.06}{12}\right)^n, \quad n = 1, 2, 3, \ldots$$

(a) Compute the first 10 terms of the sequence.

(b) Find the balance in this account after 10 years by computing the 120th term of the sequence.

In Exercises 25–28, determine whether the sequence is arithmetic. If it is, find the common difference.

**25.** $3, 8, 13, 18, \ldots$

**26.** $1, 3, 9, 27, \ldots$

**27.** $4, \frac{7}{2}, 3, \frac{5}{2}, 2, \ldots$

**28.** $3.2, 4.0, 4.8, 5.6, 6.4, \ldots$

In Exercises 29 and 30, write the first five terms of the specified sequence. Determine whether the sequence is arithmetic, and, if it is, find the common difference.

**29.** $a_n = 4 + 5n$

**30.** $a_n = 2 + (n - 3)5$

In Exercises 31 and 32, find a formula for $a_n$ for the arithmetic sequence.

**31.** $a_1 = 2, d = 3$

**32.** $a_1 = 15, d = -3$

In Exercises 33 and 34, write the first five terms of the arithmetic sequence.

**33.** $a_1 = 4, a_8 = 25$

**34.** $a_6 = 28, a_{14} = 16$

In Exercises 35–38, find the sum of the first $n$ terms of the arithmetic sequence.

**35.** $3, 9, 15, 21, 27, \ldots, n = 8$

**36.** $-8, -4, 0, 4, 8, \ldots, n = 12$

**37.** $a_1 = 27, a_{40} = 300, n = 20$

**38.** $a_1 = 9, a_{15} = 93, n = 15$

**39.** *Job Offers*    A person is offered a job by two companies. The position with Company A has a salary of $26,500 for the first year with a guaranteed annual raise of $1600 per year for the first 5 years. The position with Company B has a salary of $23,000 for the first year with a guaranteed annual raise of $2500 per year for the first 5 years.

(a) What will the salary be during the sixth year of employment at Company A? At Company B?

(b) How much will company A have paid the person at the end of 6 years?

(c) How much will Company B have paid the person at the end of 6 years?

(d) Which job should the person accept, and why?

**40.** *Seating Capacity*    Determine the seating capacity of an auditorium with 20 rows of seats if there are 25 seats in the first row, 28 seats in the second row, and so on, in arithmetic progression.

In Exercises 41–44, determine whether the sequence is geometric. If it is, find its common ratio.

**41.** $1, -3, 9, -27, \ldots$    **42.** $3, 9, 15, 21, \ldots$

**43.** $16, 8, 4, 2, 1, \frac{1}{2}, \frac{1}{4}, \frac{1}{8}, \ldots$

**44.** $1, -\frac{1}{3}, \frac{1}{9}, -\frac{1}{27}, \frac{1}{81}, -\frac{1}{243}, \ldots$

In Exercises 45–48, write the first five terms of the geometric sequence.

**45.** $a_1 = 2, r = 3$    **46.** $a_1 = 2, r = e$

**47.** $a_1 = 10, r = -\frac{1}{5}$    **48.** $a_1 = 1, r = \frac{3}{4}$

In Exercises 49–52, find the indicated term of the geometric sequence.

**49.** $a_1 = 8, r = \frac{1}{2}$, 40th term

**50.** $a_1 = 100, r = -\frac{1}{10}$, 5th term

**51.** $a_1 = 200, r = -\frac{1}{2}$, 10th term

**52.** $a_1 = 10, r = 1.05$, 20th term

In Exercises 53 and 54, find the sum.

**53.** $\displaystyle\sum_{n=1}^{10} 4(2^n)$

**54.** $\displaystyle\sum_{n=0}^{20} 2\left(\frac{2}{3}\right)^n$

In Exercises 55–58, find the sum of the infinite geometric sequence, if it exists.

**55.** $\displaystyle\sum_{n=0}^{\infty} \left(\frac{1}{3}\right)^n$

**56.** $\displaystyle\sum_{n=0}^{\infty} \left(\frac{3}{2}\right)^n$

**57.** $\displaystyle\sum_{n=0}^{\infty} \left(\frac{4}{3}\right)^n$

**58.** $\displaystyle\sum_{n=0}^{\infty} 2\left(\frac{1}{10}\right)^n$

**59.** *Compound Interest*    A deposit of $100 is made at the beginning of each month for 10 years in an account that pays 8% compounded monthly. The balance $A$ at the end of 10 years is as follows.

$$A = 100\left(1 + \frac{0.08}{12}\right)^1 + \cdots + 100\left(1 + \frac{0.08}{12}\right)^{120}$$

(a) Find the balance.

(b) Is the balance after 20 years twice what it is after 10 years? Explain your reasoning.

**60.** *Compound Interest*    Suppose you deposit $200 in an account at the beginning of each month for 10 years. The account pays 6% compounded monthly. What will your balance be at the end of 10 years? If the interest were compounded continuously, what would the balance be?

**61.** *Profit*    The annual profit for a company from 1984 to 1999 can be approximated by the model

$$a_n = 156.4 e^{0.16n}$$

where $a_n$ is the annual profit (in millions of dollars) and $n$ represents the year, with $n = 0$ corresponding to 1984.

(a) Sketch a bar graph that represents the company's profit during the 15-year period.

(b) Find the total profit during the 15-year period.

**62.** *Sales*    The annual sales for a company from 1984 to 1999 can be approximated by the model

$$a_n = 221 e^{0.146n}$$

where $a_n$ is the annual sales (in millions of dollars) and $n$ represents the year, with $n = 0$ corresponding to 1984. Use the formula for the sum of a geometric sequence to approximate the total sales earned during the 15-year period.

In Exercises 63–66, evaluate $_nC_r$.

**63.** $_7C_4$

**64.** $_{12}C_8$

**65.** $_{30}C_{30}$

**66.** $_{20}C_9$

In Exercises 67 and 68, expand the binomial using Pascal's Triangle to determine the coefficients.

**67.** $(3 - 2y)^4$

**68.** $(3x + 4y)^5$

In Exercises 69–74, use the Binomial Theorem to expand the binomial. Simplify your answer.

**69.** $(x + 1)^6$

**70.** $(x - 1)^7$

**71.** $(x + y)^5$

**72.** $(2x - y)^6$

**73.** $(x - 2y)^8$

**74.** $(x^2 + 4)^5$

In Exercises 75 and 76, find the required term in the expansion of the binomial.

| Binomial | Term |
|----------|------|
| **75.** $(x + 2)^{12}$ | $ax^7$ |
| **76.** $(2x - 5y)^9$ | $ax^4y^5$ |

In Exercises 77 and 78, use the Binomial Theorem to expand the expression. [In the study of probability, it is sometimes necessary to use the expansion of $(p + q)^n$, where $p + q = 1$.]

**77.** $\left(\frac{1}{3} + \frac{2}{3}\right)^5$

**78.** $(0.2 + 0.8)^{10}$

**79.** *Computer Systems* A customer in a computer store can choose one of five monitors, one of four keyboards, and one of five computers. If all of the choices are compatible, how many different systems can be chosen?

**80.** *Sound System* A customer in an electronics store can choose one of five CD players, one of six speaker systems, and one of three radio/cassette players to design a sound system. How many different systems can be designed?

**81.** *Roller Coaster Ride* Six people are lining up for a ride on a roller coaster, but only three of the six are willing to sit in the front two seats of the coaster car which has six seats. With that constraint, in how many ways can the six people be seated in the car?

**82.** *True-False Exam* In how many ways can a 15-question true-false exam be answered? (Assume that no questions are omitted.)

**83.** *True-False Exam* In how many ways can a 20-question true-false exam be answered? (Assume that no questions are omitted.)

**84.** *Four-Digit Numbers* How many four-digit numbers can be formed under the following conditions?

(a) Leading digits cannot be zero.

(b) Leading digits cannot be zero and no repetition of digits is allowed.

(c) Leading digits cannot be zero and the number must be divisible by 2.

**85.** *Five-Digit Numbers* How many five-digit numbers can be formed under the following conditions?

(a) Leading digits cannot be zero.

(b) Leading digits cannot be zero and no repetition of digits is allowed.

(c) Leading digits cannot be zero and the number must be odd.

In Exercises 86–89, evaluate $_nP_r$.

**86.** $_9P_3$

**87.** $_{30}P_1$

**88.** $_8P_8$

**89.** $_{15}P_2$

**90.** *Test Questions* A student may answer any 15 questions from a total of 20 questions on an exam. In how many different ways can the student select the 15 questions?

**91.** *Starting Lineup* In how many ways can the starting lineup for a coed, six-member flag football team be selected from a team of eight men and eight women? League rules require that every team have three men and three women on the field.

**92.** *Defective Units* A shipment of 30 VCRs contains three defective units. In how many ways can a vending company purchase five of these units and receive (a) all good units, (b) two good units, and (c) at least two good units?

**93.** *Job Applicants* An employer interviews 10 people for five openings in the company. Six of the 10 people are women. If all 10 are qualified, in how many ways can the employer fill the five positions if (a) the selection is random, (b) exactly three of those hired are women, and (c) all five of those hired are women?

*Drawing a Card* In Exercises 94 and 95, find the indicated probability in the experiment of selecting one card from a standard deck of 52 playing cards.

**94.** The probability of getting a ten, jack, or queen

**95.** The probability of getting a red card that is not a face card

*Tossing a Die* In Exercises 96 and 97, find the indicated probability in the experiment of tossing a six-sided die twice.

**96.** The probability that the sum is 7

**97.** The probability that the sum is 7 or 11

In Exercises 98 and 99, you are given the probability that an event will happen. Find the probability that the event will not happen.

**98.** $p = 0.57$ **99.** $p = 0.23$

**100.** *Preparing for a Test* An instructor gives a class a list of 12 study problems, from which eight will be selected to construct an exam. If a given student knows how to solve 10 of the problems, find the probabilities that the student will be able to answer (a) all eight questions on the exam, (b) exactly seven questions on the exam, and (c) at least seven questions on the exam.

**101.** *Game Show* On a game show, you are given five digits to arrange in the proper order to represent the price of a car. If you are correct, you win the car. What are the probabilities of winning under the following conditions?

(a) You must guess the position of each digit.

(b) You know the first two digits, but must guess the remaining three.

**102.** *Payroll Mix-Up* Four paychecks and envelopes are addressed to four different people. If the paychecks get mixed up and are randomly inserted into the envelopes, what are the probabilities that (a) exactly one will be inserted in the correct envelope and (b) at least one will be inserted in the correct envelope?

**103.** *Poker Hand* Five cards are drawn from an ordinary deck of 52 playing cards. What is the probability of getting two pair? (Two pair consists of two of one kind, two of another kind, and one of a third kind. For example, KK995 is two pair.)

**104.** *A Boy or a Girl?* Assume that the probability of the birth of a child of a particular gender is 50%. In a family with five children, what are the probabilities that (a) all the children are boys, (b) all the children are of the same gender, and (c) there is at least one boy?

**105.** *A Boy or a Girl?* Assume that the probability of the birth of a child of a particular gender is 50%. In a family with 10 children, what are the probabilities that (a) all the children are girls, (b) all the children are of the same gender, and (c) there is at least one girl?

**106.** *Defective Units* A shipment of 1000 radar detectors contains five defective units. A retail outlet has ordered 50 units.

(a) What is the probability that all 50 units are good?

(b) What is the probability that at least one unit is defective?

In Exercises 107–110, use mathematical induction to prove the formula for every positive integer $n$.

**107.** $1 + 4 + \cdots + (3n - 2) = \dfrac{n}{2}(3n - 1)$

**108.** $1 + \dfrac{3}{2} + 2 + \dfrac{5}{2} + \cdots + \dfrac{1}{2}(n + 1) = \dfrac{n}{4}(n + 3)$

**109.** $\displaystyle\sum_{i=0}^{n-1} ar^i = \dfrac{a(1 - r^n)}{1 - r}$

**110.** $\displaystyle\sum_{k=0}^{n-1} (a + kd) = \dfrac{n}{2}[2a + (n - 1)d]$

In Exercises 111–114, find the sum of the integers using the formulas for the sums of powers of integers.

**111.** $\displaystyle\sum_{n=1}^{30} n$ **112.** $\displaystyle\sum_{n=1}^{5} n^2$

**113.** $\displaystyle\sum_{n=1}^{8} n^3$ **114.** $\displaystyle\sum_{n=1}^{6} (n^2 + n)$

In Exercises 115 and 116, find a formula for the sum of the first $n$ terms of the sequence.

**115.** $1, \dfrac{3}{5}, \dfrac{9}{25}, \dfrac{27}{125}, \ldots$ **116.** $5, -\dfrac{5}{2}, \dfrac{5}{4}, -\dfrac{5}{8}, \ldots$

In Exercises 117 and 118, write the first five terms of the sequence where $a_1 = f(1)$. Then calculate the first and second differences of the sequence. Does the sequence have a linear model, a quadratic model, or neither?

**117.** $a_0 = 2; \ a_n = a_{n-1} + n$

**118.** $f(1) = 1; \ a_n = a_{n-1} + 2n$

## Chapter Test

Take this test as you would take a test in class. After you are done, check your work against the answers given in the back of the book.

In Exercises 1–4, write the first five terms of the sequence. (Begin with $n = 1$.)

**1.** $a_n = 2n + 1$     **2.** $a_n = (-1)^n n^2$     **3.** $a_n = (n + 1)!$     **4.** $a_n = \left(\frac{1}{2}\right)^n$

In Exercises 5–8, decide whether the sequence is arithmetic, geometric, or neither. If possible, find its common difference or common ratio.

**5.** $a_n = 5n - 2$     **6.** $a_n = 2^n + 2$     **7.** $a_n = 5(2^n)$     **8.** $a_n = 3n^2 - 2$

In Exercises 9–11, evaluate the sum.

**9.** $\displaystyle\sum_{n=1}^{50} (2n + 1)$     **10.** $\displaystyle\sum_{n=1}^{20} 3\left(\frac{3}{2}\right)^n$     **11.** $\displaystyle\sum_{n=1}^{\infty} \left(\frac{1}{3}\right)^n$

**12.** A deposit of \$10,000 is made in an account that pays 8% compounded monthly. The balance in the account after $n$ months is given by

$$A = 10,000\left(1 + \frac{0.08}{12}\right)^n, \quad n = 1, 2, 3, \ldots$$

Find the balance in the account after 10 years.

**13.** You deposit \$250 in an account at the beginning of each month for 10 years. The account pays 6% compounded monthly. What is the balance at the end of 10 years?

**14.** Use the Binomial Theorem to expand $(x + 2)^5$.

**15.** Determine the numerical coefficient of the $x^7$ term of $(x - 3)^{12}$.

**16.** A customer in an electronics store can choose one of five CD players, one of six speaker systems, and one of four radio/cassette players to design a sound system. How many different systems can be designed?

**17.** In how many ways can a 10-question true-false exam be answered? (Assume that no question is omitted.) Explain your reasoning.

**18.** How many five-digit numbers can be formed if the leading digit cannot be zero and the number must be odd?

**19.** On a game show, you are given five different digits to arrange in the proper order to represent the price of a car. You know only the first digit. What is the probability that you will arrange the other four digits correctly?

**20.** Five coins are tossed. What is the probability that all are heads?

## Cumulative Test: Chapters 6–8

Take this test as you would take a test in class. After you are done, check your work against the answers given in the back of the book.

In Exercises 1 and 2, solve the system by the indicated method.

**1.** *Elimination*

$$2x - 3y + z = 18$$
$$3x \quad - 2z = -4$$
$$x - y + 3z = 20$$

**2.** *Matrices*

$$3x - 4y + 2z = -32$$
$$2x + 3y \quad = 8$$
$$y - 3z = 19$$

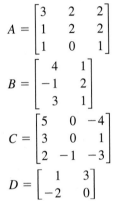

$$A = \begin{bmatrix} 3 & 2 & 2 \\ 1 & 2 & 2 \\ 1 & 0 & 1 \end{bmatrix}$$

$$B = \begin{bmatrix} 4 & 1 \\ -1 & 2 \\ 3 & 1 \end{bmatrix}$$

$$C = \begin{bmatrix} 5 & 0 & -4 \\ 3 & 0 & 1 \\ 2 & -1 & -3 \end{bmatrix}$$

$$D = \begin{bmatrix} 1 & 3 \\ -2 & 0 \end{bmatrix}$$

Matrices for 3–6

In Exercises 3–6, use the matrices to find the indicated matrix.

**3.** $2A - C$     **4.** $AB$     **5.** $BD$     **6.** $A^{-1}$

In Exercises 7 and 8, evaluate the determinant of the matrix.

**7.** $\begin{bmatrix} 1 & -2 & 4 \\ 3 & 7 & -5 \\ 6 & 1 & 4 \end{bmatrix}$     **8.** $\begin{bmatrix} 1 & 0 & 0 \\ 0 & 7 & 0 \\ 0 & 0 & -1 \end{bmatrix}$

**9.** Find the point of equilibrium for a system with demand function $p = 75 - 0.0005x$ and supply function $p = 30 + 0.002x$.

**10.** A total of $45,000 is invested at 7.5% and 8.5% simple interest. If the yearly interest is $3625, how much of the $45,000 is invested at each rate?

**11.** A factory produces three different models of a product, which are shipped to three different warehouses. The number of units $i$ that are shipped to warehouse $j$ is represented by $a_{ij}$ in matrix $A$ at the left. The price per unit is represented by matrix $B$. Find the product $BA$ and state what each entry of the product represents.

$$A = \begin{bmatrix} 1000 & 3000 & 2000 \\ 2000 & 4000 & 5000 \\ 3000 & 1000 & 1000 \end{bmatrix}$$

$$B = \begin{bmatrix} \$25 & \$32 & \$27 \end{bmatrix}$$

Matrices for 11

**12.** Write the first five terms of the sequence given by $a_n = 2 + 3n$.

**13.** Write the first five terms of the geometric sequence with $a_1 = 3$ and $r = \frac{1}{3}$.

**14.** Use the Binomial Theorem to expand $(y - 4)^7$.

**15.** In how many ways can the letters in MATRIX be arranged?

**16.** How many five-digit numbers are multiples of 5?

**17.** A shipment of 2000 microwave ovens contains 15 defective units. You order three units from the shipment. What is the probability that all three are good?

**18.** On a game show, the digits 7, 8, and 9 must be arranged in the proper order to form the price of an appliance. If the digits are arranged correctly, the contestant wins the appliance. What is the probability of winning if the contestant knows that the price is at least $800?

**19.** Write the first five terms of the sequence where $a_1 = 2$ and $a_n = a_{n-1} + n$. Then calculate the first and second differences of the sequence. Does the sequence have a linear model, a quadratic model, or neither?

## Appendix A

# Review of Fundamental Concepts of Algebra

## A.1    Real Numbers:  Order and Absolute Value

Real Numbers  ■  The Real Number Line and Ordering  ■  Absolute Value and Distance  ■  Application

## Real Numbers

The formal term that is used in mathematics to refer to a collection of objects is the word **set.** For instance, the set

$$\{1, 2, 3\}$$

contains the three numbers 1, 2, and 3. Note that a pair of braces $\{\ \ \}$ is used to enclose the members of the set. In this text, a *pair* of braces will always indicate the members of a set. Parentheses $(\ \ )$ and brackets $[\ \ ]$ are used to represent other ideas.

The set of numbers that is used in arithmetic is the **set of real numbers.** The term *real* distinguishes real numbers from *imaginary* or *complex* numbers.

A set $A$ is called a **subset** of a set $B$ if every member of $A$ is also a member of $B$. Here are two examples.

- $\{1, 2, 3\}$ is a subset of $\{1, 2, 3, 4\}$.

- $\{0, 4\}$ is a subset of $\{0, 1, 2, 3, 4\}$.

One of the most commonly used subsets of real numbers is the set of **natural numbers** or **positive integers**

$$\{1, 2, 3, 4, \ldots\}. \qquad \text{Set of positive integers}$$

Note that the three dots indicate that the pattern continues. For instance, the set also contains the numbers 5, 6, 7, and so on.

Positive integers can be used to describe many quantities that you encounter in everyday life—for instance, you might be taking four classes this term, or you might be paying $240 dollars a month for rent. But even in everyday life, positive integers cannot describe some concepts accurately. For instance, you could have a zero balance in your checking account, or the temperature could be $-10°$ (10 degrees below zero). To describe such quantities you need to expand the set of positive integers to include **zero** and the **negative integers.** The expanded set is called the set of **integers,** which can be written as follows.

$$\underbrace{\{\ldots, -3, -2, -1,}_{\text{Negative integers}} \overset{\overset{\displaystyle\text{Zero}}{\displaystyle\mid}}{0,} \underbrace{1, 2, 3, \ldots\}}_{\text{Positive integers}}$$

The set of integers is a **subset** of the set of real numbers—this means that every integer is a real number.

Even with the set of integers, there are still many quantities in everyday life that you cannot describe accurately. The costs of many items are not in whole-dollar amounts, but in parts of dollars, such as $1.19 or $39.98. You might work $8\frac{1}{2}$ hours, or you might miss the first *half* of a movie. To describe such quantities, the set of integers is expanded to include **fractions.** The expanded set is called the set of **rational numbers.** Formally, a real number is called **rational** if it can be written as the ratio $p/q$ of two integers, where $q \neq 0$. (The symbol $\neq$ means **not equal to.**) For instance,

$$2 = \frac{2}{1}, \quad 0.333\ldots = \frac{1}{3}, \quad 0.125 = \frac{1}{8}, \quad \text{and} \quad 1.126126\ldots = \frac{125}{111}$$

are rational numbers. Real numbers that cannot be written as ratios of two integers are called **irrational.** For instance, the numbers

$$\sqrt{2} = 1.4142135\ldots \quad \text{and} \quad \pi = 3.1415926\ldots$$

are irrational. The decimal representation of a rational number is either *terminating* or *repeating*. For instance, the decimal representation of $\frac{1}{4} = 0.25$ is terminating, and the decimal representation of $\frac{4}{11} = 0.363636\ldots = 0.\overline{36}$ is repeating. (The line over "36" indicates which digits repeat.)

The decimal representation of an irrational number neither terminates nor repeats. When you perform calculations using decimal representations of nonterminating decimals, you usually use a decimal approximation that has been **rounded** to a certain number of decimal places. For instance, rounded to four decimal places, the decimal approximations of $\frac{2}{3}$ and $\pi$ are

$$\frac{2}{3} \approx 0.6667 \quad \text{and} \quad \pi \approx 3.1416.$$

The symbol $\approx$ means **approximately equal to.**

Figure A.1 shows several commonly used subsets of real numbers and their relationships to each other.

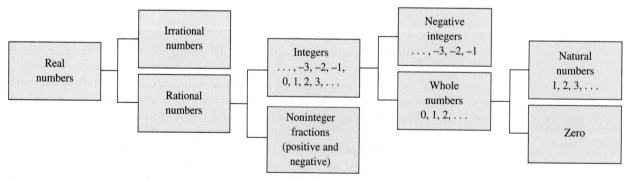

Figure A.1

# The Real Number Line and Ordering

The picture that is used to represent the real numbers is the **real number line.** It consists of a horizontal line with a point (the **origin**) labeled as 0. Points to the left of 0 are associated with **negative numbers,** and points to the right of 0 are associated with **positive numbers,** as shown in Figure A.2. The real number zero is neither positive nor negative. Thus, when you want to talk about real numbers that might be positive *or* zero, you can use the term **nonnegative real numbers.**

Origin

Negative numbers          Positive numbers

**Figure A.2**    *The Real Number Line*

Each point on the real number line corresponds to exactly one real number, and each real number corresponds to exactly one point on the real number line, as shown in Figure A.3. The number associated with a point on the real number line is the **coordinate** of the point.

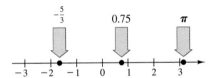

$-\frac{5}{3}$          0.75          $\pi$

**Figure A.3**    *Every real number corresponds to a point on the real number line.*

The real number line provides you with a way of comparing any two real numbers. For instance, if you choose any two (different) numbers on the real number line, one of the numbers must be to the left of the other number. The number to the left is **less than** the number to the right, and the number to the right is **greater than** the number to the left.

> ▶ **Definition of Order on the Real Number Line**
>
> If the real number $a$ lies to the left of the real number $b$ on the real number line, $a$ is **less than** $b$, which is denoted by
>
> $$a < b$$
>
> as shown in Figure A.4. This relationship can also be described by saying that $b$ is **greater than** $a$ and writing $b > a$. The symbol $a \le b$ means that $a$ is **less than or equal to** $b$, and the symbol $b \ge a$ means that $b$ is **greater than or equal to** $a$.

$a < b$

**Figure A.4**    *a is to the left of b.*

The symbols $<$, $>$, $\le$, and $\ge$ are called **inequality symbols.** Inequalities are useful in denoting subsets of real numbers, as shown in Examples 1 and 2.

**Example 1**    Interpreting Inequalities

a. The inequality $x \le 2$ denotes all real numbers that are less than or equal to 2, as shown in Figure A.5 (a).

b. The inequality $-2 \le x < 3$ means that $x \ge -2$ *and* $x < 3$. This **double inequality** denotes all real numbers between $-2$ and 3, including $-2$ but *not* including 3, as shown in Figure A.5 (b).

c. The inequality $x > -5$ denotes all real numbers that are greater than $-5$, as shown in Figure A.5 (c).

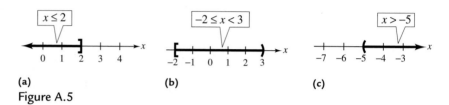

(a)                    (b)                    (c)

Figure A.5

In Figure A.5, notice that a bracket is used to *include* the endpoint of an interval and a parenthesis is used to *exclude* the endpoint.

**Example 2**    Inequalities and Sets of Real Numbers

a. "$c$ is nonnegative" means that $c$ is greater than or equal to zero, which you can write as $c \ge 0$.

b. "$b$ is at most 5" can be written as $b \le 5$.

c. "$d$ is negative" can be written as $d < 0$, and "$d$ is greater than $-3$" can be written as $-3 < d$. Combining these two inequalities produces $-3 < d < 0$.

d. "$x$ is positive" can be written as $0 < x$, and "$x$ is not more than 6" can be written as $x \le 6$. Combining these two inequalities produces $0 < x \le 6$.

The following property of real numbers is the **Law of Trichotomy.** As the "tri" in its name suggests, this law tells you that for any two real numbers $a$ and $b$, precisely one of *three* relationships is possible.

$$a < b, \qquad a = b, \qquad \text{or} \qquad a > b \qquad \text{Law of Trichotomy}$$

## Absolute Value and Distance

The **absolute value** of a real number is its *magnitude*, or its value disregarding its sign. For instance, the absolute value of $-3$ is $|-3|$, which has the value of 3.

▶ **Definition of Absolute Value**

Let $a$ be a real number. The **absolute value** of $a$, denoted by $|a|$, is

$$|a| = \begin{cases} a, \text{if } a \ge 0 \\ -a, \text{if } a < 0. \end{cases}$$

The absolute value of any real number is either positive or zero. Moreover, 0 is the only real number whose absolute value is zero. That is, $|0| = 0$.

Be sure you see from the definition on the preceding page that the absolute value of a real number is never negative. For instance, if $a = -5$, then $|-5| = -(-5) = 5$.

---

**Example 3**    Finding Absolute Value

**a.** $|-7| = 7$        **b.** $\left|\frac{1}{2}\right| = \frac{1}{2}$

**c.** $|-4.8| = 4.8$      **d.** $-|-9| = -(9) = -9$

---

**Example 4**    Comparing Real Numbers

Place the correct symbol ($<$, $>$, or $=$) between the real numbers.

**a.** $|-4|$ ⬚ $|4|$      **b.** $|-5|$ ⬚ $3$     **c.** $-|-1|$ ⬚ $|-1|$

**Solution**

**a.** $|-4| = |4|$, because both are equal to 4.

**b.** $|-5| > 3$, because $|-5| = 5$ and 5 is greater than 3.

**c.** $-|-1| < |-1|$, because $-|-1| = -1$ and $|-1| = 1$.

---

▶ **Properties of Absolute Value**

Let $a$ and $b$ be real numbers. Then the following properties are true.

**1.** $|a| \geq 0$                **2.** $|-a| = |a|$

**3.** $|ab| = |a|\,|b|$            **4.** $\left|\dfrac{a}{b}\right| = \dfrac{|a|}{|b|}$

---

Absolute value can be used to define the distance between two numbers on the real number line. To see how this is done, consider the numbers $-3$ and 4, as shown in Figure A.6. To find the distance between these two numbers, subtract *either* number from the other and then take the absolute value of the difference. For instance,

$$(\text{Distance between } -3 \text{ and } 4) = |-3 - 4| = |-7| = 7.$$

**Figure A.6**    *The distance between $-3$ and 4 is 7.*

---

▶ **Distance Between Two Numbers**

Let $a$ and $b$ be real numbers. The **distance between $a$ and $b$** is given by

$$\text{Distance} = |b - a| = |a - b|.$$

### Example 5    Finding the Distance Between Two Numbers

**a.** The distance between 2 and 7 is

$$\text{Distance} = |2 - 7| = |-5| = 5.$$

**b.** The distance between 0 and $-4$ is

$$\text{Distance} = |0 - (-4)| = |4| = 4.$$

**c.** The statement "the distance between $x$ and 2 is at least 3" can be written as

$$|x - 2| \geq 3.$$

## Application

### Example 6    Budget Variance

You work in the accounting department of a company. One of your jobs is to check whether the monthly expenses of the other departments vary too much from their budgets. Your company's definition of "too much" is that the difference between actual and budgeted expenses must be less than or equal to $500 *and* less than or equal to 5% of the budgeted expenses. By letting $a$ represent the actual expenses and $b$ the budgeted expenses, you can translate these requirements as follows.

$$|a - b| \leq 500 \qquad \text{and} \qquad |a - b| \leq 0.05b$$

For travel, the budgeted expense was $12,500 and the actual expense was $12,872.56. For office supplies, the budgeted expense was $750 and the actual expense was $704.15. For wages, the budgeted expense was $84,600 and the actual expense was $85,143.95. Are these amounts within budget restrictions?

Solution

One way to determine whether these three expenses are within budget restrictions is to create a table, as follows.

|  | *Budgeted Expense, b* | *Actual Expense, a* | $|a - b|$ | $0.05b$ |
|---|---|---|---|---|
| Travel | $12,500.00 | $12,872.56 | $372.56 | $625.00 |
| Office supplies | $750.00 | $704.15 | $45.85 | $37.50 |
| Wages | $84,600.00 | $85,143.95 | $543.95 | $4230.00 |

From this table, you can see that the expense for travel passes both tests, but that each of the other two expenses fails one of the tests. Can you see why?

## A.1    Exercises

In Exercises 1–4, determine which numbers in the set are (a) natural numbers, (b) integers, (c) rational numbers, and (d) irrational numbers.

**1.** $\left\{-9, -\frac{7}{2}, 5, \frac{2}{3}, \sqrt{2}, 0.1\right\}$

**2.** $\left\{12, -13, 1, \sqrt{4}, \sqrt{6}, \frac{3}{2}\right\}$

**3.** $\left\{3, -1, \frac{1}{3}, \frac{6}{3}, -\frac{1}{2}, \sqrt{2}, -7.5\right\}$

**4.** $\left\{25, -17, \frac{12}{5}, \sqrt{9}, \sqrt{8}, -\sqrt{8}\right\}$

In Exercises 5–8, plot the two real numbers on the real number line and place the appropriate inequality symbol (< or >) between them.

**5.** $\frac{3}{2}, 7$

**6.** $-4, -8$

**7.** $\frac{5}{6}, \frac{2}{3}$

**8.** $-\frac{8}{7}, -\frac{3}{7}$

In Exercises 9–14, describe the subset of real numbers that is represented by the inequality, and sketch the subset on the real number line.

**9.** $x \le 5$

**10.** $x \ge -2$

**11.** $x > 3$

**12.** $x \ge 4$

**13.** $-2 < x < 2$

**14.** $0 \le x \le 5$

In Exercises 15–18, use inequality notation to describe the set of real numbers.

**15.** $x$ is negative.

**16.** $y$ is greater than 5 and less than or equal to 12.

**17.** The annual rate of inflation $r$ is expected to be at least 3.5%, but no more than 6%.

**18.** The price $p$ of unleaded gasoline is not expected to go above $1.45 per gallon during the coming year.

In Exercises 19–24, write the expression without using absolute value signs.

**19.** $|-10|$

**20.** $|3 - \pi|$

**21.** $\dfrac{-5}{|-5|}$

**22.** $-3 - |-3|$

**23.** $-3|-3|$

**24.** $|-1| - |-2|$

In Exercises 25–28, place the correct symbol (<, >, or =) between the two real numbers.

**25.** $|-4|$ ▨ $|4|$

**26.** $-5$ ▨ $-|5|$

**27.** $-|-6|$ ▨ $|-6|$

**28.** $-|-2|$ ▨ $-|2|$

In Exercises 29–34, find the distance between $a$ and $b$.

**29.** $a = -1$ ⟶ $b = 3$

**30.** $a = -\frac{5}{2}$ ⟶ $b = 0$

**31.** $a = -\frac{7}{2}, b = 0$

**32.** $a = \frac{3}{4}, b = \frac{9}{4}$

**33.** $a = 126, b = 75$

**34.** $a = -126, b = -75$

In Exercises 35–38, use absolute value notation to describe the expression.

**35.** The distance between $z$ and $\frac{3}{2}$ is greater than 1.

**36.** The distance between $z$ and 0 is less than 8.

**37.** $y$ is at least 6 units from 0.

**38.** $y$ is at most 2 units from $a$.

In Exercises 39 and 40, use a calculator to order the numbers from smallest to largest.

**39.** $\frac{7071}{5000}, \frac{584}{413}, \sqrt{2}, \frac{47}{33}, \frac{127}{90}$

**40.** $\frac{26}{15}, \sqrt{3}, 1.73\overline{20}, \frac{381}{220}, \sqrt{10} - \sqrt{2}$

In Exercises 41 and 42, use a calculator to find the decimal form of the rational number. If it is a nonterminating decimal, write the repeating pattern.

**41.** $\frac{5}{8}$

**42.** $\frac{6}{11}$

*Budget Variance*    In Exercises 43–48, the accounting department of a company is checking to see whether the actual expenses of a department differ from the budgeted expenses by more than $500 or 5%. Complete the missing parts of the table and determine whether each actual expense passes the "budget variance test."

| Budgeted Expense, b | Actual Expense, a | $\|a - b\|$ | $0.05b$ |
|---|---|---|---|
| **43.** $25,000.00 | $24,872.12 | | |
| **44.** $112,700.00 | $113,356.52 | | |
| **45.** $9400.00 | $9972.59 | | |
| **46.** $7500.00 | $7104.68 | | |
| **47.** $37,640.00 | $36,968.25 | | |
| **48.** $2575.00 | $2613.15 | | |

*Median Incomes* In Exercises 49–54, use the bar graph, which shows the median income for American households that purchase equipment for skiing, tennis, golf, fishing, camping, and hunting. In each exercise you are given a household income. Find the amount by which the income differs from the median income for the particular sport. (Source: National Sporting Goods Association)

| | *Median Income, y* | *Household Income, x* | $\lvert y - x \rvert$ |
|---|---|---|---|
| **49.** Skiing | | \$65,400 | |
| **50.** Hunting | | \$75,400 | |
| **51.** Tennis | | \$37,300 | |
| **52.** Golf | | \$18,760 | |
| **53.** Camping | | \$21,300 | |
| **54.** Fishing | | \$26,370 | |

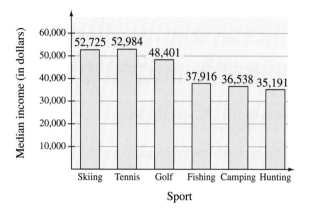

## A.2   The Basic Rules of Algebra

Algebraic Expressions ■ Basic Rules of Algebra ■ Equations ■ Calculators and Rounding

## Algebraic Expressions

One of the basic characteristics of algebra is the use of letters (or combinations of letters) to represent numbers. The letters used to represent numbers are called **variables,** and combinations of letters and numbers are called **algebraic expressions.** Here are a few examples of algebraic expressions.

$$5x, \qquad 2x - 3, \qquad \frac{4}{x^2 + 2}, \qquad 7x + y$$

▶ **Algebraic Expression**

A collection of letters (called **variables**) and real numbers (called **constants**) that are combined using the operations of addition, subtraction, multiplication, and division is an **algebraic expression.** (Other operations can also be used to form an algebraic expression.)

The **terms** of an algebraic expression are those parts that are separated by addition. For example, the algebraic expression $x^2 - 5x + 8$ has three terms: $x^2$, $-5x$, and 8. Note that $-5x$, rather than $5x$, is a term, because

$$x^2 - 5x + 8 = x^2 + (-5x) + 8.$$

The terms $x^2$ and $-5x$ are the **variable terms** of the expression, and 8 is the **constant term** of the expression. The numerical factor of a variable term is the **coefficient** of the variable term. For instance, the coefficient of the variable term $-5x$ is $-5$, and the coefficient of the variable term $x^2$ is 1.

---

**Example 1**    Identifying the Terms of an Algebraic Expression

| *Algebraic Expression* | *Terms* |
|---|---|
| **a.** $4x - 3$ | $4x, -3$ |
| **b.** $2x + 4y - 5$ | $2x, 4y, -5$ |

---

To **evaluate** an algebraic expression, substitute numerical values for each of the variables in the expression. Here are some examples.

| *Expression* | *Value of Variable* | *Substitute* | *Value of Expression* |
|---|---|---|---|
| $-3x + 5$ | $x = 3$ | $-3(3) + 5$ | $-9 + 5 = -4$ |
| $3x^2 + 2x - 1$ | $x = -1$ | $3(-1)^2 + 2(-1) - 1$ | $3 - 2 - 1 = 0$ |
| $-2x(x + 4)$ | $x = -2$ | $-2(-2)(-2 + 4)$ | $-2(-2)(2) = 8$ |
| $\dfrac{1}{x - 2}$ | $x = 2$ | $\dfrac{1}{2 - 2}$ | Undefined |

---

**Example 2**    Evaluating Algebraic Expressions

Evaluate the following algebraic expressions when $x = -2$ and $y = 3$.

**a.** $4y - 2x$     **b.** $5 + x^2$     **c.** $5 - x^2$

Solution

**a.** When $x = -2$ and $y = 3$, the expression $4y - 2x$ has a value of

$$4(3) - 2(-2) = 12 + 4 = 16.$$

**b.** When $x = -2$, the expression $5 + x^2$ has a value of

$$5 + (-2)^2 = 5 + 4 = 9.$$

**c.** When $x = -2$, the expression $5 - x^2$ has a value of

$$5 - (-2)^2 = 5 - 4 = 1.$$

---

## Basic Rules of Algebra

The four basic arithmetic operations are **addition, multiplication, subtraction,** and **division,** denoted by the symbols $+$, $\times$ or $\cdot$, $-$, and $\div$. Of these, addition and multiplication are considered to be the two primary arithmetic operations. Subtraction and division are defined as the inverse operations of addition and multiplication, as follows.

*Subtraction*                                    *Division*

$$a - b = a + (-b)$$        If $b \neq 0$, then $a \div b = a\left(\dfrac{1}{b}\right) = \dfrac{a}{b}$.

In these definitions, $-b$ is called the **opposite** (or additive inverse) of $b$, and $1/b$ is called the **reciprocal** (or multiplicative inverse) of $b$. In place of $a \div b$, you can use the fraction symbol $a/b$. In this fractional form, $a$ is called the **numerator** of the fraction and $b$ is called the **denominator.**

Be sure you see that the **basic rules of algebra,** listed below, are true for variables and algebraic expressions as well as for real numbers.

---

▶ **Basic Rules of Algebra**

Let $a$, $b$, and $c$ be real numbers, variables, or algebraic expressions.

| *Property* | *Example* |
|---|---|
| Commutative Property of Addition | |
| $a + b = b + a$ | $4x + x^2 = x^2 + 4x$ |
| Commutative Property of Multiplication | |
| $ab = ba$ | $(4 - x)x^2 = x^2(4 - x)$ |
| Associative Property of Addition | |
| $(a + b) + c = a + (b + c)$ | $(-x + 5) + 2x^2 =$ $-x + (5 + 2x^2)$ |
| Associative Property of Multiplication | |
| $(ab)c = a(bc)$ | $(2x \cdot 3y)(8) = (2x)(3y \cdot 8)$ |
| Distributive Property | |
| $a(b + c) = ab + ac$ | $3x(5 + 2x) = 3x \cdot 5 + 3x \cdot 2x$ |
| $(a + b)c = ac + bc$ | $(y + 8)y = y \cdot y + 8 \cdot y$ |
| Additive Identity Property | |
| $a + 0 = a$ | $5y^2 + 0 = 5y^2$ |
| Multiplicative Identity Property | |
| $a \cdot 1 = 1 \cdot a = a$ | $(4x^2)(1) = (1)(4x^2) = 4x^2$ |
| Additive Inverse Property | |
| $a + (-a) = 0$ | $5x^3 + (-5x^3) = 0$ |
| Multiplicative Inverse Property | |
| $a \cdot \dfrac{1}{a} = 1, \quad a \neq 0$ | $(x^2 + 4)\left(\dfrac{1}{x^2 + 4}\right) = 1$ |

---

Because subtraction is defined as "adding the opposite," the Distributive Properties are also true for subtraction. For instance, the "subtraction form" of $a(b + c) = ab + ac$ is

$$a(b - c) = a[b + (-c)] = ab + a(-c) = ab - ac.$$

**Example 3**    Identifying the Basic Rules of Algebra

Identify the rule of algebra illustrated in each of the following.

**a.** $(4x^2)5 = 5(4x^2)$

**b.** $(2y^3 + y) - (2y^3 + y) = 0$

**c.** $(4 + x^2) + 3x^2 = 4 + (x^2 + 3x^2)$

**d.** $(x - 5)7 + (x - 5)x = (x - 5)(7 + x)$

**e.** $2x \cdot \dfrac{1}{2x} = 1, \quad x \neq 0$

### Solution

**a.** This equation illustrates the Commutative Property of Multiplication.

**b.** This equation illustrates the Additive Inverse Property.

**c.** This equation illustrates the Associative Property of Addition. In other words, to form the sum $4 + x^2 + 3x^2$, it doesn't matter whether 4 and $x^2$ are added first or $x^2$ and $3x^2$ are added first.

**d.** This equation illustrates the Distributive Property in reverse order.

$$ab + ac = a(b + c) \qquad \text{Distributive Property}$$

$$(x - 5)7 + (x - 5)x = (x - 5)(7 + x)$$

**e.** This equation illustrates the Multiplicative Inverse Property. Note that it is important that $x$ be a nonzero number. If $x$ were allowed to be zero, you would be in trouble because the reciprocal of zero is undefined.

---

The following three lists summarize the basic properties of negation, zero, and fractions. When you encounter such lists, we suggest that you not only *memorize* a verbal description of each property, but that you also try to gain an *intuitive feeling* for the validity of each.

> ▶ **Properties of Negation**
>
> Let $a$ and $b$ be real numbers, variables, or algebraic expressions.
>
> | *Property* | *Example* |
> |---|---|
> | **1.** $(-1)a = -a$ | $(-1)7 = -7$ |
> | **2.** $-(-a) = a$ | $-(-6) = 6$ |
> | **3.** $(-a)b = -(ab) = a(-b)$ | $(-5)3 = -(5 \cdot 3) = 5(-3)$ |
> | **4.** $(-a)(-b) = ab$ | $(-2)(-6) = 12$ |
> | **5.** $-(a + b) = (-a) + (-b)$ | $-(3 + 8) = (-3) + (-8)$ |

Be sure you see the difference between the *negative* (or *opposite*) of a number and a *negative number*. If $a$ is already negative, then its additive inverse, $-a$, is positive. For instance, if $a = -5$, then $-a = -(-5) = 5$.

---

▶ **Properties of Zero**

Let $a$ and $b$ be real numbers, variables, or algebraic expressions. Then the following properties are true.

**1.** $a + 0 = a$     and     $a - 0 = a$

**2.** $a \cdot 0 = 0$

**3.** $\dfrac{0}{a} = 0, \quad a \neq 0$

**4.** $\dfrac{a}{0}$ is undefined.

**5.** Zero-Factor Property: If $ab = 0$, then $a = 0$ or $b = 0$.

---

The "or" in the Zero-Factor Property includes the possibility that both factors are zero. This is called an **inclusive or,** and it is the way the word "or" is always used in mathematics.

---

▶ **Properties of Fractions**

Let $a$, $b$, $c$, and $d$ be real numbers, variables, or algebraic expressions such that $b \neq 0$ and $d \neq 0$. Then the following properties are true.

**1.** *Equivalent fractions:* $\dfrac{a}{b} = \dfrac{c}{d}$ if and only if $ad = bc$.

**2.** *Rules of signs:* $-\dfrac{a}{b} = \dfrac{-a}{b} = \dfrac{a}{-b}$     and     $\dfrac{-a}{-b} = \dfrac{a}{b}$

**3.** *Generate equivalent fractions:* $\dfrac{a}{b} = \dfrac{ac}{bc}, \quad c \neq 0$

**4.** *Add or subtract with like denominators:* $\dfrac{a}{b} \pm \dfrac{c}{b} = \dfrac{a \pm c}{b}$

**5.** *Add or subtract with unlike denominators:* $\dfrac{a}{b} \pm \dfrac{c}{d} = \dfrac{ad \pm bc}{bd}$

**6.** *Multiply fractions:* $\dfrac{a}{b} \cdot \dfrac{c}{d} = \dfrac{ac}{bd}$

**7.** *Divide fractions:* $\dfrac{a}{b} \div \dfrac{c}{d} = \dfrac{a}{b} \cdot \dfrac{d}{c} = \dfrac{ad}{bc}, \quad c \neq 0$

---

In Property 1 (equivalent fractions) the phrase "if and only if" implies two statements. One statement is: If $a/b = c/d$, then $ad = bc$. The other statement is: If $ad = bc$, where $b \neq 0$ and $d \neq 0$, then $a/b = c/d$.

> **Example 4**    Properties of Zero and Properties of Fractions
>
> **a.** $x - \dfrac{0}{5} = x - 0 = x$        Properties 3 and 1 of zero
>
> **b.** $\dfrac{x}{5} = \dfrac{3 \cdot x}{3 \cdot 5} = \dfrac{3x}{15}$        Generate equivalent fractions.
>
> **c.** $\dfrac{x}{3} + \dfrac{2x}{5} = \dfrac{x \cdot 5 + 3 \cdot 2x}{15}$        Add fractions with unlike denominators.
>
> **d.** $\dfrac{7}{x} \div \dfrac{3}{2} = \dfrac{7}{x} \cdot \dfrac{2}{3} = \dfrac{14}{3x}$        Divide fractions.

If $a$, $b$, and $c$ are integers such that $ab = c$, then $a$ and $b$ are **factors** or **divisors** of $c$. For example, 2 and 3 are factors of 6 because $2 \cdot 3 = 6$. A **prime number** is a positive integer that has exactly two factors: itself and 1. For example, 2, 3, 5, 7, and 11 are prime numbers, whereas 1, 4, 6, 8, 9, and 10 are not. The numbers 4, 6, 8, 9, and 10 are **composite** because they can be written as the products of two or more prime numbers. The number 1 is neither prime nor composite. The **Fundamental Theorem of Arithmetic** states that every positive integer greater than 1 can be written as the product of prime numbers in precisely one way (disregarding order). For instance, the prime factorization of 24 is

$$24 = 2 \cdot 2 \cdot 2 \cdot 3.$$

When adding or subtracting fractions with unlike denominators, you can use Property 4 of fractions by rewriting both fractions so that they have the same denominator. This is called the **least common denominator** method.

> **Example 5**    Adding and Subtracting Fractions
>
> Evaluate $\dfrac{2}{15} - \dfrac{5}{9} + \dfrac{4}{5}$.
>
> **Solution**
>
> By prime factoring the denominators ($15 = 3 \cdot 5, 9 = 3 \cdot 3$, and $5 = 5$) you can see that the least common denominator is $3 \cdot 3 \cdot 5 = 45$. Thus, it follows that
>
> $$\frac{2}{15} - \frac{5}{9} + \frac{4}{5} = \frac{2 \cdot 3}{15 \cdot 3} - \frac{5 \cdot 5}{9 \cdot 5} + \frac{4 \cdot 9}{5 \cdot 9}$$
>
> $$= \frac{6 - 25 + 36}{45}$$
>
> $$= \frac{17}{45}.$$

## Equations

An **equation** is a statement of equality between two expressions. Thus, the statement

$$a + b = c + d$$

means that the expressions $a + b$ and $c + d$ represent the same number. For instance, because $1 + 4$ and $3 + 2$ both represent the number 5, you can write $1 + 4 = 3 + 2$. Three important properties of equality are as follows.

---

▶ **Properties of Equality**

Let $a$, $b$, and $c$ be real numbers, variables, or algebraic expressions.

**1.** Reflexive: $a = a$

**2.** Symmetric: If $a = b$, then $b = a$.

**3.** Transitive: If $a = b$ and $b = c$, then $a = c$.

---

In algebra, you often rewrite expressions by making substitutions that are permitted under the **Substitution Principle.** "If $a = b$, then $a$ can be replaced by $b$ in any expression involving $a$." Two important consequences of the Substitution Principle are the following rules.

**1.** If $a = b$, then $a + c = b + c$.          Add $c$ to both sides.

**2.** If $a = b$, then $ac = bc$.          Multiply both sides by $c$.

The first rule allows us to add the same number to both sides of an equation. The second allows us to multiply both sides of an equation by the same number. The converses of these two rules are called the **Cancellation Laws** for addition and multiplication.

**1.** If $a + c = b + c$, then $a = b$.          Subtract $c$ from both sides.

**2.** If $ac = bc$ and $c \neq 0$ then $a = b$.          Divide both sides by $c$.

When adding, subtracting, multiplying, or dividing more than two numbers, it is important to use symbols of grouping (such as parentheses) to indicate the order of operations.

---

**Example 6**    Symbols of Grouping

**a.** $7 - 3(4 - 2) = 7 - 3(2) = 7 - 6 = 1$

**b.** $(4 - 5) - (3 - 6) = (-1) - (-3) = -1 + 3 = 2$

---

## Calculators and Rounding

The specific keystrokes that are listed here are those that correspond to a standard scientific calculator and to a graphing calculator. The keystrokes listed here may not be the same as those for your graphing calculator. Consult your user's guide for specific keystrokes. Here are some comparisons between graphing calculator keys and scientific calculator keys.

**1.** The key marked $\boxed{\text{ENTER}}$ is similar to $\boxed{=}$.

**2.** The key marked $\boxed{(-)}$ is similar to $\boxed{+/-}$.

**3.** The key marked $\boxed{\wedge}$ is similar to $\boxed{y^x}$ .

**4.** The key marked $\boxed{x^{-1}}$ is similar to $\boxed{1/x}$ .

For example, you can evaluate $13^3$ on a graphing calculator or a scientific calculator as follows.

| *Graphing Calculator* | *Scientific Calculator* |
|---|---|
| 13 $\boxed{\wedge}$ 3 $\boxed{\text{ENTER}}$ | 13 $\boxed{y^x}$ 3 $\boxed{=}$ |

### Example 7    Using a Calculator

**Scientific Calculator**

| *Expression* | *Keystrokes* | *Display* |
|---|---|---|
| **a.** $7 - (5 \cdot 3)$ | 7 $\boxed{-}$ 5 $\boxed{\times}$ 3 $\boxed{=}$ | $-8$ |
| **b.** $-12^2 - 100$ | 12 $\boxed{+/-}$ $\boxed{x^2}$ $\boxed{-}$ 100 $\boxed{=}$ | $-244$ |
| **c.** $24 \div 2^3$ | 24 $\boxed{\div}$ 2 $\boxed{y^x}$ 3 $\boxed{=}$ | 3 |
| **d.** $3(10 - 4^2) \div 2$ | 3 $\boxed{\times}$ $\boxed{(}$ 10 $\boxed{-}$ 4 $\boxed{x^2}$ $\boxed{)}$ $\boxed{\div}$ 2 $\boxed{=}$ | $-9$ |
| **e.** 37% of 40 | .37 $\boxed{\times}$ 40 $\boxed{=}$ | 14.8 |

**Graphing Calculator**

| *Expression* | *Keystrokes* | *Display* |
|---|---|---|
| **a.** $7 - (5 \cdot 3)$ | 7 $\boxed{-}$ 5 $\boxed{\times}$ 3 $\boxed{\text{ENTER}}$ | $-8$ |
| **b.** $-12^2 - 100$ | $\boxed{(-)}$ 12 $\boxed{x^2}$ $\boxed{-}$ 100 $\boxed{\text{ENTER}}$ | $-244$ |
| **c.** $24 \div 2^3$ | 24 $\boxed{\div}$ 2 $\boxed{\wedge}$ 3 $\boxed{\text{ENTER}}$ | 3 |
| **d.** $3(10 - 4^2) \div 2$ | 3 $\boxed{(}$ 10 $\boxed{-}$ 4 $\boxed{x^2}$ $\boxed{)}$ $\boxed{\div}$ 2 $\boxed{\text{ENTER}}$ | $-9$ |
| **e.** 37% of 40 | .37 $\boxed{\times}$ 40 $\boxed{\text{ENTER}}$ | 14.8 |

Be sure you see the difference between the change sign keys $\boxed{+/-}$ or $\boxed{(-)}$ and the subtraction key $\boxed{-}$, as used in Example 7(b).

For all their usefulness, calculators do have a problem representing some numbers because they are limited to a finite number of digits. For instance, what does your calculator display when you compute $2 \div 3$? Some calculators simply truncate (drop) the digits that exceed their display range and display .66666666. Others round the number and display .66666667. Although the second display is more accurate, *both* of these decimal representations of $\frac{2}{3}$ contain rounding errors. When rounding decimals in this text, use the following rule: *round up on 5 or greater, round down on 4 or less.*

### Example 8    Rounding Decimal Numbers

| *Number* | *Rounded to Three Decimal Places* | |
|---|---|---|
| **a.** $\sqrt{2} = 1.4142135 \ldots$ | 1.414 | Round down. |
| **b.** $\pi = 3.1415926 \ldots$ | 3.142 | Round up. |
| **c.** $\dfrac{7}{9} = 0.7777777 \ldots$ | 0.778 | Round up. |

One of the best ways to minimize error due to rounding is to leave numbers in your calculator until your calculations are complete. If you want to save a number for future use, store it in your calculator's memory.

---

# Warm Up

The following warm-up exercises involve skills that were covered in earlier sections. You will use these skills in the exercise set for this section.

In Exercises 1–4, place the correct inequality symbol ($<$ or $>$) between the two numbers.

**1.** $-4 \quad -2$    **2.** $0 \quad -3$    **3.** $\sqrt{3} \quad 1.73$    **4.** $-\pi \quad -3$

In Exercises 5–8, find the distance between the two numbers.

**5.** $4, 6$      **6.** $-2, 2$      **7.** $0, -5$      **8.** $-1, 3$

In Exercises 9 and 10, evaluate the expression.

**9.** $|-7| + |7|$      **10.** $-|8 - 10|$

---

## A.2 Exercises

In Exercises 1–4, identify the terms of the algebraic expression.

**1.** $7x + 4$          **2.** $-5 + 3x$

**3.** $x^2 - 4x + 8$      **4.** $4x^3 + x - 5$

In Exercises 5–8, evaluate the expression for the values of $x$. (If not possible, state the reason.)

| | Expression | | Values |
|---|---|---|---|
| **5.** | $4x - 6$ | (a) $x = -1$ | (b) $x = 0$ |
| **6.** | $x^2 - 3x + 4$ | (a) $x = -2$ | (b) $x = 2$ |
| **7.** | $-x^3 + 4x$ | (a) $x = 0$ | (b) $x = 2$ |
| **8.** | $\dfrac{x + 1}{x - 1}$ | (a) $x = 1$ | (b) $x = -1$ |

In Exercises 9–23, identify the rule(s) of algebra illustrated by the equation.

**9.** $3 + 4 = 4 + 3$      **10.** $x + 9 = 9 + x$

**11.** $-15 + 15 = 0$      **12.** $2(x + 3) = 2x + 6$

**13.** $2\left(\frac{1}{2}\right) = 1$

**14.** $(5 + 11) \cdot 6 = 5 \cdot 6 + 11 \cdot 6$

**15.** $\dfrac{1}{h + 6}(h + 6) = 1, \quad h \neq -6$

**16.** $h + 0 = h$

**17.** $57 \cdot 1 = 57$

**18.** $(z - 2) + 0 = z - 2$

**19.** $1 \cdot (1 + x) = 1 + x$

**20.** $6 + (7 + 8) = (6 + 7) + 8$

**21.** $x + (y + 10) = (x + y) + 10$

**22.** $x(3y) = (x \cdot 3)y = (3x)y$

**23.** $\frac{1}{7}(7 \cdot 12) = \left(\frac{1}{7} \cdot 7\right)12 = 1 \cdot 12 = 12$

In Exercises 24 and 25, write the prime factorization of the integer.

**24.** $36$                **25.** $30$

In Exercises 26–37, perform the operations. (Write fractional answers in reduced form.)

**26.** $(8 - 17) + 3$      **27.** $-3(5 - 2)$

**28.** $(4 - 7)(-2)$      **29.** $(-5)(-8)$

**30.** $2\left(\dfrac{77}{-11}\right)$      **31.** $\dfrac{27 - 35}{4}$

**32.** $8(-6)(-2)$      **33.** $\frac{10}{11} + \frac{6}{33} - \frac{13}{66}$

**34.** $\frac{2}{5} \cdot \frac{7}{8}$      **35.** $\frac{2}{3} \cdot \frac{5}{8} \cdot \frac{3}{4}$

**36.** $\frac{2}{3} \div 8$      **37.** $\left(\frac{3}{5} \div 3\right) - \left(6 \cdot \frac{4}{8}\right)$

In Exercises 38–41, use a calculator to evaluate the expression. (Round to two decimal places.)

**38.** $3\left(-\frac{5}{12} + \frac{3}{8}\right)$      **39.** $2\left(-7 + \frac{1}{6}\right)$

**40.** $\dfrac{11.46 - 5.37}{3.91}$          **41.** $\dfrac{-8.31 + 4.83}{7.65}$

In Exercises 42 and 43, use a calculator to solve the percent problem.

**42.** 33% of 57          **43.** 121% of 34

In Exercises 44 and 45, find the percent that corresponds to the unlabeled portion of the circle graph.

**44.**

**45.**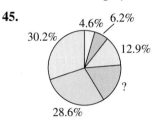

**46.** *Cookie Sales*    The circle graph shows the shares for the best-selling Girl Scout cookies for 1998. What percent of the total cookie sales did Thin Mints have? If 175 million boxes of Girl Scout cookies were sold in 1998, find the numbers of boxes of Thin Mints, Samoas, Tagalongs, Do-Si-Dos, and Trefoils that were sold. (Round your answers to the nearest tenth of a million boxes.)    (Source: Girl Scouts of the USA)

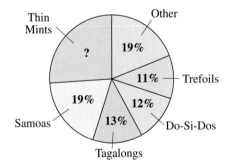

**47.** *Federal Government Expenses*    The circle graph shows the types of expenses for the federal government in 1996. What percent of the total budget was spent by Social Security? If the total expenses were 1,560,500,000,000 dollars, find the expense for each of the indicated categories. (Round your answers to the nearest billion dollars.)    (Source: Office of Management and Budget)

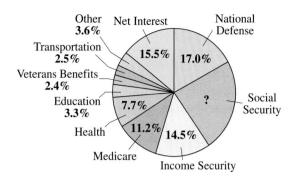

**Figure for 47**

**48.** *Calculator Keystrokes*    Write the algebraic expression that corresponds to the following keystrokes.

5 ⊠ ⎛ 2.7 ⊟ 9.4 ⎞ ⊜          Scientific

5 ⎛ 2.7 ⊟ 9.4 ⎞ ⎕ENTER⎕          Graphing

**49.** *Calculator Keystrokes*    Write the algebraic expression that corresponds to the following keystrokes.

2 ⊠ ⎛ 4 ⎕+/−⎕ ⊞ 2 ⎞ ⊜          Scientific

2 ⎛ ⎕(−)⎕ 4 ⊞ 2 ⎞ ⎕ENTER⎕          Graphing

**50.** *Calculator Keystrokes*    Write the keystrokes used to evaluate the following algebraic expression. (Base your description on either a scientific or a graphing calculator.)

$5(18 - 2^3) \div 10$

**51.** *Calculator Keystrokes*    Write the keystrokes used to evaluate the following algebraic expression. (Base your description on either a scientific or a graphing calculator.)

$-6^2 - [7 + (-2)^3]$

**52.** *College Enrollment*    The percent of students 24 years old or older at a college is 44.7%. If the college enrollment for the 1998–1999 academic year was 13,385 students, how many students were under 24 years old?

**53.** *College Enrollment*    The percent of students who attend evening classes at a college is 44.1%. If the total enrollment for the college is 11,875 students, how many students attend evening classes?

**A.3** Integer Exponents

Properties of Exponents ■ Scientific Notation ■ Applications

## Properties of Exponents

Repeated multiplication of a real number by itself can be written in **exponential form.** Here are some examples.

| *Repeated Multiplication* | *Exponential Form* |
|---|---|
| $7 \cdot 7$ | $7^2$ |
| $a \cdot a \cdot a \cdot a \cdot a$ | $a^5$ |
| $(-4)(-4)(-4)$ | $(-4)^3$ |
| $(2x)(2x)(2x)(2x)$ | $(2x)^4$ |

### Study Tip

It is important to recognize the difference between exponential forms such as $(-2)^4$ and $-2^4$. In $(-2)^4$, the parentheses indicate that the exponent applies to the negative sign as well as to the 2, but in $-2^4 = -(2^4)$, the exponent applies only to the 2. Similarly, in $(5x)^3$, the parentheses indicate that the exponent applies to the 5 as well as to the $x$, whereas in $5x^3 = 5(x^3)$, the exponent applies only to the $x$.

▶ **Exponential Notation**

Let $a$ be a real number, a variable, or an algebraic expression, and let $n$ be a positive integer. Then

$$a^n = \underbrace{a \cdot a \cdot a \cdots a}_{n \text{ factors}}$$

where $n$ is the **exponent** and $a$ is the **base.** The expression $a^n$ is read as "$a$ to the $n$th **power**" or simply "$a$ to the $n$th."

When multiplying exponential expressions with the same base, *add* exponents.

$$a^m \cdot a^n = a^{m+n} \qquad \text{Add exponents when multiplying.}$$

For instance, to multiply $2^2$ and $2^3$, you can write

$$2^2 \cdot 2^3 = \overbrace{(2 \cdot 2)}^{\substack{\text{Two} \\ \text{factors}}} \cdot \overbrace{(2 \cdot 2 \cdot 2)}^{\substack{\text{Three} \\ \text{factors}}} = \overbrace{2 \cdot 2 \cdot 2 \cdot 2 \cdot 2}^{\substack{\text{Five} \\ \text{factors}}}$$
$$= 2^{2+3} = 2^5.$$

On the other hand, when dividing exponential expressions, *subtract* exponents. That is,

$$\frac{a^m}{a^n} = a^{m-n}, \qquad a \neq 0. \qquad \text{Subtract exponents when dividing.}$$

These and other properties of exponents are summarized in the list on the following page.

> ▶ **Properties of Exponents**
>
> Let $a$ and $b$ be real numbers, variables or algebraic expressions, and let $m$ and $n$ be integers. (Assume all denominators and bases are nonzero.)
>
> | *Property* | *Example* |
> |---|---|
> | **1.** $a^m a^n = a^{m+n}$ | $3^2 \cdot 3^4 = 3^{2+4} = 3^6$ |
> | **2.** $\dfrac{a^m}{a^n} = a^{m-n}$ | $\dfrac{x^7}{x^4} = x^{7-4} = x^3$ |
> | **3.** $a^{-n} = \dfrac{1}{a^n}$ | $y^{-4} = \dfrac{1}{y^4}$ |
> | **4.** $a^0 = 1, \quad a \neq 0$ | $(x^2 + 1)^0 = 1$ |
> | **5.** $(ab)^m = a^m b^m$ | $(5x)^3 = 5^3 x^3 = 125x^3$ |
> | **6.** $(a^m)^n = a^{mn}$ | $(y^3)^{-4} = y^{3(-4)} = y^{-12} = \dfrac{1}{y^{12}}$ |
> | **7.** $\left(\dfrac{a}{b}\right)^m = \dfrac{a^m}{b^m}$ | $\left(\dfrac{2}{x}\right)^3 = \dfrac{2^3}{x^3} = \dfrac{8}{x^3}$ |
> | **8.** $\left(\dfrac{a}{b}\right)^{-n} = \left(\dfrac{b}{a}\right)^n, \quad a \neq 0, b \neq 0$ | $\left(\dfrac{3}{2}\right)^{-3} = \left(\dfrac{2}{3}\right)^3$ |
> | **9.** $|a^2| = |a|^2 = a^2$ | $|(-2)^2| = |-2|^2 = (-2)^2 = 4$ |

Be sure you see that these properties of exponents apply for *all* integers $m$ and $n$, not just positive ones. For instance, by Property 2, you have

$$\frac{3^4}{3^{-5}} = 3^{4-(-5)}$$

$$= 3^{4+5}$$

$$= 3^9.$$

**Example 1**    Using Properties of Exponents

**a.** $3^4 \cdot 3^{-1} = 3^{4-1} = 3^3 = 27$

**b.** $\dfrac{5^6}{5^4} = 5^{6-4} = 5^2 = 25$

**c.** $5\left(\dfrac{2}{5}\right)^3 = 5 \cdot \dfrac{2^3}{5^3} = 5 \cdot 5^{-3} \cdot 2^3 = 5^{-2}2^3 = \dfrac{2^3}{5^2} = \dfrac{8}{25}$

**d.** $(-5 \cdot 2^3)^2 = (-5)^2 \cdot (2^3)^2 = 25 \cdot 2^6 = 25 \cdot 64 = 1600$

**e.** $(-3ab^4)(4ab^{-3}) = -12a \cdot a(b^4)(b^{-3}) = -12a^2 b$

**f.** $3a(-4a^2)^0 = 3a(1) = 3a, \quad a \neq 0$

**g.** $\left(\dfrac{5x^3}{y}\right)^2 = \dfrac{5^2(x^3)^2}{y^2} = \dfrac{25x^6}{y^2}$

The next two examples show how expressions involving negative exponents can be rewritten using positive exponents.

### Example 2    Rewriting with Positive Exponents

**a.** $x^{-1} = \dfrac{1}{x}$                    Property 3: $a^{-n} = \dfrac{1}{a^n}$

**b.** $\dfrac{1}{3x^{-2}} = \dfrac{1(x^2)}{3} = \dfrac{x^2}{3}$             $-2$ exponent does not apply to 3.

**c.** $\dfrac{12a^3b^{-4}}{4a^{-2}b} = \dfrac{12a^3 \cdot a^2}{4b \cdot b^4} = \dfrac{3a^5}{b^5}$

### Example 3    Quotients Raised to Negative Powers

Rewrite $\left(\dfrac{3x^2}{y}\right)^{-2}$ and simplify.

Solution

$$\left(\frac{3x^2}{y}\right)^{-2} = \frac{3^{-2}(x^2)^{-2}}{y^{-2}} = \frac{3^{-2}x^{-4}}{y^{-2}} = \frac{y^2}{3^2x^4} = \frac{y^2}{9x^4}$$

### Example 4    Ratio of Volume to Surface Area

The volume $V$ and surface area $S$ of a sphere are given by

$$V = \frac{4}{3}\pi r^3 \qquad \text{and} \qquad S = 4\pi r^2$$

where $r$ is the radius of the sphere. A spherical weather balloon has a radius of 2 feet, as shown in Figure A.7. Find the quotient of the volume and the surface area.

Solution

Form the quotient $V/S$, and simplify, as follows.

$$\frac{V}{S} = \frac{\frac{4}{3}\pi r^3}{4\pi r^2}$$

$$= \frac{\frac{4}{3}\pi 2^3 \text{ ft}^3}{4\pi 2^2 \text{ ft}^2}$$

$$= \frac{1}{3}(2) \text{ feet}$$

$$= \frac{2}{3} \text{ feet}$$

$r = 2$ ft

Figure A.7

## Scientific Notation

Exponents provide an efficient way of writing and computing with very large (or very small) numbers. For instance, a drop of water contains more than 33 billion billion molecules—that is, 33 followed by 18 zeros.

33,000,000,000,000,000,000

It is convenient to write such numbers in **scientific notation.** This notation has the form $c \times 10^n$, where $1 \leq c < 10$ and $n$ is an integer. Thus, the number of molecules in a drop of water can be written in scientific notation as

$$3.3 \times 10,000,000,000,000,000,000 = 3.3 \times 10^{19}.$$

The *positive* exponent 19 indicates that the number is large (10 or more) and that the decimal point has been moved 19 places. A *negative* exponent in scientific notation indicates that the number is *small* (less than 1). For instance, the mass (in grams) of one electron is approximately

$$9.0 \times 10^{-28} = 0.00000000000000000000000000009.$$

28 decimal places

### Example 5    Converting to Scientific Notation

**a.** $0.0000572 = 5.72 \times 10^{-5}$     Number is less than 1.

**b.** $149,400,000 = 1.494 \times 10^8$     Number is greater than 10.

### Example 6    Converting to Decimal Notation

**a.** $3.125 \times 10^2 = 312.5$     Number is greater than 10.

**b.** $3.73 \times 10^{-6} = 0.00000373$     Number is less than 1.

Most calculators automatically switch to scientific notation when they are showing large (or small) numbers that exceed the display range. Try multiplying $86,500,000 \times 6000$. If your calculator follows standard conventions, its display should be

   5.19   11    or    5.19   E   11 .

This means that $c = 5.19$ and the exponent of 10 is $n = 11$, which implies that the number is $5.19 \times 10^{11}$. To *enter* numbers in scientific notation, your calculator should have an exponential entry key labeled $\boxed{\text{EE}}$.

### Example 7    The Speed of Light

The distance between earth and the sun is approximately 93 million miles, as shown in Figure A.8. How long does it take for light to travel from the sun to earth? Use the fact that light travels at a rate of approximately 186,000 miles per second.

Solution

Using the formula Distance = (rate)(time), you find the time as follows.

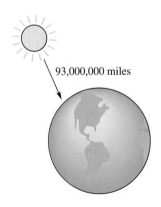

93,000,000 miles

Figure A.8

$$\text{Time} = \frac{\text{distance}}{\text{rate}} = \frac{93 \text{ million miles}}{186,000 \text{ miles per second}}$$

$$= \frac{9.3 \times 10^7 \text{ miles}}{1.86 \times 10^5 \text{ miles/second}}$$

$$= 5 \times 10^2 \text{ seconds}$$

$$\approx 8.33 \text{ minutes}$$

Note that to convert 500 seconds to 8.33 minutes, you divide by 60 because there are 60 seconds in a minute.

## Applications

One of the most useful features of a calculator is its ability to evaluate exponential expressions. Consult your user's guide for specific keystrokes.

| Example 8 | Using a Calculator to Raise a Number to a Power |

**Scientific Calculator**

| | *Expression* | *Keystrokes* | *Display* |
|---|---|---|---|
| **a.** | $13^4 + 5$ | 13 $\boxed{y^x}$ 4 $\boxed{+}$ 5 $\boxed{=}$ | 28566 |
| **b.** | $3^{-2} + 4^{-1}$ | 3 $\boxed{y^x}$ 2 $\boxed{+/-}$ $\boxed{+}$ 4 $\boxed{y^x}$ 1 $\boxed{+/-}$ $\boxed{=}$ | .3611111111 |
| **c.** | $\dfrac{3^5 + 1}{3^5 - 1}$ | $\boxed{(}$ 3 $\boxed{y^x}$ 5 $\boxed{+}$ 1 $\boxed{)}$ $\boxed{\div}$ | |
| | | $\boxed{(}$ 3 $\boxed{y^x}$ 5 $\boxed{-}$ 1 $\boxed{)}$ $\boxed{=}$ | 1.008264463 |

**Graphing Calculator**

| | *Expression* | *Keystrokes* | *Display* |
|---|---|---|---|
| **a.** | $13^4 + 5$ | 13 $\boxed{\wedge}$ 4 $\boxed{+}$ 5 $\boxed{\text{ENTER}}$ | 28566 |
| **b.** | $3^{-2} + 4^{-1}$ | 3 $\boxed{\wedge}$ $\boxed{(-)}$ 2 $\boxed{+}$ 4 $\boxed{\wedge}$ $\boxed{(-)}$ 1 $\boxed{\text{ENTER}}$ | .3611111111 |
| **c.** | $\dfrac{3^5 + 1}{3^5 - 1}$ | $\boxed{(}$ 3 $\boxed{\wedge}$ 5 $\boxed{+}$ 1 $\boxed{)}$ $\boxed{\div}$ | |
| | | $\boxed{(}$ 3 $\boxed{\wedge}$ 5 $\boxed{-}$ 1 $\boxed{)}$ $\boxed{\text{ENTER}}$ | 1.008264463 |

The following box gives a formula for finding the balance in a savings account.

---

▶ **Balance in an Account**

The balance $A$ in an account that earns an annual interest rate of $r$ (in decimal form) for $t$ years is given by one of the following.

$$A = P(1 + rt) \qquad \text{Simple interest}$$

$$A = P\left(1 + \frac{r}{n}\right)^{nt} \qquad \text{Compound interest}$$

In both formulas, $P$ is the principal (or the initial deposit). In the formula for compound interest, $n$ is the number of compoundings *per year.*

**Example 9**    Finding the Balance in an Account

**a.** If a friend borrows $250 at 8% annual interest for 6 months, the amount due after 6 months (at simple interest) is

$$A = P(1 + rt) = 250\left[1 + 0.08\left(\frac{1}{2}\right)\right] = \$260.00.$$

**b.** If you deposit $1000 at 6% annual interest compounded quarterly for 5 years, the balance after 5 years is

$$A = P\left(1 + \frac{r}{n}\right)^{nt} = 1000\left(1 + \frac{0.06}{4}\right)^{(4)(5)} = \$1346.86.$$

In addition to finding the balance in an account, the compound interest formula can also be used to determine the rate of inflation. To apply the formula, you must know the cost of an item for two different years, as demonstrated in Example 10.

**Example 10**    Finding the Rate of Inflation

In 1974, the cost of a first-class postage stamp was $0.10. By 1999, the cost had risen to $0.33, as shown in Figure A.9. Find the average annual rate of inflation for first-class postage over this 25-year period.    (Source: U.S. Postal Service)

**Solution**

To find the average annual rate of inflation, use the formula for compound interest with *annual* compounding. Thus, you need to find the value of $r$ that will make the following equation true.

$$A = P\left(1 + \frac{r}{n}\right)^{nt}$$

$$0.33 = 0.10(1 + r)^{25}$$

You can begin by guessing that the average annual rate of inflation was 5%. Entering $r = 0.05$ in the formula, you find that $0.10(1 + 0.05)^{25} \approx 0.3386$. Because this result is more than 0.33, try some smaller values of $r$. Finally, you can discover that

$$0.10(1 + 0.049)^{25} \approx 0.33.$$

So, the average annual rate of inflation for first-class postage from 1974 to 1999 was about 4.9%.

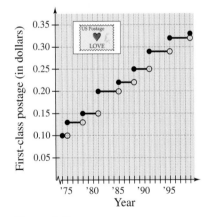

**Figure A.9**

# Warm Up

The following warm-up exercises involve skills that were covered in earlier sections. You will use these skills in the exercise set for this section.

In Exercises 1–10, perform the indicated operation(s) and simplify.

**1.** $\left(\frac{2}{3}\right)\left(\frac{3}{2}\right)$

**2.** $\left(\frac{1}{4}\right)(5)(4)$

**3.** $3\left(\frac{2}{7}\right) + 11\left(\frac{2}{7}\right)$

**4.** $\frac{1}{2} \div 2$

**5.** $\frac{1}{3} + \frac{1}{2} - \frac{5}{6}$

**6.** $\frac{1}{3} \div \frac{1}{3}$

**7.** $\frac{1}{7} + \frac{1}{3} - \frac{1}{21}$

**8.** $11\left(\frac{1}{4}\right) + \frac{5}{4}$

**9.** $\frac{1}{12} - \frac{1}{3} + \frac{1}{8}$

**10.** $\left(\frac{1}{2} - \frac{1}{3}\right) \div \frac{1}{6}$

## A.3 Exercises

In Exercises 1–16, evaluate the expression. Write fractional answers in simplified form.

**1.** $2^2 \cdot 2^4$

**2.** $\dfrac{2^6}{2^3}$

**3.** $(3^3)^2$

**4.** $(2^4)^2$

**5.** $-3^4$

**6.** $(-3)^4$

**7.** $\left(\frac{1}{2}\right)^{-3}$

**8.** $5^{-1} - 2^{-1}$

**9.** $\left(\frac{3}{4}\right)^2$

**10.** $\left(\frac{3}{4}\right)^{-2}$

**11.** $(2^3 \cdot 3^2)^2$

**12.** $\left(-\frac{3}{5}\right)^3\left(\frac{5}{3}\right)^2$

**13.** $6^5 \cdot 6^{-3}$

**14.** $6 \cdot 2^{-3} \cdot 3^{-1}$

**15.** $3^0$

**16.** $(-2)^0$

In Exercises 17–20, evaluate the expression for the indicated value of x.

| Expression | Value |
|---|---|
| **17.** $\dfrac{x^2}{2}$ | $x = 6$ |
| **18.** $4x^{-3}$ | $x = 2$ |
| **19.** $7x^{-2}$ | $x = 4$ |
| **20.** $6x^0 - (6x)^0$ | $x = 10$ |

In Exercises 21–38, simplify the expression.

**21.** $(-5z)^3$

**22.** $(8x^4)(2x^3)$

**23.** $5x^4(x^2)$

**24.** $10(x^2)^2$

**25.** $(-z)^3(3z^4)$

**26.** $\dfrac{25y^8}{10y^4}$

**27.** $\dfrac{3x^5}{x^3}$

**28.** $\left(\dfrac{4}{y}\right)^3\left(\dfrac{3}{y}\right)^4$

**29.** $\dfrac{7x^2}{x^3}$

**30.** $(2x^5)^0, \quad x \neq 0$

**31.** $(x + 5)^0, \quad x \neq -5$

**32.** $(4y^{-2})(8y^4)$

**33.** $(-2x^2)^3(4x^3)^{-1}$

**34.** $\left(\dfrac{x^{-3}y^4}{5}\right)^{-3}$

**35.** $\left(\dfrac{x}{10}\right)^{-1}$

**36.** $\dfrac{x^2 \cdot x^n}{x^3 \cdot x^n}$

**37.** $3^n \cdot 3^{2n}$

**38.** $2^m \cdot 2^{3m}$

In Exercises 39–43, write the number in scientific notation.

**39.** *Land area of earth:* 57,500,000 square miles

**40.** *Ocean area of earth:* 139,400,000 square miles

**41.** *Light year:* 9,461,000,000,000,000 kilometers

**42.** *Relative density of hydrogen:* 0.0000899

**43.** *One point (printer's measure):* 0.013837 inch

In Exercises 44–48, write the number in decimal form.

**44.** *Number of air sacs in lungs:* $3.5 \times 10^8$

**45.** *Temperature of sun:* $1.3 \times 10^7$ degrees Celsius

**46.** *Number of insect species:* $9.5 \times 10^5$

**47.** *Electron charge:* $4.8 \times 10^{-10}$ electrostatic unit

**48.** *Width of human hair:* $9.0 \times 10^{-4}$ meter

In Exercises 49 and 50, use a calculator to evaluate the expression. (Round to three decimal places.)

**49.** (a) $2400(1 + 0.06)^{20}$

(b) $750\left(1 + \dfrac{0.11}{365}\right)^{800}$

**50.** (a) $5000(1 + 0.02)^{60}$

(b) $2500\left(1 + \dfrac{0.085}{12}\right)^{60}$

**51.** *Balance in an Account*    Ten thousand dollars is deposited in an account with an annual percentage rate of 6.5% for 10 years. What is the balance in the account if the interest is compounded (a) quarterly and (b) monthly?

**52.** *Balance in an Account*    Two thousand dollars is deposited in an account with an annual percentage rate of 7.75% for 15 years. What is the balance in the account if the interest is compounded (a) quarterly and (b) monthly?

**53.** *Reasoning*    A student calculates the amount in an account after 5 years to be $10,616.78 when $2000 was invested at 6% interest compounded monthly. Determine the error the student made in the calculation if the problem was set up correctly. Explain to the student how to correct the error.

**54.** *Reasoning*    A student tells you that $1000 invested at 8% for 10 years compounded quarterly will yield $2208.04. The student then concludes that $2000 invested at 8% for 10 years compounded quarterly will yield $4416.08. Do you agree? Explain.

**55.** *College Tuition*    The bar chart shows the average tuition at private universities in the United States from 1986 to 1996. Estimate the average annual rate of inflation for tuition over this 10-year period. (Source: College Board)

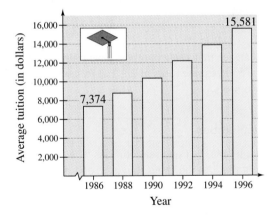

**56.** *Minimum Wage*    The bar chart shows the minimum wage in the United States from 1977 to 1997. Estimate the average annual rate of inflation for minimum wages over this 20-year period. (Source: U.S. Department of Labor)

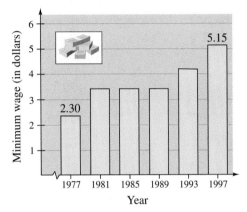

**57.** *Calculator Keystrokes*    Write the algebraic expression that corresponds to the following keystrokes.

⎛ 5.1 �| 3.6 ⎞ $y^x$ 5 =        Scientific

⎛ 5.1 �| 3.6 ⎞ ^ 5 ENTER        Graphing

**58.** *Calculator Keystrokes*    Write the algebraic expression that corresponds to the following keystrokes.

1 + 3 × 2 = $y^x$ 2 +/− =        Scientific

1 + 3 × 2 ENTER ^ (−) 2 ENTER        Graphing

**59.** *Think About It*    Because $-2^3 = -8$ and $(-2)^3 = -8$, a student concludes that $-a^n = (-a)^n$ where $n$ is an integer. Do you agree? Can you find an example where $-a^n \neq (-a)^n$?

**60.** *Think About It*    Because $2^0 = 1$ and $3^0 = 1$, a student concludes that $0^0 = 1$. Do you agree? What argument would you give to show that $0^0 \neq 1$?

## A.4   Radicals and Rational Exponents

Radicals and Properties of Radicals ▪ Simplifying Radicals ▪ Rational Exponents ▪ Radicals and Calculators

## Radicals and Properties of Radicals

A **square root** of a number is defined as one of its two equal factors. For example, 5 is a square root of 25 because 5 is one of the two equal factors of 25. In a similar way, a **cube root** of a number is one of its three equal factors. Here are some examples.

| *Number* | *Equal Factors* | *Root* | |
|---|---|---|---|
| $25 = (-5)^2$ | $(-5)(-5)$ | $-5$ | (square root) |
| $-64 = (-4)^3$ | $(-4)(-4)(-4)$ | $-4$ | (cube root) |
| $81 = 3^4$ | $3 \cdot 3 \cdot 3 \cdot 3$ | $3$ | (fourth root) |

In general, an **nth root** of a number is defined as follows.

---

▶ **Definition of nth Root of a Number**

Let $a$ and $b$ be real numbers and let $n$ be a positive integer. If

$$a = b^n$$

then $b$ is an **nth root of a.** If $n = 2$, the root is a **square root,** and if $n = 3$, the root is a **cube root.**

---

From this definition, you can see that some numbers have more than one nth root. For example, both 5 and $-5$ are square roots of 25. The following definition distinguishes between these two roots.

---

▶ **Principal nth Root of a Number**

Let $a$ be a real number that has at least one real nth root. The **principal nth root of a** is the nth root that has the same sign as $a$, and it is denoted by the **radical symbol**

$$\sqrt[n]{a}. \qquad \text{Principal } n\text{th root}$$

The positive integer $n$ is the **index** (the plural of index is *indexes* or *indices*) of the radical, and the number $a$ is the **radicand.** If $n = 2$, omit the index and write $\sqrt{a}$ rather than $\sqrt[2]{a}$.

---

**Example 1**    Evaluating Expressions Involving Radicals

**a.** The principal square root of 121 is $\sqrt{121} = 11$ because $11^2 = 121$.

**b.** The principal cube root of $\frac{125}{64}$ is $\sqrt[3]{\frac{125}{64}} = \frac{5}{4}$ because $\left(\frac{5}{4}\right)^3 = \frac{5^3}{4^3} = \frac{125}{64}$.

**c.** The principal fifth root of $-32$ is $\sqrt[5]{-32} = -2$ because $(-2)^5 = -32$.

**d.** $-\sqrt{49} = -7$ because $7^2 = 49$.

**e.** $\sqrt[4]{-81}$ is not a real number because there is no real number that can be raised to the fourth power to produce $-81$.

From Example 1, you can make the following generalizations about $n$th roots of a real number.

**1.** If $a$ is a positive real number and $n$ is a positive *even* integer, then $a$ has exactly two real $n$th roots, which are denoted by $\sqrt[n]{a}$ and $-\sqrt[n]{a}$.

**2.** If $a$ is any real number and $n$ is an *odd* integer, then $a$ has only one (real) $n$th root. It is the principal $n$th root and is denoted by $\sqrt[n]{a}$.

**3.** If $a$ is negative and $n$ is an *even* integer, then $a$ has no (real) $n$th root.

Integers such as 1, 4, 9, 16, 49, and 81 are called **perfect squares** because they have integer square roots. Similarly, integers such as 1, 8, 27, 64, and 125 are called **perfect cubes** because they have integer cube roots.

---

▶ **Properties of Radicals**

Let $a$ and $b$ be real numbers, variables, or algebraic expressions such that the indicated roots are real numbers, and let $m$ and $n$ be positive integers. Then the following properties are true.

| *Property* | *Example* |
|---|---|
| **1.** $\sqrt[n]{a^m} = \left(\sqrt[n]{a}\right)^m$ | $\sqrt[3]{8^2} = \left(\sqrt[3]{8}\right)^2 = (2)^2 = 4$ |
| **2.** $\sqrt[n]{a} \cdot \sqrt[n]{b} = \sqrt[n]{ab}$ | $\sqrt{5} \cdot \sqrt{7} = \sqrt{5 \cdot 7} = \sqrt{35}$ |
| **3.** $\dfrac{\sqrt[n]{a}}{\sqrt[n]{b}} = \sqrt[n]{\dfrac{a}{b}}, \quad b \neq 0$ | $\dfrac{\sqrt[4]{27}}{\sqrt[4]{9}} = \sqrt[4]{\dfrac{27}{9}} = \sqrt[4]{3}$ |
| **4.** $\sqrt[m]{\sqrt[n]{a}} = \sqrt[mn]{a}$ | $\sqrt[3]{\sqrt{10}} = \sqrt[6]{10}$ |
| **5.** $\left(\sqrt[n]{a}\right)^n = a$ | $\left(\sqrt{3}\right)^2 = 3$ |
| **6.** For $n$ even, $\sqrt[n]{a^n} = \lvert a \rvert$. | $\sqrt{(-12)^2} = \lvert -12 \rvert = 12$ |
| For $n$ odd, $\sqrt[n]{a^n} = a$. | $\sqrt[3]{(-12)^3} = -12$ |

A common special case of Property 6 is $\sqrt{a^2} = \lvert a \rvert$.

## Simplifying Radicals

An expression involving radicals is in **simplest form** when the following conditions are satisfied.

**1.** All possible factors have been removed from the radical.

**2.** All fractions have radical-free denominators (accomplished by a process called *rationalizing the denominator*).

**3.** The index of the radical has been reduced as far as possible.

To simplify a radical, factor the radicand into factors whose powers are multiples of the index. The roots of these factors are written outside the radical, and the "leftover" factors make up the new radicand.

### Example 2    Simplifying Even Roots

**a.** $\sqrt[4]{48} = \sqrt[4]{16 \cdot 3}$                    Find largest fourth-power factor.

$\quad\quad\quad = \sqrt[4]{2^4 \cdot 3}$                    Rewrite.

$\quad\quad\quad = 2\sqrt[4]{3}$                    Find fourth root.

**b.** $\sqrt{75x^3} = \sqrt{25x^2 \cdot 3x}$                    Find largest square factor.

$\quad\quad\quad = \sqrt{(5x)^2 \cdot 3x}$                    Rewrite.

$\quad\quad\quad = 5x\sqrt{3x}, \quad x \geq 0$                    Find root of perfect square.

**c.** $\sqrt[4]{(5x)^4} = |5x| = 5|x|$.

In Example 2(c), note that absolute value is included in the answer because $\sqrt[4]{x^4} = |x|$.

### Example 3    Simplifying Odd Roots

**a.** $\sqrt[3]{24} = \sqrt[3]{8 \cdot 3}$                    Find largest cube factor.

$\quad\quad\quad = \sqrt[3]{2^3 \cdot 3}$                    Rewrite.

$\quad\quad\quad = 2\sqrt[3]{3}$                    Find root of perfect cube.

**b.** $\sqrt[3]{24a^4} = \sqrt[3]{8a^3 \cdot 3a}$                    Find largest cube factor.

$\quad\quad\quad = \sqrt[3]{(2a)^3 \cdot 3a}$                    Rewrite.

$\quad\quad\quad = 2a\sqrt[3]{3a}$                    Find root of perfect cube.

**c.** $\sqrt[3]{-40x^6} = \sqrt[3]{(-8x^6) \cdot 5}$                    Find largest cube factor.

$\quad\quad\quad = \sqrt[3]{(-2x^2)^3 \cdot 5}$                    Rewrite.

$\quad\quad\quad = -2x^2\sqrt[3]{5}$                    Find root of perfect cube.

To **rationalize the denominator** of a fraction, multiply the numerator and denominator by an appropriate factor. To find the "appropriate factor," make use of the form $a + b\sqrt{m}$ and its **conjugate** $a - b\sqrt{m}$. The product of this conjugate pair has no radical. For instance,

$$\left(4 + \sqrt{3}\right)\left(4 - \sqrt{3}\right) = 4^2 - \left(\sqrt{3}\right)^2 = 16 - 3 = 13.$$

Therefore, to rationalize a denominator of the form $a - b\sqrt{m}$ (or $a + b\sqrt{m}$), you can multiply the numerator and denominator by the conjugate factor. If $a = 0$, the rationalizing factor of $\sqrt{m}$ is itself, $\sqrt{m}$.

| Example 4 | Rationalizing Single-Term Denominators |

**a.** To rationalize the denominator of the following fraction, multiply *both* the numerator and the denominator by $\sqrt{3}$ to obtain

$$\frac{5}{2\sqrt{3}} = \frac{5}{2\sqrt{3}} \cdot \frac{\sqrt{3}}{\sqrt{3}} = \frac{5\sqrt{3}}{2\sqrt{3^2}} = \frac{5\sqrt{3}}{2(3)} = \frac{5\sqrt{3}}{6}.$$

**b.** To rationalize the denominator of the following fraction, multiply *both* the numerator and the denominator by $\sqrt[3]{5^2}$. Note how this eliminates the radical from the denominator by producing a perfect *cube* in the radicand.

$$\frac{2}{\sqrt[3]{5}} = \frac{2}{\sqrt[3]{5}} \cdot \frac{\sqrt[3]{5^2}}{\sqrt[3]{5^2}} = \frac{2\sqrt[3]{5^2}}{\sqrt[3]{5^3}} = \frac{2\sqrt[3]{25}}{5}$$

| Example 5 | Rationalizing a Denominator with Two Terms |

$$\frac{2}{3 + \sqrt{7}} = \frac{2}{3 + \sqrt{7}} \cdot \frac{3 - \sqrt{7}}{3 - \sqrt{7}} \qquad \text{Multiple numerator and denominator by conjugate.}$$

$$= \frac{2(3 - \sqrt{7})}{3^2 - (\sqrt{7})^2} \qquad \text{Multiply fractions.}$$

$$= \frac{2(3 - \sqrt{7})}{9 - 7} \qquad \text{Simplify.}$$

$$= \frac{2(3 - \sqrt{7})}{2} \qquad \text{Cancel like factors.}$$

$$= 3 - \sqrt{7} \qquad \text{Simplify.}$$

Don't confuse an expression such as $\sqrt{2} + \sqrt{7}$ with $\sqrt{2 + 7}$. In general, $\sqrt{x + y} \neq \sqrt{x} + \sqrt{y}$.

## Rational Exponents

The following definition shows how radicals are used to define **rational exponents.** Until now, work with exponents has been restricted to integer exponents.

**Study Tip**

The numerator of a rational exponent denotes the *power* to which the base is raised, and the denominator denotes the *index* or the *root* to be taken. Moreover, it doesn't matter in which order the two operations are performed, provided the $n$th root exists. Here is an example.

$$8^{2/3} = \left(\sqrt[3]{8}\right)^2 = 2^2 = 4$$

$$8^{2/3} = \sqrt[3]{8^2} = \sqrt[3]{64} = 4$$

▶ **Definition of Rational Exponents**

If $a$ is a real number and $n$ is a positive integer such that the principal $n$th root of $a$ exists, then $a^{1/n}$ is defined to be

$$a^{1/n} = \sqrt[n]{a}.$$

If $m$ is a positive integer that has no common factor with $n$, then

$$a^{m/n} = (a^{1/n})^m = \left(\sqrt[n]{a}\right)^m$$

and

$$a^{m/n} = (a^m)^{1/n} = \sqrt[n]{a^m}.$$

The properties of exponents that were listed in Section A.3 also apply to rational exponents (provided the roots indicated by the denominators exist). Some of those properties are relisted here, with different examples.

---

▶ **Properties of Exponents**

Let $r$ and $s$ be rational numbers, and let $a$ and $b$ be real numbers, variables, or algebraic expressions. If the roots indicated by the rational exponents exist, then the following properties are true.

| *Property* | *Example* |
|---|---|
| **1.** $a^r a^s = a^{r+s}$ | $4^{1/2}(4^{1/3}) = 4^{5/6}$ |
| **2.** $\dfrac{a^r}{a^s} = a^{r-s}, \quad a \neq 0$ | $\dfrac{x^2}{x^{1/2}} = x^{2-(1/2)} = x^{3/2}$ |
| **3.** $a^{-r} = \dfrac{1}{a^r}, \quad a \neq 0$ | $4^{-1/2} = \dfrac{1}{4^{1/2}} = \dfrac{1}{2}$ |
| **4.** $\left(\dfrac{a}{b}\right)^{-r} = \left(\dfrac{b}{a}\right)^r, \quad a \neq 0, \quad b \neq 0$ | $\left(\dfrac{x}{4}\right)^{-1/2} = \left(\dfrac{4}{x}\right)^{1/2} = \dfrac{2}{x^{1/2}}$ |
| **5.** $(ab)^r = a^r b^r$ | $(2x)^{1/2} = 2^{1/2}(x^{1/2})$ |
| **6.** $(a^r)^s = a^{rs}$ | $(x^3)^{1/3} = x$ |
| **7.** $\left(\dfrac{a}{b}\right)^r = \dfrac{a^r}{b^r}, \quad b \neq 0$ | $\left(\dfrac{x}{3}\right)^{1/3} = \dfrac{x^{1/3}}{3^{1/3}}$ |

---

Rational exponents are particularly useful for evaluating roots of numbers on a calculator, for reducing the index of a radical, and for simplifying (and factoring) algebraic expressions. Examples 6 and 7 demonstrate some of these uses.

## Study Tip

Rational exponents can be tricky, and you must remember that the expression $b^{m/n}$ is not defined unless $\sqrt[n]{b}$ is a real number. This restriction produces some unusual-looking results. For instance, the number $(-8)^{1/6}$ is not defined because $\sqrt[6]{-8}$ is not a real number. And yet, $(-8)^{1/3}$ is defined because $\sqrt[3]{-8} = -2$.

**Example 6**    Simplifying with Rational Exponents

**a.** $(27)^{1/3} = \sqrt[3]{27} = 3$

**b.** $(-32)^{-4/5} = \left(\sqrt[5]{-32}\right)^{-4} = (-2)^{-4} = \dfrac{1}{(-2)^4} = \dfrac{1}{16}$

**c.** $(-5x^{2/3})(3x^{-1/3}) = -15x^{(2/3)-(1/3)} = -15x^{1/3}, \quad x \neq 0$

**Example 7**    Reducing the Index of a Radical

**a.** $\sqrt[6]{a^4} = a^{4/6} = a^{2/3} = \sqrt[3]{a^2}$

**b.** $\sqrt[3]{\sqrt{125}} = (125^{1/2})^{1/3}$     Rewrite with rational exponents.

$\qquad = (125)^{1/6}$     Multiply exponents.

$\qquad = (5^3)^{1/6}$     Rewrite base as perfect cube.

$\qquad = 5^{3/6}$     Multiply exponents.

$\qquad = 5^{1/2}$     Reduce exponent.

$\qquad = \sqrt{5}$     Rewrite as radical.

Radical expressions can be combined (added or subtracted) if they are **like radicals**—that is, if they have the same index and radicand. For instance, $2\sqrt{3x}$ and $\frac{1}{2}\sqrt{3x}$ are like radicals, but $\sqrt[3]{3x}$ and $2\sqrt{3x}$ are not like radicals. To determine whether two expressions involve like radicals, you should first simplify each radical.

| Example 8 | Simplifying and Combining Like Radicals |

**a.** $2\sqrt{48} + 3\sqrt{27} = 2\sqrt{16 \cdot 3} + 3\sqrt{9 \cdot 3}$    Find square factors.

$\qquad\qquad\qquad = 8\sqrt{3} + 9\sqrt{3}$    Find square roots.

$\qquad\qquad\qquad = 17\sqrt{3}$    Combine like terms.

**b.** $\sqrt[3]{16x} - \sqrt[3]{54x} = \sqrt[3]{8 \cdot 2x} - \sqrt[3]{27 \cdot 2x}$    Find cube factors.

$\qquad\qquad\qquad = 2\sqrt[3]{2x} - 3\sqrt[3]{2x}$    Find cube roots.

$\qquad\qquad\qquad = -\sqrt[3]{2x}$    Combine like terms.

## Radicals and Calculators

There are two methods of evaluating radicals on most calculators. For square roots, you can use the *square root key*. For other roots, you should first convert the radical to exponential form and then use the *exponential key*.

| Example 9 | Evaluating a Cube Root with a Calculator |

Use a calculator to evaluate $\sqrt[3]{25}$. Round to three decimal places.

**Solution**

First write $\sqrt[3]{25}$ as $25^{1/3}$. Then use the following keystrokes.

25 $\boxed{y^x}$ $\boxed{(}$ 1 $\boxed{\div}$ 3 $\boxed{)}$ $\boxed{=}$    Scientific

25 $\boxed{\wedge}$ $\boxed{(}$ 1 $\boxed{\div}$ 3 $\boxed{)}$ $\boxed{\text{ENTER}}$    Graphing

For either of these keystroke sequences, the calculator display should read 2.9240177. Thus, you have

$$\sqrt[3]{25} \approx 2.924.$$

| Example 10 | Evaluating Radicals with a Calculator |

Evaluate the following radicals. Round to three decimal places.

**a.** $\sqrt[3]{-4}$    **b.** $(1.4)^{-2/5}$

**Solution**

**a.** Because

$$\sqrt[3]{-4} = \sqrt[3]{(-1)(4)} = \sqrt[3]{-1} \cdot \sqrt[3]{4} = -\sqrt[3]{4}$$

you can attach the negative sign of the radicand as follows.

$$4 \boxed{y^x} \boxed{(} 1 \boxed{\div} 3 \boxed{)} \boxed{=} \boxed{+/-}$$    Scientific

$$\boxed{(-)} \boxed{(} 4 \boxed{\wedge} \boxed{(} 1 \boxed{\div} 3 \boxed{)} \boxed{)} \boxed{\text{ENTER}}$$    Graphing

The calculator display is $-1.5874011$, which implies that

$$\sqrt[3]{-4} \approx -1.587.$$

b. Using the following keystroke sequence

$$1.4 \boxed{y^x} \boxed{(} 2 \boxed{\div} 5 \boxed{+/-} \boxed{)} \boxed{=}$$    Scientific

$$1.4 \boxed{\wedge} \boxed{(} \boxed{(-)} 2 \boxed{\div} 5 \boxed{)} \boxed{\text{ENTER}}$$    Graphing

you obtain a calculator display of $0.8740751$. Thus, you have

$$(1.4)^{-2/5} \approx 0.874.$$

## Warm Up

The following warm-up exercises involve skills that were covered in earlier sections. You will use these skills in the exercise set for this section.

In Exercises 1–10, simplify the expression.

1. $\left(\frac{1}{3}\right)\left(\frac{2}{3}\right)^2$

2. $3(-4)^2$

3. $(-2x)^3$

4. $(-2x^3)(-3x^4)$

5. $(7x^5)(4x)$

6. $(5x^4)(25x^2)^{-1}, \quad x \neq 0$

7. $\dfrac{12z^6}{4z^2}, \quad z \neq 0$

8. $\left(\dfrac{2x}{5}\right)^2\left(\dfrac{2x}{5}\right)^{-4}, \quad x \neq 0$

9. $\left(\dfrac{3y^2}{x}\right)^0, \quad x \neq 0, \ y \neq 0$

10. $[(x+2)^2(x+2)^3]^2$

## A.4    Exercises

In Exercises 1–12, fill in the missing form.

| Radical Form | Rational Exponent Form |
|---|---|
| 1. $\sqrt{9} = 3$ | |
| 2. $\sqrt[3]{64} = 4$ | |
| 3. | $32^{1/5} = 2$ |
| 4. | $-(144^{1/2}) = -12$ |
| 5. | $196^{1/2} = 14$ |
| 6. $\sqrt[3]{614.125} = 8.5$ | |
| 7. $\sqrt[3]{-216} = -6$ | |
| 8. | $(-243)^{1/5} = -3$ |
| 9. | $27^{2/3} = 9$ |
| 10. $\left(\sqrt[4]{81}\right)^3 = 27$ | |
| 11. $\sqrt[4]{81^3} = 27$ | |
| 12. | $16^{5/4} = 32$ |

In Exercises 13–28, evaluate the expression.

13. $\sqrt{9}$

14. $\sqrt[3]{64}$

15. $-\sqrt[3]{-27}$

16. $\sqrt[3]{0}$

17. $\dfrac{4}{\sqrt{64}}$

18. $\dfrac{\sqrt[4]{81}}{3}$

19. $\left(\sqrt[3]{-125}\right)^3$

20. $\sqrt[4]{562^4}$

21. $16^{1/2}$

22. $27^{1/3}$

23. $36^{3/2}$

24. $16^{3/2}$

25. $\left(\frac{16}{81}\right)^{-3/4}$

26. $\left(\frac{9}{4}\right)^{-1/2}$

27. $\left(-\frac{1}{64}\right)^{-1/3}$

28. $\left(-\frac{125}{27}\right)^{-1/3}$

In Exercises 29–32, simplify the expression.

29. $\sqrt[3]{16x^5}$

30. $\sqrt[4]{(3x^2)^4}$

31. $\sqrt{75x^2y^{-4}}$

32. $\sqrt[5]{96x^5}$

In Exercises 33–40, rewrite the expression by rationalizing the denominator. Simplify the result.

**33.** $\dfrac{1}{\sqrt{3}}$

**34.** $\dfrac{5}{\sqrt{10}}$

**35.** $\dfrac{8}{\sqrt[3]{2}}$

**36.** $\dfrac{5}{\sqrt[3]{(5x)^2}}$

**37.** $\dfrac{2x}{5 - \sqrt{3}}$

**38.** $\dfrac{5}{\sqrt{14} - 2}$

**39.** $\dfrac{3}{\sqrt{5} + \sqrt{6}}$

**40.** $\dfrac{5}{2\sqrt{10} - 5}$

In Exercises 41–44, simplify the expression.

**41.** $5\sqrt{x} - 3\sqrt{x}$

**42.** $3\sqrt{x+1} + 10\sqrt{x+1}$

**43.** $2\sqrt{4y} - 2\sqrt{9y}$

**44.** $2\sqrt{80} + \sqrt{125}$

In Exercises 45–50, simplify the expression.

**45.** $\sqrt{2}\sqrt{3}$

**46.** $\sqrt{2}\sqrt{5}$

**47.** $5^{1/2} \cdot 5^{3/2}$

**48.** $\dfrac{2^{3/2}}{2}$

**49.** $\dfrac{x^2}{x^{1/2}}$

**50.** $\sqrt{\sqrt{x^4}}$

In Exercises 51 and 52, use rational exponents to reduce the index of the radical.

**51.** $\sqrt[4]{3^2}$

**52.** $\sqrt[4]{(3x^2)^4}$

In Exercises 53–56, fill in the blank with <, =, or >.

**53.** $\sqrt{5} + \sqrt{3}$ ____ $\sqrt{5+3}$

**54.** $5$ ____ $\sqrt{3^2 + 4^2}$

**55.** $\sqrt{3} \cdot \sqrt[4]{3}$ ____ $\sqrt[8]{3}$

**56.** $\sqrt{\dfrac{3}{11}}$ ____ $\dfrac{\sqrt{3}}{\sqrt{11}}$

**57.** *Geometry*   Find the measurements of a cube that has a volume of 13,824 cubic inches (see figure).

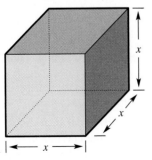

**58.** *Geometry*   Find the measurements of a square classroom that has 800 square feet of floor space (see figure).

*Declining Balances Depreciation*   In Exercises 59 and 60, find the annual depreciation rate $r$. To find the annual depreciation rate by the declining balances method, use the formula

$$r = 1 - \left(\frac{S}{C}\right)^{1/n}$$

where $n$ is the useful life of the item (in years), $S$ is the salvage value (in dollars), and $C$ is the original cost (in dollars).

**59.** A truck whose original cost is $75,000 is depreciated over an 8-year period, as shown in the bar graph.

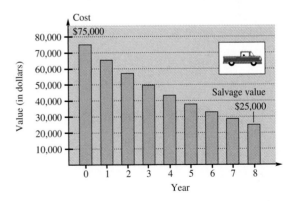

**60.** A printing press whose original cost is $125,000 is depreciated over a 10-year period, as shown in the bar graph.

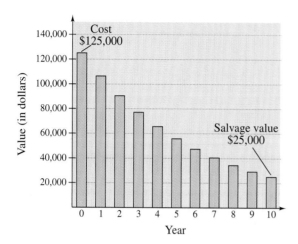

**61.** *Period of a Pendulum* The period $T$ of a pendulum in seconds is given by

$$T = 2\pi\sqrt{\frac{L}{32}}$$

where $L$ is the length of the pendulum in feet. Find the period of a pendulum whose length is 2 feet.

**62.** *Period of a Pendulum* Use the formula given in Exercise 61 to find the period of a pendulum whose length is 1.5 feet.

*Notes on a Musical Scale* In Exercises 63 and 64, find the frequency of the indicated note on a piano (see figure). The musical note A above middle C has a frequency of 440 vibrations per second. If we denote this frequency by $F_1$, then the frequency of the next higher note is given by $F_2 = F_1 \cdot 2^{1/12}$. Similarly, the frequency of the next note is given by $F_3 = F_2 \cdot 2^{1/12}$.

**63.** Find the frequency of the musical note B above middle C.

**64.** Find the frequency of the musical note C that is one octave above middle C.

**65.** *Calculator Experiment* Enter any positive real number in your calculator and repeatedly take the square root. What real number does the display appear to be approaching?

**66.** *Calculator Experiment* Square the real number $2/\sqrt{5}$ and note that the radical is eliminated from the denominator. Is this equivalent to rationalizing the denominator? Why or why not?

---

## A.5 Polynomials and Special Products

Polynomials ■ Operations with Polynomials ■ Special Products ■ Applications

## Polynomials

One of the simplest and most common types of algebraic expression is a **polynomial**. Here are some examples.

$$2x + 5, \qquad 3x^4 - 7x^2 + 2x + 4, \qquad 5x^2y^2 - xy + 3$$

The first two are *polynomials in x* and the third is a *polynomial in x and y*. The terms of a polynomial in $x$ have the form $ax^k$, where $a$ is the **coefficient** and $k$ is the **degree** of the term. Because a polynomial is defined as an algebraic sum, the coefficients take on the signs between the terms. For instance, the polynomial

$$2x^3 - 5x^2 + 1 = 2x^3 + (-5)x^2 + (0)x + 1$$

has coefficients $2, -5, 0,$ and $1.$

---

▶ **Definition of a Polynomial in $x$**

Let $a_n, \ldots, a_2, a_1, a_0$ be real numbers and let $n$ be a *nonnegative integer*. A **polynomial in $x$** is an expression of the form

$$a_n x^n + \cdots + a_2 x^2 + a_1 x + a_0$$

where $a_n \neq 0$. The polynomial is of **degree** $n$, and the number $a_n$ is the **leading coefficient.** The number $a_0$ is the **constant term.** The constant term is considered to have a degree of zero.

---

Note in the definition of a polynomial in $x$ that the polynomial is written with descending powers of $x$. This is called the **standard form** of a polynomial.

**Example 1**    Rewriting a Polynomial in Standard Form

| Polynomial | Standard Form | Degree |
|---|---|---|
| **a.** $4x^2 - 5x^3 - 2 + 3x$ | $-5x^3 + 4x^2 + 3x - 2$ | 3 |
| **b.** $4 - 9x^2$ | $-9x^2 + 4$ | 2 |
| **c.** $8$ | $8 \ (8 = 8x^0)$ | 0 |

Polynomials with one, two, and three terms are called **monomials, binomials,** and **trinomials,** respectively.

A polynomial that has all zero coefficients is called the **zero polynomial,** denoted by 0. This particular polynomial is not considered to have a degree.

**Example 2**    Identifying a Polynomial and Its Degree

**a.** $-2x^3 + x^2 + 3x - 2$ is a polynomial of degree 3.

**b.** $\sqrt{x^2 - 3x}$ is not a polynomial because the radical sign indicates a noninteger power of $x$.

**c.** $x^2 + 5x^{-1}$ is not a polynomial because of the negative exponent.

For polynomials in more than one variable, the *degree of a term* is the sum of the powers of the variables in the term. The *degree of the polynomial* is the highest degree of all its terms. For instance, the polynomial

$$5x^3 y - x^2 y^2 + 2xy - 5$$

has two terms of degree 4, one term of degree 2, and one term of degree 0. The degree of the polynomial is 4.

## Operations with Polynomials

You can **add** and **subtract** polynomials in much the same way that you add and subtract real numbers—you simply add or subtract the *like terms* (terms having the same variables to the same powers) by adding their coefficients. For instance, $-3x^2$ and $5x^2$ are like terms and their sum is given by

$$-3x^2 + 5x^2 = (-3 + 5)x^2 = 2x^2.$$

**Example 3** Sums and Differences of Polynomials

**a.** $(5x^3 - 7x^2 - 3) + (x^3 + 2x^2 - x + 8)$

$\qquad = (5x^3 + x^3) + (2x^2 - 7x^2) - x + (8 - 3)$  Group like terms.

$\qquad = 6x^3 - 5x^2 - x + 5$  Combine like terms.

**b.** $(7x^4 - x^2 - 4x + 2) - (3x^4 - 4x^2 + 3x)$

$\qquad = 7x^4 - x^2 - 4x + 2 - 3x^4 + 4x^2 - 3x$  Distribute sign.

$\qquad = (7x^4 - 3x^4) + (4x^2 - x^2) + (-3x - 4x) + 2$  Group like terms.

$\qquad = 4x^4 + 3x^2 - 7x + 2$  Combine like terms.

A common mistake is to fail to change the sign of *each* term inside parentheses preceded by a minus sign. For instance, note the following.

$\qquad -(3x^4 - 4x^2 + 3x) = -3x^4 + 4x^2 - 3x$  Correct

$\qquad \overline{-(3x^4 - 4x^2 + 3x) = -3x^4 - 4x^2 + 3x}$  Common mistake

To find the **product** of two polynomials, you can use the left and right Distributive Properties. For example, if you treat $(5x + 7)$ as a single quantity, you can multiply $(3x - 2)$ by $(5x + 7)$ as follows.

$$(3x - 2)(5x + 7) = 3x(5x + 7) - 2(5x + 7)$$

$$= (3x)(5x) + (3x)(7) - (2)(5x) - (2)(7)$$

$$= 15x^2 + 21x - 10x - 14$$

| Product of First terms | Product of Outer terms | Product of Inner terms | Product of Last terms |
| --- | --- | --- | --- |

$$= 15x^2 + 11x - 14$$

With practice you should be able to multiply two binomials without writing all of the steps above. In fact, the four products in the boxes above suggest that you can put the product of two binomials in the FOIL form in just one step. This is called the **FOIL Method.**

When multiplying two polynomials, be sure to multiply *each* term of one polynomial by *each* term of the other. The following vertical pattern is a convenient way to multiply two polynomials.

**Example 4** Using a Vertical Format to Multiply Polynomials

Multiply $(x^2 - 2x + 2)$ by $(x^2 + 2x + 2)$.

## Solution

$$x^2 - 2x + 2 \qquad\qquad \text{Standard form}$$

$$x^2 + 2x + 2 \qquad\qquad \text{Standard form}$$

$$x^4 - 2x^3 + 2x^2 \qquad\qquad \Longleftarrow \quad x^2(x^2 - 2x + 2)$$

$$2x^3 - 4x^2 + 4x \qquad\qquad \Longleftarrow \quad 2x(x^2 - 2x + 2)$$

$$2x^2 - 4x + 4 \qquad\qquad \Longleftarrow \quad 2(x^2 - 2x + 2)$$

$$x^4 + 0x^3 + 0x^2 + 0x + 4 = x^4 + 4$$

Thus, $(x^2 - 2x + 2)(x^2 + 2x + 2) = x^4 + 4$.

## Special Products

▶ **Special Products**

Let $u$ and $v$ be real numbers, variables, or algebraic expressions.

|  *Special Product*  |  *Example*  |
| --- | --- |
| **Sum and Difference of Two Terms** | |
| $(u + v)(u - v) = u^2 - v^2$ | $(x + 4)(x - 4) = x^2 - 16$ |
| **Square of a Binomial** | |
| $(u + v)^2 = u^2 + 2uv + v^2$ | $(x + 3)^2 = x^2 + 6x + 9$ |
| $(u - v)^2 = u^2 - 2uv + v^2$ | $(3x - 2)^2 = 9x^2 - 12x + 4$ |
| **Cube of a Binomial** | |
| $(u + v)^3 = u^3 + 3u^2v + 3uv^2 + v^3$ | $(x + 2)^3 = x^3 + 6x^2 + 12x + 8$ |
| $(u - v)^3 = u^3 - 3u^2v + 3uv^2 - v^3$ | $(x - 1)^3 = x^3 - 3x^2 + 3x - 1$ |

**Example 5**    Sum and Difference of Two Terms

$$(5x + 9)(5x - 9) = (5x)^2 - 9^2 = 25x^2 - 81$$

**Example 6**    Square of a Binomial

$$(6x - 5)^2 = (6x)^2 - 2(6x)(5) + 5^2$$
$$= 36x^2 - 60x + 25$$

**Example 7**    Cube of a Binomial

$$(3x + 2)^3 = (3x)^3 + 3(3x)^2(2) + 3(3x)(2)^2 + 2^3$$
$$= 27x^3 + 54x^2 + 36x + 8$$

**Example 8**   The Product of Two Trinomials

$$(x + y - 2)(x + y + 2) = [(x + y) - 2][(x + y) + 2]$$
$$= (x + y)^2 - 2^2$$
$$= x^2 + 2xy + y^2 - 4$$

## Applications

**Example 9**   A Savings Plan  

During the summers prior to your four college years you are able to save $1500, $1800, $2400, and $2600. During the summer after graduation you save $3000. This money is deposited at the end of each summer in an account that pays 7% interest, compounded annually. Find the balance in the account at the end of the summer after graduation.

### Solution

Using the formula for compound interest, for *each* deposit you have

$$\text{Balance} = P\left(1 + \frac{r}{n}\right)^{nt} = P(1 + 0.07)^t = P(1.07)^t.$$

For the first deposit, $P = 1500$ and $t = 4$. For the second deposit, $P = 1800$ and $t = 3$, and so on. The balances for the five deposits are as follows.

| Date | Deposit | Time in Account | Balance in Account |
|------|---------|-----------------|--------------------|
| First Summer | $1500 | 4 years | $1500(1.07)^4$ |
| Second Summer | $1800 | 3 years | $1800(1.07)^3$ |
| Third Summer | $2400 | 2 years | $2400(1.07)^2$ |
| Fourth Summer | $2600 | 1 year | $2600(1.07)$ |
| Fifth Summer | $3000 | 0 years | 3000 |

By adding these five balances, you can find the total balance in the account to be

$$1500(1.07)^4 + 1800(1.07)^3 + 2400(1.07)^2 + 2600(1.07) + 3000.$$

Note that this expression is in polynomial form. By evaluating the expression, you can find the balance to be $12,701.03, as shown in Figure A.10.

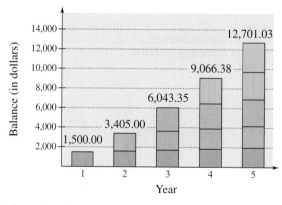

Figure A.10

**Example 10** Geometry: Volume of a Box

An open box is made by cutting squares from the corners of a piece of metal that is 16 inches by 20 inches, and turning up the sides, as shown in Figure A.11. If the edge of each cut-out square is $x$ inches, what is the volume of the box? Find the volume when $x = 1$, $x = 2$, and $x = 3$.

Solution

Figure A.11

*Verbal Model:*

Volume = Length · Width · Height

*Labels:*

| | |
|---|---|
| Height = $x$ | (inches) |
| Width = $16 - 2x$ | (inches) |
| Length = $20 - 2x$ | (inches) |

*Equation:*

$$\text{Volume} = (20 - 2x)(16 - 2x)(x)$$
$$= (320 - 72x + 4x^2)(x)$$
$$= 320x - 72x^2 + 4x^3$$

When $x = 1$ inch, the volume of the box is

$$\text{Volume} = 320(1) - 72(1)^2 + 4(1)^3 = 252 \text{ cubic inches.}$$

When $x = 2$ inches, the volume of the box is

$$\text{Volume} = 320(2) - 72(2)^2 + 4(2)^3 = 384 \text{ cubic inches.}$$

When $x = 3$ inches, the volume of the box is

$$\text{Volume} = 320(3) - 72(3)^2 + 4(3)^3 = 420 \text{ cubic inches.}$$

# Warm Up

The following warm-up exercises involve skills that were covered in earlier sections. You will use these skills in the exercise set for this section.

In Exercise 1–10, perform the indicated operation(s).

**1.** $(7x^2)(6x)$

**2.** $(10z^3)(-2z^{-1})$

**3.** $(-3x^2)^3$

**4.** $-3(x^2)^3$

**5.** $\dfrac{27z^5}{12z^2}$

**6.** $\sqrt{24} \cdot \sqrt{2}$

**7.** $\left(\dfrac{2x}{3}\right)^{-2}$

**8.** $16^{3/4}$

**9.** $\dfrac{4}{\sqrt{8}}$

**10.** $\sqrt[3]{-27x^3}$

## A.5    Exercises

In Exercises 1–4, find the degree and leading coefficient of the polynomial.

**1.** $2x^2 - x + 1$

**2.** $-3x^4 + 2x^2 - 5$

**3.** $x^5 - 1$

**4.** $3$

In Exercises 5–8, determine whether the algebraic expression is a polynomial. If it is, write the polynomial in standard form and state its degree.

**5.** $2x - 3x^3 + 8$

**6.** $2x^3 + x - 3x^{-1}$

**7.** $\dfrac{3x + 4}{x}$

**8.** $\sqrt{y^2 - y^4}$

In Exercises 9 and 10, evaluate the polynomial for the indicated values of $x$.

**9.** $4x + 3$       (a) $x = -1$   (b) $x = 0$
                  (c) $x = 1$    (d) $x = 2$

**10.** $-2x^2 + 3x + 4$   (a) $x = -2$   (b) $x = -1$
                       (c) $x = 0$    (d) $x = 1$

In Exercises 11–20, perform the indicated operations and write the resulting polynomial in standard form.

**11.** $(6x + 5) - (8x + 15)$

**12.** $-(5x^2 - 1) - (-3x^2 + 5)$

**13.** $(15x^2 - 6) - (-8x^3 - 14x^2 - 17)$

**14.** $(15x^4 - 18x - 19) - (13x^4 - 5x + 15)$

**15.** $5z - [3z - (10z + 8)]$

**16.** $(y^3 + 1) - [(y^2 + 1) + (3y - 7)]$

**17.** $3x(x^2 - 2x + 1)$

**18.** $-4x(3 - x^3)$

**19.** $(-2x)(-3x)(5x + 2)$

**20.** $(1 - x^3)(4x)$

In Exercises 21–44, find the product.

**21.** $(x + 3)(x + 4)$

**22.** $(x - 5)(x + 10)$

**23.** $(3x - 5)(2x + 1)$

**24.** $(7x - 2)(4x - 3)$

**25.** $(x + 6)^2$

**26.** $(3x - 2)^2$

**27.** $(2x - 5y)^2$

**28.** $(5 - 8x)^2$

**29.** $[(x - 3) + y]^2$

**30.** $[(x + 1) - y]^2$

**31.** $(x + 2y)(x - 2y)$

**32.** $(2x + 3y)(2x - 3y)$

**33.** $(m - 3 + n)(m - 3 - n)$

**34.** $(x + y + 1)(x + y - 1)$

**35.** $(x + 1)^3$

**36.** $(x - 2)^3$

**37.** $(2x - y)^3$

**38.** $(3x + 2y)^3$

**39.** $\left(\sqrt{x} + \sqrt{y}\right)\left(\sqrt{x} - \sqrt{y}\right)$

**40.** $\left(5 + \sqrt{x}\right)\left(5 - \sqrt{x}\right)$

**41.** $(x^2 - x + 1)(x^2 + x + 1)$

**42.** $(x^2 + 9)(x^2 - x - 4)$

**43.** $5x(x + 1) - 3x(x + 1)$

**44.** $(2x - 1)(x + 3) + 3(x + 3)$

**45.** *Savings Plan*    A person who is attending 4 years of college has a summer job. During the summers prior to each of the four college years, the person is able to save $1200, $1400, $800, and $2300. During the summer after graduation the person saves $2900. This money is deposited in an account that pays 6% interest, compounded annually. Find the balance in the account at the end of the summer after graduation.

**46.** *Savings Plan*    Suppose you have an investment that pays an annual dividend in July. Each year for 6 years, you reinvest this dividend in an account that pays 6.5% interest, compounded annually. For the dividends shown in the table, find the balance in the account at the end of 6 years. (*Note:* The first deposit will earn 5 years of interest, and the sixth deposit will earn no interest.)

| Year | 1 | 2 | 3 |
|---|---|---|---|
| Dividend | $840 | $930 | $760 |

| Year | 4 | 5 | 6 |
|---|---|---|---|
| Dividend | $1020 | $980 | $1130 |

**47.** *Compound Interest*    After 2 years, an investment of $500 compounded annually at an interest rate of $r$ will yield an amount of

$$500(1 + r)^2.$$

Write this polynomial in standard form.

**48.** *Compound Interest*    After 3 years, an investment of $1200 compounded annually at an interest rate of $r$ will yield an amount of

$$1200(1 + r)^3.$$

Write this polynomial in standard form.

**49.** *Geometry*    An open box is made by cutting squares from the corners of a piece of metal that is 18 inches by 26 inches, and folding up the sides (see figure). If the edge of each cut-out square is $x$ inches, what is the volume of the box? Find the volume when $x = 1$, $x = 2$, and $x = 3$.

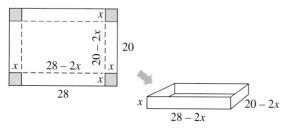

**50.** *Geometry*    An open box is made by cutting squares from the corners of a piece of cardboard that is 20 inches by 28 inches, and turning up the sides (see figure). If the edge of each cut-out square is $x$ inches, what is the volume of the box? Find the volume when $x = 1$, $x = 2$, and $x = 3$.

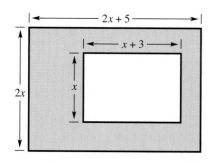

**51.** *Geometry*    Find the area of the shaded region in the figure. Write your answer as a polynomial in standard form.

**52.** *Geometry*    Find the area of the shaded region in the figure. Write your answer as a polynomial in standard form.

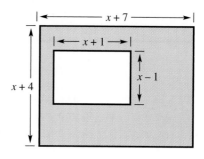

**53.** *Geometry*    Find a polynomial that represents the total number of square feet for the following floor plan.

**54.** *Geometry*    Find a polynomial that represents the total number of square feet for the following floor plan.

## A.6 Factoring

Common Factors ■ Factoring Special Polynomial Forms ■
Trinomials with Binomial Factors ■ Factoring by Grouping

## Common Factors

The process of writing a polynomial as a product is called **factoring.** It is an important tool for solving equations and reducing fractional expressions.

A polynomial that cannot be factored using integer coefficients is called **prime** or **irreducible over the integers.** For instance, the polynomial $x^2 - 3$ is irreducible over the integers. [Over the *real numbers*, this polynomial can be factored as

$$x^2 - 3 = \left(x + \sqrt{3}\right)\left(x - \sqrt{3}\right).]$$

A polynomial is **completely factored** when each of its factors is prime. For instance,

$$x^3 - x^2 + 4x - 4 = (x - 1)(x^2 + 4) \qquad \text{Completely factored}$$

is completely factored, but

$$x^3 - x^2 - 4x + 4 = (x - 1)(x^2 - 4) \qquad \text{Not completely factored}$$

is not completely factored. Its complete factorization would be

$$x^3 - x^2 - 4x + 4 = (x - 1)(x + 2)(x - 2).$$

The simplest type of factoring involves a polynomial that is the product of a monomial and another polynomial. To factor such a polynomial, you can use the Distributive Property in the *reverse* direction.

$$ab + ac = a(b + c) \qquad a \text{ is a common factor.}$$

### Example 1 Removing Common Factors

Factor each of the following.

**a.** $6x^3 - 4x$     **b.** $(x - 2)(2x) + (x - 2)(3)$

Solution

**a.** Each term of this polynomial has $2x$ as a common factor.

$$6x^3 - 4x = 2x(3x^2) - 2x(2)$$
$$= 2x(3x^2 - 2)$$

**b.** The binomial factor $(x - 2)$ is common to both terms.

$$(x - 2)(2x) + (x - 2)(3) = (x - 2)(2x + 3)$$

# Factoring Special Polynomial Forms

> ► Factoring Special Polynomial Forms
>
> | *Factored Form* | *Example* |
> |---|---|
>
> **Difference of Two Squares**
>
> $$u^2 - v^2 = (u + v)(u - v) \qquad 9x^2 - 4 = (3x + 2)(3x - 2)$$
>
> **Perfect Square Trinomial**
>
> $$u^2 + 2uv + v^2 = (u + v)^2 \qquad x^2 + 6x + 9 = (x + 3)^2$$
>
> $$u^2 - 2uv + v^2 = (u - v)^2 \qquad x^2 - 6x + 9 = (x - 3)^2$$
>
> **Sum or Difference of Two Cubes**
>
> $$u^3 + v^3 = (u + v)(u^2 - uv + v^2) \quad x^3 + 8 = (x + 2)(x^2 - 2x + 4)$$
>
> $$u^3 - v^3 = (u - v)(u^2 + uv + v^2) \quad 27x^3 - 1 = (3x - 1)(9x^2 + 3x + 1)$$

**Study Tip**

In Example 2, note that the first step in factoring a polynomial is to check for common factors. Once the common factor is removed, it is often possible to recognize patterns that were not obvious at first glance.

**Example 2**    Removing a Common Factor First

Factor $3 - 12x^2$.

Solution

$$3 - 12x^2 = 3(1 - 4x^2) \qquad \text{Common factor}$$

$$= 3[1^2 - (2x)^2] \qquad \text{Difference of squares}$$

$$= 3(1 + 2x)(1 - 2x) \qquad \text{Completely factored}$$

**Example 3**    Factoring the Difference of Two Squares

**a.** $(x + 2)^2 - y^2 = [(x + 2) + y][(x + 2) - y]$

$$= (x + 2 + y)(x + 2 - y)$$

$$= (x + y + 2)(x - y + 2)$$

**b.** Applying the difference of two squares formula twice, you have

$$16x^4 - 81 = (4x^2)^2 - 9^2$$

$$= (4x^2 + 9)(4x^2 - 9) \qquad \text{First application}$$

$$= (4x^2 + 9)[(2x)^2 - 3^2]$$

$$= (4x^2 + 9)(2x + 3)(2x - 3). \qquad \text{Second application}$$

A perfect square trinomial is the square of a binomial, and it has the following form.

$$u^2 + 2uv + v^2 = (u + v)^2 \qquad \text{or} \qquad u^2 - 2uv + v^2 = (u - v)^2$$

Same sign       Same sign

Note that the first and last terms are squares and the middle term is twice the product of $u$ and $v$.

> ### Example 4 Factoring Perfect Square Trinomials
>
> **a.** $16x^2 + 8x + 1 = (4x)^2 + 2(4x)(1) + 1^2$
>
> $= (4x + 1)^2$
>
> **b.** $x^2 - 10x + 25 = x^2 - 2(x)(5) + 5^2$
>
> $= (x - 5)^2$

The next two formulas show that sums and differences of cubes factor easily. Pay special attention to the signs of the terms.

Like signs

Like signs

$$u^3 + v^3 = (u + v)(u^2 - uv + v^2) \qquad u^3 - v^3 = (u - v)(u^2 + uv + v^2)$$

Unlike signs

Unlike signs

> ### Example 5 Factoring the Sum and Difference of Cubes
>
> Factor the following.
>
> **a.** $x^3 - 27$    **b.** $3x^3 + 192$
>
> **Solution**
>
> **a.** $x^3 - 27 = x^3 - 3^3$
>
> $= (x - 3)(x^2 + 3x + 9)$
>
> **b.** $3x^3 + 192 = 3(x^3 + 64)$    Common factor
>
> $= 3(x^3 + 4^3)$
>
> $= 3(x + 4)(x^2 - 4x + 16)$

## Trinomials with Binomial Factors

To factor a trinomial of the form $ax^2 + bx + c$, use the following pattern.

Factors of $a$

$$ax^2 + bx + c = (\quad x + \quad )(\quad x + \quad )$$

Factors of $c$

The goal is to find a combination of factors of $a$ and $c$ such that the outer and inner products add up to the middle term $bx$. For instance, in the trinomial $6x^2 + 17x + 5$, you have $a = 6$, $c = 5$, and $b = 17$. After some experimentation, you can find that the factorization is $(2x + 5)(3x + 1)$.

> **Example 6**   Factoring a Trinomial:  Leading Coefficient Is 1

Factor the trinomial $x^2 - 7x + 12$.

### Solution

For this trinomial, you have $a = 1$, $b = -7$, and $c = 12$. Because $b$ is negative and $c$ is positive, both factors of 12 must be negative. That is, $12 = (-2)(-6)$, $12 = (-1)(-12)$, or $12 = (-3)(-4)$. Therefore, the possible factorizations of $x^2 - 7x + 12$ are

$$(x - 2)(x - 6), \qquad (x - 1)(x - 12), \qquad \text{and} \qquad (x - 3)(x - 4).$$

Testing the middle term, you can find the correct factorization to be

$$x^2 - 7x + 12 = (x - 3)(x - 4).$$

> **Example 7**   Factoring a Trinomial:  Leading Coefficient Is Not 1

Factor the trinomial $2x^2 + x - 15$.

### Solution

For this trinomial, you have $a = 2$ and $c = -15$, which means that the factors of $-15$ must have unlike signs. The eight possible factorizations are as follows.

$$(2x - 1)(x + 15) \qquad (2x + 1)(x - 15)$$
$$(2x - 3)(x + 5) \qquad (2x + 3)(x - 5)$$
$$(2x - 5)(x + 3) \qquad (2x + 5)(x - 3)$$
$$(2x - 15)(x + 1) \qquad (2x + 15)(x - 1)$$

Testing the middle term, you can find the correct factorization to be

$$2x^2 + x - 15 = (2x - 5)(x + 3).$$

## Factoring by Grouping

Sometimes polynomials with more than three terms can be **factored by grouping.**

> **Example 8**   Factoring by Grouping

$$
\begin{aligned}
x^3 - 2x^2 - 3x + 6 &= (x^3 - 2x^2) - (3x - 6) && \text{Group terms.} \\
&= x^2(x - 2) - 3(x - 2) && \text{Factor groups.} \\
&= (x - 2)(x^2 - 3) && \text{Common factor}
\end{aligned}
$$

# Warm Up

The following warm-up exercises involve skills that were covered in earlier sections. You will use these skills in the exercise set for this section.

In Exercises 1–10, find the product.

**1.** $3x(5x - 2)$  **2.** $-2y(y + 1)$

**3.** $(2x + 3)^2$  **4.** $(3x - 8)^2$

**5.** $(2x - 3)(x + 8)$  **6.** $(4 - 5z)(1 + z)$

**7.** $(2y + 1)(2y - 1)$  **8.** $(x + a)(x - a)$

**9.** $(x + 4)^3$  **10.** $(2x - 3)^3$

## A.6 Exercises

In Exercises 1–4, factor out the common factor.

**1.** $3x + 6$  **2.** $4x^3 - 6x^2 + 12x$

**3.** $(x - 1)^2 + 6(x - 1)$  **4.** $3x(x + 2) - 4(x + 2)$

In Exercises 5–8, factor the difference of two squares.

**5.** $x^2 - 36$  **6.** $49 - 9y^2$

**7.** $(x - 1)^2 - 4$  **8.** $25 - (z + 5)^2$

In Exercises 9–12, factor the perfect square trinomial.

**9.** $x^2 - 4x + 4$  **10.** $x^2 + 10x + 25$

**11.** $25y^2 - 10y + 1$  **12.** $z^2 + z + \frac{1}{4}$

In Exercises 13–22, factor the trinomial.

**13.** $x^2 + x - 2$  **14.** $x^2 + 5x + 6$

**15.** $y^2 + y - 20$  **16.** $z^2 - 5z - 24$

**17.** $x^2 - 30x + 200$  **18.** $x^2 - 13x + 42$

**19.** $3x^2 - 5x + 2$  **20.** $2x^2 - x - 1$

**21.** $5x^2 + 26x + 5$  **22.** $5u^2 + 13u - 6$

In Exercises 23–26, factor the sum or difference of cubes.

**23.** $x^3 - 8$  **24.** $x^3 - 27$

**25.** $8t^3 - 1$  **26.** $27x^3 + 8$

In Exercises 27–30, factor by grouping.

**27.** $x^3 - x^2 + 2x - 2$  **28.** $x^3 + 5x^2 - 5x - 25$

**29.** $2x^3 - x^2 - 6x + 3$  **30.** $5x^3 - 10x^2 + 3x - 6$

In Exercises 31–46, completely factor the expression.

**31.** $4x^2 - 8x$  **32.** $12x^2 - 48$

**33.** $x^3 - 4x^2$  **34.** $16 + 6x - x^2$

**35.** $1 - 4x + 4x^2$  **36.** $2y^3 - 7y^2 - 15y$

**37.** $3x^3 + x^2 + 15x + 5$  **38.** $5 - x + 5x^2 - x^3$

**39.** $x^4 - 4x^3 + x^2 - 4x$  **40.** $3u - 2u^2 + 6 - u^3$

**41.** $25 - (x + 5)^2$  **42.** $(t - 1)^2 - 49$

**43.** $(x^2 + 1)^2 - 4x^2$  **44.** $(x^2 + 8)^2 - 36x^2$

**45.** $2t^3 - 16$  **46.** $5x^3 + 40$

**47.** *Geometry* The room that is shown in the figure has a floor space of $2x^2 + 3x + 1$ square feet. If the width of the room is $(x + 1)$ feet, what is the length?

**48.** *Geometry* The room that is shown in the figure has a floor space of $3x^2 + 8x + 4$ square feet. If the width of the room is $(x + 2)$ feet, what is the length?

## A.7    Fractional Expressions and Probability

Domain of an Expression    ■    Rational Expressions    ■    Complex Fractions    ■
Introduction to Probability

## Domain of an Expression

The set of all real numbers for which an algebraic expression is defined is called
the **domain** of the expression. For instance, the domain of

$$\frac{1}{x}$$

is all real numbers other than $x = 0$. Two algebraic expressions are said to be
**equivalent** if they yield the same value for all numbers in their domain. For
instance, the expressions

$$[(x + 1) + (x + 2)] \quad \text{and} \quad 2x + 3$$

are equivalent.

### Study Tip

For all algebraic expressions,
exclude from the domain all
values that could create *division
by zero* or *the square root of a
negative number.*

**Example 1**    Finding the Domain of an Algebraic Expression

**a.** The domain of the polynomial

$$2x^3 + 3x + 4$$

is the set of all real numbers. In fact, the domain of any polynomial is the set
of all real numbers (unless the domain is specifically restricted).

**b.** The domain of the polynomial

$$x^2 + 5x + 2, \qquad x > 0$$

is the set of positive real numbers, because the polynomial is specifically
restricted to that set.

**c.** The domain of the radical expression

$$\sqrt{x}$$

is the set of nonnegative real numbers, because the square root of a negative
number is not a real number.

**d.** The domain of the expression

$$\frac{x + 2}{x - 3}$$

is the set of all real numbers except $x = 3$, because the value of $x = 3$ would
produce an undefined division by zero.

## Rational Expressions

The quotient of two algebraic expressions is a **fractional expression.** Moreover,
the quotient of two *polynomials* such as

$$\frac{1}{x}, \qquad \frac{2x-1}{x+1}, \qquad \text{or} \qquad \frac{x^2-1}{x^2+1}$$

is a **rational expression.** Recall that a fraction is in reduced form if its numerator and denominator have no factors in common aside from ±1. To write a fraction in reduced form, you can apply the following **Cancellation Law.**

$$\frac{a \cdot \overset{1}{\cancel{c}}}{b \cdot \underset{1}{\cancel{c}}} = \frac{a}{b}, \qquad b \neq 0, \qquad c \neq 0$$

The key to success in simplifying rational expressions lies in your ability to *factor* polynomials. For example,

$$\frac{18x^2 - 18}{6x - 6} = \frac{3(6)(x+1)\overset{1}{\cancel{(x-1)}}}{\underset{1}{\cancel{6}}\,\underset{1}{\cancel{(x-1)}}} = 3(x+1), \quad x \neq 1.$$

Note that the original expression is undefined when $x = 1$ (because division by zero is undefined). Thus, in order to make sure that the reduced expression is *equivalent* to the original expression, we must restrict the domain of the reduced expression by excluding the value $x = 1$.

---

**Example 2**    Reducing a Rational Expression

$$\frac{x^2 + 4x - 12}{3x - 6} = \frac{(x+6)\overset{1}{\cancel{(x-2)}}}{3\underset{1}{\cancel{(x-2)}}} \qquad \text{Factor completely.}$$

$$= \frac{x+6}{3}, \quad x \neq 2 \qquad \text{Cancel common factors.}$$

---

In Example 2, do not make the mistake of trying to reduce further by canceling *terms*.

$$\frac{\cancel{x}+6}{\cancel{3}} = \frac{x+\overset{2}{\cancel{6}}}{\underset{1}{\cancel{3}}} = x + 2$$

Remember that to reduce fractions, you cancel *factors*, not terms.

When simplifying rational expressions, be sure to factor each polynomial completely before concluding that the numerator and denominator have no factors in common. Moreover, changing the sign of a factor may allow further reduction, as demonstrated in part (b) of the next example.

---

**Example 3**    Reducing Rational Expressions

**a.** $\dfrac{x^3 - 4x}{x^2 + x - 2} = \dfrac{x(x+2)(x-2)}{(x+2)(x-1)}$       Factor completely.

$$= \frac{x(x-2)}{x-1}, \qquad x \neq -2 \qquad \text{Cancel common factors.}$$

**b.** $\dfrac{12 + x - x^2}{2x^2 - 9x + 4} = \dfrac{(4 - x)(3 + x)}{(2x - 1)(x - 4)}$    Factor completely.

$= \dfrac{-(x - 4)(3 + x)}{(2x - 1)(x - 4)}$    $4 - x = -(x - 4)$

$= -\dfrac{3 + x}{2x - 1}, \quad x \neq 4$    Cancel common factors.

To multiply or divide rational expressions, use the properties of fractions (see Section A.2). Recall that to divide fractions you invert the divisor and multiply.

**Example 4**    Multiplying Rational Expressions

$\dfrac{6x^2 - 6x}{x^2 + 2x - 3} \cdot \dfrac{x^2 + x - 6}{2x}$

$= \dfrac{6x(x - 1)(x + 3)(x - 2)}{(x - 1)(x + 3)(2x)}$    Factor and multiply.

$= \dfrac{3(2x)(x - 1)(x + 3)(x - 2)}{(x - 1)(x + 3)(2x)}$    Cancel common factors.

$= 3(x - 2), \quad x \neq 0, x \neq 1, x \neq -3$    Simplify.

**Example 5**    Dividing Rational Expressions

$\dfrac{2x}{3x - 12} \div \dfrac{x^2 - 2x}{x^2 - 6x + 8} = \dfrac{2x}{3x - 12} \cdot \dfrac{x^2 - 6x + 8}{x^2 - 2x}$    Invert and multiply.

$= \dfrac{(2x)(x - 2)(x - 4)}{(3)(x - 4)(x)(x - 2)}$    Factor and multiply.

$= \dfrac{(2x)(x - 2)(x - 4)}{(3)(x - 4)(x)(x - 2)}$    Cancel common factors.

$= \dfrac{2}{3}, \quad x \neq 0, x \neq 2, x \neq 4$    Simplify.

To add or subtract rational expressions, use the least common denominator method or the following basic definition for adding two fractions.

$$\frac{a}{b} \pm \frac{c}{d} = \frac{ad \pm bc}{bd}, \qquad b \neq 0, d \neq 0$$

This definition is efficient for adding or subtracting *two* fractions that have no common factors in their denominators.

**Example 6**    Adding Rational Expressions

$$\frac{x}{x-3} + \frac{2}{3x+4} = \frac{x(3x+4) + 2(x-3)}{(x-3)(3x+4)} \qquad \frac{a}{b} + \frac{c}{d} = \frac{ad+bc}{bd}$$

$$= \frac{3x^2 + 4x + 2x - 6}{(x-3)(3x+4)} \qquad \text{Remove parentheses.}$$

$$= \frac{3x^2 + 6x - 6}{(x-3)(3x+4)} \qquad \text{Collect like terms.}$$

**Example 7**    Combining Rational Expressions

Perform the given operations and simplify.

$$\frac{3}{x-1} - \frac{2}{x} + \frac{x+3}{x^2-1}$$

Solution

Using the factored denominators $(x-1)$, $x$, and $(x+1)(x-1)$, you can see that the least common denominator is $x(x+1)(x-1)$.

$$\frac{3}{x-1} - \frac{2}{x} + \frac{x+3}{x^2-1}$$

$$= \frac{3(x)(x+1)}{x(x+1)(x-1)} - \frac{2(x+1)(x-1)}{x(x+1)(x-1)} + \frac{(x+3)(x)}{x(x+1)(x-1)}$$

$$= \frac{3(x)(x+1) - 2(x+1)(x-1) + (x+3)(x)}{x(x+1)(x-1)}$$

$$= \frac{3x^2 + 3x - 2x^2 + 2 + x^2 + 3x}{x(x+1)(x-1)}$$

$$= \frac{2x^2 + 6x + 2}{x(x+1)(x-1)}$$

$$= \frac{2(x^2 + 3x + 1)}{x(x+1)(x-1)}$$

## Complex Fractions

Problems involving the division of two rational expressions are sometimes written as **complex fractions.** The rules for dividing fractions still apply in such cases: invert the denominator and multiply, as follows.

$$\frac{\left(\dfrac{x-1}{5}\right)}{\left(\dfrac{x+3}{x}\right)} = \frac{x-1}{5} \cdot \frac{x}{x+3} = \frac{x(x-1)}{5(x+3)}$$

**Example 8**   Simplifying a Complex Fraction

$$\frac{\left(\dfrac{x^2 + 2x - 3}{x - 3}\right)}{(4x + 12)} = \frac{\left(\dfrac{x^2 + 2x - 3}{x - 3}\right)}{\left(\dfrac{4x + 12}{1}\right)} \qquad \text{Rewrite denominator.}$$

$$= \frac{x^2 + 2x - 3}{x - 3} \cdot \frac{1}{4x + 12} \qquad \text{Invert and multiply.}$$

$$= \frac{(x - 1)(x + 3)}{(x - 3)(4)(x + 3)} \qquad \text{Factor and multiply.}$$

$$= \frac{(x - 1)(x + 3)}{(x - 3)(4)(x + 3)} \qquad \text{Cancel common factors.}$$

$$= \frac{x - 1}{4(x - 3)}, \quad x \neq -3 \qquad \text{Simplify.}$$

**Example 9**   Simplifying a Complex Fraction

$$\frac{\left(\dfrac{x}{x + 1}\right)}{\left(\dfrac{2x}{x^2 - 1}\right)} = \frac{x}{x + 1} \cdot \frac{x^2 - 1}{2x} \qquad \text{Invert and multiply.}$$

$$= \frac{x(x - 1)(x + 1)}{(x + 1)(2)(x)} \qquad \text{Multiply and factor.}$$

$$= \frac{x - 1}{2}, \quad x \neq 0, x \neq -1 \qquad \text{Cancel common factors.}$$

# Introduction to Probability

As a member of a complex society, you are used to living with varying amounts of uncertainty. For example, you may be questioning the likelihood of getting a good job after graduation, or winning a state lottery, or the probability of rain tomorrow.

In assigning measurements to uncertainties in everyday life, we often use ambiguous terminology, such as "fairly certain," "probable," or "highly unlikely." In mathematics, we attempt to remove this ambiguity by assigning a number to the likelihood of the occurrence of an event. We call this measurement the **probability** that the event will occur. For example, if we toss a fair coin, we say that the probability that it will land heads up is one-half, or 0.5.

In the study of probability, we call any happening whose result is uncertain an **experiment.** The possible results of the experiment are called **outcomes,** the set of all possible outcomes of an experiment is called the **sample space** of the experiment, and any subcollection of a sample space is called an **event.**

To calculate the probability of an event, we count the number of outcomes in the event and in the sample space. The *number of outcomes* in event $E$ is denoted by $n(E)$ and the number of outcomes in the sample space $S$ is denoted by $n(S)$.

> ▶ The Probability of an Event
>
> If an event $E$ has $n(E)$ equally likely outcomes and its sample space has $n(S)$ equally likely outcomes, the **probability** of event $E$ is
>
> $$P(E) = \frac{n(E)}{n(S)}.$$

Because the number of outcomes in an event must be *less than or equal to* the number of outcomes in the sample space, we can see that the probability of an event must be a number between 0 and 1. That is, for any event $E$, it must be true that $0 \le P(E) \le 1$, as indicated in Figure A.12.

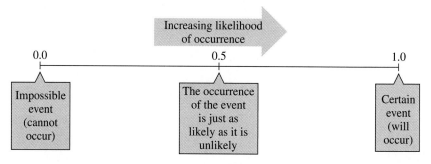

Figure A.12

> ▶ Properties of the Probability of an Event
>
> Let $E$ be an event that is a subset of a finite sample space $S$.
>
> **1.** $0 \le P(E) \le 1$
>
> **2.** If $P(E) = 0$, $E$ *cannot occur* and is called an **impossible event.**
>
> **3.** If $P(E) = 1$, $E$ *must occur* and is called a **certain event.**

**Example 10**  Finding the Probability of an Event

A card is drawn at random from a standard deck of 52 playing cards.

**a.** What is the probability that the card is the ace of hearts?

**b.** What is the probability that the card is an ace?

Solution

**a.** Because there are 52 cards in the deck, the number of possible outcomes in the sample space is 52. Moreover, because only one of these cards is the ace of hearts, the number of outcomes in the event "drawing the ace of hearts" is 1. Thus, the probability of drawing the ace of hearts is

$$P(\text{drawing the ace of hearts}) = \frac{1}{52}.$$

**b.** Because there are 52 cards in the deck, the number of possible outcomes in the sample space is 52. Moreover, because four of these cards are aces, the number of outcomes in the event "drawing an ace" is 4. Thus, the probability of drawing an ace is

$$P(\text{drawing an ace}) = \frac{4}{52} = \frac{1}{13}.$$

This probability is shown graphically in Figure A.13.

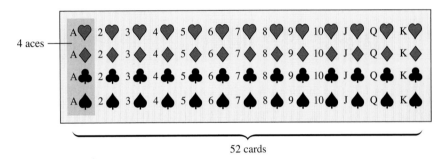

Figure A.13

---

**Example 11**   Finding the Probability of an Event

A parachutist plans to land in the middle of a football stadium during a halftime show. The parachutist is certain of landing in the stadium field, but cannot guarantee a landing in a square area in the center of the field, as shown in Figure A.14. Assuming that the probability of landing at any given point in the field is the same, what is the probability that the parachutist will land in the center square?

**Solution**

The area of the entire field is given by

$$\text{Area of field} = (\text{length})(\text{width}) = (6x)(12x + 25).$$

The area of the center square is given by

$$\text{Area of center square} = (\text{length})(\text{width}) = x^2.$$

Therefore, the probability of landing in the center square is

$$P(\text{landing in center square}) = \frac{x^2}{(6x)(12x + 25)} = \frac{x}{6(12x + 25)}.$$

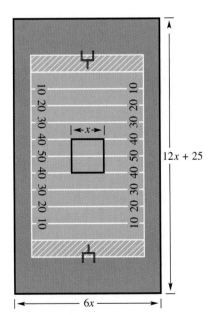

Figure A.14

# Warm Up

The following warm-up exercises involve skills that were covered in earlier sections. You will use these skills in the exercise set for this section.

In Exercises 1–10, completely factor the polynomials.

**1.** $5x^2 - 15x^3$

**2.** $16x^2 - 9$

**3.** $9x^2 - 6x + 1$

**4.** $9 + 12y + 4y^2$

**5.** $z^2 + 4z + 3$

**6.** $x^2 - 15x + 50$

**7.** $3 + 8x - 3x^2$

**8.** $3x^2 - 46x + 15$

**9.** $s^3 + s^2 - 4s - 4$

**10.** $y^3 + 64$

## A.7   Exercises

In Exercises 1–6, find the domain of the expression.

**1.** $3x^2 - 4x + 7$

**2.** $6x^2 + 7x - 9, \ x > 0$

**3.** $\dfrac{1}{x - 2}$

**4.** $\dfrac{x + 1}{2x + 1}$

**5.** $\dfrac{x - 1}{x^2 - 4x}$

**6.** $\dfrac{1}{\sqrt{x + 1}}$

In Exercises 7–12, find the missing factor in the numerator such that the two fractions will be equivalent.

**7.** $\dfrac{5}{2x} = \dfrac{5(\quad)}{6x^2}$

**8.** $\dfrac{3}{4} = \dfrac{3(\quad)}{4(x + 1)}$

**9.** $\dfrac{x + 1}{x} = \dfrac{(x + 1)(\quad)}{x(x - 2)}$

**10.** $\dfrac{3y - 4}{y + 1} = \dfrac{(3y - 4)(\quad)}{y^2 - 1}$

**11.** $\dfrac{3x}{x - 3} = \dfrac{3x(\quad)}{x^2 - x - 6}$

**12.** $\dfrac{1 - z}{z^2} = \dfrac{(1 - z)(\quad)}{z^3 + z^2}$

In Exercises 13–24, write the rational expression in reduced form.

**13.** $\dfrac{15x^2}{10x}$

**14.** $\dfrac{18y^2}{60y^5}$

**15.** $\dfrac{2x}{4x + 4}$

**16.** $\dfrac{9x^2 + 9x}{2x + 2}$

**17.** $\dfrac{x - 5}{10 - 2x}$

**18.** $\dfrac{x^2 - 25}{5 - x}$

**19.** $\dfrac{x^3 + 5x^2 + 6x}{x^2 - 4}$

**20.** $\dfrac{x^2 + 8x - 20}{x^2 + 11x + 10}$

**21.** $\dfrac{y^2 - 7y + 12}{y^2 + 3y - 18}$

**22.** $\dfrac{x + 1}{x^2 + 11x + 10}$

**23.** $\dfrac{z^3 - 8}{z^2 + 2z + 4}$

**24.** $\dfrac{x^2 - 9}{x^3 + x^2 - 9x - 9}$

In Exercises 25–46, perform the indicated operations and simplify.

**25.** $\dfrac{5}{x - 1} \cdot \dfrac{x - 1}{25(x - 2)}$

**26.** $\dfrac{x + 13}{x^3(3 - x)} \cdot \dfrac{x(x - 3)}{5}$

**27.** $\dfrac{(x - 9)(x + 7)}{x + 1} \cdot \dfrac{x}{9 - x}$

**28.** $\dfrac{(x + 5)(x - 3)}{x + 2} \cdot \dfrac{1}{(x + 5)(x + 2)}$

**29.** $\dfrac{r}{r - 1} \cdot \dfrac{r^2 - 1}{r^2}$

**30.** $\dfrac{4y - 16}{5y + 15} \cdot \dfrac{2y + 6}{4 - y}$

**31.** $\dfrac{t^2 - t - 6}{t^2 + 6t + 9} \cdot \dfrac{t + 3}{t^2 - 4}$

**32.** $\dfrac{y^3 - 8}{2y^3} \cdot \dfrac{4y}{y^2 - 5y + 6}$

**33.** $\dfrac{x^2 + x - 2}{x^3 + x^2} \cdot \dfrac{x}{x^2 + 3x + 2}$

**34.** $\dfrac{x^3 - 1}{x + 1} \cdot \dfrac{x^2 + 1}{x^2 - 1}$

**35.** $\dfrac{3(x + y)}{4} \div \dfrac{x + y}{2}$

**36.** $\dfrac{x + 2}{5(x - 3)} \div \dfrac{x - 2}{5(x - 3)}$

**37.** $\dfrac{\left[\dfrac{x^2}{(x + 1)^2}\right]}{\left[\dfrac{x}{(x + 1)^3}\right]}$

**38.** $\dfrac{\left(\dfrac{x^2 - 1}{x}\right)}{\left[\dfrac{(x - 1)^2}{x}\right]}$

**39.** $\dfrac{5}{x-1} + \dfrac{x}{x-1}$

**40.** $\dfrac{2x-1}{x+3} + \dfrac{1-x}{x+3}$

**41.** $6 - \dfrac{5}{x+3}$

**42.** $\dfrac{2x}{x-5} - \dfrac{5}{5-x}$

**43.** $\dfrac{2}{x^2-4} - \dfrac{1}{x^2-3x+2}$

**44.** $\dfrac{x}{x^2+x-2} - \dfrac{1}{x+2}$

**45.** $-\dfrac{1}{x} + \dfrac{2}{x^2+1} + \dfrac{1}{x^3+x}$

**46.** $\dfrac{2}{x+1} + \dfrac{2}{x-1} + \dfrac{1}{x^2-1}$

In Exercises 47–52, simplify the complex fraction.

**47.** $\dfrac{\left(\dfrac{x}{2}-1\right)}{(x-2)}$

**48.** $\dfrac{(x-3)}{\left(\dfrac{x}{4}-\dfrac{4}{x}\right)}$

**49.** $\dfrac{\left(\dfrac{1}{x}-\dfrac{1}{x+1}\right)}{\left(\dfrac{1}{x+1}\right)}$

**50.** $\dfrac{\left(\dfrac{5}{y}-\dfrac{6}{2y+1}\right)}{\left(\dfrac{5}{y}+4\right)}$

**51.** $\dfrac{\left(\sqrt{x}-\dfrac{1}{2\sqrt{x}}\right)}{\sqrt{x}}$

**52.** $\dfrac{\left(\dfrac{x+4}{x+5}-\dfrac{x}{x+1}\right)}{4}$

**53.** *Random Selection*   One of the 35 students in your class is chosen at random. What is the probability that it will be you?

**54.** *Misplaced Test*   Thirty-eight of the students in your class took a test. Your instructor lost two of the tests. What is the probability that yours got lost?

**55.** *Lottery*   The "Daily Number" lottery in many states involves choosing an integer from 000 to 999. Suppose that you have purchased a lottery ticket that has one number. What is the probability that your number will be selected?

**56.** *Defective Product*   Suppose that you just bought a new camera. The manufacturer of your camera made 250,000 cameras of your model, and 5000 of them had defective lenses. What is the probability that your new camera has a defective lens?

**57.** *Playing Cards*   Suppose that you draw a single card from a standard deck of playing cards. What is the probability that the card is a heart?

*Monthly Payment*   In Exercises 58 and 59, use the following formula, which gives the approximate annual percentage rate $r$ of a monthly installment loan:

$$r = \dfrac{\left[\dfrac{24(NM-P)}{N}\right]}{\left(P+\dfrac{NM}{12}\right)}$$

where $N$ is the total number of payments, $M$ is the monthly payment, and $P$ is the amount financed.

**58.** (a) Approximate the rate $r$ for a 4-year car loan of \$15,000 that has monthly payments of \$400.

   (b) Simplify the expression for the annual percentage rate $r$, and then rework part (a).

**59.** (a) Approximate the rate $r$ for a 5-year car loan of \$18,000 that has monthly payments of \$400.

   (b) Simplify the expression for the annual percentage rate $r$, and then rework part (a).

**60.** *Refrigeration*   When food (at room temperature) is placed in a refrigerator, the time required for the food to cool depends on the amount of food, the air circulation in the refrigerator, the original temperature of the food, and the temperature of the refrigerator. Consider the following model, which gives the temperature of food that is at 75°F and is placed in a 40°F refrigerator:

$$T = 10\left(\dfrac{4t^2+16t+75}{t^2+4t+10}\right),\ 0 \le t$$

where $T$ is the temperature in degrees Fahrenheit and $t$ is the time in hours. Sketch a bar graph showing the temperature of the food when $t = 0, 1, 2, 3, 4,$ and 5 hours.

**61.** *Oxygen Level*   The mathematical model

$$O = \dfrac{t^2-t+1}{t^2+1},\ 0 \le t$$

measures the percentage of the normal level of oxygen in a pond, where $t$ is the time in weeks after organic waste is dumped into the pond. Sketch a bar graph showing the oxygen level of the pond when $t = 0, 1, 2, 3, 4,$ and 5 weeks.

## Appendix B
# Conic Sections

### B.1 Conic Sections

Introduction to Conic Sections ■ Parabolas ■ Ellipses ■ Hyperbolas

## Introduction to Conic Sections

Conic sections were discovered during the classical Greek period, which lasted from 600 to 300 B.C. By the beginning of the Alexandrian period, enough was known of conics for Apollonius (262–190 B.C.) to produce an eight-volume work on the subject.

This early Greek study was largely concerned with the geometric properties of conics. It was not until the early seventeenth century that the broad applicability of conics became apparent.

A **conic section** (or simply **conic**) can be described as the intersection of a plane and a double-napped cone. Notice from Figure B.1 that in the formation of the four basic cones, the intersecting plane does not pass through the vertex of the cone. When the plane does pass through the vertex, we call the resulting figure a **degenerate conic,** as shown in Figure B.2

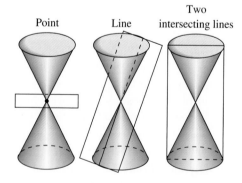

Circle   Parabola   Ellipse   Hyperbola

Point   Line   Two intersecting lines

**Figure B.1** *Conic Sections*

**Figure B.2** *Degenerate Conics*

There are several ways to approach the study of conics. We could begin by defining conics in terms of the intersections of planes and cones, as the Greeks did, or we could define them algebraically, in terms of the general second-degree equation

$$Ax^2 + Bxy + Cy^2 + Dx + Ey + F = 0.$$

However, we will use a third approach, in which each of the conics is defined as a *locus*, or collection of points satisfying a certain geometric property. For example, in Section 2.2 we saw how the definition of a circle as *the collection of all points $(x, y)$ that are equidistant from a fixed point $(h, k)$* led easily to the standard equation of a circle, $(x - h)^2 + (y - k)^2 = r^2$.

We will restrict our study of conics in this section to parabolas with vertices at the origin and ellipses and hyperbolas with centers at the origin. In the following section, we will look at the more general cases.

## Parabolas

In Section 4.1 we determined that the graph of the quadratic function $f(x) = ax^2 + bx + c$ is a parabola that opens upward or downward. The following definition of a parabola is more general in the sense that it is independent of the orientation of the parabola.

---

▶ **Definition of a Parabola**

A **parabola** is the set of all points $(x, y)$ that are equidistant from a fixed line called the **directrix** and a fixed point called the **focus** (not on the line).

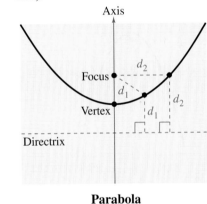

**Parabola**

---

The midpoint between the focus and the directrix is called the **vertex,** and the line passing through the focus and the vertex is called the **axis** of the parabola.

Using this definition, we can derive the following standard form of the equation of a parabola.

---

▶ **Standard Equation of a Parabola (Vertex at Origin)**

The **standard form of the equation of a parabola** with a vertex at $(0, 0)$ and directrix $y = -p$ is

$$x^2 = 4py, \quad p \neq 0. \qquad \text{Vertical axis}$$

For directrix $x = -p$, the equation is

$$y^2 = 4px, \quad p \neq 0. \qquad \text{Horizontal axis}$$

The focus is on the axis $p$ units (directed distance) from the vertex.

---

| Example 1 | Finding the Focus of a Parabola |

Find the focus of the parabola whose equation is $y = -2x^2$.

**Solution**

Since the squared term in the equation involves $x$, we know that the axis is vertical, and we should use the standard form

$$x^2 = 4py. \qquad \text{Standard form, vertical axis}$$

Writing the given equation in this form, we have

$$-2x^2 = y \qquad \text{Original equation}$$

$$x^2 = -\frac{1}{2}y \qquad \text{Divide both sides by } -2.$$

$$x^2 = 4\left(-\frac{1}{8}\right)y. \qquad \text{Standard form}$$

Thus, $p = -\frac{1}{8}$. Since $p$ is negative, the parabola opens downward and the focus of the parabola is $(0, p) = \left(0, -\frac{1}{8}\right)$, as shown in Figure B.3.

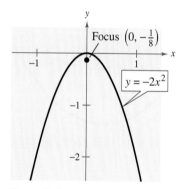

Figure B.3

| Example 2 | A Parabola with a Horizontal Axis |

Write the standard form of the equation of the parabola with the vertex at the origin and the focus at $(2, 0)$.

**Solution**

The axis of the parabola is horizontal, passing through $(0, 0)$ and $(2, 0)$, as shown in Figure B.4. So, we consider the standard form

$$y^2 = 4px. \qquad \text{Standard form, horizontal axis}$$

Since the focus is $p = 2$ units from the vertex, the equation is

$$y^2 = 4(2)x \qquad \text{Standard form}$$

$$y^2 = 8x.$$

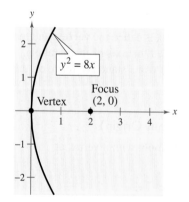

Figure B.4

The term *parabola* is a technical term used in mathematics and does not simply refer to *any* U-shaped curve.

Parabolas occur in a wide variety of applications. For instance, a parabolic reflector can be formed by revolving a parabola about its axis. The resulting surface has the property that all incoming rays parallel to the axis are reflected through the focus of the parabola—this is the principal behind the construction of the parabolic mirrors used in reflecting telescopes. Conversely, the light rays emanating from the focus of a parabolic reflector used in a flashlight are all reflected parallel to one another, as shown in Figure B.5.

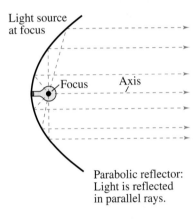

Light source at focus

Focus   Axis

Parabolic reflector:
Light is reflected
in parallel rays.

**Figure B.5**

## Ellipses

The second basic type of conic is called an **ellipse,** and it is defined as follows.

▶ **Definition of an Ellipse**

An **ellipse** is the set of all points $(x, y)$ the sum of whose distances from two distinct fixed points, called **foci,** is constant.

$(x, y)$

$d_1$   $d_2$

Focus   Focus

$d_1 + d_2$ is constant.

Major axis

Center

Vertex

Minor axis

The line through the foci intersects the ellipse at two points, called the **vertices.** The chord joining the vertices is called the **major axis,** and its midpoint is called the **center** of the ellipse. The chord perpendicular to the major axis at the center is called the **minor axis** of the ellipse.

**Figure B.6**

You can visualize the definition of an ellipse by imagining two thumbtacks placed at the foci, as shown in Figure B.6. If the ends of a fixed length of string are fastened to the thumbtacks and the string is drawn taut with a pencil, the path traced by the pencil will be an ellipse.

The standard form of the equation of an ellipse takes one of two forms, depending on whether the major axis is horizontal or vertical.

---

▶ **Standard Equation of an Ellipse (Center at Origin)**

The **standard form of the equation of an ellipse** with the center at the origin and major and minor axes of the lengths $2a$ and $2b$, where $0 < b < a$, is

$$\frac{x^2}{a^2} + \frac{y^2}{b^2} = 1 \qquad \text{or} \qquad \frac{x^2}{b^2} + \frac{y^2}{a^2} = 1, \quad 0 < b < a.$$

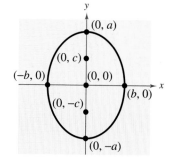

**Major axis is horizontal.**
**Minor axis is vertical.**

**Major axis is vertical.**
**Minor axis is horizontal.**

The vertices and foci lie on the major axis, $a$ and $c$ units, respectively, from the center. Moreover, $a$, $b$, and $c$ are related by the equation $c^2 = a^2 - b^2$.

---

| | Example 3 | Finding the Standard Equation of an Ellipse |

Find the standard form of the equation of the ellipse that has a major axis of length 6 and foci at $(-2, 0)$ and $(2, 0)$, as shown in Figure B.7.

**Solution**

Since the foci occur at $(-2, 0)$ and $(2, 0)$, the center of the ellipse is $(0, 0)$, and the major axis is horizontal. Thus, the ellipse has an equation of the form

$$\frac{x^2}{a^2} + \frac{y^2}{b^2} = 1. \qquad \text{Standard form, horizontal major axis}$$

Since the length of the major axis is 6, we have

$$2a = 6 \qquad \text{Length of major axis}$$

which implies that $a = 3$. Moreover, the distance from the center to either focus is $c = 2$. Finally, we have

$$b^2 = a^2 - c^2 = 3^2 - 2^2 = 9 - 4 = 5$$

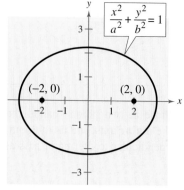

**Figure B.7**

which yields the equation

$$\frac{x^2}{9} + \frac{y^2}{5} = 1. \qquad \text{Standard form}$$

**Example 4**    Sketching an Ellipse

Sketch the ellipse given by $4x^2 + y^2 = 36$, and identify the vertices.

**Solution**

We begin by writing the equation in standard form.

$$4x^2 + y^2 = 36 \qquad \text{Original equation}$$

$$\frac{4x^2}{36} + \frac{y^2}{36} = \frac{36}{36} \qquad \text{Divide both sides by 36.}$$

$$\frac{x^2}{3^2} + \frac{y^2}{6^2} = 1 \qquad \text{Standard form}$$

Since the denominator of the $y^2$-term is larger than the denominator of the $x^2$-term, we conclude that the major axis is vertical. Moreover, since $a = 6$, the vertices are $(0, -6)$ and $(0, 6)$. Finally, since $b = 3$, the endpoints of the minor axis are $(-3, 0)$ and $(3, 0)$, as shown in Figure B.8.

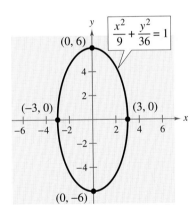

**Figure B.8**

In Example 4, note that from the standard form of the equation we can sketch the ellipse by locating the endpoints of the two axes. Since $3^2$ is the denominator of the $x^2$-term, we move 3 units to the *right and left* of the center to locate the endpoints of the horizontal axis. Similarly, since $6^2$ is the denominator of the $y^2$-term, we move 6 units *up and down* from the center to locate the endpoints of the vertical axis.

## Hyperbolas

The definition of a **hyperbola** is similar to that of an ellipse. The distinction is that, for an ellipse, the *sum* of the distances between the foci and a point on the ellipse is constant, while for a hyperbola, the *difference* of these distances is constant.

> ▶ **Definition of a Hyperbola**
>
> A **hyperbola** is the set of all points $(x, y)$, the difference of whose distances from two distinct fixed points, called **foci,** is constant.
>
>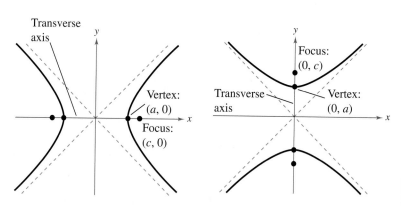
>
> $d_2 - d_1$ is constant.

The graph of a hyperbola has two disconnected parts, called **branches.** The line through the two foci intersects the hyperbola at two points, called **vertices.** The line segment connecting the vertices is called the **transverse axis,** and the midpoint of the transverse axis is called the **center** of the hyperbola.

> ▶ **Standard Equation of a Hyperbola (Center at Origin)**
>
> The **standard form of the equation of a hyperbola** with the center at $(0, 0)$ is
> $$\frac{x^2}{a^2} - \frac{y^2}{b^2} = 1 \qquad \text{or} \qquad \frac{y^2}{a^2} - \frac{x^2}{b^2} = 1.$$
>
> The vertices and foci are, respectively, $a$ and $c$ units from the center, and $b^2 = c^2 - a^2$.

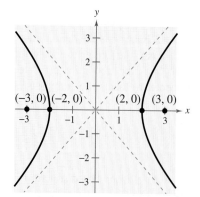

**Figure B.9**

**Example 5**   Finding the Standard Equation of a Hyperbola

Find the standard form of the equation of the hyperbola with foci at $(-3, 0)$ and $(3, 0)$ and vertices at $(-2, 0)$ and $(2, 0)$, as shown in Figure B.9.

**Solution**

It can be seen that $c = 3$ because the foci are 3 units from the center. Moreover, $a = 2$ because the vertices are 2 units from the center. So, it follows that

$$b^2 = c^2 - a^2 = 3^2 - 2^2 = 9 - 4 = 5.$$

Since the transverse axis is horizontal, the standard form of the equation is

$$\frac{x^2}{a^2} - \frac{y^2}{b^2} = 1. \qquad \text{Standard form, horizontal transverse axis}$$

Finally, substituting $a^2 = 4$ and $b^2 = 5$, we have

$$\frac{x^2}{4} - \frac{y^2}{5} = 1. \qquad \text{Standard form}$$

An important aid in sketching the graph of a hyperbola is the determination of its **asymptotes,** as shown in Figure B.10. Each hyperbola has two asymptotes that intersect at the center of the hyperbola. Furthermore, the asymptotes pass through the corners of a rectangle of dimensions $2a$ by $2b$. The line segment of length $2b$, joining $(0, b)$ and $(0, -b)$ [or $(-b, 0)$ and $(b, 0)$], is referred to as the **conjugate axis** of the hyperbola.

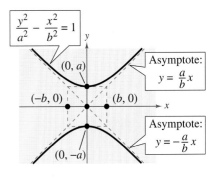

*Transverse axis is horizontal*
Figure B.10

*Transverse axis is vertical*

---

▶ **Asymptotes of a Hyperbola (Center at Origin)**

The asymptotes of a hyperbola with center at $(0, 0)$ are

$$y = \frac{b}{a}x \qquad \text{and} \qquad y = -\frac{b}{a}x \qquad \text{Transverse axis is horizontal.}$$

or

$$y = \frac{a}{b}x \qquad \text{and} \qquad y = -\frac{a}{b}x. \qquad \text{Transverse axis is vertical.}$$

---

**Example 6**    Sketching the Graph of a Hyperbola

Sketch the graph of the hyperbola whose equation is $4x^2 - y^2 = 16$.

### Solution

We begin by rewriting the equation in standard form.

$$4x^2 - y^2 = 16 \qquad \text{Original equation}$$

$$\frac{4x^2}{16} - \frac{y^2}{16} = \frac{16}{16} \qquad \text{Divide both sides by 16.}$$

$$\frac{x^2}{2^2} - \frac{y^2}{4^2} = 1 \qquad \text{Standard form}$$

Because the $x^2$-term is positive, we conclude that the transverse axis is horizontal and the vertices occur at $(-2, 0)$ and $(2, 0)$. Moreover, the endpoints of the conjugate axis occur at $(0, -4)$ and $(0, 4)$, and we are able to sketch the rectangle shown in Figure B.11. Finally, by drawing the asymptotes through the corners of this rectangle, we complete the sketch shown in Figure B.12.

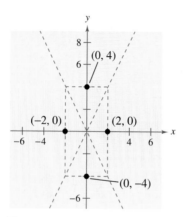

Figure B.11

Figure B.12

### Example 7    Finding the Standard Equation of a Hyperbola

Find the standard form of the equation of the hyperbola having vertices at $(0, -3)$ and $(0, 3)$ and with asymptotes $y = -2x$ and $y = 2x$, as shown in Figure B.13.

### Solution

Since the transverse axis is vertical, we have asymptotes of the form

$$y = \frac{a}{b}x \qquad \text{and} \qquad y = -\frac{a}{b}x. \qquad \text{Transverse axis is vertical.}$$

Using the given equations for the asymptotes, it follows that

$$\frac{a}{b} = 2$$

and since $a = 3$, we can determine that $b = \frac{3}{2}$. Finally, we conclude that the hyperbola has the following equation.

$$\frac{y^2}{3^2} - \frac{x^2}{(3/2)^2} = 1 \qquad \text{Standard form}$$

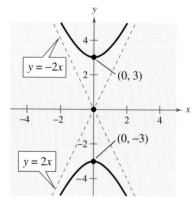

Figure B.13

# Warm Up

The following warm-up exercises involve skills that were covered in earlier sections. You will use these skills in the exercise set for this section.

In Exercises 1–4, rewrite the equations so that they have no fractions.

1. $\dfrac{x^2}{16} + \dfrac{y^2}{9} = 1$

2. $\dfrac{x^2}{32} + \dfrac{4y^2}{32} = \dfrac{32}{32}$

3. $\dfrac{x^2}{1/4} - \dfrac{y^2}{4} = 1$

4. $\dfrac{3x^2}{1/9} + \dfrac{4y^2}{9} = 1$

In Exercises 5–8, solve for $c$. (Assume $c > 0$.)

5. $c^2 = 3^2 - 1^2$

6. $c^2 = 2^2 + 3^2$

7. $c^2 + 2^2 = 4^2$

8. $c^2 - 1^2 = 2^2$

In Exercises 9 and 10, find the distance between the point and the origin.

9. $(0, -4)$

10. $(-2, 0)$

## B.1   Exercises

In Exercises 1–8, match the equation with its graph. [The graphs are labeled (a), (b), (c), (d), (e), (f), (g), and (h).]

(a)

(b)

(c)

(d)

(e)

(f)

(g)

(h)
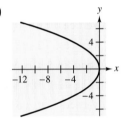

1. $x^2 = 4y$

2. $x^2 = -4y$

3. $y^2 = 4x$

4. $y^2 = -4x$

5. $\dfrac{x^2}{1} + \dfrac{y^2}{4} = 1$

6. $\dfrac{x^2}{4} + \dfrac{y^2}{1} = 1$

7. $\dfrac{x^2}{1} - \dfrac{y^2}{4} = 1$

8. $\dfrac{y^2}{4} - \dfrac{x^2}{1} = 1$

In Exercises 9–16, find the vertex and focus of the parabola and sketch its graph.

9. $y = 4x^2$

10. $y = 2x^2$

11. $y^2 = -6x$

12. $y^2 = 3x$

13. $x^2 + 8y = 0$

14. $x + y^2 = 0$

15. $y^2 - 8x = 0$

16. $x^2 + 12y = 0$

In Exercises 17–26, find an equation of the specified parabola with vertex at the origin.

17. Focus: $\left(0, -\frac{3}{2}\right)$

18. Focus: $(2, 0)$

19. Focus: $(-2, 0)$

20. Focus: $(0, -2)$

**21.** Directrix: $y = -1$

**22.** Directrix: $x = 3$

**23.** Directrix: $y = 2$

**24.** Directrix: $x = -2$

**25.** Horizontal axis; Passes through the point $(4, 6)$

**26.** Vertical axis; Passes through the point $(-2, -2)$

In Exercises 27–34, find the center and vertices of the ellipse and sketch its graph.

**27.** $\dfrac{x^2}{25} + \dfrac{y^2}{16} = 1$

**28.** $\dfrac{x^2}{144} + \dfrac{y^2}{169} = 1$

**29.** $\dfrac{x^2}{16} + \dfrac{y^2}{25} = 1$

**30.** $\dfrac{x^2}{169} + \dfrac{y^2}{144} = 1$

**31.** $\dfrac{x^2}{9} + \dfrac{y^2}{5} = 1$

**32.** $\dfrac{x^2}{28} + \dfrac{y^2}{64} = 1$

**33.** $5x^2 + 3y^2 = 15$

**34.** $x^2 + 4y^2 = 4$

In Exercises 35–42, find an equation of the specified ellipse with center at the origin.

**35.** Vertices: $(0, \pm 2)$; Minor axis of length 2

**36.** Vertices: $(\pm 2, 0)$; Minor axis of length 3

**37.** Vertices: $(\pm 5, 0)$; Foci: $(\pm 2, 0)$

**38.** Vertices: $(0, \pm 8)$; Foci: $(0, \pm 4)$

**39.** Foci: $(\pm 5, 0)$; Major axis of length 12

**40.** Foci: $(\pm 2, 0)$; Major axis of length 8

**41.** Vertices: $(0, \pm 5)$; Passes through the point $(4, 2)$

**42.** Major axis vertical; Passes through the points $(0, 4)$ and $(2, 0)$

In Exercises 43–50, find the center and vertices of the hyperbola and sketch its graph.

**43.** $x^2 - y^2 = 1$

**44.** $\dfrac{x^2}{9} - \dfrac{y^2}{16} = 1$

**45.** $\dfrac{y^2}{1} - \dfrac{x^2}{4} = 1$

**46.** $\dfrac{y^2}{9} - \dfrac{x^2}{1} = 1$

**47.** $\dfrac{y^2}{25} - \dfrac{x^2}{144} = 1$

**48.** $\dfrac{x^2}{36} - \dfrac{y^2}{4} = 1$

**49.** $2x^2 - 3y^2 = 6$

**50.** $3y^2 - 5x^2 = 15$

In Exercises 51–58, find an equation of the specified hyperbola with center at the origin.

**51.** Vertices: $(0, \pm 2)$; Foci: $(0, \pm 4)$

**52.** Vertices: $(\pm 3, 0)$; Foci: $(\pm 5, 0)$

**53.** Vertices: $(\pm 1, 0)$; Asymptotes: $y = \pm 3x$

**54.** Vertices: $(0, \pm 3)$; Asymptotes: $y = \pm 3x$

**55.** Foci: $(0, \pm 8)$; Asymptotes: $y = \pm 4x$

**56.** Foci: $(\pm 10, 0)$; Asymptotes: $y = \pm \frac{3}{4}x$

**57.** Vertices: $(0, \pm 3)$; Passes through the point $(-2, 5)$

**58.** Vertices: $(\pm 2, 0)$; Passes through the point $\left(3, \sqrt{3}\right)$

**59.** *Satellite Antenna*    The receiver in a parabolic television dish antenna is 3 feet from the vertex and is located at the focus (see figure). Find an equation of a cross section of the reflector. (Assume that the dish is directed upward and the vertex is at the origin.)

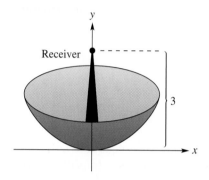

**60.** *Suspension Bridge*    Each cable of a suspension bridge is suspended (in the shape of a parabola) between two towers that are 400 feet apart and 50 feet above the roadway (see figure). The cables touch the roadway midway between the towers. Find an equation for the parabolic shape of each cable.

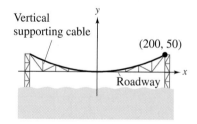

**61.** *Fireplace Arch*    A fireplace arch is to be constructed in the shape of a semiellipse. The opening is to have a height of 2 feet at the center and a width of 5 feet along the base (see figure). The contractor draws the outline of the ellipse by the method shown in Figure B.6. Where should the tacks be placed and what should be the length of the piece of string?

Figure for 61

62. *Mountain Tunnel*  A semielliptical arch over a tunnel for a road through a mountain has a major axis of 100 feet, and its height at the center is 30 feet (see figure). Determine the height of the arch 5 feet from the edge of the tunnel.

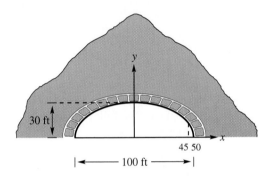

63. Sketch a graph of the ellipse that consists of all points $(x, y)$ such that the sum of the distances between $(x, y)$ and two fixed points is 15 units and the foci are located at the centers of the two sets of concentric circles, as shown in the figure.

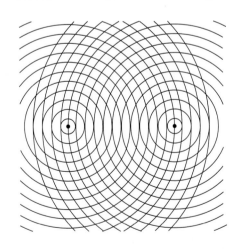

64. *Think About It*  A line segment through a focus with endpoints on the ellipse and perpendicular to the major axis is called a **latus rectum** of the ellipse. Therefore, an ellipse has two latera recta. Knowing

the length of the latera recta is helpful in sketching an ellipse because it yields other points on the curve (see figure). Show that the length of each latus rectum is $2b^2/a$.

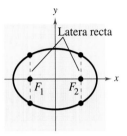

In Exercises 65–68, sketch the graph of the ellipse, making use of the latera recta (see Exercise 64).

65. $\dfrac{x^2}{4} + \dfrac{y^2}{1} = 1$

66. $\dfrac{x^2}{9} + \dfrac{y^2}{16} = 1$

67. $9x^2 + 4y^2 = 36$

68. $5x^2 + 3y^2 = 15$

69. *Loran*  Long-range navigation for aircraft and ships is accomplished by synchronized pulses transmitted by widely separated transmitting stations. These pulses travel at the speed of light (186,000 miles per second). The difference in the arrival times of these pulses at an aircraft or ship is constant on a hyperbola having the transmitting stations as foci. Assume that two stations 300 miles apart are positioned on the rectangular coordinate system at points with coordinates $(-150, 0)$ and $(150, 0)$ and that a ship is traveling on a path with coordinates $(x, 75)$ (see figure). Find the $x$-coordinate of the position of the ship if the time difference between the pulses from the transmitting stations is 1000 microseconds (0.001 second).

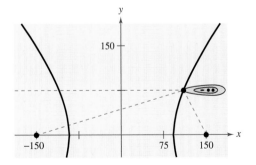

**70.** *Hyperbolic Mirror*   A hyperbolic mirror (used in some telescopes) has the property that a light ray directed at one focus will be reflected to the other focus (see figure). The focus of the hyperbolic mirror has coordinates (12, 0). Find the vertex of the mirror if its **mount** has coordinates (12, 12).

---

## B.2   Conic Sections and Translations

**Vertical and Horizontal Shifts of Conics**  ■  **Writing Equations of Conics in Standard Form**

# Vertical and Horizontal Shifts of Conics

In Section B.1 we looked at conic sections whose graphs were in *standard position*. In this section we will study the equations of conic sections that have been shifted vertically or horizontally in the plane. The following summary lists the standard forms of the equations of the four basic conics.

> ▶ **Standard Forms of Equations of Conics**
>
> Circle ($r$ = radius)
>
> $(x - h)^2 + (y - k)^2 = r^2$
>
> Ellipse ($2a$ = major axis length,
>          $2b$ = minor axis length)
>
> Center: $(h, k)$
>
>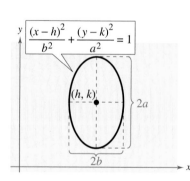

▶ **Standard Forms of Equations of Conics    (continued)**

Hyperbola ($2a$ = transverse axis length,
            $2b$ = conjugate axis length)
Center: $(h, k)$

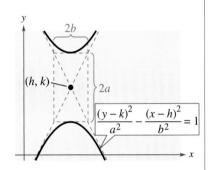

Parabola ($p$ = directed distance
          from vertex to focus)
Vertex: $(h, k)$

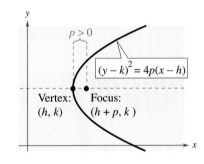

| **Example 1** | **Equations of Conic Sections** |
|---|---|

**a.** The graph of

$$(x - 1)^2 + (y + 2)^2 = 3^2 \qquad \text{Circle}$$

is a circle whose center is the point $(1, -2)$ and whose radius is 3, as shown
in Figure B.14(a) on the following page.

**b.** The graph of

$$\frac{(x - 2)^2}{3^2} + \frac{(y - 1)^2}{2^2} = 1 \qquad \text{Ellipse}$$

is an ellipse whose center is the point $(2, 1)$. The major axis of the ellipse is
horizontal with a length of $2(3) = 6$. The minor axis of the ellipse is vertical
with a length of $2(2) = 4$, as shown in Figure B.14(b) on the following page.

**c.** The graph of

$$\frac{(x-3)^2}{1^2} - \frac{(y-2)^2}{3^2} = 1 \qquad \text{Hyperbola}$$

is a hyperbola whose center is the point $(3, 2)$. The transverse axis is horizontal with a length of $2(1) = 2$. The conjugate axis is vertical with a length of $2(3) = 6$, as shown in Figure B.14(c).

**d.** The graph of

$$(x-2)^2 = 4(-1)(y-3) \qquad \text{Parabola}$$

is a parabola whose vertex is the point $(2, 3)$. The axis of the parabola is vertical. The focus of the parabola is 1 unit above or below the vertex. Moreover, since $p = -1$, it follows that the focus lies *below* the vertex, as shown in Figure B.14(d).

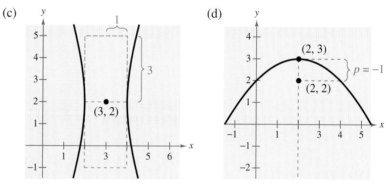

**Figure B.14**

---

# Writing Equations of Conics in Standard Form

To write the equation of a conic in standard form, we complete the square, as demonstrated in Examples 2, 3, and 4.

**Example 2**   Finding the Standard Form of a Parabola

Find the vertex and focus of the parabola given by

$$x^2 - 2x + 4y - 3 = 0.$$

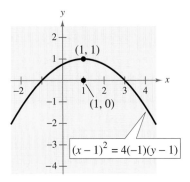

**Figure B.15**

## Solution

$$x^2 - 2x + 4y - 3 = 0 \qquad \text{Original equation}$$
$$x^2 - 2x = -4y + 3 \qquad \text{Group terms.}$$
$$x^2 - 2x + 1 = -4y + 3 + 1 \qquad \text{Add 1 to both sides.}$$
$$(x - 1)^2 = -4y + 4 \qquad \text{Completed square form}$$
$$(x - 1)^2 = 4(-1)(y - 1) \qquad \text{Standard form}$$
$$(x - h)^2 = 4p(y - k)$$

From this standard form, we see that $h = 1$, $k = 1$, and $p = -1$. Since the axis is vertical and $p$ is negative, the parabola opens downward. The vertex and focus are

*Vertex*: $(h, k) = (1, 1)$

*Focus*: $(h, k + p) = (1, 0)$.

The graph of this parabola is shown in Figure B.15.

---

In Example 1, $p$ is the *directed distance* from the vertex to the focus. Because the axis of the parabola is vertical and $p = -1$, the focus is 1 unit *below* the vertex, and the parabola opens downward.

### Example 3    Sketching an Ellipse

Sketch the graph of the ellipse whose equation is

$$x^2 + 4y^2 + 6x - 8y + 9 = 0.$$

## Solution

$$x^2 + 4y^2 + 6x - 8y + 9 = 0 \qquad \text{Original equation}$$
$$(x^2 + 6x + \quad) + (4y^2 - 8y + \quad) = -9 \qquad \text{Group terms.}$$
$$(x^2 + 6x + \quad) + 4(y^2 - 2y + \quad) = -9 \qquad \text{Factor 4 out of } y\text{-terms.}$$
$$(x^2 + 6x + 9) + 4(y^2 - 2y + 1) = -9 + 9 + 4(1) \qquad \text{Add 9 and 4 to both sides.}$$
$$(x + 3)^2 + 4(y - 1)^2 = 4 \qquad \text{Completed square form}$$
$$\frac{(x + 3)^2}{4} + \frac{(y - 1)^2}{1} = 1 \qquad \text{Standard form}$$
$$\frac{(x - h)^2}{a^2} + \frac{(y - k)^2}{b^2} = 1$$

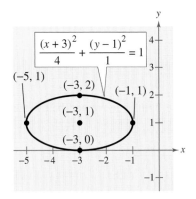

**Figure B.16**

From this standard form, we see that the center is $(h, k) = (-3, 1)$. Since the denominator of the $x$-term is $4 = a^2 = 2^2$, the endpoints of the major axis lie 2 units to the right and left of the center. Similarly, since the denominator of the $y$-term is $1 = b^2 = 1^2$, the endpoints of the minor axis lie 1 unit up and down from the center. The graph of this ellipse is shown in Figure B.16.

---

**Example 4** Sketching a Hyperbola

Sketch the graph of the hyperbola given by the equation

$$y^2 - 4x^2 + 4y + 24x - 41 = 0.$$

**Solution**

| | |
|---|---|
| $y^2 - 4x^2 + 4y + 24x - 41 = 0$ | Original equation |
| $(y^2 + 4y + \quad) - (4x^2 - 24x + \quad) = 41$ | Group terms. |
| $(y^2 + 4y + \quad) - 4(x^2 - 6x + \quad) = 41$ | Factor 4 out of $x$-terms. |
| $(y^2 + 4y + 4) - 4(x^2 - 6x + 9) = 41 + 4 - 4(9)$ | Add 4, subtract 36. |
| $(y + 2)^2 - 4(x - 3)^2 = 9$ | Completed square form |
| $\dfrac{(y + 2)^2}{9} - \dfrac{4(x - 3)^2}{9} = 1$ | Divide both sides by 9. |
| $\dfrac{(y + 2)^2}{9} - \dfrac{(x - 3)^2}{9/4} = 1$ | Change 4 to $1/(1/4)$. |
| $\dfrac{(y + 2)^2}{3^2} - \dfrac{(x - 3)^2}{(3/2)^2} = 1$ | Standard form |
| $\dfrac{(y - k)^2}{a^2} - \dfrac{(x - h)^2}{b^2} = 1$ | |

From the standard form, we see that the transverse axis is vertical and the center lies at $(h, k) = (3, -2)$. Since the denominator of the $y$-term is $a^2 = 3^2$, we know that the vertices occur 3 units above and below the center.

*Vertices:* $(3, -5)$ and $(3, 1)$

To sketch the hyperbola, we draw a rectangle whose top and bottom pass through the vertices. Since the denominator of the $x$-term is $b^2 = \left(\frac{3}{2}\right)^2$, we locate the sides of the rectangle $\frac{3}{2}$ units to the right and left of the center, as shown in Figure B.17. Finally, we sketch the asymptotes by drawing lines through the opposite corners of the rectangle. Using these asymptotes, we complete the graph of the hyperbola, as shown in Figure B.17.

$$\dfrac{(y+2)^2}{3^2} - \dfrac{(x-3)^2}{(3/2)^2} = 1$$

**Figure B.17**

To find the foci in Example 4, we first find c.

$$c^2 = a^2 + b^2 = 9 + \frac{9}{4} = \frac{45}{4} \quad \Longrightarrow \quad c = \frac{3\sqrt{5}}{2}$$

Because the transverse axis is vertical, the foci lie $c$ units above and below the center.

$$\textit{Foci:} \left(3, -2 + \frac{3\sqrt{5}}{2}\right) \quad \text{and} \quad \left(3, -2 - \frac{3\sqrt{5}}{2}\right)$$

| Example 5 | Writing the Equation of an Ellipse |

Write the standard form of the equation of an ellipse whose vertices are $(2, -2)$ and $(2, 4)$. The length of the minor axis of the ellipse is 4, as shown in Figure B.18.

### Solution

The center of the ellipse lies at the midpoint of its vertices. Thus, the center is

$$(h, k) = (2, 1). \qquad \text{Center}$$

Since the vertices lie on a vertical line and are 6 units apart, it follows that the major axis is vertical and has a length of $2a = 6$. Thus, $a = 3$. Moreover, since the minor axis has a length of 4, it follows that $2b = 4$, which implies that $b = 2$. Therefore, we can conclude that the standard form of the equation of the ellipse is as follows.

$$\frac{(x - h)^2}{b^2} + \frac{(y - k)^2}{a^2} = 1 \qquad \text{Major axis is vertical.}$$

$$\frac{(x - 2)^2}{2^2} + \frac{(y - 1)^2}{3^2} = 1 \qquad \text{Standard form}$$

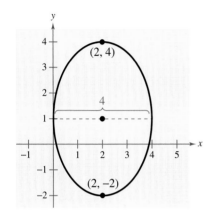

**Figure B.18**

An interesting application of conic sections involves the orbits of comets in our solar system. Of the 610 comets identified prior to 1970, 245 have elliptical orbits, 295 have parabolic orbits, and 70 have hyperbolic orbits. For example, Halley's comet has an elliptical orbit, and we can predict the reappearance of this comet every 75 years. The center of the sun is a focus of each of these orbits, and each orbit has a vertex at the point where the comet is closest to the sun, as shown in Figure B.19.

If $p$ is the distance between the vertex and the focus, and $v$ is the speed of the comet at the vertex, then the orbit is:

an *ellipse* if $\quad v < \sqrt{\dfrac{2GM}{p}}$

a *parabola* if $\quad v = \sqrt{\dfrac{2GM}{p}}$

a *hyperbola* if $\quad v > \sqrt{\dfrac{2GM}{p}}$

where $M$ is the mass of the sun and $G \approx 6.67 \times 10^{-8} \ \text{cm}^3/(\text{gm} \cdot \text{sec}^2)$.

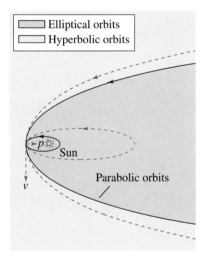

Elliptical orbits
Hyperbolic orbits

Sun

Parabolic orbits

**Figure B.19**

# Warm Up

The following warm-up exercises involve skills that were covered in earlier sections. You will use these skills in the exercise set for this section.

In Exercises 1–10, identify the conic represented by each equation.

**1.** $\dfrac{x^2}{4} - \dfrac{y^2}{4} = 1$

**2.** $\dfrac{x^2}{9} + \dfrac{y^2}{1} = 1$

**3.** $2x + y^2 = 0$

**4.** $\dfrac{x^2}{9} - \dfrac{y^2}{4} = 1$

**5.** $\dfrac{x^2}{4} + \dfrac{y^2}{16} = 1$

**6.** $4x^2 + 4y^2 = 25$

**7.** $\dfrac{y^2}{4} - \dfrac{x^2}{2} = 1$

**8.** $x^2 - 6y = 0$

**9.** $3x - y^2 = 0$

**10.** $\dfrac{x^2}{9/4} + \dfrac{y^2}{4} = 1$

## B.2    Exercises

In Exercises 1–12, find the vertex, focus, and directrix of the parabola and sketch its graph.

**1.** $(x - 1)^2 + 8(y + 2) = 0$

**2.** $(x + 3) + (y - 2)^2 = 0$

**3.** $\left(y + \frac{1}{2}\right)^2 = 2(x - 5)$

**4.** $\left(x + \frac{1}{2}\right)^2 = 4(y - 3)$

**5.** $y = \frac{1}{4}(x^2 - 2x + 5)$

**6.** $y = -\frac{1}{6}(x^2 + 4x - 2)$

**7.** $4x - y^2 - 2y - 33 = 0$

**8.** $y^2 + x + y = 0$

**9.** $y^2 + 6y + 8x + 25 = 0$

**10.** $x^2 - 2x + 8y + 9 = 0$

**11.** $y^2 - 4y - 4x = 0$

**12.** $y^2 - 4x - 4 = 0$

In Exercises 13–20, find an equation of the specified parabola.

**13.** Vertex: $(3, 2)$; Focus: $(1, 2)$

**14.** Vertex: $(-1, 2)$; Focus: $(-1, 0)$

**15.** Vertex: $(0, 4)$; Directrix: $y = 2$

**16.** Vertex: $(-2, 1)$; Directrix: $x = 1$

**17.** Focus: $(2, 2)$; Directrix: $x = -2$

**18.** Focus: $(0, 0)$; Directrix: $y = 4$

**19.**

**20.**

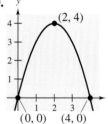

In Exercises 21–28, find the center, foci, and vertices of the ellipse and sketch its graph.

**21.** $\dfrac{(x - 1)^2}{9} + \dfrac{(y - 5)^2}{25} = 1$

**22.** $(x + 2)^2 + \dfrac{(y + 4)^2}{1/4} = 1$

**23.** $9x^2 + 4y^2 + 36x - 24y + 36 = 0$

**24.** $9x^2 + 4y^2 - 36x + 8y + 31 = 0$

**25.** $16x^2 + 25y^2 - 32x + 50y + 16 = 0$

**26.** $9x^2 + 25y^2 - 36x - 50y + 61 = 0$

**27.** $12x^2 + 20y^2 - 12x + 40y - 37 = 0$

**28.** $36x^2 + 9y^2 + 48x - 36y + 43 = 0$

In Exercises 29–36, find an equation of the specified ellipse.

**29.** Vertices: $(0, 2), (4, 2)$; Minor axis of length 2

**30.** Foci: $(0, 0), (4, 0)$; Major axis of length 8

**31.** Foci: $(0, 0), (0, 8)$; Major axis of length 16

**32.** Center: $(2, -1)$; Vertex: $\left(2, \frac{1}{2}\right)$;
Minor axis of length 2

**33.** Vertices: $(3, 1), (3, 9)$ Minor axis of length 6

**34.** Center: $(3, 2)$; $a = 3c$; Foci: $(1, 2), (5, 2)$

**35.** Center: $(0, 4)$; $a = 2c$; Vertices $(-4, 4), (4, 4)$

**36.** Vertices: $(5, 0), (5, 12)$; Endpoints of the minor axis: $(0, 6), (10, 6)$

In Exercises 37–46, find the center, vertices, and foci of the hyperbola and sketch its graph, using asymptotes as an aid.

**37.** $\dfrac{(x - 1)^2}{4} - \dfrac{(y + 2)^2}{1} = 1$

**38.** $\dfrac{(x + 1)^2}{144} - \dfrac{(y - 4)^2}{25} = 1$

**39.** $(y + 6)^2 - (x - 2)^2 = 1$

**40.** $\dfrac{(y - 1)^2}{1/4} - \dfrac{(x + 3)^2}{1/9} = 1$

**41.** $9x^2 - y^2 - 36x - 6y + 18 = 0$

**42.** $x^2 - 9y^2 + 36y - 72 = 0$

**43.** $9y^2 - x^2 + 2x + 54y + 62 = 0$

**44.** $16y^2 - x^2 + 2x + 64y + 63 = 0$

**45.** $x^2 - 9y^2 + 2x - 54y - 107 = 0$

**46.** $9x^2 - y^2 + 54x + 10y + 55 = 0$

In Exercises 47–54, find an equation of the specified hyperbola.

**47.** Vertices: $(2, 0), (6, 0)$; Foci: $(0, 0), (8, 0)$

**48.** Vertices: $(2, 3), (2, -3)$; Foci: $(2, 5), (2, -5)$

**49.** Vertices: $(4, 1), (4, 9)$; Foci: $(4, 0), (4, 10)$;

**50.** Vertices: $(-2, 1), (2, 1)$; Foci: $(-3, 1), (3, 1)$

**51.** Vertices: $(2, 3), (2, -3)$; Passes through the point $(0, 5)$

**52.** Vertices: $(-2, 1), (2, 1)$; Passes through the point $(4, 3)$

**53.** Vertices: $(0, 2), (6, 2)$;
Asymptotes: $y = \frac{2}{3}x, y = 4 - \frac{2}{3}x$

**54.** Vertices: $(3, 0), (3, 4)$;
Asymptotes: $y = \frac{2}{3}x, 4 - \frac{2}{3}x$

In Exercises 55–62, classify the graph of each equation as a circle, a parabola, an ellipse, or a hyperbola.

**55.** $x^2 + y^2 - 6x + 4y + 9 = 0$

**56.** $x^2 + 4y^2 - 6x + 16y + 21 = 0$

**57.** $4x^2 - y^2 - 4x - 3 = 0$

**58.** $y^2 - 4y - 4x = 0$

**59.** $4x^2 + 3y^2 + 8x - 24y + 51 = 0$

**60.** $4y^2 - 2x^2 - 4y - 8x - 15 = 0$

**61.** $25x^2 - 10x - 200y - 119 = 0$

**62.** $4x^2 + 4y^2 - 16y + 15 = 0$

**63.** *Satellite Orbit* A satellite in a 100-mile-high circular orbit around the earth has a velocity of approximately 17,500 miles per hour. If this velocity is multiplied by $\sqrt{2}$, then the satellite will have the minimum velocity necessary to escape the earth's gravity, and it will follow a parabolic path with the center of the earth as the focus (see figure).

(a) Find the escape velocity of the satellite.

(b) Find an equation of its path (assume the radius of the earth is 4000 miles).

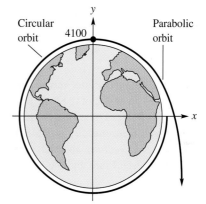

**64.** *Fluid Flow* Water is flowing from a horizontal pipe 48 feet above the ground. The falling stream of water has the shape of a parabola whose vertex $(0, 48)$ is at the end of the pipe (see figure). The stream of water strikes the ground at the point $\left(10\sqrt{3}, 0\right)$. Find the equation of the path taken by the water.

In Exercises 65–71, $e$ is called the **eccentricity** of the ellipse and is defined by $e = c/a$. It measures the flatness of the ellipse.

65. Find an equation of the ellipse with vertices $(\pm 5, 0)$ and eccentricity $e = \frac{3}{5}$.

66. Find an equation of the ellipse with vertices $(0, \pm 8)$ and eccentricity $e = \frac{1}{2}$.

67. *Orbit of the Earth* The earth moves in an elliptical orbit with the sun at one of the foci (see figure). The length of half of the major axis is $92.957 \times 10^6$ miles and the eccentricity is 0.017. Find the shortest distance (*perihelion*) and the greatest distance (*aphelion*) between the earth and the sun.

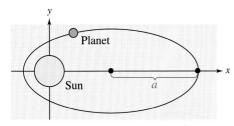

Figure for 67–69

68. *Orbit of Pluto* The planet Pluto moves in an elliptical orbit with the sun at one of the foci (see figure). The length of half of the major axis is $3.666 \times 10^9$ miles and the eccentricity is 0.248. Find the shortest distance (*perihelion*) and the greatest distance (*aphelion*) between Pluto and the sun.

69. *Orbit of Saturn* The planet Saturn moves in an elliptical orbit with the sun at one of the foci (see figure). The shortest distance and the greatest distance between the planet and the sun are $1.3495 \times 10^9$ kilometers and $1.5045 \times 10^9$ kilometers, respectively. Find the eccentricity of the orbit.

70. *Satellite Orbit* The first artificial satellite to orbit the earth was Sputnik I (launched by Russia in 1957). Its highest point above the earth's surface was 583 miles, and its lowest point was 132 miles (see figure). Assume that the center of the earth is the focus of the elliptical orbit and the radius of the earth is 4000 miles. Find the eccentricity of the orbit.

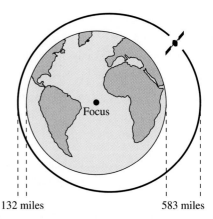

Figure for 70

71. Show that the equation of an ellipse can be written as

$$\frac{(x - h)^2}{a^2} + \frac{(y - k)^2}{a^2(1 - e^2)} = 1.$$

Note that as $e$ approaches zero, the ellipse approaches a circle of radius $a$.

# Appendix C
# Further Concepts in Statistics

## C.1   Representing Data and Linear Modeling

Stem-and-Leaf Plots   ■   Histograms and Frequency Distributions   ■   Scatter Plots   ■
Fitting a Line to Data

## Stem-and-Leaf Plots

Statistics is the branch of mathematics that studies techniques for collecting, organizing, and interpreting data. In this section, you will study several ways to organize and interpret data.

One type of plot that can be used to organize sets of numbers by hand is a **stem-and-leaf plot.** A set of test scores and the corresponding stem-and-leaf plot are shown below.

| *Test Scores* | | *Stems* | *Leaves* |
|---|---|---|---|
| 93, 70, 76, 58, 86, 93, 82, 78, 83, 86, | | 5 | 8 |
| 64, 78, 76, 66, 83, 83, 96, 74, 69, 76, | | 6 | 4 4 6 9 |
| 64, 74, 79, 76, 88, 76, 81, 82, 74, 70 | | 7 | 0 0 4 4 4 6 6 6 6 8 8 9 |
| | | 8 | 1 2 2 3 3 3 6 6 8 |
| | | 9 | 3 3 6 |

Note that the *leaves* represent the units digits of the numbers and the *stems* represent the tens digits. Stem-and-leaf plots can also be used to compare two sets of data, as shown in the following example.

### Example 1   Comparing Two Sets of Data

Use a stem-and-leaf plot to compare the test scores given above with the following test scores. Which set of test scores is better?

> 90, 81, 70, 62, 64, 73, 81, 92, 73, 81, 92, 93, 83, 75, 76,
> 83, 94, 96, 86, 77, 77, 86, 96, 86, 77, 86, 87, 87, 79, 88

**Solution**

Begin by ordering the second set of scores.

> 62, 64, 70, 73, 73, 75, 76, 77, 77, 77, 79, 81, 81, 81, 83,
> 83, 86, 86, 86, 86, 87, 87, 88, 90, 92, 92, 93, 94, 96, 96

Now that the data have been ordered, you can construct a *double* stem-and-leaf plot by letting the leaves to the right of the stems represent the units digits for the first group of test scores and letting the leaves to the left of the stems represent the units digits for the second group of test scores.

| Leaves (2nd Group) | Stems | Leaves (1st Group) |
|---:|:---:|:---|
|  | 5 | 8 |
| 4 2 | 6 | 4 4 6 9 |
| 9 7 7 7 6 5 3 3 0 | 7 | 0 0 4 4 4 6 6 6 6 6 8 8 9 |
| 8 7 7 6 6 6 6 3 3 1 1 1 | 8 | 1 2 2 3 3 3 6 6 8 |
| 6 6 4 3 2 2 0 | 9 | 3 3 6 |

By comparing the two sets of leaves, you can see that the second group of test scores is better than the first group.

---

### Example 2    Using a Stem-and-Leaf Plot

The table below shows the percent of the population of each state and the District of Columbia that was at least 65 years old in 1997. Use a stem-and-leaf plot to organize the data.    (Source: U.S. Bureau of the Census)

| | | | | | | | | | |
|---|---|---|---|---|---|---|---|---|---|
| AK | 5.3 | AL | 13.0 | AR | 14.3 | AZ | 13.2 | CA | 11.1 |
| CO | 10.1 | CT | 14.4 | DC | 13.9 | DE | 12.9 | FL | 18.5 |
| GA | 9.9 | HI | 13.2 | IA | 15.0 | ID | 11.3 | IL | 12.5 |
| IN | 12.5 | KS | 13.5 | KY | 12.5 | LA | 11.4 | MA | 14.1 |
| MD | 11.5 | ME | 13.9 | MI | 12.4 | MN | 12.3 | MO | 13.7 |
| MS | 12.2 | MT | 13.2 | NC | 12.5 | ND | 14.4 | NE | 13.7 |
| NH | 12.1 | NJ | 13.7 | NM | 11.2 | NV | 11.5 | NY | 13.4 |
| OH | 13.4 | OK | 13.4 | OR | 13.3 | PA | 15.8 | RI | 15.8 |
| SC | 12.1 | SD | 14.3 | TN | 12.5 | TX | 10.1 | UT | 8.7 |
| VA | 11.2 | VT | 12.3 | WA | 11.5 | WI | 13.2 | WV | 15.1 |
| WY | 11.3 | | | | | | | | |

### Solution

Begin by ordering the numbers, as shown below.

5.3, 8.7, 9.9, 10.1, 10.1, 11.1, 11.2, 11.2, 11.3, 11.3, 11.4,
11.5, 11.5, 11.5, 12.1, 12.1, 12.2, 12.3, 12.3, 12.4, 12.5,
12.5, 12.5, 12.5, 12.5, 12.9, 13.0, 13.2, 13.2, 13.2, 13.2,
13.3, 13.4, 13.4, 13.4, 13.5, 13.7, 13.7, 13.7, 13.9, 13.9,
14.1, 14.3, 14.3, 14.4, 14.4, 15.0, 15.1, 15.8, 15.8, 18.5

Next construct the stem-and-leaf plot using the leaves to represent the digits to the right of the decimal points.

| Stems | Leaves |
|-------|--------|
| 5. | 3         Alaska has the lowest percent. |
| 6. | |
| 7. | |
| 8. | 7 |
| 9. | 9 |
| 10. | 1 1 |
| 11. | 1 2 2 3 3 4 5 5 5 |
| 12. | 1 1 2 3 3 4 5 5 5 5 5 9 |
| 13. | 0 2 2 2 2 3 4 4 4 5 7 7 7 9 9 |
| 14. | 1 3 3 4 4 |
| 15. | 0 1 8 8 |
| 16. | |
| 17. | |
| 18. | 5         Florida has the highest percent. |

## Histograms and Frequency Distributions

With data such as those given in Example 2, it is useful to group the numbers into intervals and plot the frequency of the data in each interval. For instance, the **frequency distribution** and **histogram** shown in Figure C.1 represent the data given in Example 2.

| *Interval* | *Tally* |
|------------|---------|
| [5, 7) | I |
| [7, 9) | I |
| [9, 11) | III |
| [11, 13) | ⊔⊓⊤ ⊔⊓⊤ ⊔⊓⊤ ⊔⊓⊤ I |
| [13, 15) | ⊔⊓⊤ ⊔⊓⊤ ⊔⊓⊤ ⊔⊓⊤ |
| [15, 17) | IIII |
| [17, 19) | I |

Figure C.1

A histogram has a portion of a real number line as its horizontal axis. A **bar graph** is similar to a histogram, except that the rectangles (bars) can be either horizontal or vertical and the labels of the bars are not necessarily numbers.

Another difference between a bar graph and a histogram is that the bars in a bar graph are usually separated by spaces, whereas the bars in a histogram are not separated by spaces.

## •TECHNOLOGY•

Most graphing utilities have built-in statistical programs that can create scatter plots. Use your graphing utility to plot the points given in the table below. The data shows the number of people $P$ (in millions) in the United States who were part of the labor force from 1980 through 1990. In the table, $t$ represents the year, with $t = 0$ corresponding to 1980.
(Source: U.S. Bureau of Labor Statistics)

| $t$ | $P$ |
|-----|-----|
| 0 | 109 |
| 1 | 110 |
| 2 | 112 |
| 3 | 113 |
| 4 | 115 |
| 5 | 117 |

| $t$ | $P$ |
|-----|-----|
| 6 | 120 |
| 7 | 122 |
| 8 | 123 |
| 9 | 126 |
| 10 | 126 |

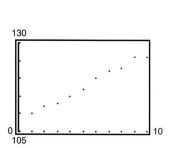

Figure C.3

| Example 3 | Constructing a Bar Graph |
|-----------|--------------------------|

The data below show the average monthly precipitation (in inches) in Houston, Texas. Construct a bar graph for these data. What can you conclude?    (Source: PC USA)

| January | 3.2 | February | 3.3 | March | 2.7 |
|---------|-----|----------|-----|-------|-----|
| April | 4.2 | May | 4.7 | June | 4.1 |
| July | 3.3 | August | 3.7 | September | 4.9 |
| October | 3.7 | November | 3.4 | December | 3.7 |

### Solution

To create a bar graph, begin by drawing a vertical axis to represent the precipitation and a horizontal axis to represent the months. The bar graph is shown in Figure C.2. From the graph, you can see that Houston receives a fairly consistent amount of rain throughout the year—the driest month tends to be March and the wettest month tends to be September.

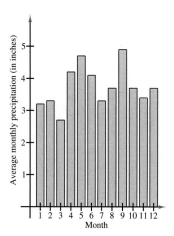

Figure C.2

## Scatter Plots

Many real-life situations involve finding relationships between two variables, such as the year and the number of people in the labor force. In a typical situation, data are collected and written as a set of ordered pairs. The graph of such a set is called a **scatter plot.**

From the scatter plot in Figure C.3, it appears that the points describe a relationship that is nearly linear. (The relationship is not *exactly* linear because the labor force did not increase by precisely the same amount each year.) A mathematical equation that approximates the relationship between $t$ and $P$ is called a *mathematical model.* When developing a mathematical model, you strive for two (often conflicting) goals—accuracy and simplicity.

Consider a collection of ordered pairs of the form $(x, y)$. If $y$ tends to increase as $x$ increases, the collection is said to have a **positive correlation.** If $y$ tends to decrease as $x$ increases, the collection is said to have a **negative correlation.** Figure C.4 shows three examples: one with a positive correlation, one with a negative correlation, and one with no (discernible) correlation.

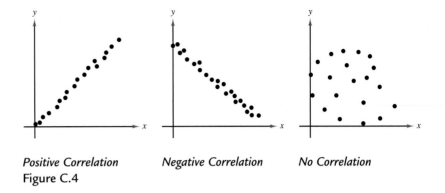

*Positive Correlation*    *Negative Correlation*    *No Correlation*
Figure C.4

## Fitting a Line to Data

Finding a linear model that represents the relationship described by a scatter plot is called **fitting a line to data.** You can do this graphically by simply sketching the line that appears to fit the points, finding two points on the line, and then finding the equation of the line that passes through the two points.

**Example 4**    Fitting a Line to Data

Find a linear model that relates the year with the number of people $P$ (in millions) who were part of the United States labor force from 1987 through 1997. In the table below, $t$ represents the year, with $t = 0$ corresponding to 1987.    (Source: U.S. Bureau of Labor Statistics)

| $t$ | 0 | 1 | 2 | 3 | 4 | 5 | 6 | 7 | 8 | 9 | 10 |
|---|---|---|---|---|---|---|---|---|---|---|---|
| $P$ | 120 | 122 | 124 | 126 | 126 | 128 | 129 | 131 | 132 | 134 | 136 |

Solution

After plotting the data from the table, draw the line that you think best represents the data, as shown in Figure C.5. Two points that lie on this line are (0, 120) and (9, 134). Using the point-slope form, you can find the equation of the line to be

$$P = \frac{14}{9}t + 120.$$    Linear model

Figure C.5

Once you have found a model, you can measure how well the model fits the data by comparing the actual values with the values given by the model, as shown in the following table.

| t | 0 | 1 | 2 | 3 | 4 | 5 | 6 | 7 | 8 | 9 | 10 |
|---|---|---|---|---|---|---|---|---|---|---|---|
| Actual → P | 120 | 122 | 124 | 126 | 126 | 128 | 129 | 131 | 132 | 134 | 136 |
| Model → P | 120 | 121.6 | 123.1 | 124.7 | 126.2 | 127.8 | 129.3 | 130.9 | 132.4 | 134 | 135.6 |

The sum of the squares of the differences between the actual values and the model's values is the **sum of the squared differences.** The model that has the least sum is called the **least squares regression line** for the data. For the model in Example 4, the sum of the squared differences is 3.16. The least squares regression line for the data is

$$P = 1.5t + 120.5. \qquad \text{Best-fitting linear model}$$

Its sum of squared differences is 2.5.

---

▶ **Least Squares Regression Line**

The least squares regression line, $y = ax + b$, for the points $(x_1, y_1)$, $(x_2, y_2)$, $(x_3, y_3)$, . . . , is given by

$$a = \frac{n\sum\limits_{i=1}^{n} x_i y_i - \sum\limits_{i=1}^{n} x_i \sum\limits_{i=1}^{n} y_i}{n\sum\limits_{i=1}^{n} x_i^2 - \left(\sum\limits_{i=1}^{n} x_i\right)^2} \qquad \text{and} \qquad b = \frac{1}{n}\left(\sum\limits_{i=1}^{n} y_i - a\sum\limits_{i=1}^{n} x_i\right).$$

---

**Example 5**   Finding a Least Squares Regression Line

Find the least squares regression line for the points $(-3, 0)$, $(-1, 1)$, $(0, 2)$, and $(2, 3)$.

**Solution**

Begin by constructing a table like the one shown below.

| $x$ | $y$ | $xy$ | $x^2$ |
|---|---|---|---|
| $-3$ | 0 | 0 | 9 |
| $-1$ | 1 | $-1$ | 1 |
| 0 | 2 | 0 | 0 |
| 2 | 3 | 6 | 4 |
| $\sum\limits_{i=1}^{n} x_i = -2$ | $\sum\limits_{i=1}^{n} y_i = 6$ | $\sum\limits_{i=1}^{n} x_i y_i = 5$ | $\sum\limits_{i=1}^{n} x_i^2 = 14$ |

Applying the formulas for the least squares regression line with $n = 4$ produces

$$a = \frac{n\sum\limits_{i=1}^{n}x_i y_i - \sum\limits_{i=1}^{n}x_i \sum\limits_{i=1}^{n}y_i}{n\sum\limits_{i=1}^{n}x_i^2 - \left(\sum\limits_{i=1}^{n}x_i\right)^2} = \frac{4(5) - (-2)(6)}{4(14) - (-2)^2} = \frac{8}{13}$$

and

$$b = \frac{1}{n}\left(\sum\limits_{i=1}^{n}y_i - a\sum\limits_{i=1}^{n}x_i\right) = \frac{1}{4}\left[6 - \frac{8}{13}(-2)\right] = \frac{47}{26}.$$

Thus, the least squares regression line is $y = \frac{8}{13}x + \frac{47}{26}$. This line is shown in Figure C.6.

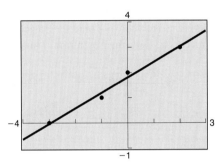

Figure C.6

Many graphing utilities have "built-in" least squares regression programs. If your calculator has such a program, try using it to duplicate the results shown in the following example.

**Example 6**    Finding a Least Squares Regression Line

The following ordered pairs $(w, h)$ represent the shoe sizes $w$ and the heights $h$ (in inches) of 25 men. Use a computer program or a statistical calculator to find the least squares regression line for these data.

(10.0, 70.0), (10.5, 71.0), (9.5, 70.0), (11.0, 72.0), (12.0, 74.0),
(8.5, 66.0), (9.0, 68.5), (13.0, 76.0), (10.5, 71.5), (10.5, 70.5),
(10.0, 72.0), (9.5, 70.0), (10.0, 71.0), (10.5, 69.5), (11.0, 71.5),
(12.0, 73.5), (12.5, 74.0), (11.0, 71.5), (9.0, 67.5), (10.0, 70.0),
(13.0, 73.5), (10.5, 72.5), (10.5, 71.0), (11.0, 73.0), (8.5, 68.0)

Solution

A scatter plot for the data is shown in Figure C.7 on the following page. Note that the plot does not have 25 points because some of the ordered pairs graph as the same point. After entering the data into a statistical calculator, you can obtain

$$a = 1.67 \quad \text{and} \quad b = 53.57.$$

Thus, the least squares regression line for the data is

$$h = 1.67w + 53.57.$$

Figure C.7

If you use a statistical calculator or graphing utility to duplicate the results of Example 6, you will notice that the program also outputs a value of $r \approx 0.918$. This number is called the **correlation coefficient** of the data. Correlation coefficients vary between $-1$ and $1$. Basically, the closer $|r|$ is to $1$, the better the points can be described by a line. Three examples are shown in Figure C.8.

$r = 0.981$     $r = -0.866$     $r = 0.190$

Figure C.8

## C.1 Exercises

*Exam Scores* In Exercises 1 and 2, use the following scores from a math class with 30 students. The scores are given for two 100-point exams.

*Exam #1:* 77, 100, 77, 70, 83, 89, 87, 85, 81, 84, 81, 78, 89, 78, 88, 85, 90, 92, 75, 81, 85, 100, 98, 81, 78, 75, 85, 89, 82, 75

*Exam #2:* 76, 78, 73, 59, 70, 81, 71, 66, 66, 73, 68, 67, 63, 67, 77, 84, 87, 71, 78, 78, 90, 80, 77, 70, 80, 64, 74, 68, 68, 68

1. Construct a stem-and-leaf plot for exam #1.

2. Construct a double stem-and-leaf plot to compare the scores for exam #1 and exam #2. Which set of test scores is higher?

3. *Insurance Coverage* The following table shows the total number of persons (in thousands) without health insurance coverage in the 50 states and the District of Columbia in 1996. Use a stem-and-leaf plot to organize the data. (Source: U.S. Bureau of the Census)

| AK | 89 | AL | 550 | AR | 566 | AZ | 1159 | CA | 6514 |
|----|----|----|-----|----|-----|----|------|----|------|
| CO | 644 | CT | 368 | DC | 80 | DE | 98 | FL | 2722 |
| GA | 1319 | HI | 101 | IA | 335 | ID | 196 | IL | 1337 |
| IN | 600 | KS | 292 | KY | 601 | LA | 890 | MA | 766 |
| MD | 581 | ME | 146 | MI | 857 | MN | 480 | MO | 700 |
| MS | 518 | MT | 124 | NC | 1160 | ND | 62 | NE | 190 |
| NH | 109 | NJ | 1317 | NM | 412 | NV | 255 | NY | 3132 |
| OH | 1292 | OK | 570 | OR | 496 | PA | 1133 | RI | 93 |
| SC | 634 | SD | 67 | TN | 841 | TX | 4680 | UT | 240 |
| VA | 811 | VT | 65 | WA | 761 | WI | 438 | WV | 261 |
| WY | 66 | | | | | | | | |

Table for 3

4. *Snowfall* The data below show the seasonal snowfall (in inches) at Lincoln, Nebraska for the years 1968 through 1997 (the amounts are listed in order by year). How would you organize these data? Explain your reasoning. (Source: University of Nebraska-Lincoln)

39.8, 26.2, 49.0, 21.6, 29.2, 33.6, 42.1, 21.1, 21.8, 31.0, 34.4, 23.3, 13.0, 32.3, 38.0, 47.5, 21.5, 18.9, 15.7, 13.0, 19.1, 18.7, 25.8, 23.8, 32.1, 21.3, 21.8, 30.7, 29.0, 44.6

**5.** *Fruit Crops*   The data below show the cash receipts (in millions of dollars) from fruit crops of farmers in 1996. Construct a bar graph for these data.   (Source: U.S. Department of Agriculture)

| | | | |
|---|---|---|---|
| Apples | 1846 | Peaches | 380 |
| Cherries | 264 | Pears | 292 |
| Grapes | 2334 | Plums and Prunes | 295 |
| Lemons | 228 | Strawberries | 770 |
| Oranges | 1798 | | |

**6.** *Travel to the United States*   The data below give the places of origin and the numbers of travelers (in millions) to the United States in 1995. Construct a bar graph for these data.   (Source: U.S. Department of Commerce)

| | | | |
|---|---|---|---|
| Canada | 14.7 | Mexico | 8.0 |
| Europe | 8.8 | Far East | 6.6 |
| Other | 5.2 | | |

*Crop Yield*   In Exercises 7–10, use the data in the table, where $x$ is the number of units of fertilizer applied to sample plots and $y$ is the yield (in bushels) of a crop.

| $x$ | 0 | 1 | 2 | 3 | 4 | 5 | 6 | 7 | 8 |
|---|---|---|---|---|---|---|---|---|---|
| $y$ | 58 | 60 | 59 | 61 | 63 | 66 | 65 | 67 | 70 |

**7.** Sketch a scatter plot of the data.

**8.** Determine whether the points are positively correlated, are negatively correlated, or have no discernible correlation.

**9.** Sketch a linear model that you think best represents the data. Find an equation of the line you sketched. Use the line to predict the yield if 10 units of fertilizer are used.

**10.** Can the model found in Exercise 9 be used to predict yields for arbitrarily large values of $x$? Explain.

*Speed of Sound*   In Exercises 11–14, use the data in the table, where $h$ is altitude in thousands of feet and $v$ is the speed of sound in feet per second.

| $h$ | 0 | 5 | 10 | 15 | 20 | 25 | 30 | 35 |
|---|---|---|---|---|---|---|---|---|
| $v$ | 1116 | 1097 | 1077 | 1057 | 1036 | 1015 | 995 | 973 |

**11.** Sketch a scatter plot of the data.

**12.** Determine whether the points are positively correlated, are negatively correlated, or have no discernible correlation.

**13.** Sketch a linear model that you think best represents the data. Find an equation of the line you sketched. Use the line to predict the speed of sound at an altitude of 27,000 feet.

**14.** The speed of sound at an altitude of 70,000 feet is approximately 971 feet per second. What does this suggest about the validity of using the model in Exercise 13 to extrapolate beyond the data given in the table?

In Exercises 15 and 16, (a) sketch a scatter plot of the points, (b) find an equation of the linear model you think best represents the data and find the sum of the squared differences, and (c) use the formulas in this section to find the least squares regression line and the sum of the squared differences.

**15.** $(-1, 0), (0, 1), (1, 3), (2, 3)$

**16.** $(0, 4), (1, 3), (2, 2), (4, 1)$

In Exercises 17–20, sketch a scatter plot of the points, use the formulas in this section to find the least squares regression line, and sketch the graph of the line.

**17.** $(-2, 0), (-1, 1), (0, 1), (2, 2)$

**18.** $(-3, 1), (-1, 2), (0, 2), (1, 3), (3, 5)$

**19.** $(1, 5), (2, 8), (3, 13), (4, 16), (5, 22), (6, 26)$

**20.** $(1, 10), (2, 8), (3, 8), (4, 6), (5, 5), (6, 3)$

In Exercises 21–24, use a graphing utility to find the least squares regression line for the data. Sketch a scatter plot and the regression line.

**21.** $(0, 23), \ (1, 20), \ (2, 19), \ (3, 17), \ (4, 15), \ (5, 11),$ $(6, 10)$

**22.** $(4, 52.8), \ (5, 54.7), \ (6, 55.7), \ (7, 57.8), \ (8, 60.2)$ $(9, 63.1), (10, 66.5)$

**23.** $(-10, 5.1), \ (-5, 9.8), \ (0, 17.5), \ (2, 25.4), \ (4, 32.8),$ $(6, 38.7), (8, 44.2), (10, 50.5)$

**24.** $(-10, 213.5), \ (-5, 174.9), \ (0, 141.7), \ (5, 119.7),$ $(8, 102.4), (10, 87.6)$

**25.** *Advertising* The management of a department store ran an experiment to determine if a relationship existed between sales $S$ (in thousands of dollars) and the amount spent on advertising $x$ (in thousands of dollars). The following data were collected.

| $x$ | 1 | 2 | 3 | 4 | 5 | 6 | 7 | 8 |
|---|---|---|---|---|---|---|---|---|
| $S$ | 405 | 423 | 455 | 466 | 492 | 510 | 525 | 559 |

(a) Use a graphing utility to find the least squares regression line. Use the equation to estimate sales if $4500 is spent on advertising.

(b) Make a scatter plot of the data and sketch the graph of the regression line.

(c) Use a computer or calculator to determine the correlation coefficient.

**26.** *School Enrollment* The table gives the preprimary school enrollments $y$ (in millions) for the years 1990 through 1995, where $t = 0$ corresponds to 1990. (Source: U.S. Bureau of the Census)

| $t$ | 0 | 1 | 2 | 3 | 4 | 5 |
|---|---|---|---|---|---|---|
| $y$ | 11.21 | 11.37 | 11.54 | 11.95 | 12.33 | 12.52 |

(a) Use a computer or calculator to find the least squares regression line. Use the equation to estimate enrollment in 1996.

(b) Make a scatter plot of the data and sketch the graph of the regression line.

(c) Use a computer or calculator to determine the correlation coefficient.

## C.2 Measures of Central Tendency and Dispersion

Mean, Median, and Mode ■ Choosing a Measure of Central Tendency ■ Variance and Standard Deviation

## Mean, Median, and Mode

In many real-life situations, it is helpful to describe data by a single number that is most representative of the entire collection of numbers. Such a number is called a **measure of central tendency.** The most commonly used measures are as follows.

1. The **mean,** or **average,** of $n$ numbers is the sum of the numbers divided by $n$.

2. The **median** of $n$ numbers is the middle number when the numbers are written in order. If $n$ is even, the median is the average of the two middle numbers.

3. The **mode** of $n$ numbers is the number that occurs most frequently. If two numbers tie for most frequent occurrence, the collection has two modes and is called **bimodal.**

### Example 1   Comparing Measures of Central Tendency

You are interviewing for a job. The interviewer tells you that the average income of the company's 25 employees is $60,849. The actual annual incomes of the 25 employees are shown below. What are the mean, median, and mode of the incomes? Was the person telling you the truth?

| | | | | |
|---|---|---|---|---|
| $17,305, | $478,320, | $45,678, | $18,980, | $17,408, |
| $25,676, | $28,906, | $12,500, | $24,540, | $33,450, |
| $12,500, | $33,855, | $37,450, | $20,432, | $28,956, |
| $34,983, | $36,540, | $250,921, | $36,853, | $16,430, |
| $32,654, | $98,213, | $48,980, | $94,024, | $35,671 |

Solution

The mean of the incomes is

$$\text{Mean} = \frac{17{,}305 + 478{,}320 + 45{,}678 + 18{,}980 + \cdots + 35{,}671}{25}$$

$$= \frac{1{,}521{,}225}{25} = \$60{,}849.$$

To find the median, order the incomes as follows.

| | | | | |
|---|---|---|---|---|
| $12,500, | $12,500, | $16,430, | $17,305, | $17,408, |
| $18,980, | $20,432, | $24,540, | $25,676, | $28,906, |
| $28,956, | $32,654, | $33,450, | $33,855, | $34,983, |
| $35,671, | $36,540, | $36,853, | $37,450, | $45,678, |
| $48,980, | $94,024, | $98,213, | $250,921, | $478,320 |

From this list, you can see the median (the middle number) is $33,450. From the same list, you can see that $12,500 is the only income that occurs more than once. Thus, the mode is $12,500. Technically, the person was telling the truth because the average is (generally) defined to be the mean. However, of the three measures of central tendency

*Mean:* $60,849      *Median:* $33,450      *Mode:* $12,500

it seems clear that the median is most representative. The mean is inflated by the two highest salaries.

---

## Choosing a Measure of Central Tendency

Which of the three measures of central tendency is the most representative? The answer is that it depends on the distribution of the data *and* the way in which you plan to use the data.

For instance, in Example 1, the mean salary of $60,849 does not seem very representative to a potential employee. To a city income tax collector who wants to estimate 1% of the total income of the 25 employees, however, the mean is precisely the right measure.

**Example 2**    Choosing a Measure of Central Tendency

Which measure of central tendency is the most representative of the data given in each of the following frequency distributions?

| a. Number | Tally | | b. Number | Tally | | c. Number | Tally |
|---|---|---|---|---|---|---|---|
| 1 | 7 | | 1 | 9 | | 1 | 6 |
| 2 | 20 | | 2 | 8 | | 2 | 1 |
| 3 | 15 | | 3 | 7 | | 3 | 2 |
| 4 | 11 | | 4 | 6 | | 4 | 3 |
| 5 | 8 | | 5 | 5 | | 5 | 5 |
| 6 | 3 | | 6 | 6 | | 6 | 5 |
| 7 | 2 | | 7 | 7 | | 7 | 4 |
| 8 | 0 | | 8 | 8 | | 8 | 3 |
| 9 | 15 | | 9 | 9 | | 9 | 0 |

Solution

a. For these data, the mean is 4.23, the median is 3, and the mode is 2. Of these, the mode is probably the most representative.

b. For these data, the mean and median are each 5 and the modes are 1 and 9 (the distribution is bimodal). Of these, the mean or median is the most representative.

c. For these data, the mean is 4.59, the median is 5, and the mode is 1. Of these, the mean or median is the most representative.

# Variance and Standard Deviation

Very different sets of numbers can have the same mean. You will now study two **measures of dispersion,** which give you an idea of how much the numbers in the set differ from the mean of the set. These two measures are called the *variance* of the set and the *standard deviation* of the set.

---

▶ Definitions of Variance and Standard Deviation

Consider a set of numbers $\{x_1, x_2, \ldots, x_n\}$ with a mean of $\bar{x}$. The **variance** of the set is

$$v = \frac{(x_1 - \bar{x})^2 + (x_2 - \bar{x})^2 + \cdots + (x_n - \bar{x})^2}{n}$$

and the **standard deviation** of the set is

$$\sigma = \sqrt{v}$$

($\sigma$ is the lowercase Greek letter *sigma*).

---

The standard deviation of a set is a measure of how much a typical number in the set differs from the mean. The greater the standard deviation, the more the numbers in the set *vary* from the mean. For instance, each of the following sets has a mean of 5.

$$\{5, 5, 5, 5\}, \qquad \{4, 4, 6, 6\}, \qquad \text{and} \qquad \{3, 3, 7, 7\}$$

The standard deviations of the sets are 0, 1, and 2.

$$\sigma_1 = \sqrt{\frac{(5 - 5)^2 + (5 - 5)^2 + (5 - 5)^2 + (5 - 5)^2}{4}} = 0$$

$$\sigma_2 = \sqrt{\frac{(4 - 5)^2 + (4 - 5)^2 + (6 - 5)^2 + (6 - 5)^2}{4}} = 1$$

$$\sigma_3 = \sqrt{\frac{(3 - 5)^2 + (3 - 5)^2 + (7 - 5)^2 + (7 - 5)^2}{4}} = 2$$

**Example 3**    Estimations of Standard Deviation

Consider the three sets of data represented by the bar graphs in Figure C.9. Which set has the smallest standard deviation? Which has the largest?

**Figure C.9**

Solution

Of the three sets, the numbers in set $A$ are grouped most closely to the center and the numbers in set $C$ are the most dispersed. Thus, set $A$ has the smallest standard deviation and set C has the largest standard deviation.

**Example 4**    Finding Standard Deviation

Find the standard deviation of each set shown in Example 3.

Solution

Because of the symmetry of each bar graph, you can conclude that each has a mean of $\bar{x} = 4$. The standard deviation of set $A$ is

$$\sigma = \sqrt{\frac{(-3)^2 + 2(-2)^2 + 3(-1)^2 + 5(0)^2 + 3(1)^2 + 2(2)^2 + (3)^2}{17}}$$

$$\approx 1.53.$$

The standard deviation of set $B$ is

$$\sigma = \sqrt{\frac{2(-3)^2 + 2(-2)^2 + 2(-1)^2 + 2(0)^2 + 2(1)^2 + 2(2)^2 + 2(3)^2}{14}}$$

$$= 2.$$

The standard deviation of set $C$ is

$$\sigma = \sqrt{\frac{5(-3)^2 + 4(-2)^2 + 3(-1)^2 + 2(0)^2 + 3(1)^2 + 4(2)^2 + 5(3)^2}{26}}$$

$$\approx 2.22.$$

These values confirm the results of Example 3. That is, set $A$ has the smallest standard deviation and set $C$ has the largest.

The following alternative formula provides a more efficient way to compute the standard deviation.

---

▶ **Alternative Formula for Standard Deviation**

The standard deviation of $\{x_1, x_2, \ldots, x_n\}$ is

$$\sigma = \sqrt{\frac{x_1^2 + x_2^2 + \cdots + x_n^2}{n} - \bar{x}^2}.$$

---

Because of messy computations, this formula is difficult to verify. Conceptually, however, the process is straightforward. It consists of showing that the expressions

$$\sqrt{\frac{(x_1 - \bar{x})^2 + (x_2 - \bar{x})^2 + \cdots (x_n - \bar{x})^2}{n}}$$

and

$$\sqrt{\frac{x_1^2 + x_2^2 + \cdots + x_n^2}{n} - \bar{x}^2}$$

are equivalent. Try verifying this equivalence for the set $\{x_1, x_2, x_3\}$ with $\bar{x} = (x_1 + x_2 + x_3)/3$.

**Example 5** Using the Alternative Formula

Use the alternative formula for standard deviation to find the standard deviation of the following set of numbers.

5, 6, 6, 7, 7, 8, 8, 8, 9, 10

**Solution**

Begin by finding the mean of the set, which is 7.4. Thus, the standard deviation is

$$\sigma = \sqrt{\frac{5^2 + 2(6^2) + 2(7^2) + 3(8^2) + 9^2 + 10^2}{10} - (7.4)^2}$$

$$= \sqrt{\frac{568}{10} - 54.76}$$

$$= \sqrt{2.04}$$

$$\approx 1.43.$$

You can use the statistical features of a graphing utility to check this result.

A well-known theorem in statistics, called *Chebychev's Theorem*, states that at least

$$1 - \frac{1}{k^2}$$

of the numbers in a distribution must lie within $k$ standard deviations of the mean. So, 75% of the numbers in a collection must lie within two standard deviations of the mean, and at least 88.9% of the numbers must lie within three standard deviations of the mean. For most distributions, these percentages are low. For instance, in all three distributions shown in Example 3, 100% of the numbers lie within two standard deviations of the mean.

### Example 6    Describing a Distribution

The following table shows the number of dentists (per 100,000 people) in each state and the District of Columbia. Find the mean and standard deviation of the numbers. What percent of the numbers lie within two standard deviations of the mean? (Source: American Dental Association)

| AK | 66 | AL | 40 | AR | 39 | AZ | 51 | CA | 62 |
|----|----|----|----|----|----|----|----|----|----|
| CO | 69 | CT | 80 | DC | 94 | DE | 44 | FL | 50 |
| GA | 46 | HI | 80 | IA | 55 | ID | 53 | IL | 61 |
| IN | 47 | KS | 51 | KY | 53 | LA | 45 | MA | 74 |
| MD | 68 | ME | 47 | MI | 62 | MN | 67 | MO | 53 |
| MS | 37 | MT | 62 | NC | 42 | ND | 47 | NE | 63 |
| NH | 59 | NJ | 77 | NM | 45 | NV | 49 | NY | 73 |
| OH | 55 | OK | 47 | OR | 70 | PA | 61 | RI | 56 |
| SC | 41 | SD | 49 | TN | 53 | TX | 47 | UT | 66 |
| VA | 54 | VT | 57 | WA | 68 | WI | 65 | WV | 43 |
| WY | 52 |  |  |  |  |  |  |  |  |

### Solution

Begin by entering the numbers into a graphing utility that has a standard deviation program. After running the program, you should obtain

$$\bar{x} \approx 56.76 \quad \text{and} \quad \sigma = 12.14.$$

The interval that contains all numbers that lie within two standard deviations of the mean is

$$[56.76 - 2(12.14), \, 56.76 + 2(12.14)] \quad \text{or} \quad [32.48, 81.04].$$

From the histogram in Figure C.10, you can see that all but one of the numbers (98%) lie in this interval—all but the number that corresponds to the number of dentists (per 100,000 people) in Washington, DC.

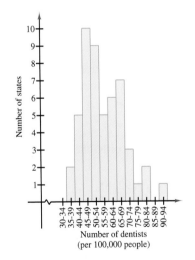

Figure C.10

## C.2 Exercises

In Exercises 1–6, find the mean, median, and mode of the set of measurements.

1. 5, 12, 7, 14, 8, 9, 7

2. 30, 37, 32, 39, 33, 34, 32

3. 5, 12, 7, 24, 8, 9, 7

4. 20, 37, 32, 39, 33, 34, 32

5. 5, 12, 7, 14, 9, 7

6. 30, 37, 32, 39, 34, 32

7. *Reasoning* Compare your answers for Exercises 1 and 3 with those for Exercises 2 and 4. Which of the measures of central tendency is sensitive to extreme measurements? Explain your reasoning.

8. *Reasoning*

   (a) Add 6 to each measurement in Exercise 1 and calculate the mean, median, and mode of the revised measurements. How are the measures of central tendency changed?

   (b) If a constant $k$ is added to each measurement in a set of data, how will the measures of central tendency change?

9. *Electric Bills* A person had the following monthly bills for electricity. What are the mean and median of the collection of bills?

   | | | | |
   |---|---|---|---|
   | January | $67.92 | February | $59.84 |
   | March | $52.00 | April | $52.50 |
   | May | $57.99 | June | $65.35 |
   | July | $81.76 | August | $74.98 |
   | September | $87.82 | October | $83.18 |
   | November | $65.35 | December | $57.00 |

10. *Car Rental* A car rental company kept the following record of the numbers of miles driven by a car that was rented. What are the mean, median, and mode of these data?

    | | | | |
    |---|---|---|---|
    | Monday | 410 | Tuesday | 260 |
    | Wednesday | 320 | Thursday | 320 |
    | Friday | 460 | Saturday | 150 |

11. *Six-Child Families* A study was done on families having six children. The table gives the number of families in the study with the indicated number of girls. Determine the mean, median, and mode of this set of data.

    | Number of girls | 0 | 1 | 2 | 3 | 4 | 5 | 6 |
    |---|---|---|---|---|---|---|---|
    | Frequency | 1 | 24 | 45 | 54 | 50 | 19 | 7 |

12. *Baseball* A baseball fan examined the records of a favorite baseball player's performance during his last 50 games. The number of games in which the player had 0, 1, 2, 3, and 4 hits are recorded in the table.

    | Number of hits | 0 | 1 | 2 | 3 | 4 |
    |---|---|---|---|---|---|
    | Frequency | 14 | 26 | 7 | 2 | 1 |

    (a) Determine the average number of hits per game.

    (b) Determine the player's batting average if he had 200 at bats during the 50-game series.

13. *Think About It* Construct a collection of numbers that has the following properties. If this is not possible, explain why it is not.

    Mean = 6, median = 4, mode = 4

14. *Think About It* Construct a collection of numbers that has the following properties. If this is not possible, explain why it is not.

    Mean = 6, median = 6, mode = 4

15. *Test Scores* A professor records the following scores for a 100-point exam.

    99, 64, 80, 77, 59, 72, 87, 79, 92, 88, 90, 42, 20, 89, 42, 100, 98, 84, 78, 91

    Which measure of central tendency best describes these test scores?

16. *Shoe Sales* A salesman sold eight pairs of a certain style of men's shoes. The sizes of the eight pairs were as follows: $10\frac{1}{2}$, 8, 12, $10\frac{1}{2}$, 10, $9\frac{1}{2}$, 11 and $10\frac{1}{2}$. Which measure (or measures) of central tendency best describes the typical shoe size for these data?

In Exercises 17–24, find the mean ($\bar{x}$), variance ($v$), and standard deviation ($\sigma$) of the numbers.

**17.** 4, 10, 8, 2

**18.** 3, 15, 6, 9, 2

**19.** 0, 1, 1, 2, 2, 2, 3, 3, 4

**20.** 2, 2, 2, 2, 2, 2

**21.** 1, 2, 3, 4, 5, 6, 7

**22.** 1, 1, 1, 5, 5, 5

**23.** 49, 62, 40, 29, 32, 70

**24.** 1.5, 0.4, 2.1, 0.7, 0.8

In Exercises 25–30, use the alternative formula to find the standard deviation of the numbers.

**25.** 2, 4, 6, 6, 13, 5

**26.** 10, 25, 50, 26, 15, 33, 29, 4

**27.** 246, 336, 473, 167, 219, 359

**28.** 6.0, 9.1, 4.4, 8.7, 10.4

**29.** 8.1, 6.9, 3.7, 4.2, 6.1

**30.** 9.0, 7.5, 3.3, 7.4, 6.0

**31.** *Reasoning*    Without calculating the standard deviation, explain why the set {4, 4, 20, 20} has a standard deviation of 8.

**32.** *Reasoning*    If the standard deviation of a set of numbers is 0, what does this imply about the set?

**33.** *Test Scores*    An instructor adds five points to each student's exam score. Will this change the mean or standard deviation of the exam scores? Explain.

**34.** *Think About It*    Consider the four sets of data represented by the histograms. Order the sets from the smallest to the largest variance.

Set A

Set B

Set C

Set D

**35.** *Test Scores*    The scores of a mathematics exam given to 600 science and engineering students at a college had a mean and standard deviation of 235 and 28, respectively. Use Chebychev's Theorem to determine the intervals containing at least $\frac{3}{4}$ and at least $\frac{8}{9}$ of the scores. How would the intervals change if the standard deviation were 16?

**36.** *Price of Gold*    The following data represent the average prices of gold (in dollars per fine ounce) for the years 1978 to 1997. Use a computer or calculator to find the mean, variance, and standard deviation of the data. What percent of the data lie within two standard deviations of the mean?    (Source: U.S. Bureau of Mines)

| | | | | |
|---|---|---|---|---|
| 194, | 308, | 613, | 460, | 376, |
| 424, | 361, | 318, | 368, | 448, |
| 438, | 383, | 385, | 363, | 345, |
| 361, | 385, | 386, | 389, | 333 |

# Answers to Warm Ups, Odd-Numbered Exercises, Quizzes, and Tests

## Chapter 1
### Section 1.1 *(page 8)*

---

**Warm Up** *(page 8)*

1. $-3x - 10$   2. $5x - 12$   3. $x$   4. $x + 26$

5. $\dfrac{8x}{15}$   6. $\dfrac{3x}{4}$   7. $-\dfrac{1}{x(x + 1)}$   8. $\dfrac{5}{x}$

9. $\dfrac{7x - 8}{x(x - 2)}$   10. $-\dfrac{2}{x^2 - 1}$

---

1. Identity   3. Conditional equation

5. Conditional equation

7. (a) No   9. (a) Yes   11. (a) Yes   13. (a) No
   (b) No      (b) Yes      (b) No       (b) No
   (c) Yes     (c) No       (c) No       (c) No
   (d) No      (d) No       (d) No       (d) Yes

15. 5   17. $-4$   19. 3   21. 9   23. $-26$

25. $-4$   27. $-\frac{6}{5}$   29. 9   31. No solution

33. 10   35. 4   37. 3   39. 5   41. No solution

43. $\frac{11}{6}$   45. No solution   47. 0

49. All real numbers   51. No solution

53. Extraneous solutions may arise when a fractional expression is multiplied by factors involving the variable.

55. $x \approx 138.889$   57. $x \approx 62.372$   59. $x \approx -2.386$

61. Use the table feature in the ASK mode or evaluate using the scientific calculator part of a graphing utility.

63. (a) $\approx 6.46$   65. (a) $\approx 1.06$   67. (a) $\approx 56.09$
    (b) $\approx 6.41$; yes   (b) $\approx 1.06$; no   (b) $\approx 56.13$; yes

69. Sometime during 1999; $t = 9.935$   71. 61.2 inches

73. 1990; $x = 10.07$   75. 0.46

### Section 1.2 *(page 19)*

---

**Warm Up** *(page 19)*

1. 14   2. 4   3. $-3$   4. 4   5. $-2$

6. 1   7. $\frac{2}{5}$   8. $\frac{10}{3}$   9. 6   10. $-\frac{11}{5}$

---

1. $x + (x + 1) = 2x + 1$   3. $50t$   5. $0.2x$   7. $6x$

9. $1200 + 25x$   11. 262, 263   13. 37, 185

15. $-5, -4$   17. Coworker's check: \$300
    Your check: \$345

19. Coworker's check: \$348.65
    Your check: \$296.35

21. $\approx 96\%$ decrease $(-96.4\%)$   23. 19,300% increase

25. (a) 1990: 7.8 years
    (b) 1995: 8.5 years

27. Guaranteed investment: 540   29. 0.63
    Money fund: 333
    Diversified stock: 311
    Aggressive stock: 303
    Company stock: 296

31. 15 feet $\times$ 22.5 feet   33. $\approx 5.7$ years (5.714 years)

35. 97 or greater   37. $\approx \$22,316.98$   39. $\approx \$1016.77$

41. 38%   43. \$361.25   45. 3 hours   47. $\frac{1}{3}$ hour

49. Family 1 (42 miles per hour): 3.8 hours
    Family 2 (50 miles per hour): 3.2 hours

51. $\approx 1.29$ seconds   53. 62.5 feet   55. \$563,952

57. \$4500 at 10.5%   59. 11.43%
    \$7500 at 13%

61. 8823 units   63. $\approx 48$ feet   65. $\approx 32.1$ gallons

67. A mathematical model should be accurate and reasonable to use. Hence, the major factors are accuracy and ease of use. If the model is very complicated, the users may make errors or choose not to use the model. So sometimes a reasonably accurate model that is easy to use is the best model.

69. $\approx 10.894$ miles per hour   71. $l = \dfrac{P - 2w}{2}$

73. $h = \dfrac{V}{\pi r^2}$   75. $L = \dfrac{S}{1 - R}$   77. $P = \dfrac{A}{\left(1 + \dfrac{r}{n}\right)^{nt}}$

79. $\theta = \dfrac{360A}{\pi r^2}$   81. $r = \dfrac{s - a}{s - L}$

83. $r = \sqrt{\dfrac{A}{4\pi}}$ or $r = \dfrac{\sqrt{A\pi}}{2\pi}$

## Section 1.3 *(page 32)*

**Warm Up** *(page 32)*

1. $\dfrac{\sqrt{14}}{10}$   2. $4\sqrt{2}$   3. $14$   4. $\dfrac{\sqrt{10}}{4}$

5. $x(3x + 7)$   6. $(2x - 5)(2x + 5)$

7. $-(x - 7)(x - 15)$   8. $(x - 2)(x + 9)$

9. $(5x - 1)(2x + 3)$   10. $(6x - 1)(x - 12)$

1. $2x^2 + 5x - 3 = 0$   3. $x^2 - 25x = 0$

5. $x^2 - 6x + 7 = 0$   7. $2x^2 - 2x + 1 = 0$

9. $3x^2 - 60x - 10 = 0$   11. $0, -\frac{1}{2}$   13. $4, -2$

15. $-5$   17. $3, -\frac{1}{2}$   19. $2, -6$   21. $-2, -5$

23. $\pm 4$   25. $\pm\sqrt{7} \approx \pm 2.65$   27. $\pm 2\sqrt{3} \approx \pm 3.46$

29. $12 + 3\sqrt{2} \approx 16.24$   31. $-2 + 2\sqrt{3} \approx 1.46$

$12 - 3\sqrt{2} \approx 7.76$     $-2 - 2\sqrt{3} \approx -5.46$

33. $\pm 5$   35. $\pm\dfrac{\sqrt{115}}{5} \approx \pm 2.14$   37. $\pm 8$   39. $1$

41. $\pm\frac{3}{4}$   43. $\frac{3}{2}$   45. $6, -12$   47. $\frac{3}{2}, -\frac{1}{2}$   49. $5, -\frac{10}{3}$

51. $9, 3$   53. $\frac{1}{5}, 1$   55. $-1, -5$   57. $-\frac{1}{2}$

59. Algebra argument:

$(x + 2)^2 = (x + 2)(x + 2)$      Definition of exponent

$\qquad\qquad = x^2 + 2x + 2x + 4$      FOIL

$\qquad\qquad = x^2 + 4x + 4$      Combine like terms.

so $(x + 2)^2 \ne x^2 + 4$.

Graphing utility argument:

(1) Let $y_1 = (x + 2)^2$ and $y_2 = x^2 + 4$. Use the table feature with an example value of $x$ (but not $x = 0$). The table will show that $y_1$ is not the same as $y_2$.

(2) Use the scientific calculator portion to show that, if $x = 5$, $(5 + 2)^2 = 49$ and $5^2 + 4 = 29$, so $(x + 2)^2$ is not the same as $x^2 + 4$.

61. $34$ feet $\times$ $48$ feet   63. Base: $2\sqrt{2}$ feet

Height: $2\sqrt{2}$ feet

65. $\dfrac{9\sqrt{13}}{4} \approx 8.11$ seconds   67. $\approx 1.43$ seconds

69. $\approx 3.54$ centimeters   71. $886$ miles

73. $\approx 1414$ feet   75. $50,000$ units

77. $2005$ $(t \approx 15.4)$

79. $1987$ $(t \approx 18.74)$. The model was a good representation through 1990.

81. The model in Exercise 80 is *not* valid for the population in 2050 because it predicts 555,909,100 people (not 390,000,000).

83. $\approx 0.4490$

## Section 1.4 *(page 42)*

**Warm Up** *(page 42)*

1. $3\sqrt{17}$   2. $2\sqrt{3}$   3. $4\sqrt{6}$   4. $3\sqrt{73}$

5. $2, -1$   6. $\frac{3}{2}, -3$   7. $5, -1$   8. $\frac{1}{2}, -7$

9. $3, 2$   10. $4, -1$

1. One real solution   3. Two real solutions

5. No real solutions   7. Two real solutions

9. $\frac{1}{2}, -1$   11. $\frac{1}{4}, -\frac{3}{4}$   13. $1 \pm \sqrt{3}$

15. $-7 \pm \sqrt{5}$   17. $-4 \pm 2\sqrt{5}$   19. $\dfrac{2}{3} \pm \dfrac{\sqrt{7}}{3}$

21. $-\dfrac{1}{3} \pm \dfrac{\sqrt{11}}{6}$   23. $-\dfrac{1}{2} \pm \sqrt{2}$   25. $\dfrac{2}{7}$

27. $2 \pm \dfrac{\sqrt{6}}{2}$   29. $6 \pm \sqrt{11}$   31. $x \approx 0.976, -0.643$

33. $x \approx 0.561, 0.126$   35. $x \approx 1.687, -0.488$

37. $-11$   39. $\pm\sqrt{10}$   41. $-\dfrac{3}{2} \pm \dfrac{\sqrt{5}}{2}$   43. $-2, 4$

45. $\pm 2$   47. $50, 50$   49. $7, 8$ or $-8, -7$

51. $200$ units   53. $653$ units   55. $9$ seats per row

57. $14$ inches $\times$ $14$ inches   59. Moon: $14.9$ seconds

Earth: $2.6$ seconds

61. Shorter period of time on earth

63. $259$ miles, $541$ miles   65. (a) $1991$ $(t = 11.523)$

(b) $2000$ $(t \approx 20)$

67. $2002$ $(t \approx 12)$

69. Southbound: $\approx 550$ miles per hour; eastbound: $\approx 600$ miles per hour

71. $3761$ units or $146,239$ units

73. In an application, one of the solutions may not be reasonable. In Example 5, the other solution is $t = -144.415$. The model is for $t \ge 0$ and not $t < 0$. Additionally, $t \approx -144$ would be about the year 1846, which is not reasonable.

## Mid-Chapter Quiz   *(page 46)*

**1.** $x = 2$   **2.** $x = 6$   **3.** $x = -2$   **4.** No solution

**5.** 180.244   **6.** 431.398

**7.** Use the table feature in ASK mode or the scientific calculator portion of the graphing utility.

**8.** $7.50x + 20,000 = 80,000$; 8000 units

**9.** 4290 units ($x = 4289.322$) or
145,711 units ($x = 145,710.678$)

**10.** $\frac{2}{3}, -5$   **11.** $x = \pm\sqrt{5}; x \approx \pm2.24$

**12.** $x = -3 \pm \sqrt{17}; x \approx -7.12, 1.12$

**13.** $x = \dfrac{-7 \pm \sqrt{73}}{6}$   **14.** $x = -1 \pm \sqrt{6}$

**15.** $x \approx 1.568, -0.068$   **16.** Two real solutions

**17.** No real solutions

**18.** Answers will vary. Use the FOIL method $[(x + 3)^2 = (x + 3)(x + 3) = x^2 + 6x + 9]$; or use the table feature of your graphing utility; or use the scientific calculator portion of your graphing utility to evaluate the solution.

**19.** $\approx 3.95$ seconds   **20.** 6 inches $\times$ 6 inches

## Section 1.5   *(page 55)*

### Warm Up   *(page 55)*

**1.** 11   **2.** $20, -3$   **3.** $5, -45$   **4.** $0, -\frac{1}{5}$

**5.** $\frac{2}{3}, -2$   **6.** $\frac{11}{6}, -\frac{5}{2}$   **7.** $1, -5$   **8.** $\frac{3}{2}, -\frac{5}{2}$

**9.** $\dfrac{3 \pm \sqrt{5}}{2}$   **10.** $2 \pm \sqrt{2}$

**1.** $0, \pm\dfrac{3\sqrt{2}}{2}$   **3.** $3, -1, 0$   **5.** $\pm3$   **7.** $-3, 0$

**9.** $3, 1, -1$   **11.** $\pm1$   **13.** $\pm3, \pm1$   **15.** $\pm2$

**17.** $\pm\frac{1}{2}, \pm4$   **19.** $1, -2$   **21.** 50   **23.** 26

**25.** $-16$   **27.** $\frac{1}{4}$   **29.** $6, 5$   **31.** $2, -5$   **33.** 0

**35.** $-59, 69$   **37.** 1   **39.** $\pm\sqrt{69}$   **41.** $\dfrac{-3 \pm \sqrt{21}}{6}$

**43.** $4, -5$   **45.** $2, -\frac{3}{2}$   **47.** $-1$   **49.** $1, -3$

**51.** $1, -3$   **53.** $3, -2$   **55.** $\sqrt{3}, -3$   **57.** $10, -1$

**59.** The quadratic equation was not written in standard form before values for $a, b$, and $c$ were substituted in the Quadratic Formula. Standard form for this equation is $3x^2 - 7x - 4 = 0$ $(a = 3, b = -7$, and $c = -4)$, and the correct solution is
$$x = \dfrac{-(-7) \pm \sqrt{(-7)^2 - 4(3)(-4)}}{2(3)}.$$

**61.** $x \approx \pm1.038$   **63.** $x \approx 16.756$   **65.** 34

**67.** 7%   **69.** 19.2%   **71.** 26,250 passengers

**73.** 61 years old   **75.** 2,566,025 units

**77.** $\approx 12.12$ feet   **79.** $t = 13\frac{1}{3}$ minutes

## Section 1.6   *(page 66)*

### Warm Up   *(page 66)*

**1.** $-\frac{1}{2}$   **2.** $-\frac{1}{6}$   **3.** $-3$   **4.** $-6$   **5.** $x \geq 0$

**6.** $-3 < z < 10$   **7.** $P \leq 2$   **8.** $W \geq 200$

**9.** $2, 7$   **10.** $0, 1$

**1.** (a) Yes  (b) No  (c) Yes  (d) No   **3.** c   **4.** h

**5.** f   **6.** e   **7.** g   **8.** a   **9.** b   **10.** d

**11.** $x \geq 3$                        **13.** $x < 12$

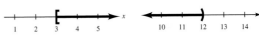

**15.** $x > -4$                        **17.** $x \geq 12$

**19.** $x < -\frac{1}{2}$                        **21.** $x \geq \frac{1}{2}$

**23.** $2 \leq x < 4$                        **25.** $-1 < x < 3$

**27.** $-\frac{9}{2} < x < \frac{15}{2}$                        **29.** $-\frac{3}{4} < x < -\frac{1}{4}$

**31.** $-8 < x < 2$                        **33.** $x < -6, x > 6$

**35.** $16 \leq x \leq 24$                        **37.** $x < -\frac{1}{2}, x > \frac{11}{2}$

**39.** $x \leq -7, x \geq 13$                        **41.** $4 < x < 5$

**43.** $x \le -\frac{29}{2}, x \ge -\frac{11}{2}$     **45.** No solution

**47.** $|x| \le 2$     **49.** $|x - 9| \ge 3$     **51.** $|x - 12| \le 10$

**53.** $|x + 3| > 5$     **55.** $m > 400$ miles     **57.** $r > 12.5\%$

**59.** 24 weeks

**61.** (a)

| $x$ | 10 | 20 | 30 | 40 | 50 |
|---|---|---|---|---|---|
| $R$ | \$1159.50 | \$2319.00 | \$3478.50 | \$4638.00 | \$5797.50 |
| $C$ | \$1700.00 | \$2650.00 | \$3600.00 | \$4550.00 | \$5500.00 |

(b) $x \ge 36$ units

**63.** $m < 24{,}062.5$     **65.** 2001 ($t \ge 11.661$)

**67.** Overcharged about \$0.12 or undercharged \$0.12

**69.** $[65.8, 71.2]$

---

**Math Matters**     *(page 69)*

| Cube | Ratio of $\dfrac{\text{surface area}}{\text{weight}}$ |
|---|---|
| 1 | 6 |
| 2 | 3 |
| 3 | 2 |
| 4 | 1.5 |

---

## Section 1.7     *(page 77)*

**Warm Up**     *(page 77)*

**1.** $y < -6$     **2.** $z > -\frac{9}{2}$     **3.** $-3 \le x < 1$

**4.** $x \le -5$     **5.** $-3 < x$     **6.** $5 < x < 7$

**7.** $-\frac{7}{2} \le x \le \frac{7}{2}$     **8.** $x < 2, x > 4$

**9.** $x < -6, x > -2$     **10.** $-2 \le x \le 6$

**1.** $-3 \le x \le 3$          **3.** $x < -2, x > 2$

---

**5.** $-7 < x < 3$          **7.** $x \le -5, x \ge 1$

**9.** $-3 < x < 2$          **11.** $x < -1, x > 1$

**13.** $-3 < x < 1$          **15.** $x < 0, 0 < x < \frac{3}{2}$

**17.** $-2 \le x \le 0, x \ge 2$          **19.** $x < -1, 0 < x < 1$

**21.** $x < -1, x > 4$          **23.** $5 < x < 15$

**25.** $-5 < x < -\frac{3}{2}, x > -1$     **27.** $-\frac{3}{4} < x < 3, x \ge 6$

**29.** $[-2, 2]$     **31.** $(-\infty, 3], [4, \infty)$     **33.** $[-4, 3]$

**35.** All real numbers     **37.** The cube root of any real number is a real number.

**39.** $-3.51 < x < 3.51$     **41.** $-0.13 < x < 25.13$

**43.** $2.26 < x < 2.39$     **45.** $4 < t < 6$

**47.** 13.8 meters $\le l \le 36.2$ meters

**49.** (a) $90{,}000 \le x < 100{,}000$     (b) $\$30 \le p \le \$32$

**51.** 9.5%     **53.** 2002 ($t > 22.824$)

**55.** 2001 ($t > 11.626$)

## Review Exercises     *(page 82)*

**1.** Conditional equation

**3.** (a) No   (b) Yes   (c) Yes   (d) No

**5.** $-\frac{1}{2}$     **7.** $-10$     **9.** $-\frac{2}{3}$     **11.** $\approx 377.778$     **13.** 12

**15.** May: \$121,833.51     **17.** 29.5 feet $\times$ 59 feet
June: \$103,558.49

**19.** \$12     **21.** \$399     **23.** 2 hours     **25.** 2.9 quarts

**27.** $-\frac{1}{2}, \frac{4}{3}$     **29.** 3, 8     **31.** $\pm\sqrt{11}, \pm 3.32$

**33.** $-4 + 3\sqrt{2} \approx 0.24$
$-4 - 3\sqrt{2} \approx -8.24$

**35.** (1) Use the table feature in ASK mode with the variable equal to a solution.

(2) Use the scientific calculator portion to evaluate the quadratic equation at a particular solution.

**37.** 15 feet × 27 feet    **39.** 200,000 units or 300,000 units

**41.** Two solutions    **43.** $6 \pm \sqrt{6}$    **45.** $\dfrac{-19 \pm \sqrt{165}}{2}$

**47.** $-3 \pm 2\sqrt{3}$

**49.** The quadratic equation was not written in standard form before values for $a, b$, and $c$ were substituted in the Quadratic Formula. Standard form for this equation is $10x^2 - 11x - 2 = 0$ ($a = 10$, $b = -11$, and $c = -2$), and the correct solution is
$$x = \frac{-(-11) \pm \sqrt{(-11)^2 - 4(10)(-2)}}{2(10)} = \frac{11 \pm \sqrt{201}}{20}.$$

**51.** 8.544, 0.162    **53.** 17 inches × 17 inches ($x = 11$)

**55.** $\pm 2, \pm 1$    **57.** $-3, 1$    **59.** 2 (9 is extraneous)

**61.** $-2, 7$    **63.** $-4, -2, 2$    **65.** $-1, 1$

**67.** $r \approx 21.18\%$

**69.** $x < 11$

**71.** $-\dfrac{13}{2} < x < \dfrac{11}{2}$

**73.** $-12 < x < -8$    **75.** $x \geq 36$ units

**77.** $-1 < x < 3$    **79.** $x < -3, 0 < x < 3$

**81.** $x < \dfrac{6}{5}, x > 4$    **83.** $-1.69 < x < 1.69$

**85.** $1.65 < x < 1.74$    **87.** $x \geq 10$

**89.** $x \leq 6$ or $x \geq 9$    **91.** 3.65 seconds $< t <$ 4.72 seconds

**93.** $r = 6.96\%$    **95.** 25,359 units $\leq x \leq$ 94,641 units

## Chapter Test    *(page 86)*

**1.** $\dfrac{17}{23}$    **2.** $x \geq \dfrac{9}{2}$    **3.** April: \$175,364.00
May: \$140,291.20

**4.** $-\dfrac{5}{3}, \dfrac{1}{2}$    **5.** $4, -\dfrac{3}{2}$    **6.** $\pm\sqrt{15}$    **7.** $\dfrac{-13 \pm \sqrt{69}}{2}$

**8.** $\dfrac{11 \pm \sqrt{145}}{6}$    **9.** 1.038, $-0.446$    **10.** $2, -\dfrac{10}{3}$

**11.** 4    **12.** $-1, 1, -3, 3$    **13.** $-6, 6$

**14.** Selling either 341,421 units or 58,579 units will produce a revenue of \$2,000,000.

**15.** $x < 3$

**16.** $x \geq \dfrac{28}{5}, x \leq -4$

**17.** $(-11, -7)$

**18.** $x \leq -2, 0 \leq x \leq 2$

**19.** $10,839.2 \leq x \leq 129,160.8$
For a profit of at least \$500,000, the company must sell more than 10,839 units but less than 129,161 units.

**20.** $r > 9.6\%$

# Chapter 2

## Section 2.1    *(page 96)*

**Warm Up**    *(page 96)*

**1.** 5    **2.** $3\sqrt{2}$    **3.** 1    **4.** $-2$
**5.** $3(\sqrt{2} + \sqrt{5})$    **6.** $2(\sqrt{3} + \sqrt{11})$    **7.** $-3, 11$
**8.** 9, 1    **9.** 11    **10.** 4

**1.**

**3.**

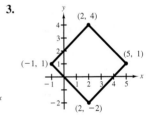

**5.** $(-3, -4)$ is shifted to $(-1, 1)$
$(1, -3)$ is shifted to $(3, 2)$
$(-2, -1)$ is shifted to $(0, 4)$

**7.** 6    **9.** 11

**11.** Answers vary. $(3, 4), (1, 4)$    **13.** 5    **15.** $\sqrt{109}$

**17.** (a)

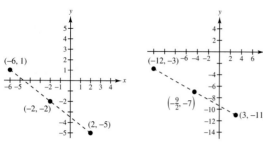

(b) 10   (c) $(-2, -2)$

**19.** (a)

(b) 17   (c) $\left(-\frac{9}{2}, -7\right)$

**21.** (a)

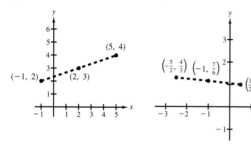

(b) $2\sqrt{10}$   (c) $(2, 3)$

**23.** (a)

(b) $\dfrac{\sqrt{82}}{3}$   (c) $\left(-1, \dfrac{7}{6}\right)$

**25.** (a)

(b) $\sqrt{34.25}$

(c) $(-0.75, 6.9)$

**27.** (a)

(b) $6\sqrt{277}$

(c) $(6, -45)$

**29.** \$630,000    **31.** $\left(\sqrt{5}\right)^2 + \left(\sqrt{45}\right)^2 = \left(\sqrt{50}\right)^2$

**33.** All sides have a length of $\sqrt{5}$.    **35.** $x = 15, -9$

**37.** $y = 9, -23$    **39.** IV    **41.** I

**43.** $\approx 33.5$ yards    **45.** $\approx 54$ years

**47.** $\approx \$610$ in 1980    **49.** $\approx 48\%$    **51.** $\approx 4000$

## Section 2.2    *(page 106)*

**Warm Up**    *(page 106)*

**1.** $y = \dfrac{3x - 2}{5}$    **2.** $y = -\dfrac{(x - 5)(x + 1)}{2}$

**3.** 2    **4.** $1, -5$    **5.** $0, \pm 3$    **6.** $\pm 2$

**7.** $y = x^3 + 4x$    **8.** $x^2 + y^2 = 4$

**9.** $y = 4x^2 + 8$    **10.** $y^2 = 3x + 4$

**1.** (a) Yes   (b) Yes    **3.** (a) No   (b) No

**5.** (a) Yes   (b) Yes    **7.** $\left(\frac{1}{2}, 0\right), (0, -1)$

**9.** $(-2, 0), (1, 0), (0, -2)$    **11.** $(0, 0), (-2, 0)$

**13.** $\left(\frac{4}{3}, 0\right), (0, 2)$

**15.** Every ordered pair on the $x$-axis has a $y$-coordinate of zero $[(x, 0)]$, so to find an $x$-intercept we let $y = 0$. Similarly, every ordered pair on the $y$-axis has an $x$-coordinate of zero $[(0, y)]$, so to find a $y$-intercept we let $x = 0$.

**17.** $y$-axis symmetry    **19.** $y$-axis symmetry

**21.** Origin symmetry    **23.** Origin symmetry

**25.**    **27.**

**29.** c    **30.** f    **31.** d    **32.** a    **33.** e    **34.** b

**35.**

Intercepts: $\left(\frac{5}{3}, 0\right), (0, 5)$

Symmetry: none

**37.**

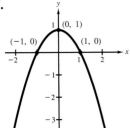

Intercepts:
$(-1, 0), (1, 0), (0, 1)$
Symmetry: $y$-axis

**39.**

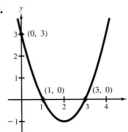

Intercepts:
$(3, 0), (1, 0), (0, 3)$
Symmetry: none

**41.**

Intercepts: $\left(-\sqrt[3]{2}, 0\right), (0, 2)$
Symmetry: none

**43.**

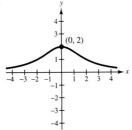

Intercepts: no $x$-intercepts;
$(0, 2)$
Symmetry: $y$-axis

**45.**

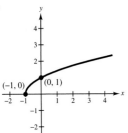

Intercepts: $(-1, 0), (0, 1)$
Symmetry: none

**47.**

Intercept: $(0, 0)$
Symmetry: origin

**49.**

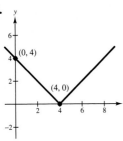

Intercepts: $(4, 0), (0, 4)$
Symmetry: none

**51.**

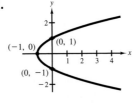

Intercepts: $(-1, 0), (0, 1),$
$(0, -1)$
Symmetry: $x$-axis

**53.**

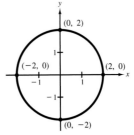

Intercepts: $(-2, 0), (2, 0), (0, 2), (0, -2)$
Symmetry: $x$-axis, $y$-axis, origin

**55.** $x^2 + y^2 = 9$    **57.** $(x + 4)^2 + (y - 1)^2 = 2$

**59.** $(x + 1)^2 + (y - 2)^2 = 5$

**61.** $(x - 1)^2 + (y - 1)^2 = 25$

**63.** $(x - 3)^2 + (y + 2)^2 = 16$

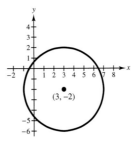

**65.** $(x - 1)^2 + (y + 3)^2 = 0$

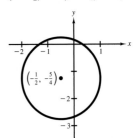

**67.** $\left(x - \frac{1}{2}\right)^2 + \left(y - \frac{1}{2}\right)^2 = 2$   **69.** $\left(x + \frac{1}{2}\right)^2 + \left(y + \frac{5}{4}\right)^2 = \frac{9}{4}$

**71.** 6    **73.** 4

**75.** $(2, -3), 4, x^2 + y^2 - 4x + 6y - 3 = 0$

**77.** (a)

| Year | Data | Model |
|------|------|-------|
| 1989 | 0.807 | 0.803 |
| 1990 | 0.766 | 0.768 |
| 1991 | 0.734 | 0.738 |
| 1992 | 0.713 | 0.712 |
| 1993 | 0.692 | 0.690 |
| 1994 | 0.675 | 0.671 |
| 1995 | 0.656 | 0.653 |
| 1996 | 0.638 | 0.639 |

(b)

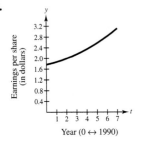

(c) $0.554 (compared with 1982 dollars)

**79.**

## Section 2.3 *(page 115)*

### Warm Up *(page 115)*

**1.** $y = 4 - 3x$  **2.** $y = x$  **3.** $y = \frac{2}{3}(1 - x)$

**4.** $y = \frac{2}{5}(2x + 1)$  **5.** $y = \frac{1}{4}(5 - 3x)$

**6.** $y = \frac{2}{3}(-x + 3)$  **7.** $y = 4 - x^2$

**8.** $y = \frac{2}{3}(x^2 - 1)$  **9.** $y = \pm\sqrt{4 - x^2}$

**10.** $y = \pm\sqrt{x^2 - 9}$

**1.** d  **2.** g  **3.** a  **4.** f  **5.** i

**6.** b  **7.** j  **8.** c  **9.** e  **10.** h

**11.**

**13.**

**15.**

**17.**

**19.**

**21.**

**23.**

**25.**

**27.**

**29.**

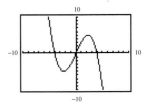

**31.** Yes. The intercepts are easily seen.

**33.** Yes. The intercepts are easily seen.

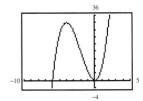

**35.** $-30$     **37.**

| Xmin = -10 |
| Xmax = 10 |
| Xscl = 1 |
| Ymin = -12 |
| Ymax = 30 |
| Yscl = 6 |
| Xres = 1 |

**39.**

| Xmin = -10 |
| Xmax = 10 |
| Xscl = 1 |
| Ymin = -10 |
| Ymax = 10 |
| Yscl = 1 |
| Xres = 1 |

**41.** (a) No *x*-intercepts

(b) Two *x*-intercepts

**43.** Three *x*-intercepts

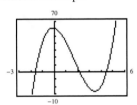

**45.** The graphs of 41(b) and 42(a) cross the *x*-axis and have two *x*-intercepts. The graphs of 41(a) and 42(b) do *not* cross the *x*-axis and have no *x*-intercepts.

**47.**

Square

**49.**

Circle

**51.** (a) $0.741 (in 1982 dollars)

(b) The model is reasonable.

**53.** The viewing window did not contain any ordered pairs of the equation. I would suggest using the table feature to find some ordered pairs for the equation, and using them to help select an appropriate viewing window.

**55.**

**57.** 65%

**59.**

**61.** $0.34 \le x \le 0.35$

## Mid-Chapter Quiz   *(page 118)*

**1.** (a)

(b) $7\sqrt{2}$

(c) $\left(\frac{1}{2}, -\frac{3}{2}\right)$

**2.** (a)

(b) $\sqrt{52.04}$

(c) $(-1.2, -1.9)$

**3.** $1,610,000. Used the Midpoint Formula because 1997 is midway between 1995 and 1999.

**4.** 250 miles     **5.** Right isosceles triangle

**6.** Parallelogram

**7.**

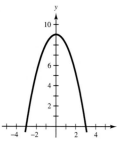

Intercepts:
$(3, 0), (-3, 0), (0, 9)$
Symmetry: $y$-axis

**8.**

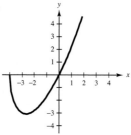

Intercepts: $(0, 0), (-4, 0)$
Symmetry: none

**9.**

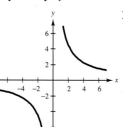

Intercepts: none
Symmetry: origin

**10.**

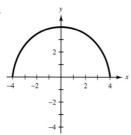

Intercepts:
$(0, 4), (4, 0), (-4, 0)$
Symmetry: $y$-axis

**11.**

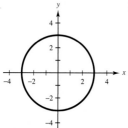

Intercepts: $(3, 0), (-3, 0)$,
$(0, 3), (0, -3)$
Symmetry:
$x$-axis, $y$-axis, origin

**12.**

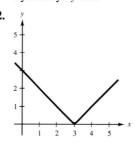

Intercepts: $(3, 0), (0, 3)$
Symmetry: none

**13.** $(x - 2)^2 + (y + 3)^2 = 16$   **14.** $x^2 + \left(y + \frac{1}{2}\right)^2 = 4$

**15.**

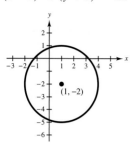

$(x - 1)^2 + (y + 2)^2 = 9$

**16.** Two $x$-intercepts   **17.** Two $x$-intercepts

**18.**

| Xmin = 0 |
| Xmax = 7 |
| Xscl = 1 |
| Ymin = 0 |
| Ymax = 2.8 |
| Yscl = 0.2 |
| Xres = 1 |

**19.**

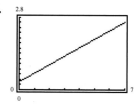

**20.** \$3.61

## Section 2.4   *(page 127)*

**Warm Up**   *(page 127)*

**1.** $-\frac{9}{2}$   **2.** $-\frac{13}{3}$   **3.** $-\frac{5}{4}$   **4.** $\frac{1}{2}$

**5.** $y = \frac{2}{3}x - \frac{5}{3}$   **6.** $y = -2x$

**7.** $y = 3x - 1$   **8.** $y = \frac{2}{3}x + 5$

**9.** $y = -2x + 7$   **10.** $y = x + 3$

**1.** 1   **3.** 0   **5.** $-3$

**7.**

**9.**

$m = 1$

**11.**

$m$ is undefined.

**13.**

$m = \frac{1}{2}$

**15.** Perpendicular    **17.** Neither parallel nor perpendicular

**19.** Parallel    **21.** $(3, -2), (-1, -2), (0, -2)$

**23.** $(6, -5), (7, -4), (8, -3)$    **25.** $(2, 3), (-4, 0), (4, 4)$

**27.** $m = 4, (0, -6)$    **29.** $m$ is undefined, no $y$-intercept

**31.** $m = -\frac{7}{6}, (0, 5)$    **33.** $x + 2y + 1 = 0$

**35.** $x + 9 = 0$    **37.** $x + 2y - 3 = 0$

**39.** $2x - 5y + 1 = 0$

**41.** $x - y - 7 = 0$    **43.** $2x + y = 0$

**45.** $x + 3y - 4 = 0$    **47.** $x - 6 = 0$

**49.** $8x - 6y - 17 = 0$

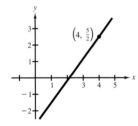

**51.** Answers may vary. You could graph a vertical line and pick two convenient points on the vertical line. Find the slope. Regardless of the vertical line selected and the points selected, the slope will have zero in the denominator. Division by zero is not possible, so the slope does not exist.

**53.** $4x - y - 4 = 0$    **55.** $12x + 3y + 2 = 0$

**57.** (a) $2x - y - 10 = 0$    **59.** (a) $4x - 6y - 5 = 0$

    (b) $x + 2y - 10 = 0$        (b) $36x + 24y + 7 = 0$

**61.** (a) $y = 0$    (b) $x + 1 = 0$    **63.** $F = \frac{9}{5}C + 32$

**65.** $A = 4.5t + 750$    **67.** $\$166{,}000$    **69.** $\$36{,}200$

**71.** $\$9553$ million; no

**73.** Rate of change per year: $\$175/3$ or $\$58\frac{1}{3}$ or $\$58.33$; Rent in 2005: $\$1208\frac{1}{3}$ or $\$1208.33$

## Section 2.5   *(page 138)*

**Warm Up**   *(page 138)*

**1.**

**2.**

**3.**

**4.**

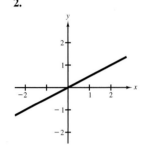

**5.** $y = x + 2$    **6.** $y = \frac{3}{2}x + 3$    **7.** $y = x + 2$

**8.** $y = \frac{6}{7}x + 4$    **9.** $5x + 40y - 213 = 0$

**10.** $29x + 60y - 448 = 0$

**1.**

The model is a "good fit" for the actual data.

**3.** $y = 4.7225 - 0.0125t$

Predicted time for 2000 Olympics: 3.97 minutes

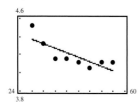

The prediction is probably too low. When the actual times are compared with the model times, it can be seen that the actual times are not decreasing as quickly as the model times. The model could be better.

**5.** $y = \dfrac{3}{8}x$     **7.** $y = 20x$     **9.** $y = \dfrac{3.2}{7}x$ or $y = \dfrac{32}{70}x$

**11.** $I = 0.075P$     **13.** (a) $y = 0.0368x$   (b) \$3128

**15.** (a) $C = \dfrac{33}{13}I$

(b)

| Inches | 5 | 10 | 20 | 25 | 30 |
|---|---|---|---|---|---|
| Centimeters | 12.69 | 25.38 | 50.77 | 63.46 | 76.15 |

**17.** $V = 125t + 2540, \quad 0 \le t \le 5$

**19.** $V = -2000t + 20,400, \quad 0 \le t \le 5$

**21.** $V = 12,500t + 154,000, \quad 0 \le t \le 5$

**23.** $V = 875 - 175t$     **25.** $S = 0.85L$

**27.** $W = 0.75x + 11.50$

**29.** (a) $h = 7000 - 20t$   (b) 2:13:50 P.M.

**31.** b; slope $= -10$; the amount owed *decreases* by \$10 per week.

**33.** a; slope $= 0.25$; the amount received *increases* by \$0.25 per mile driven.

**35.** (a) $N = 1200 + 50t$

(b) 2200 students

(c)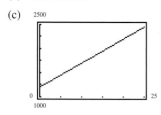

**37.** Yes     **39.** No

**41.** Answers will vary.
$y = 0.4x + 3$

**43.** Answers will vary.
$y = -0.5x + 3$

**45.** (a)

(b) $y = 0.13t + 2.50$

(c) July 1999:  4.97 hours
July 2000:  5.1 hours
Predictions seem reasonable.

**47.** (a) Answers will vary. $y = 1.13t + 124.9, \quad 0 \le t \le 96$

(b) 237.9 feet

(c) Answers will vary.

**49.** (a) $C = 16.75t + 36,500$     (b) $R = 27t$

(c) $P = 10.25t - 36,500$     (d) 3561 hours

**51.** (a) $x = -\dfrac{1}{15}p + \dfrac{226}{3}$

(b) 45 units rented when $p = \$455$

## Review Exercises   *(page 145)*

**1.** 13

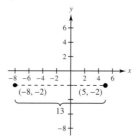

**3.** (a) 4 and 3, hypotenuse $= 5$     (b) 5

**5.** (a)

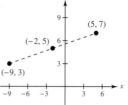

(b) $2\sqrt{53} \approx 14.56$

(c) $(-2, 5)$

**7.** (a)

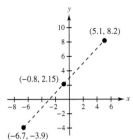

(b) $\sqrt{285.65} \approx 16.9$

(c) $(-0.8, 2.15)$

**9.** Slope of line through $(1, 1)$ and $(8, 2)$ is $\frac{1}{7}$.
Slope of line through $(8, 2)$ and $(9, 5)$ is 3.
Slope of line through $(9, 5)$ and $(2, 4)$ is $\frac{1}{7}$.
Slope of line through $(2, 4)$ and $(1, 1)$ is 3.
Consequently, opposite sides are parallel.

**11.** $-10$ or 30    **13.** 3 or $-7$    **15.** \$725,000

**17.** (a) Yes  (b) Yes    **19.** $x$-intercepts: $(-3, 0), (2, 0)$
$y$-intercept: $(0, -6)$

**21.** $x$-intercepts: $(0, 0), (-3, 0)$    **23.** Origin    **25.** None
$y$-intercept: $(0, 0)$

**27.**

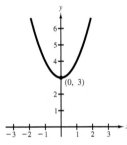

$y$-intercept: $(0, 3)$
Symmetry: $y$-axis

**29.**

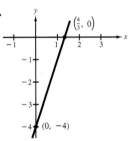

$x$-intercept: $\left(\frac{4}{3}, 0\right)$
$y$-intercept: $(0, -4)$
Symmetry: none

**31.**

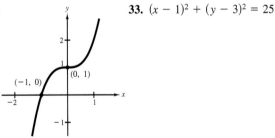

$x$-intercept: $(-1, 0)$
$y$-intercept: $(0, 1)$
Symmetry: none

**35.** $(x - 1)^2 + \left(y + \frac{1}{2}\right)^2 = \frac{25}{4}$

**37.** $(x - 3)^2 + (y + 2)^2 = 16$

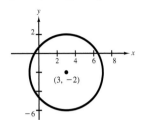

**39.** b    **40.** a    **41.** e    **42.** d    **43.** f    **44.** c

**45.**

**47.**

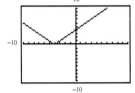

**49.** 3    **51.** Two $x$-intercepts

**53.**

Circle

**55.**

$m = \frac{11}{4}$

**57.** $8x - y - 17 = 0$

**59.** $x - 3 = 0$

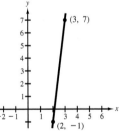

**61.** $3x - 2y - 10 = 0$

**63.** $2x + 3y - 6 = 0$

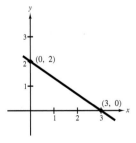

**65.** Slope: $\frac{5}{4}$; $y$-intercept: $\left(0, \frac{11}{4}\right)$

**67.** Parallel     **69.** Neither

**71.** (a) $5x - 4y - 23 = 0$   (b) $4x + 5y - 2 = 0$

**73.** $y = \frac{7}{3}x$     **75.** $y = 348x$     **77.** $I = 0.0626\,P$

**79.** $y = \frac{85}{3}x$

| Ounces | 6 | 8 | 12 | 16 | 24 |
|--------|---|---|----|----|----|
| Grams | 170 | 226.7 | 340 | 453.3 | 680 |

**81.** \$210,000

**83.** $V = 85 + 3.75t$,   $t = 0$ represents 2000
Value in 2005: \$103.75

**85.** $r = 1.25x - 25$; \$1725

**87.**

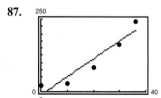

Model: $y = 6.19t - 10.98$

Predicted number of cigarettes exported in 2000:
236.62 million

## Chapter Test   *(page 149)*

**1.** Distance: $4\sqrt{5}$     **2.** Distance: 7.81
Midpoint: $(1, 0)$          Midpoint: $(0.44, 4.34)$

**3.** $x$-intercepts: $(-5, 0), (3, 0)$     **4.** $x$-intercept: $(2, 0)$
$y$-intercept: $(0, -15)$          $y$-intercepts: none

**5.** Symmetric with respect to the origin

**6.** $y - 4 = 0$     **7.** $3x - 4y - 24 = 0$

**8.** $x - 2 = 0$     **9.** $4x - 3y - 12 = 0$

**10.** (a)     **11.** (b)     **12.** (c)     **13.** (d)

**14.** $(x - 2)^2 + (y - 1)^2 = 9$

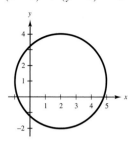

**15.** \$220,000     **16.** $W = 0.55x + 8.75$

**17.** $V = 25,000 - 4000x$     **18.** $S = L - 0.30L = 0.70L$

**19.** $y = -3.74t + 57.6$

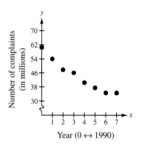

## Cumulative Test: Chapters 1–2   *(page 150)*

**1.** $-4, 4$     **2.** $\frac{12}{7}$     **3.** $\pm 1, \pm 4$     **4.** $-1, 4$

**5.** $1$     **6.** $\dfrac{-1 \pm \sqrt{41}}{4}$     **7.** \$11,663.24

**8.** $-\frac{7}{2} \le x \le \frac{17}{2}$     **9.** $-6 \le x \le -2$

**10.** $\frac{5}{4} < x < 4$     **11.** $(0, 0), (-4, 0)$

**12.** $y$-axis symmetry

**13.** $(x - 3)^2 + (y + 2)^2 = 16$

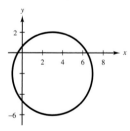

**14.** 11     **15.** $7x + 4y - 13 = 0$     **16.** $2x - 3y - 7 = 0$

**17.** $y + 3 = 0$     **18.** $y = 0.07x + 1000$     **19.** \$250,000

**20.** $x = 79,623.23$ or $x = 376.66$. Hence, revenue will equal
cost when either 79,623 units or 377 units are sold.

# Chapter 3

## Section 3.1     *(page 160)*

---

**Warm Up**   *(page 160)*

**1.** $-73$     **2.** 13     **3.** $2(x + 2)$     **4.** $-8(x - 2)$

**5.** $y = \frac{7}{5} - \frac{2}{5}x$     **6.** $y = \pm x$     **7.** $x \le -2, x \ge 2$

**8.** $-3 \le x \le 3$     **9.** All real numbers

**10.** $x \le 1, x \ge 2$

---

**1.** (a) $-6$   (b) 34   (c) $6 - 4t$   (d) $2 - 4c$

**3.** (a) $-1$   (b) $\frac{1}{15}$   (c) $\frac{1}{t^2 - 2t}$   (d) $\frac{1}{t^2 - 1}$

**5.** (a) $-1$   (b) $-9$   (c) $2x - 5$   (d) $-\frac{5}{2}$

**7.** (a) 0   (b) 3   (c) $x^2 + 2x$   (d) $-0.75$

**9.** (a) 1   (b) $-7$   (c) $3 - 2|x|$   (d) 2.5

**11.** (a) $\frac{1}{7}$   (b) $-\frac{1}{9}$   (c) Undefined   (d) $\frac{1}{y^2 + 6y}$

**13.** (a) 1   (b) $-1$   (c) 1   (d) $\frac{|x - 1|}{x - 1}$

**15.** (a) $-1$   (b) 2   (c) 4   (d) 6

**17.** 5     **19.** $\pm 3$     **21.** $\frac{10}{7}$     **23.** All real numbers

**25.** All real numbers except $t = 0$     **27.** $y \ge 10$

**29.** $-1 \le x \le 1$     **31.** All real numbers except $x = 0, -2$

**33.** All real numbers $x \ge -1$ except $x = 2$

**35.** The domain of $f(x) = \sqrt{x - 2}$ is all real numbers $x \ge 2$, since an even root of a negative number is not a real number. The domain of $g(x) = \sqrt[3]{x - 2}$ is all real numbers. $f$ and $g$ have different domains because an odd root of a negative number is a real number but an even root of a negative number is not a real number.

**37.** Not a function     **39.** Function     **41.** Function

**43.** Not a function     **45.** Function

**47.** This is a function from $A$ to $B$, because each element in $A$ is matched with an element in $B$.

**49.** This is a function from $A$ to $B$, because each element in $A$ is matched with an element in $B$.

**51.** Not a function. The relationship assigns two elements of $B$ to the element $c$ in $A$.

**53.** Not a function. The relationship defines a function from $B$ to $A$.

**55.** This is a function from $A$ to $B$, because each element in $A$ is matched with an element in $B$.

**57.** Not a function. The relationship does not match an element of $A$ with an element of $B$.

**59.** $\{(-2, 4), (-1, 1), (0, 0), (1, 1), (2, 4)\}$

**61.** $\{(-2, 0), (-1, 1), (0, \sqrt{2}), (1, \sqrt{3}), (2, 2)\}$

**63.** (a) $V = 4x(6 - x)^2$   (b) Domain: $0 < x < 6$

**65.** $h = \sqrt{d^2 - 2000^2}$; domain: $d \ge 2000$

**67.** (a) $A = 2000(1.015)^{4t}$   (b) Domain: $t \ge 0$

**69.** (a) $C = 12.30x + 98{,}000$   (b) $R = 17.98x$

   (c) $P = 5.68x - 98{,}000$   (d) $x \approx 17{,}254$ units

**71.** (a) $C = 6000 + 0.95x$   (b) $\overline{C} = \dfrac{6000}{x} + 0.95$

(c)

| $x$ | 100 | 1000 | 10,000 | 100,000 |
|-----|-----|------|--------|---------|
| $\overline{C}$ | 60.95 | 6.95 | 1.55 | 1.01 |

   (d) The average cost per unit decreases as $x$ gets larger.

**73.** (a) $R = 12.00n - 0.05n^2$,   $n \ge 80$

(b)

| $n$ | 90 | 100 | 110 | 120 | 130 | 140 | 150 |
|-----|-----|-----|-----|-----|-----|-----|-----|
| $R$ | \$675 | \$700 | \$715 | \$720 | \$715 | \$700 | \$675 |

   (c) Maximum profit of \$720 occurs when a group of 120 people charter a bus.

**75.** (a) $r(t) = 0.8t$

   (b) $A = \pi(0.8t)^2$

| Time, $t$ | 1 | 2 | 3 | 4 | 5 |
|-----------|---|---|---|---|---|
| Radius, $r$ | 0.8 ft | 1.6 ft | 2.4 ft | 3.2 ft | 4 ft |
| Area, $A$ | 2.011 ft$^2$ | 8.042 ft$^2$ | 18.096 ft$^2$ | 32.170 ft$^2$ | 50.265 ft$^2$ |

   (c) $\dfrac{A(2)}{A(1)} = \dfrac{8.042}{2.011} \approx 3.999$, $\dfrac{A(4)}{A(2)} = \dfrac{32.170}{8.042} \approx 4.000$

   Predicted area when $t = 8$: 128.68 square feet
   Calculated area when $t = 8$: 128.680 square feet

## Section 3.2     *(page 172)*

---

**Warm Up**   *(page 172)*

**1.** 2     **2.** 0     **3.** $-\dfrac{3}{x}$     **4.** $x^2 + 3$     **5.** $0, \pm 4$

**6.** $\frac{1}{2}, 1$     **7.** All real numbers except $x = 4$

**8.** All real numbers except $x = 4, 5$     **9.** $t \le \frac{5}{3}$

**10.** All real numbers

---

**1.** Domain: $[1, \infty)$; range: $[0, \infty)$

**3.** Domain: $(-\infty, -2] \cup [2, \infty)$; range: $[0, \infty)$

**5.** Domain: $[-5, 5]$; range: $[0, 5]$

**7.** Domain: $(-\infty, \infty)$; range: $(-\infty, \infty)$

**9.** Function    **11.** Not a function    **13.** Function

**15.** Increasing on $(-\infty, \infty)$; no change

**17.** Increasing on $(-\infty, 0)$ and $(2, \infty)$, decreasing on $(0, 2)$; behavior changes at $(0, 0)$ and at $(2, 4)$

**19.** Increasing on $(-1, 0)$ and $(1, \infty)$, decreasing on $(-\infty, -1)$ and $(0, 1)$; behavior changes at $(-1, -3)$, $(0, 0)$, and $(1, -3)$

**21.** Increasing on $(-2, \infty)$, decreasing on $(-3, -2)$; behavior changes at $(-2, -2)$

**23.** Even    **25.** Odd    **27.** Neither even nor odd

**29.** Even    **31.** Neither even nor odd

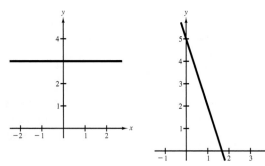

**33.** Odd    **35.** Neither even nor odd

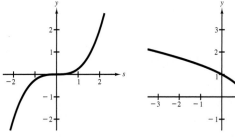

**37.** Neither even nor odd    **39.** Neither even nor odd

**41.**

**43.**

**45.**

**47.**

**49.**

**51.**

**53.**

**55.** Increasing: $(2, \infty)$
Decreasing: $(-\infty, 2)$

**57.** Increasing: $(-\infty, 0), (2, \infty)$
Decreasing: $(0, 2)$

**59.** Decreasing: 1986–1989
  Increasing: 1990–1996

**61.**

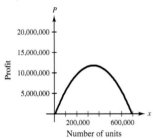

Increasing: 1990–1995
Decreasing: 1996–2000
Maximum revenue: \$2,558,200
Minimum revenue: \$891,000

**63.** Maximum or minimum values may occur at endpoints.

**65.** Approximately 350,000 units

**67.**

**69.**

**71.**

## Section 3.3   *(page 182)*

### Warm Up   *(page 182)*

**1.** 12      **2.** $\dfrac{-2x}{-x-3}$      **3.** $0, \pm\sqrt{10}$      **4.** $\dfrac{4}{3}, -2$

**5.** $f(x) = -2$                    **6.** $f(x) = -x$

**7.** $f(x) = x + 5$              **8.** $f(x) = 2 - x$

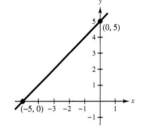

**9.** $f(x) = 3x - 4$          **10.** $f(x) = 9x + 10$

**1.**                            **3.**

**5.**

**7.**

**25.**

**27.**

**9.**

**11.**

**29.**

**31.** Common function:
$y = x^3$
Transformation: shifted
two units to the right
Equation: $y = (x - 2)^3$

**13.**

**15.**

**33.** Common function: $y = x^2$
Transformation: reflection about the $x$-axis
Equation: $y = -x^2$

**35.** Common function: $y = \sqrt{x}$
Transformation: reflection about the $x$-axis and shifted one
unit up
Equation: $y = -\sqrt{x} + 1$

**37.** (a) Vertical shift of two units

**17.**

**19.**

(b) Horizontal shift of two units

**21.**

**23.**

(c) Slope of the function changes

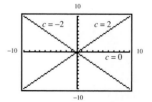

**39.** (a) $g(x) = (x - 1)^2 + 1$

(b) $g(x) = -(x + 1)^2$

**41.** (a) $y = f(x) + 2$    (b) $y = -f(x)$

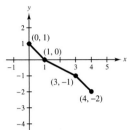

(c) $y = f(x - 2)$    (d) $y = f(x + 3)$

(e) $y = 2f(x)$    (f) $y = f(-x)$

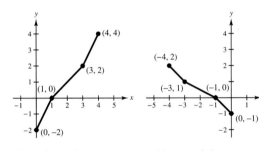

**43.** $g(x) = |x - 2| + 1$    **45.** $g(x) = -3|x|$

**47.** $h(x) = \sqrt{x + 3} + 2$    **49.** $h(x) = -4\sqrt{x}$

**51.** $g(x) = -x^3 + 3x^2 + 1$

**53.** Shifted one unit to the right and two units down

$h(x) = (x - 1)^2 - 2$

**55.** (a)

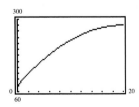

(b) $P(x) = 80 + 20x - 0.5x^2 - 25$

$P(x) = 55 + 20x - 0.5x^2, \quad 0 \le x \le 20$

Shifted 25 units down

(c) $P\left(\dfrac{x}{100}\right) = 80 + \dfrac{x}{5} - 0.00005x^2$

Vertical shrink

**57.** (a), (c), and (e) are odd functions; (b), (d), and (f) are even. Also, (a), (c), and (e) are increasing for all real numbers; (b), (d), and (f) are decreasing for all $x < 0$ and increasing for all $x > 0$.

**59.**

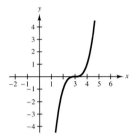

## Mid-Chapter Quiz *(page 185)*

**1.** 16    **2.** 4    **3.** 1    **4.** $2a^2 - a + 1$

**5.** $-2, 0, 2$    **6.** $\frac{4}{3}$    **7.** $P = 5.49x - 85,000$

**8.** All real numbers except 1 and 0.    **9.** $x \le 5$

**10.** $x \ge 1$

**11.** Increasing: $x < 2.5$

Decreasing: $x > 2.5$

**12.** Increasing: $x < 0, x > \frac{5}{3}$

Decreasing: $0 < x < \frac{5}{3}$

**13.** Increasing: $x > 0$
Decreasing: $x < 0$

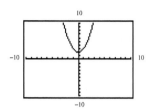

**14.** Increasing: $x > -1$

**15.** $g(x)$ is reflected about the $x$-axis and shifted two units to the right.

**16.** $g(x)$ is shifted one unit to the left.

**17.** $g(x) = (x - 2)^3 - 1$

**18.** $y = (x - 3)^2 - 2$     **19.** $y = -\sqrt{x + 2}$

       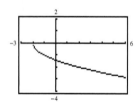

**20.** $P = x(80 - 0.001x) - (300,000 + 25x)$

$P = -0.001x^2 + 55x - 300,000$

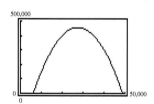

Estimated number of units to produce a profit: $\approx 27,500$

**Section 3.4**   *(page 192)*

**Warm Up**   *(page 192)*

**1.** $\dfrac{1}{x(1 - x)}$     **2.** $-\dfrac{12}{(x + 3)(x - 3)}$     **3.** $\dfrac{3x - 2}{x(x - 2)}$

**4.** $\dfrac{4x - 5}{3(x - 5)}$     **5.** $\sqrt{\dfrac{x - 1}{x + 1}}$     **6.** $\dfrac{x + 1}{x(x + 2)}$

**7.** $5(x - 2)$     **8.** $\dfrac{x + 1}{(x - 2)(x + 3)}$     **9.** $\dfrac{1 + 5x}{3x - 1}$

**10.** $\dfrac{x + 4}{4x}$

**1.**

**3.**

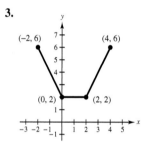

**5.** (a) $2x$   (b) $2$   (c) $x^2 - 1$

(d) $\dfrac{x + 1}{x - 1}$; domain: $(-\infty, 1) \cup (1, \infty)$

**7.** (a) $x^2 - x + 1$   (b) $x^2 + x - 1$   (c) $x^2 - x^3$

(d) $\dfrac{x^2}{1 - x}$; domain: $(-\infty, 1) \cup (1, \infty)$

**9.** (a) $x^2 + \sqrt{1 - x} + 5$   (b) $x^2 - \sqrt{1 - x} + 5$

(c) $x^2\sqrt{1 - x} + 5\sqrt{1 - x}$

(d) $\dfrac{x^2 + 5}{\sqrt{1 - x}}$; domain: $(-\infty, 1)$

**11.** (a) $\dfrac{x + 1}{x^2}$   (b) $\dfrac{x - 1}{x^2}$   (c) $\dfrac{1}{x^3}$

(d) $x, x \neq 0$; domain: $(-\infty, 0) \cup (0, \infty)$

**13.** $9$     **15.** $4t^2 - 2t + 5$     **17.** $0$     **19.** $26$

**21.** $5$     **23.** $\frac{3}{5}$     **25.** (a) $x^2 - 2x + 1$   (b) $x^4$

**27.** (a) $20 - 3x$   (b) $9x + 20$

**29.** (a) $\sqrt{x^2 + 4}$   (b) $x + 4, x \geq -4$

**31.** (a) $x - \frac{8}{3}$   (b) $x - 8$     **33.** (a) $\sqrt[4]{x}$   (b) $\sqrt[4]{x}$

**35.** (a) $|x + 6|$   (b) $|x| + 6$     **37.** (a) $3$   (b) $0$

**39.** (a) $0$   (b) $4$

**41.** (a) $x \geq 0$, or $[0, \infty)$

(b) All real numbers, or $(-\infty, \infty)$

(c) All real numbers, or $(-\infty, \infty)$

**43.** (a) All real numbers except $x = \pm 1$, or
$(-\infty, -1) \cup (-1, 1) \cup (1, \infty)$

(b) All real numbers, or $(-\infty, \infty)$

(c) All real numbers except $x = -2$ and $x = 0$, or
$(-\infty, -2) \cup (-2, 0) \cup (0, \infty)$

**45.** $f(x) = x^2$, $g(x) = 2x + 1$ (Answers will vary.)

**47.** $f(x) = \sqrt[3]{x}$, $g(x) = x^2 - 4$ (Answers will vary.)

**49.** $f(x) = \dfrac{1}{x}$, $g(x) = x + 2$ (Answers will vary.)

**51.** $f(x) = x^2 + 2x$, $g(x) = x + 4$ (Answers will vary.)

**53.** $T = \frac{3}{4}x + \frac{1}{15}x^2$

**55.** Total sales: $R_1 + R_2 = -0.8t^2 + 0.78t + 753.9$, $t = 5, 6,$
7, 8, 9, 10
Total sales have been decreasing.

**57.** (a) $N(T(t)) = 90t^2 + 590$

(b) Approximately 3.18 hours

**59.**

| Year | $P$ | $E$ | $P/E$ |
|------|-----|-----|-------|
| 1989 | $14.41 | $0.98 | 14.7 |
| 1990 | $15.51 | $1.10 | 14.1 |
| 1991 | $16.76 | $1.18 | 14.2 |
| 1992 | $22.10 | $1.30 | 17.0 |
| 1993 | $26.13 | $1.46 | 17.9 |
| 1994 | $28.73 | $1.68 | 17.1 |
| 1995 | $37.43 | $1.97 | 19.0 |
| 1996 | $47.52 | $2.21 | 21.5 |
| 1997 | $48.09 | $2.29 | 21.0 |

**61.** $(A \circ r)(t) = \pi(0.6t)^2 = 0.36\pi t^2$

$A \circ r$ represents the area of the circle at time $t$.

**63.** $(C \circ x)(t) = 2800t + 375$

$C \circ x$ represents the cost of production at $t$ hours.

## Section 3.5    (page 203)

**Warm Up**    (page 203)

**1.** All real numbers    **2.** $[-1, \infty)$

**3.** All real numbers except $x = 0, 2$

**4.** All real numbers except $x = -\frac{5}{3}$    **5.** $x$

**6.** $x$    **7.** $x$    **8.** $x$    **9.** $x = \frac{3}{2}y + 3$

**10.** $x = \dfrac{y^3}{2} + 2$

**1.**

**3.**

**5.**

**7.**

**9.**

**11.** $f$ has an inverse.

**13.** $f$ doesn't have an inverse.    **15.** $g$ has an inverse.

**17.** $h$ doesn't have an inverse.

**19.** $f$ doesn't have an inverse.

**21.** Error: $f^{-1}$ does not mean to take the reciprocal of $f(x)$.

**23.** $f^{-1}(x) = \dfrac{x+3}{2}$   **25.** $f^{-1}(x) = \sqrt[5]{x}$

**27.** $f^{-1}(x) = x^2,\ x \geq 0$   **29.** $f^{-1}(x) = \sqrt{4 - x^2}$,
$\qquad\qquad\qquad\qquad\qquad\qquad 0 \leq x \leq 2$

**31.** $f^{-1}(x) = x^3 + 1$   **33.** $f$ doesn't have an inverse.

**35.** $g^{-1}(x) = 8x$   **37.** $p$ doesn't have an inverse.

**39.** $f^{-1}(x) = \sqrt{x} - 3,\ x \geq 0$   **41.** $h^{-1}(x) = \dfrac{1}{x}$

**43.** $f^{-1}(x) = \dfrac{x^2 - 3}{2},\ x \geq 0$

**45.** $g$ doesn't have an inverse.

**47.** $f^{-1}(x) = -\sqrt{25 - x},\ x \leq 25$

**49.**

| $x$ | 0 | 1 | 2 | 3 | 4 |
|---|---|---|---|---|---|
| $f^{-1}(x)$ | $-2$ | 0 | 1 | 2 | 4 |

**51.** 32   **53.** 600

**55.** $(g^{-1} \circ f^{-1})(x) = \dfrac{x+1}{2}$   **57.** $(f \circ g)^{-1}(x) = \dfrac{x+1}{2}$

**59.** $f^{-1}(t) = \dfrac{141.919 - t}{6.05503}$   **61.** $t = \sqrt{\dfrac{y - 2594}{31.036}}$

The average monthly
bill was \$51 in 1995.

There were 7,000,000,000
transactions in 1992.

**63.** \$1.41

## Review Exercises   *(page 208)*

**1.** $y$ is a function of $x$.   **3.** $y$ is a function of $x$.

**5.** Function. Every element in $A$ is assigned to an element in $B$.

**7.** (a) $-2$   (b) $-11$   (c) $3m - 5$   (d) $3x - 2$

**9.** (a) 5   (b) 8   (c) $5\frac{1}{2}$   (d) $x^2 + 5$

**11.** $-\frac{7}{2}$   **13.** All real numbers   **15.** $[-9, \infty)$

**17.** $[3, 5) \cup (5, \infty)$

**19.** The domain of $h(x)$ is all real numbers except $x = 0$, since division by zero is undefined. The domain of $k(x)$ is all real numbers except $x = -2$ and $x = 2$, since if $x = 2$ or $x = -2$, then $x^2 - 4$ equals zero, and division by zero is undefined. Using a graphing utility and the table feature, $h(0)$ results in an error and $k(-2)$ or $k(2)$ also results in an error.

**21.** (a) $V = x(20 - 2x)(20 - 2x) = 4x(10 - x)^2$

   (b) Domain: $(0, 10)$

   (c)

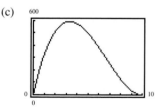

**23.** (a) 16 feet per second   (b) 1.5 seconds

   (c) $-16$ feet per second

**25.** (a) Domain: all real numbers
       Range: $[1, \infty)$

   (b) Decreasing: $(-\infty, 0)$
       Increasing: $(0, \infty)$

   (c) Even

**27.** (a) Domain: all real numbers
       Range: all real numbers

   (b) Decreasing: $\left(0, \frac{8}{3}\right)$
       Increasing: $(-\infty, 0) \cup \left(\frac{8}{3}, \infty\right)$

   (c) Neither

**29.**

**31.**

**33.**

**35.**

**37.**

**39.**

**41.** $y$ is a function of $x$.    **43.** $y$ is not a function of $x$.

**45.** $y$ is a function of $x$.

**47.**

**49.**

**51.**

**53.**

**55.**

**57.**

**59.** Common function: $y = x$
Transformation: shifted three units to the left
Equation: $y = x + 3$

**61.** Common function: $y = x^2$
Transformation: shifted down one unit
Equation: $y = x^2 - 1$

**63.** $(f + g)(x) = 2x - 1$
$(f - g)(x) = 5$
$(fg)(x) = x^2 - x - 6$
$(f/g)(x) = \dfrac{x + 2}{x - 3}$

Domain of $f/g$: all real numbers except $x = 3$, or
$(-\infty, 3) \cup (3, \infty)$

**65.** $(f + g)(x) = \sqrt{x^2 - 4} + 3$
$(f - g)(x) = \sqrt{x^2 - 4} - 3$
$(fg)(x) = 3\sqrt{x^2 - 4}$
$(f/g)(x) = \dfrac{\sqrt{x^2 - 4}}{3}$

Domain of $f/g$: $x \le -2$, $x \ge 2$, or $(-\infty, -2] \cup [2, \infty)$

**67.** 7    **69.** 40    **71.** (a) $x^2 + 6x + 9$   (b) $x^2 + 3$

**73.** (a) $|x - 5| + 2$   (b) $|x| - 3$

**75.** (a) $x^2 - 2x + 1$   (b) $x^2 - 1$

**77.** $f(x) = \dfrac{1}{x^2}$, $g(x) = x - 2$   (Answers will vary.)

**79.** $f(x) = x^2 - 6x + 9$, $g(x) = \sqrt{x}$   (Answers will vary.)

**81.** $R = R_1 + R_2$
$R = 600.52 + 0.22t - 0.3t^2$

Total sales are decreasing.

**83.** (a) $r = \dfrac{x}{2}$ (b) $A = \pi r^2$ (c) $A \circ r(x) = \dfrac{\pi x^2}{4}$

$A \circ r(x)$ is the area of the base of the tank in $x$, where $x$ is the side of the square base.

**85.** $(C \circ x)(t) = 5000t + 900$

$(C \circ x)(t)$ is the cost of producing $x$ units in $t$ hours.

**87.** $f(g)(x) = x = g(f)(x)$,

$f(g(x)) = 3\left(\dfrac{x-5}{3}\right) + 5 = x$

$g(f(x)) = \dfrac{3x + 5 - 5}{3} = x$

so $f$ and $g$ are inverse functions of each other.

**89.** $f(x)$ does not have an inverse.

**91.** $f^{-1}(x) = \dfrac{1}{x}$   **93.** $f^{-1}(x) = \dfrac{3x+5}{2}$

($f$ is its own inverse.)

**95.** (a) $f^{-1}(x) = 2(x + 3) = 2x + 6$

(b)

(c) $f \circ f^{-1}(x) = \frac{1}{2}(2x + 6) - 3$

$= x$

$f^{-1} \circ f(x) = 2\left(\frac{1}{2}x - 3\right) + 6$

$= x$

**97.** (a) $f^{-1}(x) = x^2 - 1, x \geq 0$

(b)

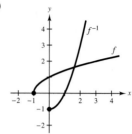

(c) $f \circ f^{-1}(x) = \sqrt{(x^2 - 1) + 1}$

$= x, x \geq 0$

$f^{-1} \circ f(x) = \left(\sqrt{x + 1}\right)^2 - 1$

$= x, x \geq -1$

**99.** (a) $f^{-1}(x) = \sqrt{x}, x \geq 0$

(b)

(c) $f \circ f^{-1}(x) = \left(\sqrt{x}\right)^2 = x, x \geq 0$

$f^{-1} \circ f(x) = \sqrt{x^2} = x, x \geq 0$

**101.** (a) $y = \frac{4}{3}x - \frac{32}{3}$

$x$ is the hourly wage. $y$ is the number of units produced.

(b)

(c) Approximately $15.50

(d) Approximately 19 units

## Chapter Test *(page 212)*

**1.** $(-\infty, \infty)$ **2.** $[7, \infty)$ **3.** $[2, 7) \cup (7, \infty)$

**4.** $(-\infty, -2) \cup (-2, 2) \cup (2, \infty)$

**5.** True. To each value of $x$ there corresponds exactly one value of $y$.

**6.** False. The $\pm$ indicates that to a given value of $x$ there correspond two values of $y$.

**7.** False. The element $-9$ is not included in set $B$.

**8.** (a) Domain: $(-\infty, \infty)$ (b) Increasing: $(0, \infty)$
   Range: $[2, \infty)$    Decreasing: $(-\infty, 0)$

   (c) Even

**9.** (a) Domain: $(-\infty, -2] \cup [2, \infty)$
   Range: $[0, \infty)$

   (b) Decreasing: $(-\infty, 2)$ (c) Even
   Increasing: $(2, \infty)$

**10.** (a) 0  (b) 2  (c) 11

**11.**

**12.**

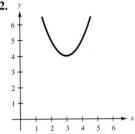

**13.** $f^{-1} = \sqrt[3]{x + 5}$  **14.** $(f + g)(x) = x^2 + 2x + 1$

**15.** $(f - g)(x) = x^2 - 2x + 3$

**16.** $(fg)(x) = 2x^3 - x^2 + 4x - 2$  **17.** $\left(\dfrac{f}{g}\right)(x) = \dfrac{x^2 + 2}{2x - 1}$

**18.** $(f \circ g)(x) = 4x^2 - 4x + 3$  **19.** $(g \circ f)(x) = 2x^2 + 3$

**20.** (a) $C = 3.17x + 75,000$  (b) $R = 5.25x$

   (c) $P = 2.08x - 75,000$

# Chapter 4

## Section 4.1 *(page 222)*

---

### Warm Up *(page 222)*

**1.** $\frac{1}{2}, -6$  **2.** $-\frac{3}{5}, 3$  **3.** $\frac{3}{2}, -1$  **4.** $-10$

**5.** $3 \pm \sqrt{5}$  **6.** $-2 \pm \sqrt{3}$  **7.** $4 \pm \dfrac{\sqrt{14}}{2}$

**8.** $-5 \pm \dfrac{\sqrt{3}}{3}$  **9.** $-\dfrac{3}{2} \pm \dfrac{\sqrt{5}}{2}$  **10.** $-\dfrac{3}{2} \pm \dfrac{\sqrt{21}}{2}$

---

**1.** g **2.** e **3.** c **4.** f **5.** b **6.** a

**7.** h **8.** d **9.** $y = -(x + 2)^2$

**11.** $y = (x - 3)^2 - 9$  **13.** $y = -2(x + 3)^2 + 3$

**15.** Intercept: $(0, 2)$   **17.** Intercepts: $(\pm 4, 0), (0, 16)$
   Vertex: $(0, 2)$    Vertex: $(0, 16)$

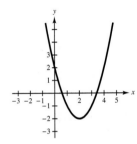

**19.** Intercepts:    **21.** Intercepts:
   $\left(-5 \pm \sqrt{6}, 0\right), (0, 19)$   $\left(2 \pm \sqrt{2}, 0\right), (0, 2)$
   Vertex: $(-5, -6)$   Vertex: $(2, -2)$

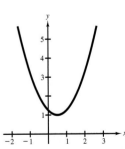

**23.** Intercepts:    **25.** Intercept: $\left(0, \frac{5}{4}\right)$
   $(1, 0), (-3, 0), (0, 3)$   Vertex: $\left(\frac{1}{2}, 1\right)$
   Vertex: $(-1, 4)$

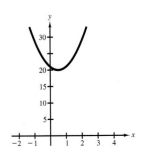

**27.** Intercepts:    **29.** Intercept: $(0, 21)$
   $\left(1 \pm \sqrt{6}, 0\right), (0, 5)$   Vertex: $\left(\frac{1}{2}, 20\right)$
   Vertex: $(1, 6)$

**31.** Intercepts: $\left(8 \pm 4\sqrt{2}, 0\right), (0, 8)$
Vertex: $(8, -8)$

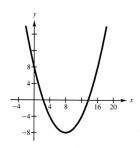

**33.** $y = -\frac{1}{2}(x - 2)^2 - 1$     **35.** $y = \frac{3}{4}(x - 5)^2 + 12$

**37.** $f(x) = x^2 - x - 2$          **39.** $f(x) = x^2 - 10x$
     $g(x) = -x^2 + x + 2$               $g(x) = -x^2 + 10x$
     (Answers are not unique.)          (Answers are not unique.)

**41.** $f(x) = 2x^2 + 7x + 3$
     $g(x) = -2x^2 - 7x - 3$
     (Answers are not unique.)

**43.** $A = 50x - x^2$; 25 feet × 25 feet

**45.** $x = 25$ feet, $y = 33\frac{1}{3}$ feet; 50 feet × $33\frac{1}{3}$ feet

**47.** 45,000 units   **49.** 20 fixtures   **51.** 14 feet

**53.** (a)

(b) $p(x) = -1.28x^2 + 21.63x + 236.49$
   (Models may vary.)

(c)

(d) Approximately 84,492 units

(e) Production costs may be growing faster than revenue, so profit decreases.

**55.** 1995

---

**Math Matters**   *(page 225)*

Both regions have the same area, $9\pi r^2$.

---

**Section 4.2**   *(page 233)*

---

**Warm Up**   *(page 233)*

**1.** $(3x - 2)(4x + 5)$     **2.** $x(5x - 6)^2$
**3.** $z^2(12z + 5)(z + 1)$     **4.** $(y + 5)(y^2 - 5y + 25)$
**5.** $(x + 3)(x + 2)(x - 2)$     **6.** $(x + 2)(x^2 + 3)$
**7.** No real solution     **8.** $3 \pm \sqrt{5}$
**9.** $-\frac{1}{2} \pm \sqrt{3}$     **10.** $\pm 3$

---

**1.** e   **2.** c   **3.** g   **4.** d   **5.** f   **6.** h
**7.** a   **8.** b   **9.** Rises to the left
                      Falls to the right
**11.** Rises to the left     **13.** Rises to the left
     Falls to the right           Rises to the right
**15.** Rises to the left     **17.** Falls to the left
     Rises to the right           Falls to the right
**19.** $\pm 4$   **21.** $-4$   **23.** $1, -2$   **25.** No real zeros
**27.** $2, 0$   **29.** $\pm 1$   **31.** $\pm\sqrt{5}$   **33.** 3

**35.**

**37.**

**39.**

**41.**

**43.**

**45.**

**47.**

**49.**

**51.**

**53.**

**55.** $a_n < 0$            $a_n > 0$

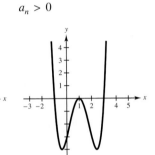

Answers will vary.

**57.** $f(x) = \frac{1}{2}(x^4 - 11x^3 + 28x^2)$

**59.** (a)

260
−9                    10
−20

(b) 1985–86 playing season

(c) No, the model predicts an average weight of 212.4 pounds for 1997–98.

**61.** (200, 320)

**63.** The functions have a common shape. The graphs are not identical.

---

## Section 4.3    *(page 243)*

**Warm Up**    *(page 243)*

**1.** $x^3 - x^2 + 2x + 3$    **2.** $2x^3 + 4x^2 - 6x - 4$

**3.** $x^4 - 2x^3 + 4x^2 - 2x - 7$

**4.** $2x^4 + 12x^3 - 3x^2 - 18x - 5$

**5.** $(x - 3)(x - 1)$    **6.** $2x(2x - 3)(x - 1)$

**7.** $x^3 - 7x^2 + 12x$    **8.** $x^2 + 5x - 6$

**9.** $x^3 + x^2 - 7x - 3$

**10.** $x^4 - 3x^3 - 5x^2 + 9x - 2$

**1.** $3x - 4$    **3.** $x + 3$    **5.** $x^3 + 3x^2 - 1$

**7.** $7 - \dfrac{11}{x + 2}$    **9.** $3x + 5 - \dfrac{2x - 3}{2x^2 + 1}$

**11.** $x^2 - 3x + 2$

**13.** $2x^3 + 4x^2 - 2x - 8 - \dfrac{10x - 7}{x^2 - 2x + 1}$

**15.** $2x^2 - 3x + 5$    **17.** $4x^2 - 9$

**19.** $-x^2 + 10x - 25$    **21.** $5x^2 + 14x + 56 + \dfrac{232}{x - 4}$

**23.** $10x^3 + 10x^2 + 60x + 360 + \dfrac{1360}{x - 6}$

**25.** $2x^4 + 8x^3 + 2x^2 + 8x - 5 - \dfrac{7}{x - 4}$

**27.** $-3x^3 - 6x^2 - 12x - 24 - \dfrac{48}{x - 2}$

**29.** $-x^2 + 3x - 6 + \dfrac{11}{x + 1}$    **31.** $4x^2 + 14x - 30$

**33.** $(x + 2)(x + 2)(x - 4)$ or $(x + 2)^2(x - 4)$

**35.** $(2x - 3)(x + 4)(x - 1)$

**37.** $(x + 2)\big(x + \sqrt{3}\big)\big(x - \sqrt{3}\big)$

**39.** $f(x) = (2x - 1)(x^2 - 4x + 3) - 6$, $f\big(\frac{1}{2}\big) = -6$

**41.** $f(x) = \big(x - \sqrt{2}\big)\big[x^2 + \big(3 + \sqrt{2}\big)x + 3\sqrt{2}\big] - 8$, $f\big(\sqrt{2}\big) = -8$

**43.** (a) $-69$  (b) $-6306$  (c) $-6$  (d) $446$

**45.** (a) $1$  (b) $-267$  (c) $-\frac{11}{3}$  (d) $-3.8$

**47.** (a) $72$  (b) $0$  (c) $37.648$  (d) $30$

**49.** b, $3$, $\dfrac{-1 \pm \sqrt{17}}{2}$    **51.** a, $-1$, $-2 \pm \sqrt{2}$

**53.** $f(x) = 3x^3 - 13x^2 + 4x + 20$

$f(x) = -3x^3 + 13x^2 - 4x - 20$

Infinitely many polynomial functions

**55.** $x^2 + x - 2$ **57.** $3x^2 + 5x + 7$ **59.** $x^2 + 3x$

**61.** $x^2 + 10x + 24$ square feet **63.** \$192,116

## Section 4.4 *(page 255)*

### Warm Up *(page 255)*

**1.** $f(x) = 3x^3 - 8x^2 - 5x + 6$

**2.** $f(x) = 4x^4 - 3x^3 - 16x^2 + 12x$

**3.** $x^4 - 3x^3 + 5 + \dfrac{3}{x + 3}$

**4.** $3x^3 + 15x^2 - 9 - \dfrac{2}{x + (2/3)}$

**5.** $\frac{1}{2}, -3 \pm \sqrt{5}$ **6.** $10, -\frac{2}{3}, -\frac{3}{2}$ **7.** $-\frac{3}{4}, 2 \pm \sqrt{2}$

**8.** $\frac{2}{5}, -\frac{7}{2}, -2$ **9.** $\pm\sqrt{2}, \pm 1$ **10.** $\pm 2, \pm\sqrt{3}$

**1.** Possible: $\pm 1, \pm 2, \pm 4$ **3.** Possible: $\pm\frac{1}{4}, \pm\frac{1}{2}, \pm 1, \pm 2, \pm 4$

Actual: $-1, \pm 2$ Actual: $\pm\frac{1}{2}, \pm 2$

**5.** $1, 2, 3$ **7.** $-1, -10$ **9.** $\frac{1}{2}, -1$ **11.** $\pm 1, \pm\sqrt{2}$

**13.** $-1, 2$ **15.** $-6, \frac{1}{2}, 1$ **17.** $-2, 0, 1$

**19.** (a) $\pm 1, \pm 3, \pm\frac{1}{2}, \pm\frac{3}{2}, \pm\frac{1}{4}, \pm\frac{3}{4}, \pm\frac{1}{8}, \pm\frac{3}{8}, \pm\frac{1}{16}, \pm\frac{3}{16}, \pm\frac{1}{32}, \pm\frac{3}{32}$

(b)  (c) $1, \frac{3}{4}, -\frac{1}{8}$

**21.** $\pm 2, \pm\frac{3}{2}$ **23.** Real zero $\approx 0.7$ **25.** Real zero $\approx 3.3$

**27.** e; $-1.769$ **28.** c; $0.755$ **29.** d; $0.206$

**30.** a; $0.266, 1.175, 2.559$ **31.** f; $2.769$

**32.** b; $-1.675, -0.539, 2.21$

**33.** $-1.164, 1.453$ **35.** $0.529$ **37.** $-1.453, 1.164$

**39.** $-2.177, 1.563$

**41.** (a) $V = x(12 - 2x)(10 - 2x)$

Domain: $0 \le x \le 5$

(b)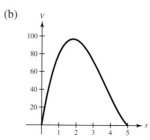

Approximate measurements:

2 in. $\times$ 8 in. $\times$ 6 in. or 1.81 in. $\times$ 8.38 in. $\times$ 6.38 in.

(c) $x \approx 1.628, 2, \approx 7.372$

$x \approx 7.372$ inches is physically impossible because $10 - 2(7.372)$ and $12 - 2(7.372)$ would yield a negative length and width.

**43.** 18 in. $\times$ 18 in. $\times$ 36 in.

**45.** 1.5 in. $\times$ 5 in. $\times$ 7 in. or 1.445 in. $\times$ 5.11 in. $\times$ 7.11 in.

**47.** (a)

(b) $y = 0.45x^3 - 11.03x^2 + 111.78x + 15.97$

(Models may vary.)

(c) 900

Yes, the model seems to be a good fit.

(d) 1994

**49.** 4.5 hours

**51.** \$384,000 or \$461,000

## Mid-Chapter Quiz  *(page 259)*

**1.** Vertex: $(-1, -2)$
Intercepts: $\left(-1 \pm \sqrt{2}, 0\right), (0, -1)$

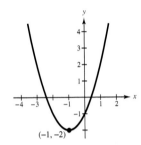

**2.** Vertex: $(0, 16)$
Intercepts: $(\pm 4, 0), (0, 16)$

**3.** Falls to the left    **4.** Falls to the left
Rises to the right            Falls to the right

**5.** $(x + 2)(x - 3)(2x + 1)$

**6.** 13

**7.** $f(x) = (x - 1)(x^3 + x^2 - 4x - 4) + 0; f(1) = 0$

**8.** $f(x) = (x + 3)(x^2 + 2x - 8) + 0; f(-3) = 0$

**9.** $2x^2 + 5x - 12$    **10.** $P = \$2,534,375, x \approx 33.76$

**11.** $\pm\sqrt{5}, -\frac{7}{2}$    **12.** $\pm 3, \pm\frac{1}{2}$    **13.** $1, -\frac{4}{3}$

**14.** $\frac{3}{2}$    **15.** $x \approx -1.485$    **16.** $x \approx -1.414, 1.414$

**17.** $160 = x(12 - 2x)(14 - 2x)$

**18.** 2 in. $\times$ 8 in. $\times$ 10 in. or
$\left(\dfrac{11 - \sqrt{41}}{2}\right)$ in. $\times \left(1 + \sqrt{41}\right)$ in. $\times \left(3 + \sqrt{41}\right)$ in.

**19.** 2 in. $\times$ 8 in. $\times$ 10 in. or 2.298 in. $\times$ 7.404 in. $\times$ 9.404 in.

**20.** $\dfrac{11 + \sqrt{41}}{2} \approx 8.702$ results in $l = -5.404$ and

$w = -3.404$.

## Section 4.5  *(page 268)*

> ### Warm Up  *(page 268)*
>
> **1.** $2\sqrt{3}$    **2.** $10\sqrt{5}$    **3.** $\sqrt{5}$    **4.** $-6\sqrt{3}$
>
> **5.** 12    **6.** 48    **7.** $\dfrac{\sqrt{3}}{3}$    **8.** $\sqrt{2}$
>
> **9.** $-\dfrac{1}{2} \pm \dfrac{\sqrt{5}}{2}$    **10.** $-1 \pm \sqrt{2}$

**1.** $i, -1, -i, 1, i, -1, -i, 1, i, -1, -i, 1, i, -1, -i, 1$

**3.** $a = 3, b = 12$    **5.** $a = 4, b = -3$

**7.** $9 + 4i, 9 - 4i$    **9.** $-3 - 2\sqrt{3}i, -3 + 2\sqrt{3}i$

**11.** $-21, -21$    **13.** $-1 - 6i, -1 + 6i$    **15.** $-5i, 5i$

**17.** $-3, -3$    **19.** $2 + i$    **21.** $5 + 6i$

**23.** $3 - 3\sqrt{2}i$    **25.** $\frac{1}{6} + \frac{7}{6}i$    **27.** $-2\sqrt{6}$    **29.** $-10$

**31.** $5 + i$    **33.** 41    **35.** $21 + 42i$    **37.** 8

**39.** $\left(16 + 4\sqrt{3}\right) + \left(-16\sqrt{2} + 2\sqrt{6}\right)i$    **41.** $-9 + 40i$

**43.** $\frac{3}{5} + \frac{4}{5}i$    **45.** $1 + \frac{1}{2}i$    **47.** $-7 - 6i$    **49.** $\frac{1}{8}i$

**51.** $-\frac{5}{4} - \frac{5}{4}i$    **53.** $\frac{35}{29} + \frac{595}{29}i$    **55.** $-10$

**57.** Error: $(3 - 2i)(3 + 2i) = 9 - 4i^2 = 9 + 4 = 13$
(not $9 - 4 = 5$)

**59.** $1 \pm i$    **61.** $-2 \pm \frac{1}{2}i$    **63.** $-\frac{3}{2}, -\frac{5}{2}$

**65.** $\dfrac{1}{8} \pm \dfrac{\sqrt{11}}{8}i$    **67.**

**69.**                    **71.**

**73.** The complex number 0 is in the Mandelbrot Set because, for $c = 0$, the corresponding Mandelbrot sequence is $0, 0, 0, 0, 0, 0$, which is bounded.

**75.** The complex number $\frac{1}{2}i$ is in the Mandelbrot Set because, for $c = \frac{1}{2}i$, the corresponding Mandelbrot sequence is $\frac{1}{2}i$, $-\frac{1}{4} + \frac{1}{2}i$, $-\frac{3}{16} + \frac{1}{4}i$, $-\frac{7}{256} + \frac{13}{32}i$, $-\frac{10,767}{65,636} + \frac{1957}{4096}i$, $-\frac{864,513,055}{4,294,967,296} + \frac{46,037,845}{134,217,728}i$, which is bounded.

**77.** The complex number 1 is not in the Mandelbrot Set because, for $c = 1$, the corresponding Mandelbrot sequence is 1, 2, 5, 26, 677, 458,330, which is unbounded.

## Section 4.6   *(page 276)*

---
**Warm Up**   *(page 276)*

**1.** $4 - \sqrt{29}i, 4 + \sqrt{29}i$     **2.** $-5 - 12i, -5 + 12i$

**3.** $-1 + 4\sqrt{2}i, -1 - 4\sqrt{2}i$     **4.** $6 + \frac{1}{2}i, 6 - \frac{1}{2}i$

**5.** $-13 + 9i$     **6.** $12 + 16i$     **7.** $26 + 22i$

**8.** 29     **9.** $i$     **10.** $-9 + 46i$
---

**1.** $\pm 4i, (x + 4i)(x - 4i)$

**3.** $\dfrac{5 \pm \sqrt{5}}{2}, \left(x - \dfrac{5 + \sqrt{5}}{2}\right)\left(x - \dfrac{5 - \sqrt{5}}{2}\right)$

**5.** $\pm 4, \pm 4i, (x - 4)(x + 4)(x - 4i)(x + 4i)$

**7.** $0, \pm \sqrt{5}i, x\left(x - \sqrt{5}i\right)\left(x + \sqrt{5}i\right)$

**9.** $2, 2 \pm i, (x - 2)(x - 2 + i)(x - 2 - i)$

**11.** $-5, 4 \pm 3i, (t + 5)(t - 4 + 3i)(t - 4 - 3i)$

**13.** $-10, -7 \pm 5i, (x + 10)(x + 7 - 5i)(x + 7 + 5i)$

**15.** $-\frac{3}{4}, 1 \pm \frac{1}{2}i, (4x + 3)(2x - 2 + i)(2x - 2 - i)$

**17.** $-2, 1 \pm \sqrt{2}i, (x + 2)\left(x - 1 + \sqrt{2}i\right)\left(x - 1 - \sqrt{2}i\right)$

**19.** $-\frac{1}{5}, 1 \pm \sqrt{5}i, (5x + 1)\left(x - 1 + \sqrt{5}i\right)\left(x - 1 - \sqrt{5}i\right)$

**21.** $2, \pm 2i, (x - 2)^2(x + 2i)(x - 2i)$

**23.** $\pm i, \pm 3i, (x + i)(x - i)(x + 3i)(x - 3i)$

**25.** $-4, 3, \pm i, (t + 4)^2(t - 3)(t + i)(t - i)$

**27.** $x^3 + 3x^2 + 36x + 108$     **29.** $x^3 - 5x^2 + 9x - 5$

**31.** $x^5 + 4x^4 + 13x^3 + 52x^2 + 36x + 144$

**33.** $x^4 + 8x^3 + 9x^2 - 10x + 100$

**35.** $16x^4 + 36x^3 + 16x^2 + x - 30$

**37.** (a) $(x^2 - 8)(x^2 + 1)$

(b) $\left(x - 2\sqrt{2}\right)\left(x + 2\sqrt{2}\right)(x^2 + 1)$

(c) $\left(x - 2\sqrt{2}\right)\left(x + 2\sqrt{2}\right)(x - i)(x + i)$

**39.** (a) $(x^2 - 2x - 2)(x^2 - 2x + 3)$

(b) $\left(x - 1 + \sqrt{3}\right)\left(x - 1 - \sqrt{3}\right)(x^2 - 2x + 3)$

(c) $\left(x - 1 + \sqrt{3}\right)\left(x - 1 - \sqrt{3}\right)$
$\left(x - 1 + \sqrt{2}i\right)\left(x - 1 - \sqrt{2}i\right)$

**41.** $\pm 4i, \frac{5}{3}$     **43.** $\pm 6i, 1$     **45.** $-3 \pm i, \frac{1}{4}$

**47.** $1, 2, -3 \pm \sqrt{2}i$     **49.** $\dfrac{3}{4}, \dfrac{1}{2} \pm \dfrac{\sqrt{5}}{2}i$

**51.** (Answers will vary.)
$f(x) = x^4 - 7x^3 + 17x^2 - 17x + 6$
4 real zeros: 1, 1, 2, 3

$g(x) = x^4 - 3x^3 + 3x^2 - 3x + 2$
2 real zeros: 1, 2; 2 complex zeros: $\pm i$

$h(x) = x^4 + 5x^2 + 4$
4 complex zeros: $\pm i, \pm 4i$

Similarities: $f, g,$ and $h$ all rise to the left and rise to the right, and are similar in shape.

Differences: differences in the number of intersections with the $x$-axis.

**53.** Solving the equation for $P = 9,000,000$ and graphing shows no real solutions.

**55.** No. The conjugate pair statement specifies polynomials with *real* coefficients. $f(x)$ has imaginary coefficients.

**57.** No. $\left(x + \sqrt{3}\right)$ and $\left(x - \sqrt{3}\right)$ are not rational factors.

## Section 4.7   *(page 286)*

---
**Warm Up**   *(page 286)*

**1.** $x(x - 4)$     **2.** $2x(x^2 - 3)$     **3.** $(x - 5)(x + 2)$

**4.** $(x - 5)(x - 2)$     **5.** $x(x + 1)(x + 3)$

**6.** $(x^2 - 2)(x - 4)$

**7.**      **8.**

**9.**     **10.**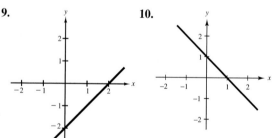
---

**1.** Domain: all $x \neq -2$    **3.** Domain: all $x \neq 3$

**5.** Domain: all reals    **7.** Domain: all reals    **9.** f

**10.** e    **11.** a    **12.** b    **13.** c    **14.** d

**15.** $g(x)$ shifts down 2 units.    **17.** $g(x)$ is a reflection about the x-axis.

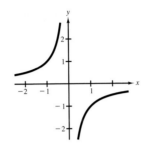

**19.** $g(x)$ shifts up 3 units.    **21.** $g(x)$ is a reflection about the x-axis.

**23.** $g(x)$ shifts up 5 units.    **25.** $g(x)$ is a reflection about the x-axis.

**27.**

**29.**

**31.**

**33.**

**35.**

**37.**

**39.**

**41.**

**43.**

**45.**

**47.**

**49.**

**51.** Asymptotes: none.
Graph is equivalent to the graph of $y = x - 4$, $x \neq 3$.

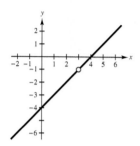

**53.** Asymptotes: vertical, $x = -1$; slant, $y = x - 1$

**55.** $f(x) = \dfrac{1}{x^2 + 1}$

(Answers will vary.)

**57.** $f(x) = \dfrac{x}{x^2 - 2x - 3}$

(Answers will vary.)

**59.** (a) $176 million
(b) $528 million
(c) $1584 million
(d) No. The model would generate a zero in the denominator.

**61.** (a) 167, 250, 400    (b) 750

**63.** (a)

(b)

| $x$ | 1000 | 10,000 | 100,000 |
|---|---|---|---|
| $\overline{C}$ | $150.25 | $15.25 | $1.75 |

Eventually, the average cost per unit will approach the horizontal asymptote of $0.25.

**65.** (a)

| $n$ | 1 | 2 | 3 | 4 | 5 |
|---|---|---|---|---|---|
| $P$ | 0.50 | 0.74 | 0.82 | 0.86 | 0.89 |

| $n$ | 6 | 7 | 8 | 9 | 10 |
|---|---|---|---|---|---|
| $P$ | 0.91 | 0.92 | 0.93 | 0.94 | 0.95 |

(b) 100%

**67.** (a)

(b)

| $x$ | 1000 | 10,000 | 100,000 |
|---|---|---|---|
| $\overline{C}$ | $355 | $40 | $8.5 |

Eventually, the average recycling cost will approach the horizontal asymptote of $5.

**69.** The sale of long-playing albums dropped after cassette tapes and compact discs were introduced.

**71.** (a) $\approx 1.56$ hours

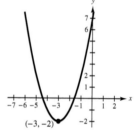

(b) Yes, the function has a horizontal asymptote of $y = 0$, which means that eventually the concentration of the medication will be untraceable.

## Review Exercises    *(page 292)*

**1.**

Vertex: $(-3, -2)$
Intercepts: $(0, 7), \left(-3 + \sqrt{2}, 0\right)$

**3.** Vertex: $(2, -1)$

Intercepts:

$(0, 11)$, $\left(\dfrac{6 \pm \sqrt{3}}{3}, 0\right)$

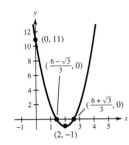

**5.** $f(x) = -3(x + 4)^2 - 2$

**7.** $A(x) = -(x - 50)^2 + 2500$, 50 feet × 50 feet

**9.** (a)

(b) $\approx 363$ to $364$ units

(c) Calculate the vertex from the equation

$\left(\dfrac{4000}{11}, \dfrac{1{,}540{,}000}{121}\right)$.

The cost is minimum when $x = \frac{4000}{11} \approx 364$ units.

**11.** (a)

(b) $d(x) = -0.077x^2 + 6.590x + 2.353$
(Models may vary.)

(c)

(d) $\approx 42.8°$

**13.** Falls to the left      **15.** Rises to the left
Rises to the right            Falls to the right

**17.** $\pm 5$     **19.** $0, 2, 5$     **21.** $x^2 - 3x + 1 - \dfrac{1}{2x + 1}$

**23.** $x^2 + 11x + 24$     **25.** $x^2 - 3x + 3$

**27.** $(x - 5)(x - 2)(x + 3)$     **29.** (a) $-10$   (b) $11$

**31.** $x^2 + 11x + 28$ square feet

**33.** $\pm 1, \pm 3, \pm 5, \pm 15, \pm\frac{1}{2}, \pm\frac{3}{2}, \pm\frac{5}{2}, \pm\frac{15}{2}, \pm\frac{1}{4}, \pm\frac{3}{4}, \pm\frac{5}{4}, \pm\frac{15}{4}$

From the graph: $x \approx 2.357$

**35.** $x \approx -2.3$     **37.** $-3, -1, 2$     **39.** $\pm 2, \pm\sqrt{5}$

**41.** $-2, 1, \pm\dfrac{\sqrt{3}}{3}$     **43.** $-1.321, -0.283, 1.604$

**45.** $\$212{,}000$ or $\$509{,}000$

**47.** $4\sqrt{2}i, -4\sqrt{2}i$     **49.** $-3 + 4i, -3 - 4i$

**51.** $5 + i$     **53.** $11 + 9\sqrt{3}i$     **55.** $89$     **57.** $-10 - 8i$

**59.** $-7 + 24i$     **61.** $3 - 2i$     **63.** $-3 - 4i$     **65.** $10$

**67.** $\dfrac{1 \pm \sqrt{23}i}{4}$     **69.** $\dfrac{-11 \pm \sqrt{73}}{8}$

**71.**

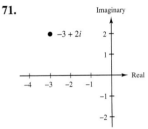

**73.** $(x - 3)(x + 3)(x - 3i)(x + 3i)$

**75.** $(t + 5)\left(t - \sqrt{3}i\right)\left(t + \sqrt{3}i\right)$

**77.** $(x - 2)(2x - 3i)(2x + 3i)$     **79.** $x^3 - 3x^2 + 16x - 48$

**81.** (a) $(x^2 + 8)(x^2 - 3)$

(b) $(x^2 + 8)\left(x - \sqrt{3}\right)\left(x + \sqrt{3}\right)$

(c) $\left(x + 2\sqrt{2}i\right)\left(x - 2\sqrt{2}i\right)\left(x - \sqrt{3}\right)\left(x + \sqrt{3}\right)$

**83.** $\pm 4i, \frac{1}{4}$     **85.** $-1 \pm 3i, -1, -4$

**87.** Domain: all $x \neq 6$
Vertical asymptote: $x = 6$
Horizontal asymptote: $y = 0$

**89.** Domain: all $x \neq \pm 2$
Vertical asymptotes: $x = -2, x = 2$
Horizontal asymptote: $y = 1$

**91.**

**93.**

**95.**

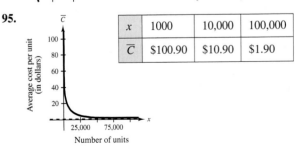

| $x$ | 1000 | 10,000 | 100,000 |
|---|---|---|---|
| $\overline{C}$ | $\$100.90$ | $\$10.90$ | $\$1.90$ |

**97.**

| $n$ | 1 | 2 | 3 | 4 | 5 |
|---|---|---|---|---|---|
| $P$ | 0.5 | 0.737 | 0.821 | 0.865 | 0.891 |

| $n$ | 6 | 7 | 8 | 9 | 10 |
|---|---|---|---|---|---|
| $P$ | 0.909 | 0.922 | 0.932 | 0.939 | 0.945 |

**99.** (a)

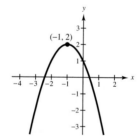

Number of pounds of waste

(b)

| $x$ | 1000 | 10,000 | 100,000 |
|---|---|---|---|
| $\overline{C}$ | $456 | $51 | $10.50 |

Eventually, the average recycling cost will approach the horizontal asymptote of $6.

## Chapter Test    *(page 296)*

**1.** Vertex: $(-1, 2)$
Intercepts: $\left(-1 \pm \sqrt{2}, 0\right), (0, 1)$
Line of symmetry: $x = -1$

**2.** (a) Falls to the left    (b) Rises to the left
     Rises to the right        Rises to the right

**3.** $\pm 1, \pm 5, \pm \frac{1}{2}, \pm \frac{5}{2}$
$f\left(\frac{5}{2}\right) = 0$
$f(x) = (2x - 5)(x - i)(x + i)$

**4.** $x^2 + 7x + 12$    **5.** $16 - 3i$    **6.** $7 - 9i$

**7.** $27 - 4\sqrt{3}i$    **8.** $23 - 14i$    **9.** $-5 - 12i$

**10.** $21 + 20i$    **11.** $i$    **12.** $-2 - 5i$

**13.** $x = \dfrac{-5 \pm \sqrt{3}i}{2}$    **14.** $x = \dfrac{5 \pm 3\sqrt{7}i}{4}$

**15.** $\pm \sqrt{5}i, -2$    **16.** $x^4 - 7x^3 + 19x^2 - 63x + 90$

**17.**

Intercept: $(0, 0)$
Vertical asymptote:
$x = -1$
Horizontal asymptote:
$y = 1$
Domain: all $x \neq -1$

**18.** $x = 2500$ units

**19.**

| $x$ | 10,000 | 100,000 | 1,000,000 |
|---|---|---|---|
| $\overline{C}$ | $50 | $9.50 | $5.45 |

Conclusion: High volume yields lower average cost of recycling.

**20.** $200,000 $(x = 20)$ and $777,250 $(x = 77.725)$ both result in profits of $222,000. The company should spend $200,000, which is the smaller amount.

## Chapter 5
### Section 5.1    *(page 306)*

---

**Warm Up**    *(page 306)*

**1.** $5^x$    **2.** $3^{2x}$    **3.** $4^{3x}$    **4.** $10^x$    **5.** $4^{2x}$

**6.** $4^{10x}$    **7.** $\left(\frac{3}{2}\right)^x$    **8.** $4^{3x}$    **9.** $2^{-x}$    **10.** $16^{x/4}$

---

**1.** 3.463    **3.** 95.946    **5.** 0.079    **7.** 54.598

**9.** 1.948    **11.** g    **12.** e    **13.** b    **14.** h    **15.** d

**16.** a    **17.** f    **18.** c

**19.**

**21.**

**23.**

**25.**

**27.**

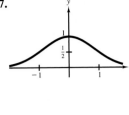

Each graph passes through (0, 1).

**29.**

**31.**

**33.**

**35.**

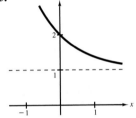

**37.** The graphs have similar shapes and are increasing as $x$ gets larger. The larger the base, the more quickly the $y$-values increase. The graphs have similar shapes and are increasing as $x$ gets larger, but the larger the base, the smaller the $y$-values and the closer the graph is to the $x$-axis.

**39.**

| $n$ | 1 | 2 | 4 |
|---|---|---|---|
| $A$ | $7346.64 | $7401.22 | $7429.74 |

| $n$ | 12 | 365 | Continuous compounding |
|---|---|---|---|
| $A$ | $7449.23 | $7458.80 | $7459.12 |

**41.**

| $n$ | 1 | 2 | 4 |
|---|---|---|---|
| $A$ | $24,115.73 | $25,714.29 | $26,602.23 |

| $n$ | 12 | 365 | Continuous compounding |
|---|---|---|---|
| $A$ | $27,231.38 | $27,547.07 | $27,557.94 |

**43.**

| $t$ | 1 | 10 | 20 |
|---|---|---|---|
| $P$ | $91,393.12 | $40,656.97 | $16,529.89 |

| $t$ | 30 | 40 | 50 |
|---|---|---|---|
| $P$ | $6720.55 | $2732.37 | $1110.90 |

**45.**

| $t$ | 1 | 10 | 20 |
|---|---|---|---|
| $P$ | $90,521.24 | $36,940.70 | $13,646.15 |

| $t$ | 30 | 40 | 50 |
|---|---|---|---|
| $P$ | $5040.98 | $1862.17 | $687.90 |

**47.** $222,822.57

**49.** (a) $472.70

   (b) $298.29

   (c)
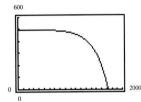

   (d) No, the graph shows only the demand curve, not the cost curve.

**51.** (a) 100  (b) 300  (c) 900

**53.** (a)                                        (b) 0.29 pound

   (c) On the graph, when $P = 2.5$, $t \approx 24,364$.

**55.** (a)

| Year | 1990 | 1992 | 1995 |
|---|---|---|---|
| Estimated shares (in millions) | 42,000 | 55,000 | 90,000 |

   (b)

| Year | 1990 | 1992 | 1995 |
|---|---|---|---|
| Shares (in millions) | 42,567 | 54,213 | 88,790 |

**57.** (a)

| Year | 1980 | 1990 | 1997 |
|------|------|------|------|
| Estimated age (in years) | 22 | 24 | 24.9 |

(b)

| Year | 1980 | 1990 | 1997 |
|------|------|------|------|
| Age (in years) | 22.1 | 24.0 | 24.9 |

**59.** Women tend to marry about 2 years younger than men do. The average ages of both have been rising.

**61.** For a person who will be 21 years old on his or her next birthday,

(a) $20,700.76  (b) $103,503.78  (c) $207,007.55

# Section 5.2   *(page 317)*

## Warm Up   *(page 317)*

**1.** 3    **2.** 0    **3.** $-1$    **4.** 1    **5.** 7.389

**6.** 0.368    **7.** Graph is shifted 2 units to the left.

**8.** Graph is reflected about the $x$-axis.

**9.** Graph is shifted down 1 unit.

**10.** Graph is reflected about the $y$-axis.

**1.** 2    **3.** $-4$    **5.** $\frac{1}{3}$    **7.** 1    **9.** $-4$

**11.** 1    **13.** $-4$    **15.** 5    **17.** $\log_4 256 = 4$

**19.** $\log_{81} 3 = \frac{1}{4}$    **21.** $\log_6 \frac{1}{36} = -2$    **23.** $\ln e = 1$

**25.** $\ln 4 = x$    **27.** 2.538    **29.** 2.161    **31.** 2.913

**33.** 0.549    **35.** $t \approx 23.7$ years

**37.**

**39.**

**41.** d    **42.** e    **43.** a    **44.** c    **45.** f    **46.** b

**47.** Domain: $(0, \infty)$
Asymptote: $x = 0$
$x$-Intercept: $(1, 0)$

**49.** Domain: $(-4, \infty)$
Asymptote: $x = -4$
$x$-Intercept: $(-3, 0)$

**51.** Domain: $(0, \infty)$
Asymptote: $x = 0$
$x$-Intercept: $(1, 0)$

**53.** Domain: $(-\infty, 0)$
Asymptote: $x = 0$
$x$-Intercept: $(-1, 0)$

**55.** (a) 80.0    (b) 68.1    (c) 62.3

**57.** (a)

| $K$ | 1 | 2 | 4 | 6 | 8 | 10 | 12 |
|-----|---|---|---|---|---|----|----|
| $t$ | 0 | 7.3 | 14.6 | 18.9 | 21.9 | 24.2 | 26.2 |

(b)

**59.**

The domain $0 \leq t \leq 12$ covers the period of 12 months or 1 year. The range $0 \leq f(t) \leq 100$ covers the possible scores.

**61.** $\approx 29.7$ years    **63.** 1980: $\approx 29.7$ years
2000: $\approx 35.0$ years
Change of $\approx 5.3$ years

**65.** $\approx 20$ years    **67.** $\approx \$280,178.40$

## Section 5.3   *(page 324)*

### Warm Up   *(page 324)*

**1.** 2   **2.** −5   **3.** −2   **4.** −3   **5.** $e^5$

**6.** $\dfrac{1}{e}$   **7.** $e^6$   **8.** 1   **9.** $x^{-2}$   **10.** $x^{1/2}$

**1.** $\dfrac{\log_{10} 8}{\log_{10} 5}$   **3.** $\dfrac{\log_{10} n}{\log_{10} 3}$   **5.** $\dfrac{\ln 8}{\ln 5}$   **7.** $\dfrac{\ln n}{\ln 3}$

**9.** 2.585   **11.** −0.683   **13.** 2.633   **15.** −1.661

**17.** $\log_3 4 + \log_3 n$   **19.** $\log_5 x - \log_5 25$ or $\log_5 x - 2$

**21.** $4 \log_2 x$   **23.** $\frac{1}{2} \ln z$   **25.** $\ln x + \ln y + \ln z$

**27.** $\frac{1}{2} \ln(a - 1)$   **29.** $2 \ln(z - 1) - \ln z$

**31.** $2 \log_b x - 3 \log_b y$   **33.** $\frac{1}{3}(\ln x - \ln y)$

**35.** $\dfrac{1}{2} \ln y - 5 \ln z$   **37.** $\log_3 5x$   **39.** $\ln\left(\dfrac{8}{x}\right)$

**41.** $\log_{10}(x + 4)^2$   **43.** $\ln\left(\dfrac{x^5}{3^2}\right)$   **45.** $\ln(5x)^{1/3}$

**47.** $\ln\left(\dfrac{x - 2}{x + 2}\right)$   **49.** $\ln \dfrac{3^2}{\sqrt{x^2 + 1}}$

**51.** $\ln\left[\dfrac{x}{(x + 2)(x - 2)}\right]$   **53.** 1.1833   **55.** −0.2084

**57.** 1.0686   **59.** 0.1781   **61.** 1.8957   **63.** −2.7124

**65.** 0.9136   **67.** 3   **69.** 2.4   **71.** 4.5   **73.** $\frac{3}{2}$

**75.** $\frac{1}{2} + \frac{1}{2} \log_7 10$   **77.** $-3 - \log_5 2$   **79.** $6 + \ln 5$

**81.** $y = \sqrt[4]{x}$   **83.** $y = 2.5x^{-1/4}$

**85.** The two graphs are the same.

$\log_a \dfrac{u}{v} = \log_a u - \log_a v$

**87.** Graph the two functions in the same window. Functions and graphs vary.

**89.** $y = 0.07x^{2/3}$

**91.** Let $\log_a u = x$ and $\log_a v = y$.

$a^x = u$ and $a^y = v$

$uv = a^x \cdot a^y = a^{x+y}$

$\log_a uv = x + y$

$\log_a uv = \log_a u + \log_a y$

## Mid-Chapter Quiz   *(page 328)*

**1.**

**2.**

**3.**   **4.**

**5.** $4439.28, $4451.08   **6.** 0.393, 0.632

**7.** (a) 100   (b) 609   (c) 3713

**8.** $591.97   **9.** 2   **10.** 4   **11.** −2   **12.** 0

**13.**

Domain of $f(x)$: $(-\infty, \infty)$
Domain of $g(x)$: $(0, \infty)$

Graphs are reflections about the line $y = x$.

**14.** $\frac{1}{3}(\log x + \log y - \log z)$   **15.** $\ln(x^2 + 3) - 3 \ln x$

**16.** $\ln\left(\dfrac{xy}{3}\right)$   **17.** $\log_{10}\left(\dfrac{1}{64x^3}\right)$   **18.** $\dfrac{3}{2}$   **19.** $\dfrac{6}{5}$

**20.** $y^3 = x$ or $\log_y x = 3$ or $y = \sqrt[3]{x}$

## Section 5.4   *(page 336)*

### Warm Up   *(page 336)*

**1.** $\dfrac{\ln 3}{\ln 2}$   **2.** $1 + \dfrac{2}{\ln 4}$   **3.** $\dfrac{e}{2}$   **4.** $2e$

**5.** $2 \pm i$   **6.** $\frac{1}{2}, 1$   **7.** $x$   **8.** $2x$

**9.** $2x$   **10.** $-x^2$

**1.** 3     **3.** −2     **5.** 2     **7.** 64     **9.** $\frac{1}{10}$

**11.** $x^2$     **13.** $-8 + x^3$     **15.** $x^2$     **17.** $\ln 20 \approx 2.996$

**19.** $\ln 14 \approx 2.639$     **21.** $\ln\left(\frac{23}{3}\right) \approx 2.037$

**23.** $\ln 15 \approx 2.708$     **25.** $\ln\left(\frac{31}{3}\right) \approx 2.335$     **27.** 0

**29.** $\frac{1}{4}\ln 20 \approx 0.749$     **31.** $\ln\left(\frac{5}{28}\right) \approx -1.723$

**33.** $\ln\left(\frac{5}{3}\right) \approx 0.508$     **35.** $\frac{1}{0.06}\ln 3 \approx 18.310$

**37.** $\ln 2 \approx 0.693$     **39.** $\frac{1}{2}\left(-1 + \ln\frac{962}{25}\right) \approx 1.325$

**41.** $\ln 6 \approx 1.792$     **43.** $-\frac{1}{2}\ln(0.05) \approx 1.498$

**45.** $e^7 \approx 1096.633$     **47.** $e^{-8} \approx 0.000335$

**49.** $\frac{1}{2}e \approx 1.359$     **51.** $e^{7/2} \approx 33.115$

**53.** $\frac{1}{4}$     **55.** $\frac{1}{2}e^{-1} \approx 0.184$

**57.** $\pm\sqrt{\dfrac{e^5}{2}} \approx \pm 8.614$     **59.** $2e^5 \approx 296.826$

**61.** $1 + \sqrt{1 + e} \approx 2.928$

**63.** 8.15 years     **65.** 16.28 years

**67.** (a) 1426 units   (b) 1498 units

(c)

**69.** (a) 29.3 years   (b) 39.8 years

**71.**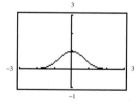

**73.** (a) 69.41 inches

(b) The model is an increasing function with horizontal asymptotes at 0 and 1.

**75.** (a)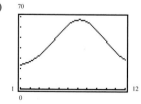

(b) January, February, March, December

(c) The model is bell-shaped and peaks.

**77.** $\approx 35.649$

## Section 5.5   *(page 346)*

**Warm Up**   *(page 346)*

**1.**
**2.**

**3.**
**4.**

**5.**
**6.**

**7.** $\frac{1}{2}\ln\frac{7}{3} \approx 0.424$     **8.** $\frac{1}{5}e^{7/2} \approx 6.623$

**9.** $-5\ln 0.001 \approx 34.539$     **10.** $\frac{1}{2}e^2 \approx 3.695$

| | Initial Investment | Annual % Rate | Time to Double | Amount After 10 Years |
|---|---|---|---|---|
| **1.** | $5000 | 7% | 9.90 yr | $10,068.76 |
| **3.** | $500 | 6.93% | 10 yr | $999.85 |
| **5.** | $1000 | 8.25% | 8.40 yr | $2281.88 |
| **7.** | $6392.79 | 11% | 6.30 yr | $19,205.00 |
| **9.** | $5000 | 8% | 8.66 yr | $11,127.70 |

| | Half-Life (Years) | Initial Quantity | Amount After 1000 years |
|---|---|---|---|
| **11.** | 1620 | 4 g | 2.61 g |
| **13.** | 5730 | 3.95 g | 3.5 g |
| **15.** | 24,360 | 1.65 g | 1.6 g |

**17.** Growth   **19.** Decay   **21.** $\frac{1}{4}\ln 10 \approx 0.5756$

**23.** $\frac{1}{4} \ln \frac{1}{4} \approx -0.3466$    **25.** 2014

**27.** $k \approx 0.0447$
Population in 2010:  61,122

**29.** 3.03 hours to double
4.80 hours to triple

**31.** 95.8%    **33.** (a) $-0.05$    (b) $\approx 36$ days

**35.** (a)

(b) Sales of snowmobiles in millions of dollars increased at a higher rate in 1992.

**37.** (a)

(b) 64.9 inches

**39.** (a)

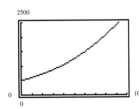

(b) 1252
(c) 7.8 months

**41.** (a) $k \approx 0.1656$    (b) 203 animals

**43.** Competition and lower production costs cause the price to fall over time. Additionally, compact disks appeared in the market. (Answers will vary.)

**45.** (a) 7.91    (b) 7.68    (c) 7.60    (d) 7.30

**47.** (a) 20    (b) 70    (c) 95    (d) 120

**49.** 74.9% decrease    **51.** 7:30 A.M.

**53.**

Model: $P = 39,887(1.102)^t$ or $P = 39,887e^{0.0969t}$
Population in 2000:  $\approx 105,353$
Population in 2002:  $\approx 127,941$

## Review Exercises    *(page 353)*

**1.** d    **2.** f    **3.** a    **4.** b
**5.** c    **6.** e    **7.** h    **8.** g

**9.**

**11.**

**13.**

**15.**

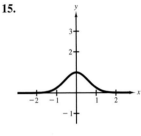

**17.**

| $n$ | 1 | 2 | 4 |
|---|---|---|---|
| $A$ | \$9499.28 | \$9738.91 | \$9867.22 |

| $n$ | 12 | 365 | Continuous compounding |
|---|---|---|---|
| $A$ | \$9956.20 | \$10,000.27 | \$10,001.78 |

**19.**

| $t$ | 1 | 10 | 20 |
|---|---|---|---|
| $P$ | \$184,623.27 | \$89,865.79 | \$40,379.30 |

| $t$ | 30 | 40 | 50 |
|---|---|---|---|
| $P$ | \$18,143.59 | \$8152.44 | \$3663.13 |

**21.** 34th and 35th birthdays    **23.** 5    **25.** 7
**27.** $-\frac{1}{2}$    **29.** $\log_4 64 = 3$
**31.**

**33.** Domain: $(4, \infty)$
Vertical asymptote: $x = 4$
$x$-intercept: $(5, 0)$

**35.** Average score decreased from 85 to about 75.

**37.**

Model: $S = 83.7(0.9568)^t$ or
$S = 83.7e^{-0.0442t}$

**39.** Median age in 1995: 34.18 years
Median age in 2010: 36.35 years

**41.** 2.096    **43.** 2.132    **45.** $\log_{10} x - \log_{10} y$

**47.** $\ln x + \frac{1}{2}\ln(x - 3)$    **49.** $3\log_{10}(y - 5)$

**51.** $\log_4 6$    **53.** $\ln\sqrt{x}$    **55.** $\ln\dfrac{x}{(x - 3)(x + 1)}$

**57.** 0.9208    **59.** 0.2823    **61.** 2    **63.** 3.2

**65.** $\ln 8 \approx 2.079$    **67.** $-\frac{1}{3}\ln\left(\frac{73}{2}\right) \approx -1.199$

**69.** $\frac{1}{3}e^{8.2} \approx 1213.650$    **71.** $\frac{1}{5}e^2 \approx 1.478$

**73.** $3e^2 \approx 22.167$    **75.** (a) 1162 units   (b) 1300 units

**77.** Amount after 1000 years: 10.63 grams

**79.** (a)

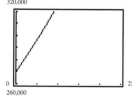

(b) 1995

**81.** $y = x^{4/3}$

**83.** (a) $N = 50(1 - e^{-0.04838t})$   (b) 48 days

**85.** (a) $k \approx 0.3615$   (b) 302 deer

**87.** Yes    **89.** 9.81 hours

## Chapter Test    *(page 357)*

**1.** $y = 2^x$

**2.** $y = 3^{-x}$

**3.** $y = \ln x$

**4.** $y = \log_3(x - 1)$

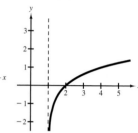

**5.** The fund will be greater than \$100,000 when 16.6 years have elapsed.

**6.** $2\ln x + 3\ln y - \ln z$

**7.** $\log_{10} 3 + \log_{10} x + \log_{10} y + 2\log_{10} z$

**8.** $\ln x + \dfrac{1}{3}\ln(x - 2)$    **9.** $\ln\dfrac{yz^2}{x^3}$    **10.** $\log_{10}\sqrt[3]{x^2 y^2}$

**11.** $\frac{1}{4}\ln 21 \approx 0.761$    **12.** $\ln 6 \approx 1.792$
$\ln 2 \approx 0.693$

**13.** $\frac{1}{4}e^3 \approx 5.021$    **14.** $e^6 - 2 \approx 401.429$

**15.** 87    **16.** 79.8, 76.5

**17.** After 12 months: 70.3
After 18 months: 67.8
Slowly over time, some (but not all) memory fades.

**18.** About 2015. Since the exponent for $e$ is positive, the population of the city is growing.

**19.** 3.4 hours    **20.** 2.98 grams, 0.89 gram

## Cumulative Test: Chapters 3–5
*(page 358)*

**1.** $x^2 + 3x - 4$    **2.** $x^2 - 3x + 6$

**3.** $3x^3 - 5x^2 + 3x - 5$    **4.** $\dfrac{x^2 + 1}{3x - 5}, x \neq \dfrac{5}{3}$

**5.** $9x^2 - 30x + 26$    **6.** $3x^2 - 2$

**7.** Domain: $(-\infty, \infty)$
Range: $[3, \infty)$

**8.** Domain: $(-\infty, 3) \cup (3, \infty)$
Range: $(-\infty, 0) \cup (0, \infty)$

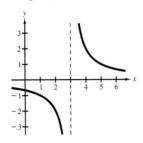

**9.** Domain: $(-\infty, \infty)$
Range: $(0, \infty)$

**10.** Domain: $(1, \infty)$
Range: $(-\infty, \infty)$

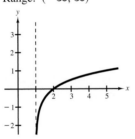

**11.** Domain: $(-\infty, \infty)$
Range: $(-\infty, \infty)$

**12.** $x = 5000$ units
$P = \$25{,}200$

**13.** $38 - 34i$    **14.** $-9 + 40i$

**15.** $\dfrac{1 + 7i}{10}$ or $\dfrac{1}{10} + \dfrac{7}{10}i$    **16.** $\dfrac{5 \pm \sqrt{59}\,i}{6}$

**17.** $-3i, 3i, i, -i$. Since a given zero is $3i$ and $f(x)$ is a polynomial with real coefficients, then $-3i$ is also a zero. Dividing by $3i$ and then by $-3i$ yields a quadratic that can easily be solved.

**18.** $\ln 3 \approx 1.099$    **19.** $e^8 + 3 \approx 2983.958$    **20.** 2006
$\ln 8 \approx 2.079$

# Chapter 6

## Section 6.1    *(page 367)*

### Warm Up    *(page 367)*

**1.**

**2.**

**3.**

**4.**
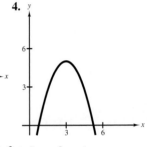

**5.** $x$    **6.** $-37v$    **7.** $2x^2 + 9$    **8.** $-1$

**9.** $x = 6$    **10.** $y = 1$

**1.** (a) No  (b) Yes    **3.** $(2, -3)$    **5.** $(-1, 2), (2, 5)$

**7.** $(-2, -2), (0, 0), (2, 2)$    **9.** $(0, 4), \left(\frac{12}{5}, -\frac{16}{5}\right)$

**11.** $(-1, 0), (1, 0)$    **13.** $(-1, 1)$    **15.** $\left(\frac{1}{2}, 3\right)$

**17.** $(1.5, 0.3)$    **19.** $\left(\frac{20}{3}, \frac{40}{3}\right)$    **21.** $(10, 3)$    **23.** $(1, 2)$

**25.** $\left(\frac{29}{10}, \frac{21}{10}\right), (-2, 0)$    **27.** $(-1, -2), (2, 1)$

**29.** $(0, 1), (\pm 1, 0)$    **31.** $(2, 1)$    **33.** $\left(\frac{1}{2}, \frac{3}{4}\right), (-3, -1)$

**35.** $(1, 4), (4, 7)$    **37.** $(0, 1)$

**39.** No points of intersection    **41.** 192 units

**43.** 233,333 units    **45.** 6400 units

**47.** 2002    **49.** No

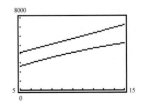

**51.** $12,000 at 8.5%; $13,000 at 8%

**53.** $8333.33  **55.** 1997

## Section 6.2  *(page 378)*

### Warm Up  *(page 378)*

**1.**  **2.**

**3.** $x - y + 4 = 0$  **4.** $5x + 3y - 28 = 0$

**5.** $-\frac{1}{2}$  **6.** $\frac{7}{4}$  **7.** Perpendicular  **8.** Parallel

**9.** Neither parallel nor perpendicular

**10.** Perpendicular

**1.** $(2, 2)$  **3.** $(2, 0)$  **5.** Inconsistent

**7.** $(2a, 3a - 3)$ where $a$ is any real number

**9.** $\left(-\frac{1}{3}, -\frac{2}{3}\right)$  **11.** $(-7, -13)$  **13.** $(4, -5)$

**15.** $(4, -1)$  **17.** $(40, 40)$  **19.** Inconsistent

**21.** $\left(\frac{18}{5}, \frac{3}{5}\right)$  **23.** $\left(\frac{19}{7}, -\frac{2}{7}\right)$

**25.** $\left(a, \frac{5}{6}a - \frac{1}{2}\right)$ where $a$ is any real number  **27.** $\left(\frac{90}{31}, -\frac{67}{31}\right)$

**29.** $\left(-\frac{6}{35}, \frac{43}{35}\right)$  **31.** No, lines intersect at $(79{,}400, 398)$.

**33.** 550 miles per hour, 50 miles per hour

**35.** $6\frac{2}{3}$ gallons at 20%  **37.** $8000 at 12%
$3\frac{1}{3}$ gallons at 50%  $4000 at 10.5%

**39.** 231 adult tickets
519 children's tickets

**41.** $x = 309{,}091$ units, $p = \$25.09$

**43.** $x = 2{,}000{,}000$ units, $p = \$100.00$

**45.** (a) $x = 3$ corresponds to 1993.
Exports: $y = 55.12t + 298.76$
Imports: $y = 71.22t + 374.21$
(b) No

**47.** 1800 packages, 600 packages  **49.** $y = 0.97x + 2.1$

**51.** $y = 0.318x + 4.061$

**53.** Models may vary.

(a) $y = 4.15t + 73.72$, $t = 2$ corresponds to 1992.

(b) $106.92 billion

(c) Same model: $y = 4.15t + 73.72$, $t = 2$ corresponds to 1992.

## Section 6.3  *(page 390)*

### Warm Up  *(page 390)*

**1.** $(15, 10)$  **2.** $\left(-2, -\frac{8}{3}\right)$  **3.** $(28, 4)$

**4.** $(4, 3)$  **5.** Not a solution  **6.** Not a solution

**7.** Solution  **8.** Solution  **9.** $5a + 2$

**10.** $a + 13$

**1.** $(4, -2, -2)$  **3.** $(2, -3, -2)$  **5.** $(3, -4, 2)$

**7.** Inconsistent  **9.** $\left(1, -\frac{3}{2}, \frac{1}{2}\right)$

**11.** $(-3a + 10, 5a - 7, a)$

**13.** $\left(-4a + 13, -\frac{15}{2}a + \frac{45}{2}, a\right)$  **15.** $(-a, 2a - 1, a)$

**17.** $\left(-\frac{3}{2}a + \frac{1}{2}, -\frac{3}{2}a + 1, a\right)$  **19.** $(1, 1, 1, 1)$

**21.** Inconsistent  **23.** $(-3, 4, 2)$  **25.** $\left(-\frac{3}{5}a, \frac{4}{5}a, a\right)$

**27.** The solution $(-a, 2a - 1, a)$ is generated by solving for $x$ and $y$ in terms of $z$. The solution $(b, -2b - 1, -b)$ is generated by solving for $y$ and $z$ in terms of $x$. Both yield equivalent solutions.

**29.** $a = 1$: $(3, 4, 1)$
$a = 0$: $(0, 5, 0)$
$a = 2$: $(6, 3, 2)$
Answers may vary.

**31.** $y = 2x^2 + 3x - 4$  **33.** $x^2 + y^2 - 4x = 0$

**35.** $4000 at 5%  **37.** $75,000 at 10%
$5000 at 6%  $400,000 at 9%
$7000 at 7%  $300,000 at 8%

**39.** 25 pounds of grade A paper
15 pounds of grade B paper
10 pounds of grade C paper

**41.** 10 gallons of spray X  **43.** One possible solution:
5 gallons of spray Y
12 gallons of spray Z  $C = -\dfrac{a}{2} + 250{,}000$

$M = \dfrac{a}{2} + 125{,}000$

$B = -a + 125{,}000$

$G = a$

**45.** $y = 0.079x^2 + 0.63x + 2.943$

**47.** $y = -0.207x^2 - 0.887x + 5.083$

**49.** Models may vary.

$y = 0.0714285714x^2 + 0.686x + 5.077142857;$

10.29 million personal computers in 2000

**51.** (a)

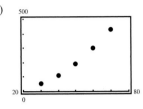

(b) $y = 0.115x^2 - 2.17x + 15.8$

(c) $y = 0.115x^2 - 2.17x + 15.8$

Models may vary.

(d) Models are the same. Estimated stopping distance when $x = 75$ mph: $\approx 500$ feet

**53.** $s = -16t^2 - 20t + 220$

$(g = -32, v_0 = -20, s_0 = 220)$

## Mid-Chapter Quiz   *(page 395)*

**1.** Answers may vary.

$3x + y = 11$
$x - 2y = -1$

**2.** Answers may vary.

$x + 2y + z = 4$
$2x - y - z = 9$
$3y + 2z = 0$

**3.** $(1, 3)$    **4.** $(0.927, 2.854), (-1.727, -2.454)$

**5.** 1342 units    **6.** 500,000 units    **7.** $(2, -1)$

**8.** $\left(1, \frac{3}{2}\right)$    **9.** $x = 18,333$ units

$p = \$26.67$

**10.** $x = 71,429$ units    **11.** $(1, -2, 3)$

$p = \$80.71$

**12.** Answers may vary. $(a + 6, a + 6, a)$, $a$ is any real number

**13.** Inconsistent

**14.** Answers may vary.

$a = 1: (1, -1, 3)$
$a = 0: (0, -2, 0)$
$a = 3: (3, 1, 9)$

**15.** Answers may vary.

$a = 0: (0, 5, 0)$
$a = 2: (4, 7, 2)$
$a = 1: (2, 6, 1)$

**16.** $s = -16t^2 + 10t + 150$    **17.** Infinitely many

$(g = -32, v_0 = 10, s_0 = 150)$

**18.** $3\frac{1}{3}$ gallons of 25% and $6\frac{2}{3}$ gallons of 40%

**19.** True    **20.** False

## Section 6.4   *(page 403)*

**Warm Up**   *(page 403)*

**1.** Line    **2.** Line    **3.** Parabola    **4.** Parabola

**5.** Circle    **6.** Ellipse    **7.** $(1, 1)$    **8.** $(2, 0)$

**9.** $(2, 1), \left(-\frac{5}{2}, -\frac{5}{4}\right)$    **10.** $(2, 3), (3, 2)$

**1.** d    **2.** b    **3.** a    **4.** c    **5.** f    **6.** e

**7.**

**9.**

**11.**

**13.**

**15.**

**17.**

**19.**

**21.**

**23.**

**25.**

**27.**

**29.**

**31.**

**33.**

**35.**

**37.**

**39.**

**41.** $1 \leq x \leq 5, 1 \leq y \leq 5$

**43.** $y \leq \frac{5}{3}x, y \geq 0, y \leq -\frac{5}{4}x + \frac{35}{4}$

**45.** $x^2 + y^2 \leq 16, x \geq 0, y \geq 0$

**47.** $x + \frac{3}{2}y \leq 12, \frac{4}{3}x + \frac{3}{2}y \leq 15, x \geq 0, y \geq 0$

**49.** (a) $x + y \leq 20,000$

   $x \geq 5000$

   $y \geq 5000, y \geq 2x$

(b)

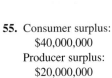

**51.** (a) $20x + 10y \geq 300$

   $15x + 10y \geq 150$

   $10x + 20y \geq 200$

   $x \geq 0, y \geq 0$

(b)

**53.** Consumer surplus:
   $4,777,001.41
   Producer surplus:
   $477,545.60

**55.** Consumer surplus:
   $40,000,000
   Producer surplus:
   $20,000,000

**57.** (a)

(b) 49,000

## Section 6.5 *(page 413)*

### Warm Up *(page 413)*

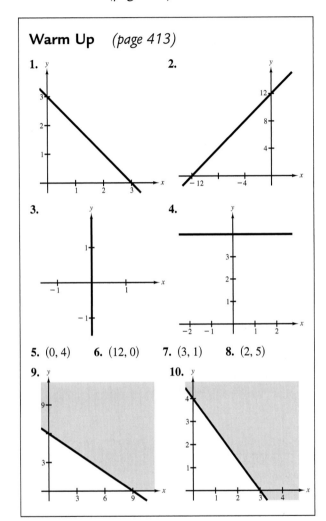

1.

2.

3.

4.

5. $(0, 4)$    6. $(12, 0)$    7. $(3, 1)$    8. $(2, 5)$

9.

10.

1. Minimum value at $(0, 0)$: 0
   Maximum value at $(6, 0)$: 36

3. Minimum value at $(0, 0)$: 0
   Maximum value at $(6, 0)$: 48

5. Minimum value at $(0, 0)$: 0
   Maximum value at $(3, 4)$: 17

7. Minimum value at $(0, 0)$: 0
   Maximum value at $(4, 0)$: 20

9. Minimum value at $(0, 0)$: 0
   Maximum value at
   $(5, 0)$ or $(0, 3)$: 30

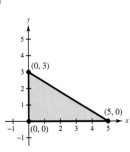

11. Minimum value at $(0, 0)$: 0
    Maximum value at $(5, 0)$: 45
    Same graph as in Answer 9

13. Minimum value at $(5, 3)$: 35
    No maximum value

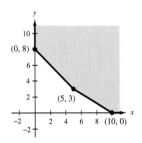

15. Minimum value at $(10, 0)$: 20
    No maximum value
    Same graph as in Answer 13

17. Minimum value at $(24, 8)$: 136
    Maximum value at $(40, 0)$: 200

19. Minimum value at $(36, 0)$: 72
    Maximum value at $(24, 8)$: 88
    Same graph as in Answer 17

21. Maximum value at $(3, 6)$: 12

23. Maximum value at $(0, 10)$: 10

25. Maximum value at $(4, 4)$: 28

27. Maximum value at $(5, 3)$: 71

29. Answers may vary.
    $z = 2x + 11y$

31. Answers may vary.
    $z = 5x + y$

33. \$1100 model:  200 units
    \$1500 model:  50 units

35. Brand X:  3 bags
    Brand Y:  6 bags
    Minimum cost:  \$195

37. Model A:  750 units
    Model B:  1000 units

39. Eight audits
    Eight tax returns

41. Television:  \$540,000
    Newspaper:  \$60,000

43. Type A:  \$62,500
    Type B:  \$187,500

**45.** $z$ is maximum at any point on the line segment connecting the vertices $(2, 0)$ and $\left(\frac{20}{19}, \frac{45}{19}\right)$.

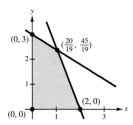

**47.** The constraint $x \leq 10$ is extraneous. The maximum value of $z$ occurs at $(0, 7)$.

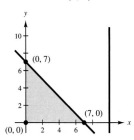

**49.** The constraint $2x + y \leq 4$ is extraneous. The maximum value of $z$ occurs at $(0, 1)$.

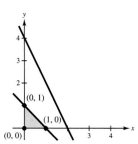

**51.** Model A: 929 units; model B: 77 units

# Review Exercises   *(page 419)*

**1.** $(-2, 4)$     **3.** $(8, -10)$     **5.** $(8, 6), (0, 10)$

**7.** $(1, 9), (1.5, 8.75)$        **9.** 5406 units

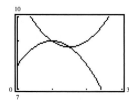

**11.** $15,000 at 6.75%
     $35,000 at 7.5%

**13.** $(3, -5)$

**15.** Infinitely many solutions of the form $\left(a, \frac{5}{8}a - \frac{7}{4}\right)$

**17.** Inconsistent     **19.** Graph is a triangle.
                   Solution: $(-1, 1)$

**15.** Infinitely many solutions     **17.** Inconsistent

**21.** (a) $0.1x + 0.5y = 0.25(12)$
          $x + \quad y = 12$

(b) 7.5 gallons of 10% solution
     4.5 gallons of 50% solution

**23.** $x = 71,429$ units     **25.** $(3, 5, 2)$
     $p = \$22.71$

**27.** $(2, -1, 3)$     **29.** $\left(\frac{1}{5}a + \frac{8}{5}, -\frac{6}{5}a + \frac{42}{5}, a\right)$

**31.** Inconsistent     **33.** $y = 2x^2 + x - 6$

**35.** $x^2 + y^2 - 4x + 2y - 4 = 0$

**37.** $2571.43 at 6%
     $15,428.57 at 7%
     $3571.43 at 8%

**39.** Certificates of deposit: $200,000
     Municipal bonds: $100,000
     Blue-chip stocks: $190,000
     Growth stocks: $10,000

**41.** $y = 0.114x^2 + 0.8x + 1.59$

**43.**

**45.**

**47.**

**49.** $x - 4y \leq -2$
     $2x + 3y \geq \quad 29$
     $3x - \quad y \geq \quad 5$

**51.**
$$x + y \geq 16,000$$
$$20x + 30y \geq 400,000$$
$$x \geq 9,000$$
$$y \geq 4,000$$

**53.** August

**55.** Consumer surplus: $4,000,000
Producer surplus: $6,000,000

**57.** Minimum value at $(0, 0)$: 0
Maximum value at $(3, 3)$: 81

**59.** Minimum value at $(0, 300)$: 18,000
Maximum value at any point on line segment connecting $(0, 500)$ and $(600, 0)$: 30,000

**61.** Minimum value at $(0, 0)$: 0
Maximum value at $(2, 5)$: 50

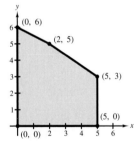

**63.** Minimum value at $(3, 0)$: 30
Maximum value at $(5, 4)$: 94

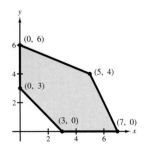

**65.** 50 of the $300 model
50 of the $500 model

**67.** 369 of model A
538 of model B

**69.** $\frac{1}{2}$ of A
$\frac{1}{2}$ of B

**71.** Seven audits
Ten tax returns

## Chapter Test   *(page 424)*

**1.** $(2, 4)$    **2.** $(-2, 5), (3, 0)$    **3.** $(5, -3)$

**4.** $(1.68, 2.38), (-4.18, -26.89)$    **5.** $(3.35, -1.32)$

**6.** $(2, -1, 3)$    **7.** $(2, -3, 4)$

**8.** The system is inconsistent.

**9.** $(50,000, 30)$    **10.** The system is inconsistent.

**11.** The system has infinitely many solutions.

**12.** $y = 2x^2 - 5x - 2$

**13.** $20,000 at 8%; $30,000 at 8.5%

**14.**

**15.**

**16.**

**17.**

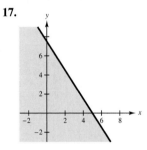

**18.**

**19.** Maximum value at $(3, 3)$: 39
Minimum value at $(0, 0)$: 0

**20.** Model A: 297; model B: 569

# Chapter 7

## Section 7.1    *(page 436)*

---

### Warm Up    *(page 436)*

**1.** $-3$    **2.** $30$    **3.** $6$    **4.** $-\frac{1}{9}$    **5.** Solution

**6.** Not a solution    **7.** $(5, 2)$    **8.** $\left(\frac{12}{5}, -3\right)$

**9.** $(40, 14, 2)$    **10.** $\left(\frac{15}{2}, 4, 1\right)$

---

**1.** $2 \times 3$    **3.** $4 \times 3$    **5.** $4 \times 2$

**7.** Reduced row-echelon form

**9.** Not in row-echelon form

**11.** (a) $\begin{bmatrix} 1 & 1 & 2 \\ 0 & 1 & -9 \\ 2 & -1 & 1 \end{bmatrix}$ (b) $\begin{bmatrix} 1 & 1 & 2 \\ 0 & 1 & -9 \\ 0 & -3 & -3 \end{bmatrix}$

(c) $\begin{bmatrix} 1 & 1 & 2 \\ 0 & 1 & -9 \\ 0 & 0 & -30 \end{bmatrix}$ (d) $\begin{bmatrix} 1 & 1 & 2 \\ 0 & 1 & -9 \\ 0 & 0 & 1 \end{bmatrix}$

**13.** $\begin{bmatrix} 1 & 2 & -1 & 5 \\ 0 & 1 & -1 & 1 \\ 0 & 0 & 1 & -2 \end{bmatrix}$    **15.** $\begin{bmatrix} 1 & -1 & -1 & 1 \\ 0 & 1 & 6 & 3 \\ 0 & 0 & 0 & 0 \end{bmatrix}$

**17.** $\begin{bmatrix} 1 & 0 & 0 \\ 0 & 1 & 0 \\ 0 & 0 & 1 \end{bmatrix}$    **19.** $\begin{bmatrix} 1 & 2 & 0 & 0 \\ 0 & 0 & 1 & 0 \\ 0 & 0 & 0 & 1 \\ 0 & 0 & 0 & 0 \end{bmatrix}$

**21.** $\begin{aligned} 2x + 4y &= 6 \\ -x + 3y &= -8 \end{aligned}$    **23.** $\begin{aligned} x \quad + 2z &= -10 \\ 3y - z &= 5 \\ 4x + 2y \quad &= 3 \end{aligned}$

**25.** $\begin{aligned} x - 5y &= 6 \\ y &= -2 \\ (-4, -2) \end{aligned}$    **27.** $\begin{aligned} x + 3y - z &= 15 \\ y + 4z &= -12 \\ z &= -5 \\ (-14, 8, -5) \end{aligned}$

**29.** $(-4, 6)$    **31.** $(-4, -8, 2)$

**33.** $\begin{bmatrix} 2 & -1 & \vdots & 3 \\ 5 & 7 & \vdots & 12 \end{bmatrix}$    **35.** $\begin{bmatrix} 1 & 10 & -3 & \vdots & 2 \\ 5 & -3 & 4 & \vdots & 0 \\ 2 & 4 & 0 & \vdots & 6 \end{bmatrix}$

**37.** $(3, 2)$    **39.** $(4, -2)$    **41.** $\left(\frac{1}{2}, -\frac{3}{4}\right)$

**43.** Inconsistent    **45.** $\left(-\frac{3}{2}a + \frac{3}{2}, \frac{1}{3}a + \frac{1}{3}, a\right)$

**47.** $(2a + 1, 3a + 2, a)$    **49.** $(5a + 4, -3a + 2, a)$

**51.** $(0, 2 - 4a, a)$    **53.** $\left(\frac{141}{71}, -\frac{10}{71}, \frac{784}{71}, -\frac{312}{71}\right)$

**55.** $(0, 0)$    **57.** $(-2a, a, a)$    **59.** $100{,}000$ at $9\%$
$\phantom{59.}$ $\$250{,}000$ at $10\%$
$\phantom{59.}$ $\$150{,}000$ at $12\%$

**61.** Both are correct. Since there are infinitely many ordered triples that are solutions to this system, a solution can be written many different ways. If $a = 3$, the ordered triple would be $(3, -3, 5)$, which is the same triple as when $b = 5$.

**63.** $y = 54.2897 + 2.04061$; $\$70{,}614.58$

## Section 7.2    *(page 450)*

---

### Warm Up    *(page 450)*

**1.** $-5$    **2.** $-7$    **3.** Not in reduced row-echelon form

**4.** Not in reduced row-echelon form

**5.** $\begin{bmatrix} -5 & 10 & \vdots & 12 \\ 7 & -3 & \vdots & 0 \\ -1 & 7 & \vdots & 25 \end{bmatrix}$

**6.** $\begin{bmatrix} 10 & 15 & -9 & \vdots & 42 \\ 6 & -5 & 0 & \vdots & 0 \end{bmatrix}$

**7.** $(0, 2)$    **8.** $(2 + a, 3 - a, a)$

**9.** $(1 - 2a, a, -1)$    **10.** $(2, -1, -1)$

---

**1.** $x = -3, y = 2$    **3.** $x = -2, y = 5$

**5.** (a) $\begin{bmatrix} 8 & -1 \\ 1 & 7 \end{bmatrix}$ (b) $\begin{bmatrix} 2 & -3 \\ 5 & -5 \end{bmatrix}$

(c) $\begin{bmatrix} 15 & -6 \\ 9 & 3 \end{bmatrix}$ (d) $\begin{bmatrix} 9 & -8 \\ 13 & -9 \end{bmatrix}$

**7.** (a) $\begin{bmatrix} 7 & 3 \\ 1 & 9 \\ -2 & 15 \end{bmatrix}$ (b) $\begin{bmatrix} 5 & -5 \\ 3 & -1 \\ -4 & -5 \end{bmatrix}$

(c) $\begin{bmatrix} 18 & -3 \\ 6 & 12 \\ -9 & 15 \end{bmatrix}$ (d) $\begin{bmatrix} 16 & -11 \\ 8 & 2 \\ -11 & -5 \end{bmatrix}$

**9.** (a) $\begin{bmatrix} 3 & 3 & -2 & 1 \\ -2 & 5 & 7 & -6 \\ 1 & -8 & 11 & 4 \end{bmatrix}$ (b) $\begin{bmatrix} 1 & 1 & 0 & -1 \\ 4 & -3 & -11 & 6 \\ 1 & 6 & -5 & 0 \end{bmatrix}$

(c) $\begin{bmatrix} 6 & 6 & -3 & 0 \\ 3 & 3 & -6 & 0 \\ 3 & -3 & 9 & 6 \end{bmatrix}$ (d) $\begin{bmatrix} 4 & 4 & -1 & -2 \\ 9 & -5 & -24 & 12 \\ 3 & 11 & -7 & 2 \end{bmatrix}$

**11.** (a) $\begin{bmatrix} 0 & 15 \\ 6 & 12 \end{bmatrix}$ (b) $\begin{bmatrix} -2 & 2 \\ 31 & 14 \end{bmatrix}$ (c) $\begin{bmatrix} 9 & 6 \\ 12 & 12 \end{bmatrix}$

**13.** (a) $\begin{bmatrix} 3 & 5 \\ 1 & 6 \end{bmatrix}$ (b) $\begin{bmatrix} 11 & 5 & 0 \\ -7 & -4 & -1 \\ 14 & 8 & 2 \end{bmatrix}$ (c) Not possible

**15.** (a) $[11]$ (b) $\begin{bmatrix} -4 & 2 & 3 \\ 0 & 0 & 0 \\ -20 & 10 & 15 \end{bmatrix}$ (c) Not possible

**17.** $\begin{bmatrix} -7 & 10 & -12 & 17 \\ 6 & 21 & 7 & 15 \\ -3 & 3 & -5 & 6 \end{bmatrix}$   **19.** $\begin{bmatrix} -1 & 19 \\ 4 & -27 \\ 0 & 14 \end{bmatrix}$

**21.** $\begin{bmatrix} 1 & 0 & 0 \\ 0 & 1 & 0 \\ 0 & 0 & \frac{7}{2} \end{bmatrix}$   **23.** $\begin{bmatrix} 60 & 72 \\ -20 & -24 \\ 10 & 12 \\ 60 & 72 \end{bmatrix}$

**25.** $\begin{bmatrix} -6 & -9 \\ -1 & 0 \\ 17 & -10 \end{bmatrix}$   **27.** $\begin{bmatrix} 3 & 3 \\ -\frac{1}{2} & 0 \\ -\frac{13}{2} & \frac{11}{2} \end{bmatrix}$

**29.** $A = \begin{bmatrix} -1 & 1 \\ -2 & 1 \end{bmatrix}, X = \begin{bmatrix} x \\ y \end{bmatrix}, B = \begin{bmatrix} 4 \\ 0 \end{bmatrix}; x = 4, y = 8$

**31.** $A = \begin{bmatrix} 1 & 2 \\ 3 & -1 \end{bmatrix}, X = \begin{bmatrix} x \\ y \end{bmatrix}, B = \begin{bmatrix} 3 \\ 2 \end{bmatrix}; x = 1, y = 1$

**33.** $A = \begin{bmatrix} 1 & -2 & 3 \\ -1 & 3 & -1 \\ 2 & -5 & 5 \end{bmatrix}, X = \begin{bmatrix} x \\ y \\ z \end{bmatrix}, B = \begin{bmatrix} 9 \\ -6 \\ 17 \end{bmatrix};$

$x = 1, y = -1, z = 2$

**35.** $AC = BC = \begin{bmatrix} 12 & -6 & 9 \\ 16 & -8 & 12 \\ 4 & -2 & 3 \end{bmatrix}$

**37.** $\begin{bmatrix} 110 & 99 & 77 & 33 \\ 44 & 22 & 66 & 66 \end{bmatrix}$

**39.**
$$\begin{matrix} \text{Hotel} & \text{Hotel} & \text{Hotel} \\ x & y & z \end{matrix}$$
$$\begin{bmatrix} 86.9 & 93.5 & 119.9 \\ 95.7 & 102.3 & 132 \\ 104.5 & 121 & 141.9 \\ 104.5 & 137.5 & 148.5 \end{bmatrix} \begin{matrix} \text{Single} \\ \text{Double} \\ \text{Triple} \\ \text{Quadruple} \end{matrix} \Bigg\} \text{Occupancy}$$

**41.** (a) $18,300   (b) $21,260

(c)
$$\begin{matrix} \text{Wholesale} & \text{Retail} \end{matrix}$$
$$\begin{bmatrix} \$15,770 & \$18,300 \\ \$26,500 & \$29,250 \\ \$21,260 & \$24,150 \end{bmatrix} \begin{matrix} 1 \\ 2 \\ 3 \end{matrix} \Bigg\} \text{Outlet}$$

This represents the wholesale and retail prices of the inventory at each outlet.

**43.** $\begin{bmatrix} -2 & 3 \\ 1 & -1 \end{bmatrix}$   **45.** $\begin{bmatrix} 0.40 & 0.15 & 0.15 \\ 0.28 & 0.53 & 0.17 \\ 0.32 & 0.32 & 0.68 \end{bmatrix}$

Matrix is unique.

**47.** Player A: $6800
Player B: $4500
Player C: $6400

## Section 7.3   (page 460)

### Warm Up   (page 460)

**1.** $\begin{bmatrix} 4 & 24 \\ 0 & -16 \\ 48 & 8 \end{bmatrix}$   **2.** $\begin{bmatrix} \frac{11}{2} & 5 & 24 \\ \frac{1}{2} & 0 & 8 \\ 0 & 1 & 4 \end{bmatrix}$

**3.** $\begin{bmatrix} -5 & -2 & -13 \\ 4 & -13 & -2 \end{bmatrix}$   **4.** $\begin{bmatrix} -13 & 11 \\ -19 & 21 \end{bmatrix}$

**5.** $\begin{bmatrix} 1 & 0 \\ 0 & 1 \end{bmatrix}$   **6.** $\begin{bmatrix} 6 & 5 \\ 3 & -2 \end{bmatrix}$   **7.** $\begin{bmatrix} 1 & 0 & 0 \\ 0 & 1 & 0 \\ 0 & 0 & 1 \end{bmatrix}$

**8.** $\begin{bmatrix} 1 & 0 & 0 \\ 0 & 1 & 0 \\ 0 & 0 & 1 \end{bmatrix}$   **9.** $\begin{bmatrix} 1 & 0 & \vdots & 3 & -2 \\ 0 & 1 & \vdots & 4 & -3 \end{bmatrix}$

**10.** $\begin{bmatrix} 1 & 0 & 0 & \vdots & -6 & -4 & 3 \\ 0 & 1 & 0 & \vdots & 11 & 6 & -5 \\ 0 & 0 & 1 & \vdots & -2 & -1 & 1 \end{bmatrix}$

**1.** Answers may vary.   **3.** Answers may vary.

**5.** Answers may vary.   **7.** Answers may vary.

**9.** Answers may vary.   **11.** $\begin{bmatrix} \frac{1}{4} & \frac{1}{2} \\ -\frac{1}{4} & -1 \end{bmatrix}$

**13.** Does not exist   **15.** $\begin{bmatrix} 1 & -1 \\ 2 & -1 \end{bmatrix}$   **17.** $\begin{bmatrix} \frac{1}{2} & -\frac{1}{3} \\ \frac{1}{4} & 0 \end{bmatrix}$

**19.** Does not exist   **21.** $\begin{bmatrix} 1 & 1 & -1 \\ -3 & 2 & -1 \\ 3 & -3 & 2 \end{bmatrix}$

**23.** $\begin{bmatrix} -175 & 37 & -13 \\ 95 & -20 & 7 \\ 14 & -3 & 1 \end{bmatrix}$   **25.** $\frac{1}{2}\begin{bmatrix} -3 & 3 & 2 \\ 9 & -7 & -6 \\ -2 & 2 & 2 \end{bmatrix}$

**27.** $\begin{bmatrix} \frac{1}{3} & 0 & 0 \\ 0 & -\frac{1}{2} & 0 \\ 0 & 0 & \frac{1}{4} \end{bmatrix}$   **29.** $\begin{bmatrix} 1 & 0 & 0 \\ -\frac{3}{4} & \frac{1}{4} & 0 \\ \frac{7}{20} & -\frac{1}{4} & \frac{1}{5} \end{bmatrix}$

**31.** Does not exist   **33.** $\begin{bmatrix} -\frac{1}{8} & 0 & 0 & 0 \\ 0 & 1 & 0 & 0 \\ 0 & 0 & \frac{1}{4} & 0 \\ 0 & 0 & 0 & -\frac{1}{5} \end{bmatrix}$

**35.** $\begin{bmatrix} -24 & 7 & 1 & -2 \\ -10 & 3 & 0 & -1 \\ -29 & 7 & 3 & -2 \\ 12 & -3 & -1 & 1 \end{bmatrix}$   **37.** (4, 8)   **39.** (10, 30)

**41.** $(-2, 3)$   **43.** $(2, 0)$   **45.** $\left(-5, -\frac{81}{2}, 48\right)$

**47.** $(2, 1, 0, 0)$

**49.** (a) $\begin{bmatrix} 0.2571 & -0.0286 \\ -0.0286 & 0.0044 \end{bmatrix}$   **51.** $\begin{aligned} 2x + y + 3z &= 16 \\ 4x \quad\;\; - 2z &= -2 \\ 3y + 2z &= 1 \end{aligned}$

$y = 1.01 + 0.24t$

(b) \$5.09 billion

**53.** \$9000 at 8%     **55.** \$23,750 at 8%
\$9500 at 6%          \$4750 at 6%
\$19,000 at 7%        \$9500 at 7%

**57.** 15 computer chips   **59.** 70 computer chips
10 resistors              30 resistors
10 transistors            10 transistors

**61.** 75 muffins
200 cookies
100 brownies

**63.** $AB = \begin{bmatrix} 13 & 4 \\ 1 & 8 \end{bmatrix}$, $BA = \begin{bmatrix} 8 & 1 \\ 4 & 13 \end{bmatrix}$

Row 1 of $AB$ is Row 2 of $BA$ with reversed entries. Row 2 of $AB$ is Row 1 of $BA$ with reversed entries.

**65.** If $k = 3$, then

$\begin{bmatrix} 4 & 3 \\ -2 & 3 \end{bmatrix}^{-1} = \begin{bmatrix} \frac{1}{6} & -\frac{1}{6} \\ \frac{1}{9} & \frac{2}{9} \end{bmatrix}.$

If $k = -\frac{3}{2}$, the matrix is singular.

## Mid-Chapter Quiz   *(page 464)*

**1.** Any matrix with three rows and four columns

**2.** Any matrix with one row and three columns

**3.** $\begin{bmatrix} 3 & 2 & \vdots & -2 \\ 5 & -1 & \vdots & 19 \end{bmatrix}$   **4.** $\begin{bmatrix} 1 & 0 & 3 & \vdots & -5 \\ 1 & 2 & -1 & \vdots & 3 \\ 3 & 0 & 4 & \vdots & 0 \end{bmatrix}$

**5.** $(2.769, -5.154)$   **6.** $(4, -2, -3)$   **7.** $\begin{bmatrix} 2 & -10 \\ 15 & 11 \end{bmatrix}$

**8.** $\begin{bmatrix} -5 & 2 & -13 \\ 5 & 6 & 11 \end{bmatrix}$   **9.** $\begin{bmatrix} 1 & 2 \\ -3 & 2 \end{bmatrix}$

**10.** $\begin{bmatrix} -6 & -2 \\ 3 & -5 \end{bmatrix}$   **11.** $\begin{bmatrix} \frac{2}{5} & \frac{1}{5} \\ -\frac{3}{10} & \frac{1}{10} \end{bmatrix}$   **12.** Not possible

**13.** $\begin{bmatrix} 2 & 2 \\ -3 & 5 \end{bmatrix}$   **14.** $\begin{bmatrix} \frac{3}{2} & -2 \\ 3 & \frac{11}{2} \end{bmatrix}$

**15.** $A = \begin{bmatrix} 1 & -3 \\ -2 & 1 \end{bmatrix}$, $X = \begin{bmatrix} x \\ y \end{bmatrix}$, $B = \begin{bmatrix} 10 \\ -10 \end{bmatrix}$;

$x = 4, y = -2$

**16.** $A = \begin{bmatrix} 2 & -1 & 1 \\ 3 & 0 & -1 \\ 0 & 4 & 3 \end{bmatrix}$, $X = \begin{bmatrix} x \\ y \\ z \end{bmatrix}$, $B = \begin{bmatrix} 3 \\ 15 \\ -1 \end{bmatrix}$;

$x = 4, y = 2, z = -3$

**17.** \$39.40   **18.** \$83.60   **19.** \$114.60

**20.** $\begin{bmatrix} 20.4 & 19 \\ 44.2 & 39.4 \\ 59 & 55.6 \end{bmatrix}$

Total labor for each model listed by plant.

## Section 7.4   *(page 472)*

### Warm Up   *(page 472)*

**1.** $\begin{bmatrix} 3 & 5 \\ 4 & 0 \end{bmatrix}$   **2.** $\begin{bmatrix} -2 & 8 \\ 2 & -4 \end{bmatrix}$

**3.** $\begin{bmatrix} 9 & -12 & 6 \\ 3 & 0 & -3 \\ 0 & 3 & -6 \end{bmatrix}$   **4.** $\begin{bmatrix} 0 & 8 & 12 \\ -4 & 8 & 12 \\ -8 & 4 & -8 \end{bmatrix}$

**5.** $-22$   **6.** $35$   **7.** $-15$   **8.** $-\frac{1}{8}$

**9.** $-45$   **10.** $-16$

**1.** $-2$   **3.** $1$   **5.** $3$   **7.** $0$   **9.** $4$   **11.** $0$

**13.** $-0.002$   **15.** $0$   **17.** $0$   **19.** $-6$

**21.** $-16$   **23.** $120$

**25.** (a) $M_{11} = -5, M_{12} = 2, M_{21} = 4, M_{22} = 3$

(b) $C_{11} = -5, C_{12} = -2, C_{21} = -4, C_{22} = 3$

**27.** (a) $M_{11} = 30, M_{12} = 12, M_{13} = 11, M_{21} = -36,$

$M_{22} = 26, M_{23} = 7, M_{31} = -4,$

$M_{32} = -42, M_{33} = 12$

(b) $C_{11} = 30, C_{12} = -12, C_{13} = 11, C_{21} = 36,$

$C_{22} = 26, C_{23} = -7, C_{31} = -4,$

$C_{32} = 42, C_{33} = 12$

**29.** $-99$   **31.** $-145$   **33.** $170$   **35.** $-58$

**37.** $-30$   **39.** $-108$   **41.** $0$   **43.** $412$

**45.** Matrices will vary. The product of the main diagonals equals $-18$.

**47.** Rows 2 and 4 are identical.

## Section 7.5   *(page 481)*

### Warm Up   *(page 481)*

**1.** 1    **2.** 0    **3.** $-8$    **4.** $x^2$    **5.** 8    **6.** 60

**7.** $\begin{bmatrix} 7 & -3 \\ -2 & 1 \end{bmatrix}$    **8.** $\left[\begin{array}{ccc:ccc} 1 & 0 & 0 & 1 & 2 & -1 \\ 0 & 1 & 0 & -1 & -2 & 2 \\ 0 & 0 & 1 & 2 & 5 & 0 \end{array}\right]$

**9.** $\begin{bmatrix} 0.16 \\ 0.36 \\ 0.48 \end{bmatrix}$    **10.** $\begin{bmatrix} 23 & 54 & 8 \end{bmatrix}$

---

**1.** 11    **3.** 28    **5.** $\frac{33}{8}$    **7.** $\frac{5}{2}$    **9.** 28

**11.** $x = \frac{16}{5}$ or $x = 0$    **13.** 250 square miles

**15.** Collinear    **17.** Not collinear    **19.** Not collinear

**21.** $3x - 5y = 0$    **23.** $x + 3y - 5 = 0$

**25.** $2x + 3y - 8 = 0$    **27.** $x = -3$

**29.** $\begin{bmatrix} 3 & 15 \end{bmatrix}\begin{bmatrix} 13 & 5 \end{bmatrix}\begin{bmatrix} 0 & 8 \end{bmatrix}\begin{bmatrix} 15 & 13 \end{bmatrix}\begin{bmatrix} 5 & 0 \end{bmatrix}\begin{bmatrix} 19 & 15 \end{bmatrix}\begin{bmatrix} 15 & 14 \end{bmatrix}$
48, 81, 28, 51, 24, 40, 54, 95, 5, 10, 64, 113, 57, 100

**31.** $\begin{bmatrix} 3 & 1 & 12 \end{bmatrix}\begin{bmatrix} 12 & 0 & 13 \end{bmatrix}\begin{bmatrix} 5 & 0 & 20 \end{bmatrix}\begin{bmatrix} 15 & 13 & 15 \end{bmatrix}$
$\begin{bmatrix} 18 & 18 & 15 \end{bmatrix}\begin{bmatrix} 23 & 0 & 0 \end{bmatrix}$
$-68, 21, 35, -66, 14, 39, -115, 35, 60, -62, 15, 32,$
$-54, 12, 27, 23, -23, 0$

**33.** $1, -25, -65, 17, 15, -9, -12, -62, -119,$
27, 51, 48, 43, 67, 48, 57, 111, 117

**35.** $-5, -41, -87, 91, 207, 257, 11, -5, -41,$
40, 80, 84, 76, 177, 227

**37.** HAPPY NEW YEAR    **39.** NEWS AT ELEVEN

**41.** SEND MORE MONEY

**43.** MEET ME TONIGHT RON

### Review Exercises   *(page 486)*

**1.** $2 \times 5$    **3.** $\begin{bmatrix} 1 & 3 & 0 & 2 \\ 0 & 1 & 1 & 2 \\ 0 & 0 & 1 & 2 \end{bmatrix}$

**5.** $\begin{bmatrix} 1 & 0 & 0 \\ 0 & 1 & 0 \\ 0 & 0 & 1 \end{bmatrix}$    **7.** $(3, -2)$    **9.** $(3, 2, -1)$

**11.** $(10 - 4a, a, a)$    **13.** Inconsistent

**15.** There are three variables but only two equations.

**17.** $100,000 at 9%
$300,000 at 10%
$200,000 at 12%

**19.** (a) $\begin{bmatrix} 3 & 7 \\ -4 & 4 \end{bmatrix}$ (b) $\begin{bmatrix} -5 & 3 \\ 8 & -2 \end{bmatrix}$

(c) $\begin{bmatrix} -4 & 20 \\ 8 & 4 \end{bmatrix}$ (d) $\begin{bmatrix} -16 & 14 \\ 26 & -5 \end{bmatrix}$

**21.** (a) $\begin{bmatrix} 3 & 4 & 2 & 1 \\ 3 & -5 & 6 & 0 \end{bmatrix}$ (b) $\begin{bmatrix} -1 & 2 & -6 & 11 \\ -3 & 7 & 0 & 4 \end{bmatrix}$

(c) $\begin{bmatrix} 4 & 12 & -8 & 24 \\ 0 & 4 & 12 & 8 \end{bmatrix}$ (d) $\begin{bmatrix} -2 & 9 & -20 & 39 \\ -9 & 22 & 3 & 14 \end{bmatrix}$

**23.** (a) $\begin{bmatrix} 4 \end{bmatrix}$ (b) $\begin{bmatrix} 2 & -6 & 8 \\ -2 & 6 & -8 \\ -1 & 3 & -4 \end{bmatrix}$ (c) Not possible

**25.** (a) $\begin{bmatrix} 12 & 8 & -4 \\ -3 & -10 & 5 \\ 8 & 2 & -1 \end{bmatrix}$ (b) $\begin{bmatrix} 2 & 0 & 4 \\ -4 & -1 & 7 \\ 15 & 2 & 0 \end{bmatrix}$

(c) $\begin{bmatrix} 3 & 2 & 4 \\ 4 & -1 & 2 \\ 5 & 2 & 1 \end{bmatrix}$

**27.** $\begin{bmatrix} 8 \\ 5 \\ -6 \end{bmatrix}$    **29.** $\begin{bmatrix} 1 & 0 & 0 \\ 0 & 1 & 0 \\ 0 & 0 & 1 \end{bmatrix}$    **31.** Not possible

**33.** Answers will vary.

$A = \begin{bmatrix} 2 & 0 \\ 4 & -1 \\ 2 & 3 \end{bmatrix}, B = \begin{bmatrix} 3 & 4 & 0 \\ 6 & -5 & 9 \end{bmatrix}$

$AB = \begin{bmatrix} 6 & 8 & 0 \\ 6 & 21 & -9 \\ 24 & -7 & 27 \end{bmatrix}, BA = \begin{bmatrix} 22 & -4 \\ 10 & 32 \end{bmatrix}$

**35.** $\begin{bmatrix} 5 & -13 \\ -3 & 2 \\ 1 & 0 \end{bmatrix}$    **37.** $\begin{bmatrix} 96 & 84 & 108 & 48 \\ 60 & 36 & 96 & 24 \\ 108 & 72 & 120 & 60 \end{bmatrix}$

**39.** $ST = \begin{bmatrix} 5100 & 7950 \\ 4700 & 7450 \\ 4700 & 7550 \end{bmatrix}$

This represents the wholesale and retail prices of the inventory at each outlet.

**41.** $BA$ is the $3 \times 3$ identity matrix, and so is $AB$.

**43.** $\begin{bmatrix} -5 & 3 \\ 2 & -1 \end{bmatrix}$    **45.** $\begin{bmatrix} -1 & 0 & 0 & 0 \\ 0 & \frac{1}{2} & 0 & 0 \\ 0 & 0 & \frac{1}{4} & 0 \\ 0 & 0 & 0 & \frac{1}{6} \end{bmatrix}$

**47.** $(3, 4)$    **49.** $\left(2, \frac{1}{2}, 3\right)$    **51.** 160 of model A
140 of model B
80 of model C

**53.** $29,000 at 8%
$7000 at 6%
$14,000 at 7%

**55.** 4    **57.** 0    **59.** 44

**61.** $-12$   **63.** 60   **65.** 36   **67.** 10

**69.** Not collinear   **71.** Collinear

**73.** $x + 15y - 38 = 0$

**75.** $[20\ 18][1\ 14][19\ 13][9\ 20][0\ 14][15\ 23]$

94, 132, 44, 59, 77, 109, 78, 107, 42, 56, 99, 137

**77.** SEIZE THE DAY

## Chapter Test   *(page 490)*

**1.** $\begin{bmatrix} 3 & 1 & \vdots & 1 \\ 1 & -1 & \vdots & 7 \end{bmatrix}$   **2.** $\begin{bmatrix} 3 & 4 & 2 & \vdots & 4 \\ 2 & 3 & 0 & \vdots & -2 \\ 0 & 2 & -3 & \vdots & -13 \end{bmatrix}$

**3.** $\left(\frac{7}{3}a - \frac{10}{3}, -\frac{8}{3}a + \frac{29}{3}, a\right)$   **4.** $(-12, -16, -6)$

**5.** $(2, -3, 1)$   **6.** $\begin{bmatrix} 2 & 4 \\ 7 & 13 \end{bmatrix}$   **7.** $\begin{bmatrix} 14 & -1 & 6 \\ 20 & -2 & 10 \end{bmatrix}$

**8.** $\begin{bmatrix} 1 \\ 11 \end{bmatrix}$   **9.** $\begin{bmatrix} 7 & 15 \\ 10 & 22 \end{bmatrix}$   **10.** $\frac{1}{5}\begin{bmatrix} 4 & 1 \\ 3 & 2 \end{bmatrix}$

**11.** $\begin{bmatrix} 1 & 0 \\ 0 & 1 \end{bmatrix}$   **12.** $\frac{1}{5}\begin{bmatrix} -9 & 16 & -6 \\ 6 & -9 & 4 \\ 4 & -6 & 1 \end{bmatrix}$   **13.** 25

**14.** $-23$   **15.** $-20$   **16.** $(2, -2, 3)$

**17.** Matrices will vary.   **18.** Yes, collinear   **19.** $\frac{15}{2}$

$\begin{bmatrix} 2 & -2 \\ -2 & 2 \end{bmatrix} \times \begin{bmatrix} 3 & 3 \\ 3 & 3 \end{bmatrix}$

**20.** $[260,000 \quad 514,000]$

Total expense of stocking each warehouse

# Chapter 8

## Section 8.1   *(page 500)*

**Warm Up**   *(page 500)*

**1.** $\frac{4}{5}$   **2.** $\frac{1}{3}$   **3.** $(2n + 1)(2n - 1)$

**4.** $(2n - 1)(2n - 3)$   **5.** $(n - 1)(n - 2)$

**6.** $(n + 1)(n + 2)$   **7.** $\frac{1}{3}$   **8.** 24

**9.** $\frac{13}{24}$   **10.** $\frac{3}{4}$

**1.** 1, 4, 7, 10, 13   **3.** 2, 4, 8, 16, 32   **5.** $\frac{1}{2}, \frac{1}{4}, \frac{1}{8}, \frac{1}{16}, \frac{1}{32}$

**7.** $1, \frac{2}{3}, \frac{1}{2}, \frac{2}{5}, \frac{1}{3}$   **9.** $-1, \frac{1}{2}, -\frac{1}{3}, \frac{1}{4}, -\frac{1}{5}$   **11.** $\frac{5}{2}, \frac{11}{4}, \frac{23}{8}, \frac{47}{16}, \frac{95}{32}$

**13.** $3, \frac{9}{2}, \frac{9}{2}, \frac{27}{8}, \frac{81}{40}$   **15.** 0, 1, 2, 3, 4

**17.** $-1, \frac{1}{4}, -\frac{1}{9}, \frac{1}{16}, -\frac{1}{25}$   **19.** 3, 6, 9, 12, 15

**21.** $5, \frac{19}{4}, \frac{43}{9}, \frac{77}{16}, \frac{121}{25}$   **23.** 56   **25.** 19,656

**27.** $n + 1, 1, 2$   **29.** $\frac{1}{2n(2n + 1)}$, undefined, $\frac{1}{6}$

**31.** $a_n = 2n - 1$   **33.** $a_n = \frac{2}{n^2}$   **35.** $a_n = \frac{n + 1}{n + 2}$

**37.** $a_n = \frac{(-1)^{n+1}}{2^n}$   **39.** $a_n = 1 + \frac{1}{n}$   **41.** $a_n = \frac{1}{n!}$

**43.** $a_n = (-1)^{n+1}$   **45.** 72   **47.** $-84$   **49.** 40

**51.** 30   **53.** $\frac{29}{40}$   **55.** 14   **57.** 4   **59.** $\sum_{i=1}^{9} \frac{1}{3i}$

**61.** $\sum_{i=1}^{8} \left[2\left(\frac{i}{8}\right) + 3\right]$   **63.** $\sum_{i=1}^{6} (-1)^{i+1}3i$

**65.** $\sum_{i=1}^{20} \frac{(-1)^{i+1}}{i^2}$   **67.** $\sum_{i=1}^{5} \frac{2^i - 1}{2^{i+1}}$

**69.** (a) $A_1 = \$5100.00$, $A_2 = \$5202.00$, $A_3 = \$5306.04$,
$A_4 = \$5412.16$, $A_5 = \$5520.40$, $A_6 = \$5630.81$,
$A_7 = \$5743.43$, $A_8 = \$5858.30$

(b) $A_{40} = \$11,040.20$

(c) No, the balance is $24,377.20.

**71.** (a) $a_1 = 1.0092$   (b)
$a_2 = 0.9848$
$a_3 = 0.9668$
$a_4 = 0.9552$
$a_5 = 0.9500$
$a_6 = 0.9512$
$a_7 = 0.9588$

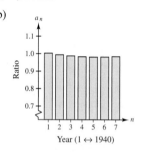

(c) 128.1 million women, 121.9 million men

**73.** $a_0 = \$578.10$
$a_1 = \$639.39$
$a_2 = \$699.27$
$a_3 = \$757.75$
$a_4 = \$814.83$
$a_5 = \$870.50$
$a_6 = \$924.77$
$a_7 = \$977.63$

$n = 12$ represents 2000. So, $a_{12} = \$1220.90$.

**75.** $9.75

## Section 8.2   *(page 509)*

**Warm Up**   *(page 509)*

**1.** 36   **2.** 240   **3.** $\frac{11}{2}$   **4.** $\frac{10}{3}$   **5.** 18

**6.** 4   **7.** 143   **8.** 160   **9.** 430   **10.** 256

**1.** Arithmetic sequence, $d = 4$

**3.** Not an arithmetic sequence

**5.** Arithmetic sequence, $d = -\frac{1}{2}$

**7.** Not an arithmetic sequence

**9.** Arithmetic sequence, $d = 0.4$

**11.** 8, 11, 14, 17, 20; arithmetic sequence, $d = 3$

**13.** 1, 1, 1, 1, 1; arithmetic sequence, $d = 0$

**15.** 2, 8, 24, 64, 160; not an arithmetic sequence

**17.** $\frac{1}{2}, \frac{1}{3}, \frac{1}{4}, \frac{1}{5}, \frac{1}{6}$; not an arithmetic sequence

**19.** $a_n = 5n - 1$     **21.** $a_n = -12n + 62$

**23.** $a_n = 5n + 2$     **25.** $a_n = 4n - 4$     **27.** $a_n = 6n - 4$

**29.** 10, 14, 18, 22, 26     **31.** 5, 14, 23, 32, 41

**33.** 2, 6, 10, 14, 18     **35.** $-2, 2, 6, 10, 14$     **37.** 84

**39.** 12     **41.** 23     **43.** 4000     **45.** 1275

**47.** 25,250     **49.** 44,625     **51.** 520     **53.** 2725

**55.** 10,000     **57.** (a) \$35,000   (b) \$187,500

**59.** 2625 seats     **61.** (a)

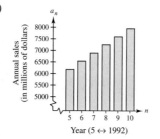

Year (5 ↔ 1992)

(b) \$42,395.10 million

**63.** 1024 feet     **65.** 126 logs     **67.** \$150,000

## Section 8.3   *(page 518)*

---

### Warm Up   *(page 518)*

**1.** $\frac{64}{125}$   **2.** $\frac{9}{16}$   **3.** $\frac{1}{16}$   **4.** 324   **5.** $6n^3$

**6.** $27n^4$   **7.** $4n^3$   **8.** $n^2$   **9.** $\displaystyle\sum_{n=1}^{4} 2(3)^{n-1}$

**10.** $\displaystyle\sum_{n=1}^{4} 3(2)^{n-1}$

---

**1.** Geometric sequence, $r = 4$, $a_n = 4^{n-1}$

**3.** Geometric sequence, $r = -3$, $a_n = 4(-3)^{n-1}$

**5.** Geometric sequence, $r = -\frac{1}{2}$, $a_n = \left(-\frac{1}{2}\right)^{n-1}$

**7.** Not a geometric sequence

**9.** Not a geometric sequence     **11.** 4, 8, 16, 32, 64

**13.** $1, \frac{1}{3}, \frac{1}{9}, \frac{1}{27}, \frac{1}{81}$     **15.** $7, -\frac{7}{5}, \frac{7}{25}, -\frac{7}{125}, \frac{7}{625}$

**17.** $1, e, e^2, e^3, e^4$     **19.** $16\left(\frac{1}{4}\right)^4 = \frac{1}{16}$

**21.** $6\left(-\frac{1}{3}\right)^{11} = -\frac{2}{3^{10}}$     **23.** $100e^8$

**25.** $500(1.02)^{39} \approx 1082.37$     **27.** $16\left(\frac{3}{4}\right)^2 = 9$

**29.** $54\left(-\frac{1}{3}\right)^5 = -\frac{2}{9}$     **31.** 16,368

**33.** $\frac{349,525}{32,768} \approx 10.67$     **35.** 511     **37.** $\approx 29,921.31$

**39.** 2     **41.** $\frac{2}{3}$     **43.** $\frac{16}{3}$

**45.** Sum does not exist; $|r| > 1$

**47.** (a) \$7011.89   (b) \$189.16   (c) \$17,409.45

**49.** \$13,812.43; \$13,833.34     **51.** $\approx$ \$2425.70 million

**53.** $\approx$ \$4910.61 million     **55.** \$5,368,709.11 for 29 days
\$10,737,418.23 for 30 days
\$21,474,836.47 for 31 days

**57.** 126 square inches

## Mid-Chapter Quiz   *(page 522)*

**1.** 3, 5, 7, 9, 11     **2.** $\frac{1}{4}, \frac{2}{9}, \frac{3}{16}, \frac{4}{25}, \frac{5}{36}$     **3.** 20

**4.** 24,360     **5.** $\frac{2}{n^2}$     **6.** $\frac{1}{n+1}$     **7.** $5n - 3$

**8.** $3n + 4$     **9.** Arithmetic, common difference is 3

**10.** Neither; no common ratio, no common difference

**11.** Geometric, common ratio is $\frac{1}{2}$

**12.** Geometric, common ratio is 2     **13.** 25     **14.** 60

**15.** 650     **16.** 126     **17.** $\frac{3}{2}$     **18.** $\frac{25}{4}$

**19.** (a) \$38,500   (b) \$201,000     **20.** \$39,297.40

## Section 8.4   *(page 528)*

---

### Warm Up   *(page 528)*

**1.** $5x^5 + 15x^2$   **2.** $x^3 + 5x^2 - 3x - 15$

**3.** $x^2 + 8x + 16$   **4.** $4x^2 - 12x + 9$   **5.** $\frac{3x^3}{y}$

**6.** $-32z^5$   **7.** 120   **8.** 336   **9.** 720   **10.** 20

---

**1.** 15   **3.** 1   **5.** 15,504   **7.** 4950   **9.** 4950

**11.** $x^4 + 4x^3 + 6x^2 + 4x + 1$     **13.** $x^3 + 6x^2 + 12x + 8$

**15.** $x^4 - 8x^3 + 24x^2 - 32x + 16$

**17.** $x^5 + 5x^4y + 10x^3y^2 + 10x^2y^3 + 5xy^4 + y^5$

**19.** $s^4 + 8s^3 + 24s^2 + 32s + 16$

**21.** $64 + 576s + 2160s^2 + 4320s^3 + 4860s^4$
$+ 2916s^5 + 729s^6$

**23.** $x^5 - 5x^4y + 10x^3y^2 - 10x^2y^3 + 5xy^4 - y^5$

**25.** $1 - 6x + 12x^2 - 8x^3$

**27.** $x^{10} - 5x^8y^2 + 10x^6y^4 - 10x^4y^6 + 5x^2y^8 - y^{10}$

**29.** $\dfrac{1}{x^5} + \dfrac{5y}{x^4} + \dfrac{10y^2}{x^3} + \dfrac{10y^3}{x^2} + \dfrac{5y^4}{x} + y^5$

**31.** $6, 7, 8, 9, n + 1$ **33.** $-4$ **35.** $-7488 - 2816i$

**37.** $1$ **39.** $32t^5 - 80t^4 + 80t^3 - 40t^2 + 10t - 1$

**41.** $81 + 216z + 216z^2 + 96z^3 + 16z^4$ **43.** $3360x^6$

**45.** $180x^8y^2$ **47.** $-326{,}592x^4y^5$ **49.** $70x^8$

**51.** $\dfrac{1}{128} + \dfrac{7}{128} + \dfrac{21}{128} + \dfrac{35}{128} + \dfrac{35}{128} + \dfrac{21}{128} + \dfrac{7}{128} + \dfrac{1}{128} = 1$

**53.** $\dfrac{1}{6561} + \dfrac{16}{6561} + \dfrac{112}{6561} + \dfrac{448}{6561} + \dfrac{1120}{6561}$
$+ \dfrac{1792}{6561} + \dfrac{1792}{6561} + \dfrac{1024}{6561} + \dfrac{256}{6561} = 1$

**55.** $0.07776 + 0.2592 + 0.3456 + 0.2304 + 0.0768$
$+ 0.01024 = 1$

**57.** Fibonacci sequence

**59.** Rows 8, 9, 10 Rows 6, 7, 8

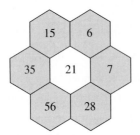

Rows 10, 11, 12 Rows 7, 8, 9

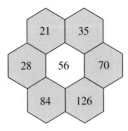

## Section 8.5 *(page 537)*

### Warm Up *(page 537)*

**1.** 6656 **2.** 291,600 **3.** 7920 **4.** 13,800

**5.** 792 **6.** 2300 **7.** $n(n - 1)(n - 2)(n - 3)$

**8.** $n(n - 1)(2n - 1)$ **9.** $n!$ **10.** $n!$

**1.** 20 **3.** 240 **5.** 6,760,000 **7.** 64

**9.** (a) 900 (b) 648 (c) 180 **11.** 64,000

**13.** (a) 40,320 (b) 384 **15.** 24 **17.** 336

**19.** 11,880 **21.** 9900 **23.** 720 **25.** 120

**27.** 11,880 **29.** 14,950 **31.** 3,838,380

**33.** 3,921,225 **35.** 1800 **37.** (a) 210 (b) 90

**39.** 3744 **41.** 5 **43.** 20 **45.** 30 **47.** 105

**49.** 2520

**51.**

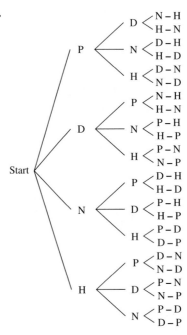

There are 24 possible arrangements.

**53.**

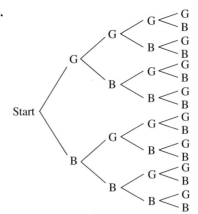

There are 16 possible arrangements.

## Section 8.6  *(page 549)*

> **Warm Up**  *(page 549)*
>
> **1.** $\dfrac{9}{16}$  **2.** $\dfrac{8}{15}$  **3.** $\dfrac{1}{6}$  **4.** $\dfrac{1}{80{,}730}$
>
> **5.** $\dfrac{1}{804{,}722{,}688{,}000}$  **6.** $\dfrac{1}{24}$  **7.** $\dfrac{1}{12}$  **8.** $\dfrac{135}{323}$
>
> **9.** 0.366  **10.** 0.997

**1.** $\dfrac{3}{8}$  **3.** $\dfrac{7}{8}$  **5.** $\dfrac{1}{9}$  **7.** $\dfrac{7}{12}$  **9.** $\dfrac{1}{3}$  **11.** $\dfrac{3}{13}$

**13.** $\dfrac{5}{13}$  **15.** $\dfrac{3}{5}$  **17.** $\dfrac{2}{5}$  **19.** 0.6  **21.** 0.8

**23.** Answers will vary.  **25.** 0.22

**27.** (a) $\dfrac{139}{198}$  (b) $\dfrac{59}{198}$  (c) $\dfrac{1}{33}$

**29.** (a) $\dfrac{1}{4}$  (b) $\dfrac{1}{2}$  (c) $\dfrac{9}{100}$  (d) $\dfrac{1}{30}$  **31.** $\dfrac{3}{28}$

**33.** (a) $\dfrac{1}{169}$  (b) $\dfrac{1}{221}$  **35.** (a) $\dfrac{1}{120}$  (b) $\dfrac{1}{24}$  (c) $\dfrac{1}{6}$

**37.** (a) $\dfrac{3}{8}$  (b) $\dfrac{19}{30}$  **39.** $\dfrac{6}{4165}$  **41.** (a) 0.922  (b) 0.078

**43.** (a) 0.9702  (b) 0.9998  (c) 0.0002

**45.** (a) $\dfrac{1}{81}$  (b) $\dfrac{16}{81}$  (c) $\dfrac{65}{81}$  **47.** (a) $\dfrac{1}{64}$  (b) $\dfrac{1}{32}$  (c) $\dfrac{63}{64}$

**49.** $\dfrac{7}{32}$  **51.** $\dfrac{76{,}545}{524{,}288} \approx 0.1459980011$

**53.** 0.1024  **55.** 0.004

## Section 8.7  *(page 562)*

> **Warm Up**  *(page 562)*
>
> **1.** 24  **2.** 40  **3.** $\dfrac{77}{60}$  **4.** $\dfrac{7}{2}$  **5.** $\dfrac{2k+5}{5}$
>
> **6.** $\dfrac{3k+1}{6}$  **7.** $8 \cdot 2^{2k} = 2^{2k+3}$  **8.** $\dfrac{1}{9}$
>
> **9.** $\dfrac{1}{k}$  **10.** $\dfrac{4}{5}$

**1.** $\dfrac{5}{(k+1)(k+2)}$  **3.** $\dfrac{(k+1)^2(k+2)^2}{4}$

**5.–17.** Answers will vary.

**19.** 210  **21.** 91  **23.** 979  **25.** 70  **27.** $-3402$

**29.** $S_n = n(2n-1)$  **31.** $S_n = 10 - 10\left(\dfrac{9}{10}\right)^n$

**33.** $S_n = \dfrac{n}{2(n+1)}$  **35.–45.** Answers will vary.

**47.** See page 554.  **49.** 1, 3, 5, 7, 9

**51.** $4, 2, -2, -4, -2$

**53.** 0, 3, 6, 9, 12
First differences: 3, 3, 3, 3
Second differences: 0, 0, 0
Linear

**55.** $3, 1, -2, -6, -11$
First differences: $-2, -3, -4, -5$
Second differences: $-1, -1, -1$
Quadratic

**57.** 0, 1, 3, 6, 10
First differences: 1, 2, 3, 4
Second differences: 1, 1, 1
Quadratic

**59.** 2, 4, 6, 8, 10
First differences: 2, 2, 2, 2
Second differences: 0, 0, 0
Linear

**61.** 1, 2, 6, 15, 31
First differences: 1, 4, 9, 16
Second differences: 3, 5, 7
Neither

**63.** $a_n = n^2 - n + 3$  **65.** $a_n = \frac{1}{2}n^2 + n - 3$

## Review Exercises  *(page 566)*

**1.** 5, 9, 13, 17, 21  **3.** $\dfrac{2}{3}, \dfrac{4}{5}, \dfrac{6}{7}, \dfrac{8}{9}, \dfrac{10}{11}$

**5.** $1, -\dfrac{4}{3}, 2, -\dfrac{16}{5}, \dfrac{16}{3}$  **7.** 43,680  **9.** $a_n = 2^{n-1}$

**11.** $a_n = \dfrac{1}{n}$  **13.** $a_n = (-1)^{n+1}3^n$  **15.** 24  **17.** 100

**19.** $\displaystyle\sum_{k=1}^{12} \dfrac{3}{1+k}$  **21.** $\displaystyle\sum_{k=1}^{13} (-1)^{k-1}2^k$

**23.** (a) $A_1 = \$2040.00, A_2 = \$2080.80, A_3 = \$2122.42,$
$A_4 = \$2164.86, A_5 = \$2208.16, A_6 = \$2252.32,$
$A_7 = \$2297.37, A_8 = \$2343.32$

 (b) $A_{40} = \$4416.08$

 (c) $A_{120} = \$21{,}530.33$

**25.** Arithmetic, $d = 5$  **27.** Arithmetic, $d = -\frac{1}{2}$

**29.** 9, 14, 19, 24, 29; arithmetic, $d = 5$

**31.** $a_n = 3n - 1$  **33.** 4, 7, 10, 13, 16

**35.** 192  **37.** 1870

**39.** (a) Company A: \$34,500
   Company B: \$35,500

 (b) \$183,000

 (c) \$175,500

 (d) For the short term, one should pick Company A, because at the end of 6 years Company A has paid \$7500 more in salary.

**41.** Geometric, $r = -3$  **43.** Geometric, $r = \frac{1}{2}$

**45.** 2, 6, 18, 54, 162   **47.** $10, -2, \frac{2}{5}, -\frac{2}{25}, \frac{2}{125}$

**49.** $a_{40} = \dfrac{1}{2^{36}}$   **51.** $a_{10} = -\dfrac{25}{64}$   **53.** 8184   **55.** 1.5

**57.** The sum does not exist; $\left|\frac{4}{3}\right| > 1$

**59.** (a) $18,416.57

(b) No; it will be more because the interest is compound.

**61.** (a)

(b) $9034.71 million

**63.** 35   **65.** 1

**67.** $81 - 216y + 216y^2 - 96y^3 + 16y^4$

**69.** $x^6 + 6x^5 + 15x^4 + 20x^3 + 15x^2 + 6x + 1$

**71.** $x^5 + 5x^4y + 10x^3y^2 + 10x^2y^3 + 5xy^4 + y^5$

**73.** $x^8 - 16x^7y + 112x^6y^2 - 448x^5y^3 + 1120x^4y^4$
$\quad - 1792x^3y^5 + 1792x^2y^6 - 1024xy^7 + 256y^8$

**75.** $25,344x^7$

**77.** $\frac{1}{243} + \frac{10}{243} + \frac{40}{243} + \frac{80}{243} + \frac{80}{243} + \frac{32}{243}$

**79.** 100   **81.** 144   **83.** 1,048,576

**85.** (a) 90,000   (b) 27,216   (c) 45,000

**87.** 30   **89.** 210   **91.** 3136

**93.** (a) 252   (b) 120   (c) 6   **95.** $\frac{5}{13}$   **97.** $\frac{2}{9}$

**99.** 0.77   **101.** (a) $\frac{1}{120}$   (b) $\frac{1}{6}$   **103.** $\frac{396}{4165}$

**105.** (a) $\frac{1}{1024}$   (b) $\frac{1}{512}$   (c) $\frac{1023}{1024}$   **107.** Answers will vary.

**109.** Answers will vary.   **111.** 465   **113.** 1296

**115.** $S_n = \frac{5}{2}\left[1 - \left(\frac{3}{5}\right)^n\right]$

**117.** 2, 3, 5, 8, 12

| $n$: | 0 | 1 | 2 | 3 | 4 |
|---|---|---|---|---|---|
| $a_n$: | 2 | 3 | 5 | 8 | 12 |

*1st difference:*   1   2   3   4

*2nd difference:*   1   1   1

Quadratic model

## Chapter Test   *(page 570)*

**1.** 3, 5, 7, 9, 11   **2.** $-1, 4, -9, 16, -25$

**3.** 2, 6, 24, 120, 720   **4.** $\frac{1}{2}, \frac{1}{4}, \frac{1}{8}, \frac{1}{16}, \frac{1}{32}$

**5.** Arithmetic, $d = 5$   **6.** Neither

**7.** Geometric, $r = 2$   **8.** Neither   **9.** 2600

**10.** $\approx 29,918.311$   **11.** $\frac{1}{2}$   **12.** $22,196.40

**13.** $41,174.69

**14.** $x^5 + 10x^4 + 40x^3 + 80x^2 + 80x + 32$

**15.** $-192,456$   **16.** 120   **17.** $2^{10} = 1024$

**18.** 45,000   **19.** $\frac{1}{24}$   **20.** $\frac{1}{32}$

## Cumulative Test: Chapters 6–8
*(page 571)*

**1.** $x = 2, y = -3, z = 5$   **2.** $x = -2, y = 4, z = -5$

**3.** $\begin{bmatrix} 1 & 4 & 8 \\ -1 & 4 & 3 \\ 0 & 1 & 5 \end{bmatrix}$   **4.** $\begin{bmatrix} 16 & 9 \\ 8 & 7 \\ 7 & 2 \end{bmatrix}$   **5.** $\begin{bmatrix} 2 & 12 \\ -5 & -3 \\ 1 & 9 \end{bmatrix}$

**6.** $\begin{bmatrix} \frac{1}{2} & -\frac{1}{2} & 0 \\ \frac{1}{4} & \frac{1}{4} & -1 \\ -\frac{1}{2} & \frac{1}{2} & 1 \end{bmatrix}$   **7.** $-39$   **8.** $-7$

**9.** $x = 18,000$   **10.** $20,000 at 7.5%
$\quad p = 66$   $\qquad\qquad$ $25,000 at 8.5%

**11.** $BA = \begin{bmatrix} 170,000 & 230,000 & 237,000 \end{bmatrix}$
$170,000 is the value of the inventory in warehouse 1.
$230,000 is the value of the inventory in warehouse 2.
$237,000 is the value of the inventory in warehouse 3.

**12.** 5, 8, 11, 14, 17   **13.** $3, 1, \frac{1}{3}, \frac{1}{9}, \frac{1}{27}$

**14.** $y^7 - 28y^6 + 336y^5 - 2240y^4 + 8960y^3$
$\quad - 21,504y^2 + 28,672y - 16,384$

**15.** 720   **16.** 18,000   **17.** $\approx 0.978$   **18.** $\frac{1}{4}$

**19.** 2, 4, 7, 11, 16

| $n$: | 1 | 2 | 3 | 4 | 5 |
|---|---|---|---|---|---|
| $a_n$: | 2 | 4 | 7 | 11 | 16 |

*1st difference:*   2   3   4   5

*2nd difference:*   1   1   1

Quadratic model

# Appendix A

## Section A.1    *(page A7)*

**1.** (a) Natural: $\{5\}$

(b) Integer: $\{-9, 5\}$

(c) Rational: $\left\{-9, -\frac{7}{2}, 5, \frac{2}{3}, 0.1\right\}$

(d) Irrational: $\{\sqrt{2}\}$

**3.** (a) Natural: $\left\{3, \frac{6}{3}\right\}$ $\left(Note: \frac{6}{3} = 2\right)$

(b) Integer: $\left\{3, -1, \frac{6}{3}\right\}$

(c) Rational: $\left\{3, -1, \frac{1}{3}, \frac{6}{3}, -7.5\right\}$

(d) Irrational: $\left\{-\frac{1}{2}\sqrt{2}\right\}$

**5.** $\frac{3}{2} < 7$

**7.** $\frac{5}{6} > \frac{2}{3}$

**9.** $x \le 5$ denotes all real numbers less than or equal to 5.

**11.** $x > 3$ denotes all real numbers greater than 3.

**13.** $-2 < x < 2$ denotes all real numbers greater than $-2$ and less than 2.

**15.** $x < 0$    **17.** $3.5\% \le r \le 6\%$    **19.** 10    **21.** $-1$

**23.** $-9$    **25.** $|-4| = |4|$    **27.** $-|-6| < |-6|$

**29.** 4    **31.** $\frac{7}{2}$    **33.** 51    **35.** $\left|z - \frac{3}{2}\right| > 1$

**37.** $|y - 0| \ge 6 \Rightarrow |y| \ge 6$    **39.** $\frac{127}{90}, \frac{584}{413}, \frac{7071}{5000}, \sqrt{2}, \frac{47}{33}$

**41.** 0.625

| | | Passes Budget |
|---|---|---|
| $\|a - b\|$ | $0.05b$ | Variance Test |
| **43.** $127.88 | $1250 | Yes |
| **45.** $572.59 | $470 | No |
| **47.** $671.75 | $1882 | No |

| | Median Income, $y$ | Household Income, $x$ | $\|y - x\|$ |
|---|---|---|---|
| **49.** | $52,725 | $65,400 | $12,675 |
| **51.** | $52,984 | $37,300 | $15,684 |
| **53.** | $36,538 | $21,300 | $15,238 |

## Section A.2    *(page A16)*

### Warm Up    *(page A16)*

**1.** $-4 < -2$    **2.** $0 > -3$    **3.** $\sqrt{3} > 1.73$

**4.** $-\pi < -3$    **5.** $|6 - 4| = 2$    **6.** $|2 - (-2)| = 4$

**7.** $|0 - (-5)| = 5$    **8.** $|3 - (-1)| = 4$

**9.** $|-7| + |7| = 7 + 7 = 14$

**10.** $-|8 - 10| = -|-2| = -2$

**1.** $7x, 4$    **3.** $x^2, -4x, 8$    **5.** (a) $-10$    (b) $-6$

**7.** (a) 0    (b) 0    **9.** Commutative (addition)

**11.** Inverse (addition)    **13.** Inverse (multiplication)

**15.** Inverse (multiplication)    **17.** Identity (multiplication)

**19.** Identity (multiplication)    **21.** Associative (addition)

**23.** $\frac{1}{7}(7 \cdot 12) = \left(\frac{1}{7} \cdot 7\right)12$ Associative (multiplication)

$= 1 \cdot 12$    Inverse (multiplication)

$= 12$    Identity (multiplication)

**25.** $2 \cdot 3 \cdot 5$    **27.** $-9$    **29.** 40    **31.** $-2$    **33.** $\frac{59}{66}$

**35.** $\frac{5}{16}$    **37.** $-\frac{14}{5}$    **39.** $-13.67$    **41.** $-0.45$

**43.** 41.14    **45.** 17.5%

**47.** Social Security spent 23.4% of the total budget.

National Defense: 266 billion dollars

Social Security: 350 billion dollars

Income Security: 226 billion dollars

Medicare: 174 billion dollars

Health: 119 billion dollars

Education: 52 billion dollars

Veterans Benefits: 37 billion dollars

Transportation: 40 billion dollars

Net Interest: 241 billion dollars

Other: 56 billion dollars

**49.** $2(-4 + 2)$

**51.** Scientific: 6 $\boxed{x^2}$ $\boxed{+/-}$ $\boxed{-}$ $\boxed{(}$ 7 $\boxed{+}$ $\boxed{(}$ 2 $\boxed{+/-}$ $\boxed{)}$ $\boxed{x^y}$ 3 $\boxed{)}$ $\boxed{=}$

Graphing: $\boxed{(-)}$ 6 $\boxed{x^2}$ $\boxed{-}$ $\boxed{(}$ 7 $\boxed{+}$ $\boxed{(}$ $\boxed{-}$ 2 $\boxed{)}$ $\boxed{\wedge}$ 3 $\boxed{)}$ $\boxed{ENTER}$

**53.** $\approx 5237$ students

## Section A.3    *(page A24)*

### Warm Up    *(page A24)*

**1.** 1    **2.** 5    **3.** 4    **4.** $\frac{1}{4}$    **5.** 0    **6.** 1

**7.** $\frac{3}{7}$    **8.** 4    **9.** $-\frac{1}{8}$    **10.** 1

**1.** 64   **3.** 729   **5.** $-81$   **7.** 8   **9.** $\frac{9}{16}$
**11.** 5184   **13.** 36   **15.** 1   **17.** 18   **19.** $\frac{7}{16}$
**21.** $-125z^3$   **23.** $5x^6$   **25.** $-3z^7$   **27.** $3x^2$   **29.** $\frac{7}{x}$
**31.** $1, x \neq -5$   **33.** $-2x^3$   **35.** $\frac{10}{x}$   **37.** $3^{3n}$
**39.** $5.75 \times 10^7$ square miles   **41.** $9.461 \times 10^{15}$ kilometers
**43.** $1.3837 \times 10^{-2}$ inch   **45.** 13,000,000 °C
**47.** 0.00000000048 electrostatic unit
**49.** (a) 7697.125   (b) 954.448
**51.** (a) \$19,055.59   (b) \$19,121.84
**53.** Insert parentheses around the exponent.

$$A = 2000\left(1 + \frac{0.06}{12}\right)^{(12 \cdot 5)}$$

**55.** About 7.77%   **57.** $(5.1 - 3.6)^5$
**59.** No. Let $a = -1$ and $n = 2$, then $-1^2 = -1$ and $(-1)^2 = 1$.

## Section A.4   *(page A32)*

**Warm Up**   *(page A32)*

**1.** $\frac{4}{27}$   **2.** 48   **3.** $-8x^3$   **4.** $6x^7$
**5.** $28x^6$   **6.** $\frac{1}{5}x^2$   **7.** $3z^4$   **8.** $\frac{25}{4x^2}$   **9.** 1
**10.** $(x + 2)^{10}$

**1.** $9^{1/2} = 3$   **3.** $\sqrt[5]{32} = 2$   **5.** $\sqrt{196} = 14$
**7.** $(-216)^{1/3} = -6$   **9.** $\sqrt[3]{27^2} = 9$   **11.** $81^{3/4} = 27$
**13.** 3   **15.** 3   **17.** $\frac{1}{2}$   **19.** $-125$   **21.** 4
**23.** 216   **25.** $\frac{27}{8}$   **27.** $-4$   **29.** $2x\sqrt[3]{2x^2}$
**31.** $\frac{5|x|\sqrt{3}}{y^2}$   **33.** $\frac{\sqrt{3}}{3}$   **35.** $4\sqrt[3]{4}$   **37.** $\frac{x(5 + \sqrt{3})}{11}$
**39.** $3(\sqrt{6} - \sqrt{5})$   **41.** $2\sqrt{x}$   **43.** $-2\sqrt{y}$   **45.** $\sqrt{6}$
**47.** 25   **49.** $x^{3/2}, x \neq 0$   **51.** $3^{1/2} = \sqrt{3}$
**53.** $\sqrt{5} + \sqrt{3} > \sqrt{5 + 3}$   **55.** $\sqrt{3} \cdot \sqrt[4]{3} > \sqrt[8]{3}$
**57.** 24 in. × 24 in. × 24 in.   **59.** $\approx 12.8\%$
**61.** 1.57 seconds   **63.** $\approx 494$ vibrations per second
**65.** 1

## Section A.5   *(page A40)*

**Warm Up**   *(page A39)*

**1.** $42x^3$   **2.** $-20z^2$   **3.** $-27x^6$   **4.** $-3x^6$
**5.** $\frac{9}{4}z^3, z \neq 0$   **6.** $4\sqrt{3}$   **7.** $\frac{9}{4x^2}$   **8.** 8
**9.** $\sqrt{2}$   **10.** $-3x$

**1.** Degree: 2   **3.** Degree: 5
Leading coefficient: 2   Leading coefficient: 1
**5.** Polynomial, $-3x^3 + 2x + 8$, degree 3
**7.** Not a polynomial
**9.** (a) $-1$   (b) 3   (c) 7   (d) 11
**11.** $-2x - 10$   **13.** $8x^3 + 29x^2 + 11$   **15.** $12z + 8$
**17.** $3x^3 - 6x^2 + 3x$   **19.** $30x^3 + 12x^2$
**21.** $x^2 + 7x + 12$   **23.** $6x^2 - 7x - 5$
**25.** $x^2 + 12x + 36$   **27.** $4x^2 - 20xy + 25y^2$
**29.** $x^2 + 2xy + y^2 - 6x - 6y + 9$   **31.** $x^2 - 4y^2$
**33.** $m^2 - n^2 - 6m + 9$   **35.** $x^3 + 3x^2 + 3x + 1$
**37.** $8x^3 - 12x^2y + 6xy^2 - y^3$   **39.** $x - y$
**41.** $x^4 + x^2 + 1$   **43.** $2x^2 + 2x$   **45.** $\approx \$9419.27$
**47.** $500r^2 + 1000r + 500$
**49.** Volume $= 4x^3 - 88x^2 + 468x$
$x = 1$ inch: $V = 384$ cubic inches
$x = 2$ inches: $V = 616$ cubic inches
$x = 3$ inches: $V = 720$ cubic inches
**51.** $3x^2 + 7x$   **53.** $x^2 + 42x + 360$

## Section A.6   *(page A46)*

**Warm Up**   *(page A46)*

**1.** $15x^2 - 6x$   **2.** $-2y^2 - 2y$   **3.** $4x^2 + 12x + 9$
**4.** $9x^2 - 48x + 64$   **5.** $2x^2 + 13x - 24$
**6.** $-5z^2 - z + 4$   **7.** $4y^2 - 1$   **8.** $x^2 - a^2$
**9.** $x^3 + 12x^2 + 48x + 64$
**10.** $8x^3 - 36x^2 + 54x - 27$

**1.** $3(x + 2)$ **3.** $(x - 1)(x + 5)$ **5.** $(x + 6)(x - 6)$

**7.** $(x + 1)(x - 3)$ **9.** $(x - 2)^2$ **11.** $(5y - 1)^2$

**13.** $(x + 2)(x - 1)$ **15.** $(y + 5)(y - 4)$

**17.** $(x - 20)(x - 10)$ **19.** $(3x - 2)(x - 1)$

**21.** $(5x + 1)(x + 5)$ **23.** $(x - 2)(x^2 + 2x + 4)$

**25.** $(2t - 1)(4t^2 + 2t + 1)$ **27.** $(x - 1)(x^2 + 2)$

**29.** $(2x - 1)(x^2 - 3)$ **31.** $4x(x - 2)$ **33.** $x^2(x - 4)$

**35.** $(1 - 2x)^2$ **37.** $(3x + 1)(x^2 + 5)$

**39.** $x(x - 4)^2(x + 1)^2$ **41.** $-x(x + 10)$

**43.** $(x + 1)^2(x - 1)^2$ **45.** $2(t - 2)(t^2 + 2t + 4)$

**47.** $(2x + 1)$ feet

## Section A.7 *(page A54)*

**Warm Up** *(page A54)*

**1.** $5x^2(1 - 3x)$ **2.** $(4x + 3)(4x - 3)$

**3.** $(3x - 1)^2$ **4.** $(2y + 3)^2$ **5.** $(z + 3)(z + 1)$

**6.** $(x - 5)(x - 10)$ **7.** $(3 - x)(1 + 3x)$

**8.** $(3x - 1)(x - 15)$ **9.** $(s + 1)(s + 2)(s - 2)$

**10.** $(y + 4)(y^2 - 4y + 16)$

**1.** All real numbers **3.** All real numbers except $x = 2$

**5.** All real numbers except $x = 0$ and $x = 4$

**7.** $3x$ **9.** $x - 2$, $x \neq 2$ **11.** $x + 2$, $x \neq -2$

**13.** $\dfrac{3x}{2}$, $x \neq 0$ **15.** $\dfrac{x}{2(x + 1)}$ **17.** $-\dfrac{1}{2}$, $x \neq 5$

**19.** $\dfrac{x(x + 3)}{x - 2}$, $x \neq -2$ **21.** $\dfrac{y - 4}{y + 6}$, $y \neq 3$ **23.** $z - 2$

**25.** $\dfrac{1}{5(x - 2)}$, $x \neq 1$ **27.** $-\dfrac{x(x + 7)}{x + 1}$, $x \neq 9$

**29.** $\dfrac{r + 1}{r}$, $r \neq 1$ **31.** $\dfrac{t - 3}{(t + 3)(t - 2)}$, $t \neq -2$

**33.** $\dfrac{x - 1}{x(x + 1)^2}$, $x \neq -2$ **35.** $\dfrac{3}{2}$, $x \neq -y$

**37.** $x(x + 1)$, $x \neq -1, 0$ **39.** $\dfrac{x + 5}{x - 1}$ **41.** $\dfrac{6x + 13}{x + 3}$

**43.** $\dfrac{x - 4}{(x + 2)(x - 2)(x - 1)}$ **45.** $\dfrac{2 - x}{x^2 + 1}$, $x \neq 0$

**47.** $\dfrac{1}{2}$, $x \neq 2$ **49.** $\dfrac{1}{x}$, $x \neq -1$ **51.** $\dfrac{2x - 1}{2x}$

**53.** $\dfrac{1}{35} \approx 0.029$ **55.** $\dfrac{1}{1000} = 0.001$

**57.** $\dfrac{13}{52} = \dfrac{1}{4} = 0.25$ **59.** (a) 12% (b) $\dfrac{288\,(NM - P)}{N(12P + NM)}$

**61.**

## Section B.1 *(page A65)*

**Warm Up** *(page A65)*

**1.** $9x^2 + 16y^2 = 144$ **2.** $x^2 + 4y^2 = 32$

**3.** $16x^2 - y^2 = 4$ **4.** $243x^2 + 36y^2 = 9$

**5.** $c = \pm 2\sqrt{2}$ **6.** $c = \pm\sqrt{13}$

**7.** $c = \pm 2\sqrt{3}$ **8.** $c = \pm\sqrt{5}$

**9.** 4 **10.** 2

**1.** a **3.** d **5.** g **7.** e

**9.** Vertex: $(0, 0)$ **11.** Vertex: $(0, 0)$

Focus: $\left(0, \dfrac{1}{16}\right)$ Focus: $\left(-\dfrac{3}{2}, 0\right)$

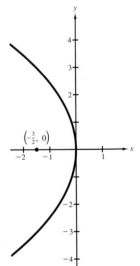

**13.** Vertex: $(0, 0)$
Focus: $(0, -2)$

**15.** Vertex: $(0, 0)$
Focus: $(2, 0)$

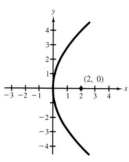

**43.** Center: $(0, 0)$
Vertices: $(\pm 1, 0)$

**45.** Center: $(0, 0)$
Vertices: $(0, \pm 1)$

**17.** $x^2 = -6y$     **19.** $y^2 = -8x$     **21.** $x^2 = 4y$

**23.** $x^2 = -8y$     **25.** $y^2 = 9x$

**27.** Center: $(0, 0)$
Vertices: $(\pm 5, 0)$

**29.** Center: $(0, 0)$
Vertices: $(0, \pm 5)$

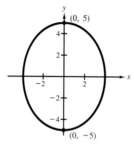

**47.** Center: $(0, 0)$
Vertices: $(0, \pm 5)$

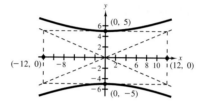

**49.** Center: $(0, 0)$
Vertices: $\left(\pm \sqrt{3}, 0\right)$

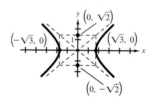

**51.** $\dfrac{y^2}{4} - \dfrac{x^2}{12} = 1$     **53.** $\dfrac{x^2}{1} - \dfrac{y^2}{9} = 1$

**55.** $\dfrac{y^2}{1024/17} - \dfrac{x^2}{64/17} = 1$     **57.** $\dfrac{y^2}{9} - \dfrac{x^2}{9/4} = 1$

**59.** $x^2 = 12y$

**61.** Tacks should be placed at $\frac{3}{2}$ feet to the left and right of the center along the base. The string should be 5 feet long.

**31.** Center: $(0, 0)$
Vertices: $(\pm 3, 0)$

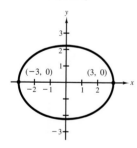

**33.** Center: $(0, 0)$
Vertices: $\left(0, \pm \sqrt{5}\right)$

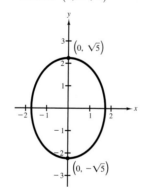

**35.** $\dfrac{x^2}{1} + \dfrac{y^2}{4} = 1$     **37.** $\dfrac{x^2}{25} + \dfrac{y^2}{21} = 1$

**39.** $\dfrac{x^2}{36} + \dfrac{y^2}{11} = 1$     **41.** $\dfrac{x^2}{400/21} + \dfrac{y^2}{25} = 1$

**63.**

**65.**

**67.**

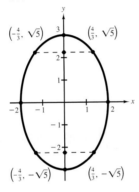

**69.** $x \approx 110.3$ miles

## Section B.2    *(page A74)*

**Warm Up**    *(page A74)*

1. Hyperbola    2. Ellipse    3. Parabola
4. Hyperbola    5. Ellipse    6. Circle
7. Hyperbola    8. Parabola    9. Parabola
10. Ellipse

**1.** Vertex: $(1, -2)$
   Focus: $(1, -4)$
   Directrix: $y = 0$

**3.** Vertex: $\left(5, -\frac{1}{2}\right)$
   Focus: $\left(\frac{11}{2}, -\frac{1}{2}\right)$
   Directrix: $x = \frac{9}{2}$

**5.** Vertex: $(1, 1)$
   Focus: $(1, 2)$
   Directrix: $y = 0$

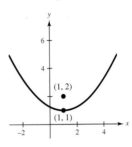

**7.** Vertex: $(8, -1)$
   Focus: $(9, -1)$
   Directrix: $x = 7$

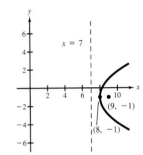

**9.** Vertex: $(-2, -3)$
   Focus: $(-4, -3)$
   Directrix: $x = 0$

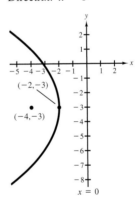

**11.** Vertex: $(-1, 2)$
   Focus: $(0, 2)$
   Directrix: $x = -2$

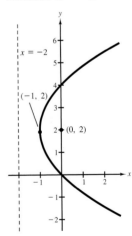

**13.** $(y - 2)^2 = -8(x - 3)$    **15.** $x^2 = 8(y - 4)$

**17.** $(y - 2)^2 = 8x$    **19.** $x^2 = -(y - 4)$

**21.** Center: $(1, 5)$

Foci: $(1, 9), (1, 1)$

Vertices: $(1, 10), (1, 0)$

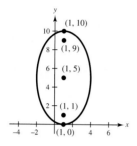

**23.** Center: $(-2, 3)$

Foci: $\left(-2, 3 - \sqrt{5}\right), \left(-2, 3 + \sqrt{5}\right)$

Vertices: $(-2, 0), (-2, 6)$

**25.** Center: $(1, -1)$

Foci: $\left(\frac{1}{4}, -1\right), \left(\frac{7}{4}, -1\right)$

Vertices: $\left(-\frac{1}{4}, -1\right), \left(\frac{9}{4}, -1\right)$

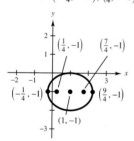

**27.** Center: $\left(\frac{1}{2}, -1\right)$

Foci: $\left(\frac{1}{2} - \sqrt{2}, -1\right), \left(\frac{1}{2} + \sqrt{2}, -1\right)$

Vertices: $\left(\frac{1}{2} - \sqrt{5}, -1\right), \left(\frac{1}{2} + \sqrt{5}, -1\right)$

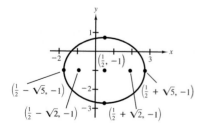

**29.** $\dfrac{(x - 2)^2}{4} + \dfrac{(y - 2)^2}{1} = 1$    **31.** $\dfrac{x^2}{48} + \dfrac{(y - 4)^2}{64} = 1$

**33.** $\dfrac{(x - 3)^2}{9} + \dfrac{(y - 5)^2}{16} = 1$    **35.** $\dfrac{x^2}{16} + \dfrac{(y - 4)^2}{12} = 1$

**37.** Center: $(1, -2)$

Vertices: $(-1, -2), (3, -2)$

Foci: $\left(1 - \sqrt{5}, -2\right), \left(1 + \sqrt{5}, -2\right)$

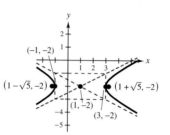

**39.** Center: $(2, -6)$

Vertices: $(2, -7), (2, -5)$

Foci: $\left(2, -6 - \sqrt{2}\right), \left(2, -6 + \sqrt{2}\right)$

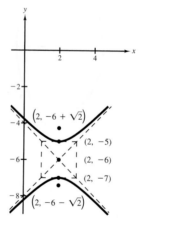

**41.** Center: $(2, -3)$

Vertices: $(3, -3), (1, -3)$

Foci: $\left(2 - \sqrt{10}, -3\right), \left(2 + \sqrt{10}, -3\right)$

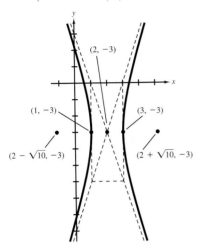

**43.** Center: $(1, -3)$

Vertices: $\left(1, -3 - \sqrt{2}\right), \left(1, -3 + \sqrt{2}\right)$

Foci: $\left(1, -3 - 2\sqrt{5}\right), \left(1, -3 + 2\sqrt{5}\right)$

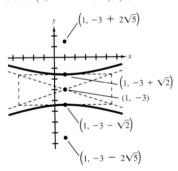

**45.** Center: $(-1, -3)$

Vertices: $\left(-1 - 3\sqrt{3}, -3\right), \left(-1 + 3\sqrt{3}, -3\right)$

Foci: $\left(-1 - \sqrt{30}, -3\right), \left(-1 + \sqrt{30}, -3\right)$

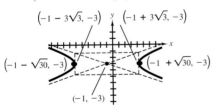

**47.** $\dfrac{(x - 4)^2}{4} = \dfrac{y^2}{12} = 1$     **49.** $\dfrac{(y - 5)^2}{16} - \dfrac{(x - 4)^2}{9} = 1$

**51.** $\dfrac{y^2}{9} - \dfrac{(x - 2)^2}{9/4} = 1$     **53.** $\dfrac{(x - 3)^2}{9} - \dfrac{(y - 2)^2}{4} = 1$

**55.** Circle     **57.** Hyperbola     **59.** Ellipse

**61.** Parabola     **63.** (a) 24,748.74 miles per hour

(b) $x^2 = -16,400(y - 4100)^2$

**65.** $\dfrac{x^2}{25} + \dfrac{y^2}{16} = 1$

**67.** Perihelion: $9.138 \times 10^7$ miles

Aphelion: $9.454 \times 10^7$ miles

**69.** $e \approx 5.431 \times 10^{-2}$

**71.**

| | |
|---|---|
| $\dfrac{(x - h)^2}{a^2} + \dfrac{(y - k)^2}{b^2} = 1$ | Standard form of equation of ellipse |
| $\dfrac{(x - h)^2}{a^2} + \dfrac{(y - k)^2}{a^2 - c^2} = 1$ | In an ellipse, $c^2 = a^2 - b^2$. |
| $\dfrac{(x - h)^2}{a^2} + \dfrac{(y - k)^2}{a^2/a^2(a^2 - c^2)} = 1$ | Multiply denominator by $a^2/a^2$. |
| $\dfrac{(x - h)^2}{a^2} + \dfrac{(y - k)^2}{a^2(1 - c^2/a^2)} = 1$ | Distribute $1/a^2$ into $(a^2 - c^2)$. |
| $\dfrac{(x - h)^2}{a^2} + \dfrac{(y - k)^2}{a^2[1 - (c/a)^2]} = 1$ | Simplify. |
| $\dfrac{(x - h)^2}{a^2} + \dfrac{(y - k)^2}{a^2(1 - e^2)} = 1$ | Definition of $e$ |

## Section C.1  *(page A77)*

**1.**

| Stems | Leaves |
|---|---|
| 7 | 0 5 5 5 7 7 8 8 8 |
| 8 | 1 1 1 1 2 3 4 5 5 5 5 7 8 9 9 9 |
| 9 | 0 2 8 |
| 10 | 0 0 |

**3.**

| Stems | Leaves |
|---|---|
| 0 | 89 66 67 65 80 98 62 93 |
| 1 | 09 01 46 24 96 90 |
| 2 | 92 55 40 61 |
| 3 | 68 35 |
| 4 | 12 96 80 38 |
| 5 | 81 18 50 70 66 |
| 6 | 44 00 34 01 |
| 7 | 61 66 00 |
| 8 | 11 57 41 90 |
| 9 | |
| 10 | |
| 11 | 60 59 33 |
| 12 | 92 |
| 13 | 19 17 37 |
| | |
| 27 | 22 |
| | |
| 31 | 32 |
| | |
| 46 | 80 |
| | |
| 65 | 14 |

**5.**

**7.**

**9.** $y = 57.49 + 1.43x$; 71.8

**11.**

**13.** $v = 1117.3 - 4.1h$; 1006.6  **15.** (a)

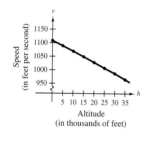

(b) $y = 1.1x + 1.2$; 0.7

(c) $y = \frac{11}{10}x + \frac{6}{5}$; 0.7

**17.** $y = \frac{16}{35}x + \frac{39}{35}$

**19.** $y = \frac{30}{7}x$

**21.** $y = -2.179x + 22.964$

**23.** $y = 2.378x + 23.546$

**25.** (a) $S = 384.1 + 21.2x$; \$479.50

(b)

(c) $r \approx 0.996$

## Section C.2    *(page A86)*

**1.** Mean: 8.86; median: 8; mode: 7

**3.** Mean: 10.29; median: 8; mode: 7

**5.** Mean: 9; median: 8; mode: 7

**7.** The mean is sensitive to extreme values.

**9.** Mean: \$67.14; median: \$65.35

**11.** Mean: 3.07; median: 3; mode: 3

**13.** One possibility: $\{4, 4, 10\}$

**15.** The median gives the most representative description.

**17.** $\bar{x} = 6$, $v = 10$, $\sigma = 3.16$

**19.** $\bar{x} = 2$, $v = \frac{4}{3}$, $\sigma = 1.15$    **21.** $\bar{x} = 4$, $v = 4$, $\sigma = 2$

**23.** $\bar{x} = 47$, $v = 226$, $\sigma = 15.03$    **25.** 3.42

**27.** 101.55    **29.** 1.65

**31.** $\bar{x} = 12$ and $|x_i - 12| = 8$ for all $x_i$.

**33.** It will increase the mean by 5, but the standard deviation will not change.

**35.** With $\bar{x} = 235$ and $\sigma = 28$:

At least $\frac{3}{4}$ of the scores in $[179, 291]$.

At least $\frac{8}{9}$ of the scores in $[151, 319]$.

With $\bar{x} = 235$ and $\sigma = 16$:

At least $\frac{3}{4}$ of the scores in $[203, 267]$.

At least $\frac{8}{9}$ of the scores in $[187, 283]$.

# Index of Applications

## CHEMISTRY AND PHYSICS APPLICATIONS

## CONSTRUCTION APPLICATIONS

## CONSUMER APPLICATIONS

## MISCELLANEOUS APPLICATIONS

# Index